ARTIFICIAL DIETS FOR INSECTS, MITES, AND SPIDERS

ARTIFICIAL DIETS FOR INSECTS, MITES, AND SPIDERS

Pritam Singh

Entomology Division
Department of Scientific and Industrial Research
Auckland, New Zealand

IFI/PLENUM · NEW YORK · WASHINGTON · LONDON

Library of Congress Cataloging in Publication Data

Singh, Pritam, 1935-
Artificial diets for insects, mites, and spiders.

Bibliography: p.
Includes indexes.
1. Insect rearing. 2. Insects–Feeding and feeds. 3. Mites–Feeding and feeds. 4. Spiders–Feeding and feeds. I. Title.
SF518.S56 638'.5 76-55694

ISBN 978-1-4684-8351-2 ISBN 978-1-4684-8349-9 (eBook)
DOI 10.1007/978-1-4684-8349-9

A Division of Plenum Publishing Corporation
227 West 17th Street, New York, N.Y. 10011

To my parents

and

my two little joys

Sona and Seema

Preface

This is a handbook of artificial diets for rearing insects, mites and spiders. Its origin goes back to my association with Dr H.L. House at the Research Institute of the Canadian Department of Agriculture at Belleville, Ontario. In 1967, Dr House published a bulletin *Artificial Diets for Insects: A Compilation of References with Abstracts*. We jointly updated this work in 1971 to include papers published from 1967-1970 and to take special note of the use of antimicrobial agents. In 1970 I left Belleville to join the Department of Scientific and Industrial Research in New Zealand where I continued to catalogue insect diets, and in 1972 compiled a *Bibliography on Artificial Diets for Insects and Mites* covering the period of 1900-1970. This was followed in 1974 by another publication entitled *Artificial Diets for Insects* containing insect diets published during May 1970-December 1972. These publications laid the foundations for the present volume.

The aim of this book is to give complete but concise coverage of existing knowledge of insect dietetics. It brings together in one volume the most complete survey of artificial diets for insects, mites and spiders to date. I hope its content will aid all those concerned with insect rearing programmes.

I would like to thank the Department of Scientific and Industrial Research for permission to publish this book. I am particularly grateful to Dr J.M. Hoy, Director of the Entomology Division, for his encouragement. I wish to express my sincere thanks to all those who assisted in the production of this book, and also to those who made the task easier by regularly supplying reprints of their papers. In particular, I am indebted to the staff of Entomology Division's Insect Rearing Section for their assistance, the Library for literature search and advice on bibliographic problems, and the Systematics Section for clarification of many taxonomic anomalies. I would like to express my special gratitude to Dr P.S. Corbet for his continued interest and several encouraging discussions about this book; Dr T.K. Crosby for advice on format and editing; Mr J.G. Charles for editing, abstracting and translation of French papers; Mrs E.M. Jerram for literature search and some abstracting; and Mr W.W. Batch for the translation of German papers. Special thanks are due to Mrs G.L. Manning for typing drafts and to Mrs R.L. Harding for typing on camera-ready forms. I would also like to thank my wife, Harminder, whose encouragement was a constant source of needed strength.

Auckland, New Zealand — Pritam Singh

October 1976

Contents

Introduction

The use of artificial diets for rearing insects is a phenomenon of this century which has been developed mainly since the 1950's in order to meet the demand for large numbers of insects required for fundamental research in the fields of physiology, ecology and genetics, and for insect control techniques such as male sterilization, pathogen production, hormone and pheromone manipulations, and biological and integrated control programmes.

The diversity of natural foods of insects is extraordinarily large including plants, grain, blood, feathers, refuse, wood and so on. Brues (1946) has given a detailed description of the dietetics of natural foods of insects.

Over the past 20 years or so, much effort has been expended in attempting to combine the 40-50 nutrients common to most foodstuffs into an acceptable food for insects. This has resulted in papers describing diets for more than 750 species of insect, most having been published within the past 10 years (see Fig. 1). Though the first reference to an insect artificial diet dates back to Bogdanov (1908), there are only 154 pre-1950 references. In the following pages I have abstracted or referred to 1961 publications in more than 100 scientific journals and periodicals covering the period from 1900 to the beginning of 1976.

The taxonomic distribution of the 754 species so far reared is summarized in Fig. 2 and detailed in Table 1. The majority of species reared are from the orders of Lepidoptera, Coleoptera and Diptera. Though there are 42 species listed under Hymenoptera, 32 of these are ants from one paper and the remaining 10 species have been reared only in the last 5-6 years. Very little work has been done on this economically important order. Only 23 species are listed under Acarina (mites) and 2 species under Aranedia (spiders). Compared with insects, these groups have received little attention, and although this is perhaps understandable in view of their lesser economic importance, spiders, for example, may be predators and thus could warrant greater attention. On the whole it seems that diets for predators and parasites and blood feeding insects have been neglected in comparison with diets for plant pests.

In this book, I have made no attempt to explain, criticize or otherwise evaluate the relative performances of the diets listed. I have assumed that each diet, as it was reported, fulfilled reasonably well the purpose for which it was developed. As the success of these diets depends on the techniques and procedures used in their preparation and methods of handling the insects, it is possible that special (but unspecified) precautions may have been lost in abbreviating the paper. It is therefore advisable, and

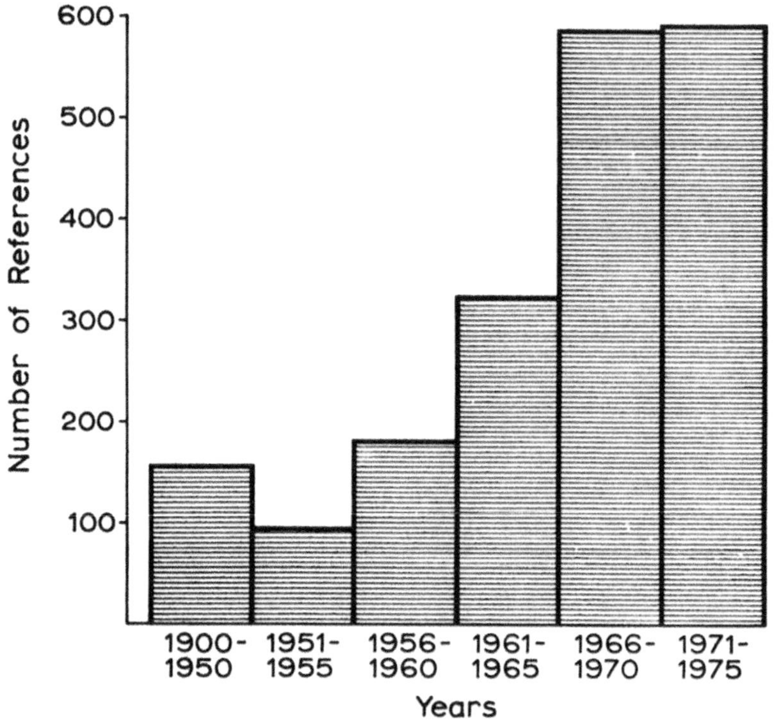

Fig. 1. Distribution of literature on artificial diets from 1900 - 1975.

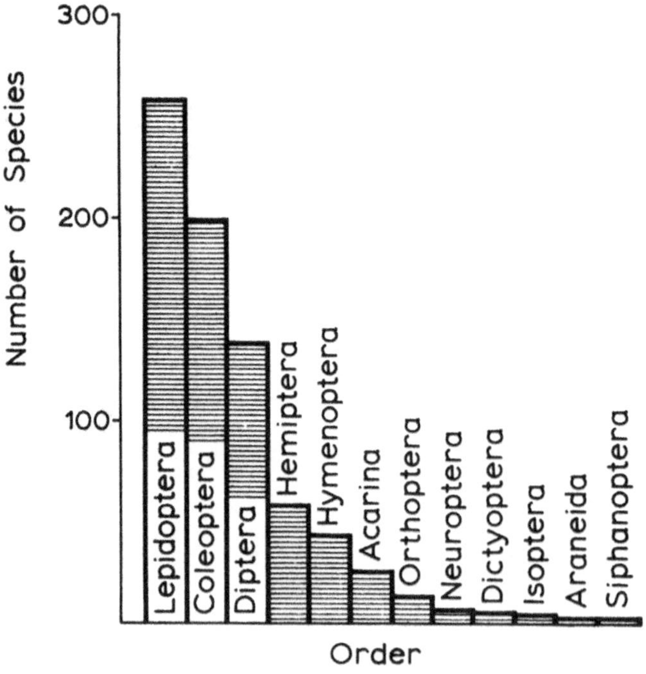

Fig. 2. Number of species reared on artificial diets.

TABLE 1

Taxonomic distribution of species reared on artificial diets

Order/Family	No. Species Reared	
Coleoptera	204	
Anobiidae		3
Bostrichidae		2
Bruchidae		1
Cerambycidae		60
Chrysomelidae		10
Coccinellidae		37
Cucujidae		3
Curculionidae		20
Dermestidae		8
Dolichopodidae		1
Elateridae		4
Lyctidae		1
Meloidae		3
Nitidulidae		1
Ptinidae		1
Scarabaeidae		15
Scolytidae		25
Tenebrionidae		7
Trogositidae		2
Dictyoptera	4	
Blattellidae		1
Blattidae		3
Diptera	138	
Anthomyiidae		7
Calliphoridae		12
Ceratopogonidae		15
Chironomidae		3
Chloropidae		4
Culicidae		15
Cuterebridae		1
Drosophilidae		12
Lycoriidae		1
Muscidae		10
Oestridae		1
Phoridae		2
Piophilidae		1
Psilidae		1
Psychodidae		3
Sarcophagidae		5
Scatopsidae		1
Sciaridae		22
Simuliidae		3
Sphaeroceridae		1
Syrphidae		3
Tachinidae		1
Tephritidae		13
Borboridae		1
Hemiptera	58	
A. Heteroptera	12	
Anthocoridae		5
Lygaeidae		1
Miridae		4
Pentatomidae		1
Reduviidae		1
B. Homoptera	46	
Aphididae		30
Cercopidae		1
Cicadellidae		7
Coccidae		1
Delphacidae		6
Pseudococcidae		1
Hymenoptera	42	
Aphelinidae		1
Apidae		1
Cephidae		1
Formicidae		32
Ichneumonidae		4
Megachilidae		1
Pteromalidae		1
Trichogrammatidae		1
Isoptera	3	
Kalotermitidae		1
Rhinotermitidae		2
Lepidoptera	258	
Arctiidae		6
Bombycidae		2
Carposinidae		1
Cochylidae		1
Cossidae		2
Gelechiidae		6
Geometridae		12
Hepialidae		2
Lasiocampidae		2
Limacodidae		1
Lymantriidae		4
Megalopygidae		1
Megathymidae		6
Noctuidae		92
Notodontidae		1
Nymphalidae		4
Oecophoridae		2
Olethreutidae		12
Papilionidae		2
Pieridae		6
Pyralidae		43
Saturniidae		6
Sesiidae		2
Sphingidae		4
Tineidae		1
Tortricidae		35
Yponomeutidae		2
Neuroptera	5	
Chrysopidae		5
Orthoptera	13	
Acrididae		8
Gryllidae		5
Siphonaptera	2	
Pulicidae		2
Acarina	25	
Acaridae		10
Ceratozetidae		2
Oppiidae		1
Phytoseiidae		5
Pyroglyphidae		2
Tetranychidae		3
Tydeidae		2
Aranedia	2	
Clubionidae		1
Lycosidae		1
Total Species		754

sometimes necessary, to read the original publication carefully before proceeding to make any diet. However, I have made every attempt to abstract sufficient information to make this book a practical guide for those interested in using artificial diets for insect breeding programmes.

Historical Background

The first insect to be reared from egg to adult axenically on an artificial diet (consisting of peptone, meat extract, starch and minerals) was *Calliphora vomitoria* by Bogdanov in 1908. Loeb (1915a) was successful in rearing *Drosophila* sp. for 5 generations on the diet of grape sugar, cane sugar, ammonium tartrate, citric acid, dipotassium hydrogen phosphate, magnesium sulphate and water. Guyénot (1917) successfully raised *Drosophila ampelophila* on a completely artificial diet. Żabiński (1926, 1928a) reared the cockroaches *Periplaneta orientalis* and *Blatella germanica* on ovalbumine, starch, saccharose, agar, and salt mixture; and several stored product pests were reared by Fraenkel and his associates in the early 1940's on a casein based diet. The steps leading up to these initial accomplishments were reviewed by Uvarov (1928), Trager (1941, 1947, 1953), and Fraenkel (1943, 1959b).

Schultz, St. Lawrence and Newmeyer (1946) were the first to rear a multicellular organism (*Drosophila melanogaster*) axenically on a chemically defined diet composed of pure amino acids, salts, vitamins, etc.

Rearing of a phytophagous insect (*Ostrinia nubilalis*) on an artificial diet was first reported by Bottger (1942); the diet was compounded from casein, sugar, fats, salts, vitamins, cellulose, agar and water. This work was followed by Beck *et al.* (1949), their diet being formulated from highly purified natural products, but containing an unidentified growth factor later identified as ascorbic acid by Chippendale & Beck (1964). Many subsequent investigators modelled their test diets on these two pioneer reports. Ishii (1952) reared *Chilo suppressalis* and Matsumoto (1954) reared *Grapholitha molesta* in Japan by including extracts of host plants in the diet developed by Beck *et al.* (1949) for *Ostrinia nubilalis*. In Japan, Ishii promoted the use of artificial diets, and in 1959 published *Methods of Experiments with Insects* (in Japanese) giving detailed procedures for rearing insects on artificial diets. Vanderzant and Reiser (1956a, b) reared *Pectinophora gossypiella* aseptically on a diet with no plant extracts. Since then, a number of phytophages have been reared on diets compounded from pure chemicals and other nutritive substances.

Rearing of an hemipteran on an artificial diet was not achieved until 1957 when Scheel *et al.* succeeded with the large milkweed bug, *Oncopeltus fasciatus*, and the one spot stink bug, *Euschistus variolorius*. Aphids were first reared by two groups of workers independently: Mittler and Dadd (1962) reared *Myzus persicae* in the United States and Auclair and Cartier (1963) reared *Acyrthosiphon pisum* in Canada.

Only recently a breakthrough has been reported by Yazgan & House (1970) and Yazgan (1972) on the rearing of an hymenopterous parasite, *Itoplectis conquistor* on a chemically defined diet under aseptic conditions.

One of the most important advances in rearing lepidopterous and other phytophagous insects in the laboratory was the use of wheatgerm in the formulation of diets as reported by Adkisson *et al.* (1960) for *Pectinophora gossypiella* and Berger (1963) for *Heliothis* species. Their formulations, with some modification, are the basis of diets for many other insects.

Mites on artificial diets have been reared only in recent years. Brickhill (1958) cultured two species of tydeid mites on a paste of enzymatic protein hydrolysate, fructose and water. Rodriguez and his associates reared the plant feeding mite (*Tetranychus urticae*) from egg to adult in about 1968. Acarid mites were cultured by Bot & Meyer (1967), Matsumoto (1968), Pillai & Winston (1968) and Rodriguez & Lasheen (1971). The house dust mite (*Dermatophagoides farinae*) was reared only 3 years ago on equal parts of fish and dried yeast in Japan (Waki & Matsumoto, 1973a). The predatory mite, *Phytoseiulus persimilis* has been artificially reared for only one generation (Shehata & Weisman, 1972).

Literature Review

The literature on artificial diets is voluminous. Singh (1972) compiled a bibliography of artificial diets for insects and mites covering the period of 1900-1970, which lists 980 references; insect diets have been catalogued in 3 "recipe" books by House (1967c), House *et al.* (1971) and Singh (1974d). Wyniger (1974) in his book on methods of rearing and maintaining insects and mites in the laboratory (in German) has listed some artificial diets. Boness (1968, 1969, 1970), McKinley (1971) and Vanderzant (1966, 1969a, 1974) reviewed various aspects concerning diets. Reviews on particular aspects are given by Meynadier (1971) on aseptic rearing for tissue culture; Helms & Raun (1971) and Vago (1972) on perennial laboratory cultures of disease free insects; Boller (1972), Gast (1968), Beck & Chippendale (1968), Chippendale & Beck (1968), and Yushima (1970), on behavioural aspects of mass rearing of insects; Finney & Fisher (1964) on culturing of entomophagous insects; MacKauer (1972, 1976) and Davidson (1974) on genetical aspects of insect production. Two books for general use in artificial diet work are *Insect Colonization and Mass Production* (Smith, 1966) and *Radiation, Radioisotopes and Rearing Methods in the Control of Insect Pests* (Proceedings I.A.E.A., 1968). Earlier work was edited by Needham *et al.* (1937) under the title *Culture Methods for Invertebrate Animals.*

Rodriguez (1964) reviewed mite nutrition and Audy & Lavoipierre (1964) laboratory rearing of parasitic Acarina; Arthur (1965) emphasized feeding in ectoparasitic ticks.

Insect nutrition, a closely related subject, has been reviewed by Chapman (1969b), House (1961a, 1962, 1963, 1965a, 1972a), Patton (1963), Gordon (1968b), David (1967), Dadd (1970a, 1973), Guennelon (1967), and Pant (1973a). Specialized reviews are also available: on phytophagous insects, Friend (1958), McGinnis & Kasting (1964a), Beck & Chippendale (1968), Pant (1973b); on silkworm, (Ito, 1967, 1972a, b); on honeybees, (Haydak, 1968, 1970); on locust and grasshoppers, (Dadd, 1963); on aphids, (Auclair, 1963; Yushima, 1968; Massonié, 1971a); on Diptera, (Friend, 1968), and muscid flies, (Spiller, 1964b); on reproduction, (Hurprin, 1962; Johansson, 1964; Wigglesworth, 1960); on *Drosophila*, (Sang, 1976); on plant sucking insects, (Auclair, 1969); on the nutrition of entomophagous insects, (Hagen, 1964); and on stored product pests, (Punj & Girish, 1968; Misra, 1973). Extensive tabular summaries of qualitative requirements are presented in a compendium by Altman & Dittmer (1968); and a similar work on diets is compiled by Singh (1976c). Rodriguez (1972b) edited *Insect and Mite Nutrition* detailing the significance and implications in ecology and pest management covering the latest development in this field.

Terminology

Various terms have been synonymously and often confusingly used to describe insect diets such as 'artificial', 'synthetic', 'semi-synthetic', 'pure', 'defined', 'semi-defined', 'chemically defined', and 'completely chemically defined'. Thus, one can find a diet containing starch, casein or wheatgerm described as 'chemically defined'. To some authors, a 'synthetic' diet is a mixture of nutritive substances, with perhaps a plant preparation with yeast, or vitamins or sugar added; to others it is a mixture of pure chemicals only. 'Artificial' usually refers to any diet that is not the natural food for insects.

To avoid this confusion Dougherty (1959) proposed and defined terminology based on the composition and purity of dietary ingredients. Thus holidic media are those in which the constituents, other than purified inert materials are of exactly known chemical structure before compounding. Meridic media are composed of a holidic base to which is added at least one substance or preparation of unknown structure or of uncertain purity. In oligidic media crude organic materials supply most dietary requirements.

This nomenclature can be applied to 'axenic', 'synxenic' or to any 'xenic' cultivation; when axenic, the species being reared is the only living member of the system; when synxenic, two or more species are reared together; and when xenic, a known species has unidentified companions. The term 'gnotobiotic' refers to a system in which any or all species present are identified. The words 'axenic' and 'aseptic', though often used synonymously, do not necessarily imply a chemically defined medium or diet, but merely indicate that the organism being reared is the only living member of the system. The term 'aseptic' should be reserved for describing techniques, and means the exclusion of contaminating microorganisms from the field of operation (in this case insect). The term 'medium' and 'diet' are synonymous when used in the context of insect cultures. In this book 'diet' is used in preference to 'medium' which is reserved for referring to the culture of microorganisms.

Some other common terms used in papers on dietetics and nutrition also require comment. 'Nutritional requirements' refer to the chemical factors of ingested food essential for normal metabolism and development of the insect. This term was originally defined by Beck (1956a) and redefined by House (1965a). 'Chemical feeding requirements' and 'physical feeding requirements' are the chemical and physical factors important to normal feeding behaviour (Beck, 1956a). These factors include taste, dietary texture, light intensities, etc. Dadd (1970a) clarified the use of the terms 'essential', 'required' and 'indispensable' in the identification of chemicals that may be utilized for growth or energy. Essential constituents are those which cannot be synthesized by either the metabolic system of the insect or its normal complement of symbionts. Non-essential constituents include materials that have to be consumed to produce energy and that can be converted into a form in which they can be utilized through metabolic processes. In the absence of essential nutrients, growth eventually ceases in one, or several consecutive generations. It is sometimes assumed that if a particular nutrient is synthesized by the insect, then it is non-essential, but this will be true only if the rate of synthesis is high enough to provide the minimal requirement for survival and growth. Nutrients which are required for optimal performance, though not necessarily, essential may be termed 'required'. In these cases synthesis of the nutrient presumably occurs, but at so slow a rate that without augmentation it becomes the limiting factor in overall performance. The terms 'essential', 'indispensable', 'required', or 'necessary' are often applied indiscriminately when their omission produces any adverse effect whatever, but it is preferable when the status of a supposed nutrient is not clear to use a noncommittal term, such as 'useful'.

Several definitions of 'artificial diet' are found in the literature. Ishii (1959) refers to a food which is not the natural diet of the insect but has been synthesized or processed. Vanderzant (1966) defines it as any diet that is not the natural food of the insect. McKinley (1971) applies the term artificial diet to one which does not include the insect's natural food except in minute traces.

In this book, the term 'artificial diet' has been adopted irrespective of diet purity or rearing conditions, and is defined as an unfamiliar food which has been formulated, synthesized, processed, and/or concocted by man, on which an insect in captivity can develop through all or part of its life cycle.

The Distinction between Insect Nutrition and Dietetics

As nutrition and dietetics are closely related subjects, it is necessary to distinguish these terms. Nutrition is the study of the food requirements of organisms (in this case insects), whereas dietetics may be called the study of practical nutrition. The two are intimately related subjects and where analysis allows some chemical definition of natural foods, nutritional inferences may be gleaned. Such inferences are usually the basis for formulating chemically defined diets for experimental use; advances in nutritional studies depend on the development of artificial diets of increasing chemical definition. Knowledge of nutrition is essential but not necessary in the development of practical artificial diets.

Types of Diet, Advantages and Usage

Insect diets can be conveniently classified in 3 categories. The first group is primarily used for studies of nutrition and metabolic pathways and includes the chemically defined (holidic) diets. The ingredients of these diets can be represented by chemical formulae, but they are expensive to prepare and, in general, aseptic techniques are required with their use. They may be dry, aqueous or agar-based depending on the insects' feeding habits.

The second group of diets contains one or more unrefined plant or animal substance such as liver powder or extract, yeast, wheatgerm or other ingredients of a similar nature. In some cases, the ingredient(s) may provide only one nutrient or class of nutrients, e.g. vitamins or lipids; or sometimes it may contribute to a wide spectrum of nutrients. The main characteristic of these diets is that most of the nutrients are provided as pure or refined substances. Most diets for laboratory rearing of insects are included in this group.

The third type of diet is composed mainly of crude materials. They are either designed to imitate the natural food, or are composed of substances of high nutrient content. These diets are assumed to have all the required nutrients, but may also contain nutrients that are not used as well as inert, indigestible materials. The efficiency and utilization of these diets can be improved with an increased understanding of the nutrition of the species. These diets are economical and are used for mass rearing of insects.

The advantages of artificial diets over natural foodstuffs can be summarized. Rearing is generally easier and life histories and behaviour can be precisely studied with less effort. The logistics of rearing insects is facilitated and large numbers may be reared simultaneously and economically in a limited space. Insects can often be reared throughout the year irrespective of food source and season giving uniform specimens of known age whose nutrition and metabolism can be regulated and manipulated. Thus insects with special traits can be developed and the "quality" of test insects can be controlled.

The use of laboratory reared insects has been dealt with in detail by Knipling (1966) in the Introductory Chapter of *Insect Colonization and Mass Production*. It is obvious that the uses of artificial diets in entomological research programmes are varied. Of these, mass-rearing has become a major weapon against insect pests. It has been used in the combat of pest species to produce parasites, predators, host material for pathogen production, attractants to lure pest species into traps, and pheromones and hormones. One highly effective use of mass-reared insects has been in sterile-male release programmes. Another area to which the mass-rearing of insects could be beneficial is as cheap food for animals (e.g. in zoos) and birds (e.g. in poultry). Mass rearing on artificial diets also has its own place in helping endangered species by rearing them in the laboratory and releasing them in their own natural habitat.

Formulation of Diets

For detailed reviews of this aspect, see Friend (1955, 1958), Edelman (1961), Patton (1967), Vanderzant (1966, 1969a, 1974), Davis (1972b) and Pant (1973b).

The successful formulation of an artificial diet depends on a sound, basic understanding of nutrition, the chemical composition of the insect and its natural food, and knowledge of the habitat and feeding behaviour of the species. Four principle requirements in formulating a diet are that it must be physically and/or chemically attractive so that it induces and stimulates the insect to feed on an unfamiliar food; it must possess all the essential nutrients in balanced proportions needed for normal growth, development and reproduction; and it must be free from microbial contamination. These topics are discussed below:-

(1) *Physical Factors*

The consistency and structure required for a suitable diet are governed largely by the feeding habits and mouth-part structure of the insect. Thus for species which live on pollen or debris, e.g. stored grain pests, crickets, and cockroaches, powdery or granular materials are suitable; phytophagous insects or borers require solid masses with high water content; fly larvae usually thrive in gelatinous or broth-like diets; for mosquito larvae the food must be suspended or dissolved in water, and for plant suckers (aphids and leafhoppers) it is usual to provide liquid food through a membrane. Vanderzant (1966, 1969a, 1974) has strongly emphasized the physical aspects of artificial diets, and the following paragraphs are based on her reviews.

(a) *Texture*

The physical properties of the diet such as hardness, texture, homogeneity, water content, and other factors that influence them should be considered. Physical modification of a diet may be accomplished by adding cellulose because it is not digested by insects. Its main purpose is to add texture to the diet so that insects will feed. It also provides roughage which aids the passage of food material through the gut (Neville & Luckey, 1962). However, as the amount of cellulose is increased, more of the thus diluted diet must be consumed to obtain the same amount of nutrients (Dadd, 1960d; McGinnis & Kasting, 1960, 1966; Mukaiyama & Ito, 1962). A number of other inert substances such as polyethylene beads, clays and other insoluble materials have been used in diets as diluents and to provide texture and roughage.

The hardness of a diet may be difficult to regulate because most plant feeding insects require a high water content, but need a firm surface against which they can press their mouth-parts. The polysaccharide, agar, is the preferred substance for controlling hardness because it is compatible with dietary ingredients, and in highly purified form is satisfactory for use in nutritional experiments where its only disadvantage is in the traces of minerals it contains. This may make mineral requirements determination difficult. It forms a firm gel at concentrations of 1.5% or more depending on the nature of the dietary ingredients and the total water content. Most diets use about 3% agar. Tanton (1965) observed that the amount of agar eaten by the mustard beetle, *Phaedon cochleariae*, increased as the agar concentration increased from 1.5% to 6%. In fact from 3% to 6% the intake of agar tripled. This resulted from physical texture alone because the test substances contained only water and agar with sinigrin as a feeding stimulant. On the other hand when agar concentration in boll weevil diets was doubled no marked changed in the development occurred, but adults were smaller than those from the control. Other substitutes for agar - polysaccharide gums from seaweeds, plant seeds, and roots - are used in the food industry, but have limited use here because of higher water content of insect diets. Sodium alginate (a thickening agent and stabilizer) improved the diet for *Pectinophora gossypiella* but was deleterious to that for *Anthonomus grandis*.

Ground plant tissues added primarily as a source of nutrient may also influence texture and hardness because of the cellulose, starch, pectins or other natural polysaccharides present. It is possible that in some instances the improvements gained by adding natural products is mainly due to their effects on the physical properties of diets rather than nutrients they contain.

(b) *Water Content*

Plant-feeding insects generally obtain their high water requirement from their food, although many insects can drink water, which is advantageous during periods of food shortage or when food is desiccated. For example, large larvae of the salt marsh caterpillar, *Estigmene acrea* fed readily on dry diet and obtained water by eating water-soaked filter paper (Vanderzant *et al.*, 1962). In laboratory rearing of Orthoptera, it is common practice to feed the insects relatively dry food and provide water separately. Nayar (1964a, b) found that nymphs of *Melanoplus bivittatus* grew better when water was present in the rearing containers. Adults of *Tenebrio molitor* are also fed water through cotton wick in our laboratory. However, for newly hatched insects or those that cannot move easily, the presence of free water is a disadvantage because they may drown in it. The optimum water content of a diet is difficult to ascertain. A change in the water content affects the physical properties of the diet and changes the concentrations of the nutrients. Also, it is difficult to maintain a constant water content because of loss by evaporation.

(c) *Preparation of Diet*

The method of preparation of the diet may influence the physical and chemical characteristics of the final product. The quality of the ingredients, the mixing of insoluble materials and the presence of trace substances in diets have been discussed by Vanderzant (1966). All ingredients must be thoroughly mixed to make the diet homogeneous. As far as possible fresh material should be used in the preparation of the diet. Several dry chemicals (e.g. cholesterol and proline) deteriorate slowly at room temperature resulting in loss of their original concentration. Fatty acids of wheatgerm change in composition during prolonged storage at room temperature (Schaefer, 1968). Chemicals sensitive to light (e.g. riboflavine) should be stored in the dark.

Insufficient heating may be just as detrimental as no heating at all. It is common practice in the preparation of diets to heat the water and agar mixture just long enough to stabilize the agar. This is then combined with the other ingredients without lowering the temperature of the diet to the solidifcation point of the agar (39° C) so that the mixture can be dispensed into the rearing containers. Complete diets and the portions of diet containing large molecular weight compounds are heat sterilized under pressure at 121°C for 15 minutes in an autoclave. Heat-labile small molecules and those that react chemically such as amino- and aldehyde-groups from sugars are sterilized by filtration (Vanderzant, 1965). The advantage of a sterile diet is that microorganisms are excluded. These may cause changes in the diet that result in misinterpretation of nutritional data or even disease or death of the insect. In addition to sterilizing the diets, autoclaving has several other advantages:- it stops enzyme action, helps dissolve the ingredients, and ruptures plant cells and stored granules. This makes the mixture more homogeneous and also makes some nutrients more readily available to insects. It is a misconception that heat treatment completely destroys heat-sensitive vitamins. The pH of the mixture and the protective action of other dietary components influence the amount of decomposition of heat labile substances.

The homogeneity of solid diets can be regulated by including gums and other stabilizers to keep insoluble particles suspended. Free sterols and fatty acids do not mix without a solubilizing agent such as a lipoprotein, phospholipid or some product, such as Tween 80, manufactured for the purpose. Soybean lecithin has been used to dissolve cholesterol and provide fatty acids for lygus bugs (Vanderzant, 1967b). Trace metals may precipitate as insoluble salts unless some chelating agent is present to keep them in solution.

(d) *Shape of Diet*

The physical properties of diets and some factors involved in their preparation are generally similar for most insects. However, there are a number of specific conditions required for particular insects such as presence of cracks and holes, raised areas on the surface, and curvature of the surface that could also be considered as physical aspects. One example is the form of the diet for the adult boll weevil. Adults feed and oviposit in the diet, but maximum egg production is obtained only if the surface of the diet has a curvature similar to that of the bud of the cotton plant (Vanderzant & Davich, 1961). *Prodenia eridania* caterpillars would not feed on their diet unless it was cut in narrow strips and laid out in a criss-cross manner, so that they could feed by placing themselves in gaps between the diet (Elliot, 1955). First instar larvae of *Lymantria dispar* would only feed on their diet if they were attached to the inner wall of the test tube in the form of a plate, imitating a leaf lamina (Edelman, 1961). Poor growth is not necessarily due to nutritional deficiencies - the physical properties of the diet may be to blame.

(2) *Chemical Factors*

(a) *Nutrients*

Phagostimulants are compounds that release feeding behaviour and induce insects to ingest (Dethier *et al.*, 1960). Continued feeding depends on the presence of feeding stimulants but not necessarily on the same ones in each stage of the insects. Phagostimulant compounds may be of nutritional value or may only provide token stimuli. Carbohydrates, proteins, amino acids, lipids, sterols, salts, vitamins, cellulose, agar, guanine monophosphate, terpenes and several organic aldehydes have shown phagostimulatory properties for various insects as reported in literature. Sugars are perhaps the most important feeding stimulants for phytophagous insects. Beck (1956d) demonstrated that the feeding of *Ostrinia nubilalis* larvae was stimulated mainly be glucose, and to a lesser extent by fructose and sucrose. *Bombyx mori* larvae, however, were stimulated by sucrose and there was little if any stimulation from glucose (Ito, 1960). Adults of *Anthonomus grandis* also preferred sucrose to glucose (Vanderzant & Davich, 1961). Sugars also affected feeding in locusts. Sucrose was definitely phagostimulatory to aphids (Mittler & Dadd, 1963a). For *Heteronychus arator* maltose was found to be the preferred phagostimulant (Sutherland & Hillier, 1976). Scheel *et al.* (1957) found that starch stimulated the feeding of *Oncopeltus fasciatus* whereas glucose, fructose and sucrose did not.

Proteins and amino acids are another important class of nutrients that stimulate feeding. The effect of proteins is probably related to their amino acid content, although impurities might exert an effect. Beck (1956a) found that casein was slightly stimulatory to feeding of *Ostrinia nubilalis.* Beck & Hanec (1958) reported that feeding of *O. nubilalis* was increased by *l*-alanine, *dl*-α-amino-*n*-butyric acid, *l*-serine, and *l*-threonine.

Lipids may also stimulate feeding in some insects. For instance, the plant sterols, β-sitosterol, stigmasterol and campesterol are powerful feeding stimulants for larvae of *Bombyx mori* (Ito *et al.*, 1964). Wheatgerm oil and soybean oil increased feeding in locusts (Dadd, 1960d) and *B. mori* (Ito, 1961a). Plant phospholipids isolated from wheatgerm oil, particularly lecithin and phosphatidyl inositol, evoked strong feeding responses in the grasshoppers, *Melanoplus bivittatus* and *Camnula pellucida* (Thorsteinson & Nayar, 1963).

(b) *Token Stimuli*

Other phagostimulants are not of obvious nutritional value and may function as token feeding stimuli. For example, the mustard oil glycoside sinigrin, found in cruciferous plants, stimulate feeding in *Plutella maculipennis* larvae (Thorsteinson, 1953). Phagostimulant effects of sinigrin have also been demonstrated in other Lepidoptera (David & Gardiner, 1966c); aphids (Wensler, 1962; Wearing, 1968), and in coleopterous larvae (Tanton, 1965).

Specific token chemicals are known for only a few insects besides those restricted to Brassicae: methyl chavicol and other components of essential oils of Umbelliferae are necessary for feeding by *Papilio* spp. (Dethier, 1953); hypericin for *Chrysomela brunsvicensis* feeding on *Hypericum* spp. (Schoonhoven, 1968); catalposides for *Ceratomia catalpae* (Nayar & Fraenkel, 1963a); curcurbitacins for the cucumber beetle (Chambliss & Jones, 1966); and the cyanogenic glycosides phaseolunatin and lotaustrin for the Mexican bean beetle feeding principally on *Phaseolus* (Nayar & Fraenkel, 1963b).

(c) *Organic Compounds*

Other organic compounds have also been found to stimulate feeding, for example, guanine monophosphate in houseflies (Robbins *et al.*, 1965), terpenes in *Leperesinus fraxini* (Chararas, 1960) and ethanol in *Xyleborus ferrugineus* (Norris & Baker, 1969).

The intentional use of phagostimulants in artificial diets for insects has not been widespread. A knowledge of phagostimulants is not necessary but can be very helpful in developing a successful artificial diet. The first inclusion of a specific and identified phagostimulant in an artificial diet was for the silkworm (Ito, 1961c). Recent nutritional investigations with aphids (Mittler & Dadd, 1965) and with large white butterfly (David & Gardiner, 1966c) have also made use of the knowledge of phagostimulants in the formulation of artificial diets. This was recently demonstrated by us after Sutherland & Hillier (1976) reported maltose as a dominant phagostimulant for larvae and adults of *Heteronychus arator*. Earlier attempts to rear this insect on a general purpose artificial diet developed by Singh (1974c) containing 3% sucrose and 0.5% glucose as the only sugars, were frustrating; incorporation of maltose with sucrose and glucose into the diet made it much more attractive and both larvae and adults now feed vigorously on it and can be satisfactorily reared.

The role of phagostimulants in the development of artificial diets and their use in nutritional studies has been reviewed by Vanderzant (1966), Wardojo (1967), Davis (1968c) and Dadd (1970a).

(3) *Nutrient Balance*

Qualitatively at least, the nutritional requirements of most insects are remarkably similar, and the choice of a particular food by a particular species is determined largely by non-nutritional factors such as physical properties and phagostimulants. The similarity of nutrient requirement is such that some generalisations can be made; for example, ascorbic acid is required for normal development of phytophagous insects and unsaturated fatty acids (linoleic or linolenic acids) are necessary for normal wing development.

It is, therefore, the quantitative balance of the nutrients that is the dominant factor in the success of a diet. Proteins, carbohydrates, lipids and vitamins must all be present in adequate supply and suboptimal amounts of one nutrient can lower the rate of use of others and decrease the efficiency with which the food is utilized. An unsatisfactory nutrient balance may lead to nutritional diseases affecting growth, development, reproduction or other life processes (Friend *et al.*, 1959; Gordon, 1959; House, 1959, 1961, 1963, 1965a). An insect's quantitative requirement is likely to be best fulfilled by its 'natural' foodstuffs since selection pressures favour genotypes that can use a given food supply with the maximum efficiency.

House (1966b) generalized knowledge of nutrient requirements and proposed three principles applicable to insect nutrition. (i) The rule of sameness:- the qualitative nutritional requirements of all insects are similar irrespective of their feeding habits and systematic position. (ii) The principle of nutrient proportionality:- metabolically suitable proportions of nutrients are required for normal nutrition. This however does not rule out the possibility of a number of balances between nutrients each equally satisfactory for optimal growth and development. (iii) The principle of cooperating supplements:- supplementary sources of nutrients, as provided by nutrient resources and sometimes symbionts may play an important role in the nutrition of almost any insect.

(4) *Microbial Contamination*

Generally the presence of microorganisms in an artificial diet causes spoilage of the diet and is detrimental to insects (Singh & Bucher, 1971). The most usual microbial contaminants encountered in artificial diets are yeasts, bacteria and moulds belonging to the genera *Aspergillus*,

Rhizopus and *Penicillium*, and several species may often be found on the same diet. It is common practice to use one or more antimicrobial agents to combat these organisms; most commonly used are formaldehyde, methyl *p*-hydroxybenzoate, sodium benzoate, potassium sorbate, sorbic acid, Streptomycin, Penicillin and Aureomycin.

Singh & House (1970b) reviewed the literature on the effects of antimicrobial agents on insects. The safe level of a compound may be defined as the concentration that does not reduce the yield of pupae and adults and does not prolong the time for larval development more than 25% compared with that of the control. Above the safe level, antimicrobial agents exert detrimental effects in proportion to the concentration used; the size of the insect may be reduced, larval life prolonged and mortality in larval and pupal stages increased (Singh & House, 1970a, c). The severity of these effects is influenced by the concentration of the nutrients in the diet (Singh & House, 1970b). Based on this work Singh & Bucher (1971) concluded that antimicrobial compounds must be chosen carefully with due regard for the species of microorganisms that are responsible for the most serious contamination, for the tolerance of the insect species concerned to the effective concentrations of antimicrobials, and for the nutritional level of the diet. The ideal antimicrobial food additive would suppress a wide variety of microorganisms at a concentration safe for the insect. Too often investigators have added antimicrobials to artificial diets for insects before determining the concentration at which they suppressed contamination without harming the insect.

Alternative methods of checking microbial contamination in diets, such as by adjusting the pH or by sterilizing the diet and equipment should be considered. House *et al.* (1971) took special notice of antimicrobial compounds used in insect diets. It is generally recognized that Coleoptera and Diptera are more sensitive to the effects of antimicrobials than Lepidoptera. Greenberg (1970) dealt with sterilizing procedures and agents, antibiotics and inhibitors in mass rearing of insects. Further information on the use of antimicrobial agents in insect diets can be gained from:- Afrikian (1960), Chawla *et al.* (1967), Clark *et al.* (1961), Dunkel & Boush (1968), Gochnauer & Furgala (1962), Golanski (1961), Henry (1968), Ignoffo (1963), Ignoffo & Garcia (1968), Lewis & Lynch (1969, 1970), McLaughlin & Keller (1964), Ouye (1962), Power & Singh (1974), Riddiford (1967), and Vail *et al.* (1968).

The Role of Symbionts

Insects may have symbiotic relationships with bacteria, fungi, yeasts, or even protozoa. A species which harbours such organisms is described as symbiotic, and one lacking them as asymbiotic. If a symbiotic species is artificially deprived of its symbiotic organism(s), it is termed aposymbiotic. Symbiotic bacteria living in the gut, usually in specialized cells at the edges of the fat body adjacent to the mid gut, or deep in the fat body, are called mycetocytes or mycetomes. These are transferred from one generation of the host insect to the next by contaminated foodstuff, or eggs, or by specialized processes during oogenesis. Symbionts may be eliminated by surface sterilization of eggs, centrifugation, thermal treatment, microsurgery or chemotherapy. It is generally supposed that the relationship between the host insect and the microorganism(s) is mutually beneficial.

Microorganisms are important in the nutrition of some insects. They may be a principle source of food (as in the fungus growing ants and beetles, in mosquito larvae and in *Drosophila*) or aid digestion by secreting enzymes in the intestine or by converting food internally from a nonusable to a usable form (e.g. in termites and cockroaches), or provide accessory growth factors such as vitamins or perform essential biochemical functions to enable an insect to survive and grow on what would otherwise be an inadequate diet.

In general, arthropods possess symbiotic microorganisms only if they feed on nutritionally inadequate diets during their entire life (Brooks, 1964). Examples of such inadequate foodstuffs are wood and stored grain (rich in cellulose and deficient in protein); wool, hair and feathers (rich in keratin and deficient in vitamins); plant juices (deficient in nitrogen); and blood or serum (deficient in water soluble vitamins). Whatever their role, the fact remains that any symbionts associated with insects and other microorganisms associated with foodstuffs make it difficult to determine the actual nutritional requirements of the insects concerned. However, although knowledge of symbiotic relationships is important it is not essential in the development of practical diets for insects.

The role of microorganisms and symbionts in insect nutrition has been reviewed by Brooks (1960, 1964), Brooks & Richards (1955, 1956), Büchner (1953), Henry (1962), Fraenkel (1959a, b), Koch (1954), Lipke & Fraenkel (1956), Pant (1973c) and Wigglesworth (1952).

Evaluation of Diets

Methods for evaluating artificial diets or conversely, determining the nutritional requirements of insects have been described by Friend (1955), Dougherty (1959), House (1959), Gordon (1959), Huprin (1962), David (1967), Patton (1963) and Vanderzant (1974). The primary method used is the deletion experiment which first requires the development of a satisfactory artificial diet of sufficient definition to allow exclusion of chemicals whose nutritional functions are then assayed in growth studies. The effects of variations in diet are measured in terms of rate of growth, development, reproduction, mortality, longevity, and morphological abnormalities such as wing deformity. These should be measured over several generations, since body reserves of some nutrients are large enough to permit normal development for more than one generation in spite of deficiencies in the diet.

Gordon (1968b) pointed out that many possible factors, singly or in combination may account for an insect failing to grow well on a diet. (1) Food intake may be abnormally low because (a) phagostimulants essential for normal feeding are deficient or (b) phagodeterrents present inhibit the feeding reflexes. (2) Ingested food may be poorly digested because (a) the animal lacks the requisite lytic enzymes or (b) normal secretion or action of the enzymes may be inhibited by an antimetabolite. (3) The absorption of one or more of the nutrients is blocked by an antimetabolite acting at any of many possible transfer sites. (4) Absorbed food may not be efficiently converted into body substance because (a) one or more essential nutrients is deficient or (b) one or more antimetabolites are present and interfere with efficient conversion. An excess of nutrient can also act as antimetabolite in various ways.

Physical aspects of diets and rearing methods (see appropriate sections) can also play an important part in the acceptance of a diet.

In our laboratory the information pertaining to all rearing work is recorded on three evaluation forms, data recording, calculating and evaluation (see Table 2, 3 and 4) which with slight modifications could be used for most situations.

Table 2

Data Recording Sheet for Evaluation of Diet

Species

Experiment No.

Generation No.

Date larvae or eggs inoculated

Date experiment terminated

Date diet prepared

Test diet No.

Replicate No.

Temperature

Relative humidity

Light

Remarks

PURPOSE:

COMPOSITION:

LARVAL NO. DATE	Larval Instars									Pre-Pupa	Pupa	Pupal wt (mg)	Adult Emergence	Sex
1														
2														
3														
4														
5														
6														
7														
8														
9														
10														
11														
12														
13														
14														
15														
16														
17														
18														
19														
20														
21														
22														
23														
24														
25														
26														
27														
28														
29														
30														
n														
Dead														

TABLE 3

Calculating Sheet

Experiment No.

Diet No.

Species

Larva No.	Date Pre-pupa	Date Pupa	Days to Pupa	Wt. of Pupa (mg)	Date Adult	Days to Adult	Sex	Pair No.	Date Paired	Longevity (days) ♂ ♀	Oviposition Period (days)	No. Egg Batches	Eggs/ Batch	Remarks
1														
2														
3														
4														
5														
6														
7														
8														
9														
10														
11														
12														
13														
14														
15														
16														
17														
18														
19														
20														
21														
22														
23														
24														
25														
26														
27														
28														
29														
30														
n														

TABLE 4

Final Evaluation of Test Diet

Species

Test Diet No.

Generation No.

Date inoculated

Date terminated

No. of larvae/or eggs inoculated

No. of larvae established after 24 h.

Temperature

Relative humidity

Light

Observations		No.	Mean	Minimum	Maximum	Standard Deviation	Standard Error
Larval Period (days)	♂ ♀						
Pre-pupal & Pupal Period (days)	♂ ♀						
Weight of Pupae (mg)	♂ ♀						
Total Life Cycle (days) Egg to Adult	♂ ♀						
Longevity, Adults (days)	♂ ♀						
Fecundity/ pair							

% Pupae

% Adults

Laboratory Rearing

Rearing has been described as the propagation for one or more generations of field collected material (in this case insects) in the laboratory (MacKauer, 1972). Peterson (1964) divided rearing into 2 categories: (a) complete rearing: including all stages between the egg stage of one generation to the production of fertile eggs in the succeeding generations; (b) incomplete or partial rearing: usually including one or two stages without completing a life cycle, e.g. larvae or pupae held until adult emergence.

For complete rearing it is essential to have a detailed knowledge of the life cycle of the species, especially of habitats and habits of the various stages, of optimum conditions of temperature, light, moisture, wind currents, etc. Knowledge of mouth part structure, feeding behaviour and food preferences is most helpful in laboratory rearing and subsequently in the development of artificial diets. Knowledge of adult behaviour - especially mating and oviposition - may also be useful when designing cages.

Insects can be either reared on their natural food or an artificial diets. Brues (1946) gave an excellent account of the food habits of insects. Uses of artificial diet for rearing insects have been given by Smith (1966), House (1967c), House *et al.* (1971) and Singh (1974d) to mention only a few.

Successful rearing of insects also depends on the prevention or suppression of diseases. Selection of clean disease-free rearing stocks and careful sanitation are essential to maintain continuous cultures of insects. Infected colonies should be isolated and quarantined. Non-infectious diseases may be caused by improper conditions of temperature, light and humidity in rearing laboratories. Each species has its own requirements which must be considered in order to produce healthy insects. Too high or too low humidity may cause abnormalities, e.g. low humidity affects emergence of adults from the pupa and subsequent wing expansion; and some epizootic conditions occur under high humidity conditions. Temperature has a profound effect on the development of insects. Some bacterial infections can be eliminated by raising or lowering the rearing temperature. Non-sporeforming bacterial infection of *Phthorimaea operculella* was checked by lowering the temperature and some studies have demonstrated that viruses can be eliminated from cultures by raising the temperature to 35°C or higher. Quality and quantity of light, as well as day-length, is also important. Photoperiod manipulation may be used to control diapause or adult mating behaviour.

Factors influencing non-infectious diseases also indirectly affect infectious diseases. Infectious maladies can be checked by preventing the causative organism from reaching the insects. This is best achieved by continuous sanitation, and reducing air-borne contamination by "clean air" or microbiologically filtered air under positive pressure in rearing rooms. Flow of personnel should be controlled in the rearing laboratory. A "No smoking" sign should be strictly observed as some insects are quite sensitive to nicotine. The egg is usually the best stage to start a colony because it is least likely to carry disease and is most easily cleaned. Many pathogens can be easily eliminated by surface sterilization of eggs in germicidal solution. Steinhaus (1963) and Vago (1972) reviewed insect diseases and House (1963) reviewed nutritional diseases. Sonleitner (1964) has reviewed the stress factors in insect colonies; some important factors are physical environment, food, disease and crowding. In a healthy colony factors causing undue stress should be avoided or minimized. In an optimum environment the population increase of the insect is greatest.

Some useful generalizations on insect rearing in the laboratory were given by Singh & Charles (1975). They stated that the importance of hygiene in insect rearing cannot be overemphasized. Dirty or unsanitary laboratory conditions will inevitably lead to contamination from other arthropods, or microorganisms varying from annoying moulds to pathogenic diseases which will rapidly destroy the whole colony. Bench space should be cleaned regularly with an antiseptic solution, such as Savlon or zephiran chloride. Glassware, metal cages and artificial diets when possible should be sterilized in an autoclave (15 lb for 20 minutes is most satisfactory). Plastic ware can be sterilized in a weak solution of Clorox or Janola. The addition of antimicrobial agents to artificial food media also helps to suppress microorganisms. Arthropod pests such as mites and psocids can be discouraged by regular cleaning out of cages, and removing faeces, corpses and rotting food. Stored product pests can be kept in jars placed in a tray with a layer of paraffin oil. Food can be kept in a deepfreezer before use, to kill any living organisms. If a pathogenic disease does occur in the colony then often the only way to remedy the situation is to destroy the population, sterilize all the equipment and start again. It is generally futile just to remove those individuals that "look ill".

On the biological side, problems may arise with populations that have been kept going for several generations. If the population is to be kept as similar as possible to a wild one, then care must be taken to avoid genetic selection and inbreeding resulting from too small a gene pool which may give rise to sterility, or affect fecundity or larval behaviour. The addition of wild stock every so often may help to alleviate these problems (but not always), and there is the risk of introducing disease. If it is intended to keep the species for a long time, then as many individuals as possible should be used to start the colony.

Care should be taken to avoid overcrowding in cages as this often leads to cannibalism or reduction in size and fecundity. When possible one stage (generally the egg) should be sterilized, again to reduce microorganic contamination. When working with more than one species, care must be taken to see that the populations are kept separate, preferably in different rooms, especially when they are likely to interact as, for example, during the study of predator/prey relationships.

Factors affecting rearing were dealt with exhaustively by Finney & Fisher (1964), while Helms & Raun (1971) and Meynadier (1971) recently reviewed the whole subject of insect rearing. Peterson (1964) reported that out of 85000 known species described in North America, only about 300 have been completely reared in the laboratory. In this book only 754 species listed are reared on artificial diets, thus indicating the enormous scope for future work.

Mass Rearing

The main distinction between laboratory rearing and mass rearing of insects is one of scale and economics. In laboratory rearing, it is important to obtain the maximum yield of insects, and costs of up to 15 to 20 cents/insect may not be excessive; on the other hand, mass rearing has a simple objective - to produce large numbers of 'acceptable' insects at the lowest possible cost.

The usefulness of mass rearing of insects on artificial diets has been recognized since the eradication of the screw worm, *Cochliomyia hominivorax* from the Southeastern United States by the sterile-male release technique in the 1960's. Some insects are easier to mass rear than others. An ideal insect for mass production should have three basic traits; a short life cycle, a high biotic potential and simple food requirements. In the past the goal of mass rearing has been to produce the maximum number of insects, in as short a time and as economically as possible, using the minimum possible labour and space. In other words emphasis has been on "quantity" rather than "quality". This concept has recently been questioned by Boller (1972) and Chambers (1975). Thus standards for mass rearing programmes must be established, depending on the ultimate use of the insects.

Artificial diets developed for mass rearing should not only be nutritionally efficient and meet the species' behavioural requirements, but also be inexpensive and simple to prepare. The economics of rearing should be the main consideration keeping in mind the standards required for the programme; this may involve automation. Different diets may be required for larval and adult stages.

The main problems encountered in mass rearing are contamination of diets by microorganisms, pathogenic diseases, and undesirable genetic or physiological changes, such as loss of pheromone production or ability to fly or to attack natural hosts. In addition, diapause and cannibalism may pose a problem in continuous mass rearing. Other problems may be similar to those encountered with ordinary laboratory rearing of the species. Methods of combating these problems vary with the particular species.

The subject of mass rearing of insects is vast and complex. There are some excellent general reviews: Smith (1966) edited a book *Insect Colonization and Mass Production*. Gast (1968) reviewed concept, methods and problems of mass rearing; Boller (1972) and Beck & Chippendale (1968) behavioural aspects; MacKauer (1972, 1976) the genetical aspect; Vago (1972) pathological problems; Vanderzant (1974) dietary aspects; Finney & Fisher (1964) mass production of entomophagous insects; Smith (1967) mass rearing procedures and Chambers (1975) the quality of mass produced insects. There are also useful publications by the International Atomic Energy Agency, Vienna which are listed in the bibliography.

Smith (1967) outlined areas in which research is necessary for developing mass production: (i) development and standardization of natural foods or artificial diets which can be economically prepared from abundant new materials; (ii) techniques that will enable insects to tolerate crowded conditions thereby economizing space; (iii) techniques for extracting various stages from the diet; (iv) understanding the chemical and physical stimuli controlling mating and oviposition; and (v) automated systems for handling all phases of the process. To these can be added problems concerning the quality and standardization of mass produced insects. In this regard, attention is particularly drawn to the *Report on the Workshop on Genetics of Insect Behaviour* held in Gainesville, Florida in 1974 which identified problems in mass rearing and made recommendations on standardization and on quality control.

Axenic Insects

Axenic insects are used primarily for studies of tissue culture, precise nutritional requirements, microorganism/symbiont relationships and pathogenic diseases. Success in producing axenic insects depends on (i) preparation of aseptic food, (ii) setting up sterile rearing units and (iii) a reliable method of surface sterilization of eggs. All procedures involving preparation of diets, sterilization of eggs and transfer of larvae should be carried out in a sterile chamber.

Some of the commonly used egg disinfectants are different forms and brands of sodium hypochlorite, potassium permanganate, hydrogen peroxide, mercuric chloride, formalin, alcohol, Hyamine, Zephiran and White's solution. The rates of hatching and disinfecting curves should be determined at various combinations of disinfectant concentration and duration of treatment, so that the best combination can be found (Vago *et al.*, 1961). Oothecae of Dictyoptera can be sterilized like eggs by dipping in disinfectant. Oviviviparous insects, such as the dipterous parasitoid, *Agaria housei*, may be disinfected by dipping gravid females in a 2.5% formalin solution before using the larvae (House 1954a).

After disinfection, rearing should be accompanied by bacteriological and mycological testing. Greenberg (1970) and Vago (1961) have reviewed the subject of disinfection; and Greenberg (1967) and Meynadier (1971) reviewed the techniques for producing axenic insects.

Standardization of Insect Cultures; Concept and Definition of Strains

Entomologists in the past have often neglected to standardize insect cultures. Standardization is of special use when studying genetics, resistance to insecticides, or pathogen, hormone or pheromone production. Special strains may be devised for a specific purpose (the word "strain" is used here as defined in *Terminology of Malaria and Malaria Eradication*, World Health Organization, Geneva, 1963, p. 120). Coker in collaboration with others (1967) in *Genetics of Insect Vectors of Disease* has written an excellent chapter detailing the necessity for development and use of standardized strains of insects. He gives a list of existing and required strains and also discusses rearing, maintenance and dispatch of strains. A major advantage of using standard strains is that results from different laboratories can be compared.

Format

Abstracts of diets have been collated and are arranged alphabetically within the taxonomic categories of order, family, genus and species. Thus, for example, abstracts on the Coleoptera precede those on the Diptera and Lepidoptera; and within the Coleoptera abstracts of diets for the Anobiidae are to be found earlier than those for the Tenebrionidae. Finally, individual species within families are also arranged alphabetically, and where there are several diets for one species, they are listed either chronologically or alphabetically by author's name as is most convenient. When several species or genera are covered within one paper, then this paper is positioned at the end of the family or order under the title "Miscellaneous". The orders Acarina (mites) and Araneida (spiders) are positioned after the insect orders.

The generic and specific names cited are generally those in common usage as approved by the Entomological Society of America (Anderson, 1975); if a name is not found in this list then it is taken from the Australian list (Callan, 1973). Otherwise the name cited in the original paper is used.

The names given to chemicals and dietary ingredients are those from the original paper. Where the source or a description of dietary ingredients is significant it is included; for example, if it is found that a particular brand of chemical is necessary for development, then the trading company is listed. Many authors, however, have given full particulars of the ingredients used, and the addresses of the suppliers without stipulating their importance, if any, to the success of the diet. Also, many substances listed are known by various synonymic names, and so it may be helpful to refer to an encyclopaedia of chemicals, such as the Merck Index (Merck & Co. Inc., Rahway, N.J.) to clarify the names and properties of many dietary materials.

American and English spellings have not been standardized, and so pyridoxin or pyridoxine, and thiamin or thiamine, for example, are given as in the original paper.

A. *Abstract Information*

As far as possible each abstract is divided into five sections. A missing section generally indicates lack of information in the original paper; but occasionally may be due to translation difficulties, in which case the original paper should be consulted.

1. REFERENCE A complete citation giving author(s), year of publication, title of paper, name of journal, volume and pages is given. The journals and periodicals are abbreviated according to *Bibliographic Guide for Editors and Authors* (1974).

2. COMPOSITION Quantities of all ingredients used in the preparation or compounding of the diet(s) are detailed. No attempt has been made to standardize the units which are given as in the original citation, e.g. (g), (mg) or (%), (cm) or (inches), etc.

3. PREPARATION Generally brief but complete instructions are given. Attention is drawn to any important points, such as pH, order of mixing, sterilization, or any special precautions required in compounding the diet.

4. REARING Techniques adopted for rearing the species are described. Whenever possible, details of temperature, relative humidity and photoperiod are given, followed by descriptions of cages or containers, or methods of sterilization of eggs or equipment, and other relevant information.

5. DEVELOPMENT Rates of growth and development are given. Life cycle information (larval and pupal periods, larval and/or pupal weights, adult longevity, fecundity, etc.), numbers of generations reared and, when available, data on natural food and artificial diet comparisons are included.

B. *Other Categories*

Corollary: Sometimes the main reference (abstract) is followed by "Corollary". This deals with supplementary reference(s) supporting the previous work, e.g. a substantial improvement in dietary composition or modification in rearing technique. Under this section only the modifications are detailed, the remainder being similar to that in the main reference. The advancement made in a corollary is generally of greater significance than that provided in the following category.

See also: Papers listed here are also supporting references; (i) they detail progressive, minor changes to an original recipe reflecting the advancement of nutritional knowledge of the species, either by the same or different authors. (ii) For some species (such as silkworm, housefly, fruit fly, cockroaches and other common insects) many successful diets, almost similar in composition have been reported in the literature. In such cases, only those references emphasizing significant differences have been selected for abstracting; the remainder are listed as "See also". The reader should consult these if further or detailed information is required.

C. *Bibliography*

The bibliography is arranged alphabetically and chronologically by author. References indicated by one asterisk (*) are abstracted and include corollaries; those indicated by two asterisks (**) are references in the introductory chapter, and mainly represent review papers. References with no asterisk are those under "See also". When an auxillary reference in the text of the abstracts is indicated by one asterisk (*) it means that that reference is abstracted elsewhere under an appropriate section.

D. *Addendum*

The addendum includes references abstracted after the completion of the manuscript; however, authors and species from these are included in the indices.

E. *Appendix*

The appendix lists the composition of various salts and vitamin mixtures commonly used in insect diets which are referred to in the abstracts.

Finally, author and species indices are provided. Italicized page numbers indicate references occurring only in the bibliography.

REFERENCES

Anderson, D.M. (Chairman, Committee on Common Names of Insects). 1975. Common Names of Insects (1975 Revised). Approved by the Entomological Society of America. Special Publication Entomological Society of America 75-1: 37 pp.

Bibliographic Guide for Editors. 1974. The American Chemical Society, Washington D.C., 362 pp.

Callan, E.McC. (Convenor, Common Names List Subcommittee of the Commonwealth and States Entomology Committee). 1973. Scientific and Common Names of Insects and Allied Forms Occurring in Australia. AUST CSIRO BULL 287: 47 pp.

General Purpose Diets

→ Fraenkel, G.; Blewett, M. 1943a. The basic food requirements of several insects. J EXP BIOL 20: 28-34.

COMPOSITION (parts): casein (casein E, ashless and vitamin-free, Glaxo Lab), 50; carbohydrates (starch or glucose), 50; McCollum's salt mixture No. 185, 1; sterol (usually cholesterol), 3, (or, later, 1); yeast (dried brewers', debittered, Glaxo Lab), or roughly equivalent quantity of yeast extract and insoluble residue, 5; water, 15.

PREPARATION Ingredients except salts mixed dry. Water added to salt mixture and poured into diets; thoroughly mixed. Left overnight and mixed again.

REARING At 25° C and 70% RH in 2 x 1" glass tubes with 2 g diet/tube.

DEVELOPMENT Species reared: *Tribolium confusum* Jacquelin duVal (Col., Tenebrionidae), *Oryzaephilus surinamensis* (Linnaeus) [previously *Silvanus*], (Col., Cucujidae), *Stegobium paniceum* (Linnaeus) [previously *Sitodrepa panicea* Linnaeus], (Col., Anobiidae), *Lasioderma serricorne* (Fabricius) (Col., Anobiidae), *Ptinus tectus* Boieldiue (Col., Ptinidae), and *Anagasta kühniella* (Zeller) [previously *Ephestia kühniella* Zeller] (Lep., Phycitidae). With glucose substituted for starch, efficiency of diet unaltered for *Stegobium*; growth of *Tribolium, Lasioderma, Silvanus* and *Ptinus* retarded. But on other occasions, development of *Tribolium, Lasioderma* and *Silvanus* on glucose diet equal to that on wholemeal flour. *Ptinus* growth better on starch than on glucose. Poor development of *Anagasta* on glucose with unaccountable difficulties. *Tribolium* to pupae in 30 days; *Oryzaephilus* to adults in 24 days; *Lasioderma* to adults in 35 days; *Stegobium* to adults in 35-40 days; *Ptinus* to adults in 46-50 days.

See also: Fraenkel (1953a); Fraenkel & Blewett (1942a b, 1943b c d e, 1945, 1947a).

→ George, B.W.; Raun, E.S.; Peters, D.C.; Mendoza, C. 1960. Artificial medium for rearing some lepidopterous corn insects. J ECON ENTOMOL 53: 318-319.

COMPOSITION (mg/g): water, 833.10; agar, 22.16; cellulose, 11.08; glucose, 34.7; vitamin-free casein, 34.7; cholesterol, 2.72; corn oil + 1% α-tocopherol, 1.36; salt mixture U.S.P. XIII, 4.18; choline chloride, 0.39; brewers' yeast, 27.82; ground alfalfa, 27.82.

PREPARATION See Beck *et al.** (1949).

REARING At 85° F.

DEVELOPMENT *Ostrinia nubilalis* (Hübner) [previously *Pyrausta*] (Lep., Pyralidae) larval period, 14-17 days. *Heliothis zea* (Boddie) (Lep., Noctuidae). From 2nd instar to pupation, 15.5 days. *Agrotis ipsilon* (Hufnagel) (Lep., Noctuidae). Egg to pupation, 31 days. *Pseudaletia unipuncta* (Haworth) (Lep., Noctuidae). Larval growth but no pupation. *Laphygma frugiperda* (J.E. Smith) (Lep., Noctuidae). Complete development.

Colias philodice eurytheme (Boisduval) (Lep., Pieridae) and *Simyra henrici* (Grote) (Lep., Noctuidae) not reared successfully.

→ Grisdale, D.G. 1963. Rearing insects on artificial diet. Interim Research Report, Insect Pathology Research Institute, Sault Ste Marie, Ontario, Canada.

COMPOSITION & PREPARATION See Berger* (1963).

REARING At 72° F and 70% RH. Eggs sterilized by immersion in NaClO solution (0.5 ml "Javex" in 100 ml distilled water) for 3 minutes. Field collected larvae immersed for 30-40 seconds before placing on diet. Larvae reared individually or in small groups in 6-dram vials filled with ½-¾" diet and plugged with non-absorbent cotton. Amount sufficient for development, but for prolonged larval life extra diet added or changed every 2-3 weeks as necessary.

DEVELOPMENT Lepidoptera species successfully reared:- *Hyphantria cunea* (Drury) (Arctiidae) - 3rd instar to pupa; *Malacosoma disstria* Hübner (Lasiocampidae) - 1st instar to adult and egg; *Orgyia antiqua* (Linnaeus) (Liparidae) - early instar to adult and egg; *Orgyia leucostigma* J.E. Smith (Liparidae) - early instar to adult and egg, 2 generations; *Orthosia revicta* Morrison (Noctuidae) - 1st instar to pupa; *Depressaria groteella* Robinson (Oecophoridae) - mid instar to pupa; *Ecdytolopha insiticiana* Zeller (Olethreutidae) - mid instar to adult; *Pseudexentera cressoniana* Clemson (Olethreutidae) - mid instar to pupa; *Tetralopha aplastella* Hulst (Pyralidae) - mid instar to overwintering larva; *Tetralopha robustella* Zeller (Pyralidae) - mid instar to pupa; *Archips cerasivorana* (Fitch) (Tortricidae) 1st instar to adult; *Argyrotaenia pinatubana* (Kearfott) (Tortricidae) - mid instar to adult; *Argyrotaenia quercifoliana* Fitch (Tortricidae) - mid instar to adult; *Choristoneura fumiferana* Clemson (Tortricidae) - 2nd instar to adult, 3 generations; *Choristoneura pinus* Freeman (Tortricidae) - mid instar to adult, 2 generations; *Choristoneura rosaceana* (Harris) - mid instar to adult; *Sparganothis unifasciana* Clemson (Tortricidae) - mid instar to adult.

→ Haydak, M.H. 1936a. A food for rearing laboratory insects. J ECON ENTOMOL 29: 1026.

COMPOSITION (parts) (a): cornflour, 4; whole-wheat flour, 2; skim milk powder, 2; dried, powdered yeast, 1; wheat middlings or bran, 2.

(b): honey, 1; glycerine, 1.

PREPARATION Equal parts of (a) and (b) mixed. Allowed to stand for 24 h then dispensed.

REARING At 32°C and 75% RH. Food placed in clumps in mason jars with screen covers. Proportions of (a) and (b) changed for different species. Liquid eliminated for stored-products insects. Grasshoppers given water on cotton or cheesecloth.

DEVELOPMENT Species reared:- greater wax moth, *Galleria mellonella* (Linnaeus) (Lep., Pyralidae); lesser wax moth, *Achoria grisella* (Fabricius) (Lep., Pyralidae); saw-toothed beetle, *Oryzaephilus surinamensis* (Linnaeus) (Col., Cucujidae); Indian meal moth, *Plodia interpunctella* (Hübner) (Lep., Pyralidae); *Trogoderma* sp. (Col., Dermestidae); grasshoppers (Orth., Acrididae).

→ Luckey, T.D. 1961. A study in comparative nutrition. COMP BIOCHEM PHYSIOL 2: 100-124.

Best of three diets tested.

COMPOSITION (g/kg): casein, 300; corn starch, 300; Alphacel, 120; sucrose, 120; corn oil, 80; vitamin A, 10000 I.U.; vitamin D, 2000 I.U.; vitamin E, 0.3; vitamin K (menadione), 0.010; vitamin C, 10; thiamin, 0.020; riboflavin, 0.020; pyridoxine, 0.020; calcium pantothenate, 0.050; niacin, 0.100; biotin, 0.001; folic acid, 0.020; choline chloride, 2; vitamin B_{12}, 0.005; *i*-inositol, 2; corn starch (as vitamin carrier), 5; salts L-II, 60.

salts: $CaCO_3$, 18.0; $CaHPO_4$, 3.3; K_2HPO_4, 13.5; $NaHPO_4$, 12.0; NaCl, 3.0; KI, 0.045; $MgSO_4.7H_2O$, 4.5; $MnSO_4.4H_2O$, 0.75; $Fe(C_6H_5O_7)_2$, 4.5; $CuSO_4.5H_2O$, 0.23; $CoCl_2.6H_2O$, 0.03; $ZnSO_4.7H_2O$, 0.06; $Na_2B_4O_7.10H_2O$, 0.03; AlK $(SO_4)_2.12H_2O$, 0.045.

PREPARATION Ingredients mixed and powdered to 300-800 mesh. Fed as powder, or with water or agar depending on species.

REARING Adapted to species and phyla.

DEVELOPMENT Universal diet suitable for 13 phyla. Representatives of Insecta eat diet readily. Three successive generations of *Blattella germanica* (Dict., Blattellidae) reared on powdered diet and water. Slow growth of *Acheta domesticus* (Linnaeus) (Orth., Gryllidae) achieved.

→ Patana, R. 1969. Rearing cotton insects in the laboratory. U S DEP AGRIC PROD RES REP 108: 6 pp.

Modified diet of Shorey* (1963) for noctuids.

COMPOSITION (g): soaked baby lima beans, 1200; brewers' yeast, 120; methyl *p*-hydroxybenzoate, 12; ascorbic acid, 12; Gelcarin, 30; and (ml): water, 2400; formaldehyde, 4.

Species of Coleoptera reared on diet of Sterling & Adkisson* (1966) used for *Anthonomus grandis*.

PREPARATION 600 g dry beans soaked in 1860 ml water for 18-20 h. Gelcarin added to 1200 ml hot water, blended and heated to boiling, then added to beans and remaining components. Blended for 5 minutes. Hot diet dispensed with air pressure into pans. Left overnight and shredded through ⅛" mesh. Left 24 h before use.

REARING Lepidopteran species except *P. gossypiella* adults kept in 1-gal glass jars lined with plastic bag. Sterilized pupae added and 2 strips paper towelling hung inside bag. Top of jar covered with 5" square paper towelling. Reared at 75-80°F. Water vial placed on top of jar fitted with 5% sugar solution and inverted through towelling top. Eggs laid on towelling. Bags renewed daily. Eggs surface sterilized as in Ignoffo* (1963). Pupae of pink bollworm, *Pectinophora gossypiella* (Saunders) surface sterilized in 0.3% NaClO and 100 placed in 1' cube screen cage at 80° F. Two cotton sponges for water and oviposition placed on cage; one moistened with 5% sugar solution, another with water. Cage top and cotton sponges covered with wax paper. Strips of absorbent cotton placed between cages for oviposition. Pads and eggs placed in Buchner funnel and soaked in 1:200 formaldehyde solution for 5 minutes. Air dried. Larvae individually reared in inverted 9/16-oz cup half full of diet capped with plastic lid. Mass rearing in 1-gal carton with shredded diet. After 10 days diet separated from cotton and placed in Berlese funnels. Mature larvae drop into jar with tissue paper for pupation. *Heliothis zea* (Boddie) reared at 82° F in 1-oz plastic cup $\frac{2}{3}$rd full of diet @ 2 neonate larvae/cup. *Heliothis virescens* Fabricius, reared in 16-oz cup @ 30/cup at 82° F. Pupae washed from cup using hose. *Spodoptera exigua* Hübner, reared at 82° F @ 50/16-oz cup, $\frac{1}{3}$rd full of diet. Pupae separated by washing with host. 25 pupae placed in plastic bags in emergence jars. *Trichoplusia ni* Hübner, reared at 82° F @ 35/16-oz cup. Pupae hand picked from cocoons. 20 placed in emergence cages. *Estigmene acrea* (Drury), reared at 82° F @ 25/16-oz cup placed on its side. 14 day old larvae placed in 1-oz diet cups. Pupae surface sterilized and 15-20 placed in cage.

COLEOPTERA

Anthonomus grandis Boheman, and *Anthonomus grandis thurberiae* Pierce, reared at 80° F in petri dishes. Diet scoured and eggs dropped on to surface. Lid replaced. Adults fed on paraffin-waxed pellets of diet of Gast (1965) in plastic shoe boxes. Bottom lined with paper towels and eggs recovered by agitating in water. Diet washed through series of screens and eggs collected from diet in NaCl solution. Eggs sterilized in 1:50 formaldehyde solution and then in 1:200 formaldehyde solution. Pipetted on to diet.

DEVELOPMENT *H. zea:* larval period, 14 days. *H. virescens:* larval period, 12-14 days. *S. exigua:* larval period, 10-12 days; pupal period, 6 days. *T. ni:* larval period, 12-14 days. *E. acrea:* larval period, 20 days; pupal period, 8-10 days. *P. gossypiella:* larval period, 14 days. *A. grandis* and *A. thurberiae:* egg to adult, 17-20 days.

Poitout, S.; Bues, R. 1970. Élevage de plusíeurs espèces de Lépidoptère Noctuidae sur milieu artificiel riche et sur milieu artificel simplifié. ANN ZOOL ECOL ANIM 2: 79-91.

Seventeen species reared on two artificial diets.

COMPOSITION Diet 1 (g): water, 170.0; agar, 5.0; cellulose, 6.0; pieces of desiccated cabbage leaf, 10.0; casein proteolysate, 7.0; glucose, 7.0; cholesterol, 0.60; Wesson's salt mixture, 2.0; ascorbic acid, 1.0; brewers' yeast, 6.0; Vanderzant's fortification vitamin mixture, 4.0; benzoic acid, 0.25; Nipagin, 0.20.

Diet 2 (g): water, 170.0; agar, 5.0; maize semolina, 28.0; wheat germ, 7.0; brewers' yeast, 7.5; ascorbic acid, 1.0; benzoic acid, 0.25; Nipagin, 0.20.

PREPARATION Agar and benzoic acid stirred in water, heated to boiling point and cooled to 56° C. Mixture transferred to mixing bowl; other ingredients added singly, with 2-3 minutes mixing for each. Warm mixture poured into plastic container, allowed to set, and refrigerated at 7° C.

REARING At 25° C with 18 h photoperiod, or 20° C with 16 h photoperiod.

DEVELOPMENT Noctuid species reared on either diet; (numbers of generations reared in parentheses): *Scotia segetum* Shiffermiller (3); *S. exclamationis* Linné (1); *S. ipsilon* Hufnagel (4); *Noctua pronuba* Linné (1); *Peridroma saucia* (Hübner) (9); *Amathes nigrum* Linné (13); *Mamestra brassicae* Linné (2); *M. oleracea* Linné (2); *Phlogophora meticulosa* Linné (1); *Spodoptera exigua* (Hübner) (30); *Pseudaletia unipuncta* (Haworth) (1); *Chloridea peltigera* Shiffermiller (1); *Helicoverpa armigera* Hübner (17); *Autographa gamma* Linné (3); *Chrysodeixis chalcites* Esper (1); *Macdunnoughia confusa* Stephens (1); *Trichoplusia ni* (Hübner) (4).

→ Shumakov, E.M.; Edelman, N.M.; Borisova, A.E. 1967. Propagation of phytophagans in artificial media. ZASHCH RAST 6: 21-22. (In Russian)

COMPOSITION For *Heliothis armigera* (Hübner) (Lep., Noctuidae); *Platyedra malvella* (Hübner) (Lep., Gelechidae); and *Loxostege sticticalis* Linnaeus (Fam., Pyralidae) (g/100 g): agar, 4.000; cellulose, 5.000; casein, 8.000; tyrosine, 0.015; valine, 0.015; arginine, 0.015; leucine, 0.015; lysine, 0.015; aspartic acid, 0.004; alanine, 0.010; methionine, 0.010; serine, 0.010; cystine, 0.150; glycine, 0.150; vegetable oil, 1.000; cholesterol, 0.100; sucrose, 6.000; vitamin B_1, 0.0012; vitamin B_2, 0.0018; vitamin $B_?$ (numeral in original text indistinct), 0.0016; vitamin B_{12}, 0.00006; pantothenic acid, 0.004; folic acid, 0.001; nicotinic acid, 0.024; ascorbic acid, 0.030; vitamin E (tocopherol), 0.036; brewers' yeast (autolysate), 8.0; mixture of mineral salts, 2.0; and water to 100 g.

Two semi-artificial diets (one wet and one dry) for migratory locust, *Locusta migratoria migratoria* (Linnaeus) (Orth., Acrididae), also developed; wet diet (g/100 g): agar, 3.00; casein, 4.00; wheat bran, 4.00; sucrose, 8.00; mineral salt, 2.50; dry reed, 2.00; ascorbic acid, 0.15; cholesterol, 0.30; vitamin D, 1.00; brewers' yeast, 1.60; water to 100 g.

dry diet (g): dry reed, 2.00; dry cows' milk, 2.00; bran, 2.00; sucrose, 3.00; ascorbic acid, 0.15; dry brewers' yeast, 1.00; and cholesterol, 0.30.

Continuous development of several generations possible.

Original paper gives no specific results, techniques, or procedures. Components of mineral salt mixture, etc., not given.

→ Singh, P. 1974c. Improvements in or relating to artificial dietary mixtures for insects. N.Z. Patent, Application No. 174449. Provisional Application filed 4 June 1974. Complete specification filed 4 September 1975.

This diet is licensed by Diga International Company as PS-diet (see commercial diets).

COMPOSITION (g/kg): part 1: agar, 25; casein, 35; cellulose powder, 100; Wesson's salt mixture, 10; wheat germ, 30.

part 2: cholesterol, 0.5; linoleic acid, 2.5; and (ml): methylene chloride, 50.

part 3: sucrose, 30; glucose, 5; inositol, 0.4; ascorbic acid, 4.0; choline chloride, 2.0; and (mg): niacinamide, 20; calcium pantothenate, 20; thiamine hydrochloride, 5; riboflavin, 10; pyridoxine hydrochloride, 5; folic acid, 5; biotin, 0.4; B_{12}, 0.04.

part 4 (ml): 4 N KOH, 5.0; mould inhibitor, 15.0; streptomycin sulphate, 10.0 (0.1 g/10 ml): and water to 1000.

mould inhibitor (g): methyl *p*-hydroxybenzoate, 15; sorbic acid, 20; and 95% alcohol, 170 ml.

PREPARATION Ingredients prepared in bulk and refrigerated until use. Items in part 1 finely blended. Cholesterol and linoleic acid dissolved in methylene dichloride and mixed well with powdered ingredients of part 1. Solvent evaporated at room temperature. Sugar and vitamins in part 3 dissolved in distilled water to make 100 ml. To prepare 1 kg diet, 200 g dry mixture blended thoroughly with 670 ml distilled water in 2 l stainless steel beaker. pH adjusted to ca. 6.5 with 4 N KOH. Diet autoclaved for 20 minutes at 105 kPa pressure, cooled to 60° C and sugar and vitamin solution added and mixed. Finally 10 ml streptomycin sulphate solution and 15 ml mould inhibitor added and completed diet poured into stainless steel trays to ca. 2.5 cm depth. When set cut into slabs and refrigerated for up to 3-4 weeks; can also be frozen and used when required.

REARING At 25° C and 50-60% RH. Rearing containers and light requirement adapted to each species.

DEVELOPMENT Species reared from egg to adult over 1-3 generations; *Epiphyas postvittana* for >30 generations.

Species reared:-

COLEOPTERA:

Family Curculionidae

Hyperodes bonariensis Kuschel

DIPTERA:

Family Drosophilidae

Drosophila melanogaster Meigen
Drosophila immigrans Sturtevant

LEPIDOPTERA:

Family Arctiidae

Nyctemera annulata (Boisduval)

Family Carposinidae

Carposina adreptella (Walker)

Family Geometridae

Austrocidaria similata (Walker)
Cleora scriptaria (Walker)
Gellonia dejectaria (Walker)
Helastia lucidata (Walker)
Helastia rosearia (Walker)
Microdes quadristrigata (Walker)
Phrissogonus laticostatus (Walker)
Poecilasthena subpurpureata (Walker)
Pseudocoremia rudisata (Walker)
Scopula rubraria (Doubleday)

Family Noctuidae

Agrotis ipsilon (Hufnagel)
Chrysodeixis eriosoma (Doubleday)
Dasypodia cymatoides Guenée
Graphania mutans mutans (Walker)
Graphania mutans erebia (Hudson)
Graphania ustistriga (Walker)
Graphania insignis (Walker)
Graphania scutata (Meyrick)
Graphania lignana (Walker)
Heliothis armigera conferta (Walker)
"Melanchra" alycone Hudson
"Melanchra" ochthistis Meyrick
Mythimna loreyi (Duponchel)
Persectania aversa Walker
Plusia chalcites Esper
Plusia argentifera Guenée
Pseudaletia separata (Walker)
Spodoptera litura (Fabricius)

Family Oecophoridae

Eutorna phaulacosma Meyrick

Family Pyralidae

Orocrambus flexuosellus (Doubleday)
Mecyna flavidalis (Doubleday)
Salebria sublignalis (Walker)
Scoparia halopis Meyrick
Uresiphita polygonalis maorialis Felder

Family Tortricidae

"Batodes" jactatana Walker
Cydia molesta Busck
Cydia pomonella (Linnaeus)
Cryptophlebia pallifimbria Bradley
Ctenopseustis obliquana (Walker)
Epalxiphora axenana Meyrick
Epiphyas postvittana (Walker)
Harmologa oblongana (Walker)
Merophyas leucaniana (Walker)
Planotortrix excessana (Walker)
"Dipterina" incessana (Walker)
Planotortrix conditana (Walker)
Planotortrix notophaea (Turner)
Pyrgotis plagiatana (Walker)
Sperchia intractana Walker
Strepsicrates macropetana (Meyrick)
'Tortrix' flavescens Butler

Partially reared species:

Coleoptera: *Graphognathus leucoloma* (Boheman) (Curculionidae); Scarabaeidae, *Adoryphorus couloni* Burmeister, *Costelytra zealandica* White, *Heteronychus arator* (Fabricius); Lepidoptera: *Declana leptomera* Walker (Geometridae); *Wiseana cervinata* Meyrick (Hepialidae); *Persectania atristriga* Walker (Noctuidae); *Plusia albostriata* Bremer & Grey (Noctuidae).

→ Wilkinson, R.S. 1966. The David-Gardiner method of feeding lepidopterous larvae on a semi-synthetic diet. MICH ENTOMOL 1: 26-28.

COMPOSITION & PREPARATION See David & Gardiner* (1965).

REARING Larvae introduced on to sides of jars. Jars closed with filter paper until 5th instar when gauze used. 15, 5th instar/jar. Smaller species reared in specimen tubes with larger density of insects/jar.

DEVELOPMENT Species reared: Lepidoptera: *Philosamia cynthia ricini* Boisduval (Saturniidae), *Arctia caja* Linnaeus (Arctiidae), *Pieris brassicae* (Linnaeus), *P. napi* (Linnaeus), *P. rapae* Linnaeus (Pieridae), *Panaxia dominula* Linnaeus (Arctiidae), *Vanessa atalanta* (Linnaeus) (Nymphalidae); and *Carausius morosis* Brunher (Phasmida; Phasmatidae).

Commercial Diets

This chapter is intended to provide information on commercially available insect diets, most of which are readily available and easily prepared. The list here does not give a complete coverage of commercial diets, but is based on information received from the 7 companies which responded to enquiries and granted permission for this material to be used. The companies are arranged in alphabetical order. The listing of diets in this chapter is in no way an endorsement of the product.

Bio-Serv, Inc.
P.O. Box 100
Frenchtown, N.J. 08825
U.S.A.

Phone (201) 996-2129

Insect Diets (ready to feed)

Bio-Serv, Inc. has designed, in cooperation with present users, effective ways of supplying insect diets.

For ease of rearing certain Lepidoptera and other insects, Bio-Serv can supply a specific diet (except amino acid defined diets) in ready to feed containers. The finished product is low in bacteria and mould. Storage life under 40° F is 5-7 weeks.

Diets can also be custom made on order for specific insects and other laboratory animals. Diets for fruit flies available.

Dietary ingredients are also available. For further details see catalogue of Biological Test Diets and Equipment, from which the following list of complete insect diets is reproduced.

Achaea catella Guevin-Meneville (Lep., Noctuidae)
Achaea finita Guenée (Lep., Noctuidae)
Agrotis ipsilon (Hufnagel) (Lep., Noctuidae)
Anastrepha suspensa (Loew) (Dipt., Tephritidae)
Anthonomus grandis Boheman (Col., Curculionidae)
Anua tirrhaca Cramer (Lep., Noctuidae)
Cadra cautella (Walker) (Lep., Pyralidae)
Carpocapsa pomonella (Linnaeus) (Lep., Olethreutidae)
Ceratitis capitata (Wiedemann) (Dipt., Tephritidae)
Chilo zonellus Swinhoe (Lep., Pyralidae)

Ctenicera destructor (Brown) (Col., Elateridae)
Diabrotica balteata Le Conte (Col., Chrysomelidae)
Diatraea saccharalis Dyar (Lep., Pyralidae)
Estigmene acrea (Drury) (Lep., Arctiidae)
Galleria mellonella (Linnaeus) (Lep., Pyralidae)
Gretchena bolliana (Slingerland) (Lep., Olethreutidae)
Heliothis armigera (Hübner) (Lep., Noctuidae)
Heliothis virescens Fabricius (Lep., Noctuidae)
Heliothis zea (Boddie) (Lep., Noctuidae)
Hemerocampa pseudotsugata McDunnough (Lep., Lymantriidae)
Herpetogramma licarsisalis (Walker) (Lep., Pyralidae)
Leptinotarsa decemlineata (Say) (Col., Chrysomelidae)
Pectinophora gossypiella (Saunders) (Lep., Gelechiidae)
Plagiohammus spinipennis Thomson (Col., Cerambycidae)
Porthetria dispar (Linnaeus) (Lep., Lymantriidae)
Rhagoletis pomonella (Walsh) (Dipt., Tephritidae)
Scolytus multistriatus (Marsham) (Col., Scolytidae)
Spodoptera exigua (Hübner) (Lep., Noctuidae)
Trichoplusia ni (Hübner) (Lep., Noctuidae)
Zeadiatraea grandisosella Dyar (Lep., Pyralidae)

Carolina Biological Supply Company
Burlington, N.C. 27215
U.S.A.

Instant Drosophila Medium (Formula 4-24)

Contents poured into vial, 15 ml water and pinch of viable yeast added, vial plugged. Adult flies added after 1 minute. Medium preferably allowed to set for 5 minutes before introduction of flies.

Carolina Instant Housefly Medium (67-5050)

Medium for growing housefly and blowfly larvae (species not specified). Clean and easy to use. One litre of medium will support 2000-3000 housefly larvae or 500 to 750 blowfly larvae.

Also kit sets available (containing diet) for *Manduca sexta* (Linnaeus) (Lep., Sphingidae) (L907), *Vanessa cardui* (Linnaeus) (Lep., Nymphalidae) (L912), and *Junonia coenia* Hübner (Lep., Nymphalidae).

Diga International Company
Biological and Chemical Division
9233 Colorado Street
P.O. Box 7004
Riverside, California 92503
U.S.A.

Telephone (714) 687-7927

PS - insect diet. Details and list of species under General Purpose diets.

ELPA lepidopterous diet. ELPA lepidopterous Larval Rearing Diets with and without alfalfa available for following species:-

Agrotis ipsilon (Hufnagel) (Lep., Noctuidae)
Agrotis subterranea (Fabricius) (Lep., Noctuidae)
Autographa californica (Speyer) (Lep., Noctuidae)
Choristoneura fumiferana (Clemens) (Lep., Tortricidae)
Estigmene acrea (Drury) (Lep., Arctiidae)
Heliothis phloxiphaga (Grote & Robinson) (Lep., Noctuidae)
Heliothis virescens Fabricius (Lep., Noctuidae)
Heliothis zea (Boddie) (Lep., Noctuidae)
Malacosoma fragile (Stretch) (Lep., Lasiocampidae)
Manduca quinquemaculata (Haworth) (Lep., Sphingidae)
Manduca sexta (Linnaeus) (Lep., Sphingidae)
Nomophila noctuella Schiffermüller (Lep., Pyralidae)
Ostrinia nubilalis (Hübner) (Lep., Pyralidae)
Peridroma saucia (Hübner) (Lep., Noctuidae)
Pieris rapae (Linnaeus) (Lep., Pieridae)
Portheria dispar (Linnaeus) (Lep., Lymantriidae)
Prodenia ornithogalli Guenée (Lep., Noctuidae)
Prodenia praefica Grote (Lep., Noctuidae)
Pseudaletia unipuncta (Haworth) (Lep., Noctuidae)
Schizura concinna (J.E. Smith) (Lep., Notodontidae)
Spodoptera mauritia acronyctoides (Boisduval) (Lep., Noctuidae)
Spodoptera exigua (Hübner) (Lep., Noctuidae)
Spodoptera frugiperda (J.E. Smith) (Lep., Noctuidae)
Trichoplusia ni (Hübner) (Lep., Noctuidae)
Vanessa cardui (Linnaeus) (Lep., Nymphalidae)
Vanessa virginiensis (Drury) (Lep., Nymphalidae)

...... and many other species (supplied on request).

Diets for other Orders.

Acheta assimilis (Fabricius) (Orth., Gryllidae)
Blattella germanica (Linnaeus) (Dic., Blattellidae)
Hypera brunneipennis (Boheman) (Col., Curculionidae)
Melanoplus sp. (Orth., Acrididae)

Diets also custom made on order for other species.

The Drosophila Supply Company
2412 W. Rollins Road
Columbia, Miss. 65201
U.S.A.

Diets for culturing *Drosophila*.

ICN Pharmaceuticals, Inc.
Life Sciences Group
26201 Miles Road
Cleveland, Ohio 44128
U.S.A.

Phone (216) 831-3000

Special wheat germ diet for insects (Vanderzant-Adkisson) (Cat. No. 102942).

COMPOSITION (%): wheat germ, 24; vitamin-free casein, 28; sucrose, 28; salt mixture, 8; Alphacel, 12.

Vanderzant modification vitamin mixture for insect diets (Cat. No. 103244): (amount/kg): α-tocopherol, 8 g; ascorbic acid, 270 g; biotin, 20 mg; calcium pantothenate, 1 g; choline chloride, 50 g; folic acid, crystalline,

250 mg; inositol, 20 g; niacinamide, 1 g; pyridoxine hydrochloride, 250 mg; riboflavin, 520 mg; thiamine hydrochloride, 250 mg. Vitamin B_{12} titrated in mannitol, 2 g.

Directions for use for Vanderzant-Adkisson wheat germ diet and Vanderzant vitamin mixture

1. Add 353.27 g of NBCo dry Vanderzant vitamin mixture to 1000 ml graduate cylinder.
2. Q.S. to 1000 ml with distilled water.
3. Mix thoroughly.
4. Store under refrigeration in closed amber bottle until ready to use.
5. Add 25 g granulated NBCo agar to 840 ml distilled water.
6. Bring to boil.
7. Add 125 g NBCo Vanderzant-Adkisson wheat germ diet to agar and water in a large blender.
8. Add antimicrobial agent if desired. The inhibitor solution can consist of sorbic acid, methyl *p*-hydroxybenzoate, butyl *p*-hydroxybenzoate. The inhibitor solution can be prepared in 95% ethyl alcohol. The use of inhibitors, plus clean equipment and careful technique allows large numbers of insects to be reared non-aseptically with little contamination.
9. Cool to 50°C before adding vitamin mixture.
10. Add 15 ml prepared vitamin solution.
11. Add 10 ml 10% KOH.
12. Blend entire mixture thoroughly before dispensing into rearing containers.

<u>Fruit fly diet</u> (Cat. No. 101735).

COMPOSITION (g): agar (granulated kabe No. 1), 114; brewers' yeast, 192; sucrose, 324.

PREPARATION Add 630 g powdered mixture and 30 ml propionic acid to 7200 ml warm water. Heat gently until completely dissolved. Store at 37°C. Mixture prepares 7860 ml of liquid diet.

Diets can be prepared on order for specific insects and other laboratory animals. Dietary ingredients also available.

Teklad Test Diets
P.O. Box 4220
2826 Lantham Drive
Madison, Wisconsin 53711
U.S.A.

Phone (608) 221-2211

<u>Insect Diet</u> (Untor). Test Insect: Fruit fly. Catalogue No. 170330.

<u>Vanderzant-Adkisson larval diet</u>. Test Insect: Pink bollworm. Catalogue No. 170987.

Diets can be custom made on order for specific insects and other laboratory animals. Dietary ingredients also available.

Turtox/Cambosco Macmillian Science Co. Inc.
8200 South Hoyne Ave
Chicago, Illinois 60620
U.S.A.

Phone (312) 488-4100

9V150 Drosophila culture medium

Freshly prepared culture medium furnished in 2-oz, wide mouth bottles, for individual student use in genetics experiments. The vials, wooden sticks and cotton plugs are sterilized before the culture medium (ca. ¾"/vial) is added.

9V149 Mould Inhibitor

Used in preparation of culture media to prevent the growth of moulds.

9V155 Turtox Banana Agar Concentrate

A new, convenient, easy to use concentrated medium for the culture of *Drosophila*. The set consists of 6 units, supplying all the necessary ingredients, with the exception of water. The quantity is sufficient to make enough Banana Agar Medium for 30-35 pint bottles.

9V1555 As above but with an anaesthetizer.

9V151 Turtox Instant Drosophila Medium

This special Turtox formulation becomes a complete, mould-resistant nutrient gel which may be inoculated with fruit flies just one minute after the simple addition of tap water. No weighing, heating or stirring is necessary. One unit makes over a litre of ready-to-use medium. Complete with instructions.

Live insects also available.

Coleoptera

Anobiidae

Anobium punctatum De Geer, furniture beetle

Cymorek, S. 1965. Verfahren zur Beschleunigten Züchtung des *Anobium punctatum* im laboratorium. PROC 12 INT CONGR ENTOMOL LONDON (1964) 687-689. (In German)

COMPOSITION (parts): wood powder, 32; cellulose powder, 9; starch, 11; blood protein, 11; bakers' yeast, 32; water, to give various dough consistencies.

PREPARATION Dry ingredients mixed with water to form "dough". 8 x 8 x 3 cm block pressed around funnel and oven dried at 60° C.

REARING At 22° C and 95% RH. Both ends of funnel shape covered with perforated plexiglass discs; perforations covered with gauze. Inner surface of synthetic wood scored for oviposition.

DEVELOPMENT Reared for 2½ years; 3 generations. In natural wood, life cycle 2-4 years; in synthetic wood, 8-11 months.

Lasioderma serricorne Fabricius, cigarette beetle

Jurzitza, G. 1969a. Die Rolle der hefeartigen Symbionten von *Lasioderma serricorne* F. (Coleoptera, Anobiidae) im Proteinmetabolismus ihrer Wirte. I. Das Wachstum normaler und aposymbiontischer Larven in Diäten mit Proteinen, Proteinderivaten und Aminosäuregemischen als N-Quellen. Z VGL PHYSIOL 63: 165-181. (In German with English summary)

COMPOSITION (%): McCollum's salt mixture, 4; cholesterol, 1; vitamin-free casein, 20; glucose-cellulose 1:1, to 100%.

McCollum's salt mixture (g): NaCl, 5; $MgSO_4$, oven-dried, 8; NaH_2PO_4, oven-dried, 10; K_2HPO_4, oven-dried, 30; $Ca(H_2PO_4)_2$, 14; $C_6H_5O_7Fe.5H_2O$, 3.6; $Ca[CH_3CH(OH)COO]_2.5H_2O$, 40.

vitamins (mg/50 ml distilled water): thiamine, 25; lactoflavin, 25; pyridoxine, 25; nicotinic acid, 25; calcium pantothenate, 25; biotin, 0.01; folic acid, 8; and choline chloride 1.0 g/5 ml distilled water.

PREPARATION Dry ingredients ground for 30 minutes in mortar-mill. Then for each 10 g of diet, 0.5 ml vitamin solution and 0.25 ml choline chloride solution added; diet reground before use.

REARING Stock culture at 30°C and 80% RH in 100 ml narrow neck Erlenmeyer flasks containing mixture of 27 g wheat flour and 3 g brewers' yeast. Wheat flour heat-sterilized at 120°C, cooled, and yeast added. Mixed, moistened with ca. 3 ml distilled water, and eggs or larvae introduced. Several hundred adults transferred to oviposition site of shredded black filter paper in 6 cm diam petri dish. Eggs surface sterilized for 3 minutes in 70% methanol or ethanol solution with 5% chloramin T. Rinsed in alcohol and several times with water. 10 g diet in 2 x 3.5 cm glass vials; 30 neonate larvae/vial.

DEVELOPMENT Basic diet qualitatively and quantitatively complete. Developmental period, 4½ weeks. Various components omitted in diet and substituted with others to study effects of different protein sources, amino acids, and insect's requirement of tryptophane.

See also: Jurzitza (1966, 1969b c, 1971a b, 1972, 1974); Jurzitza & Wolters (1967).

Lasioderma serricorne (Fabricius), cigarette beetle

Stegobium paniceum (Linnaeus) [previously *Sitodrepa*], drug store beetle

→ Pant, N.C.; Fraenkel, G. 1950. The function of the symbiotic yeasts of two insect species, *Lasioderma serricorne* F. and *Stegobium (Sitodrepa) paniceum* L. SCIENCE 112: 498-500.

Basic diet adapted from Fraenkel *et al.** (1950) for *Tenebrio molitor*.

COMPOSITION (parts): casein, 20; glucose, 80; cholesterol, 1; and McCollum's salt mixture, 2. B-complex vitamins added; (μg/g diet): thiamin, 25; riboflavin, 12.5; nicotinic acid, 25; pyridoxin, 12.5; pantothenic acid, 25; choline, 500; inositol, 250; pteroylglutamic acid, 2.5; and biotin, 0.1.

PREPARATION See Fraenkel *et al.** (1950).

REARING At 27°C and 70% RH in 2 x 1" shell vials plugged by holed corks with muslin tops. 10 neonate larvae from sterilized eggs placed in each tube with 2 g diet.

DEVELOPMENT Larval period of *L. serricorne* and *S. paniceum*, 26-35 and 42-55 days respectively.

See also: Blewett & Fraenkel (1944); Pant & Fraenkel (1954b); Pant & Kapoor (1963).

Stegobium paniceum (Linnaeus), drug store beetle

→ Pant, N.C.; Gupta, P.; Nayar, J.K. 1960b. Physiology of intracellular symbiotes of *Stegobium paniceum* L. with special reference to amino acid requirements of the host. EXPERIENTIA 16: 311-312.

COMPOSITION (parts): amino acid mixture, 15; starch, 85; salt mixture, 1; cholesterol, 2; nucleic acid, 0.1; and vitamin solution.

vitamins (μg/g diet): thiamin, 25; riboflavin, 25; nicotinic acid, 25; pyridoxin, 25; pantothenic acid, 25; choline, 500; biotin, 0.01; folic acid, 4; inositol, 250.

"Various combinations of amino acids" used, but not identified. Salt mixture not identified, although reference made to related work of Pant & Fraenkel* (1950) in which McCollum's salt mixture used.

PREPARATION See Pant & Fraenkel (1954b).

REARING In small shell vials containing 2 g diet.

DEVELOPMENT Larval period, 31-57 days.

Stegobium paniceum (Linnaeus), drug store beetle

Oryzaephilus surinamensis Linnaeus, saw toothed grain beetle (Fam, Cucujidae)

→ Lemonde, A.; Bernard, R. 1953. Aspects nutritifs des larves de *Stegobium paniceum* L. (Anobiidae) et d'*Oryzaephilus surinamensis* L. (Cucujidae). NAT CAN 80: 125-142. (In French)

Two of several experimental diets tested. Adapted from Lemonde & Bernard* (1951a) for *Tribolium confusum*.

COMPOSITION For *S. paniceum* (parts): casein, 15; glucose, 82; cholesterol, 1; salt mixture, No. 2, U.S.P. XII, 2; and (μg/g diet): thiamin hydrochloride, 25.0; biotin, 0.1; riboflavin, 12.5; nicotinic acid, 25.0; calcium pantothenate, 25.0; pyridoxine hydrochloride, 12.5; choline chloride, 500.0; folic acid, 0.2. Vitamins replaced by 5 parts brewers' yeast in similar diet.

For *O. surinamensis* (parts): casein, 10; glucose, 87; cholesterol, 1; salt mixture, No. 2., U.S.P. XII, 2; and (μg/g diet): thiamin hydrochloride, 50; biotin, 0.1; riboflavin, 50; nicotinic acid, 50; calcium pantothenate, 50; pyridoxine hydrochloride, 50; choline chloride, 500; folic acid, 0.2; inositol, 500; *p*-aminobenzoic acid, 50. Vitamins replaced by 3 parts brewers' yeast in similar diet.

PREPARATION AND REARING See Lemonde & Bernard* (1951a).

DEVELOPMENT *S. paniceum*. Optimum casein content of ca. 10% gave development similar to that on whole wheat flour of 30-32 days. Ten glucosides and 14 sugars utilized to various extent by larvae. Biotin and thiamine essential for growth.

O. surinamensis. Protein requirements minimal. 1.38% yeast sufficient for both protein and vitamin needs. Six glucosides utilized in presence of casein. Five indispensable vitamins for growth.

Bostrichidae

Bostrychus capucinus Linnaeus

→ Pranter, W. 1960. Untersuchungen über die Ernährungsphysiologie von *Bostrychus capucinus* L. (Coleopt., Fam. Bostrychidae). ANGEW ZOOL: 385-430. (In German with English summary)

COMPOSITION (%): starch, 36; flour (from which malt-extract obtainable), 60; peptone (*Peptonum siccum probacteriologia*), 4.

PREPARATION Starch mixed to paste with hot water and flour and peptone added to make dough. Cut into small portions and dried at 100° C; vitamins and enzymes present in flour thus lost leaving only sugars, starch, traces of protein, and peptone. Cooled when hard.

REARING At 30° C and 90% RH. Eggs or larvae placed on medium.

DEVELOPMENT Optimal development on acorns, chestnuts and oak roots. Development on artificial diet in one month resulting in normal adults.

Rhyzopertha dominica (Fabricius), lesser grain borer

→ Cline, L.D. 1973. Rearing the lesser grain borer on ground dog biscuits to obtain mature larvae and pupae. ANN ENTOMOL SOC AM 66: 551-552.

COMPOSITION Milk-bone dog biscuits.

PREPARATION Biscuits ground to No. 20 U.S. Standard sieve (841 μm openings); 200 g placed in 0.95-1 capacity mason jars and lightly compressed.

REARING Each jar seeded with 1.5 cm^3 (ca. 550) 1-2 week old adults; after 3 days sifted from medium which again lightly compressed and held at 27 ± 2° C, 60 ± 10% RH and 12 h photoperiod.

DEVELOPMENT First pupation at 6 weeks; highest percentage (21.5%) of pupae at 8 weeks. Little difference between development period on diet and on wheat, but both ♀♀ and ♂♂ weighed less on diet.

Bruchidae

Callosobruchus chinensis Linnaeus, cowpea beetle

→ Podoler, H.; Applebaum, S.W. 1971. Basic nutritional requirements of larvae of the bruchid beetle, *Callosobruchus chinensis* L. J STORED PROD RES 7: 187-193.

Best of 12 diets tested.

COMPOSITION (%): chickpea flour, (*Cicer avietinum* L. var. California), 58; glucose, 10; sucrose, 10; chickpea starch, 10; casein, 10; yeast extract, 2.

PREPARATION Chickpea flour diluted with commercial or isolated components to form other diets.

REARING At 28° C and 70% RH.

DEVELOPMENT 71% adult emergence in mean of 27 days.

Cerambycidae

Dectes texanus (Le Conte)

→ Hatchett, J.H.; Jackson, R.D.; Barry, R.M. 1973. Rearing a weed cerambycid, *Dectes texanus*, on an artificial medium, with notes on biology. ANN ENTOMOL SOC AM 66: 519-522.

Diet modified from Adkisson *et al.** (1960) for *Pectinophora gossypiella*.

COMPOSITION (g): agar, 90; casein, 126; sucrose, 135; Wesson's salt mixture, 36; wheat germ, 175; Alphacel, 75; potassium sorbate, 8; vitamin mixture (NBC), 36; Aureomycin, 1.0; and (ml): distilled water, 2600; 45% KOH, 9; methyl *p*-hydroxybenzoate (15% in 95% ethanol), 36; 40% formaldehyde, 3.0; wheat germ oil, 26.

PREPARATION Agar dissolved in 1500 ml water, autoclaved for 25 minutes and cooled to ca. 70°C. KOH, casein, sucrose, salt mixture, wheat germ and Alphacel blended in remaining 1000 ml water. Agar solution added, and after several minutes remaining ingredients. Whole blended for further 2 minutes and dispensed into metal trays to depth of 2 cm. Plugs cut and placed in 25 ml (1-oz) plastic cups.

REARING At 27 ± 1°C and 65% RH in darkness individually in diet cups to avoid cannibalism. Cups capped with plastic snap-on lids. Pupae transferred to 25 ml plastic cups with moist vermiculite. Adults in oviposition cages of 1-pint clear plastic containers, with copper screen base. Fresh green beans provided daily for food and oviposition site. Cages at 27 ± 1°C, 75% RH with 16 h photoperiod. Pods split open, eggs removed with fine camel hair brush, and placed in 50 ml plastic cups, 10/cup, on moist filter paper. Neonate larvae transferred to 25 ml cups with 1 diet plug.

DEVELOPMENT Average time from egg to adult, 43.5 days, (egg, 4.0; larva, 29.5, pupa, 10.0): adult survival, 34.4 days. Adult yield 40.4% in F_1, and 86.1% in F_4, reflecting increase in number of (selected for) non-diapausing larvae. Adults reared on diet comparable with natural population.

Enaphalodes rufulus (Haldeman), red oak borer

→ Galford, J.R. 1974. Some physiological effects of temperature on artificially reared red oak borers. J ECON ENTOMOL 67: 709-710.

Modified from Galford* (1969a) for wood boring insects.

COMPOSITION (g): agar, 40.0; sucrose, 20.0; vitamin diet fortification mixture, 15.0; torula yeast, 25.0; Wesson's salt mixture, 5.0; wheat germ, 100.0; riboflavin, 0.025; folic acid, 0.015; sorbic acid, 1.5; methyl paraben, 1.50; chloramphenicol, 0.75; Alphacel, 375; and water, 1100 ml.

PREPARATION 800 ml distilled water added to agar and sucrose in beaker; boiled and poured into 4-l blender. Sorbic acid, methyl paraben and chloramphenicol added and blended at high speed for 1 minute. Other ingredients plus ½ of Alphacel added and blended at low speed till mixture too viscous for blender. Remainder of Alphacel stirred in with spatula. Stored in fridge. Before use, excess water squeezed out with a laboratory press to give final water content of ca. 55%.

REARING Diet pressed into 21 g plastic cups. Hole punched in centre of diet, neonate larva introduced and cup capped with plastic lid. Diet replaced after 2 weeks. When larvae >10 mm long, placed in cups with diced diet. When larvae 15 mm long, diet renewed every 3 days, to avoid microbial contamination. (Dietary microbial inhibitors had previously affected larval growth). Pupation after cold treatment to break diapause. Prepupae transferred to individual 20 x 150 mm glass tubes partly lined with filter paper. Tubes corked, but corks slit to allow gaseous exchange. Adults sexually mature 8-10 days from emergence. Then transferred to 4-1 glass jars @ 1 pair/jar. Bolts of oak wrapped with bias tape provided for oviposition.

DEVELOPMENT Between 26.7 and 32.2°C, survival to adult, 86-88%. Mean days to heaviest larval weight: ♂, 105; ♀, 113-116. Greatest mean wt (g): ♂, 1.043-1.206; ♀, 1.280-1.328. Average number eggs laid/♀, 45-57, and mean egg hatch, 82-88%.

Megacyllene robiniae (Forster), locust borer

Wollerman, E.H.; Adams, C.; Heaton, G.C. 1969. Continuous laboratory culture of the locust borer, *Megacyllene robiniae*. ANN ENTOMOL SOC AM 62: 647-649.

COMPOSITION (g): agar, 12.5; cellulose, powdered (wood chips ground in Wiley Mill with 2 mm screen), 50.0; wheat germ, 15.0; sucrose, 17.5; vitamin-free casein, 17.5; Wesson's salt mixture, 5.0; ascorbic acid, 2.0; choline chloride, 0.5; methyl paraben, 1.0; butyl paraben, 1.0; sorbic acid, 1.0; and (ml): Vanderzant's vitamin mixture, 5.0; water, 375.0.

PREPARATION Agar and water mixed and heated; about half of hot agar mixture blended at low speed with half of dry ingredients (cellulose to choline chloride), followed by mould inhibitors dissolved in 10 ml of 80% ethanol, and vitamin mixture. Remaining half of hot agar then added and after a few minutes of blending, remaining dry ingredients. Complete mixture blended at high speed until uniform, when pressed into rearing containers as rapidly as possible to prevent crumbling.

REARING At 26.6°C and 45% RH. Larvae cannibalistic so reared individually. Two containers used:- clear plastic $\frac{7}{8}$ x $\frac{7}{8}$ x $\frac{3}{4}$" cages with friction fitting cover; and wide mouthed, round 1-oz glass jars with metal screw cap. Plastic containers used in early larval stages with $\frac{3}{8}$" thick layer of diet. Knife slits made in diet into which larvae placed. Diet changed once in 1st 4-5 weeks. Glass jars packed tightly with warm diet, and 5 week old larva placed in cavity in diet surface. Oviposition on 1" diam black locust branches wrapped spirally with cloth tape. Eggs laid in crevice between tape and bark. Eggs removed from tape with small brush dipped in 3000 ppm solution benzalkonium chloride, and placed on filter paper wetted with 1 mm Ringer's solution in petri dish.

DEVELOPMENT Time for each life stage, average (range) in days, respectively: egg, 7 (4-11); larva, 60 (41-115); pupa, 13 (5-17); adult, 34 (14-55). One generation completed in minimum of 50 and maximum of 143 days compared with 1 year under field conditions. First generation survival of 52% increased to 76% in 6th generation.

Plagiohammus spinipennis Thomson

→ Harley, K.L.S.; Willson, B.W. 1968. Propagation of a cerambycid borer on a meridic diet. CAN J ZOOL 46: 1265-1266.

COMPOSITION (g): cellulose, 52.80; sucrose, 12.00; soluble starch, 12.00; dextrose, 12.00; Wesson's salt mixture, 3.60; casein, 24.00; cholesterol, 0.48; linoleic acid, 0.48; lecithin, 1.20; agar, 25.00; and (ml): water, 840.00; antimicrobial mixture, 15.00; Vanderzant's fortification vitamin mixture, 20.00; 10% KOH, 10.00. Ingredients 1-9 (as listed above) also formulated by Nutritional Biochemicals Corporation, Cleveland, Ohio, as a "Special Diet Harley 66516". Antimicrobial mixture (g): sorbic acid, 20.0; methyl *p*-hydroxybenzoate, 15.0; and 95% ethyl alcohol, 170.0 ml.

REARING Each larva transferred to 15-20 g diet in small petri dish or squat glass jar; diet replaced monthly or as necessary. Adults left in larval container for several days then transferred to oviposition cages with stems and foliage of *Lantana camara* as oviposition sites and food.

DEVELOPMENT Neonate larva to pupation, 3 months; length of mature larvae, 4.5 cm; larval mortality, 4%; fecundity, 84 eggs/♀. Larval growth on diet superior, with greater larval survival and adult fecundity than on natural host.

See also: Hadlington & Johnston (1973).

Prionus imbricornis Linnaeus, tilehorned prionus

→ Payne, J.A.; Lowman, H.; Pate, R.R. 1975. Artificial diets for rearing the tilehorned *Prionus*. ANN ENTOMOL SOC AM 68: 680-682.

Best of five diets tested.

COMPOSITION (g): agar, 50; Alphacel, 307; wheat germ, 60; sucrose, 70; vitamin-free casein, 70; Wesson's salt mixture, 20; ascorbic acid, 8; choline chloride, 2; methyl paraben, 4; sorbic acid, 4; Vanderzant's vitamin mixture, 20; and (ml): water, 1500; 10% formaldehyde, 16.

PREPARATION 1000 ml distilled water boiled with agar in beaker. Sorbic acid and methyl paraben dissolved in warm 95% ethanol and added to agar solution. Remaining ingredients and water blended for 30 seconds; hot agar solution added and further blended till uniform. One pint hot diet dispensed into 7 x 21 x 6 cm plastic tray with 4 rectangular compartments and dried in a fume hood for 6-12 h.

REARING At 20°C, 65% RH in dark. Single, neonate larva placed in each compartment. Larvae transferred to fresh diet every 30 days.

DEVELOPMENT Larval survival, 92% after 210 days; survival to adult, 88%. Larval wt, 6-7 g in 200 days; 14 g in 400 days. One generation completed in 3-5 years.

Miscellaneous - Cerambycids

→ Gardiner, L.M. 1970. Rearing wood-boring beetles (Cerambycidae) on artificial diet. CAN ENTOMOL 102: 113-117.

Diet developed by McMorran* (1965) for *Choristoneura fumiferana* used to rear forty-nine identified and a number of unidentified Mexican species of Cerambycidae; 17 spp from egg. Dried ground plant tissue added to diet.

COMPOSITION See McMorran* (1965).

PREPARATION Ingredients blended. Nutrient agar dissolved in cold water liquified over steam, added while hot, and whole mixture then blended for 2 or 3 minutes. About 6 ml of warm diet dispensed to vial. Diet stored at 34°F for up to 3 weeks. When necessary, mixture modified to rear wood boring beetles by addition of dried, pulverized plant material - principally phloem tissue of host - as final step at rate of 5 ml to 100 ml diet. Tissue chopped into small pieces, oven-dried, and pulverized in hammer mill.

REARING On modified diet in 60 mm glass petri dishes in dark at room temperature. Blocks of artificial diet also used successfully as adult food, and one species, *Graphisurus fasciatus* (De Geer), oviposited in them.

DEVELOPMENT Most species reared completed development in less than half normal time; in some cases, less than one third.

Species reared:- (* Reared from egg stage)

PARANDRINAE

Parandra brunnea (Fabricius)

PRIONINAE

Orthosoma brunneum (Forster)

ASEMINAE

**Arhopalus foveicollis* (Haldeman)
**Asemum striatum* (Linnaeus)
**Tetropium cinnamopterum* Kirby
Atimia confusa (Say)

LEPTURINAE

**Pidonia ruficollis* (Say)
Grammoptera subargentata (Kirby)
**Typocerus velutinus* (Olivier)
**Anoplodera mutabilis* (Newman)
**A. sexmaculata* (Linnaeus)
A. sanguinea (Le Conte)
A. canadensis (Olivier)
A. minnesotana (Casey)
Rhagium inquisitor (Linnaeus)
Centrodera decolorata (Harris)
Anthophilax attenuatus (Haldeman)
A. malachiticus (Haldeman)
**Acmaeops proteus* (Kirby)
**Evodinus monticola* (Rondani)

CERAMBYCINAE

**Semanotus litigiosus* (Casey)
**S. ligneus* (Fabricius)
Meriellum proteus (Kirby)
Callidium frigidum Casey
Sarosesthes fulminans (Fabricius)
Xylotrechus undulatus (Say)
**X. sagittatus* (Germar)
X. colonus (Fabricius)
Neoclytus muricatulus (Kirby)
Clytus ruricola (Olivier)

LAMIINAE

**Monochamus scutellatus* (Say)
M. notatus (Drury)
M. oregonensis Le Conte
Monochamus sp.
**Microgoes oculatus* (Le Conte)
Aegoschema modestum (Gyllenhal)
Hyperplatys aspersa (Say)
**Graphisurus fasciatus* (De Geer)
Neacanthocinus obsoletus (Olivier)
N. pusillus (Kirby)
Pogonocherus penicillatus Le Conte
P. mixtus Haldeman
**Oplosia nubila* (Le Conte)
Saperda calcarata Say
S. discoidea Fabricius
S. vestita Say
Oberea schaumii Le Conte
O. tripunctata (Swederus)
**Tetraopes tetrophthalmus* (Forster)

Miscellaneous - Cerambycids and Scolytids

→ Galford, J.R. 1969a. Artificial rearing of 10 species of wood-boring insects. U S FOR SER RES NOTE NE-102, 6 pp.

COMPOSITION Diet A (g): agar, 32.0; sucrose, 16.0; fructose, 8.0; glucose, 8.0; Vanderzant's fortification mixture, 12.0;

brewers' yeast, 40.0; soybean protein, 16.0; Wesson's salt mixture, 20.0; cholesterol, 0.80; Kretschmer wheat germ, 20.0; vegetable lecithin, 0.80; vitamin B_t, 0.08; sorbic acid, 2.0; methyl paraben, 1.00; Alphacel, hydrolyzed, 320.0; and (ml): wheat germ oil, 4.0; water, 800.0.

Diet B (g): agar, 32.0; Kretschmer wheat germ, 120; sorbic acid, 2.00; methyl paraben, 1.00; Alphacel, hydrolyzed, 280; and (ml): water, 800.

PREPARATION See Galford* (1967a). Diet pressed to moisture content of 50-55%. Larvae inoculated to cavity in diet surface.

REARING At 30 ± 2° C and 50 ± 20% RH in darkness. Scolytid species to adults in glass plate rearing devices; larvae of cerambycid species started in glass plate rearing devices and transferred to 60 x 15 mm petri dishes full of unpressed diet at 3-4 weeks; then transferred to fresh diet at monthly intervals until pupation.

DEVELOPMENT Survival to adults, 50-94%. Species with life cycles of up to 2 years in nature reared on diet in less than 1 year. Adult appearance normal but some infertility. *M. robiniae* reared for 3 generations on diet B with no apparent decline. Days to adult emergence and % adult yield for each species as follows: CERAMBYCIDAE: *Chion cinctus* (Drury), 214, 50%; *Megacyllene caryae* (Gahan), 68, 89%; *Megacyllene robiniae* (Forster), 60, 80%; *Neoclytus caprea* Say, 92, 80%; *Neoclytus acuminatus* (Fabricius), 49, 94%; *Romaleum rufulum* (Haldeman), 205, 80%; *Xylotrechus colonus* (Fabricius), 47, 93%. SCOLYTIDAE: *Scolytus muticus* Say, 29, 72%; *Leperisinus aculeatus* Say, 25, 58%; *Orthotomicus caelatus* (Eichhoff), 20, 57%.

Chrysomelidae

Diabrotica balteata Le Conte, banded cucumber beetle

→ Creighton, C.S.; Cuthbert, E.R. Jr 1968. A semisynthetic diet for adult banded cucumber beetles. J ECON ENTOMOL 61: 337-338.

COMPOSITION Adult diet (g): Alphacel, 1.5; agar, granulated, 6.0; ascorbic acid, 1.2; brewers' yeast, 3.3; vitamin-free casein, 12.0; cornmeal, 10.0; dextrose, 6.0; sorbic acid, 0.4; soyflour, 3.3; streptomycin sulphate, 0.1; sucrose, 6.0; Wesson's salt mixture, 1.0; wheat germ, 3.3; and (ml): carotene, β (1% solution in anhydrous ether), 2.0; chlorophyllin [(Na-K-Cu) 10% solution in 95% ethanol], 0.5; 10% formaldehyde, 3.0; 4M KOH, 2.0; water, 285.0.

PREPARATION Casein dissolved in 80 ml distilled water by blending with KOH. All other solids, except streptomycin sulphate, then added and blended. Solutions of carotene, chlorophyll, and formaldehyde added, and further blended. Agar dissolved in 200 ml of distilled water in boiling-water bath, added to other ingredients, and blended again. When cooled to <70° C, streptomycin sulphate and ascorbic acid stirred into 5 ml of distilled water and added. Entire mixture then briefly blended again. Warm diet poured into petri-plates and allowed to solidify.

REARING From Cuthbert *et al.* (1968). Adults at 24° C in 30 x 30 x 61 cm wooden cages; wire screen on 5 sides, plywood sliding door on sixth. Blocks of diet on top of cage, collard foliage and sliced sweet potatoes supplied. Oviposition devices of 100 x 15 mm Nalgene petri dish bottoms filled with wet beach sand covered with filter paper and 4 layers black cheesecloth, placed @ 2/cage. Plastic tray containing collard foliage and sweet potato roots placed over dishes, changed 3 times/week.

Eggs collected daily and incubated in plastic box between layers of wet paper towelling for 3-5 days. Cheesecloth agitated in water to remove eggs which strained through unbleached calico and immersed in 0.05% NaClO for 15 minutes. 5 minutes rinse then muslin placed in crisper box on top 2.5 cm layer damp sand sterilized at 126° C and 21 lb pressure. Kept at 24° C. Before egg hatch 1 cup sprouting corn placed on sand. 5-day-old larvae washed from corn and sand, mixed with little water and poured into cheesecloth cups. 600-800 larvae in each cup. Larvae placed in food crisper with tight fitting ventilated top. Mixture of corn and vermiculite supplied in one half, damp sand in the other. 9 or 10 holes are drilled in side of flat next to sand for access by mature larvae.

DEVELOPMENT Adult survival, longevity, and reproduction on diet equal to or better than that on natural food; survival and growth of progeny on diet appeared normal. No significant change in sex attractant production by ♀♀ when fed on artificial diet. Pupation, 9 days after in rearing containers at 30° C. Pupae left 8-9 days at 24° C or 7 days at 30° C. Adults remain in soil 1-2 days before surfacing.

Diabrotica undecimpunctata howardi Barber, spotted cucumber beetle, and *Acalymma vittatum* (Fabricius), striped cucumber beetle also reared using this method.

See also: Cuthbert *et al*, (1968).

Diabrotica undecimpunctata howardi Barber, southern corn rootworm, spotted cucumber beetle

→ Guss, P.L.; Krysan, J.L. 1973. Maintenance of the southern corn rootworm on a dry diet. J ECON ENTOMOL 66: 352-353.

No significant difference between 3 diets tested.

COMPOSITION Adult diet (g/500 g diet): wheat germ, 54; casein, 63; sucrose, 48; Alphacel, 9; Wesson's salt mixture, 18; vitamin mixture (NBC Vanderzant's modification), 18; cholesterol, 0.125; then 290 g of pollen substitute or corn meal or Wheast (a fermented whey protein concentrate) added to basal ingredients.

PREPARATION Ingredients thoroughly blended in rotating table for ca. 1 h and stored at 20° C till use.

REARING In large cages. Diet on watch glasses covered with coarse wire mesh screen to prevent entrapment of insects. Replaced weekly. Water supplied in vials with cotton wicks.

DEVELOPMENT Increased oviposition from that on bean foliage. Average eggs/♀/day on *Vicia faba* foliage, 16.6 (range 0.2-45.9); on pollen substitute diet, 23.2 (8.9-39.1); on corn meal diet, 20.2 (1.1-51.2); and on Wheast diet, 18.1 (0.9-45.6).

→ Sutter, G.R.; Krysan, J.L.; Guss, P.L. 1971. Rearing the southern corn rootworm on artificial diet. J ECON ENTOMOL 64: 65-67.

COMPOSITION Larval diet, part A: distilled water, 3360 ml; agar (Difco) 58 g .

part B (g): wheat germ, 110; casein, 129; sucrose, 129; Alphacel, 55; Wesson's salt mixture, 37; methyl paraben, 4; sorbic acid, 2.5; cholesterol, 0.25; and (ml): raw linseed oil, 1.0.

part C (g): Streptomycin, 0.5; Achromycin, 0.5; and (ml): Vanderzant's vitamin mixture, 60.0; 10% KOH, 40.0; formalin, 4.0.

PREPARATION Agar dissolved in hot water and solution blended at low speed while 'B' ingredients added. Then blended for 1 minute at high speed, cooled to 40-45° C, and 'C' ingredients added. Mixed for 2 minutes, then dispensed to 1-oz plastic cups as described by Burton *et al.* (1966). Low concentration of diet necessary to prevent separation of components.

REARING Oviposition on moist cotton gauze in petri dish. Eggs washed from gauze with fine water spray on to 60 mesh screen and transferred to filter paper pad. Kept moist at 26.5° C for 6 days. 5-10 neonate larvae then placed in each 1-oz cup with fine brush. Cups capped with unlined paper lids, placed in paper trays at 26.5 ± 0.5° C and 75 ± 5% RH in total darkness for 13 days. Paper lids removed and cups placed on clay bulb pots containing 2:1 autoclaved mixture of damp vermiculite and sand. Cups discarded after 24 h and ventilated plexiglass lids placed over pots. Kept at 26.5° C and 75% RH. 400 ml water added to pot saucer every other day until emergence completed.

DEVELOPMENT Larval development completed in 13 days with 92% survival. Adult emergence 28 days after hatching, with ca. 75% survival. Five generations reared with no apparent change in morphology, egg production, or viability. Cost of rearing 1000 larvae about US$2.50, adults US$3.25, excluding labour.

See also: Hamilton (1972).

→ Rose, R.I.; McCabe, J.M. 1973. Laboratory rearing techniques for the southern corn rootworm. J ECON ENTOMOL 66: 398-400.

COMPOSITION Adult diet (g): Vanderzant-Adkisson special wheat germ diet for insects, 40.0; agar, 8.0; sliced sweet potato, 40.0; alfalfa meal, 10.0; Vanderzant modification vitamin mixture, 2.0; Fleischmann's active dry yeast, 2.0; ascorbic acid, 0.5; and (ml): 10% KOH, 5.0; Mazola corn oil, 3.0; distilled water, 390.0.

Larval diet (g): Vanderzant-Adkisson special wheat germ diet, 25.0; agar, 3.1; methyl paraben, 0.20; sorbic acid, 0.1; Streptomycin, 0.030; Tetracycline, 0.030; Penicillin "G", 0.075; Bacitracin, 0.030; Vanderzant modification vitamin mixture, 2.0; and (ml): distilled water, 215.0; 10% KOH, 2.5; formalin, 0.15; Mazola corn oil, 1.0.

PREPARATION For both diets, agar dissolved in boiling water. All ingredients added, except for vitamin mixtures, antibiotics, yeast and ascorbic acid, and blended at high speed for 5 minutes. Mixture cooled to 50° C and remaining ingredients added and blended for 2-3 minutes. Adult diet then poured into blocks and when jelled, cut into ½" squares. Larval diet squeezed into ½-oz containers at rate of 5-7 ml diet/container and allowed to harden. Stored in fridge at 4° C till use.

REARING Adults: at 24-26.7° C with 16 h photoperiod in 1' sq cages covered with 18-mesh aluminium screen. 1-4 precut sections of diet/cage, changed every 1-2 days. Number of adults/cage not given.

Oviposition: substrate of aquarium gravel 2-3 mm in diam washed in 1 mm sieve, placed in plastic dish and covered with round piece of waxed card placed so as to allow adults access to gravel, but also providing ample shade - a factor influencing oviposition. Gravel dampened daily. Eggs removed by washing with tap water through 1 mm sieve held directly above 150-mesh sieve. Eggs surface sterilized in 0.5% NaClO for 30 seconds, then immersed for 1 minute in water suspension of fungicide at 4 g Captan/2000 ml water. Transferred from sieve to Buchner funnel, rinsed with distilled water on to filter paper in funnel and washed with further 100 ml water. Filter paper and eggs transferred to covered petri dish, sealed and held at 100% RH at 24-26.7° C.

Larvae: 5 neonate larvae transferred to each larval cup. Lids with small holes for limited ventilation provided. Held in dark or heavy shade at 24-26.7° C.

Pupae: 5 in clay pot-saucers filled to within " of top with dampened mixture of 2 parts Vermiculite and 1 part clean silica sand. Autoclaved and cooled. Clean 5½" diam polyethylene funnels placed over top and small tubular screen collection cages fixed to funnel opening for adults. Mature larvae seeded on to top of vermiculite/sand mixture in batches as required. Mixture kept wet by standing pot in tray of water to depth of about ½". Held below fluorescent lighting.

DEVELOPMENT Incubation period, 9 days (range 3-10). Larval development, 20-21 days. Average pupal development, 20 days. Pre-oviposition period, 7-10 days (average 9). Generation completed in ca. 59 days. About 40% larval survival to pupation and 60% of pupal emergence, giving 25% survival rate from neonate larvae to adult.

See also: Rose (1971).

Diabrotica virgifera Le Conte, western corn rootworm

→ Singh, Z.; Howe, W.L. 1971. Feeding, longevity, and fecundity of the adult western corn rootworm fed artificial diets. J ECON ENTOMOL 64: 1136-1137.

Adult diet modified from Berger* (1963).

COMPOSITION Basal diet (ml): water, 880; 4M KOH, 18; 10% formaldehyde, 13; (g): vitamin-free casein, 126; Wesson's salt mixture, 36; sucrose, 96; methyl *p*-hydroxybenzoate, 5.49; whole wheat germ, 108; Alphacel, 18; Vanderzant's vitamin mixture, 36; agar (dissolved in 2000 ml of boiling water), 70; ascorbic acid, 13; Aureomycin, 0.5; kanamycin sulphate, 0.5; wheat germ oil, 7.09.

Three variations prepared by adding the following to 200 g preformulated basal diet: (i) 6 g corn leaf factor; (ii) 3 g corn leaf factor + 3 g corn kernel factor; (iii) 6 g kernel factor. Corn leaf factor; green air-dried field corn leaves. Corn kernel factor; commercial corn meal. Both mechanically powdered and sifted through fine screen.

PREPARATION See Berger* (1963). Ingredients blended; hot agar solution added, blended again and suspended at 10 ml/1-oz plastic cup. Refrigerated till use.

REARING Adults in 4" x 10" plastic cylinders with base of 20 x 20 mesh fibreglass screen, and stoppered by large cork. Cages placed on dampened cellucotton pads held in glass dishes at 26 ± 2° C. Diets offered in pie-shaped sections in 1" diam petri dishes. Fresh diet supplied every 24 h.

DEVELOPMENT Survival on basal diet, ca. 8 days. 5.1 eggs laid/♀/day. Addition of leaf kernel factors increased attractiveness, but oviposition response poor; further modifications needed to diet.

→ Branson, T.F.; Guss, P.L.; Krysan, J.L.; Sutter, G.R. 1975. Corn rootworms: laboratory rearing and manipulation. U S DEP AGRIC AGRIC RES SERV ARS-NC-28, 18 pp.

Three species reared:- *Diabrotica longicornis* (Say), northern corn rootworm, *D. virgifera* Le Conte, *D. undecimpunctata howardi* Barber.

COMPOSITION For larval diet for *D. undecimpunctata howardi*. See Sutter *et al.** (1973).

Agar-based artificial diets for *D. virgifera* adults described by Singh and Howe* (1971) and Cuthbert *et al.* (1968).

PREPARATION Larval diet: see Sutter *et al.** (1973).

Adult diet: for *D. virgifera* adults; See Guss & Krysan* (1973); physical properties of diet modified; ingredients milled overnight, then spread thinly on waxed paper sheets and moistened lightly by fine water mist and air dried. Diet also fed in pill form by subjecting ball milled dry diet to pressure of 2000 lbs/in^2 in pill die.

REARING Adults at 24-27° C and 30% RH with 12 h photoperiod from subdued and indirect lighting; held in 16-18 mesh cages @ 750-1000/ft^3. Fed 2 ears corn/1000 beetles, renewed twice weekly, or one leaf bundle/day. Egg production improved by feeding mixture of pollen substitute and honey. Dead adults removed daily. Water from cotton wick in glass vials. Flat beer substituted for water to prevent bacterial contamination, but wicks changed regularly to prevent fungal growth. Oviposition on crumped, damp gauze, or damp, 80 mesh soil in 100 x 15 mm plastic petri dishes @ 1/1000 adults. Fecundity reduced by subjecting adults to strong, direct light, and by keeping oviposition sites uncovered. Eggs of *D. undecimpunctata* separated from oviposition medium by washing over 60-mesh screen, sterilized for 5 minutes in 1% aqueous solution of peracetic acid containing 0.075% sodium alkylaryl-sulphonate, and incubated on damp filter paper at 22 to 25° C for 2 weeks, 5° C for 120 days, then 20 -25° C until eclosion. 5-10 1st instar larvae added to each diet cup and capped with unlined paper lids. Incubated in total darkness for 13 days at 26.5 ± ½° C and 75 ± 5% RH. At 3rd instar, cups placed on clay bulb pots containing 2:1 autoclaved mixture on damp Vermiculite and sand for pupation. After 24 h, each pot covered with ventilated plexiglass lid and incubated at 26.5° C and 75% RH on alternate days until adult emergence completed.

DEVELOPMENT *D. virgifera:* average ♀ longevity, 3 months; fecundity, 1000 eggs. 95% emergence within 90 days of oviposition; pre-oviposition period 2-4 weeks (cf. 10-20 days in *D. undecimpunctata howardi). D. howardi:* eclosion ca. 100% in 7-10 days.

Lema trilineata daturaphila Kogan & Goeden

→ Kogan, M.; Goeden, R.D. 1971. Feeding and host-selection behaviour of *Lema trilineata daturaphila* larvae (Coleoptera: Chrysomelidae). ANN ENTOMOL SOC AM 64: 1435-1448.

Basal diet for testing feeding responses of 4th instar larvae.

COMPOSITION (g): protein-free casein, 5.0; sucrose, 3.0; glucose, 2.0; Wesson's salt mixture, 1.0; cholesterol acetate, 0.05; choline chloride, 0.05; ascorbic acid, 0.1; *meso*-inositol, 0.02; agar, 3.0; ethyl cellulose, 3.0; KOH, 0.15; and (ml): casein hydrolyzate, 10.0; corn oil (Mazola), 0.50; B-vitamins, 1.0; water to 100.0.

B-vitamins (mg/ml): niacin, 6.0; calcium pantothenate, 6.0; riboflavin, 3.0; thiamine, 1.5; pyridoxine, 1.5; folic acid, 1.5; biotin, 0.12; B_{12}, 0.012.

PREPARATION Agar and cellulose dispersed in 50 ml of hot water, and slowly cooled while stirring. Other ingredients mixed in remaining 50 ml of water, and stirred into agar/cellulose mixture until homogeneous suspension obtained, then dispensed to petri dishes and refrigerated. Ethyl cellulose powder of basic diet replaced by equal amounts of whole-leaf powder or extraction residues. Cellulose powder used to prepare meridic diet incorporating aliquots of various fractions of extracts in amounts corresponding to 3 g leaf powder plus 10% to replace losses occurring during fractionation process.

REARING Specialized according to experimentation. 10 x 2 cm petri dishes used; lid with plaster of Paris layer soaked with distilled water to maintain saturated environment. Base with filter paper disc in centre on which diet disc placed on waxed paper. Five 4th instar larvae placed on each diet disc tested.

DEVELOPMENT Complete larval development achieved on basal artificial diet containing whole-leaf powder of *Datura metetloides* de Candolle.

Leptinotarsa decemlineata (Say), Colorado potato beetle

→ Hsiao, T.H.; Fraenkel, G. 1968. The role of secondary plant substances in the food specificity of the Colorado potato beetle. ANN ENTOMOL SOC AM 61: 485-493.

COMPOSITON Basal diet (g/100 ml): vitamin-free casein, 5.00; sucrose, 3.00; glucose, 2.00; corn oil containing 5.0 mg α-tocopherol, 0.50; Wesson's salt mixture, 1.00; cholesterol, 0.05; choline chloride, 0.05; ascorbic acid, 0.10; *meso*-inositol, 0.02; bacto agar, 3.00; cellulose powder, 4.00; KOH, 0.15; and (ml/100 ml diet): casein hydrolyzate, enzymatic, 10.00; B-vitamins, 1.00; water to 100 ml.

B-vitamins (mg): niacin, 1.0; calcium pantothenate, 1.0; thiamine hydrochloride, 0.25; riboflavin, 0.50; pyridoxine hydrochloride, 0.25; folic acid, 0.25; biotin, 0.02; and vitamin B_{12}, 0.002. Folic acid and biotin dissolved in alkali.

PREPARATION Agar dissolved in 50 ml hot water. Casein dissolved with KOH in remaining water, and added with other ingredients except B-vitamins, choline chloride and ascorbic acid, to hot agar solution. Micro constituents added when cooled. Diet poured into petri dishes and refrigerated. In some tests, cellulose powder replaced by 4% leaf powder or its equivalent in extracts or residues: leaf powder from fresh leaves air- or oven-dried at ca. 60° C, pulverized in Wiley Mill, sifted through 40-mesh screen, and stored in brown glass bottles at room temperature. Leaf extracts and residues, from 10 g dried leaf powder extracted in Soxhlet extractor with 150 ml petroleum ether (Skelly, F. bp 30-60°) for 24 h; repeated if extraction incomplete. Residue dried at 60° C; extract concentrated to near dryness in vacuo and refrigerated. Ether-extracted leaf powder then mixed with 50 ml of distilled water and heated in water bath for 2 h at 80° C. Mixture filtered through Buchner funnel using Whatman No. 1 filter paper. Residue re-extracted twice as above. Combined water extracts concentrated in vacuo to 25 ml and refrigerated. Residue of water extract dried at 60° C. Ether extract first mixed with cellulose powder portion of basic diet to form consistent paste. Ether evaporated and remaining ingredients added. One ml of water extract added to 9 ml basic diet. Residue from water extract calculated in dry weight of 4% leaf powder equivalent. Cellulose powder added to make up total of 4% dry powder in diet. Leaf-homogenate diet contained homogenate of 20 g fresh leaves in 100 ml of basic diet.

REARING From 4th instar larvae. Individually in 5 x 2.5 cm vial with 1 cm diam plug of diet. Kept in moist chamber at 26.5° C. Diet changed every 12 h.

DEVELOPMENT Best growth on fresh potato leaves, but addition of leaf homogenate to basal diet also gave satisfactory growth response. Growth on leaf-powder diet not as good as on fresh leaf material but superior to basal diet. Larvae on basal diet alone generally smaller with longer developmental period; but most pupated. Adults normal and produced progeny when fed on potato leaves.

See also: Hsiao (1967).

Corollary

→ Hsiao, T.H. 1972. Chemical feeding requirements of oligophagous insects. pp. 225-240. *in* Rodriguez, J.G. (Ed.), Insect and Mite Nutrition. North Holland Publishing Company, Amsterdam.

COMPOSITION (g/100 ml): vitamin-free casein, 4; glucose, 4; cellulose, 4; bacto agar, 3; corn oil, 0.5; salt mixture, 0.5; ascorbic acid, 0.1; choline chloride, 0.05; cholesterol, 0.05; KOH, 0.15; B-vitamin mixture.

salt mixture of Beck *et al.** (1968).

vitamins (mg): niacin, 1; calcium pantothenate, 1; thiamine hydrochloride, 0.24; riboflavin, 0.5; pyridoxine hydrochloride, 0.25; folic acid, 0.25; biotin, 0.02; *meso*-inositol, 20; vitamin B_{12}, 0.002.

phagostimulant mixture (% in diet): sucrose, 0.342; *l*-alanine, 0.045; γ-aminobutyric acid, 0.053; vegetable lecithin, 0.25; KCl, 0.298.

DEVELOPMENT Larval weight gain greatest when 3-4% phagostimulants in diet. Growth suboptimal compared with growth on potato leaves.

→ Wardojo, S. 1969a. Some factors relating to the larval growth of the Colorado potato beetle, *Leptinotarsa decemlineata* Say (Coleoptera: Chrysomelidae), on artificial diets. MEDED LANDBOUWHOGESCH WAGENINGEN 69-16: 75 pp.

Best of 10 diets tested.

COMPOSITION Larval diet (g): vitamin-free casein, 3.00; egg albumen, 4.00; *l*-alanine, 0.06; glycine, 0.04; *l*-proline, 0.04; *l*-serine, 0.08; sucrose, 2.00; potato starch (Scholten, Foxhol, Holland), 5.00; vegetable lecithin, 0.50; β-sitosterol, 0.036; stigmasterol, 0.027; cholesterol, 0.009; choline chloride, 0.10; *meso*-inositol, 0.04; ascorbic acid, 0.40; vitamin-A palmitate, 0.01; menadione, 0.005; Wesson's salt mixture, 0.50; agar, granulated, 2.50; Alphacel, 2.00; KOH, 0.56; inhibitor, 0.27; and (ml): linolenic acid, 0.225; linoleic acid, 0.075; oleic acid, 0.450; and (mg): B-vitamin mixture, 2.76; distilled water to 100 g.

B-vitamins (mg): niacinamide, 1; calcium pantothenate, 0.5; pyridoxine hydrochloride, 0.25; thiamine hydrochloride, 0.25; riboflavin, 0.25; folic acid, 0.25; biotin, 0.01; *p*-aminobenzoic acid, 0.25.

inhibitor solution (mg): sorbic acid, 150; methyl *p*-hydroxybenzoate, 100; streptomycin sulphate, 100.

PREPARATION Amino acids dissolved in distilled water, then heated in water bath at 100° C. B-vitamins, choline chloride, inositol, and vitamin-A palmitate separately dissolved in 1 ml distilled water. These stock solutions refrigerated for no longer than 3 weeks. Sterols and α-tocopherol acetate dissolved in mixture of oleic acid, linoleic acid, and linolenic acid by heating in water bath at 100° C. Ascorbic acid dissolved in 1 ml distilled water each time before preparing diet.

Casein dissolved in distilled water with KOH in 375 ml glass vial. Other solid ingredients added to casein solution and thoroughly stirred. Fatty acid mixture, amino acid solution, and vitamin solution added and stirred; vial then covered with aluminium foil and autoclaved at 121° C under 103 kPa for 15 minutes giving viscous dietary mass of distinctive layers. Uppermost, thin, brown layer absent on omission of ascorbic acid. Hot diet stirred continuously with electric hand mixer till temperature to ca. 60° C when poured into petri dish. Cooled to room temperature and refrigerated.

REARING At 25 ± 0.5° C and 50-70% RH with 18 h photoperiod in phytotron. 1 g cube of diet/2 cm diam glass tube. 6-8 h old larva placed on diet with fine brush. Tube plugged with non-absorbent cellulose. Transferred to clean glass tube with fresh diet daily. Pupation in moistened sand. Adults reared in pairs on same diet in petri dish.

DEVELOPMENT 50% to prepupae in 19 days. 13 adults (9 ♀, 4 ♂) obtained from 20 larvae. Mean wt ♀ and ♂, 128.4 ± 6.5 and 97.5 ± 3.2 mg respectively. Two generations reared. Aseptic rearing unsuccessful.

See also: Wardojo (1969b).

Oulema melanopus (Linnaeus), cereal leaf beetle

→ Wellso, S.G.; Connin, R.V.; Hoxie, R.P.; Cobb, D.L. 1970. Storage and behaviour of plant and diet-fed adult cereal leaf beetle, *Oulema melanopus* (Coleoptera: Chrysomelidae). MICH ENTOMOL 3: 103-107.

Modification of adult boll weevil (*Anthonomus grandis*) diet of Sterling *et al.* (1965).

COMPOSITION Adult diet (g): chlorophllin chlorophyll, 0.1; wheat germ, 19.6; vitamin-free casein, 22.8; sucrose, 32.8; Wesson's salt mixture, 6.5; brewers' yeast, 7.5; Alphacel, 10.8; egg albumen, 5.0; cholesterol, 1.5; agar, 5.0; gelatin, 3.5; and (ml): water, 430.0; Wesson's oil, 5.6; wheat germ oil, 2.5; inhibitor, 5.0; 10% KOH, 10.0; 10% formaldehyde, 5.0; Vanderzant's vitamin mixture, 8.0.

inhibitor solution; 10 g sorbic acid, 10 g methyl *p*-hydroxybenzoate, 100 ml of 95% ethanol.

PREPARATION For details see Sterling *et al.* (1965). Mixture of water, oils, KOH, inhibitor, and agar heated to 90° C, then cooled to 63° C when vitamin mixture added and blended. Dry ingredients added slowly during blending. Mixture dispensed into petri dishes, and refrigerated in plastic bags. Used within 2 weeks of preparation.

REARING Adults in 6 x 12 x 6 ft screened cage, with barley and corn seedlings as food. Transferred to bioclimatic cabinet, at 27.6° C and 60 ± 10% RH with 16 h photoperiod, in screen-topped lantern globes with diet filled petri dish as base. 150 adults in each globe with fresh diet every 2nd or 3rd day.

DEVELOPMENT Survival on plants greater than on diet under same conditions; mortality of diet-fed and plant-fed adults after 90 days, 57.8% and 15% respectively. Diet suitable for long-term maintenance of adults only.

Coccinellidae

Adonia undecimnotata Schneider

Ferran, A.; Laforge, J.P. 1975. L'alimentation artificielle des larves de la coccinelle aphidiphage *Adonia 11 notata* Schn. (Col. Coccinellidae). ANN ZOOL ECOL ANIM 7: 1-12. (In French with English summary)

COMPOSITION (g/100 ml diet): glucose, 2.5; sucrose, 2.5; and (mg): meat extract, 657; yeast hydrolysate, 657; hydrolysate acid of casein, 657; debitterized brewers' yeast, 657.

amino acids: alanine, 100; arginine, 120; cysteine, 50; glutamic acid, 220; glycine, 200; histidine, 80; isoleucine, 150; leucine, 220; lysine, 100; *dl*-methionine, 80; phenylalanine, 210; proline, 120; serine, 210; *dl*-threonine, 160; tryptophan, 50; tyrosine, 100; valine, 200.

lipids: palmitic acid, 80; stearic acid, 50; oleic acid, 200; cholesterol, 65; carotene, 100; lecithin, 100; and (ml): Tween, 1.

vitamins (mg): vitamin diet fortification mixture, 200.

salts: Wesson's salts, 500 mg; oligoelement solution with Ag, Cu, Zn, Co, Mn, Li, S, 2 ml.

miscellaneous: 70% ethanol, 2ml; 1% Nipagin solution, 1 ml; 1% formalin solution, 0.2 ml; agar, 2.5 g.

PREPARATION Stock solutions of various fractions made up, and required quantities mixed; sugars, salts, oligoelements, Nipagin, formalin, colouring agent or dye and lipids mixed respectively into nitrogenous fractions. pH adjusted to 7 with 4M KOH, then mixture heated to 45° C and agar added and dissolved. Dispensed to sterilized boxes, cooled and stored at 8° C.

REARING At 21° C and 60-70% RH with constant lighting. Diet changed daily.

DEVELOPMENT Incomplete. Adults only from 4th instar if instars 1-3 fed on *Aphis fabae* (Hom., Aphididae). These adults smaller than on control; fertile but with reduced fecundity. 1st instar larvae on diet lead to larger 2nd instars than normal, with decreased mortality at moult.

Coleomegilla maculata De Geer

Atallah, Y.H.; Newsom, L.D. 1966. Ecological and nutritional studies on *Coleomegilla maculata* De Geer (Coleoptera: Coccinellidae). I. The development of an artificial diet and a laboratory rearing technique. J ECON ENTOMOL 59: 1173-1179.

Best of 16 diets tested.

COMPOSITION Water, 100 ml; and (g): vitamin-free casein, 6.00; Wesson's salt mixture, 1.25; sucrose, 3.00; wheat germ, 4; and (ml): vitamin solution, 3.0; and (g): ascorbic acid, 0.50; Tegosept, 0.20; Aureomycin, 0.06; agar, 4.00; and water, 100 ml; and (g): soy bean hydrolyzate, 3.00; butter fat, 2.00; glycogen, 2.00; sorbic acid, 0.20; liver fraction I-20 (Nutritional Biochemicals Corporation), 3.00; liver fraction I (Nutritional Biochemicals Corporation), 3.00; and corn oil, 2.00 ml; and (g): brewers' yeast, 1.00; dextrose, 1.00; and cotton leaves extract (carotenoids and sterols), 1.00 g. Total, 237 g.

vitamins (mg/100 ml water): niacin, 100; calcium pantothenate, 100; riboflavin, 50; thiamine hydrochloride, 25; pyridoxine hydrochloride, 25; folic acid, 25; biotin, 2; and (ml): B_{12}, 0.2.

cotton-leaf extract prepared according to procedure described by Paech & Tracey (1955).

PREPARATION 50% of water and all ingredients except agar blended in above order for 3-4 minutes. Agar dissolved in remaining boiling water, and added to first mixture, blended at high speed for 30 seconds and dispensed into containers.

REARING Larvae at 27 ± 2° C and 25% RH with 24 h photoperiod in 1.6 x 6.0 cm vials with diet 1-1.5 cm deep. 20 adults/1-pint carton with food vial and distilled water vial plugged with absorbent cotton. For oviposition one pair or mated female/box. Eggs laid on side of feeding tube, removed before eaten. Neonate larvae placed singly in feeding vials plugged with absorbent cotton. Vials inverted to reduce contamination, decay and desiccation of diet; also in this position faeces accumulate on cotton plug, instead of diet. Pupation in vial and 90% survival to adult in 21-37 days.

DEVELOPMENT Eight generations reared with no reduction in culture viability after 2nd generation.

See also: Attallah & Killebrew (1967).

Epilachna varivestis Mulsant, Mexican bean beetle

→ Kogan, M. 1971. Feeding and nutrition of insects associated with soybeans. I. Growth and development of the Mexican bean beetle, *Epilachna varivestis*, on artificial media. ANN ENTOMOL SOC AM 64: 1044-1050.

Best of 3 tested.

COMPOSITION Larval and adult diet (g): bacto-agar, 0.2; Alphacel, 3.0; vitamin-free casein, 5.0; Wesson's salt mixture, 1.0; choline chloride, 0.05; *meso*-inositol, 0.10; cholesterol acetate, 0.05; Vanderzant's B-vitamin mixture, 2.0; ascorbic acid, 0.10; methyl *p*-hydroxybenzoate, 0.4; Aureomycin, 0.004; and (ml): casein hydrolysate (suspension of 10 g hydrolysate in 100 ml water), 10; corn oil, (Mazola), 1; 10% formaldehyde, 1; water, 100. Three concentrations of sucrose tested, 0.1 M giving best results. pH adjusted to 7.0 with KOH.

PREPARATION See Berger* (1963) and Shorey & Hale* (1965).

REARING At 27 ± 2° C and ca. 50% RH. Rearing units illustrated. Prepared from fifteen, 55 x 15 mm plastic petri dishes glued to 20 x 30 cm plates of 4 mm thick Perspex in 3 rows of 5 dishes each. Plates stacked so that cellucotton wadding layer sandwiched between rims of rows of dishes and flat bottom of adjacent Perspex plate. Drops of diet placed on wadding in centre of area limited by each of 15 dishes of unit, and larvae placed in each cell thus formed.

DEVELOPMENT Complete development to normal adults, which laid viable eggs when fed soybean leaves. Adults maintained on diet for several weeks, but oviposition rate dropped sharply after few days.

Harmonia axyridis (Pallas)

→ Okada, I.; Matsuka, M. 1973. Artificial rearing of *Harmonia axyridis* on pulverized drone honey bee brood. ENVIRON ENTOMOL 2: 301-302.

COMPOSITION & PREPARATION Larvae or pupae of drone *Apis mellifera* blended and freeze-dried in vacuo, then pulverized. This lyophilized diet stored at room temperature till use.

REARING At 25°C. Diet scattered in 26 x 32 x 4.5 cm plastic containers; water supplied on moist sponge. 100 1st instars introduced and container covered with cardboard. Water and food changed when needed. Rate of cannibalism low so possible to mass rear.

DEVELOPMENT Total developmental period, 17-32 days; average adult emergence, 35%; average duration of egg laying, 19 days. Many generations reared.

See also: Okada (1970); Okada *et al.* (1971, 1972, 1973).

→ Matsuka, M.; Okada, I. 1975. Nutritional studies of an aphidophagous coccinellid, *Harmonia axyridis*. (I). Examination of artificial diets for the larval growth with special reference to drone honeybee powder. BULL FAC AGR TAMAGAWA UNIV 15: 1-9.

Best of several diets tested.

COMPOSITION Larval diet (parts): liver, yeast and sugar, 1:1:1.

REARING At 25 ± 2°C in darkness, in 9 cm petri dishes.

DEVELOPMENT Developmental period, 21.0 days; pupal wt, 33.3 mg.

See also: Matsuka *et al.* (1972).

Miscellaneous - Coccinellids

→ Smirnoff, W.A. 1958. An artificial diet for rearing coccinellid beetles. CAN ENTOMOL 90: 563-565.

COMPOSITION (g): agar, 1.3; cane sugar, 16; honey, 6; hot water, 100; royal jelly, 4.5; alfalfa flour yeast, 0.5; pulverized dry insects (natural prey of the species), 2.

PREPARATION Agar dissolved with sugar and honey in hot water, cooled to 35-38°C. Royal jelly added to 20 ml of this solution, and stirred until homogeneous emulsion; alfalfa flour, yeast, and pulverized dry insects added, stirred vigorously and cooled to 5°C for storage. For some species supplementation with 3 parts beef jelly and 1 part royal jelly required.

REARING At 27-30°C and 60-80% RH in petri dishes. Food supplied as ca. 5 cm diam pellets on white paper. Gravid females need larger cages with twig of natural plant for oviposition.

DEVELOPMENT 20 species reared.

Exochomus nigromaculatus Erichson: 3 generations, adults lived 20 days; *Thea vigintiduopunctata* Linnaeus: 2 generations, adults lived 150 days; *Scymnus pallidivestis* Mulsant: 2-3 generations, adults lived 120 days; *Scymnus kiesenwetteri* Mulsant: 3 generations, adults lived 180-210 days; *Pharoscymnus ovoideus* Sicard: 3 generations, adults lived 15-20 days; *Mycetaea tafilaletica* Smirnoff: 2 generations, adults lived 120-140 days.

Also reared: *Coccinella septempunctata* Linnaeus; *Harmonia doublieri* Mulsant; *Harmonia conglobata* Linnaeus; *Rhizobius lophantae* Blaisdell; *Rhizobius litura* Fabricius; *Rodolia cardinalis* (Mulsant); *Exochomus anchorifer* Allard; *Exochomus nigromaculatus* var. *nigripennis* Erichson; *Exochomus quadripustulatus* Linnaeus var. *floralis* Notsch; *Scymnus suturalis* Thungberg; *Stethorus punctillum* Weise; *Chilocorus bipustulatus* Linnaeus; *Clitostethus arcuatus* Rossi; *Pharoscymnus numidicus* Pic.

All species developed more rapidly than under natural conditions, and life span for many, much longer.

→ Smith, B.C. 1965. Effects of food on the longevity, fecundity and development of adult coccinellids (Coleoptera: Coccinellidae). CAN ENTOMOL 97: 910-919.

Many moist and dry foods tested.

COMPOSITION Moist foods: diet 1. fresh, minced pork liver.

Diet 2. Banana diet (%): banana, 90; casein, 9; wheat germ, 0.9; B-vitamins, 0.1.

B-vitamins (mg/10 g diet): niacinamide, 5.0; calcium pantothenate, 2.5; thiamin hydrochloride, 1.5; riboflavin, 1.2; pyridoxine hydrochloride, 1.2; folic acid, 0.1; biotin, 0.02; vitamin B_{12}, 0.002.

Dry foods: diet 1. (%): Difco egg-meat, 40; yeast, 30; sucrose, 30.

Diet 2 (%): casein, 40; yeast, 10; sucrose, 50.

Diet 3 (%): Difco liver, 40; yeast, 10; sucrose, 50.

Diet 4: protein diets (a): (%): desiccated beef, 40; sucrose, 55; cholesterol, 0.3; McCollum and Davis' salt mixture, 1.5; ribose nucleic acid, 1.5; wheat germ oil, 1.5; choline chloride, inositol chloride and B-vitamins, less than 1.

(b): as (a) except desiccated beef replaced by egg albumen with 5% isoleucine.

(c): as (a) except desiccated beef replaced by desiccated liver.

(d): as (a) except desiccated beef replaced by yeast.

(e): as (a) except desiccated beef replaced by Alphacel.

REARING At 21.9 ± 0.7°C and 65% RH with constant lighting. Adults in petri dishes, dry food and water provided separately.

DEVELOPMENT *Coleomegilla maculata lengi* Timberlake only species to lay many eggs. Survival, 400 days; oviposition when fed fresh liver, banana diet, and dry mixtures 1 and 3.

Following 12 species more selective feeders.

Adalia bipunctata (Linnaeus) accepted fresh liver and dry mix 1. Eggs laid on banana diet and dry mix 3; longevity, 66.4 days.

Adalia frigida (Schöenherr) accepted banana diet, dry mix 1 and 3. Eggs laid on dry mix 2; longevity, 121.4 days.

Anatis mali (Say) accepted dry mix 1 and 2; eggs laid on fresh liver, banana, dry mix 3; longevity, 52.6 days.

Coccinella novemnotata Herbst accepted banana diet; longevity, 21.8 days.

Coccinella transversoguttata richardsoni Brown accepted banana diet; longevity, 66.8 days.

Coccinella trifasciata perplexa Mulsant accepted all diets; longevity, 55.1 days.

Cycloneda munda (Say) accepted fresh liver; eggs laid on banana diet; longevity, 53.7 days.

Hippodamia convergens (Guerin-Meneville) accepted dry mix 1; eggs laid on fresh liver, banana, dry mix 2; longevity, 134.3 days.

Hippodamia glacialis glacialis (Fabricius) accepted dry mix 1; eggs laid on banana diet; longevity, 120.6 days.

Hippodamia parenthesis (Say) accepted all diets; longevity, 40.8 days.

Hippodamia tredecimpunctata tibialis (Say) accepted all diets; eggs laid on banana diet; longevity, 55.3 days.

Mulsantina sp. accepted banana diet, dry mix 2 and 3; longevity, 134.3 days.

Cucujidae

Ahasverus advena (Waltl), foreign grain beetle

→ Hill, S.T. 1964. Axenic culture of the foreign grain beetle, *Ahasverus advena* (Waltl) (Coleoptera, Silvanidae), and the role of fungi in its nutrition. BULL ENT RES 55: 681-690.

COMPOSITION (parts): rolled oats, 9; wheat feed, 9; dried yeast powder, 1.

PREPARATION Ingredients mixed well, autoclaved at 10 lb/sq in and 115°C for 20 minutes in 3 x 1" specimen tubes with aluminium caps, then dried for 1 h at 45°C in oven. Transferred aseptically to test tubes.

REARING At 30°C and 80% RH. Special apparatus for axenic rearing described and illustrated. Apparatus made from 350 ml glass jar with loosely fitting metal cap for aeration. Porous sintered polythene base of culture tube containing food, and larvae placed in jar over KOH solution to maintain humidity. Jars sterilized by dry heat and culture tubes by propylene oxide vapours; diet also sterilized by propylene oxide. Jars assembled aseptically and kept for 2 days at constant temperature before eggs, sterilized in 1% CaOCl for 1 minute and rinsed 5 times in sterile distilled water, introduced @ 100 eggs/jar.

DEVELOPMENT Egg to adult, 21-25 days under axenic conditions. Six generations reared. Development not improved by addition of vitamins to diet.

Oryzaephilus mercator (Fauvel), merchant grain beetle

Kaul, S.; Saxena, S.C. 1975a. Pantothenic acid sparing activity of ascorbic acid in the larval nutrition of *Oryzaephilus mercator* (Coleoptera). ACTA ENTOMOL BOHEMOSLOV 72: 236-238.

Basic diet of Kolya & Pant* (1961) developed for *Oryzaephilus surinamensis*.

COMPOSITION (parts): glucose, 50; vitamin-free casein, 50; cholesterol, 1; McCollum's salt mixture, 2; vitamin solution.

vitamin solution (µg/g): nicotinic acid, 25; pantothenic acid, 25; thiamine, 25; riboflavin, 25; pyridoxin, 25; *p*-amino benzoic acid, 25; inositol, 150; choline chloride, 500; folic acid, 2; biotin, 0.1.

PREPARATION See Kolya & Pant* (1961).

REARING At 29 ± 2°C and 70-75% RH in jam jars. Stock culture maintained on mixture of wheat flour, wheat bran and yeast, (50:50:5). 12-h-old larvae used in experiments.

DEVELOPMENT 94% pupation in mean of 17.82 ± 0.24 days; emergence, 88%. Total developmental period, 23.66 ± 0.31 days. Average adult wt, 0.52 mg.

See also: Kaul & Saxena (1974a b, 1975b); Saxena & Kaul (1974, 1975).

Oryzaephilus surinamensis (Linnaeus), saw toothed grain beetle

G.R.F. Davis has published several papers on the development of *O. surinamensis* on artificial or synthetic diets. His first paper appeared in 1956 giving a diet that was improved in 1959. This basic diet was then used for nutritional studies and improvements in dietary composition which have been reported over the years. In 1972 he published a further improved diet which is now in use. These 3 diets are abstracted; other similar or slightly improved diets are listed in the bibliography.

Davis, G.R.F. 1956. Amino acid requirements of *Oryzaephilus surinamensis* (L.) (Coleoptera: Silvanidae) for pupation. CAN J ZOOL 34: 82-85.

COMPOSITION (g): *dl*-alanine, 0.146; *l*-arginine hydrochloride, 0.057; *l*-aspartic acid, 0.081; *l*-cystine, 0.004; *l*-glutamic acid, 0.309; glycine, 0.066; *l*-histidine hydrochloride, 0.040; *dl*-isoleucine, 0.169; *l*-leucine, 0.128; *l*-lysine hydrochloride, 0.101; *dl*-methionine, 0.090; *dl*-phenylalanine, 0.133; *l*-proline, 0.103; *dl*-serine, 0.204; *dl*-threonine, 0.101; *l*-tryptophan, 0.008; *l*-tyrosine, 0.089; *dl*-valine, 0.172; dextrin, 8.000; inositol, 0.005; choline chloride, 0.005; thiamine hydrochloride, 0.0005; riboflavin, 0.0005; nicotinic acid, 0.0005; pyridoxine hydrochloride, 0.0005; calcium pantothenate, 0.0005; *p*-aminobenzoic acid, 0.0005; cholesterol, 0.042; ergosterol, 0.042; nucleic acid, 0.050; folic acid, 0.00008; McCollum-Davis' salt mixture, 0.168; powdered cellulose, 0.100; linoleic acid, 0.060; α-tocopherol, 0.0031; and biotin, 0.025 µg.

Pablum mixed cereal (Mead Johnson and Company of Canada Ltd, Belleville, Ontario) used as diet for control larvae.

PREPARATION Components mixed in mortar and ground in ball mill for 12-18 h to ensure thorough mixing. Diet sprinkled in thick layer on moist filter paper clamped in embroidery loops and dried. Circles large enough to fit bottoms of 3 x 1½" diam rearing jars cut from dried preparation.

REARING At 32° C and 75% RH. 20 larvae placed on food in each jar. Mouth covered with unbleached cotton. Adults kept in flour, enriched with 3% wheat germ. Eggs removed every 24 h, sifted free of all flour and frass and put into covered petri dishes to hatch.

DEVELOPMENT Poor larval survival on diet compared with Pablum. Omission of racemic mixture of alanine from synthetic diet, however, resulted in significant increase in larval survival. Survival on Pablum ca. 86%; on synthetic diet without *dl*-alanine, 65%; and on complete synthetic diet, 9.3%. With no alanine, pupation took about twice as long as on Pablum.

→ Davis, G.R.F. 1959a. Alanine and proline in the diet of larvae of *Oryzaephilus surinamensis* (L.) (Coleoptera: Silvanidae). ANN ENTOMOL SOC AM 52: 164-167.

COMPOSITION (mg/g): *l*-amino acids: alanine, 0.7; arginine hydrochloride, 6.7; aspartic acid, 11.4; cystine, 0.6; glutamic acid, 37.4; glycine, 3.5; histidine hydrochloride, 5.2; isoleucine, 12.1; leucine, 16.2; lysine, 13.7; methionine, 5.6; phenylalanine, 10.2; proline, 0.7; serine, 11.0; threonine, 7.3; tryptophan, 2.1; tyrosine, 10.4; valine, 12.5; dextrin, 771.5; cholesterol, 8.1; McCollum-Davis' salt mixture, 16.2; ribose nucleic acid, 4.8; folic acid, 0.2; inositol, 0.5; choline chloride, 0.5; thiamine hydrochloride, 0.05; riboflavin, 0.05; nicotinic acid, 0.05; pyridoxine hydrochloride, 0.05; calcium pantothenate, 0.05; *p*-aminobenzoic acid, 0.05; linoleic acid, 5.2; α-tocopherol, 0.5; and (μg): biotin, 0.096; vitamin B_{12}, 0.116; and (ml): distilled water, 0.10.

PREPARATION Amino acids, dextrin, salt mixture, RNA and folic acid mixed in powdered form in mortar. Vitamin-B components added in distilled water and α-tocopherol mixed with linoleic acid and added to other components. All ground in mortar until finely granular.

REARING At 32° C and 60 ± 5% RH. Small amount of diet placed in 75 x 10 mm diam test tube. Neonate larva added, tube stoppered with cotton, and incubated.

DEVELOPMENT 66% larval survival to pupation in 569 h. Similar to that on casein diet.

→ Davis, G.R.F. 1972c. Quantitative requirements of the saw-toothed grain beetle, *Oryzaephilus surinamensis* for dietary glycine and *l*-threonine. J INSECT PHYSIOL 18: 1287-1294.

COMPOSITION (mg/g): *l*-amino acids: arginine (free base), 9.0; aspartic acid, 11.0; cystine, 2.0; glutamic acid, 30.0; histidine (free base), 6.0; isoleucine, 12.0; leucine, 14.0; lysine mono-hydrochloride, 13.0; methionine, 6.0; phenylalanine, 5.0; threonine (allo free), 9.0; tryptophan, 2.0; tyrosine, 10.0; valine, 11.0; dextrin (bacteriological), 725.0; McCollum-Davis' salt mixture, No. 185, 35.0; cholesterol, 8.1; guanine, synthetic, M.A., 0.4; cytosine, synthetic, M.A., 0.4; putrescine dihydrochloride, M.A., 0.2; folic acid, 0.2; choline chloride, 1.5; *i*-inositol, 1.5; *p*-aminobenzoic acid, 1.5; niacin, 0.15; calcium pantothenate, 0.15; pyridoxine hydrochloride, 0.15; riboflavin, 0.15; thiamine hydrochloride, 0.15; biotin, 1.0×10^{-4}; vitamin B_{12}, 1.2×10^{-4}; corn oil, 0.01 ml/g.

PREPARATION Ingredients mixed in pestle and mortar.

REARING Larvae reared individually from hatching to adult at 32 ± 0.5° C and 75% RH.

DEVELOPMENT Development not arrested by lack of dietary glycine; but graded concentration of glycine up to 32.0 mg/g diet increased rate of development. Dietary threonine necessary for survival and growth between 2.0 to 12.0 mg/g. Average time to pupation with 32 mg/g glycine, 512.3 ± 8.4 h, and to emergence, 603.3 ± 7.3 h. Similarly with 2 to 12 mg/g threonine, average time to pupation and emergence, 519.6 ± 12.7 to 539.4 ± 8.8 and 639.5 ± 24.3 to 650.1 ± 10.0 h respectively.

See also: Davis (1961c d, 1962, 1964, 1966a b c e, 1967a b c, 1968a d e, 1969a, 1973a b).

➔ Pant, N.C.; Fraenkel, G. 1954a. On the function of the intracellular symbionts of *Oryzaephilus surinamensis* L. (Cucujidae, Coleoptera). J ZOOL SOC INDIA 6: 173-177.

COMPOSITION (parts): casein, 50; glucose, 50; cholesterol, 1; McCollum's salt mixture, 2; and water, 10. Also (µg/g dry diet): thiamin, 25; riboflavin, 12.5; nicotinic acid, 25; pyridoxin, 12.5; pantothenic acid, 25; choline chloride, 500; inositol, 250; biotin, 0.1; folic acid, 2.5. In some tests biotin replaced by 2% water insoluble yeast.

REARING Adults reared at 29° C on wholemeal flour with 10% dried brewers' yeast. At 36° C to eliminate symbionts from culture.

DEVELOPMENT Delayed growth with all vitamins present due to lack of symbionts. Larval and pupal periods from 19-26 days.

➔ Kolya, A.K.; Pant, N.C. 1961. Effect of qualitative and quantitative vitamin deficiencies in larval development of *Oryzaephilus surinamensis* L. (Coleoptera: Cucujidae). INDIAN J ENTOMOL 23: 285-292.

Modified from Pant & Fraenkel* (1954a).

COMPOSITION (parts): casein, 50; glucose, 50; cholesterol, 1; McCollum-Davis' salt mixture, No. 185, 2; vitamin solution.

salts (g): $Ca(C_3H_5O_3)_2$, 35.15; $Ca(H_2PO_4)_2$, 14.60; K_2HPO_4, 25.78; $NaH_2PO_4.H_2O$, 9.38; NaCl, 4.67; $MgSO_4$, 7.19; $FeC_6H_5O_7$, 3.19.

vitamins (ppm): thiamine hydrochloride, 25; riboflavin, 25; pyridoxin hydrochloride, 25; nicotinic acid, 25; calcium pantothenate, 25; *p*-aminobenzoic acid, 25; folic acid, 2.0; inositol, 250; choline chloride, 500; biotin, 0.1.

PREPARATION Casein, glucose and salt mixture mixed thoroughly. Cholesterol dissolved in few ml acetone and added; acetone evaporated and vitamin solution added; mixed thoroughly, spread on petri dishes, and dried at 34 C. Thoroughly ground in mortar and pestle, and conditioned at 29° C and 70-75% RH for 48 h.

REARING At 29° C and 72-75% RH in 2" x 1" glass vials with pierced corks covered with muslin tops. 25 neonate larvae/vial with 3 g diet.

DEVELOPMENT Larval period, 20-22 days; adult emergence, 16-18%.

Curculionidae

Anthonomus eugenii Cano, pepper weevil

Toba, H.H.; Kishaba, A.N.; Pangaldan, R.; Riggs, S. 1969. Laboratory rearing of pepper weevils on artificial diets. J ECON ENTOMOL 62: 257-258.

COMPOSITION (g): agar, 23.7; casein, 33.2; sucrose, 33.2; alfalfa meal, 14.2; wheat germ, 28.5; Wesson's salt mixture, 9.5; Alphacel, 4.8; ascorbic acid, 4.0; methyl *p*-hydroxybenzoate, 1.5; sorbic acid, 1.0; Aureomycin, soluble powder, 0.13; and (ml): 10% choline chloride, 9.5; 4M KOH, 4.8; 10% formalin, 4.0; vitamin stock, 3.2; and water, 825.0.

vitamins (mg): niacin, 600.0; calcium pantothenate, 600.0; riboflavin, 300.0; thiamine, 150.0; pyridoxine, 150.0; folic acid, 150.0; biotin, 12.0; B_{12}, 1.2; in 100 ml of water.

PREPARATION See Henneberry & Kishaba* (1966a). Diets dispensed to 6-oz Lily paper cups and allowed to cool.

REARING At 26.5°C and 70% RH with 14 h photoperiod in Lily cups. 50 eggs/cup placed in or near punctures made in diet and cups covered. Emerging adults to 30 cm cube organdie covered cages with pepper seedlings wrapped in parafilm. Adults feed on seedlings and oviposit inside wrapping. Eggs collected by unwrapping parafilm, and rinsing, with seedlings, in large beaker of water. Water and eggs filtered, and eggs transferred with sable hair brush.

DEVELOPMENT Egg to adult survival, 69%. Total development, ca. 1-2 days shorter than on other diets.

Anthonomus grandis Boheman, boll weevil

Earle, N.W.; Walker, A.B.; Burks, M.L. 1966. An artificial diet for the boll weevil, *Anthonomus grandis* (Coleoptera: Curculionidae), based on the analysis of amino acids in cotton squares. ANN ENTOMOL SOC AM 59: 664-669.

COMPOSITION Larval diet (g): agar, 3.0; distilled water, 80.0; sucrose, 2.0; vitamin mixture, 0.0033; inositol, 0.02; choline chloride, 0.1; salt mixture B, 0.67; KOH, 0.15; amino acid mixture, 0.444; soy protein, 1.0; acetone powder of cotton squares, 10.0; cholesterol, 0.05; wheat germ triglycerides, 0.3; ascorbic acid, 0.1; sorbic acid, 0.1; methyl *p*-hydroxybenzoate, 0.1.

Adult diet (g): agar, 1.5; distilled water, 80; sucrose, 2.0; glucose, 1.0: fructose, 1.0; vitamin mixture, 0.0066; inositol, 0.02; choline chloride, 0.05; salt mixture B, 0.67; KOH, 0.15; amino acid mixture, 0.444; soy protein, 2.5; acetone powder of cotton squares, 10.0; cholesterol, 0.05; wheat germ triglycerides, 0.3; ascorbic acid, 0.2.

vitamins (mg/100 g diet): niacinamide, 1.0; calcium pantothenate, 1.0; riboflavin, 0.5; folic acid, 0.25; pyridoxine hydrochloride, 0.25; thiamine hydrochloride, 0.25; biotin, 0.02.

salts (mg/100 g diet): KH_2PO_4, 500; $MgSO_4$, 120; NaCl, 30; $(CH_3COO)_2Mn.4H_2O$, 5.0; $FePO_4.2H_2O$, 4.0; $(CH_3COO)_2Zn.2H_2O$, 3.0; $(CH_3COO)_2Cu.H_2O$, 1.5; $Co(NO_3)_2.6H_2O$, 0.1.

amino acids (g/100 g diet): asparagine, 0.225; threonine, 0.010; serine, 0.019; glutamine, 0.070; proline, 0.017; glycine, 0.005; alanine, 0.031; valine, 0.010; methionine, 0.004; isoleucine, 0.005; leucine, 0.013; tyrosine, 0.004; phenylalanine, 0.004; lysine, 0.011; histidine, 0.005; arginine, 0.006; tryptophan, 0.005.

PREPARATION Powder prepared by grinding whole cotton squares including bracteoles in 1-gal blender with acetone at rate of 500 g cotton/l acetone. Coarse mixture poured into muslin bag and acetone expressed. Remainder then blended twice more with 1 l portions of acetone. Acetone mostly removed by filtration and resulting powder dried at room temperature over $CaCl_2$ then deep frozen till use. Agar dissolved in hot distilled water. All other ingredients, except vitamins, lipids and cotton square powder, stirred in. Hot mixture poured into blender jar and remaining ingredients added. Square powder first impregnated with triglycerides and cholesterol by mixing to slurry in ethyl ether. Ether evaporated completely before adding to diet. Blended for 1 minute, dispensed into small glass vials or plastic petri dishes. Adult diet poured into 8 x 10 mm paraffin-coated cylinders.

REARING Larvae: eggs surface sterilized by treatment for 5 minutes in 1:3 ethanol-water solution with 0.02% $HgCl_2$ and placed individually in diet tubes. Aluminium foil caps applied; maintained at 27° C with 13 h photoperiod.

Adults: newly emerged adults placed in 1-pint icecream cartons covered with thin polyethylene sheet. Fed twice daily with 1 diet cylinder/3 weevils and maintained at 21-34° C with 13 h photoperiod. Eggs recovered by breaking open cylinders and washing contents through series of sieves. Eggs spread on black filter paper for counting.

DEVELOPMENT 80% survival to adult from eggs in vials; 50% when reared in plastic petri dishes. Mean developmental time from egg to adult, 18 days at 27° C; average weight of adults, 15 mg.

See also: Earle *et al.* (1959, 1965, 1967, 1970); Lambremont *et al.* (1967); Nettles & Betz (1966).

→ Gast, R.T.; Davich, T.B. 1966. Boll weevils. pp. 405-418. *in* C.N. Smith (Ed.) Insect Colonization and Mass Production. Academic Press, New York and London.

COMPOSITION Larval diet (g): sucrose, 7.0; agar, 10.5; brewers' yeast, 3.5; Wesson's salt mixture, 1.0; soy bean protein, 10.5; acetone-extracted square powder, 17.5; solvent extracted cottonseed meal, 5.0; choline chloride, 0.25; cholesterol, 0.3; methyl *p*-hydroxybenzoate, 0.3; potassium sorbate, 0.4; and (ml): B-vitamin solution, 3.5; water, 350.

mould inhibitor prepared by dissolving potassium sorbate in 20 ml of sterile water and methyl *p*-hydroxybenzoate in 10 ml of ethanol.

Adult diet (g): acetone extracted cotton squares, 4.00; soybean protein, 4.00; sucrose, 4.00; agar, 3.00; Wesson's salt mixture, 0.75; ascorbic acid, 0.40; cholesterol, 0.05; germinated cotton seed, 10.00; water, 100; and (ml): B-vitamin mixture, 1.00.

B-vitamins (g): niacinamide, 1.00; calcium pantothenate, 1.00; riboflavin, 0.50; pyridoxine hydrochloride, 0.25; thiamine hydrochloride, 0.25; folic acid, 0.25; biotin, 0.02; B_{12}, 0.002; choline chloride, 120.00; and water, 1000.00 ml.

PREPARATION Cotton seeds prepared as follows: seeds soaked for 24 h in two changes of water. Small numbers of seeds placed on wet paper towelling and covered with wet cloth. For ≥50 lb, seeds placed on screen-bottom trays in layers ½" deep and soaked. Trays then stacked to allow air circulation, and covered with plastic sheet to keep RH near saturation. At 85° F, seed develops 1" long root within 24 h. Germinated seeds autoclaved for 3 minutes at 20 lb pressure. Roots, (slightly toxic to weevils) removed by placing autoclaved seed on 6-mesh screen and washing roots away with strong water jet. Paste-like pulp removed from seed hulls by means of two 6" diam rollers; pulp sticks to roller surface and scraped off. Hull pieces discarded. Paste frozen until needed. Cotton square powder prepared by grinding fresh cotton squares in acetone (see Earle *et al.** 1966). For adult diet, all ingredients, except ascorbic acid and vitamins, blended and autoclaved at 240° F; then cooled to 200° F, vitamins and ascorbic acid added, mixed, and dispensed to mould to harden. Fecundity influenced by shape of hardened diet; short ¼ - ⅜" diam cylinder, most satisfactory. Pellets coated with wax to prevent desiccation. For larval diet all dry ingredients, except cholesterol and mould inhibitors added to boiling water in blender. Cholesterol dissolved in hot ethanol, and added. Diet autoclaved for 10 minutes at 20 lb pressure. Inhibitors and B-vitamin mixture added, stirred, and dispensed to sterile vials or petri dishes to solidify.

REARING Adults at 85-87° F and 50-60% RH in 13" square x 3" high stainless steel oviposition cages with sliding door and 14 mesh stainless steel floor. Fresh diet added twice daily in screen mesh food container. Underfloor lighting to keep adults on food. Eggs dissected from diet with scalpel, or diet washed through sieve into NaCl solution. Eggs surface sterilized for 2 h in 0.75% formaldehyde, then 0.2% formaldehyde until inoculation. Best results from hand inoculation into hole punched in 1 ml diet in 2-dram vials. Vials capped with aluminium foil. Or, inoculated on to 9 cm petri dishes with diet ¼" deep. Automated methods also described. Incubated at 85° F and 50-60% RH. Larval dishes opened and placed in large, shallow, light-tight emergence box, with several screen containers at one end. Adults move to screen traps.

DEVELOPMENT Adult emergence, 13-14 days from inoculation. Adult yield from vials, 70-80%; from petri dishes, 40-60%. Maximum oviposition, 7-8 days from emergence.

See also: Brazzel *et al.* (1959); Gast, R.T. (1961, 1966a b); Gast & Landin (1966); Griffin & Lindig (1973, 1975); Gueldner *et al.* (1975); Hudsepeth *et al.* (1969); Lindig & Malone (1973); Stewart *et al.* (1972); Tingle *et al.* (1971).

➡ McKibben, G.H.; Hedin, P.A.; Davich, T.B.; Daum, R.J.; Laseter, M.W. 1971. Addition of food acidulants to increase attractiveness to boll weevils of bait containing cottonseed oil. J ECON ENTOMOL 64: 583-585.

Bait formulation for attracting adults.

COMPOSITION (% by weight): sucrose, 10.00; cotton oil "foots", 20.50; Dacagin, 0.75; hydroxethyl cellulose (QP-4400), 0.50; glycerol, 4.00; Calco Oil Red N-1700 dye, 2.30; $Na_3PO_4.12H_2O$, 1.00; water, 60.95.

"Foots" are byproducts of cottonseed oil manufacture. Hydroxethyl cellulose and Dacagin are both water soluble thickeners.

PREPARATION Glycerol, sugar, hydroxethyl cellulose, Dacagin, $Na_3PO_4.12H_2O$, and dye added to water and agitated. When well dispersed, thickened "foots" added. "Foots" added before water solution thickens to ensure thorough blending and emulsification. Mixed for 5-10 minutes until paste formed which does not run on plants but is easily dispensed in field by spray equipment. Calco Oil Red N-1700 dye added to bait giving simple way of measuring ingestion.

BIOASSAY Drop of dyed bait applied with 5 ml syringe to each corner of 2 x 2" glass slides. 2 h later each slide placed in petri dish with 20, 3-6-day-old ♂ weevils for 3 h. Dishes placed in freezer to stop feeding activity. Addition of food acidulants (fumaric, malic, phosphoric, citric, adipic, tartaric, lactic, and succinic acids) caused significant increase (41-67%) in ingested food compared with standard unfortified formulation.

See also: Lloyd *et al.* (1968); McKibben *et al.* (1972).

→ Moore, R.F. Jr; Whisnant, F.F.; Taft, H.M. 1967. A laboratory diet containing egg albumin for larval and adult boll weevils. J ECON ENTOMOL 60: 237-241.

COMPOSITION Adult diet (g/100 g diet) group I: egg albumin, 3.00; Alphacel, 4.00; Wesson's salt mixture, 0.75; ascorbic acid, 0.10; sugar, 5.00; group II: cholesterol, 0.05; lipid mixture, 1.00; cotton leaf powder, 0.50; agar, 3.00; and (ml/100 g diet): vitamin mixture, 2.00; inhibitors, 0.5; water, 85.0.

lipids added (g): tripalmitin, 30; tristearin, 18; and triolein, 60 to 150 g of 1:1 mixture of Wesson's oil and linseed oil. 1:1 mixture of oils alone equally satisfactory.

leaf powder prepared by drying cotton leaves in sun or in mechanical convection oven at 60-70° C and then grinding through 1 mm Wiley Mill screen.

vitamin mixture described by Vanderzant* (1959). However, sufficient choline chloride added to give 0.1 g/100 g of diet.

inhibitors: 15 g methyl parasept and 20 g potassium sorbate in 170 ml of aqueous solution of 25% acetone.

Larval diet: as adult diet except 0.5 g, not 1.0 g of lipid mixture, and 1.0, not 0.5 ml of microbial inhibitor solution used.

PREPARATION Ingredients of group I mixed thoroughly. Group II ingredients dissolved in ethyl ether (10 ml/100 g of diet), and added to group I components. After evaporation of ether, vitamins and 20 ml of water/100 g diet added to dry ingredients forming slurry necessary to prevent lumping when warm agar added. Agar heated to boiling in remaining water, cooled to 75° C, added to slurry, and stirred in kitchen mixer; respective amounts of inhibitor for larval and adult diets added. For larvae, ca. 25 g diet poured into sterile plastic petri dishes, capped and dried for 7 days at room temperature before eggs implanted. Adult diet poured into moulds.

REARING At 28° C and 50 ± 5% RH with 24 h photoperiod. 400 adults in clear styrene, plastic, cylindrical containers, 7 x 8" diam with 4" screened opening in base and ¾" legs. Adult diet poured into ½ x 3 x 16" moulds from aluminium bar (illustrated). Two trays contained 150 pellets sufficient for 400 weevils in cage for 24 h. Eggs laid in diet, removed by forcing diet through 14 mesh sieve and washing with water through 20, 30 and 40 mesh sieves; eggs placed in 18% $CuSO_4.5H_2O$ solution and transferred to 250 ml separatory funnel. Floating eggs sterilized with 0.3% Roccal, then transferred to 4-5-days-old diet in petri dishes.

DEVELOPMENT Adult emergence, 16-18 days from inoculation giving 45 ± 1.5% yield and weighing average of 12 mg.

See also: Ludvik (1972).

→ Sterling, W.L.; Adkisson, P.L. 1966. An artificial diet for laboratory cultures of boll weevil larvae and adults. J ECON ENTOMOL 59: 1074-1077.

COMPOSITION (g/100 g): cottonseed meal, 13.8; egg albumen, 0.8; sucrose, 1.7; Alphacel, 1.3; brewers' yeast, 1.3; cholesterol, 0.01; Polyotic (contains as active ingredient 25 g/lb of tetracycline hydrochloride), 0.7; agar, 1.3; and (ml/100 g of diet): water, 70.8; 1% KOH, 2.4; special vitamins, 2.6; inhibitor, 1.7; wheat germ oil, 0.4; Wesson's oil, 0.9.

vitamins: special vitamins made by adding following to 60 ml of Vanderzant* (1959) vitamin solution: distilled water, 90.0 ml; and (g): ascorbic acid, 20.0; choline chloride, 5.0; and α-tocopherol, 0.6.

inhibitor: 10 g of sorbic acid and 10 g of methyl *p*-hydroxybenzoate in 100 ml of 95% ethyl alcohol.

PREPARATION Mixture of water, oils, KOH, inhibitor, and agar heated to 90° C, and cooled to 63° C; vitamin mixture added and blended. Dry ingredients added slowly during blending.

REARING At 80 ± 5° F with 24 h photoperiod. 300 adults/6" cube oviposition cage. Larval diet made into pellets for adult food and oviposition. Pellets 10 mm diam x 15 mm long with 1 rounded end. Eggs laid through paraffin shell of pellets, and removed by washing through series of sieves. Finally eggs and small debris placed in separatory funnel filled with NaCl solution. Eggs then separated and collected for inoculation.

DEVELOPMENT Six generations reared. Mean adult wt, 18 mg; within 4 generations fecundity increased from average of 196 to 751 eggs/♀. Necessity of preparing separate diets for adult and larval rearing eliminated on this diet.

See also: Sterling *et al.* (1965).

→ Vanderzant, E.S.; Davich, T.B. 1958. Laboratory rearing of the boll weevil: a satisfactory larval diet and oviposition studies. J ECON ENTOMOL 51: 288-291.

COMPOSITION (g/100 g): soybean protein (Drackett assay C-1), 5; *l*-cysteine hydrochloride, 0.05; glycine, 0.1; sucrose, 5; corn oil (Mazola), 0.5; cholesterol, 0.05; Wesson's salt mixture, 1; choline chloride (in 1 ml water), 0.1; powdered cellulose, 4; sodium alginate (Kelcosol), 0.5; agar, 3; bacto yeast extract (Difco), 1; and (ml): vitamin mixture, 1; water to 100 g.

vitamins (mg): niacinamide, 1; calcium pantothenate, 1; thiamin hydrochloride, 0.25; riboflavin, 0.5; pyridoxin hydrochloride, 0.25; folic acid, 0.25; biotin, 0.02; vitamin B_{12}, 0.002. Folic acid and biotin dissolved in alkali.

PREPARATION Constituents thoroughly mixed in Waring Blendor; pH to 6.5 with 2M KOH. Diet heated in boiling water bath until agar dissolved, then allowed to cool and harden. Plugs put in 17 x 60 mm rearing vials, or diet dispensed while hot from polyethylene squeeze bottle.

Vials plugged with cotton and sterilized at 15 lb pressure for 20 minutes. Diet cooled and hole for egg cut in diet.

REARING At 86 ± 1° F and 80-90% RH. Eggs obtained from cotton squares in dishes of wet sand. Eggs washed for 5 minutes in beaker of sterilizing solution (g): $HgCl_2$, 0.25; NaCl, 6.5; $CH_2(OH).COONa$, 0.5; and (ml): conc. HCl, 1.25; C_2H_5OH, 250; distilled water, 750. Rinsed once in sterile water and transferred aseptically with sable brushes to near hole in diet. Adults removed to 6" cube screened feeding and oviposition cages @ 30/cage.

DEVELOPMENT Five generations reared aseptically from egg to adult. Adults oviposited on cotton squares, with average production of 3 eggs/♀/day, with 7 eggs maximum. Cotton plant extracts necessary in diet for oviposition.

See also: Vanderzant (1963a b).

Corollary

→ Vanderzant, E.S. 1959. Inositol: an indispensable dietary requirement for the boll weevil. J ECON ENTOMOL 52: 1018-1019.

COMPOSITION Larval diet (g): vitamin-free casein, 5.0; *l*-cysteine hydrochloride, 0.1; sucrose, 5.0; corn oil (Mazola), 0.25; cholesterol, 0.05; Wesson's salt mixture, 1.0; choline chloride, 0.1; inositol, 0.1; sodium alginate, 0.5; agar, 3.0; and (mg): $Zn(C_2H_3O_2)_2.2H_2O$, 0.5; $CoCl_2.6H_2O$, 0.25; $Na_2MoO_4.2H_2O$, 0.25; niacinamide, 1.0; calcium pantothenate, 1.0; thiamine hydrochloride, 0.25; riboflavin, 0.5; pyridoxine hydrochloride, 0.25; folic acid, 0.25; biotin, 0.02; vitamin B_{12}, 0.002. pH to 6.5 with 2M KOH and water added to 100 g.

DEVELOPMENT Best when 60 mg inositol added/100 g diet, giving larval period of 10 days; adult average wt, 9.1 mg; and adult yield, 74%.

→ Vanderzant, E.S.; Richardson, C.D.; Davich, T.B. 1959. Feeding and oviposition by the boll weevil on artificial diets. J ECON ENTOMOL 52: 1138-1143.

Two adult diets.

COMPOSITION Diet A (g): enzymatic hydrolyzate casein, 2; glucose, 2; corn oil, 0.5; cholesterol, 0.1; Wesson's salt mixture, 0.5; choline chloride, 0.05; ascorbic acid, 0.2; agar, 2; (ml): water, 100; vitamin solution, 2.

Diet B (g): enzymatic hydrolyzate of casein, 3; glucose, 2; corn oil (Mazola), 0.25; cholesterol, 0.05; Wesson's salt mixture, 0.75; choline chloride, 0.1; ascorbic acid, 0.05; agar, 2; and (ml): water, 100; vitamin solution, 1.

vitamins: see Vanderzant* (1957).

Wheat germ diet (g): wheat germ, 4; enzymatic hydrolysate casein, 2; glucose, 1; Wesson's salt mixture, 0.75; choline, 0.1; cholesterol, 0.05; agar, 2; and (ml): water, 100; vitamin solution, 1.

PREPARATION Diet A: constituents thoroughly mixed; pH to 6.5; sterilized in autoclave at 121° C for 15 minutes. Poured while hot into ½ x 1" diam aluminium foil dishes @ 2 ml diet/dish. When solidified, dishes dipped in molten paraffin. Paraffin punched with hot needle before placing in cage.

Diet B: basal mixture heated in water bath until agar dissolved. Test substances added and heated for further 10 minutes. Diets poured aseptically into sterile aluminium dishes and refrigerated.

Wheat germ diet: constituents blended, heated in boiling water bath for 10-15 minutes and poured into dishes.

REARING At 86°F with 12-14 h photoperiod. Eggs collected, washed and transferred to rearing vials with diet as in Vanderzant & Davich* (1958). 20-30 newly emerged weevils/cage. 1 dish diet/cage changed daily.

DEVELOPMENT 2 g mixed pollen/100 g adult basal diet B required for oviposition and average adult wt, 10 mg. On wheat germ diet, adult wt, 9 mg. If larval diet supplemented with 2 g pollen or wheat germ or 10 g of germinated cottonseed, adult wt, 12 mg.

See also: Vanderzant & Richardson (1964).

→ Vanderzant, E.S. 1963a. Nutrition of the adult boll weevil: oviposition on defined diets and amino acid requirements. J INSECT PHYSIOL 9: 683-691.

Three diets developed for adult (oviposition included).

COMPOSITION Diet A (g): casein, vitamin-free, 2.0; amino acid mixture, 1.89; fructose, 3.0; glucose, 1.0; corn oil (Mazola), 0.50; Tween 80, 0.05; cholesterol, 0.05; α-tocopherol, 0.01; vitamin D_3, 0.0005; β-carotene-protein complex (Hoffman-LaRoche), 0.025; inositol, 0.04; choline, 0.10; ascorbic acid, 0.20; Wesson's salt mixture, 0.75; agar, 3.0; and (ml): B-vitamins, 2.0; cobalt, molybdenum, and zinc salts, 0.5; water to 100.0; pH to 6.2 with 10% KOH.

Diet B: as above, except that casein omitted, corn oil reduced to 0.25 g and 2.0 g of dextrin added.

Diet C as B, except that amino acids doubled to 3.78 g.

l-amino acids (g/100 g): arginine, 0.150; histidine hydrochloride, 0.075; isoleucine (allo-free), 0.113; leucine, 0.175; lysine hydrochloride, 0.113; methionine, 0.075; phenylalanine, 0.150; threonine, 0.100; tryptophan, 0.075; valine, 0.138; alanine, 0.150; aspartic acid, 0.100; glutamic acid, 0.125; glycine, 0.150; proline, 0.050; serine, 0.050; cysteine hydrochloride, 0.050; tyrosine, 0.050.

For minerals and B-vitamin mixture see also Vanderzant* (1959) (mg): $Zn(C_2H_3O_2)_2.2H_2O$, 0.5; $CoCl_2.6H_2O$, 0.25; $Na_2MoO_4.2H_2O$, 0.25; niacinamide, 1.0; calcium pantothenate, 1.0; thiamine hydrochloride, 0.25; riboflavin, 0.5; pyridoxine hydrochloride, 0.25; folic acid, 0.25; biotin, 0.02; vitamin B_{12}, 0.002.

PREPARATION Amino acids mixed in aqueous solution to 25 ml/100 ml diet and frozen till use. All ingredients except sugars and ascorbic acid mixed and placed in boiling water bath until agar dissolved. Mixture cooled to 40-45°C, sugar and ascorbic acid added, and diet poured immediately into mould.

REARING At 29 ± 1°C with 13.5 h dim light in 6" cube screened cages, @ 8 pairs/cage. Fed twice daily with diet cylinders, 1 plug/6-8 weevils.

DEVELOPMENT Nine generations reared on Diet A, average number of eggs laid after 20 days, 124/♀.

See also: Vanderzant & Chremos (1971).

→ Vanderzant, E.S. 1965. Axenic rearing of the boll weevil on defined diets: amino acid, carbohydrate, and mineral requirements. J INSECT PHYSIOL 11: 659-670.

Best of two larval diets tested.

COMPOSITION (mg/100 ml): *l*-amino acids: see Vanderzant* (1963a); sucrose, 3500; linolenic acid, 20; choline chloride, 50; inositol, 20; cholesterol, 50; nicotinamide, 1.0; riboflavin, 0.50; calcium pantothenate, 1.0; thiamine hydrochloride, 0.25; pyridoxine hydrochloride, 0.25; folic acid, 0.25; biotin, 0.02; vitamin B_{12}, 0.002; K_2HPO_4, 160; $NaH_2PO_4.H_2O$, 80; NaCl, 20; $MgSO_4.7H_2O$, 50; $FeSO_4.7H_2O$, 10; $CuSO_4.5H_2O$, 0.5; $MnSO_4.H_2O$, 2.5; $Zn(C_2H_3O_2)_2.2H_2O$, 0.5; $CoCl_2.6H_2O$, 0.25; $Na_2MoO_4.2H_2O$, 0.25; KI, 0.5; $CaCO_3$, 120; and: 3.0 g agar, KOH to pH of 6.2, and water to 100 ml.

PREPARATION See: Vanderzant* (1957). Amino acids dissolved in water. Vitamin mixture, Na, K, Mg salt solutions and sugar added and pH and volume adjusted. Solution filtered through Seitz or membrane filter. Fe and other trace mineral salt solutions added just before filtration completed. Filtered portion of diet added to heat sterilized portion containing water, agar, calcium carbonate, cholesterol, and fat or fatty acids. Sometimes fatty acids also filter sterilized and added to agar portion. Diet dispensed from heated funnel into 17 x 60 mm rearing vials.

REARING Eggs to vials. From stock weevils fed casein diet and germinated cottonseed diet. See: Vanderzant & Davich (1961).

DEVELOPMENT Average weight of ♀ and ♂, 8.1 and 7.1 mg respectively; survival from larva to adult, 64%.

See also: Vanderzant (1963b, 1973a); Vanderzant & Richardson (1964).

→ Vanderzant, E.S. 1967a. Wheat-germ diets for insects: rearing the boll weevil and the salt-marsh caterpillar. ANN ENTOMOL SOC AM 60: 1062-1066.

COMPOSITION (g): wheat germ, 3.0; vitamin-free casein, 3.5; sucrose, 3.5; Wesson's salt mixture, 1.0; agar, 2.5; cholesterol, 0.05; choline chloride, 0.1; inositol, 0.04; and (ml): ascorbic acid, 0.4; B-vitamin mixture, 1.0; water, 85.0.

B-vitamin mixture according to Vanderzant* (1959).

PREPARATION Ingredients mixed and heat- and filter-sterilized. Ascorbic acid added, and diet dispensed to sterile, larval rearing vials. For adult boll weevils, diet moulded to cylindrical plugs.

DEVELOPMENT Survival of boll weevil from egg to adult with and without cholesterol, 84-91% and 47-55% respectively. Maturation of *Estigmene acrea* (Drury) (Lep; Aegeriidae) further improved by addition of 10 mg α-tocopherol, and 250 mg of linseed oil/100 g of diet giving 85% pupation with 89% adult emergence. Five consecutive generations of salt-marsh caterpillars reared.

Corollary 1

→ Childress, D.; Williams, P.P. 1973. Control of a bacterial contaminant of boll weevil diet. J ECON ENTOMOL 66: 554-555.

Modification of Vanderzant & Davich* (1958) diet for *A. grandis*.

COMPOSITION Adult diet (%): agar, 2.6; cholesterol, 0.1; cottonseed meal, 8.7; linseed oil, 0.2; methyl parasept, 1.3; potassium sorbate, 0.1; soybean protein, 2.0; sucrose, 3.6; salts, Sure No. 2, 0.4; fortification vitamin mixture (NBC), 1.8; and water, 79.2.

Bacterial medium (%): agar, 1.5; yeast extract, 1.0; ferric ammonium citrate, 0.5; glucose, 1.0; phytone, 1.0; $CH_3.COONa.3H_2O$, 0.2; Tween 80, 0.1; yeast extract, 0.5; $MgSO_4.7H_2O$, 0.02; $MnSO_4$, 0.005; and glass-distilled water. pH 6.5 with 0.5 N HCl before sterilizing. Bacterial medium minus glucose and agar served as broth medium for incorporation of alcohols and carbohydrates at 1% concentration.

Corollary 2

→ Griffin, J.G.; Lindig, O.H.; McLaughlin, R.E. 1974. Flash sterilizers: sterilizing artificial diets for insects. J ECON ENTOMOL 67: 689.

PREPARATION Flash-type sterilizer used for sterilizing large volumes of artificial diet for *A. grandis*. Processing temperatures of 130, 138 and 144° C for 30 seconds each and cooling to ca. 57° C gave satisfactory sterilization of diet; weevils reared comparable with usual standard method of laboratory preparation.

See also: Griffin & Lindig (1975).

Conotrachelus nenuphar (Herbst), plum curculio

→ Yonce, C.E.; Gentry, C.R.; Pate, R.R. 1971. Artificial diets for rearing larvae of the plum curculio. J ECON ENTOMOL 64: 1111-1112.

Based on Burton* (1969) diet for mass rearing *Heliothis zea*. Best of 6 tested.

COMPOSITION (g): wheat germ, 25; ascorbic acid, 1.63; methyl *p*-hydroxybenzoate, 1.0; sorbic acid, 0.5; agar, 6.4; dry pinto beans, 52.5; torula yeast, 16.0; and (ml): distilled water, 320; 10% formaldehyde, 4.0.

PREPARATION See Burton* (1969). Beans soaked overnight; if dry beans used amount of water adjusted in final preparation. Prepared in 10-gal blender and processed using machine, cups, and trays.

REARING At 27° C and 60% RH with 24 h photoperiod in 4-compartmented rectangular 7 x 21 x 6 cm plastic trays with bases replaced by ¼" mesh hardware cloth. Diet supported by wax paper placed under cloth. One pint diet dispensed/plastic tray and left overnight at room temperature. 1st instar larvae from sliced thinning apples placed on diet. When mature, wax paper removed from beneath tray, and larvae fall through screen into soil-Vermiculite mixture for pupation.

DEVELOPMENT Larval life, 2.3 days shorter on diet than on natural food of apples. Pupation time also shorter. In total, life-cycle 5-7 days shorter. Survival to adult on diet and on matured apple comparable at ca. 60%.

→ Yonce, C.E.; Gentry, C.R.; Pate, R.R. 1973. A complete artificial diet for rearing the plum curculio. J ECON ENTOMOL 66: 362-363.

COMPOSITION As in Yonce *et al.** (1971) except, instead of 320 ml distilled water, 160 ml apple juice, and 160 ml distilled water used.

PREPARATION Ingredients heated and blended; then dispensed while hot to 2 cm diam glass tubes. Sealed at both ends with masking tape; and held at 4-5° C for 1 h to cool. Hardened cores forced out of tubes with air pressure device and cut into 35 mm long plugs. Plugs placed on hardware cloth drying racks inside fume hood and warm air passed over diet for 3 h to give 32% moisture loss.

REARING In cages, 16 mated ♀♀ in each with 14 diet plugs, changed twice weekly.

DEVELOPMENT Improved with addition of 0.125 g potassium sorbate. Average larval wt, 0.019 g. Life cycle, 21 days. 65% adult yield if 1 ♀/diet plug. Increased emergence with progressive generations i.e., 71% emergence after 5 generations.

Graphognathus sp., white fringed beetle/weevil

→ Barnes, E.E.; Bass, M.H. 1973. Whitefringed beetles: an artificial diet for adults. J ECON ENTOMOL 66: 581.

COMPOSITION Adult diet (g): ascorbic acid, 4; brewers' yeast, 14; cellulose, 5; vitamin diet fortification mixture (NBC), 15; casein, 15; alfalfa leaf meal, 14; cane sugar, 10; agar, 11; and (ml): corn oil (Mazola), 2; distilled water, 430.

PREPARATION Agar boiled with water and blended with dry ingredients and oil; poured into containers, allowed to solidify and refrigerated.

REARING As in Bass & Barnes (1969b). 10 adults held in ½-pint cardboard cartons and fed fresh alfalfa leaves or diet. Eggs collected daily, cleaned of debris and placed on damp filter paper in petri dishes. Newly hatched larvae transferred to 10½ x 14½ x 3¼" plastic storage boxes @ 100/box filled to depth of 3" with sterile soil mixture (equal parts peat moss and sand) and with small, clean Irish potatoes with well developed root systems. Stored at 80° F; water added as necessary. After 3 months transferred to fresh containers and food. Pupae collected and placed individually in small plastic cups with autoclaved rearing soil.

DEVELOPMENT Adults fed artificial diet surpassed those fed green alfalfa leaves: longevity, 105 vs 84 days; egg production, 783 vs 590 eggs/♀; egg viability, 93 vs 60%.

See also: Bass & Barnes (1969a b).

Hylobius abietis (Linnaeus), large pine weevil

→ Malphettes, C.B. 1968. Premiers essais d'élevage des larves d'*Hylobius abietis* sur milieu synthétique. ANN SCI FOR 25: 189-192. (In French)

COMPOSITION (g): crushed pine bark (*Pinus sylvestris*), 50.0; dried brewers' yeast, 2.0; glucose, 3.0; levulose, 3.0; sucrose, 3.0; soluble starch, 2.0; hydrolyzed casein, 2.0; Wesson's salt mixture, 1.0; agar, 6.0; corn germ, 5.0; and (ml): distilled water, 200.0; inhibitor, 5.5.

inhibitor: 5 g *p*-methyl hydroxybenzoate in 100 ml 95% ethanol.

PREPARATION Bark from selected areas of pine or spruce cut into strips and crushed to give final particle size of ca. ½ mm x 2-3 mm, with 65% moisture content and pH of ca. 5. Refrigerated until use. Water boiled; sugars and agar added and again boiled for 5 minutes. Cooled to 50° C and Wesson's salt mixture, yeast, casein, corn germ and inhibitor added. Stirred. Bark added and stirred again in blender. Poured into plastic dishes, allowed to set and refrigerated until use.

REARING In rectangular, lidded containers, filled to depth of 6 mm with diet. Grooves cut in surface; larvae placed in grooves and covered. Larvae burrow into diet. Containers large enough for only 1 or 2 larvae to avoid cannabalism. Lids placed firmly on to surface of diet to provide purchase for apodous larvae when they reach it during burrowing. Cage size increased according to larval growth. After 15-21 days, larvae reared individually at 20° C.

DEVELOPMENT Neonate larvae to adults in ca. 2½ months. Adults comparable with wild ones.

Hylobius pales (Herbst), pales weevil

→ Clark, E.W. 1973. *Hylobius pales:* a meridic larval diet using sitosterol. J ECON ENTOMOL 66: 841-843.

COMPOSITION Group 1 (g): Wesson's salt mixture, 24.0; fructose, 35.0; dextrose, 32.0; sucrose, 75.0; raffinose, 25.0; casein, 100.0; β-sitosterol, 2.0; inositol, 0.8; linoleic acid, 8.0; oleic acid, 5.0; linolenic acid, 1.0; palmitic acid, 1.0; powdered cellulose, 70.0; methyl *p*-hydroxybenzoate, 4.5; and (ml): distilled water, 660.0; 5M KOH 15.0; 10% aqueous choline chloride, 100.0; 55% formaldehyde, 1.5.

group 2: agar, 48.0 g; and distilled water, 1850.0 ml.

group 3 (g): ascorbic acid, 0.014; folic acid, 0.003; α-tocopherol, 0.020; and vitamin solution, 15.0 ml; water, 100.0 ml; vitamin solution (mg): nicotinamide, 100.0; calcium pantothenate, 100.0; riboflavin, 50.0; thiamine hydrochloride, 27.0; pyridoxine hydrochloride, 25.0; biotin, 2.0; vitamin B_{12}, 0.2.

β-sitosterol: 2 g of commercial β-sitosterol (45.5% pure) or 0.91 g of pure sterol used; other ingredients at least 95% pure.

PREPARATION Group 1 ingredients mixed in 1-gal Waring Blendor at low speed for 1-2 minutes. Group 2 heated and mixed together, added to Group 1 combination, blended at medium speed for 2-4 minutes, autoclaved for 15 minutes at 1054 g/cm^2, reblended briefly, and cooled to 60° C. Vitamin solution then added; completed diet finally mixed at high speed for 5 minutes and poured into 15 x 100 mm plastic petri dishes. Covered and held for 12 h at room temperature to cool and to lose excess moisture. pH of diet, 6.5.

REARING See: Thomas* (1969). Larvae reared individually in petri dishes in dark cabinet at 24° C.

DEVELOPMENT Larval establishment, 94%; pupation, 26.5%; pupal wt, 58 mg; days to pupation, 56; 23.5% to adult in 68 days. Three generations reared.

→ Thomas, H.A. 1969. A meridic diet and rearing technique for the pales weevil larva. J ECON ENTOMOL 62: 1491-1494.

COMPOSITION (ml): distilled water, 660.0; (g): Wesson's salt mixture, 24.0; fructose, 32.0; dextrose, 32.0; sucrose, 40.0; raffinose, 3.0; casein, 100.0; phloem powder, 30.0; (ml): 4M KOH, 15.0; (g): wheat germ, 120.0; (ml): 10% aqueous choline chloride, 100.0; 55% formalin, 1.5; (g): methyl *p*-hydroxybenzoate, 4.5; Alphacel, 70.0; wheat germ oil, 4.0; agar, 48.0; distilled water, 1860.0 ml.

phloem powder prepared from dried fresh phloem ground in Wiley Mill to pass 20 mesh screen.

vitamins (mg): nicotinamide, 100; calcium pantothenate, 100; riboflavin, 50; thiamine hydrochloride, 25; pyridoxine hydrochloride, 25; folic acid, 25; biotin, 2.0; vitamin B_{12}, 0.2.

PREPARATION Agar mixed with 1860 ml distilled water. Remaining ingredients blended in 1-gal Waring Blendor. Ingredients mixed for 10-20 seconds after each addition. Agar solution added. Mixture then autoclaved for 18 minutes at 15 lb pressure, remixed, and cooled to 60° C. Six of ascorbic acid and 15 ml aqueous vitamin solution added. After final mixing, diet dispensed to plastic petri dishes, covered and allowed to stand at room temperature for 12 h to cool and lose excess moisture. pH of medium, 5.5.

REARING Larvae at 25° C in plastic petri dishes in darkness. Adults kept at room temperature in 13 x 29 x 12 cm plastic bread boxes. Moist paper towelling provided with 7 x 10 x 10 cm sections of fresh loblolly pine phloem for food and oviposition. Eggs removed every 4-5 days with dissecting needle, rinsed in water, immersed in 1:750 aqueous zephiran chloride and incubated at room temperature.

DEVELOPMENT Duration of larval to pupal stages on diet at 25° C and on natural food at 27° C; 47.2 and 47.0 days, respectively. Mean body weight of diet-reared adults 27% greater (71 and 78 mg) than that of field collected adults (57 and 61 mg). Callow adults obtained in about 55 days.

See also: Richmond & Thomas (1965); Speers & Cody (1975); Thomas (1971).

Hypera postica (Gyllenhal), alfalfa weevil

→ Hsiao, T.H.; Hsiao, C. 1974c. Feeding requirements and artificial diets for the alfalfa weevil. ENTOMOL EXP APPL 17: 83-91.

COMPOSITION (g/100 ml): vitamin-free casein, 3.0; egg albumin, 3.0; lactoalbumin hydrolysate, 1.0; sucrose, 5.0; lecithin (vegetable), 0.50; yeast extract, 1.0; Wesson's salt mixture, 1.0; choline chloride, 0.10; cholesterol, 0.150; β-sitosterol, 0.050; corn oil (Mazola), 0.50; ascorbic acid, 0.40; B-vitamins, 0.024; adenine hydrochloride, 0.10; *l*-cysteine hydrochloride, 0.10; KOH, 0.30; agar-bacto, 2.50; cellulose powder, 4.0; inhibitors, 0.35; water to 100 ml.

vitamins (mg): niacin, 1.0; calcium pantothenate, 1.0; thiamine hydrochloride, 0.25; riboflavin, 0.50: pyridoxine hydrochloride, 0.25; folic acid, 0.25; biotin, 0.02; *meso*-inositol, 20.0; B_{12}, 0.002.

inhibitors (mg): sorbic acid, 150; methyl *p*-hydroxybenzoate, 100; streptomycin sulphate, 50; Benlate (Benomyl fungicide from Du Pont), 50.

PREPARATION Casein dissolved in portion of water with KOH. Remaining water and other ingredients except yeast, vitamins, ascorbic acid and inhibitors mixed into this solution. Autoclaved in beaker for 15 minutes at 121° C and 15 lb pressure. Diet then blended until temperature at 60° C, when remaining ingredients added, and final volume adjusted with additional water. Wrapped in foil and stored in fridge.

REARING At 26.5 ± 0.5° C and 60-80% RH with 15 h photoperiod. Rearing cage of 2.5 x 3.0 cm diam plastic vial with nylon netting folded to 5 mm in height fitted to bottom to serve as support for diet. 0.5 g diet cut into sections and placed on netting; 10 neonate larvae introduced. Vial covered with fine mesh cloth to prevent escape, and to allow air circulation. Numbers of larvae reduced to 5 third and 3 fourth instars/cage to avoid overcrowding.

DEVELOPMENT Adult yield on basal diet, 72%. Average developmental time (larva - adult), 21.9 days, and average pupal wt, 7.3 mg. Developmental time reduced to 17.7 days, and pupal weight increased to 9.1 mg with addition of 10% acetone extracted alfalfa leaf powder. This compares with 16.3 days and 9.3 mg on alfalfa leaves alone.

→ Hsiao, T.H.; Hsiao, C. 1974b. A practical artificial diet for laboratory rearing of the alfalfa weevil, *Hypera postica* (Gyllenhal). ANN ENTOMOL SOC AM 67: 149-150.

COMPOSITION (g/100 ml): vitamin-free casein, 3.0; lactoalbumin hydrolysate, 1.0; sucrose, 5.0; yeast extract, 1.0; Wesson's salt mixture, 1.0; choline chloride, 0.10; cholesterol, 0.15; corn oil (Mazola), 0.50; ascorbic acid, 0.40; KOH, 0.30; agar-bacto, 3.0; wheat germ powder, 10.0; alfalfa leaf powder (acetone extracted), 10.0; sorbic acid, 0.15; methyl *p*-hydroxybenzoate, 0.10; streptomycin sulphate, 0.05; Benlate (Benomyl fungicide from Du Pont), 0.05; and water to 100 ml.

PREPARATION AND REARING See: Hsiao & Hsiao* (1974c).

DEVELOPMENT Compared with those reared on fresh alfalfa leaves in parentheses: % adult yield, 86 (78); mean developmental time from neonate larva to adult (days), 18.7 ± 0.11 (16.3 ± 0.15); mean pupal wt (mg) 9.0 ± 0.12 (9.3 ± 0.11).

See also: Hsiao & Hsiao (1974a).

→ Nash, R.R.; Tombes, A.S. 1966. Evaluation of five artificial diets for laboratory rearing of alfalfa weevil larvae. J ECON ENTOMOL 59: 220-221.

Best of 5, adapted from Berger* (1963) for *Heliothis* sp.

COMPOSITION (g): acetone-extracted alfalfa, 5.0; wheat germ, 3.0; sucrose, 3.5; Alphacel, 0.5; casein, 3.5; Wesson's salt mixture, 1.0; choline chloride, 0.1; vitamin mixture, 1.1; ascorbic acid, 0.4; agar, 2.5; (ml): water, distilled, 120; 22.5% KOH, 0.5; 10% formaldehyde, 0.36.

PREPARATION Agar dissolved in 60 ml water, cooled to 60° C and blended with remaining ingredients except KOH and formaldehyde in 30 ml water. Remaining ingredients added and diet mixed thoroughly.

REARING At 30° C and 57% RH with 10 h photoperiod. 1st and early 2nd instar larvae in petri dishes with 3 g diet. 10 larvae/dish.

DEVELOPMENT Average larval development, 28.3 days; pupal period, 8 days; survival to adult, 17.7%.

Hyperodes bonariensis Kuschel, stem weevil

Power, R.J.B.; Singh, P. 1974. Laboratory rearing method for the stem weevil, *Hyperodes bonariensis* (Coleoptera: Curculionidae). N Z J ZOOL 1: 531-536.

COMPOSITION & PREPARATION See: Singh* (1974c).

REARING At 26 ± 0.5°C and 55 ± 5% RH with 12 h photoperiod. Two g diet dispensed into sterile 12 x 75 mm glass tubes, plugged with cotton wool and slanted. Tubes left for 24 h to allow evaporation of excess moisture. Matured eggs sterilized with 5% formalin for 10 minutes, washed 3 times in sterile distilled water treated with 0.1% Javex for 5 minutes, rewashed 3 times, dried and then inoculated on to diet. Larvae removed from diet after 2 weeks and placed in moist vermiculite to pupate. Adults anaesthetised with CO_2, sexed and paired. One pair/ oviposition cage (illustrated) with ryegrass for feeding and oviposition.

DEVELOPMENT Larval period, 14 days, followed by prepupal and pupal stage of further, 14 days; sometimes prolonged for up to 70 days depending on conditions. Pre-oviposition period, ca. 2 weeks; adults long lived. Fecundity up to 592 eggs/♀ in 15 weeks; 101 eggs in first 2 weeks. Eight generations reared with no deleterious effects.

Otiorhynchus sulcatus Fabricius, black vine weevil

Shanks, C.H. Jr; Finnigan, B. 1973. An artificial diet for *Otiorhynchus sulcatus* larvae. ANN ENTOMOL SOC AM 66: 1164-1166.

Best of 5 diets tested. Adapted from Shorey & Hale* (1965) for noctuid larvae.

COMPOSITION (g): dry pinto beans, 60; brewers' yeast, 20; *l*-ascorbic acid, 2.0; methyl *p*-hydroxybenzoate, 0.9; sorbic acid, 0.4; agar, 15; deionized water, 425 ml.

PREPARATION Pinto beans placed in 250 ml water, boiled and allowed to stand at room temperature for 3 h until beans soft. Water and softened beans mixed in Waring Blendor; other cold ingredients added; pH to 6-6.2 with 22.5% KOH. Agar dissolved in 175 ml water in autoclave, added to cold ingredients and blended for 1 minute. 7-dram plastic snap-cap vials poured half filled with diet. Cooled until condensation gone. Surface scored lightly with dissecting needle.

REARING At 70°F in darkness, individually in vials. 3 month old larvae placed in peat moss to pupate. Adults held at 70°F and 76% RH with 16 h photoperiod in cloth covered jars in closed tank over saturated NaCl. Fed on strawberry leaves. Eggs collected twice/week, soaked for 30 minutes in 0.025% NaClO solution and 0.02% Triton X-100 and rinsed with 70% ethanol.

DEVELOPMENT Average of 34.3 days to pupation; pupation, 36.6%. Survival of mature larvae to adult, 34.1%.

Pissodes strobi (Peck), white pine weevil

Zerillo, R.T.; Odell, T.M. 1973. White pine weevil: a rearing procedure and artificial medium. J ECON ENTOMOL 66: 593-594.

Adapted from Harley & Willson* (1968) diet for *Plagiohammus spinipennis*.

COMPOSITION (g): pine bark, 20; wheat germ, 30; sucrose, 12; dextrose, 12; starch, 6; Wesson's salt mixture, 3.6; casein, 24; cholesterol, 0.48; agar-agar, granular, 25; 55% linoleic acid, 0.48; and (ml): distilled water, 790; antimicrobial solution, 15; vitamin solution, 20; 10% KOH, 10.

antimicrobial solution (g): sorbic acid, 20; methyl *p*-hydroxybenzoate, 15; and 95% ethyl alcohol, 170 ml.

vitamins (mg/100 ml): *i*-inositol, 2000; calcium pantothenate, 100; niacin, 100; riboflavin, 50; pyridoxine hydrochloride, 25; thiamine hydrochloride, 25; folic acid, 25; biotin, 2; vitamin B_{12}, 0.2.

PREPARATION Pine bark dried at 176.5° C for 1 h, then ground in Wiley Mill through 20 mesh screen. Agar dissolved in water. First 8 components listed added to hot agar and blended; cooled to 55° C, remaining ingredients added and blended again. Mixture poured into 15 x 100 mm plastic petri dishes and allowed to solidify.

REARING In Sherer Gillett mobil greenhouse at 68-70° F with 16 h photoperiod. 24-48 h old eggs laid on 9 cm white pine stick, washed in 1:840 solution of zephiran chloride for 4 minutes and dried on filter paper. Eggs placed individually in small holes drilled in diet and covered with small diet plug and dishes sealed with masking tape. Adults placed @ 10-15/petri dish with 2.5 cm diet for 7-10 weeks for diet acceptance.

DEVELOPMENT Eclosion, 74%. Mortality from egg to pupal stage, 20%; survival from egg to adult, 58%. Diet-reared weevils 10% larger than controls reared on white pine. Development completed in 8 weeks.

Rhabdoscelus obscurus (Boisduval), New Guinea sugarcane weevil

→ Chang, V.C.S.; Jensen, L. 1972. A diet for studying clonal resistance of sugarcane to the New Guinea sugarcane weevil. J ECON ENTOMOL 65: 1197-1199.

COMPOSITION (g): Alphacel, 5.0; agar, 5.5; and (ml): sugarcane juice, 110; water, 137; Terramycin, 10.8 (200 ppm actual); Benomyl (methyl 1-(butylcarbamoyl)-2-benzimidazolecarbamate), 2.7 (25 ppm actual).

PREPARATION Sugarcane juice obtained by extraction from three tops of sugarcane stalks (4th to 11th internode) in hydraulic, motorised, rotary-type mill. Mixture of water and agar heated to 92° C, warm (45° C) sugarcane juice, antimicrobial agents, and Alphacel added. Blended at high speed for 2 minutes. Temperature of mixture never to exceed 60° C. pH 5.1. Poured into 1-oz containers previously sterilized with 1.25% NaClO. One day old larva, surface sterilized with 0.6% NaClO, introduced to each container after diet cooled to room temperature.

REARING In plastic containers in climatic chamber for 14 days with alternating temperatures of 25° C for 10 h and 30.6° C for 14 h.

DEVELOPMENT Larvae reared on diet with 80% sugarcane juice weighed more than larvae reared on cut sugarcane stalks, and had lower mortality. Same diet with 40% sugarcane juice produced larvae of similar weights to those reared on cut sugarcane stalks. Use of 9 antimicrobial compounds discussed; only Terramycin and Benomyl effected any microorganism control without inhibiting larval growth.

Note: Personal Communication: published diet modified by reducing Benomyl to 13 ppm and adding O-phenylphenol 25 ppm in 95% ethanol to cut down microbial contamination.

→ Napompeth, B.; Nishida, T.; Mitchell, W.C. 1972. Biology and rearing methods of the New Guinea sugarcane weevil, *Rhabdoscelus obscurus*. HAWAII AGRIC EXP STN TECH BULL 85: 51 pp.

Best of several diets tested.

COMPOSITION (g): shredded coconut husk, 140; agar, 50; Wesson's salt mixture, 3; and (ml): methyl paraben in 95% ethyl alcohol, 0.5; sorbic acid, 0.5; Vanderzant's vitamin mixture, 15; 10% KOH, 10; water, 1000.

PREPARATION Coconut husk sliced and ground in Waring Blendor. Air dried in sun for 8-10 h. Agar dissolved in some of water and boiled for 5 minutes. Poured into blender with coconut husk and antimicrobial agents and cooled to 50° C; remaining ingredients added. Blended for 3 minutes and dispensed to sterilized rearing containers.

REARING Adults maintained at 80 ± 5% humidity in 1-gal jars; layer of plaster 10 cm above bottom of jar with hole in centre. Water and 25 ml 18% $CuSO_4$ poured in up to plaster level and hole plugged. Adults fed on sugarcane stalks and eggs laid in these. Stalks removed after 3 days and placed vertically in plastic containers with moist paper towels for 1 week. Larvae transferred to diet in 8-dram shell vials, 5 cm cube plastic boxes with snap covers, or 28 x 17 x 4 cm plastic boxes with 18 compartments.

DEVELOPMENT Adult yield, 61.7% in 45-61 days (mean 54.3 ± 3.7 days).

Sitophilus granarius (Linnaeus), granary weevil

→ Baker, J.E.; Mabie, J.M. 1973b. Growth and development of larvae of the granary weevil, *Sitophilus granarius* (Coleoptera: Curculionidae), on natural and meridic diets. CAN ENTOMOL 105: 249-256.

Three diets tested.

COMPOSITION Diet 1 modified from Haydak* (1936a). (g): whole wheat flour, 30.7; yellow cornmeal, 42.2; dog meal (prepared from Gaines pellets), 16.1; brewers' yeast, 8.5; wheat germ, 2.4.

Diet 2 (g): casein, 5.0; glucose, 5.0; brewers' yeast, 0.5; cholesterol, 0.1; Wesson's salt mixture, 0.2; wheat germ oil, 0.11; and water, 1.5 ml.

Diet 3 (g): casein, 1.0; corn starch (Argo), 7.7; brewers' yeast, 0.75; cholesterol, 0.1; wheat germ, 0.25; wheat germ oil, 0.11; and water, 1.4 ml.

PREPARATION Diets prepared by grinding individual components and passing through No. 50 sieve. All dry ingredients thoroughly mixed and reground with a mortar and pestle. For diets 2 and 3, wheat germ oil added to dry components in 5.0 ml of diethyl ether. Prepared diets equilibrated in desiccator over water for 24 h prior to use.

REARING Oviposition in mixture of 100 g Argo "gloss" laundry starch and 1.0 g brewers' yeast in ½-pint glass jars at 29 ± 1.0° C and 65 ± 5% RH in dark over 4-6 day period. After 5-6 days, eggs and first instar larvae removed by sieving with water and placed in petri dishes held at 29 ± 1.0° C and 65 ± 5% RH until use. Larvae reared individually in glass tubes prepared from 9" disposable glass pipettes as follows: pipettes cut 1.0-1.5 cm above shoulder and fire-polished. Each tube sealed off with small flame 4.0-5.0 cm below shoulder. Small quantity of diet packed into end of each tapered tube to depth of 3-5 mm by tapping tube on hard surface, or, for starch-based diets, by tapping diet with small wire. One first instar larva placed in each tube and diet added to within 1.0 cm of open end (total of ca. 100 mg in each tube). Larvae also reared at 29 ± 0.5° C and 65 ± 3% RH in darkness.

DEVELOPMENT Larval growth rates and survival on diet 1 inferior to those on whole wheat flour (natural food). Maximum mean larval wt, 3.03 ± 2.2 mg on wheat and 2.47 ± 0.24 mg on diet 1. Larval growth rates and adult emergence on diet 3 better than on whole wheat flour. After 14 days, mean larval wt, 2.51 ± 0.40 mg on diet 3 and 1.25 ± 0.16 mg on whole wheat flour. Growth rates slow on diet 2, with zero survival to pupal stage.

See also: Baker & Mabie (1973a).

Sitophilus granarius (Linnaeus), granary weevil

Sitophilus oryzae (Linnaeus), rice weevil

Sitophilus zeamais Motschulsky, maize weevil

→ Baker, J.E.; Mabie, J.M. 1973c. Growth responses of larvae of the rice weevil, maize weevil, and granary weevil on a meridic diet. J ECON ENTOMOL 66: 723-726.

Modified from Baker & Mabie (1973a) diet for *S. oryzae*.

COMPOSITION (g): vitamin-free casein, 1.0; corn starch, 7.99; brewers' yeast, 0.75; wheat germ, 0.25; wheat germ oil, 0.025; cholesterol, 0.01; and distilled water, 1.4 ml.

REARING At 32° ± 0.5° C and 65 ± 3% RH in darkness.

DEVELOPMENT Mean larval weight of *S. zeamais* after 11 days on diet, 4.36 ± 1.07 mg compared with 3.01 ± 0.59 mg for *S. oryzae:* development of different life stages nearly identical for both species. Adult emergence for both species from 17-23 days with 93.5% of *S. oryzae* larvae and 80% of *S. zeamais* larvae emerging as adults. Larvae of *S. oryzae* more uniform in developmental times, reaching maturity and pupating at average of 8.5 days earlier than larvae of *S. granarius*. Survival of both species to adult stage, 90%.

→ Baker, J.E. 1974c. Differential sterol utilization by larvae of *Sitophilus oryzae* and *Sitophilus granarius*. ANN ENTOMOL SOC AM 67: 591-594.

COMPOSITION Basal diet (%): casein, 10.0; corn starch (lipid free), 80.9; brewers' yeast (lipid free), 7.0; wheat germ (lipid free), 2.0; sterol, 0.1; and distilled water to moisture content of 15-16%. In addition, 25 mg of glycerol added to each 10 g of diet to improve texture. When sterol concentration altered from 0.1% amount of corn starch adjusted to make total weight of dry components equal to 10 g.

REARING At 30° C and 65% RH.

DEVELOPMENT *S. oryzae:* mean days to pupation, 14.7; adult survival, 100%. *S. granarius:* mean days to pupation, 22.4; adult survival, 97%.

See also: Baker (1974b).

Sitophilus oryzae Linnaeus, rice weevil

Baker, J.E. 1975b. Vitamin requirements of larvae of *Sitophilus oryzae*. J INSECT PHYSIOL 21: 1337-1342.

Two diets tested.

COMPOSITION Casein diet (g/10 g): vitamin-free casein, 2.0; cornstarch, 7.89; cholesterol, 0.01; mineral mixture, 0.10; wheat germ oil, 0.025; and (ml): vitamin solution, 0.5; distilled water, 0.9.

Amino acid diet (g/10 g): arginine, 0.055; aspartic acid, 0.07; cystine, 0.015; glutamic acid, 0.19; glycine, 0.1; histidine, 0.04; isoleucine, 0.08; leucine, 0.09; lysine hydrochloride, 0.085; methionine, 0.04; phenylalanine, 0.032; threonine, 0.058; tryptophane, 0.015; tyrosine, 0.06; valine, 0.07; cornstarch, 8.89; cholesterol, 0.01; mineral mixture, 0.1; wheat germ oil, 0.025; (ml): vitamin solution, 0.5; distilled water, 0.9.

minerals: See Medici & Taylor* (1966).

vitamins (μg/g): choline chloride, 100; thiamine hydrochloride, 12.5; nicotinic acid amide, 12.5; pyridoxine hydrochloride, 12.5; calcium pantothenate, 12.5; riboflavin, 2.5; *d*-biotin, 0.5; folic acid, 0.5; *dl*-carnitine hydrochloride, 0.5.

PREPARATION Standard vitamin solution freshly prepared by adding 90 ml distilled water to weighed vitamins, adjusting pH to 6.5-6.8 with 5% KOH and bringing to 100 ml as clear, sediment-free solution; 0.5 ml solution added to diet. For amino acid diet, amino acids ground in mortar and pestle and stored at -20° C (90 mg $NaHCO_3$ added/1 g of amino acid mixture). Other dry components added with thorough mixing. For details see Baker & Mabie* (1973c).

REARING Neonate larvae reared individually in glass tubes at 30° C and 55% RH.

DEVELOPMENT On casein diet, mean days to larval-pupal ecdysis, 24.1 ± 0.9; survival to adult, 70%. On amino acid diet, mean days to larvae-pupal ecdysis, 22.5 ± 0.5; survival to adult, 84%.

S. granarius failed to develop on casein diet, but amino acid diet satisfactory.

Chippendale, G.M. 1972. Dietary carbohydrates: role in survival of the adult rice weevil, *Sitophilus oryzae*. J INSECT PHYSIOL 18: 949-957.

COMPOSITION Adult diet (g/100 g): corn starch, 72.0; vitamin-free casein, 10.0; wheat germ powder, 5.0; vitamin mixture, 2.0; cholesterol, 0.5; salt mixture, 0.5; glycerol, 10.0.

vitamin mixture: 50% vitamin-dextrose mixture and 50% corn starch; supplied by Nutritional Biochemicals Corporation, Cleveland, Ohio.

salts according to Medici & Taylor* (1966).

PREPARATION Dry powdered components mixed with glycerol in mortar and pestle. About 5 g diet dispensed/2-oz clear plastic cup.

REARING At 30° C and ca. 60% RH in darkness. Stock cultures in gallon jars partially filled with moist maize kernels (ca. 15-20% water content) in incubator. On experimental diets, in plastic cups under similar conditions as stock culture.

DEVELOPMENT Effects of various carbohydrates on feeding behaviour, food consumption, and survival of adult reported. Egg to adult, 6 weeks; adult survival, 6 months.

Corollary

Sitophilus zeamais Motschulsky, maize weevil

Brown, J.J.; Chippendale, G.M. 1975. Survival of the adult maize weevil, *Sitophilus zeamais:* role of nutrients, larval reserves and symbionts. COMP BIOCHEM PHYSIOL 50A: 83-90.

Diet modified from Chippendale* (1972) for *S. oryzae*.

COMPOSITION Adult diet (g/100 g diet): corn starch, 72; glycerol, 20; wheat germ, 5; vitamin mixture, 2; cholesterol, 0.5; salt mixture, 0.5.

DEVELOPMENT Adult survival, ca. 80% in 30 weeks.

Sphenophorus venatus vestitus Chittenden, zoysiagrass billbug

Kovitvadhi, K.; Kerr, S.H. 1968. Artificial diet for the zoysiagrass billbug, *Sphenophorus venatus vestitus* (Coleoptera: Curculionidae), and notes on its biology. FLA ENTOMOL 51: 247-250.

Diet modified from Berger* (1963) for *Heliothis* species.

COMPOSITION (g): casein, 29.2; Wesson's salt mixture, 8.5; sucrose, 22.7; wheat germ, 25.6; Alphacel, 4.3; vitamin fortification mixture, 8.5; agar, 21.3; ascorbic acid, 3.4; and (mg): streptomycin sulphate, 118.0; and (ml): water, 729.5; 22.5% KOH, 4.3; 10% formaldehyde, 3.1; 7 g methyl *p*-hydroxybenzoate + 7 g sorbic acid in 50 ml 95% ethyl alcohol, 12.5.

PREPARATION All ingredients except vitamins, ascorbic acid, agar and streptomycin blended for 2-3 minutes with 208.5 ml of water. Agar dissolved in remaining 521 ml hot water, and autoclaved at 15 p.s.i. for 30 minutes, allowed to cool briefly. Blended ingredients mixed with agar and remaining ingredients added. Diet poured into 8-dram shell vials to within 1.5" of top. Vials stoppered with cellucotton absorbent wadding and refrigerated until use.

REARING At 27°C or 22°C. Neonate larvae placed individually in shell vial with diet.

DEVELOPMENT Percent egg hatch at 22°C, 1½ times that at 27°C. Larval mortality high. 7 out of 42 at 27°C reached adult stage. No oviposition on diet.

Dermestidae

Anthrenus flavipes Le Conte, furniture carpet beetle

→ Baker, J.E.; Schwalbe, C.P. 1975. Food utilization by larvae of the furniture carpet beetle, *Anthrenus flavipes.* ENTOMOL EXP APPL 18: 213-219.

COMPOSITION (mg): wool fabric discs, 100; cholesterol, 0.1; mineral mixture, 1.0; vitamin solution, 0.5 ml.

vitamins (mg/100 ml): choline chloride, 50; thiamine hydrochloride, 12.5; nicotinic acid amide, 12.5; pyridoxine hydrochloride, 12.5; calcium pantothenate, 12.5; riboflavin, 5; folic acid, 1; *d*-biotin, 1; cyanocobalamin, 0.5. 0.5 ml of 1:25 dilution of vitamin stock solution applied to each disc. pH to 6.5 with 5% KOH.

minerals: see Medici & Taylor* (1966).

PREPARATION 1.7 cm diam punch used for 100 mg discs of wool. Cholesterol applied in 0.5 ml $CHCl_3$. Mineral mixture added to vitamin solution and applied to disc.

REARING At 30° C and 55% RH in dark. Five, 3-week-old larvae placed on each disc in 15 x 60 mm petri dish.

DEVELOPMENT After 21 days, 95% food consumed; wt gain, 90%. Not as effective as brewers' yeast dusted on to wool discs. Used as maintenance diet.

See also: Sweetman (1956).

→ Wal, Y.C.; Agarwal, P.N. 1966. Nutritional requirements of woolly bear. Part I. A new synthetic medium for rapid growth and development of *Anthrenus flavipes* Lec. LABDEV PART B 4: 71.

COMPOSITION (parts): casein, 7.0; glucose, 3.0; cholesterol, 0.2; brewers' yeast, 3.0; McCollum's salt mixture, 0.1.

PREPARATION Ingredients mixed and blended for 5 minutes. 1.5 g placed in 5 x 2.5 cm diam glass tubes plugged with perforated rubber cap.

REARING 25, 1-7-day-old larvae placed on sterilized filter paper in diet tube. Adults maintained on same diet for egg laying.

DEVELOPMENT Three generations completed in 230 days; 270 days for 1 generation on control of woollen fabric.

Corollary

→ Wal, Y.C.; Agarwal, P.N. 1968. Nutritional requirements of the woolly bear. Part II. Some B-vitamins requirements of the larvae of *Anthrenus flavipes* Lec. LABDEV PART B 6: 51-52.

COMPOSITION As in Wal & Agarwal* (1966) except half of cholesterol and yeast substituted by B-vitamins (µg/g diet): thiamine, 25; riboflavin, 25; nicotinic acid, 25; pantothenic acid, 25; pyridoxine, 25; *p*-aminobenzoic acid, 25; choline chloride, 500; inositol, 250; folic acid, 2; biotin, 0.1.

REARING At 27 ± 1° C and 70 ± 5% RH. Adults laid eggs on fragments of hog bristles smeared in solution of 20% glucose and 20% egg albumin. Eggs kept in petri dish on black paper for 7-10 days. 25 neonate larvae placed on 1.5 g diet in glass tube with filter paper.

DEVELOPMENT Larval survival after 106-130 days, 92%. Larval period, 106-121 days; pupation, 92%; adult emergence, ♂ 42% and ♀ 52%.

Attagenus sp.

→ Moore, W. 1943. Some vitamin requirements of black carpet beetles, *Attagenus* (?) spp. (Coleoptera: Dermestidae). ANN ENTOMOL SOC AM 36: 483-485.

COMPOSITION (mg): vitamin-free casein, 350; corn starch, 150; mineral mixture, 15; extracted fish meal, 30; glycerol, 70; choline chloride, 1.000; niacin, 0.050; thiamin chloride, 0.007; riboflavin, 0.005; pyridoxin hydrochloride, 0.006; calcium pantothenate, 0.020; Mead's oil of pencomorpheum, 5.0; cystine, 2.0; glucosanine hydrochloride, 2.0; urea, 2.0; phytin, 1.0; β-alanine, 0.012; *p*-aminobenzoic acid, 0.020; and benzene extract of liver, 0.025 ml.

REARING At 86-87° F and 60% RH in shell vials with 220 mg food/25 larvae.

DEVELOPMENT Slow growth for 7 weeks; increased by addition of 15% glycerol.

Corollary

→ Moore, W. 1946. Nutrition of *Attagenus* (?) sp. (Coleoptera: Dermestidae). ANN ENTOMOL SOC AM 39: 513-521.

Best diet tested.

COMPOSITION (mg): lactalbumin, 400; glycerol, 60; cooked corn starch, 100; folic acid, 1 µg/tube.

REARING At 90° F and 60% RH. 40 neonate larvae reared/tube.

DEVELOPMENT Weight at 7 weeks, 5.9 mg.

Attagenus megatoma (Fabricius), black carpet beetle

→ Baker, J.E. 1974a. Influence of nutrients on the utilization of woollen fabric as a food for larvae of *Attagenus megatoma* (F.) (Coleoptera: Dermestidae). J STORED PROD RES 10: 155-160.

Best of 4 diets tested.

COMPOSITION & PREPARATION Cholesterol (0.1 mg in 0.5 ml chloroform) and then 0.5 ml vitamin solution and 1.0 mg mineral mixture applied to each of 2.6 x 2.6 cm squares weighing 200 g of undyed, scoured, moth test fabric (100% wool). Finally wetted with Triton X-100.

vitamins (µg): choline chloride, 100; thiamine hydrochloride, 25; nicotinic acid amide, 25; pyridoxine hydrochloride, 25; calcium pantothenate, 25; riboflavin, 10; biotin, 2.

minerals: see Medici & Taylor* (1966).

REARING At 30° C and 55% RH: individually in 18 x 150 mm petri dishes.

DEVELOPMENT After 60 days, mean larval weight change (mg) and larval survival for control 0.22 ± 0.15, 80%; and for diet 0.91 ± 0.35, 95%.

Baker, J.E. 1975a. Protein utilization by larvae of the black carpet beetle, *Attagenus megatoma*. J INSECT PHYSIOL 21: 613-621.

Best of 4 diets tested.

COMPOSITION (g): amino acid mixture, 9.75; dextrose, 39.0; mineral mixture, 1.0; cholesterol, 0.25; wheat germ oil, 0.25; and vitamin solution, 3.90 ml.

l-amino acids: as in Davis (1968a). (mg): arginine, 9.0; aspartic acid, 11.0; cystine, 2.0; glutamic acid, 30.0; glycine, 53.0; histidine, 6.0; isoleucine, 12.0; leucine, 14.0; lysine monohydrochloride, 13.0; methionine, 6.0; phenylalanine, 5.0; threonine, 9.0; tryptophan, 2.0; tyrosine, 10.0; valine, 11.0.

minerals: as in Medici & Taylor* (1966).

vitamins (μg/g): choline chloride, 50; thiamine hydrochloride, 20; nicotinic acid amide, 20; pyridoxine hydrochloride, 20; calcium pantothenate, 20; riboflavin, 10; folic acid, 5; *d*-biotin, 5; *dl*-carnitine hydrochloride, 5.

Stock culture diet: 5% brewers' yeast in finely ground Purina Laboratory Chow.

PREPARATION Ingredients ground in ball mill prior to use. Diet left for 48 h at 30° C and 50-60% RH.

REARING At 30° C and 50-60% RH in 15 x 60 mm petri dishes.

Attagenus megatoma (Fabricius), black carpet beetle

Trogoderma variabile Ballion [previously *T. parbile* Beal]

Burkholder, W.E.; Schwalbe, C.P.; Boush, G.M. 1973. Antimicrobial food additives and their effects on *Trogoderma variabile* and *Attagenus megatoma* (Coleoptera, Dermestidae). J STORED PROD RES 9: 205-211.

COMPOSITION For *A. megatoma* see: Boush *et al.* (1968). Purina Dog Chow fortified with 5% brewers' yeast.

For *Trogoderma variabile* Ballion: Purina dog food, wheat germ, dry milk, brewers' yeast, and mixed meat and bone meal mixed in rates of 8:3:3:1:1 by weight.

REARING At 27 ± 1° C and 60 ± 10% RH with 16 h photoperiod in screw cap, 2-oz, 35 x 85 mm jars. 1 cm diam hole in caps covered with 0.22 μm Millipore filter, 20 mesh copper screen and filter paper.

DEVELOPMENT Complete on untreated diets. Significant toxicity to exposed eggs and larvae of both spp by addition of 2% sorbic acid, sodium benzoate, n-heptyl *p*-hydroxybenzoate and propyl *p*-hydroxybenzoate.

Dermestes maculatus De Geer [previously *D. vulpinus* Fabricius], hide beetle

Applebaum, S.W.; Konijn, A.M.; Menco, B. 1971. Growth and biochemical adaptation of larvae of the beetle *Dermestes maculatus* to carbohydrate-free diets. INSECT BIOCHEM 1: 1-13.

COMPOSITION & PREPARATION Adult diet (g): fishmeal, previously heated to 60° C overnight and sieved to 40 mesh, 95; brewers' yeast, 5; and distilled water, 9 ml.

Larval diet (%): α-cellulose, 8.0; salt mixture, USP No. 2, 2.0; yeast extract, 1.0; cholesterol, 0.1; distilled water, 8.9. Total, 20.0%.

Varying amounts of casein (73.07% protein) added; also soybean oil as neutral fat, and soybean oil free fatty acids (Drew Chemical Co. (Welcoline S)). Mixtures of soybean oil and free fatty acids prepared with α-cellulose to be equicaloric to glucose, wt/wt. All diets of same total caloric density.

REARING At 28° C and 75% RH.

DEVELOPMENT 10-50% dietary casein and either neutral fat or glucose, utilized during larval growth. Growth reduced by free fatty acids.

Fraenkel, G.; Reid, J.A.; Blewett, M. 1941. The sterol requirements of the larva of the beetle, *Dermestes vulpinus* Fabr. BIOCHEM J 35: 712-720.

Three diets tested.

COMPOSITION Diet 1 (g): ashless vitamin-free casein (casein E, Glaxo Lab.), 3.0; dried debittered brewers' yeast (Glaxo Lab.), 0.3 or 0.15; cholesterol, 0.1; cystine, 0.02; McCollum's salt mixture, 0.03; distilled water to 50% by weight.

Diet 2 (g): dried debittered brewers' yeast (Glaxo Lab.), 3.0; sterol, 0.1; McCollum's salt mixture, 0.03; distilled water to 50% by weight.

Diet 3 (g): dried debittered brewers' yeast (Glaxo Lab.), 1.5; fructose, 1.5; sterol, 0.05; McCollum's salt mixture (not essential and usually omitted), 0.03; and water, 0.4 ml.

REARING At 27° C and 70% RH. Initially reared individually in glass tubes; later at 5 larvae/tube. Food changed daily. Hygroscopic properties (giving water content of 20-25%) and high sugar content of diet 3 gave effective control of mould.

DEVELOPMENT Minimum requirement for cholesterol; 1 mg/3 g diet; for 7-dehydrocholesterol, 0.5 mg/3 g diet. On diets 2 and 3, 5th instar after 11 days. On diet 1, 4th instar after 11 days. Average time to pupation, 26 days; pupal stage, 7 days.

See also: Bloch *et al.* (1956); Clark & Bloch (1959a b); Lall (1958); Mushi & Chiang (1974).

Gay, F.J. 1938. A nutritional study of the larva of *Dermestes vulpinus* F. J EXP ZOOL 79: 93-107.

COMPOSITION (g): vitamin-free casein, 3.0; debittered dried yeast, 0.3; cholesterol, 0.1; cystine, 0.02; McCollum's salt mixture, 0.03; water, to 46% moisture content.

PREPARATION Dry ingredients thoroughly mixed. Salts dissolved in water and added to dry mix. Mixed until moist, crumbly mash. 1 g placed in 2 x ¾" glass vials.

REARING At 27° C and 70% RH individually in glass vials. Diet changed daily.

DEVELOPMENT Pupation from 25-32 days, average, 27.3; pupation, 75%; emergence, 100%. Adults fertile.

Katz, M.; Budowski, P.; Bondi, A. 1971. The effect of phytosterols on the growth and sterol composition of *Dermestes maculatus*. J INSECT PHYSIOL 17: 1295-1303.

Diet modified from Budowski *et al.* (1967) for *Dermestes maculatus*.

COMPOSITION (%): casein (lipid freed), 44.1; maize starch, 25.4; fructose, 17.8; salt mixture BS 2407, 2.6; dry yeast extract, 1.0; distilled water, 9.1. Sterols added in ethyl ether, solvent allowed to evaporate at room temperature.

salt mixture from Glaxo Laboratories Ltd, Greenford, Middlesex, England.

PREPARATION 150 g casein refluxed with 1200 ml chloroform-ethanol 2:1 (v/v) for 30 minutes. Solvent removed by suction through Buchner funnel; residue extracted once more under same conditions. Extracted casein rinsed with chloroform, dried under vacuum in rotary evaporator, ground in blender and sifted through 20-40 mesh sieve. Spread on trays to allow traces of solvent to evaporate.

Casein, starch, fructose, and salt mixed; yeast extract dissolved in water, and added; homogenized in Waring Blendor; further mixed in shaker for 30 minutes. Sterols and oil supplements added as solutions in ethyl ether. Solvent allowed to evaporate at room temperature, and mixture again thoroughly stirred.

REARING At 30° C and 65% RH. See Budowski *et al.* (1967). Stock culture reared at 30° C with fish meal diet enriched with 15% dried yeast and piece of hog adipose tissue. Eggs collected after 2 days by removing beetles and sieving diet. Eggs placed on moistened filter paper in covered petri dishes. 2 newly hatched larvae placed in small, cloth covered plastic vials with 3 g diet.

DEVELOPMENT Average time to pupation on basal diet containing 0.5% cholesterol, 36.2 days.

Levinson, Z.H.; Barelkovsky, J.; Bar Ilan, A.R. 1967. Nutritional effects of vitamin omission and antivitamin administration on development and longevity of the hide beetle *Dermestes maculatus* Deg. (Coleoptera, Dermestidae). J STORED PROD RES 3: 345-352.

COMPOSITION Larval and adult diet (%): fat- and vitamin-free casein, 44.0; maize starch, 17.6; fructose, 26.3; salt mixture, 2.6; cholesterol, 0.3; and water, 9.1; and (mg/g diet): thiamine, 0.023;

riboflavin, 0.011; nicotinic acid, 0.045; pyridoxine, 0.011; calcium pantothenate, 0.023; choline chloride, 0.450; *i*-inositol, 0.225; biotin, 0.0002; and ammonium folate, 0.0005.

salts (%): NaCl, 7.0; $Ca_3(PO_4)_2$, 41.6; $K_3C_6H_5O_7.H_2O$, 40.0; $MgSO_4.H_2O$, 9.6; $FeC_6H_5O_7$, 1.6; KI, 0.1; NaF, 0.08; $MnSO_4$, 0.02; CuI, 0.008; $ZnSO_4$, 0.008; and $K_2SO_4.Al_2(SO_4)_3.H_2O$, 0.008.

REARING At 29 ± 0.5°C and 60-70% RH.

DEVELOPMENT Average of 76 days from egg to death of adult on complete diet, and 37 days on diet deficient in all B-vitamins.

See also: Levinson (1963).

Corollary

→ Levinson, Z.H.; Cohen, E. 1973. The action of overdosed biotin on reproduction of the hide beetle, *Dermestes maculatus*. J INSECT PHYSIOL 19: 551-558.

COMPOSITION (%): dried yeast powder, 5; casein, 10; fish meal, 85; and bimonthly supplement of fresh bacon.

Trogoderma glabrum (Herbst)

→ Beck, S.D. 1971. Growth and retrogression in larvae of *Trogoderma glabrum* (Coleoptera: Dermestidae). I. Characteristics under feeding and starvation conditions. ANN ENTOMOL SOC AM 64: 149-155.

COMPOSITION (parts): Purina Lab Meal, 9; nonfat dried milk, 3; wheat germ, 3; meat and bone meal, 1; brewers' yeast powder, 1.

PREPARATION Ingredients ground together to pass through 20-mesh sieve. Autoclaved at 103 kPa for 10 minutes and stored in sealed fruit jars until use; jars ca. ⅓ filled.

REARING Eggs incubated in darkness at 30°C and 65% RH. Experimental rearing in 10 x 75 mm test tubes. Stock cultures maintained in pint fruit jars containing filter papers.

DEVELOPMENT Larval growth period: ♂♂ 26-30 days; ♀♀ 32-36 days.

Trogoderma granarium Everts, khapra beetle

→ Pant, N.C. 1957. Nutritional studies on *Trogoderma granarium* Everts. Basic food and vitamin requirements. INDIAN J ENTOMOL 18: 259-266.

COMPOSITION (parts): vitamin-free casein, 50; glucose or starch, 50; cholesterol, 1; McCollum's salt mixture, 2; water, 10.

vitamins (μg/g dry diet): thiamin, 25.0; riboflavin, 12.5; nicotinic acid, 25.0; pantothenic acid, 25.0; pyridoxin, 12.5; choline chloride, 500.0; *p*-aminobenzoic acid, 25.0; folic acid, 2.5; biotin, 0.1.

PREPARATION Adapted from Fraenkel & Blewett* (1943a). Small quantity of salt heated in water, added to casein and mixed thoroughly. Dried overnight at 50°C, ground and mixed with other ingredients. Diet thoroughly ground in mortar and pestle and kept for 48 h at 36°C and 50-60% RH.

REARING At 36° C and 50-60% RH. Neonate larvae transferred to glass vials, with perforated corks containing 2 g diet.

DEVELOPMENT Development range, 20-27 days; control on wholemeal flour of wheat + 5% yeast, 20-23 days. Growth index on diet, 4.04; on control, 4.41.

➞ Pant, N.C.; Nayar, J.K.; Gupta, P. 1958. On the significance of amino acids in the larval development of khapra beetle, *Trogoderma granarium* Everts (Coleoptera: Dermestidae). EXPERIENTIA 14: 176-177.

COMPOSITION (parts): amino acid mixture, 1; maize starch, 4; cholesterol, 0.021; lard, 0.025; nucleic acid, 0.025; Osborn-Mandel salt mixture, 0.084; and B-vitamins.

amino acids: as in Lemonde & Bernard* (1951a) for *Tribolium confusum*.

vitamins: as in Pant* (1957). Used @ 25 µg/g diet, except for choline chloride and biotin used @ 500 and 0.01 µg/g diet respectively.

DEVELOPMENT Larval weight on diet with all amino acids after 22 days, 1.92 mg.

See also: Agarwal (1970); Bhattacharya & Pant (1968); Pant (1953); Pant & Pant (1960); Punj (1971); Sehgal & Agarwal (1973).

Elateridae

Ctenicera aeripennis destructor (Brown), prairie grain wireworm

➞ Davis, G.R.F. 1959b. A method for rearing larvae of *Ctenicera aeripennis destructor* (Brown) (Coleoptera: Elateridae) aseptically in test tubes. ANN ENTOMOL SOC AM 52: 173-175.

Best of 11 diets tested.

COMPOSITION (parts): vitamin-free casein, 0 to 100; dextrin, 100 to 0; brewers' yeast powder, 5; cholesterol, 2; McCollum-Davis salt mixture No. 195, 1.

PREPARATION Ingredients ball milled for 24 h. Distilled water added and 5 mm diam pellets formed then dried from resulting paste. 1 g absorbent cotton and 3 ml distilled water placed in 147 x 15 mm diam test tubes. Pellet placed on cotton and tube plugged with non-absorbent cotton. Autoclaved at 15 lb for 20 minutes. Stored under UV light until use.

REARING At 68 ± 2° F and 50 ± 5% RH. Larvae weighing 13-28 mg immersed in 1:1000 solution of $HgCl_2$ and 80% alcohol for 2-5 minutes. Rinsed 2 times in distilled water and placed in test tubes; transferred to fresh diet after 2 months.

DEVELOPMENT Survival after 2 months, 86%. Weight gain of 0.8 mg/mg initial weight. Weight gain on pellets 3 times that on powdered diet.

→ Davis, G.R.F. 1966d. Growth and development of prairie grain wireworm *Ctenicera destructor* Brown (Coleoptera: Elateridae), on synthetic diets. CAN J ZOOL 44: 995-1001.

COMPOSITION Diet 1. See Berger* (1963) diet devised for *Heliothis* spp.

Basal diet supplemented by (as required) (μg/g): putrescine dihydrochloride, 7 and spermidine phosphate.$6H_2O$, 23; putrescine dihydrochloride, 4 and spermidine phosphate.$6H_2O$, 23; hot water extract of germinating rye seed, 19; hot water extract of tea leaves, 19; 70% ethanol extract of tea leaves, 19; and residue from hot-water extraction of tea leaves, 19. Last 4 additions resulting in dietary levels of 1.85% of each substance. Tea extracts made according to Tamaki* (1964).

Diet 2. Chemically defined diet (%): *l*-amino acids: alanine, 0.80; arginine (free base), 1.82; aspartic acid, 1.65; cystine, 0.27; glutamic acid, 5.43; glycine, 0.49; histidine (free base), 1.07; isoleucine, 2.44; leucine, 3.29; lysine hydrochloride, 2.80; methionine, 0.98; phenylalanine, 1.86; proline, 2.06; serine, 1.60; threonine, 1.99; tryptophan, 0.43; tyrosine, 2.05; valine, 2.44; sucrose, 19.16; glucose, 19.16; cellulose powder, 4.89; Wesson's salt mixture, 9.35; palmitic acid, 0.30; stearic acid, 0.04; oleic acid, 0.58; linoleic acid, 1.15; linolenic acid, 0.13; vitamin B solution, 2.82; ascorbic acid, 6.77; cholesterol, 0.71; ribonucleic acid, 0.41; and folic acid, 0.02.

vitamins: as in Davis (1961). (%): folic acid, 0.02; choline chloride, 0.05; *i*-inositol, 0.05; *p*-aminobenzoic acid, 0.005; calcium pantothenate, 0.005; niacin, 0.005; pyridoxine hydrochloride, 0.005; riboflavin, 0.005; thiamine hydrochloride, 0.005; biotin, 1×10^{-5}; vitamin B_{12}, 12×10^{-5}.

amino acid levels of diet based on analysis of amino-acid content of the prairie grain wireworm described by Kasting *et al.* (1962).

Diet modified to form second chemically defined diet by supplementing with 1.85% of material extracted from germinating rye seed with hot water.

PREPARATION Dry ingredients blended. 2.5 g nutrient agar dissolved in 62 ml distilled water, added and further blended. Dispensed to petri dishes.

REARING At 68 ± 2°F and 75 ± 5% RH in petri dishes with diet and piece of moistened filter paper. All diets described above tested over 6 month period.

DEVELOPMENT Greater larval survival on diets supplemented with putrescine and spermidine, and with hot water or ethanol extracts of tea leaves, and on unsupplemented chemically defined diet with putrescine and spermidine; survival, 85%; weight gain, >100%.

→ Davis, G.R.F. 1968b. Effects of 2,4-D, maleic acid hydrazide, and 2,4,5-T on survival and growth of the prairie grain wireworm, *Ctenicera destructor* Brown (Coleoptera: Elateridae). CAN J ZOOL 46: 747-750.

Modification of diet by Ito & Horie (1962) for *Bombyx mori*.

COMPOSITION (mg): wheat gluten, 435.6; potato starch, 96.9; glucose 48.4; sucrose, 48.4; brewers' yeast, 48.4; cellulose powder, 290.4; soybean sterols, 1.9; ascorbic acid, 19.4; β-carotene, 1.0; McCollum-Davis salt mixture No. 185, 9.7; methyl *p*-hydroxybenzoate, 0.1.

PREPARATION Components mixed and autoclaved at 15 lb pressure for 20 minutes in petri dishes. Diet sliced when cool, wrapped in aluminium foil and frozen until use.

REARING At 68 ± 2° F and 75 ± 5% RH. Diet renewed weekly.

DEVELOPMENT Initial larval weight of 57 mg increased to 86.3 mg after 6 months. Survival, 50%. Adult also feed on diet.

See also: Davis (1961, 1965).

Limonius dubitans Le Conte

Melanotus depressus (Melsheimer)

→ Keaster, A.J.; Chippendale, G.M.; Pill, B.A. 1975. Feeding behaviour and growth of the wireworms *Melanotus depressus* and *Limonius dubitans*: effect of host plants, temperature, photoperiod and artificial diets. ENVIRON ENTOMOL 4: 591-595.

Artificial diet modified from Jacob & Chippendale (1971) for *Diatraea grandiosella*. Best of natural and artificial diets tested.

COMPOSITION (g/100 g): powdered cellulose, 3.5; vitamin-free casein, 4.0; wheat germ, 3.5; agar, 3.0; sucrose, 2.9; Wesson's salt mixture, 1.1; vitamin mixture, 1.1; *l*-ascorbic acid, 0.6; 4M KOH, 0.5; wheat germ oil, 0.3; methyl *p*-hydroxybenzoate, 0.2; cholesterol, 0.2; sorbic acid, 0.1; 37% formaldehyde, 0.1; distilled water, 78.9.

PREPARATION See Jacob & Chippendale (1971). Agar dissolved in boiling water and blended with other ingredients. Diet dispensed in plastic cups.

REARING Larvae weighing 30-60 mg held individually in 28 g plastic cups at 20° C. Artificial diet given on moist cellucotton and replaced every 14 days.

DEVELOPMENT 35% larval mortality compared with 0.2% mortality on control of wheat-loam substrate. Diet incomplete.

Lyctidae

Lyctus planicollis Le Conte, southern lyctus beetle

→ Mauldin, J.K.; Lambremont, E.N.; Graves, J.B. 1971. Principal lipid classes and fatty acids synthesized during growth and development of the beetle *Lyctus planicollis*. INSECT BIOCHEM 1: 316-326.

COMPOSITION (g): whole wheat flour, 500; corn meal (Masa Harina), 400; Difco Bacto yeast extract, 96; methyl *p*-hydroxybenzoate, 2; ascorbic acid, 2.

REARING At 27° C.

DEVELOPMENT After inoculation with first instar larvae, pharate adults and emerged adults collected at 60 and 75 days respectively. Larval development incomplete on low-fat diet. Four generations reared.

Meloidae

Nemognatha lurida apicalis Le Conte

Nemognatha nigripennis Le Conte

Nemognatha nitidula (Fabricius)

→ Leppla, N.C.; Standifer, L.N.; Erickson, E.H. Jr 1974. Culturing larvae of blister beetles on diets containing different pollens collected by honeybees. J APIC RES 13: 243-247.

COMPOSITION Adult diet (g): honey, 1; methyl *p*-hydroxybenzoate, 0.1/ 100 ml 6% sucrose solution.

Larval diet: 1000 g pollen (*Tavaracum officinale*) best with little honey, few ml demineralized water and 0.5 g methyl *p*-hydroxybenzoate.

PREPARATION Ingredients well blended to obtain thick paste; diet stored in tube from trimmed, folded and elongated heavy duty aluminium foil. Procedure illustrated.

REARING At 26 ± 3°C and 20 ± 5% RH with 14 h photoperiod until triungulin stage, then at 31 ± 2°C and >90% RH for diapausing coarctate stage. Diapause terminated at 6 ± 3°C for 75 days. Then pairs placed in 16-oz waxed paper cups with transparent, perforated plastic lids. Diet provided in cotton filled 30 ml cup feeders. Oviposition sites of rolled paper towels and cups, made to resemble thistle blossoms or sunflowers. Fresh food daily. Egg masses left in adult cages until triungulins. Triungulins held in 2 ml glass vials ¾ full of diet @ 5 triungulins/vial, and sealed with cotton plug. Cannibalism and drowning reduces larval survival to one adult/cell.

DEVELOPMENT *N. nitidula:* egg to coarctate, 28.4 days and egg to adult, 158 days; survival, 71.4%.

N. lurida apicalis: egg to coarctate, 33.5 days and egg to adult, 164 days; survival, 40.0%.

N. nigripennis: egg to coarctate, 34.7 days and egg to adult, 71 days; survival, 44.0%. Shows greatest potential for establishment of laboratory colony.

→ Erickson, E.H.; Werner, F.G. 1974. Bionomics of nearctic bee-associated Meloidae (Coleoptera); life histories and nutrition of certain Meloinae. ANN ENTOMOL SOC AM 67: 394-400.

Six species reared.

COMPOSITION pollen, honey (amounts not specified), mould inhibitor, 1.5 g Moldex/1000 g medium.

PREPARATION Homogenous paste prepared so that surface moisture appeared within 30 seconds of formation of 'pellet', yet not in sufficient quantity to permit pool to form in feeding chamber.

REARING Adults at 25°C. Paired adults kept in ½-pint waxed paper cartons with perforated clear plastic lids. Fresh host plant given daily for oviposition. Beetles and debris removed and eggs left; carton sealed with unperforated lid and incubated. Neonate larvae to 5 x 1 cm diam, sterile glass tube rearing chamber with diet, plugged with cotton at both ends. Tube changed if mould or fresh diet needed. 5th instars transferred to 1.3 cm diam tube. Clean tube for coarctates. Larvae reared at 31 ± 0.5° C, and >90% RH.

DEVELOPMENT Species reared: *Lytta magister* Horn, *Lytta mutilata* (Horn), *Pleurospasta mirabilis* (Horn), *Pyrota akhurstiana* Horn, *Pyrota postica* Le Conte, *Tegrodera erosa aloga* Skinner. Large adults of uniform size of both sexes produced. Larvae of all species entered diapause. Viable eggs laid.

Nitidulidae

Carpophilus hemipterus Linnaeus

→ Stride, G.O. 1953. On the nutrition of *Carpophilus hemipterus* L. (Coleoptera: Nitidulidae). TRANS R ENTOMOL SOC LONDON 104: 171-194.

COMPOSITION Stock diet. Boiled dates sprinkled with dried yeast.

Artificial diet (g): mixture T, 10; dried yeast, 10; charcoal, 5; Cellophas B, 2.5; and (ml): saturated choline chloride solution, 0.1; α-tocopherol acetate solution (3.4 mg/ml), 0.25; vitamin mixture, 10; water, 12.5; and (mg): ergosterol, 25; yeast nucleic acid, 400; biotin in 1 ml water, 100 μg.

mixture T (parts): casein, 19; glucose, 41; cholesterol, 4; McCollum's salt mixture No. 185, 4.

vitamins (mg/50 ml solution): thiamin, 5; riboflavin, 5; nicotinic acid, 5; pantothenic acid, 5; pyridoxin, 5; *p*-aminobenzoic acid, 5; inositol, 500; folic acid, 5; thymine, 10; tryptophane, 250.

PREPARATION Mixture T components mixed, dried at 80° C and ground overnight in ball mill. Yeast, charcoal, yeast nucleic acid and alcoholic extracts added to mixture T with excess alcohol, evaporated by drying overnight. Wet fraction of "cellophas B", warm vitamin solutions (with few drops NaOH) and water mixed and left overnight. Wet and dry fractions mixed and kneaded by hand for 10 minutes. Small pad of food flattened in bottom of 2 x ½" diam tube, remainder rolled into long cylinders by hand and 1" lengths added to tubes; stoppered with rubber bungs with hole in centre plugged with cotton wool. Autoclaved for 30 minutes at 14 lb pressure.

REARING Stock culture at 25° C and 75% RH in 3-oz culture bottles with organdie tops and bakelite caps. Eggs laid between organdie and cap. Pupation in clear glass dishes. Aseptic rearing in 2 x ½" tubes. Eggs washed in tap water for 5 minutes, then White's sterilizing fluid and little bile salt in same tube for 25 minutes. Eggs rinsed in tube of distilled water then squirted on to filter paper in sterile petri dishes and transferred aseptically to tubes using brush at rate of 8 eggs/tube. Tubes kept at 25° C and 95% RH.

DEVELOPMENT Eggs hatch in 48 h; on artificial diet, wt of 14 day old larvae, 3 mg. High larval mortality if water in diet less than 50%. On stock diet, larval survival only when large number of microorganisms and dried yeast present. Adult emergence within one week of pupation.

Scarabaeidae

Aphodius constans Duftschmid

Charpentier, R. 1968. Élevage aseptique d'un Coléoptère coprophage : *Aphodius constans* Duft. (Col., Scarabaeidae). ANN EPIPHYT 19: 533-538. (In French with English summary)

COMPOSITION (g): cattle dung (40% water), 250; casein, 15; soya flour, 10; dried brewers' yeast, 10; and (ml): exchanged water, 200.

PREPARATION Dung dried, sieved and other ingredients added. RH fixed at 70%; dispensed to 25 ml tubes, slanted and sterilized at 120° C for 20 minutes.

REARING Eggs submerged in Teepol for several minutes and any debris removed under microscope. Rinsed in water and sterilized in 4% NaClO solution for 5 minutes at 18-20° C. Finally rinsed thoroughly in sterile water and dispensed at rate of 2-4 eggs/diet tube. Tubes sealed with cotton plugs and maintained at 18° C.

DEVELOPMENT Eclosion, ca. 63%; larval mortality, 59-66%; pupation, after 4-5 weeks; mean pupal wt, 35.3 mg compared with 17.7 mg for mean septic pupal weight. One generation reared.

Popillia japonica Newman, Japanese beetle

Goonewardene, H.F.; McKay, J.E. 1969. An artificial diet for the adult Japanese beetle. J ECON ENTOMOL 62: 964.

COMPOSITION Adult diet (g): apple sauce, 22.16; Alphacel, 9.49; Log Cabin syrup, 28.48; ascorbic acid, 3.16; wheat germ, 22.16; pectin, 2.53; protein (1 g tablets) (Hudson Vitamin Products, New York, N.Y.), 9.49; pet vitamins (1.5 g tablets) (Hudson Vitamin Products, New York, N.Y.), 2.53; and (ml): Hyamine 3500 (1:2500 v/v) (Rohm & Hass Co., Philadelphia, Pa.), 3.8; methyl *p*-hydroxybenzoate (0.2 g/ml in 95% ethyl alcohol), 2.5; sorbic acid solution (5% w/v in 95% ethyl alcohol), 1.7; distilled water, 158.0.

PREPARATION Apple sauce and Alphacel mixed together in heavy duty household mixer at low speed. Wheat germ, syrup and water added during mixing, and other ingredients added individually when mixture homogenous. Pectin added last, and diet mixed for further 5 minutes before storing in sealed jars at 2° C until use.

REARING Equal volumes of sassafras sandy loam soil, peat moss, Vermiculite and sand placed in quart polystyrene jars with screw caps. 15 ml pill cups lined with 2 pieces dental wick moistened with distilled water and filled with diet. 20 beetles/cup placed in jars with oviposition medium. Food and oviposition medium changed weekly; eggs collected every 4 days.

DEVELOPMENT Fecundity of 16 eggs/♀ on diet; 2 eggs/♀ on apple slices. 93% egg hatch from diet fed ♀♀.

→ Goonewardene, H.F.; Zepp, D.B.; McGuire, J.U. Jr 1974. Evaluation of artificial diets for 3rd-stage larvae of Japanese beetles. ANN ENTOMOL SOC AM 67: 413-415.

COMPOSITION 10 g acetone extracted grass-clover powder from *Lolium perenne*, *Agrostis alba* and *Trifolium repens*. Then various levels of protein source (g): tablets - (Hudson Vitamin Products, N.Y.) (1.5-2.5); cholesterol (0.4-0.6); choline chloride (0.7-0.12); Wesson's salt mixture (0.25-0.30); and (ml): linoleic acid (0.5-0.10); linolenic acid (0.7-0.12) and vitamin mixture (1.75-2.25) as supplements, alone and in combinations.

vitamins: 1 part vitamin B stock solution (NBC), 1 part inositol (10 mg/ml) and 2 parts ascorbic acid (10% aqueous solution).

PREPARATION Requisite ingredients ground and blended with grass-clover powder. Blended mixture refrigerated until required.

REARING About 1 g diet spread evenly on moistened filter paper and inverted over third instar larvae placed on 150 g mixture of moist sphagnum/peat-soil in plastic container; covered with further 150 g peat-soil. Container sealed, held at 25°C and 70% RH, and examined every 7 days.

DEVELOPMENT About 70 diets tested on field and laboratory larvae; most successful, giving about 70% pre-pupation and 40-50% adults, suggests that grass-clover powder must be supplemented with choline chloride, linoleic acid and linolenic acid.

Scolytidae

Dendroctonus frontalis Zimmerman, southern pine beetle

→ Clark, E.W. 1965. An artificial diet for the southern pine beetle and other bark beetles. U S FOR SERV RES NOTE SE-45: 3 pp.

COMPOSITION (g): fine pine bark or cellulose, 40.0; brewers' yeast, 2.0; glucose, 3.0; fructose, 3.0; sucrose, 3.0; starch, 2.0; casein hydrolysate, 2.0; Wesson's salt mixture, 0.25; agar, 8.0; and (ml): wheat germ oil, 0.5; cottonseed oil, 0.5; sorbic acid (20% ethanol w/v), 5.0; distilled water, 200.

PREPARATION Inner bark homogenized with 100 ml distilled water in blender with special blade assembly until particle size <2 mm. Ingredients, less agar, blended. Sorbic acid dissolved in ethanol and added in solution. Agar dissolved in 100 ml hot water and blended into mixture. pH adjusted to 4.8 with HCl or KOH. Diet dispensed to 15 x 100 mm petri dishes to $\frac{1}{8}$-¼" thickness and left for 24 h to allow evaporation; refrigerated for up to one month.

REARING At 27°C in dark; humidity not controlled. 24 eggs or larvae seeded under section cut in diet.

DEVELOPMENT Within range found under natural conditions; survival to adult, 20%. No oviposition on diet. Other species reared: *Ips calligraphus* (Germar); roundheaded pine beetle, *Dendroctonus adjunctus* Blanford, reared individually because of cannibalism and dietary requirements. Pales weevils, *Hylobius pales* (Herbst), reared at 22°C; oviposition; eggs removed and hatched on damp cotton.

Dendroctonus ponderosae Hopkins, mountain pine beetle

➡ Jones, R.G.; Brindley, W.A. 1970. Test of eight rearing media for the mountain pine beetle, *Dendroctonus ponderosae* (Coleoptera: Scolytidae), from lodgepole pine. ANN ENTOMOL SOC AM 63: 313-316.

Two best of 8 diets tested.

COMPOSITION & PREPARATION Diet 1: see Schmidt* (1966); and Yearin & Wilkinson* (1965) except phloem of lodgepole pine, *Pinus contorta*, dissected from fresh, uninfested logs, lopped in spring. Phloem cut into strips and 100 g macerated with 387 ml sterile water in Omni Mixer.

pH of diet adjusted with 2N HCl to 4.5 to prevent growth of mould. After mixing, filtered under vacuum with a Buchner funnel.

REARING Diet poured into petri dishes to depth of 6 mm; perforated plastic disc used to compact diet and squeeze out excess moisture. Covered with 3 layers of filter paper and petri dish cover. Diet kept at room temperature used within 7 days, or refrigerated and used up to 35 days later.

DEVELOPMENT Two diets successful for production of normal adults from 1st stage larvae on "Y" and "S" diets respectively, days to pupa 19, 19; days to adult 23, 23; pupation, 48.3%, 35.7%; adult emergence, 40.0, 21.4%.

Dendroctonus pseudotsugae Hopkins, Douglas-fir beetle

➡ Schmidt, F.H. 1966. Two artificial (oligidic) media for the Douglas-fir beetle, *Dendroctonus pseudotsugae* Hopkins (Coleoptera: Scolytidae). CAN ENTOMOL 98: 1050-1055.

Two diets described.

COMPOSITION Diet 1 (%): fresh-frozen inner bark of Douglas-fir, 16.0; brewers' yeast, 2.0; sodium benzoate, 0.1; sorbic acid, 0.05; distilled water, 81.25.

Diet 2 (%): lyophilized powder of inner bark of Douglas-fir, 10.0; brewers' yeast, 3.0; glucose, 1.5; Alphacel, 1.85; agar, 3.5; sodium benzoate, 0.1; sorbic acid, 0.05; distilled water, 80.0. pH adjusted to 4.5.

PREPARATION Diet 1. Components blended for 5 minutes and pH adjusted to 4.5. Diet transferred to petri dishes @ 100 g/dish and dried in vacuum oven at 110°C with 15" vacuum for 9-10 h. Then stored at 25 ± 1°C.

Diet 2. Dry powdered components, except agar, mixed thoroughly. Agar heated with some distilled water in steam cone to 98°C. Remaining water heated in steam cone to 95-98°C, added to dry components and mixed. Hot agar added and diet thoroughly mixed; pH adjusted to 4.5; transferred to petri dishes @ 100 g/dish. Cooled, and shallow holes drilled in diet. Stored at 25 ± 1°C.

REARING At 25 ± 1°C. 10 neonate larvae/dish. Mature adults introduced to 2 ft bole sections of Douglas fir. Held at 25 ± 3°C for 3 days then sections debarked and eggs removed. Surface sterilized in 1% NaClO for 7 minutes, rinsed with sterile distilled water and incubated at 25 ± 3°C and 100% RH.

DEVELOPMENT Both diets satisfactory. On diet 1, larval period, 20 days; adults in 24 days. On diet 2, larval period, 20 days; adults in 28 days. Control larval period, 29-30 days; adults in 38-41 days.

See also: McGhehey & Nagel (1966).

Hylastinus obscurus (Marsham), clover root borer

Byers, R.A. 1974. Artificial diets for maintaining the adult clover root borer. J ECON ENTOMOL 67: 806.

Three adult diets tested.

COMPOSITION Diet 1 (g): Vanderzant's insect diet (NBC), 30; agar, 7.2; methyl paraben, 0.4; sorbic acid, 0.4; torula yeast, 0.5; ascorbic acid, 0.7; streptomycin sulphate, 0.5; and (ml): distilled water, 190.0.

Diet 2 (g): CSM diet (Krause Milling Co., Milwaukee, WI), 20.0; agar, 5.7; methyl paraben, 0.2; sorbic acid, 0.2; torula yeast, 0.25; ascorbic acid, 0.35; streptomycin sulphate, 0.25; and (ml): distilled water, 150.0.

Diet 3 (g): Alphacel, 31.2; agar, 7.2; methyl paraben, 0.4; sorbic acid, 0.4; streptomycin sulphate, 0.5; and (ml): distilled water, 190.0.

PREPARATION Agar boiled in 120 or 130 ml water then cooled to 60°C and blended with other dry ingredients and mixed in remaining water. 1-oz plastic cups half filled with diet and left to cool. Then sealed with tight fitting lids.

REARING Holes punched in diet surface and 5 adult borers placed/cup each head down in hole. Cups held at 20-27°C in darkness for 5 months. Then in refrigerator at 5°C for 1 month in attempt to stimulate oviposition. Finally at 20-25°C for 12 h photoperiod.

DEVELOPMENT Mean survival in days on diet 1, 196.8; on diet 2, 222.2; and on diet 3, 72.9. No eggs laid. Useful maintenance diet.

Ips calligraphus (Germar)

Richeson, J.S.; Wilkinson, R.C.; Nation, J.L. 1970. Development of *Ips calligraphus* on foliage-based diets. J ECON ENTOMOL 63: 1797-1799.

Twelve diets tested for rearing larvae to adult stage. Best diet contained previous year's (mature) or current year's (immature) foliage of *Pinus elliotti* Engelman var. *elliotti*.

COMPOSITION (g): agar, granular, 7.0; brewers' yeast, 6.0; Alphacel, 3.7; glucose, 3.0; methyl *p*-hydroxybenzoate, 0.2; sorbic acid, 0.1; and (ml): distilled water, 130.0. Foliage at 10, 20, 30, 40, 50 or 60 g/diet.

PREPARATION Agar dissolved in boiling water, added to combined dry ingredients and mixed thoroughly. Poured into glass petri dishes to depth of 8-10 mm; plastic disposable petri dish pressed on to surface of hot diet and left in place; dishes stored at 7°C in plastic bags. Diets with petri dish partially removed, allowed to warm in rearing incubator to loose excess moisture for 12 h prior to use.

REARING At 30° C and 50-65% RH in darkness. 20-25 1st stage larvae transferred from egg incubation dishes to individual depressions on diet surface. Cover replaced. Eggs obtained from inner bark of artificially infested typical slash pine logs, and transferred to moist filter paper in petri dishes.

DEVELOPMENT For diets containing 11.8-28.6% foliage, 19-44% larvae reared to adults on previous year's foliage versus 1-9% on that of current year. Larval and pupal development slower on current year's foliage diets; normal pigmentation on adults not developed. Adults reared on both ages of foliage able to reproduce normally on inner bark of pine logs (natural food and habitats) for 3 generations.

→ Richeson, J.S.; Nation, J.L.; Wilkinson, R.C. 1971. Fatty acid composition in *Ips calligraphus* (Coleoptera: Scolytidae) during post-embryonic development. ANN ENTOMOL SOC AM 64: 251-254.

Dietary modification based on Bedard* (1966).

COMPOSITION (g): brewers' yeast, 10; methyl *p*-hydroxybenzoate, 0.8; fine powdered phloem tissue, 100.

PREPARATION Fresh bark phloem, from *Pinus elliotti* Engelmann var. *elliotti*, frozen at -61° C, then homogenized in Waring Blendor for 15 seconds. Resultant semipowder refrozen at -61° C, and homogenized to fine powder prior to use. Ingredients mixed at speed 3 in 6-speed Nutone Food Center Unit for about 10 minutes. About 20 g pressed firmly into glass petri dish using slightly smaller plastic petri dish base. Stored at 7° C with plastic dish cover in place and covered with glass petri dish top, or used immediately.

REARING As in Richeson *et al.** (1970).

DEVELOPMENT Average time to prepupa, ca. 8.5 days; average duration of pupal stage, ca. 3 days.

→ Yearian, W.C.; Wilkinson, R.C. 1963. An artificial rearing medium for *Ips calligraphus* Germ. FLA ENTOMOL 46: 319-320.

COMPOSITION (g): powdered cellulose, 10.0; sucrose, 4.0; brewers' yeast, 7.0; choline chloride, 0.13; cholesterol, 0.13; glycine, 0.13; vitamin diet fortification mixture (NBC), 0.50; methyl *p*-hydroxybenzoate, 0.15; sorbic acid, 0.2; agar, 4.0; and (ml): peanut oil, 0.35; water, 73.5; pH of medium about 5.

PREPARATION Agar dissolved in hot water and blended with peanut oil. Combined dry ingredients slowly added and homogenized at low speed. Hot diet dispensed to 20 x 100 mm petri dishes, @ 40 ml/dish and cooled.

REARING At 80° F and 30-40% RH. 5 neonate larvae seeded to depressions in diet surface. Eggs laid in phloem of infested pine bolts; teased out with dissecting needle and washed for 5 minutes in sterilizing solution of: $HgCl_2$, 0.25 g; NaCl, 6.5 g; HCl, 1.25 ml; ethanol, 250 ml; distilled water, 750 ml. Rinsed for 3 minutes in sterile water, transferred aseptically to petri dish lined with moist filter paper and incubated at 80° F.

DEVELOPMENT Average larval survival on diet, 10%; adult emergence from pupae, 99%. Total developmental period on diet comparable to that under natural conditions.

→ Yearian, W.C.; Wilkinson, R.C. 1965. Two larval rearing media for *Ips* bark beetles. FLA ENTOMOL 48: 25-27.

Three species reared: *Ips calligraphus* Germar, *Ips grandicollis* (Eichhoff), fine spined bark beetle, *Ips avulsus* (Eichhoff), small southern pine engraver.

COMPOSITION Diet 1 (g): Alphacel, 45.0; agar, 2.0; sucrose, 1.0; fructose, 1.0; vitamin diet fortification mixture (NBC), 1.0; brewers' yeast, 1.0; soybean protein, 1.0; choline chloride, 0.05; cystine, 0.05; glycine, 0.05; cholesterol, 0.05; Wesson's salt mixture, 0.25; sorbic acid, 0.2; methyl *p*-hydroxybenzoate, 0.15; streptomycin sulphate, 0.01; and (ml): peanut oil, 0.35; water, 47.0.

Diet 2 (g): pine phloem (fresh weight), 46.0; agar, 2.0; sucrose, 1.0; fructose, 1.0; vitamin diet fortification mixture (NBC), 0.5; brewers' yeast, 1.0; soybean protein, 1.0; Wesson's salt mixture, 0.1; sorbic acid, 0.2; methyl *p*-hydroxybenzoate, 0.15; streptomycin sulphate, 0.01; and 47.0 ml water.

PREPARATION See Yearian & Wilkinson* (1963). Agar in diet 2 dissolved in heated water phloem filtrate and macerated phloem added with dry ingredients. In both cases, blotting paper discs placed over diet when in petri dish.

REARING At 80°F and 30-40% RH. 10 neonate larvae transferred to each dish through ten small slits cut in blotting paper cover. Slits then sealed with agar. Eggs collected from phloem, washed for 2 minutes in 50% ethanol containing 1 part/1000 $HgCl_2$, rinsed for 3 minutes in sterile water then transferred aseptically to petri dish lined with moist filter paper for incubation.

DEVELOPMENT Both diets satisfactory for all three *Ips* spp, but larval survival greater on phloem-based diet. Survival on artificial diet averaged, 30% (0-50%); on phloem-based diet, 50% (30-90%). Mean larval development period for *I. calligraphus* and *I. grandicollis*, ca. 15 days (10-20 days). Pupal survival of both species, ca. 100%.

Ips cembrae Heer

→ Balogun, R.A. 1969. Rearing of *Ips cembrae* larvae on artificial diets: some essential amino acids and sugars. J INSECT PHYSIOL 15: 141-148.

Composition of diets based on qualitative amounts of amino acids and sugars obtained from analysis of larch bark tissue.

COMPOSITION & PREPARATION Amino acids: valine, leucine, arginine, threonine, histidine, glutamic acid, aspartic acid, serine, α-alanine and glycine. Sugars: raffinose, sucrose, glucose, and fructose. Unspecified amount of powdered pith of larch bark. Each nutrient prepared as 2% aqueous solution, and each autoclaved in separate container at 15 lb pressure for 15 minutes. Amino acids and sugars autoclaved separately with minimum autoclaving pressure and time. Sterilized powder pith used to provide necessary supporting matrix for tunnelling of larvae. Control diet prepared by mixing equal quantities of each of amino acids and sugars together. Diets poured into sterile vials.

REARING At 22 ± 0.5°C and 75% RH, individually and aseptically in 2 x 0.25" vials.

DEVELOPMENT Pupation from neonate larvae, 86.6%.

Ips confusus (Le Conte), California five-spined ips

Bedard, W.D. 1966. A ground phloem medium for rearing immature bark beetles (Scolytidae). ANN ENTOMOL SOC AM 59: 931-938.

COMPOSITION (g): inner bark, 100; dried brewers' yeast, 10; sodium benzoate, 0.8; and distilled water, 500 ml.

PREPARATION Inner bark ('ponderosa' pine) removed with scalpel from 20-40 year old trees, cut and stored for ≤2 months in cool, moist conditions. Then blended with water, brewers' yeast, and sodium benzoate for 2 minutes; pH to 3.7-4.0 with 2N HCl. Dispensed to petri dishes at ca. 100 g/dish. Excess liquid removed by squeezing pulp in dishes with perforated plastic disc. Stored up to 5 days at 10°C and 100% RH.

REARING Eggs removed from infested logs with moistened brush. Incubated in moist petri dishes at 26.7°C. Neonate larvae placed on diet every 10 h. Kept at 26.7°C and 100% RH for 2 days, then transferred to lower humidity.

DEVELOPMENT At 25°C, 9 adults/dish in 13 days.

Other species reared on diet: *Ips pini* (Say) [previously *I. oregonis* (Eichhoff)] adults in 13 days; *Ips plastographus* (Le Conte), 13 days; *Ips mexicanis* (Hopkins) [previously *I. radiatae*], 24 days; *Dendroctonus brevicomis* Le Conte, 46 days; *Dendroctonus pseudotsugae* Hopkins, 32 days; *Dendroctonus ponderosae* Hopkins [previously *D. monticole*], 22 days; *Dendroctonus valens* Le Conte, 50 days; *Pseudotsugae menziesii* (Mirb) on Douglas fir. Diet not suitable for *D. valens* Le Conte.

Pseudopityophthorus minutissimus (Zimmerman)

Pseudopityophthorus pruinosus (Eichhoff)

Rexrode, C.O.; Krause, C.R. 1968. Rearing larva of *Pseudopityophthorus pruinosus* and *P. minutissimus* on ground phloem media. ANN ENTOMOL SOC AM 61: 814-816.

Two ground phloem diets developed.

COMPOSITION Diet 1 (g): ground phloem of red oak, 100.0; agar, 10.0; sorbic acid, 0.5; methyl paraben, 0.5; and (ml): distilled water, 200; 70% ethyl alcohol, 10.

Diet 2 (g): ground phloem of red oak, 100; agar, 10; and (ml): distilled water, 210.

PREPARATION Diet 1: agar dissolved in boiling water; sorbic acid and methyl paraben dissolved in warm ethyl alcohol, added to agar solution and heated to evaporate alcohol. Ground phloem added and blended at 85°C; pH of ca. 4.5. About 8 g of hot diet placed in centre of 4 x 4" glass plate and sandwiched to thickness of 1 mm by another glass plate. "Sandwich" refrigerated at 4°C for 1 h, then dried for 10 minutes at 122°C with 1 glass plate removed; and cooled under culture hood at room temperature. Moisture content approximately 45% after cooling process.

Diet 2: as for diet 1, omitting sorbic acid and methyl paraben. To help prevent microbial contamination open-face sandwiches treated with ultraviolet light (2737A°) for 30 minutes.

Ground phloem prepared as follows: 3" diam bole and limb sections of 15-25 yr old black, scarlet and red oak (*Quercus velutine, Q. coccinea, Q. rubra*), debarked. Fresh phloem strips immediately placed in drying oven at 122°C for 18 h, then removed and ground in Wiley Mill using 40-mesh sieve. Ground phloem frozen and stored at -20° ± 1°C until used. Entire bark macerated; no attempt made to separate inner and outer bark.

REARING 2nd and 3rd instar larvae from infested bolts collected by sterile brush. One larva placed in each of 10 chambers in sandwich. Chambers made with dissecting needle. Top glass plate replaced, fastened with two clamps and sealed along edges with 1" masking tape. Kept in dark at 26°C and 50% RH.

DEVELOPMENT Both diets suitable for both species. All adults reared from larvae active and of normal size.

Scolytus multistriatus (Marsham), smaller European elm bark beetle

→ Galford, J.R. 1967a. A technique for rearing larvae of the smaller European elm bark beetle on an artificial medium. J ECON ENTOMOL 60: 1192.

COMPOSITION (g): agar, 32.00; sucrose, 16.00; fructose, 8.00; glucose, 8.00; vitamin diet fortification mixture, 12.00; brewers' yeast, 40.00; soybean protein, 16.00; Wesson's salt mixture, 20.00; cholesterol, 0.80; Kretschmer wheat germ, 20.00; vegetable lecithin, 0.80; vitamin B_t, 0.08; sorbic acid, 2.00; methyl paraben, 1.00; Alphacel (hydrolyzed), 320.00; and (ml): wheat germ oil, 4.00; water, 800.00.

PREPARATION Agar and sugars boiled with 800 ml distilled water. Sorbic acid and methyl paraben dissolved in 15 ml of warm 95% alcohol added to agar/sugar solution, and further boiled until alcohol evaporated. Solution then blended with remaining ingredients, except Alphacel, for 2 minutes at high speed. Alphacel then added slowly until mixture too viscous for blender, then hand blended. Diet sandwiched thinly between two 4" square glass plates.

REARING Eggs removed from artificially infested bolts and incubated on moist filter paper in petri dishes in dark at 82°F and 80% RH. Neonate larvae transferred singly with brush into each of 15 holes made in diet with dissecting needle. Plates clamped together and edges sealed with masking tape. Kept at 82°F and 80% RH in darkness.

DEVELOPMENT Rate of development on diet similar to that of laboratory culture maintained on elm bolts. Adults of normal appearance and capable of flight. Production of egg galleries in elm bolts normal. Eggs viable. After 4 generations of rearing larvae on diet, and returning diet-reared beetles to elm bolts for oviposition, larval survival to adult stage, 80%; no apparent decline in vigour of adults. Developmental period, ca. 35 days.

See also: Galford (1967b).

→ Galford, J.R. 1969b. A larval and ovipositional medium for the smaller European elm bark beetle. J ECON ENTOMOL 62: 1166-1168.

Two diets tested.

COMPOSITION Diet 1 (g): wheat germ (Kretschmer), 30.0; Alphacel (hydrolyzed), 70.0; granulated agar, 8.0; sorbic acid, 0.4; methyl paraben, 0.3; and distilled water, 200.0 ml.

Diet 2: as diet 1, except sorbic acid and methyl paraben replaced by 0.5 g sodium benzoate; and 0.1 g of Vanillin added.

PREPARATION Agar dissolved in boiling distilled water and blended at high speed with methyl paraben and sorbic acid (or sodium benzoate). Wheat germ (plus Vanillin in diet 2) added slowly and blended for 5 minutes. Speed reduced to low setting and Alphacel added slowly until mixture too viscous for blender; then remaining Alphacel added and mixed with spatula. Moisture content of diets important for successful rearing of larvae and oviposition; <56% recommended; method for achieving this described. Methyl paraben and sorbic acid toxic to eggs, so use of diet 2 with sodium benzoate recommended for oviposition.

REARING Eggs dissected from elm wood and neonate larvae placed in holes in diet. Rearing plates clamped together, and edges sealed with 1" masking tape. Kept at 30 ± 2° C and 50 ± 20% RH in darkness. Adults transferred to new diet plate for oviposition.

DEVELOPMENT Two generations reared successfully. Larval survival, 88% and 83% respectively. Mean adult development periods, 23 ± 1 and 22 ± 1 days. Viable eggs laid in both generations, with 80-93% hatch.

→ Galford, J.R. 1972. Some basic nutritional requirements of smaller European elm bark beetle larvae. J ECON ENTOMOL 65: 681-684.

Two diets tested.

COMPOSITION Diet 1 (g): Alphacel (hydrolyzed), 80; agar granulated, 8; sucrose, 8; cholesterol, 0.4; Wesson's salt mixture, 3; egg albumen, 10; sorbic acid, 0.5; methyl paraben, 0.3; choline chloride, 0.5; inositol, 0.4; and (ml): vitamin B solution, 20; water, 200.

Diet 2: as diet 1, except egg albumen replaced by 10 g vitamin-free casein.

vitamin B solution (g/100 ml water): niacinamide, 1.0; calcium pantothenate, 1.0; thiamine hydrochloride, 0.25; riboflavin, 0.5; pyridoxine hydrochloride, 0.25; folic acid, 0.25; biotin, 0.25.

PREPARATION Agar and sugar dissolved in 200 ml boiling distilled water. Sorbic acid and methyl paraben dissolved in 2 ml hot absolute alcohol and added; boiled for 1 minute to remove most of alcohol. Remaining ingredients, except vitamin solution and Alphacel, added and blended at high speed for 2 minutes. Vitamin solution added and blended for 1 minute, Alphacel added and blended by hand when too viscous for blender. Diet refrigerated in sealed petri dishes.

REARING See Galford* (1967a, 1969b). In plastic petri dishes. Neonate larvae from laboratory infested elm bolts or from laboratory culture. Eggs kept on black, moist filter paper in glass petri dish. Rearing plates kept at 30 ± 2° C and 50 ± 20% RH in darkness.

DEVELOPMENT On both diets 1 and 2, 65-78% adult emergence. Effects on adult emergence by deletion of various nutrients, lipids, sterols, salts, and vitamins discussed.

Corollary

→ Galford, J.R. 1971. Improved techniques for rearing the smaller European elm bark beetle on artificial media. J ECON ENTOMOL 64: 1327-1328.

REARING Improved technique described. Oviposition plates sealed with masking tape on 2 sides and some of other 2 sides; heated for 30 minutes at 90° C, then cooled to room temperature. Adults introduced, and ends sealed with plastic screen and masking tape. Kept in darkness at 30 ± 2°C and 50 ± 20% RH. Eggs extracted as in Gast (1961) except blender blades covered with rubber bands. Adults left in diet for 2-3 days then diet broken up, soaked for 2-3 h in water and blended at low speed for 30 seconds; eggs sieved and sterilized; incubated on black filter paper in petri dish.

➡ Gouger, R.J.; Bray, D.F. 1968. Two semiartificial oviposition media for the smaller European elm bark beetle. J ECON ENTOMOL 61: 564-565.

COMPOSITION Diet 1 (ml): water, 45.00; aged water-extract of elm phloem, 50.00; and (g): powdered elm phloem, 20.00; brewers' yeast, 5.00; sucrose, 10.00; glycine, 0.20; cholesterol, 0.20; choline chloride, 0.13; sorbic acid, 0.60; agar, 4.00; vitamin diet fortification mixture, 1.00.

Diet 2 (ml): water, 373.00; aged water-extract of elm phloem, 150.00; and (g): powdered elm phloem, 473.00; brewers' yeast, 10.00; sucrose, 5.00; dextrose, 5.00; fructose, 5.00; soybean protein, 10.00; sorbic acid, 2.00; Wesson's salt mixture, 2.00; agar, 20.00.

PREPARATION Agar dissolved in hot water, blended with dry ingredients, dispensed to petri dishes to depth of ½" and cooled. Preparation of aged, water-extract of elm phloem as follows: phloem peeled from boles of 30-year old American elm trees, *Ulmus americana*, cut into small strips and put into glass jars with 500 ml of distilled water/300 g of phloem; held at 74 ± 2°F for 20 ± 2 days (selection of aging period based on greatest response of adult beetles); water extract filtered through cheesecloth and incorporated into diet.

Powdered elm bark prepared by cutting air-dried elm phloem into small pieces and grinding finely in blender.

REARING At 72 ± 2°F and 55 ± 5% RH in darkness. In breeding chambers of 6 x 7" block of exterior masonite with ¾" hole in centre. 6 x 7 x ½" foam-rubber sponge pad with 3½" diam hole in centre placed on masonite. Plug of diet placed in hole in sponge; 6 x 7" glass plate placed over diet and sponge and held with clips. ¾" diam plug of diet removed through hole in masonite and 8 beetles introduced into cavity. Confined by 1" square of mesh screen taped into place.

DEVELOPMENT Oviposition response to diet related to water-extract of elm phloem. Adults on diets with water-extract laid eggs. On diet 1, 17 eggs on 18th day; not hatched in 26 days; on diet 2, 29 eggs on 12th day, not hatched in 26 days.

Scolytus ventralis Le Conte, fir engraver

➡ Scott, B.A. Jr; Berryman, A.A. 1971. Laboratory rearing techniques for *Scolytus ventralis* (Coleoptera: Scolytidae). BULL WASH AGRIC EXP STN 741: 9 pp.

COMPOSITION (g): phloem, 100; dried brewers' yeast, 10; sodium benzoate, 0.8; sorbic acid, 0.2; and distilled water, 500 ml.

PREPARATION Ingredients homogenized and pH adjusted to 3.7-4.0 with 2N HCl. Homogenate poured into funnel containing folded cheesecloth. Excess moisture (400-425 ml) squeezed out by hand. Resulting pulp autoclaved for 15 minutes at 5 lb pressure; or 20 ppm Benlate fungicide added instead.

REARING At 25-28°C and 80-90% RH in sterilized petri dishes with 30-35 g diet. Mites removed from field collected adults which then allowed to attack aged bolts in rearing cans for 10-20 days after which bark removed and eggs collected. Eggs either incubated on water/agar plates or transferred directly to rearing diet. Neonate larvae transferred to diet 3 times daily.

DEVELOPMENT 1st instar larval mortality, 11%; total larval mortality, 56%. Adult emergence, 36%; adults active and healthy. Eggs viable, developing to normal adults.

Xyleborus ferrugineus (Fabricius), ambrosia beetle

→ Norris, D.M.; Baker, J.M. 1968. A minimal nutritional substrate required by *Fusarium solani* to fulfill its mutualistic relationship with *Xyleborus ferrugineus*. ANN ENTOMOL SOC AM 61: 1473-1475.

Best of six diets tested.

COMPOSITION (g): Wesson's salt mixture, 2.5; sodium nitrate, 0.5; dextrose, 5.0; pure powdered cellulose, 15.0; fibrous cellulose, 4.0; bacto agar, 10.0; Streptomycin, 0.075; and (ml): Wood vitamins, 20; distilled water, 230.

Wood vitamin solution (g/l): choline chloride, 10.3; pyridoxine hydrochloride, 0.037; folic acid, 0.011.

PREPARATION Ingredients except powdered cellulose, blended for 1 minute. Cellulose stirred in by hand. pH adjusted to 6 with 1N HCl. Sterilized in autoclave for 20 minutes at 15 psi and 121°C; dispensed while hot to sterile test tubes. Tubes left at room temperature for 1 day then diet surface scored with sterile needle.

REARING At 28°C in darkness. Pupae surface sterilized with 0.1% $HgCl_2$ and resulting fungus-free adult ♀ introduced to each tube using sterile techniques. One day later each tube inoculated with small plug of potato-dextrose-agar medium taken from edge of actively growing culture of *Fusarium solani*.

DEVELOPMENT Progeny after 5 weeks/♀/tube on diet: 91 eggs, 244 larvae, 2♂ pupae, 69♀ pupae, 49♂ adults and 10♀ adults.

→ Norris, D.M.; Baker, J.M. 1969. Nutrition of *Xyleborus ferrugineus*. I. Ethanol in diets as a tunneling (feeding) stimulant. ANN ENTOMOL SOC AM 62: 592-594.

Three diets tested.

COMPOSITION Diet 1 (g): sucrose, 10.0; vitamin-free casein, 10.0; Wesson's salt mixture, 2.5; sorbic acid, 0.4; Streptomycin (20% wettable), 0.35; agar, 10.0; powdered cellulose, 15.0; fibrous cellulose, 4.0; cholesterol, 0.2; and (ml): distilled water, 230.0; vitamin solution, 20.0.

Diet 2 (g): sucrose, 20.0; casein, 20.0; Wesson's salt mixture, 5.0; sorbic acid, 0.4; Streptomycin, 0.35; agar, 10.0; powdered cellulose, 15.0; fibrous cellulose, 4.0; cholesterol, 0.4; and (ml): distilled water, 230.0; vitamin solution, 20.0.

Diet 3 (g): sucrose, 3.75; casein, 2.5; Wesson's salt mixture, 0.4; sorbic acid, 0.4; Streptomycin, 0.35; agar, 10.0; powdered cellulose, 20.0; fibrous cellulose, 4.0; bacto yeast extract, 2.5; soluble starch, 2.5; wheat germ, 1.75; and (ml): distilled water, 250.0; corn oil, 1.0.

vitamin solution (g/1 water): choline chloride, 10.3; inositol, 4.6; calcium pantothenate, 0.85; niacin, 0.33; *p*-aminobenzoic acid, 0.155; riboflavin, 0.048; pyridoxine hydrochloride, 0.37; thiamine hydrochloride, 0.028; folic acid, 0.011; biotin, 0.008; cobalamine (B_{12}), 0.002.

Additions to basic diets: diet 1, 2.5 ml 95% ethanol, 100 mg methyl linoleate + 2.5 ml 95% ethanol, 100 mg methyl linoleate, 10 g wheat germ, or 60 g (wet wt) *Fusarium solani*, respectively; diet 2, 2.5 ml 95% and diet 3, 1.0, 5.0, 10.0, or 15.0 ml 80% ethanol, respectively.

F. solani added to diet 1, grown for 6 days (at 28°C in darkness) in 100 ml quantities of Neutral-Dox-Yeast solution. Composition of this medium (g): dextrose, 20.0; K_2HPO_4, 0.5; KH_2PO_4, 0.5; KCl, 0.5; $MgSO_4$, 0.5; $FeSO_4$, 0.01; bacto yeast extract, 0.5; $NaNO_3$, 3.0; deionized water, 1000 ml.

REARING Pupae from stock cultures sterilized by submergence in 0.1% $HgCl_2$ for 4 minutes, rinsed in sterile distilled water and aseptically placed on surface of potato-dextrose-agar medium for emergence. One fungus-free adult ♀ transferred aseptically to each sterile test tube.

DEVELOPMENT Ethanol necessary for beetle tunnelling.

Progeny/♀ after 6 weeks on diet 1 with *F. solani*, 2.3 adults. On diet 1 with 95% ethanol, 1.3 adults.

See also: Kok *et al.* (1970); Norris *et al.* (1969).

Corollary

→ Chu, H.M.; Norris, D.M.; Kok, L.T. 1970. Pupation requirement of the beetle, *Xyleborus ferrugineus:* sterols other than cholesterol. J INSECT PHYSIOL 16: 1379-1387.

COMPOSITION Basic diet (g): sucrose, 10.0; vitamin-free casein, 10.0; Wesson's salt mixture, 2.5; Streptomycin, 0.1; sorbic acid, 0.4; linolenic acid, as methyl ester, 0.1; powdered cellulose, 15.0; fibrous cellulose, 4.0; bacto-agar, 10.0; and (ml): 95% ethanol, 2.5; vitamin solution, 20.0; de-ionized water, 230.0.

vitamins (g/l): choline chloride, 10.3; inositol, 4.6; calcium pantothenate, 0.85; niacin, 0.33; *p*-aminobenzoic acid, 0.155; riboflavin, 0.048; pyridoxine hydrochloride, 0.037; thiamine hydrochloride, 0.028; folic acid, 0.011; biotin, 0.008; cobalamine (B_{12}), 0.002.

DEVELOPMENT Effects of various sterols or related compounds as sole sources of sterol or sterol-substitutes on growth, development, and reproduction discussed. 0.4 g of cholesterol, 7-dehydrocholesterol, ergosterol, lanosterol, lumisterol, or vitamin D_2 or D_3 added to basic diet.

See also: Norris & Chu (1970).

Saunders, J.L.; Knoke, J.K. 1967. Diets for rearing the ambrosia beetle *Xyleborus ferrugineus* (Fabricius) *in vitro*. SCIENCE 157: 460-463.

Five diets tested.

COMPOSITION Diet 1 (g): sucrose, 15.0; yeast extract, 10.0; casein, 10.0; starch, 10.0; wheat germ, 5.0; Wesson's salt mixture, 1.3; agar, 40.0; fresh cacao sawdust, 200.0; and (ml): cottonseed oil, 5.0; distilled water, 1000.0.

Diet 2 (g): sucrose, 30.0; yeast extract, 20.0; casein, 20.0; starch, 20.0; wheat germ, 10.0; Wesson's salt mixture, 1.3; agar, 80.0; fresh cacao sawdust, 550.0; and (ml): cottonseed oil, 10.0; distilled water, 1000.0.

Diet 3 (g): sucrose, 13.4; yeast extract, 8.9; casein, 8.9; starch, 8.9; wheat germ, 4.5; Wesson's salt mixture, 1.1; agar, 36.0; dry cacao sawdust, 179.0; and (ml): cottonseed oil, 4.5; distilled water, 1000.0.

Diet 4 (g): sucrose, 18.8; yeast extract, 12.5; casein, 12.5; starch, 12.5; wheat germ, 6.3; Wesson's salt mixture, 1.6; agar, 50.0; cellulose powder, 250.0; and (ml): cottonseed oil, 6.3; cacao bark extract, 845.0; distilled water, 155.0.

Diet 5: as diet 4, except 845.0 ml of cacao bark extract omitted, and distilled water increased from 155.0 to 1000.0 ml.

PREPARATION Ingredients of diet 1 blended, dispensed to 20 x 150 mm culture tubes at 15 ml/tube, plugged with cotton, and autoclaved. For other diets all ingredients, except water and aqueous extract of cacao bark, mixed dry then liquid ingredients added and stirred to give uniform slurry. Extract of cacao bark (diet 4) prepared by macerating 100 g of air-dried cacao bark in 1-l of distilled water in Waring Blendor and filtering resulting liquid through six layers of cheesecloth. Diets 2 to 5 dispensed at 12 ml/tube and autoclaved. Diets used immediately when cool.

REARING On diet 1 at 24.5°C in uncontrolled light conditions or in continuous darkness; on diets 2 to 5 at 22°C in darkness.

DEVELOPMENT Mean yield (progeny/tube), 10.9 (diet 1, uncontrolled light); or 17.1 (diet 1, continuous darkness); 33.8 (diet 2); 12.7 (diet 3); 41.7 (diet 4); and 35.0 (diet 5). 55% of ♀♀ produced brood on diet 1, on which eggs first observed 7 days after introduction of ♀♀; larvae, pupae and adults observed in 14, 17 and 22 days respectively. Diets 2 to 5 each satisfactory for completion of brood development. 12 ml diet/tube adequate for broods of two ♀♀. Highest percentage (94%) of successful cultures (produced progeny) on diet 2; greatest mean number of adult progeny/tube on diet 4. Cacao tissue or extract not required in diet 5. Initiation of gallery construction stimulated by addition of cacao bark to diet 4. Eggs first observed 4 days after ♀♀ on diet. By 4th week from initiation of insect cultures all stages of progeny observed in galleries. Adults first seen during 4th and 5th weeks from establishment of cultures. More than 95% of individuals, mature, normal and healthy. Rearing trials with *Xyleborus posticus* Eichhoff also successful on diets containing cacao sawdust, but survival and progeny production less than *X. ferrugineus*.

See also: Norris (1972).

Tenebrionidae

Alphitobius diaperinus (Panzer), lesser mealworm

→ Preiss, F.J.; Davidson, J.A. 1971. Adult longevity, pre-oviposition period and fecundity of *Alphitobius diaperinus* in the laboratory (Coleoptera: Tenebrionidae). J GA ENTOMOL SOC 6: 105-109.

COMPOSITION Semi-synthetic diet consisting of two parts; A: distilled water, 200 ml; methyl *p*-hydroxybenzoate, 15 ml; Streptomycin, 0.1 g; Terramycin, 0.1 g; chicken starter mash, 110 g; and B: distilled water, 490 ml; granulated agar, 20 g.

PREPARATION Group "A" ingredients blended. Agar dissolved in hot water, cooled to 70°C, then blended with "A" ingredients. Dispensed to 1-oz plastic cups, allowed to solidify, and capped.

REARING At 70-75°F. Eggs laid in tightly folded 5 cm sq pieces of white paper placed in rearing cup. Adults held at room temperature, and uncontrolled photoperiod of ca. 10 h/day. Late-instar larvae introduced to cups for pupation and emergence. One pair maintained in each cup, diet changed every 10-14 days.

DEVELOPMENT Adult survival >400 days; average pre-oviposition period, 12.7 days (4-30 days range); average fecundity, 3.5; viable eggs.

Gnathocerus cornutus (Fabricius), broadhorned flour beetle

→ Leclercq, J.; Magis, N.; Rey, C. 1954. Sur les besoins nutritifs du *Gnathocerus cornutus* F. (Coléoptère, Ténébrionidae). Recherche de l'optimum glucidique et de l'optimum protidique dans un régime alimentaire artificial. ARCH INT PHYSIOL 62: 264-271. (In French)

Best of 5 diets tested.

COMPOSITION (%): casein, 90; yeast, 10.

PREPARATION Blended.

REARING At 30°C and 60-70% RH, in groups of 20 adults, each group with 20 g of diet.

DEVELOPMENT Larval mortality, 40%; mean development period, 42 days; mean wt, 1.6 mg; compared with development data on whole wheat flour of 22.5%, 37 days and 2.25 mg respectively.

Palorus ratzeburgi (Wissmann), small eyed flour beetle

→ Cooper, M.I.; Fraenkel, G. 1952. Nutritive requirements of the small-eyed flour beetle, *Palorus ratzeburgi* Wissmann (Tenebrionidae, Coleoptera). PHYSIOL ZOOL 25: 20-28.

COMPOSITION Basal diet (parts): vitamin-free casein (Labco), 20; anhydrous glucose, 80; cholesterol, 1; McCollum-Davis salt mixture No. 185, 2; water, 10. Supplemented by dried brewers' yeast or vitamin mixture.

vitamins (μg/g diet): thiamin, 25.0; riboflavin, 12.5; nicotinic acid, 50; pyridoxin, 12.5; calcium pantothenate, 25.0; choline chloride, 1000; *i*-inositol, 250; folic acid, 2.5; (biotin, 0.5, omitted at times); 3 μg B_t added when required.

PREPARATION Casein and salts mixed in water, dried, and ground. Cholesterol and glucose added. Dried brewers' yeast or vitamin mixture finally added and 4 g diet dispensed to 1 x 2" glass vials. When vitamins used - dissolved in water or dilute ammonia water, pipetted directly into vials along with added water. Mixture mixed with spatula. Diets held at constant temperature and humidity for 3 days, to reduce water content to about 10% by weight. Diets then ground and divided into 2 vials.

REARING At 30°C and 65-70% RH. 10-20 neonate larvae/vial. Pupae removed to absorbent paper in vial.

DEVELOPMENT Larval weight at 22 days, 1.10 mg. Pupation in 27.9 days. Pupal period, 5 days.

See also: Fraenkel (1951a b); Fraenkel *et al.* (1948).

Tenebrio molitor Linnaeus, yellow mealworm

→ Fraenkel, G.; Blewett, M.; Coles, M. 1950. The nutrition of the mealworm, *Tenebrio molitor* L. (Tenebrionidae, Coleoptera). PHYSIOL ZOOL 23: 92-108.

COMPOSITION (parts): vitamin-free casein, 20; glucose (anhydrous), 80; cholesterol, 1; salt mixture (McCollum's No. 185), 2; water-insoluble fraction of yeast, 2.5; water, 10; and (μg/g diet): thiamin, 25; riboflavin, 12.5; nicotinic acid, 25; pyridoxin, 12.5; pantothenic acid, 25; pteroylglutamic acid, 2.5; choline, 500; inositol, 250; aminobenzoic acid (in earlier tests only), 25.

PREPARATION Glucose, casein, salt mixture and cholesterol mixed and 3-6 g weighed into test bottles. Mixtures of vitamin solutions, B_t concentrates and other water soluble additions pipetted into dry basic diet. Kept at constant temperature and humidity for 2 days, then ground in mortar.

REARING At 25°C and 70% RH. In 2-oz wide bottle with 3 or 6 g diet; 10 or 20 neonate larvae/bottle.

DEVELOPMENT Rate of growth dependent on water content of food; fast on food in equilibrium with 70% RH, slow at 30% RH, and minimal at 13% RH. Larval weight after 10-12 weeks, 60 mg; pupation at 120-150 mg.

See also: Leclercq (1954).

Corollary

→ Fraenkel, G. 1953c. Studies on the distribution of vitamin B_t (carnitine). BIOL BULL 104: 359-371.

Larval growth and survival improved by addition of vitamin B_t or carnitine at 0.35 μg/g diet.

See also: Bieber & Monore (1969); Fraenkel (1951a b, 1954, 1955, 1957, 1958); Fraenkel & Blewett (1947a b); Fraenkel & Chang (1953, 1954); Fraenkel & Leclercq (1956); Ito & Fraenkel (1966); Leclercq (1954).

→ Leclercq, J.; DeBast, D. 1965. Projet d'utilisation des larves de "*Tenebrio molitor*" pour comparer la valeur nutritive des proteinés. ANN NUTR ALIMENT 19: 19-25. (In French)

COMPOSITION (g): glucose, anhydrous, 80 to 94; vitamin-free casein, 6-20; McCollum's salt mixture, 2; cholesterol, 1; vitamins (in µg/g diet) in 15% aqueous ethanol solution: thiamine, 25; riboflavin, 12.5; nicotinic acid, 50; calcium pantothenate, 25; pyridoxine, 12.5; folic acid, 2.5; choline chloride, 500; biotin, 0.25; *dl*-carnitine, 6; and $ZnCl_2$, 40.

PREPARATION Ingredients mixed and left for 24 h in ventilated rearing room for excess moisture and alcohol to evaporate. Then ground and dispensed to containers @ 10 g diet/101 larvae.

REARING At 27°C and ca. 75% RH.

See also: Leclercq & Lopez-Francos (1966, 1967).

Corollary 1

→ Davis, G.R.F.; Leclercq, J. 1969. Protein nutrition of *Tenebrio molitor* L. IX. Replacement caseins for the reference diet and a comparison of the nutritional values of various lactalbumins and lactalbumin hydrolysates. ARCH INT PHYSIOL BIOCHEM 77: 687-693.

COMPOSITION As in Leclercq & DeBast* (1965) except protein concentration decreased to 3 g/100 g diet, previously shown to be optimal for this insect; glucose concentration maintained at 94 g/100 g diet.

Four caseins tested for suitability in reference diet. (i) vitamin-free casein (Difco Laboratories); (ii) vitamin-free test casein (General Biochemicals, Inc.); (iii) vitamin-free casein (Nutritional Biochemicals Corporation); (iv) vitamin-free casein No. 5408 (Djac Laboratories). Similarly lactalbumin and lactalbumin hydrolysate investigated.

REARING At 27°C and 65% RH; larvae in plastic egg cups with 10 g diet.

DEVELOPMENT Larval weight gains similar on diets with 4 different caseins and with lactoalbumin. With enzymatic hydrolysates of lactalbumin growth poor than on diet containing unhydrolyzed lactalbumin.

Corollary 2

→ Davis, G.R.F. 1969b. Protein nutrition of *Tenebrio molitor* L. X. Improvement of the nutritional value of lactalbumin by supplementation with amino acids. ARCH INT PHYSIOL BIOCHEM 77: 741-748.

COMPOSITION As in Davis & Leclercq* (1969), except *l*-amino acids added (mg/g of lactalbumin): arginine, 12.5; glycine, 31.0; histidine, 10.6; methionine, 7.9; phenylalanine, 26.6; proline, 41.5; threonine, 10.6; tyrosine, 41.5; and valine, 36.0.

DEVELOPMENT Nutritional value of lactalbumin related to amino acid imbalance; utilization improved by addition of arginine or tyrosine.

Corollary 3

→ Davis, G.R.F. 1970a. Protein nutrition of *Tenebrio molitor* L. XI. Effect on growth of partial replacement of dietary casein by lactalbumin or lactalbumin hydrolysate and of partial replacement of these two proteins by casein. ARCH INT PHYSIOL BIOCHEM 78: 29-36.

COMPOSITION Larval diet: as in Davis* (1969b) and Davis & Leclercq* (1969) in which vitamin-free casein replaced by lactalbumin or lactalbumin hydrolysate at concentrations of 0.003, 0.006, 0.030, 0.150, and 0.300 g/100 g diet. Similarly, test diets prepared in which lactalbumin or lactalbumin hydrolysate replaced by casein at same concentrations.

DEVELOPMENT Larval growth on diets in which casein partially replaced by various concentrations of lactalbumin or lactalbumin hydrolysate equal to that on reference diet, containing 3 g of casein/100 g diet.

Corollary 4

→ Davis, G.R.F. 1970b. Protein nutrition of *Tenebrio molitor* L. XII. Effects of dietary casein concentration and of dietary cellulose on larvae of race F. ARCH INT PHYSIOL BIOCHEM 78: 37-41.

COMPOSITION Experimental method as in Davis and Leclercq* (1969). Four diets containing vitamin-free casein tested: two with 3 g casein/100 g diet and two with 20 g/100 g diet. In one of each diets, powdered cellulose included at rate of 1.9 g/100 g diet, with glucose concentration reduced by similar amount.

DEVELOPMENT Larval weight gain after 4 weeks on diets with 20 g casein, greater (average initial weight 6.16 mg) than those reared on diets with 3 g casein/100 g diet (4.08 mg). Also greater weight gain when cellulose in diet (average initial weight 6.74 mg on 20 g casein/100 g of diet and 5.53 mg on 3 g casein/100 g of diet).

Corollary 5

→ Davis, G.R.F. 1971b. Protein nutrition of *Tenebrio molitor* L. XV. Amino acid mixtures as replacements for protein of the artificial diet. ARCH INT PHYSIOL BIOCHEM 79: 11-17.

Five diets tested.

COMPOSITION Reference diet (g): vitamin-free casein (Difco Laboratories, Detroit, Michigan), 3.0; glucose, 94.0; McCollum's salt mixture, 2.0; cholesterol, 1.0; and (ml): vitamin solution, 10; vitamin solution prepared according Leclercq & DeBast* (1965).

In 4 test diets, casein replaced by: in diet 1, amino acid mixture at 85% protein level; diet 2, lactalbumin at 75% protein level; diet 3, pupal protein replaced at 65% level; diet 4, pupal protein replaced at 100% level.

l-amino acid mixtures (g/100 g diet):

Amino acid	1	2	3	4
Alanine	0.070	0.115	0.120	0.184
Arginine	0.090	0.058	0.127	0.196
Aspartic acid	0.166	0.247	0.168	0.258
Cystine	0.007	0.090	-	-
Glutamic acid	0.580	0.379	0.262	0.404
Glycine	0.040	0.048	0.110	0.169
Histidine	0.070	0.035	0.085	0.131
Isoleucine	0.128	0.121	0.089	0.136
Leucine	0.226	0.374	0.147	0.226
Lysine	0.208	0.201	0.141	0.217
Methionine	0.060	0.038	0.025	0.039
Phenylalanine	0.121	0.082	0.095	0.145
Proline	0.249	0.094	0.095	0.145
Serine	0.113	0.084	0.075	0.106
Threonine	0.095	0.111	0.075	0.106
Tryptophan	0.030	0.067	0.019	0.030
Tyrosine	0.128	0.077	0.180	0.276
Valine	0.163	0.128	0.135	0.208
Total	2.544	2.349	1.948	2.976

In each test diet, mineral salt mixture doubled to compensate for minerals normally added by casein or lactalbumin. Adjustments made to glucose content to maintain total weight of diet at 100 g.

REARING At 27°C and 65% RH. Larvae reared for 4 weeks on diets.

DEVELOPMENT Larval growth on reference diet nearly five fold over 4-weeks; only two fold on amino acid diets.

See also: Davis (1970c, 1971a).

Davis, G.R.F.; Sosulski, F.W. 1974a. Protein nutrition of *Tenebrio molitor* L. XVIII. Growth of larvae of race F fed protein blends of cereals and legumes. ARCH INT PHYSIOL BIOCHEM 82: 861-871.

COMPOSITION Control diet (parts): whole ground wheat, 90; brewers' yeast, 10. 14 test diets prepared in which rice or wheat flour blended with legume meals (peas, lentils, beans, broadbeans, fababeans and soybeans) or with casein so that each of two protein sources supplied 5% of protein. Cornstarch added to this such that cereal-legume-cornstarch contributed 80% diet, corn oil and antioxidant 10%, cellulose 5%, minerals 4% and vitamin mixture 1%.

salts (%): salt mix A, 1.62; $Ca_3(C_6H_5O_7)_2$, 30.83; $CaH_4(PO_4)_2$, 11.28; $MgCO_3$, 3.52; $MgSO_4$, 3.83; KCl, 12.47; KH_2PO_4, 21.88; NaCl, 7.71.

salt mix A (%): $CuSO_4.5H_2O$, 0.48; ferric ammonium citrate, 94.33; $MgSO_4$, 1.24; ammonium alum, 0.57; KI, 0.25; NaF, 3.13.

vitamins (g): thiamine, 0.01; riboflavin, 0.02; pantothenic acid, 0.10; niacin, 0.10; biotin, 0.002; pyridoxine, 0.01; 0.1% cobalamin, 0.05; folacin, 0.005; inositol, 0.04; choline, 8.00; vitamin A, (325000 I.U./g), 0.03; vitamin D_3 (2000000 I.U./g), 0.0125; vitamin E (27.5 I.U./g), 1.00; menadione, 0.40; and corn starch, 90.5.

Further 16 diets using soybeans, peas, fababeans and casein giving 10% protein tested. Supplemented with 0.2% *l*-methionine, 0.4% *l*-lysine, 0.2% *l*-methionine and 0.1% *l*-threonine.

REARING In test tubes at 27 ± 0.25°C and 65 ± 5% RH.

DEVELOPMENT Larval weight gain on wheat diets supplemented with soybeans, field beans or peas greater than with other legumes. Weight gains on casein diets intermediate to these. Larval weight gain maximum on control diets and minimum on rice-legume diets. Best diet with rice-soybean blend.

See also: Davis (1972a b, 1974); Davis & Sosulski (1973a b, 1974b).

Tribolium castaneum (Herbst), red flour beetle

→ Applebaum, S.W.; Lubin, Y. 1967. The comparative effects of vitamin deficiency on development and on adult fecundity of *Tribolium castaneum*. ENTOMOL EXP APP 10: 23-30.

COMPOSITION Basal diet (parts): vitamin-free casein, 20; glucose, 80; cholesterol, 1; and salt mixture USP No. 2, 1.

vitamins (μg/g diet): thiamine hydrochloride, 25.0; riboflavin, 12.5; nicotinic acid, 50.0; calcium pantothenate, 25.0; pyridoxine hydrochloride, 12.5; folic acid, 2.5; choline chloride, 500.0; and biotin, 0.6.

In control diets containing dried yeast replaced by 5 parts glucose.

REARING At 28°C. 5 pairs newly emerged adults/5 g diet in 7 x 5 cm plastic vials. Eggs collected by screening through 60 mesh and kept in separate vials. 10 neonate larvae on 1 g diet.

DEVELOPMENT Larval survival, 90% after 40 days when all vitamins provided. Average larval weight after 36 days, 1.2 mg. Pupation after 36 days, or 20 days if supplemented by yeast. Egg hatch, 76%; fecundity, 1170 eggs/♀ on all vitamin diet.

See also: Applebaum & Konijn (1966).

→ Hogan, G.R. 1972. Development of *Tribolium castaneum* (Coleoptera: Tenebrionidae) in media supplemented with pyrimidines and purines. ANN ENTOMOL SOC AM 65: 631-636.

COMPOSITION (%): autoclaved cornstarch, 76.0; cholesterol, 1.0; salt mixture, 2.0; vitamin mixture, 1.0; vitamin-free casein, 20.

vitamins (g): biotin, 0.005; carnitine, 0.1; choline chloride, 10.0; folic acid, 0.05; inositol, 5.0; nicotinic acid, 1.0; pantothenic acid, 0.5; pyridoxine, 0.25; riboflavin, 0.25; thiamine, 0.5; and cornstarch, 82.5.

salts: not specified.

PREPARATION Ingredients mixed in ball mill.

REARING At 31°C and ca. 70% RH in capped vials.

DEVELOPMENT Egg viability, 70-100%; larval period, 55 ± 6.3 days; survival to pupae, 78 ± 4.6%; mean pupal period, 6 ± 1.7 days; adult emergence, 95 ± 2%. When 1% yeast added to basal diet, developmental data comparable with those on flour-yeast rearing diet.

Magis, N. 1954b. Nutrition comparée des *Tribolium* III.--Action vitaminique de la *dl*-carnitine et de la *dl*-dicarnitine chez *Tribolium castaneum* Herbst (Insecte, Coléoptère). ARCH INT PHYSIOL 62: 505-511.

COMPOSITION (g/100 g): vitamin-free casein, 20.00; starch, 75.50; McCollum's salt mixture, 2.00; wheat germ oil, 1.50; cholesterol, 1.00; and (μg/g): thiamine hydrochloride, 1.00; riboflavin, 2.00; pyridoxine hydrochloride, 1.00; nicotinic acid, 8.00; calcium pantothenate, 4.00; biotin, 0.05; pteroylglutamic acid, 0.20; choline chloride, 500.00; *dl*-dicarnitine or *dl*-carnitine, 0.8. Substitution of brewers' yeast at 0.5 g/100 g for carnitine acceptable.

PREPARATION 0.5 ml vitamin solution added with 0.5 ml distilled water containing carnitine to 5 g of pulverized dry ingredients. Dried at room temperature for 48 h, ground in mortar and dispensed to rearing tubes.

REARING Individually at 27° ± 1°C and 70 ± 5% RH.

DEVELOPMENT Mean developmental periods to adult, pupal weights and larval mortalities when reared on *dl*-carnitine without yeast, and yeast without carnitine: 59.0 ± 2.9 and 42.5 ± 1 days; 2.50 ± 0.09 and 2.30 ± 0.07 mg; and 40.0 and 5.3% respectively.

Diet also satisfactory for *Tribolium confusum* and *Tribolium destructor*.

See also: Magis (1954a).

Pant, N.C.; Gabrani, K. 1963. Qualitative vitamin B requirements of *Tribolium castaneum* Herbst. INDIAN J ENTOMOL 25: 110-115.

COMPOSITION (parts): vitamin-free casein, 50; glucose, 50; McCollum's salt mixture, 2; cholesterol, 1; and (mg/g diet): thiamin, 25; riboflavin, 25; nicotinic acid, 25; pyridoxine hydrochloride, 25; pantothenic acid, 25; *p*-aminobenzoic acid, 25; inositol, 250; folic acid, 2.0; choline chloride, 500; biotin, 0.1.

PREPARATION Diet dried at 36°C for 12 h and thoroughly ground in mortar and pestle. Stored for 24 h at 29°C and 75-80% RH. 2 g diet placed in 2 x 1" vials.

REARING At 29°C and 75-80% RH. Neonate larva placed in each vial. Stock culture maintained on wholemeal flour supplemented with 5% dried yeast. Diet sieved for eggs; transferred to 2 x 1" tube for hatching.

DEVELOPMENT Egg hatch in 2-3 days. Larval period, 27.31 days; pupation, 78%.

See also: Pant & Dang (1965, 1967); Roy & Pant (1972).

Tribolium confusum Jacquelin du Val, confused flour beetle

Chirigos, M.A.; Meiss, A.N.; Pisano, J.J.; Taylor, M.W. 1960. Growth response of the confused flour beetle, *Tribolium confusum* (Duval) to six selected protein sources. J NUTR 72: 121-130.

COMPOSITION (%): crude protein, 18.0; vitamin mixture, 0.63; cholesterol, 1.0; corn oil, 3.0; mineral mixture, 1.0; and corn starch, dependent upon proetin samples.

crude protein and corn starch content for optimum results, 19.2% dried defatted beef muscle and 75.2% corn starch. Other protein source materials and corn starch levels, respectively; (%): casein, 20.4, 73.9; egg albumin, 22.5, 71.8; wheat gluten, 21.6, 72.8; peanut flour, 29.1, 65.3; and whole egg, 22.2, 72.1.

minerals (Drackett Mineral Mixture No. 446), modified to contain trace materials (mg/g diet): NaCl, 6.1; $K_3C_6H_5O_7.2H_2O$, 13.3; $CaHPO_4$, 20.0; KCl, 3.8; $Ca_3(PO_4)_2$, 13.8; $MgCO_3$, 2.3; Fe $(NH4)_3$ $(C_6H_5O_7)_2$, 0.9; and (µg/g diet): $CuSO_4.5H_2O$, 11.0; $MnSO_4.H_2O$, 66.2; KAl $(SO_4)_2.12H_2O$, 49.6; KI, 27.6; $CoCl_2.6H_2O$, 49.6; $Zn(C_3H_5O_3)_2.3H_2O$, 49.6; and NaF, 0.055.

vitamins added in 1:2 ethanol-water solution at rate of 2.0 ml to 10 g finished diet. (µg/g diet): thiamine hydrochloride, 12; riboflavin, 18; nicotinic acid, 100; calcium pantothenate, 40; pyridoxine hydrochloride, 16; inositol, 2000; biotin, 0.6; folic acid, 5; menadione, 1; *p*-aminobenzoic acid, 50; cobalamine, 0.5; ascorbic acid, 10; and choline chloride, 4000.

PREPARATION Protein source, corn starch and mineral mixture ball milled for 10-12 h. Vitamins added, thoroughly mixed, and air dried; corn oil and cholesterol in ether solution added. Diet stored in cold for 2-3 days until use.

REARING At 28°C and 70 ± 5% RH. 10 neonate larvae/2 x 5 cm shell vial with 1-2 g diet.

DEVELOPMENT Best growth on diet with whole egg, beef, or casein. Decreasing growth rates on peanut flour, wheat gluten, and egg albumin. Best larval wt, 2.75-3.25 mg after 18 days.

See also: Michael *et al.* (1960).

→ Huot, L.; Bernard, R.; Lemonde, A. 1957. Aspects quantitatifs des besoins en minéraux des larves de *Tribolium confusum* Duval. I. Pourcentage optimum d'un mélange salin. CAN J ZOOL 35: 513-518. (In French with English summary)

COMPOSITION (%): casein, 20; cholesterol, 1; sucrose, 79.

vitamins (µg/g diet): thiamine hydrochloride, 1; riboflavin, 2; pyridoxine hydrochloride, 1; nicotinic acid, 8; calcium pantothenate, 4; choline, 1500; biotin, 0.05; folic acid, 0.2.

salts: salt mixture U.S.P. XIII No. 2 in varying quantities.

PREPARATION Casein, cholesterol, sucrose and salts finely mixed; then vitamins added in aqueous solution. After 24 h, diet mixed in ball mill for 6 h.

REARING At 28 ± 1°C and 70 ± 5% RH individually in 2½ x 1½" diam tubes containing 3 g diet.

DEVELOPMENT With 3-6% salt mixture, ca. 70% survival from larva to adult in 43-46 days.

See also: Huot *et al.* (1958).

Fraenkel, G.; Printy, G.E. 1954. The amino acid requirements of the confused flour beetle, *Tribolium confusum* Duval. BIOL BULL 106: 149-157.

COMPOSITION (parts): casein or amino acid mixture, 15; corn starch, 85; cholesterol, 1; McCollum's salt mixture No. 185, 2.

vitamins (µg/g dry diet): thiamin, 25; riboflavin, 12.5; nicotinic acid, 50; pyridoxin, 12.5; pantothenic acid, 25; choline chloride, 500; inositol, 250; pterolyglutamic acid, 2.5; and biotin, 0.25.

amino acids (%): *dl*-alanine, 2.54; *l*-arginine hydrochloride, 3.18; *dl*-aspartic acid, 2.54; *l*-cystine, 1.27; *l*-glutamic acid, 12.69; glycine, 0.64; *l*-histidine hydrochloride, 6.03; *l*-hydroxyproline, 0.63; *dl*-isoleucine, 10.16; *l*-leucine, 7.62; *l*-lysine, 9.53; *dl*-methionine, 5.07; *dl*-phenylalanine, 7.62; *l*-proline, 1.27; *dl*-serine, 1.27; *dl*-threonine, 8.89; *l*-tryptophane, 2.54; *l*-tyrosine, 3.81; *dl*-valine, 12.70; $NaHCO_3$, 8.07; total, 108.07.

PREPARATION Amino acids ground with equal starch; cholesterol and salts added and ground, and then balance of starch. Vitamins added in solution and mixed with spatula. Diets left for 2 days at 30°C and 60-70% RH, then ground in mortar.

REARING At 29-30°C and 60-70% RH in 1 x 2" shell vials with 1 g diet/vial. 10 first instar larvae placed in each vial.

DEVELOPMENT Diets tested for larvae of three beetles, *T. confusum*, *Tenebrio molitor* Linnaeus and *Dermestes maculatus* DeGeer, (previously *D. vulpinus* Fabricius). Diets for *Tribolium* and *Tenebrio* identical, except for addition of 1.5 µg of carnitine/g of diet for *Tenebrio*. *Tenebrio* and *Tribolium* received 15% amino acids but *Dermestes* received 30% amino acids. *D. maculatus* failed to grow on these diets. *T. molitor* developed very poorly. For *T. confusum*:- average larval weight after 20 days on casein diet, 0.73 mg; and on 19 amino acid diet, 1.11 mg. Pupation after 30.1 days on amino acid diet.

See also: Ellinger *et al.* (1947); Fraenkel & Blewett (1942a b, 1944); French & Fraenkel (1954); Street & Palmer (1935).

Lemonde, A.; Bernard, R. 1951a. Nutrition des larves de *Tribolium confusum* Duval. I. Recherche d'un régime synthétique basal satisfaisant leurs besoins nutritifs. CAN J ZOOL 29: 70-79. (In French with English summary)

COMPOSITION (%): amino acid mix, 20.0; corn starch, 75.5; cholesterol, 1.0; wheat germ oil, 1.5; salt mixture No. 2 USP XII, 2.0; yeast, 0.5; and vitamin mix.

amino acids (g/100 g diet): *dl*-alanine, 0.49; *l*-arginine hydrochloride, 0.59; *dl*-aspartic acid, 0.49; *l*-glutamic acid, 2.44; *l*-cystine, 0.24; glycine, 0.12; *l*-histidine hydrochloride, 1.05; *l*-hydroxyproline, 0.12; *dl*-isoleucine, 1.95; *l*-leucine, 1.47; *l*-lysine hydrochloride, 1.83; *dl*-methionine, 0.98; *dl*-phenylalanine, 1.47; *l*-proline, 0.24; *dl*-serine, 0.24; *dl*-threonine, 1.71; *l*-tryptophane, 0.23; *l*-tyrosine, 0.73; *dl*-valine, 2.44.

vitamins (µg/g diet): thiamine hydrochloride, 1.0; riboflavin, 2.0; pyridoxine hydrochloride, 1.0; nicotinic acid, 8.0; calcium pantothenate, 4.0; biotin, 0.05; pteroylglutamic acid, 0.2; choline chloride, 500.0.

PREPARATION Amino acids weighed and blended with salt mixture. Wheat germ and cholesterol added, then vitamins in aqueous solution. Mix dried and finally blended.

REARING In jars at 28 ± 1°C and 70 ± 5% RH.

DEVELOPMENT Mean period to pupation, 22 days; compared with control of 19 days. Without yeast, period 30 days.

→ Chaudhary, K.D.; Lemonde, A. 1962. Phosphorus in the nutrition of *Tribolium confusum* Duval. CAN J ZOOL 40: 375-380.

Best of 4 diets described.

COMPOSITION (%): amino acids mixture, 20.0; starch, 74.5; cholesterol, 1.0; salt mixture, 4.0; brewers' yeast, 0.5; and vitamin mixture.

l-amino acids (g/100 g): α-alanine, 0.345; arginine (free base), 0.832; aspartic acid, 0.345; glutamic acid, 3.440; cystine, 0.338; glycine, 0.169; histidine (free base), 1.480; hydroxyproline, 0.169; isoleucine, 1.480; leucine, 2.074; lysine hydrochloride, 2.580; methionine, 0.691; phenylalanine, 1.036; proline, 0.339; serine, 0.169; threonine, 1.205; tryptophane, 0.691; tyrosine, 1.029; valine, 1.720; and $NaHCO_3$ (to neutralize hydrochloride of lysine), 1.190.

salts (g/100 g salt mixture): $Ca(C_5H_5O_3)_2.5H_2O$, 42.18; $FeC_6H_5O_7.5H_2O$, 3.82; $MgSO_4.7H_2O$, 17.63; NaCl, 12.59; KCl, 23.86.

vitamins (μg/g diet): thiamin hydrochloride, 25.0; riboflavin, 12.5; pyridoxin hydrochloride, 12.5; nicotinic acid, 50.0; calcium pantothenate, 25.0; biotin, 0.25; folic acid, 2.50; choline chloride, 1500.0.

REARING At 28 ± 1°C and 70 ± 5% RH in 1.5 x 2.5" vials with 3 g diet/vial.

DEVELOPMENT Larval period, 22 days; adult emergence normal.

See also: Charbonneau & Lemonde (1960a b); Lemonde & Bernard (1951b, 1955).

→ Medici, J.C.; Taylor, M.W. 1966. Mineral requirements of the confused flour beetle, *Tribolium confusum* (Duval). J NUTR 88: 181-186.

COMPOSITION Mineral control diet (%): cornstarch, 74.0; amino acid mixture, 20.0; reagent grade mineral salts, 2.0; corn oil, 3.0; vitamin mixture, 0.63; and cholesterol, 0.37.

l-amino acids (% diet): arginine, 0.9; histidine, 0.6; isoleucine, 1.2; leucine, 1.4; lysine hydrochloride, 1.3; methionine, 0.6; phenylalanine, 0.5; threonine, 0.9; tryptophan, 0.2; valine, 1.1; glutamic acid, 3.0; glycine, 5.3; cystine, 2.0; and tyrosine, 1.0.

salts (%): KH_2PO_4, 67.75; $MgSO_4$, 16.00; $Ca(H_2PO_4)_2.H_2O$, 8.00; NaCl, 5.00; $FeSO_4.7H_2O$, 2.00; $ZnCl_2$, 0.50; $MnSO_4.H_2O$, 0.50; and $CuSO_4.5H_2O$, 0.25.

vitamins (μg/g diet): choline chloride, 4000; nicotinic acid, 100; calcium *d*-pantothenate, 40; riboflavin, 18; pyridoxine hydrochloride, 16; thiamine hydrochloride, 12; folic acid, 5; biotin, 0.6; carnitine, 10; inositol, 2000; *p*-aminobenzoic acid, 500; *l*-ascorbic acid, 10; menadione, 1; and cobalamine (B_{12}), 0.5.

PREPARATION Ingredients finely ground and homogeneously mixed.

REARING At 28°C and 75 ± 5% RH. Eggs sifted from stock culture reared on 95% unbleached wheat flour and 5% dried yeast 24 h before experiment. 10 neonate larvae placed in small vials with diet.

DEVELOPMENT Larval wt, from 2.5 to 2.7 mg in 18 days. Three successive generations reared.

See also: Medici (1964); Medici & Taylor (1967); Taylor & Medici (1966).

→ Naylor, A.F. 1964. Possible value of casein, gluten, egg albumin, or fibrin as whole proteins in the diet of two strains of the flour beetle, *Tribolium confusum* (Tenebrionidae). CAN J ZOOL 42: 1-9.

Best of 7 diets described.

COMPOSITION Diet with nil protein content. (mg/g diet): *l*-alanine, 6.0; *l*-arginine hydrochloride, 8.3; *l*-aspartic acid, 13.3; *l*-cystine, 0.6; *l*-glutamic acid, 35.6; glycine, 3.7; *l*-histidine hydrochloride, 5.7; *dl*-isoleucine, 20.2; *l*-leucine, 12.5; *l*-lysine hydrochloride, 17.6; *dl*-methionine, 7.9; *dl*-phenylalanine, 14.6; *l*-proline, 11.9; *dl*-serine, 20.0; *dl*-threonine, 15.5; *l*-tryptophane, 2.5; *l*-tyrosine, 12.8; *dl*-valine, 19.9; $NaHCO_3$, 20.0; Wesson's salt mixture, 38.0; cornstarch, 562.7; sugar, 140.7; and cholesterol, 10.0; and vitamins.

Diet with 5% protein (mg/g diet): *l*-alanine, 4.0; *l*-arginine hydrochloride, 5.5; *l*-aspartic acid, 8.9; *l*-cystine, 0.4; *l*-glutamic acid, 23.7; glycine, 2.4; *l*-histidine hydrochloride, 3.8; *dl*-isoleucine, 13.5; *l*-leucine, 8.3; *l*-lysine hydrochloride, 11.7; *dl*-methionine, 5.3; *dl*-phenylalanine, 9.7; *l*-proline, 7.9; *dl*-serine, 13.3; *dl*-threonine, 10.3; *l*-tryptophane, 1.7; *l*-tyrosine, 8.6; *dl*-valine, 13.3; $NaHCO_3$, 20.0; casein, 50.0; Wesson's salt mixture, 38.0; cornstarch, 583.8; sugar (sucrose), 145.8; cholesterol, 10.0.

All media received vitamins and zinc as follows (μg/g diet): thiamin hydrochloride, 25.0; riboflavin, 12.5; pyridoxine hydrochloride, 12.5; nicotinic acid, 50.0; calcium pantothenate, 25.0; folic acid, 2.5; choline chloride, 500; inositol, 250; biotin, 0.5; *dl*-carnitine, 6.0; and $ZnCl_2$, 200.

PREPARATION Amino acid fraction ground with mortar and pestle and pulverized in micro paddle mill. Other dry ingredients similarly treated and 2 fractions mixed in small ball mill for 30-40 minutes. Zinc and vitamins added in aqueous solution @ 0.1 ml/g diet. Diet allowed to dry for 2 days in covered petri dish; then dish scraped clean and diet pulverized in the paddle mill. Dispensed to ½-oz creamer jars @ 2 g/jar.

REARING 30 larvae placed in each jar; left uncapped for 2 days at 29 ± 0.3°C and 70 ± 5% RH. Content of jars sifted every few days to check on development. Sifted daily at pupation. Live pupae removed from diet.

DEVELOPMENT Pupation on 5% casein diet in 907 h.

Frobrich, G.; Offhaus, K. 1953. Der qualitative Vitamintest mit dem Reismehilkäfer *Tribolium confusum* Duv. (Tenebrionidae, Coleoptera) als Testorganismus. Z VITAM HORM FERMENTFORSCH 5: 358-369. (In German with English summary).

COMPOSITION (g): casein (Hoffman-LaRoche or Merck), 50; glucose, 45; McCollum's salt mixture, 5; and (mg/g diet): ergosterol, 1; choline chloride, 2; and (µg/g diet): vitamin B_1, 7; vitamin B_2, 7; vitamin B_6, 35; nicotinic acid, 35; calcium pantothenate, 70; *d*-biotin, 0.4; folic acid, 10.

PREPARATION Casein and salts mixed by grinding. Ergosterol dissolved in solvent and stirred in with vitamins, dried and ground. Glucose finally added and diet reground.

REARING At 31-32°C and 50-60% RH.

DEVELOPMENT On complete basal diet; larval period, 22 days; comparable to control diet of wheat and yeast. Effect of vitamin deficiencies described.

Corollary

Schwarz, I.; Koch, A. 1962. Der Tribolium-Test als quantitativer Test für 8 B-Vitamine. I. Z VITAM HORM FERMENTFORSCH 12: 291-306. (In German with English summary)

Standard diet adapted from Frobrich & Offhaus* (1953).

COMPOSITION vitamins (µg/1 g diet): vitamin B_1, 6; vitamin B_2, 4; nicotinic acid amide, 20; calcium pantothenate, 20; *d*-biotin, 1.0; vitamin B_6, 9; folic acid, 8; choline chloride, 2000.

DEVELOPMENT From eclosion to pupa only.

See also: Schwarz & Koch (1963, 1966).

Tribolium confusum Jacquelin du Val, confused flour beetle

Acanthoscelides obtectus (Say), bean weevil (Fam., Curculionidae)

Chiu, S.F.; McCay, C.M. 1939. Nutritional studies of the confused flour beetle (*Tribolium confusum* Duval) and the bean weevil (*Acanthoscelides obtectus* (Say)). ANN ENTOMOL SOC AM 32: 164-170.

Best 2 diets of many tested.

COMPOSITION Diet 1 (g): casein, 40; cooked corn starch, 47; yeast, 5.

Diet 2 (g): casein, 20; cooked corn starch, 67; yeast, 5.

REARING At 28-35°C and 50-70% RH in darkness. 20 g diet in 4.7 x 12 cm bottles covered with muslin. 10 beetles/bottle.

DEVELOPMENT *T. confusum:* 122 adults after 49 days on diet 1; 132 adults on diet 2, and 292 adults on control diet of whole wheat flour.

A. obtectus: slow development on ether-extracted normal diet if 3% cholesterol added; average life cycle 41 days on diet; on kidney beans, 34.5 days.

Tribolium destructor Uyttenboogaart, dark flour beetle

→ Naton, E., Von 1960. Über die Entwicklung des schwarzbraunen Mehlkäfers, *Tribolium destructor* Uyttenb. Teil II. Beitrag zur Metamorphose des Darmtraktus bei natürlicher und "synthetischer" Ernährung. Z ANGEW ENTOMOL 47: 256-274. (In German with English summary)

Adapted from Fraenkel *et al.** (1950) for *Tenebrio molitor*.

COMPOSITION (g): vitamin-free casein, 35; vitamin-free rice starch, 65; McCollum's salt mixture, 5; and (mg): cholesterol, 10; and (µg): aneurin, 50; lactoflavin, 25; pyridoxine, 50; nicotinic acid amide, 100; calcium pantothenate, 100; folic acid, 2.5; *meso*-inositol, 100; *p*-amino benzoic acid, 10; choline chloride, 250; biotin, 0.5; B_{12}, 1. To each g diet either 10 mg yeast extract, 25 mg casein extract or 3 µg carnitine added.

REARING At 27°C and 80% RH.

DEVELOPMENT Equal to that on natural diet of wheat flour with 8% torula yeast.

See also: Chapman (1924); Mansingh & Pratt (1975); Naton (1961).

Trogositidae

Tenebroides mauritanicus (Linnaeus)

→ Bond, E.J.; Monro, H.A.U. 1954. Rearing the cadelle *Tenebroides mauritanicus* (L.) (Coleoptera: Ostomidae) as a test insect for insecticidal research. CAN ENTOMOL 86: 402-408.

COMPOSITION Adult diet 1 (mouldy oatmeal culture) (g): fine oatmeal, 150; water, 75 ml.

Diet 2 (yeast oatmeal culture) (g): oatmeal, 100; brewers' yeast, 5; agar, 2; water, 30 ml.

Larval diet. Whole wheat flour most successful. Oatmeal, corn meal, pablum, powdered whole milk, dog meal, peanuts, wheat germ cereal, wheat flour, pig starter, also used, and diet (parts): whole wheat, 2; corn meal, 4; powdered whole milk, 2; dried brewers' yeast, 1.

PREPARATION Diet 1: jars specially designed with diet platform made by gluing circular piece of 4" plastic screen to cylindrical piece of 1" high pliable sheet plastic; holds diet from bottom of jar and allows drainage of excess moisture. Screen tops to jars to allow air exchange yet prevent contamination. Oatmeal poured into jar, and firmly packed. Water poured in so that centre of diet soaked and outside dry. Developing moulds grow outwards from moist central part to form solid mass of mycelium throughout. Mould is ready in 2 days for feeding.

Diet 2: agar heated in water and mixed into oatmeal evenly. 3 g diet placed on 120 x 20 mm petri dish.

REARING At 77°F and 70% RH on diet 1; 5 pairs adults/jar. Diet replaced every 2 weeks. On diet 2; 3 pairs adults/dish. Diet replaced twice weekly. Oviposition on 'block' made from 2, 1" sq. 3/16" thick plastic sheets and 2, 3/8" squares of paper glued to centre of one sheet. Other sheet placed on top of paper squares. Two blocks thus held 0.3 mm apart; eggs laid in crevice. Eggs removed from block daily. Placed in small petri dishes or directly on to whole wheat flour for hatching @ 100 neonate larvae/250 g diet. Pupation within ½" cubes of coarse building cork on diet.

DEVELOPMENT Incubation period, 7-8 days; larval period, 8 weeks; prepupa to adult, 10-20 days. Maximum eggs/♀, 3581 with 72.1% viability. Maximum longevity, 218 days. Mean of 84.8% larvae and 65.7% adults reared on whole wheat flour from jars incubated with 100 larvae.

Dictyoptera

Blattellidae

Blattella germanica (Linnaeus), German cockroach

→House, H.L. 1949a. Nutritional studies with *Blattella germanica* (L.) reared under aseptic conditions. II. A chemically defined diet. CAN ENTOMOL 81: 105-112.

Five diets tested.

COMPOSITION Diet 1 (%): whole wheat flour, 40; dried skim milk, 40; dried brewers' yeast, 20.

Diet 2 (%): vitamin-free casein, 30; glucose, 65; cholesterol, 0.5; ergosterol, 0.5; mineral mixture (U.S.P. No. 1), 3; soybean oil, 1; total, 100.0; and (mg/g diet): thiamine hydrochloride, 0.05; riboflavin, 0.05; niacin, 0.05; adermine hydrochloride, 0.05; calcium pantothenate, 0.05; *p*-aminobenzoic acid, 0.05; inositol, 0.50; choline chloride, 0.50; folic acid, 0.002; and biotin, 0.25.

Diet 3: as for diet 2, except glucose replaced by 65% dextrin and amount of vitamin/g, except for biotin, doubled.

Diet 4: as for diet 3, except casein omitted; replaced by 30% amino acid mix (g): *dl*-alanine, 0.066; *l*-arginine monohydrochloride, 0.027; *dl*-aspartic acid, 0.077; *l*-cystine, 0.009; *l*-glutamic acid, 0.137; glycine, 0.018; *l*-histidine monohydrochloride, 0.018; *l*-hydroxyproline, 0.012; *dl*-isoleucine, 0.077; *l*-leucine, 0.071; *l*-lysine monohydrochloride, 0.044; *dl*-methionine, 0.048; *dl*-β-phenylalanine, 0.07; *l*-proline, 0.051; *dl*-serine, 0.089; *dl*-threonine, 0.048; *l*-tryptophane, 0.012; *l*-tyrosine, 0.042; *dl*-valine, 0.083; total, 1.000.

Diet 5: as for diet 4, but ribonucleic acid added at 5 mg/g diet.

PREPARATION For diets 2-5, glucose or dextrin, minerals, and casein or amino acids mixed first, then vitamins, sterols, soybean oil and 2 ml water mixed and blended with others. Ground with mortar and pestle, dried, reground, poured into glass bottles, autoclaved, and stored in cool place until use.

REARING Individually at 29.5 ± 0.5°C and 70% RH. Equipment and methods described in House & Patton (1949). Aseptic rearing in cages constructed from Kolle flasks. Entrance of cage plugged with removable stopper made of non-absorbant cotton wrapped in gauze. Food dish cut from end of 2" shell vial. Water supply from 50-ml Erlenmeyer flask; two tubes from top-air in through one and water out of the other protruding through back wall of Kolle flask into cage. Transfer chamber from " lumber, 36" long, 24" wide and 24" high with light fitting hinged glass-window at each side. Air locked of 2 sliding metal doors at one end to allow equipment into chamber. Long rubber gloves inserted through each side of chamber below window. Chamber fitted with ultra-violet lamp, disinfected and sterilized before use. Both rearing cage and chamber cage illustrated.

DEVELOPMENT From House (1949b). On diet 1: mortality, 8.2%; nymphal stage, 7 weeks; adult wt, 54.9 ± 3.1 mg; 20.4% adult emergence at 8 weeks. Development on diet 2-4, inferior to diet 1. On diet 5: mortality, 4.3%; 43.5% adult emergence in 12 weeks; adult wt, 48.1 ± 3.1 mg. First adult after 8 weeks; some adults reproduced.

See also: House (1949b); House & Patton (1949).

→Noland, J.L.; Lilly, J.H.; Baumann, C.A. 1949a. A laboratory method for rearing cockroaches and its application to dietary studies on the German roach. ANN ENTOMOL SOC AM 42: 63-70.

Four diets tested.

COMPOSITION Diet 1 (%): pulverized dog biscuit (Friskies brand, Albers Milling Co., Peoria, Illinois), 100.

Diet 2 (%): pulverized dog biscuit (Friskies), 95; corn oil, 5; cholesterol, 0.1.

Diet 3 (%): glucose monohydrate, 74; casein (alcohol extracted), 15; corn oil, 5; Wesson's salt mixture, 4; cholesterol, 1; *l*-cystine, 0.6; vitamin mixture, 0.4.

Diet 4 (%): glucose monohydrate, 61.6; casein (alcohol extracted), 30; corn oil, 3; Wesson's salt mixture, 4; cholesterol, 1; vitamin mixture, 0.4.

vitamins (in µg/g): thiamine hydrochloride, 12; riboflavin, 18; pyridoxine hydrochloride, 16; nicotinic acid, 100; calcium pantothenate, 40; *p*-aminobenzoic acid, 50; choline chloride, 1000; inositol, 2000; vitamin K (menadione), 1; biotin, 0.6; folic acid, 5.

PREPARATION Details in Noland *et al.* (1949b). Prepared in parts:- casein and salt mixture ground in Wiley mill to pass 100-mesh screen, then further pulverized in ball mill for 24 h, and carbohydrate added. Vitamins dissolved in water; 3 ml vitamin solution added to 20 g portion of dry ingredients, blended and dried at room temperature. Fat and cholesterol dissolved in 3 ml chloroform, mixed with other ingredients and dried as before. Diet refrigerated.

REARING Apparatus illustrated. Individual cages of 3 test tubes:- 200 x 25 mm, horizontal tube with base replaced by 50 mesh/inch wire cloth sealed with adhesive of 1:1 beeswax:resin mixture. 50 x 10 mm tube half filled with diet placed within large tube with open end facing screen. Water-filled 150 x 18 mm tube inverted and connected to horizontal tube by 12 mm length of 8 mm glass tubing angled to 120° at centre. Proximal end of tube narrowed to 3 mm diam by heating to hold water column. Individual cages grouped on wooden frameworks with vertical watering tubes spaced by furniture nails and horizontal feeding chambers by foil strips.

Three modifications of this basic design used. Rearing at about 30° C (26-33° C) with uncontrolled RH. One mating pair/cage or 5 one-day-old nymphs/cage; diet and water renewed weekly.

DEVELOPMENT Mature nymphs of about 40 and 52 days after introduction to crude and synthetic diets respectively.

Corollary

→ Noland, J.L. 1956. An improved method for rearing cockroaches. J ECON ENTOMOL 49: 411-412.

An improvement of rearing method and equipment described. Cage from 200 x 25 mm test tube with base replaced by 60-100 mesh screen affixed by 50:50 molten resin and beeswax mixture, or by perforated 1" rubber cap, removable for cleaning. Diet tubes 50 x 10 mm; watering device:- rubber stopper holding 45 x 15 mm vial with water and cotton wick.

See also: Noland (1954a b); Noland & Baumann (1949, 1951); Noland *et al.* (1949b); Brooks (1960); Brooks & Richards (1955, 1956).

→ Van Herrewege, C.; David, J. 1969. Une nouvelle méthode pour mesurer la prise de nourriture et l'utilisation digestive des aliments chez *Blattella germanica* (L.). ANN NUTR ALIMENT 23: 253-268. (In French)

Three diets used. Diets 1 and 2 for maintenance of stock colony; diet 3 modified after Gordon (1959).

COMPOSITION Diet 1 (%): powdered wheat germ, 75; skim milk, 20; dried brewers' yeast, 5.

Diet 2: Dog biscuit and fresh carrot.

Diet 3 (g): vitamin-free casein, 300; anhydrous glucose, 360; alumina or silica, 254.687; cholesterol, 10.0; inositol, 3.6; betaine, 2.4; choline chloride, 3.0; linoleic acid, 8.4; maize oil, 30.0; NaCl, 1.2; $CaCO_3$, 2.0; $MgSO_4$, 4.8; K_2SO_4, 4.4; KH_2PO_4, 13.6; and (mg): nicotinic acid, 175; calcium pantothenate, 170; thiamine hydrochloride, 118; folic acid, 77; pyridoxine, 72; biotin, 10; riboflavin, 130; cyanocobalamine, 1; acetates of Mn, Cu, Zn, and Fe, 370, 100, 220, and 470 respectively.

PREPARATION See Gordon (1959). Salts individually ground and mixed. Vitamins mixed by grinding; some dissolved in water, impregnated in celluflour and vacuum dried. Choline dissolved in hot ethanol and impregnated in celluflour. Cholesterol and linoleic acid dissolved in ether and impregnated with diet. Various mixes blended and refrigerated until use.

REARING At 30° C and 75% RH in darkness. Insects held in transparent 200 ml plastic cages with side door for addition of food and water. About 100 mg diet; replaced every 2-3 days.

DEVELOPMENT Nymphal wt: ♂, 43.47 ± 0.16 mg after 43 days; ♀, 62.96 ± 0.33 mg after 45 days. Comparable with biscuit diet, but greater than on wheatgerm based diet. Three consecutive generations reared on synthetic diet without affecting fecundity.

See also: Brooks & Kringen (1972); Clayton (1959); Gordon (1959, 1968a); Hilchey (1953); McCay (1938); McCay & Melampy (1937); Melampy & Maynard (1937); Van Herrewege (1971, 1974b); Woodruf (1939).

Blattidae

Periplaneta americana (Linnaeus), American cockroach

→ Forgash, A.J. 1958. The effect of inositol on growth, survival, and maturation in *Periplaneta americana* (L.). ANN ENTOMOL SOC AM 51: 406-409.

COMPOSITION (g/kg): dextrose, 314; vitamin-free casein, 300; non-nutritive fibre, 300; raw linseed oil, 30; Wesson's salt mixture, 40; cholesterol, 10; choline chloride, 4; and (mg/kg): biotin, 6; *p*-aminobenzoic acid, 300; nicotinic acid, 100; thiamine hydrochloride, 70; calcium pantothenate, 40; pyridoxine hydrochloride, 18; riboflavin, 18; *dl*-methionine, 18; folic acid, 6. Inositol added at rate of 500 mg/kg to improve growth, survival, and maturation.

PREPARATION Dextrose, fibre, casein and salts ground together in ball mill for 24 h, then blended with remaining ingredients (except cholesterol and linseed oil) dissolved in water. Dried in shallow trays and reground, then blended with cholesterol and linseed oil dissolved in chloroform. After drying, diet reground and refrigerated in amber bottles till use.

REARING Newly hatched nymphs reared in gallon battery jars containing several layers of surgical gauze, several masonite platforms, shallow dish of water (with pad of cotton to prevent drowning), and shallow dish of food. Dishes ringed with adhesive tape to provide traction for young nymphs, and jars coated with thin film of vaseline on inside upper 2" and capped with screen-covered embroidery hoop. Water replenished as needed and food changed twice weekly.

DEVELOPMENT On complete diet, 78% survival to maturity; on inositol deficient diet, 3% maturity.

Corollary

→ Forgash, A.J.; Moore, R.F. 1960. Dietary inositol requirement of *Periplaneta americana* (L.). ANN ENTOMOL SOC AM 53: 91-94.

Minimal levels of inositol between 70 and 80 mg/kg diet.

Haskins, K.P.F. 1962. A new system for rearing the American cockroach. ENTOMOLOGIST LONDON 95: 27-29.

COMPOSITION (g): rolled oats, 9; wheatfeed, 9; fishmeal, 1; dried yeast powder, 1.

PREPARATION Mixed. Sterilized at 50° C for 15 h.

REARING At 27.5 ± 1° C and 30% RH. Adults in 10-lb jar with perforated metal lid and paraffin smear to prevent escape. Water supplied in 2-lb jar inverted on to cotton wool in petri dish within adult cage. Egg cases collected weekly by sieving food through 0.5 cm mesh, and washed; dried between blotting paper. Incubated in 7-lb jars with little food at 30° C and 70% RH for 30-35 days. Then 400 neonate nymphs/10-lb jar. After 4 months ♂♂ and ♀♀ separated after anaesthetising with CO_2 and kept apart till adult. Then 20 ♂♂ and 30 ♀♀/10-lb jar with food in 12.5 mm diam crystallising dishes.

DEVELOPMENT Life cycle, 5-6 months from emergence to adult.

Corollary

Rochford, P.J. 1972. The American cockroach. pp. 575-581. *in* Universities Federation for Animal Welfare (Ed.), The UFAW Handbook on the Care and Management of Laboratory Animals. Fourth Edition. Churchill Livingstone, Edinburgh and London.

REARING Tolworth system: nymphs in 40 x 20 cm plastic container with metal base and perforated lid. Ten 15 x 10 cm hardboard stages separated by 1.8 cm wooden blocks placed in cage. Food and water on top stage. Hole in cage for anaesthetizing. Cages rest in Technical White Oil; 70% RH and 27.7°C.

Slough system: at 30% RH with free water available. Eggs laid in food and collected by sieving; washed, dried and incubated at 70% RH.

Rochford system: false floor of 0.6 cm wire mesh through which egg cases deposited. Colony on plywood hanging from frame. Nymphs into separate chambers at 1 week. Temperature gradient from 25°C at bottom to 28°C at top.

Other methods also referred to.

DEVELOPMENT 9-13 moults depending on method and temperature. Full development in 520 days at 25.5°C; 195 days at 28.3°C.

Saksena, C.; Perti, S.L. 1970. Development of a chemically defined diet for studies on the nutritional requirements of the cockroach, *Periplaneta americana* Linn. LABDEV PART B8: 155-160.

Best of 8 diets tested.

COMPOSITION (g): casein, 30.5; glucose, 66.0; cholesterol, 1.0; salt mixture, 3.0; and (ml): vitamins, 1.0.

salts (g): $Ca[CH_3CH(OH)COO]_2$, 35.15; $CaH_4(PO_4)_2$, 14.60; KH_2PO_4, 25.78; NaCl, 4.67; $MgSO_4.7H_2O$, 7.19; $FeC_6H_5O_7.5H_2O$, 3.19; $NaH_2PO_4.H_2O$, 9.38.

vitamins (mg/g diet): thiamine hydrochloride, 0.1; riboflavin, 0.1; niacin, 0.1; pyridoxine, 0.1; calcium pantothenate, 0.1; *p*-aminobenzoic acid, 0.1; inositol, 1.0; choline chloride, 1.0; folic acid, 0.004; and biotin, 0.25 µg.

PREPARATION Casein and mineral salts powdered and blended with vitamin solution. Cholesterol dissolved in chloroform and added to blend. Stirred to uniform paste and dried at room temperature. Glucose mixed in and diet refrigerated until use.

REARING At 27 ± 1°C and 70 ± 5% RH.

DEVELOPMENT 66% survival to adult in mean of 252 days. Adults normal and fertile.

See also: Saksena & Perti (1971).

→ Sieburth, J.F.; McLaren, B.A. 1953. Growth studies with the cockroach *Periplaneta americana* (Linn.), fed vitamin-deficient diets substituted with corresponding anti-vitamins. ANN ENTOMOL SOC AM 46: 43-48.

COMPOSITION (%): vitamin-free casein, 30; dextrose, 30; Wesson's salt mixture, 4; corn oil, 3; cholesterol, 1; cornstarch, 32.

vitamins (mg/g basal diet): thiamine hydrochloride, 0.012; riboflavin, 0.015; niacinamide, 0.100; pantothenic acid, 0.030; pyridoxine hydrochloride, 0.016; inositol, 2.0; biotin, 0.0005; folic acid, 0.005; ascorbic acid, 0.050; *p*-aminobenzoic acid, 0.050; choline, 4.0; vitamin E, 0.200; vitamin K, 0.025; vitamin A, 0.500 USP units; vitamin D, 0.400 I.U.

PREPARATION Not described. Prepared diet packed in short lengths of glass tubing sealed at one end.

REARING See Sieburth *et al.* (1951). At 30°C in ½-pint jar covered with 100 mesh wire screen. Cones of filter paper for shelter, water filled vials with wick and diet within jar. Jars cleaned and food and water renewed weekly.

DEVELOPMENT At 30, 60 and 90 days, nymphal wts: 25.0, 62.5, and 105.6 mg respectively; nymphal mortality, 6.7, 16.5, and 2.72%.

→ Wientjens, W.H.J.M. 1972. De Omzetting van β-sitosterol in cholesterol bij drie soorten kakkerlakken. MEDED CENT LAB DELFT 452: 35-46. (In Dutch).

COMPOSITION In 2% agar (g): fructose, 31; vitamin-free casein, C grade, 30; corn starch, 30; salt mixture, USP XIV, 4; sodium oleate, 3; vitamin mixture, 1.

vitamins (mg): folic acid, 5; biotin, 0.6; pyridoxine hydrochloride, 16; calcium *d*-pantothenate, 40; niacin, 100; riboflavin, 18; thiamine hydrochloride, 12; and (g): *i*-inositol, 2; choline chloride, 4.

REARING Aseptically at 30°C. Eggs surface sterilized by immersion in 70% alcohol for 1 minute and 0.2% $HgCl_2$ for 2 minutes, rinsed in 70% alcohol and twice in distilled water.

DEVELOPMENT Three species reared: German cockroach, *Blattella germanica* (Linnaeus) (Fam., Blattellidae); American cockroach, *Periplaneta americana* (Linnaeus) (Fam., Blattidae); and *Eurycotis floridana* Walker (Fam., Blattidae).

→ Żabiński, J. 1929. The growth of black beetles and of cockroaches on artificial and on incomplete diets. Part 1. J EXP BIOL 6: 360-385.

Best of many diets tested.

COMPOSITION (%): starch, 61; saccharose, 20; McCollum & Simonds mineral salts, 4; glycocoll, 15.

PREPARATION Ingredients mixed and heated at 100-110°C for 24 h; distilled water added and resulting paste dried in thin layers at 60-70°C. Then powdered and stored in sealed jars.

REARING At 25-30°C in glass jars with moistened sand-, sawdust- or filter-paper covered base. Food on small watch glass in jar changed every 3 days. Free water provided if humidity too low.

DEVELOPMENT *Blattella germanica* (Linnaeus): complete development in 3 months. *Periplaneta orientalis* Linnaeus: development incomplete.

See also: Bodenstein & Fales (1969); Gier (1947); Ramazzotto & Ludwig (1967); Żabiński (1926, 1928a b).

Diptera

Anthomyiidae

Delia antiqua (Meigen) [previously *Hylemya*], onion maggot

→ Allen, W.R.; Askew, W.L. 1970. A simple technique for mass-rearing the onion maggot (Diptera: Anthomyiidae) on an artificial diet. CAN ENTOMOL 102: 1554-1558.

COMPOSITION (g): gelatine, 32.0; sucrose, 60.0; yeast hydrolysate enzymatic, 40.0; wheat embryo, 40.0; cellulose powder, 40.0; Aureomycin, 0.3; methyl *p*-hydroxybenzoate, 1.25; and (ml): *n*-propyl disulphide, 0.2; evaporated milk, 200; water, 1000.

PREPARATION Gelatine first softened by soaking for 5 minutes in evaporated milk and 200 ml water. Remaining water added at 65°C to complete dispersion. Yeast hydrolysate, wheat embryo, and cellulose powder ground in Waring Blendor to pass through bolting cloth (16 mesh/cm), and Aureomycin, methyl *p*-hydroxybenzoate, and *n*-propyl disulphide suspended in 10 ml 95% ethyl alcohol, mixed with hot mix, and blended for 3 minutes. Final diet poured into containers to depth of 1.3 cm, covered, and stored at 2°C until required. Storage for >1 month possible.

REARING At 23 ± 1°C and 40 ± 10% RH with continuous fluorescent lighting. Adults reared in 60 cm^3 screen covered cages. Food from 9.5 cm diam petri dishes containing cotton wool pads soaked in sugar, 5 g; evaporated milk, 10 ml and 35 ml water. Yeast hydrolysate sprinkled on paper towels folded to fan-shape and inserted in 125 ml Erlenmeyer flasks filled with water also provided. Two feed dishes and paper fans replenished 3 times/week for 400 flies. 13.7 x 3.7 cm oviposition dishes with moist silica sand, 1.3 cm deep, and 1 ml mixture of *n*-propyl disulphide and evaporated milk 2.98 (v/v) emulsified with 1.0 g of saponin added. Covered with 1.3 cm layer clean moist gravel (3-5 mm diam). Left for 40 h, contents stirred in beaker of water and floating eggs decanted on to fine bolting silk. Eggs incubated on moist blotting paper in sealed plastic containers for 48 h at 23.0 ± 1°C and 35 ± 5% RH. Larval rearing in 15 cm diam x 7 cm plastic containers with 6 cm diam ventilation holes covered with bolting cloth cut in lids. 1 cm layer cellulose powder and silica sand (20-40 mesh) added to containers then moistened. 44 cm^2 x 1 cm block of diet laid on sand and 400-600 eggs or neonate larvae placed on diet surface. Diet covered with sand with 4-5% water content and dish closed. After 1 week more diet added plus either water or dry sand to adjust moisture level. After 6 days maggots recovered by placing contents of box on 18 mesh sieve and dipping repeatedly in water. Sand washed away leaving pupae and larval diet. Pupation completed after 14 days and removed by decanting on to 12 mesh sieve.

DEVELOPMENT Pupation after 14 days from inoculation. Pupal wt, 13.4 mg; average longevity: ♂, ca. 25 days; ♀, 37 days; mean fecundity, 343 eggs/♀. Three consecutive generations reared with pupal weights, percentages of pupation, adult emergence, and egg hatch equal to those reared on onion bulbs.

➔ Friend, W.G.; Patton, R.L. 1956. Studies on vitamin requirements of larvae of the onion maggot, *Hylemya antiqua* (Mg.), under aseptic conditions. CAN J ZOOL 34: 152-162.

COMPOSITION (mg/ml): agar, 20; ribonucleic acid, 1.0; inosine, 0.03; thymine, 0.004; cholesterol, 0.1; dextrose, 15.0; U.S.P. XIII No. 2 mineral mixture, 2.0; amino acid mixture, 24.21 and (μg/ml): vitamin mixture, 77.96; and distilled water, 1000 ml.

minerals (%): $CaH_4(PO_4)_2$, 13.58; $Ca[CH_3CH(OH)COO]_2$, 32.70; $C_6H_5O_7Fe.5H_2O$, 2.97; $MgSO_4.7H_2O$, 13.70; K_2HPO_4, 23.98; $Na_2HPO_4.2H_2O$, 8.72; NaCl, 4.35.

l-amino acids (mg): alanine, 1.09; arginine (free base), 0.80; aspartic acid, 1.22; cysteine, (free base), 0.48; glutamic acid, 4.42; glycine, 1.75; histidine (free base), 0.48; hydroxyproline, 0.38; isoleucine, 1.26; leucine, 2.35; lysine hydrochloride, 1.34; methionine, 0.34; phenylalanine, 1.01; proline, 1.68; serine, 0.88; threonine, 0.38; tryptophan, 1.75; tyrosine, 1.24; valine, 1.36.

vitamins (μg): biotin (aqueous solution), 0.020; B_{12}, 0.04; calcium pantothenate, 6.0; choline chloride, 20.0; coenzyme A, 1.5; folic acid, 6.0; nicotinic acid, 10.0; pyridoxine hydrochloride, 30.0; riboflavin, 2.4; thiamine hydrochloride, 1.5; thioctic acid, 0.5.

PREPARATION Amino acids except aspartic acid, glutamic acid, tryptophan and tyrosine ground in mortar. Tyrosine dissolved in hot water and concentrated HCl. Tryptophan dissolved in warm water. Glutamic and aspartic acids each dissolved in warm water. Solutions all combined and dry amino acids added and dissolved. RNA dissolved in 2N NaOH and water. Inosine and thymine dissolved in hot water and concentrated HCl; solution warmed over steam bath for 1 h. Salt mixture dissolved in hot water. Dextrose dissolved in warm water. Stock solution of cholesterol made by dissolving 500 mg in 20 ml hot 95% ethyl alcohol, pipetting into 100 ml cold water and adding 0.3 ml Tween 80; reduced to 60 ml by evaporation. Vitamins and cofactors dissolved individually; folic acid in 20% ethyl alcohol, thioctic acid in 95% ethyl alcohol, and riboflavin in 0.02 N warm acetic acid; remaining vitamins dissolved in distilled water. Salt solution decanted into large beaker and amino acid solution added. RNA and inosine and thymine solutions added and pH adjusted to 6.1 with 2N NaOH. Agar added and mixture autoclaved for 2 minutes at 15 lb pressure. Cholesterol and dextrose solutions added. Hot mixture poured into graduated flasks containing vitamin mixtures; held at 80°C. 1 ml diet dispensed into 1-dram screw-capped vials. Vials marked, loosely capped and autoclaved for 15 minutes at 15 lb pressure. Tubes slanted after autoclaving.

REARING Eggs immersed in 6.5% formalin solution for 30 minutes and washed in sterile water. Transferred to diet tube by sterile loop and held at 26°C. Adults caged in 18 x 10" diam glass jar; fed honey and brewers' yeast with growing onion plant in the cage. Plant inspected every 3 days for eggs.

DEVELOPMENT About 70% (aseptic) development to adult. Adults normal. Larval period, 14.8 days 2/3rd time taken under field conditions.

Corollary 1

→ Friend, W.G.; Backs, R.H.; Cass, L.M. 1957. Studies on amino acid requirements of larvae of the onion maggot, *Hylemya antiqua* (Mg.), under aseptic conditions. CAN J ZOOL 35: 535-543.

Vitamin mixture improved (µg/ml diet): biotin (aqueous solution), 0.02; B_{12}, 0.05; calcium pantothenate, 7.0; choline chloride, 23.0; coenzyme A (equals 246 x 10^{-6} Lipman units), 1.7; folic acid, 7.0; nicotinic acid, 12.0; pyridoxine hydrochloride, 34.0; riboflavin, 2.7; thiamine hydrochloride, 1.7; *dl*-6-thioctic acid, 0.6.

DEVELOPMENT Superior. Survival to adult, 87.9%. Larval period, 13.3 days. Eggs viable.

Corollary 2

→ Friend, W.G.; Salkeld, E.H.; Stevenson, I.L. 1959. Nutrition of onion maggots, larvae of *Hylemya antiqua* (Meig.), with reference to other members of the genus *Hylemya*. ANN N Y ACAD SCI 77: 384-393.

REARING At 25 ± 1°C in 1-dram screw-capped vials. Pupae removed from tubes and buried in moist sand in new vials.

DEVELOPMENT *Hylemya antiqua:* 93% 3rd instar larvae after 7.6 days; 65% pupae after 27.4 days. If living *Bacillus* sp. added to diet then 100% 3rd instar after 4.1 days; 94% pupae after 10.1 days; 92% adults after 22 days.

Other species: *Hylemya cilicrura* (Rondani): if 3% agar in diet, then 100% 3rd instar after 6.6 days; 75% pupae after 18.1 days; 13% adults after 26.5 days.

Hylemya brassicae (Bouché): out of 249 larvae, 3 pupated, but none emerged.

→ McClanahan, R.J. 1966. A synthetic diet for the onion maggot, *Hylemya antiqua* (Meigen) (Diptera: Anthomyiidae). CAN J ZOOL 44: 1089-1090.

COMPOSITION (g): macerated onion bulb, 100; lima bean flour, 100.0; brewers' yeast, 25.0; sucrose, 12.5; ascorbic acid, 5.0; corn oil, 2.5; methyl *p*-hydroxybenzoate, 2.0; choline chloride, 1.5; Wesson's salt mixture, 0.5; agar, 12.5; and water, 550 ml.

PREPARATION Agar dissolved in water, added to other ingredients and blended. Diet cooled and stored at 5°C in corked sections of 3.5 cm diam glass tubing.

REARING At 20°C with 14 h photoperiod. Eggs or larvae placed on diet slices on moist sand. Fresh diet provided every 3 days. Adults fed brewers' yeast, yeast hydrolysate and honey.

DEVELOPMENT 75 adults from 118 larvae. Mean days to adult, 35.7 ± 0.44; 98% of eggs fertile; second generation adults in 34 days. Comparable with growth on onions.

Ticheler, J. 1971. Rearing of the onion fly, *Hylemya antiqua* (Meigen), with a view to release of sterilized insects. pp. 341-346. *in* Sterility Principle for Insect Control or Eradication. Proceedings of a panel organized by the joint FAO/IAEA Division of Atomic Energy in Food and Agriculture.

COMPOSITION Adult diet (parts): skimmed milk powder, 10; sugar, 10; brewers' yeast, 10; soya peptone, 1.

Larval diet (g): carrot powder, 30; brewers' yeast, 12; cellulose powder, 2; methyl *p*-hydroxybenzoate, 0.25; potassium sorbate, 0.25; and (ml): 1N HCl, 11; water, 200.

PREPARATION Carrot powder from commercially available air-dried carrots ground in laboratory hammer mill to 0.8 mm particle size. All ingredients blended at high speed for ca. 5 minutes until homogenous with no visible free water. Dispensed to depth of 1.5 cm in 11 x 17 cm polythene boxes.

REARING 2000 adults in 60 x 40 x 40 cm cages with gauze top and perspex walls in greenhouse, at 20 ± 4° C and 60% RH with 16 h photoperiod. Food changed 2 times/week; water in petri dish with wet vermiculite. Few pieces of onion placed in cage to stimulate oviposition. Two oviposition devices used: 1) petri dish filled with moist silver sand with onion on top. Eggs washed out of sand and collected on sieve. 2) Bottom half of 8 cm clay flower pot, with 1-2 mm notches scratched on rim. Small basket with onion inside. Pot inverted on filter paper in clean petri dish and eggs laid under notches. Eggs removed 3 times/week and fresh onion provided. Larvae reared at 25 ± 0.5° C, and 75 ± 2% RH with continuous light. Diet seeded with 0.09 g (ca. 1250) eggs and covered with similar box. After 14 days diet placed on 1.2 mm mesh sieve under water jet to remove remaining diet from mature larvae, which remain on sieve. Larvae stored in jar in dark and high humidity. Pupae separated from larvae by putting dietary mixture on sieve (mesh 1.8 mm) through which larvae crawl. Pupae stored at 3° C and 90% RH.

DEVELOPMENT Larval stage, 14-18 days. Pupal weight, 14 mg. Fecundity, ca. 100 eggs/♀.

See also: Ticheler & Noordink (1968).

Rawlins, W.A. 1967. Oviposition by onion maggot adults fed on a chemically defined diet. J ECON ENTOMOL 60: 1747-1748.

COMPOSITION Adult diet "Nutrico Commercial" (NBC), containing 20 amino acids, 16 vitamins, 13 salts, glucose, and ethyl linoleate.

FEEDING Liquid diet dispensed to 1.5-dram vials with 2" dental roll wick/stopper proturding ½" above vial. Vials replaced once or twice/week.

DEVELOPMENT In comparative egg-laying tests, synthetic diet equal to yeast extract-honey mixture.

See also: McLeod (1964).

Delia brassicae (Bouché) [previously *Hylemya*], cabbage maggot

➡ Dambre-Raes, H. 1974. Nutrition of the cabbage root fly *Hylemya brassicae* (Bouché). I. An artificial diet and axenic culture method. MEDED FAC LANDBOUWWET RIJKSUNIV GENT 39: 1603-1610.

COMPOSITION (g): sucrose, 4; casein, 4; inositol, 0.2; RNA (from yeast), 0.1; ash-free paper pulp, 18; and (ml): torula yeast extract, 100; vitamin solution, 1.5; salt solution, 0.4.

vitamin solution (mg): biotin, 6; calcium pantothenate, 80; choline chloride, 12500; folic acid, 60; nitocinic acid, 2000; pyridoxine hydrochloride, 3000; riboflavin, 450; thiamine hydrochloride, 175; and distilled water, 1000 ml.

salt solution (g): K_2HPO_4, 22; $Na_2HPO_4.12H_2O$, 8.38; $CaCl_2$, 4.88; and distilled water, 1000 ml.

PREPARATION Torula-extract prepared from 40 g dried *Torula utilus*, boiled for 3 h with continuous stirring in 1000 ml distilled water; cooled; centrifuged at 5000 rpm for 5 minutes; supernatant decanted; residue washed with distilled water and centrifuged again; supernatant added to first and distilled water added to 1000 ml. Deep frozen till use, when adjusted to pH 7.5 with 1N NaOH. Ingredients added at 70°C. Solution poured over paper-pulp in mortar and thoroughly mixed. Mixture dispensed to screw-capped tubes or petri dishes and autoclaved at 120°C for 20 minutes.

REARING At 25 ± 1°C in darkness; 50 eggs/25 mm diam screw-capped tube filled with ca. 10 g diet, or 300 eggs in petri dishes with ca. 35 g diet. On 18th day placed in covered trays filled with moist sand for pupation. Eggs surface sterilized with White's solution using disposable syringe device (described) and placed aseptically on petri dish with sterile nutrient agar, incubated for 3 days at 20°C, then removed to diet in vials.

DEVELOPMENT Eclosion, 80%; 1st instar, 3-6 days; 2nd instar, 5-10 days, 3rd instar, 6-11 days. 78% pupation from 3rd instar. Adult emergence from 33 days after inoculation; oviposition from 8 days later; 70% eggs viable. Adult emergence, 20% - possibly due to dry conditions.

See also: Swailes (1967).

➡ Dambre-Raes, H. 1975. Axenic rearing method of cabbage root maggots *Hylemya brassicae* (Bouché) on an artificial diet. MEDED FAC LANDBOUWWET RIJKSUNIV GENT 40: 751-758.

COMPOSITION Larval diet (g): Hammarsten casein, 4.0; sucrose, 4.0; RNA (mg): β-sitosterol, 4.0; β-carotene, 1.0; $CaCl_2$, 80.0; and (ml): torula yeast extract, 100.0; 1N KOH, 5.0; 1N NaOH, 0.5; vitamin solution, 1.5.

vitamins (mg/ml): biotin, 0.002; calcium pantothenate, 0.60; pyridoxine hydrochloride, 3.0; riboflavin, 0.240; nicotinic acid, 1.0; choline chloride, 2.0; folic acid, 0.60; thiamine hydrochloride, 0.150.

PREPARATION As in Dambre-Raes (1974), except that pH of yeast solution adjusted by KOH and NaOH. $CaCl_2$ dissolved in yeast extract together with other nutrients; β-sitosterol and β-carotene absorbed on to paper pulp in ether solution prior to mixing with nutrient solution after ether evaporated.

REARING At 20°C in darkness in screw-capped tubes with 10 g diet inoculated with ca. 70 surface-sterilized eggs. On 16th day from eclosion, tubes inverted in beakers containing lightly packed moist sand. Pupae collected daily from sand, kept moist in small petri-dishes. Newly emerged adults collected daily, sexed, then transferred to 16 x 8.5 x 13 cm rearing cages @ 10 pairs/cage.

DEVELOPMENT Total development from larva to adult, 35-36 days; larval period, ca. 18 days; survival to pupa, 74.83%; pupal wt, 13.7 mg; adult emergence, 67%; sex ratio, 1:1; fertility and daily fecundity normal; mean adult longevity on diet: ♀, 10.74; ♂, 6.74 days; on rutabaga: ♀, 23.7; ♂, 17.6 days.

→ Vereecke, A.; Hertveldt, L. 1971. Laboratory rearing of the cabbage maggot. J ECON ENTOMOL 64: 670-673.

COMPOSITION Adult diet (parts): water, 5; sucrose, 3; brewers' yeast extract, 2. Neutralized with 1N NaOH.

REARING At 20° C and 68% RH. 600 adults in 56 x 56 x 50 cm cages with 3 walls of transparent plastic sheeting and 4th of glass and wooden panel. Top of nylon screen with feeding tubes. Larvae reared on rutabaga (*Brassica napus napobrassica*). Rearing cage and pupal container illustrated.

DEVELOPMENT Larval period, 19.2 days; survival from egg to pupa, 72%. Fecundity, 13.7 eggs/♀/day. Longevity, 17.1 days.

Delia platura (Meigen) [previously *Hylemya cilicrura* (Rondani)], seed corn maggot

→ Friend, W.G.; Salkeld, E.H.; McClanahan, R.J. 1958. A chemically defined diet and axenic rearing method for larvae of the seed corn maggot, *Hylemya cilicrura* (Rond.) (Diptera: Anthomyiidae). CAN J ZOOL 36: 931-936.

Two diets tested.

COMPOSITION Low level diet (mg/ml): agar, 30; ribonucleic acid, 0.50; dextrose, 7.5; U.S.P. XIII No. 2 mineral mixture, 1.0; amino acid mixture, 11.30; and (μg/ml): inosine, 15; thymine, 2.0; cholesterol, 50; vitamin mixture, 30.36.

minerals: see Friend & Patton* (1956).

l-amino acids (mg/ml): alanine, 0.55; arginine free base, 0.40; aspartic acid, 0.61; cysteine free base, 0.24; glutamic acid, 1.11; glycine, 0.88; histidine free base, 0.24; hydroxyproline, 0.19; *iso*-leucine, 0.63; leucine, 1.18; lysine hydrochloride, 0.67; methionine, 0.17; phenylalanine, 0.51; proline, 0.84; serine, 0.44; threonine, 0.19; tryptophan, 0.88; tyrosine, 0.62; valine, 0.68.

vitamins (μg/ml): biotin (aqueous), 0.01; B_{12}, 0.025; calcium pantothenate, 0.35; choline chloride, 0.12; coenzyme A, 0.85; folic acid, 3.5; nicotinic acid, 6.0; pyridoxine hydrochloride, 17.0; riboflavin, 1.35; thiamin hydrochloride, 0.85; *dl*-6-thioctic acid, 0.3.

High level diet (mg/ml): agar, 30; ribonucleic acid, 0.67; dextrose, 10.0; U.S.P. XIII No. 2 mineral mixture, 1.3; amino acid mixture, 14.69; and (μg): inosine, 20.0; thiamine, 2.7; cholesterol, 67.0; vitamin mixture, 40.70.

l-amino acids (mg/ml): alanine, 0.73; arginine free base, 0.53; aspartic acid, 0.81; cysteine free base, 0.32; glutamic acid, 1.48; glycine, 1.17; histidine free base, 0.32; hydroxyproline, 0.25; *iso*-

leucine, 0.84; leucine, 1.57; lysine hydrochloride, 0.89; methionine, 0.23; phenylalanine, 0.68; proline, 1.12; serine, 0.59; threonine, 0.25; tryptophan, 1.17; tyrosine, 0.83; valine, 0.91.

vitamins (μg/ml): biotin (aqueous), 0.013; B_{12}, 0.033; calcium pantothenate, 0.46; choline chloride, 0.16; coenzyme A, 1.13; folic acid, 4.60; nicotinic acid, 8.00; pyridoxine hydrochloride, 23.0; riboflavin, 1.80; thiamine hydrochloride, 1.10; *dl*-6-thioctic acid, 0.40.

PREPARATION Amino acids except aspartic acid, glutamic acid, tryptophan and tyrosine weighed for 1 l diet and mixed. Stock solutions of remaining 4 amino acids made; aspartic acid, 80 ml; glutamic acid, 100 ml; tryptophan, 70 ml; and tyrosine, 50 ml; each with correct dry weight of amino acid for 1 l diet. These solutions mixed and added, with 200 ml water to dry amino acids; stirred violently till dissolved. 6 g salt mixture mixed with 300 ml water and 100 ml decanted into amino acid mixture after settling. Stock solution of 90 mg inosine and 12 mg thymine made by adding 300 ml hot water and warming in water bath for 1 h. 100 ml solution added to diet mixture. RNA treated carefully to prevent 'burning' in high pH. 70 ml water poured over RNA and stirred rapidly; 3 ml 2N NaOH added in drops. When completely dissolved, water added to 100 ml. 5 ml of 3N KOH and 5 ml 3N NaOH added to amino acid mixture before addition of RNA to prevent precipitation. Aqueous stock solutions of all vitamins prepared except folic acid (20% ethyl alcohol); riboflavin (warm 0.02 N acetic acid) and *dl*-6-thioctic acid (4% ethyl alcohol). 20 ml riboflavin solution and 10 ml of other stock solutions prepared. 120 ml solution added to diet to bring volume to 930 ml. Cholesterol stock solution made by dissolving 500 mg in 20 ml hot 95% ethyl alcohol; pipetted into 100 ml cold distilled water and 0.3 ml Tween 80 to give milky suspension. Volume reduced to 100 ml by evaporation in vacuum. 20 ml added to diet in drops while stirring. 15 g glucose added to diet mixture, warmed on water bath, and stirred until suspension obtained. pH to 6.1 by 6.3 ml mixture of 50% 3N KOH and 50% 3N NaOH. Volume of diet to 1000 ml with distilled water. Aliquots for high level diet measured out, diluted to 500 ml and 10 g agar added to each. Diets heated in autoclave to melt agar, and dispensed into diet tubes at rate of 1.5-2 ml diet/1 dram screw-capped vial. Tubes marked, loosely capped and autoclaved for 15 minutes at 15 lb pressure. Tubes slanted while cooling.

REARING Eggs less than 24-h-old immersed in 20% formalin solution for 1 h and washed in sterile water. Transferred to diet surface with sterile platinum loop. Tubes kept at 25 ± 1°C. Pupation on diet surface; pupae removed and kept on moist sand or moss until emergence.

DEVELOPMENT Eclosion to pupation, 12-19 and 12-27 days on high and low diets respectively; adult emergence, 22-39 and 22-30 days; pupation, 90%; adults, 83%. Larval development ca. half as fast as on pea seedlings.

→ McClanahan, R.J.; Miller, L.A. 1958. Laboratory rearing of the seed corn maggot, *Hylemya cilicrura* (Rond.) (Diptera: Anthomyiidae). CAN ENTOMOL 90: 372-374.

COMPOSITION Adult diets: diet 1 distilled water; diet 2 (%): egg white, 10; sucrose, 1; diet 3 (%): yeast suspension, 5; diet 4 (parts): molasses, 1; evaporated milk, 1; water, 2.

FEEDING From four 3" cotton wicks in 25 ml vials containing above solutions.

REARING Pupae placed in 10 x 24" emergence cage covered with clear plastic sheet. Adults transferred to oviposition cage. Oviposition cage of 6" diam gallon jar with base removed and cheesecloth covered top placed in 9" terracotta flower pot saucers 2" deep containing 3 cupfuls of soil. Soil prepared by sifting through 15 mesh screen and adding water to give 25% moixture content. 20 peas, soaked for 2 h, planted in soil to feed developing larvae. Pots placed in 1" deep metal tray and water added periodically to tray to keep soil damp. Soil surface temperature 65° ± 4° F. Four 20 watt fluorescent lights held 10" above tops of jars to give 5 h photoperiod from 5-10 pm. 50 flies in each oviposition jar. Saucer and feeding units replaced every 3-4 days. ½" layer moist soil then added to saucer; kept at 80° F, 80-85% RH for 10 days. Contents of saucers sifted and pupae placed in small dish inside emergence cage.

DEVELOPMENT Preoviposition period, 10-20 days; adult life span up to 6 weeks, with mean of 4 weeks; incubation period, 3 days; at 80° F, larval period, 8-10 days. At 68°-72° F, pupal stage, 4-6 days. Flies reared continuously for 2 years.

Leptohylemyia coarctata (Fallén), wheat bulb fly

→ Bardner, R.; Kenten, J. 1957. Notes on the laboratory rearing and biology of the wheat bulb fly, *Leptohylemyia coarctata* (Fall.). BULL ENTOMOL RES 48: 821-831.

Best of four liquid diets tested.

COMPOSITION 1 teaspoon honey and condensed milk each in 50 ml water with uncitrated beef blood mixed in little water.

REARING At 20° C and 60% RH with 16 h photoperiod. Cages of hurricane-lamp glass 6" high and 4¾" diam containing 20 adults. Bottom covered with 2 mm mesh perforated zinc. Lamp glass supported by wire ring resting on petri dish bottom containing sheet of black filter paper giving 1-2 mm clearance between zinc and paper. Eggs laid through perforations on to filter paper. Top of cage covered by 3" square perspex sheet, perforated by 1½" diam central hole. Food from two 1½" dental wicks fitted so as to slide into position beneath perspex cover over central hole. By pushing another slide containing fresh food, old food in its slide removed without any escapes. Cage illustrated.

DEVELOPMENT Larval development, 16-24 days; pupal period, 22-27 days; fecundity, 28.1 eggs/♀.

→ Harris, C.R.; Svec, H.J.; Begg, J.A. 1966. Mass rearing of root maggots under controlled environmental conditions: seed corn maggot, *Hylemya cilicrura*; bean seed fly, *H. liturata; Euxesta notata;* and *Chaetopsis* sp. J ECON ENTOMOL 59: 407-410.

COMPOSITION (g): soya bean flour, 200; whole wheat flour, 200; brewers' yeast, 100; crushed potato or turnip, 100; condensed milk, 60; honey, 40; corn oil, 10; cholesterol, 2; formic acid, 1; distilled water, 325 ml.

REARING At 74° F (day) for 16 h and 70° F (night) for 8 h with 16 h photoperiod, 50-75% RH, and air flow at 88 ft/minute. 200-500 adults/cage fed water, honey, yeast hydrolysate, brewers' yeast and soya flour as separate components. Larvae reared in 2 plastic pots 5½ x 7½" diam and 4 x 7½ diam. ½" hole bored in bottom of smaller pot and 2" plastic tube, to hold 2, 6" dental wicks. 1" water poured into pot and smaller pot placed on top. Holes bored in bottom and covered with pottery fragments. Soil placed in 2" layer and moistened slightly. Two pieces of diet, 1½" diam, covered top and bottom with potato slices placed on soil and covered with soil layer (1:1 sand:muck mixture). Peas soaked for 24 h placed on top and also covered with thin soil layer. Larval containers placed in oviposition cages for 2-3 days, removed and ½ teaspoon fish meal added. Eggs covered with fine soil layer and kept at 70-74°F for 3 days. Pupae separated from soil by washing and sieving, and placed in pot with soil for emergence. Rearing container illustrated.

Euxesta notata (Wiedemann) (Otitidae), and *Chaetopsis* sp. (Otitidae) require temperature of 80°F. Short piece of cornstalk added to pot encourages oviposition by *Chaetopsis*.

DEVELOPMENT *Hylemya cilicrura* Rondani: life cycle, 19-30 days; larval stage, 9-15 days; pupal stage, 8-12 days; fecundity, 268 eggs/♀; 81% emergence to adults at 72°F.

No viable eggs produced by *Hylemya liturata* Meigen.

No data for *E. notata* and *Chaetopsis* sp. given, but reared at 80°F.

Calliphoridae

Aldrichina grahami Aldrich

➞ Miura, K.; Takaya, T.; Koshiba, K. 1967. The effect of biotin deficiency on the biosynthesis of the fatty acids in a blowfly, *Aldrichina grahami* during metamorphosis under aseptic conditions. ARCH INT PHYSIOL BIOCHIM 75: 65-76.

COMPOSITION See Sedee* (1956) diet for *Calliphora erythrocephala*.

REARING At 25°C in darkness. Adults supplied water and sugar cakes with 20% dry milk powder fortified with vitamins. Larvae reared aseptically.

DEVELOPMENT Pupation in 6.8 days from oviposition; 3rd instar larval wt, 77 mg.

Calliphora erythrocephala (Meigen) [previously *C. vicina* Robineau-Desvoidy], brown blow fly

➞ Sedee, J.W. 1952. Qualitative vitamin requirements for growth of larvae of *Calliphora erythrocephala* (Meig.). EXPERIENTIA 9: 142-143.

COMPOSITION (mg): vitamin-free casein, 100; *l*-cystine, 4; Osborne & Mendel salt mixture, 6; cholesterol, 10; and (µg): thiamin-chloride, 3; riboflavin, 6; nicotinic acid, 60; calcium pantothenate, 6; pyridoxin, 4; folic acid, 6; choline chloride, 460; biotin, 0.1; inositol, 180; vitamin B_{12}, 0.015; distilled water to 1000 mg; and cotton wool.

PREPARATION All components except vitamins homogenized; pH to 7.0 with 33% NaOH solution. 40 g portions dispensed to 200 ml Erlenmeyer flasks; vitamins added from stock solutions; contents absorbed on to cotton wool. Flasks plugged with clean cotton wool and autoclaved for 1 h at 120°C.

REARING At 25°C in darkness. Eggs laid on fresh meat, collected after 4-6 h and transferred to tube containing 0.9% NaCl solution and Rogipon-T4. Eggs separated with hair brush, sterilized for 40 minutes with Chloramine-T and alcohol and rinsed in sterile water; transferred aseptically in groups of 50 to flasks.

DEVELOPMENT Larval period, 7 days.

→ Sedee, J.W. 1954. Qualitative amino acid requirements of larvae of *Calliphora erythrocephala* (Meigen). ACTA PHYSIOL PHARMACOL NEÉRL 3: 262-269.

Two diets.

COMPOSITION Diet 1 (mg/g): cholesterol, 6.1; Osborne & Mendel salt mixture, 5; wheat-germ oil (Organon), 25; linseed oil, 70; nucleic acid from yeast (B.D.H.), 12.5; Amberite IR4B-treated casein hydrolysate, 62; *dl*-tryptophane, 3.5; *l*-cystine, 4; *l*-glutamic acid, 13; *l*-aspartic acid, 4.

vitamins (μg/g): thiamin chloride, 4; riboflavin, 7; nicotinic acid, 60; pyridoxin, 6; calcium pantothenate, 8; folic acid, 6; choline chloride, 500; biotin, 0.2; inositol, 180; vitamin B_{12}, 0.0025. Distilled water to 1000 mg.

Diet 2 (mg/g): cholesterol, 6.1; Osborne & Mendel salt mixture, 6.9; wheat-germ oil (Organon), 25; linseed oil (commercial), 70; nucleic acid from yeast (B.D.H.), 12.5.

amino acids (mg/g): arginine hydrochloride, 4.2; histidine hydrochloride, 3.3; lysine hydrochloride, 9.2; tyrosine, 3.5; *dl*-tryptophane, 3.5; *dl*-phenylalanine, 3.6; *dl*-methionine, 2.0; *dl*-threonine, 3.4; *l*-leucine, 5.7; *dl*-isoleucine, 4.3; *dl*-valine, 4.2; *l*-glutamic acid, 13.0; *l*-aspartic acid, 4.0; glycine, 4.8; *dl*-alanine, 1.9; *l*-proline, 6.6; *l*-cystine, 4.2.

vitamins as in diet 1.

PREPARATION Casein hydrolysate prepared by boiling casein for 17 h under reflux with 20% HCl and filtering to remove insoluble humin. Bulk of HCl removed by evaporation under reduced pressure. Hydrolysate treated with Amberlite IR4B to remove glutamic acid, aspartic acid and remaining HCl. Solution concentrated under reduced pressure till viscous, then evaporated to dryness and powdered in ball mill.

Hydrolysate, glutamic acid, aspartic acid and tryptophan mixed and dissolved in warm water with little NaOH. Nucleic acid in dilute NaOH and cystine in diluted HCl added from stock solutions. pH to 7.3. B-vitamins added from stock solution. 20 g portions poured into 100 ml Erlenmeyer flasks containing cholesterol and salts. Linseed oil and wheat germ oil added. Contents of flasks heated to boiling, absorbed on to cotton wool and autoclaved for 30 minutes at 120° C.

Diet 2 as above but without Amberlite IR4B-treated casein hydrolysate. 12 g portions poured into 50 ml Erlenmeyer flasks.

REARING 25-40 larvae in 100 ml flasks; 15-25 in 50 ml flasks.

DEVELOPMENT Larval period, 7 days; dry weight of 7-day-old larvae on diet 1, 16.6-26.8 mg; on diet 2, 8.8-15.5 mg.

Corollary

→ Sedee, J.W. 1956. Dietetic requirements and intermediary protein metabolism of the larva of *Calliphora erythrocephala* (Meig.). Van Gorcum & Co., Ltd, Assen, Netherlands. 130 pp.

Two best of several diets.

COMPOSITION Diet 1 (mg/g): vitamin-free casein, 100; *l*-cystine, 4; Osborne & Mendel salt mixture, 6; cholesterol, 10; and (μg/g) thiamin chloride, 3; riboflavin, 6; nicotinic acid, 60; calcium pantothenate, 6; pyridoxine, 4; folic acid, 6; choline chloride, 460; biotin, 0.1; inositol, 180; vitamin B_{12}, 0.02; and distilled water to 1000 mg.

Diet 2 (mg/g): cholesterol, 6.1; Osborne & Mendel salt mixture, 6.9; vegetable oils (at first, mixture of linseed and wheatgerm oils, 3:1; later, mixture of sesame and sunflower seed oils, 1:1, used), 70; yeast-nucleic acid, 12.5; glycogen, 8.3; *l*-arginine hydrochloride, 3.3; *l*-lysine hydrochloride, 9.2; *l*-tyrosine, 3.5; *l*-tryptophane, 1.8; *l*-phenylalanine, 1.8; *l*-methionine, 1; *l*-threonine, 1.7; *l*-leucine, 5.7; *l*-isoleucine, 2.1; *l*-valine, 2.1; *l*-glutamic acid, 13; *l*-aspartic acid, 4; glycine, 4.8; *l*-alanine, 1; *l*-proline, 6.6; *l*-cystine, 4.2; *l*-serine, 1.1; and (μg/g): thiamin chloride, 4; riboflavin, 7; nicotinic acid, 60; pyridoxine, 6; calcium pantothenate, 9; folic acid, 6; choline chloride, 500; biotin, 0.2; vitamin B_{12}, 0.025; and distilled water to 1000 mg.

DEVELOPMENT Larval growth on diet 1 about equal to that on natural food; growth on diet 2 with natural amino acid forms better than that on diets with racemic forms of amino acids as well.

See also: Sedee (1958, 1961).

Calliphora vomitoria (Linnaeus)

→ Bogdanow, E.A. 1908. Über die Abhängigkeit des Wachstums der Fliegenlarven von Bakterien und Fermenten und über Variabilitat und Vererbung bei den Fleischfliegen. ARCH ANAT PHYSIOL ABT SUPPL: 173-200.

Several diets tested: fresh meat; sterilized meat; sterile meat inoculated with *Bacillus megatherium* De Bary, or *Bacillus subtilis* Ebrenberg, or *Bacillus mesentericus vulgatus* Flügge, or *Bacillus mycoides* Flügge, or various bacteria closely related to *Bacterium pseudanthracis* (Dr S. Severin's "Culture No. 1"); peptone solution; nutrient gelatine; sterile congealed egg albumin; meridic diet; sterile meat with addition of trypsin. Best results achieved, beside fresh meat, with sterile meat containing trypsin, and meridic diet.

COMPOSITION Meat diet, sterilized meat; few ml of 1% trypsin solution (with 3/1000 parts of Na_2CO_3).

Meridic diet (g/25 ml of 5% peptone solution): NaOH, 0.25; meat extract (Liebigsche Fleischextrakt), 2; rice starch, 3; and salts (mg/100 ml distilled water): $Ca(NO_3)_2$, 20; $MgSO_4$, 30; $MgCl_2$, 30; H_2KPO_4, 140; KCl, 20; KOH, 70; NaOH, 920; and one drop of 5% Fe_2Cl_6 solution.

PREPARATION Meat diet: fresh meat autoclaved at 2 atmospheres (29.4 lb/sq ") twice for 25 minutes; 1% trypsin solution sterilized by filtration through chamberland filter.

Meridic diet: salts dissolved in distilled water; peptone added in quantity to give 5% peptone content. NaOH added in solution, followed by meat extract and rice starch; whole mixture boiled.

DEVELOPMENT Meat diet: larvae in trypsin treated sterile meat approximately twice as large as those in untreated sterile meat.

Meridic diet: satisfactory, providing necessary bacteria present; but slower larval growth and smaller larvae than on fresh meat.

Cochliomyia hominivorax (Coquerel) [previously *C. americana* Cushing & Patton], screw worms

Baumhover, A.H.; Husman, C.N.; Graham, A.J. 1966. Screw worms. pp. 533-554. *in* Insect Colonization and Mass Production. Smith, C.N. (Ed.) Academic Press, New York and London.

COMPOSITION (ml): water, 1000; formalin, 6; citrate beef blood, 500; lean ground meat, 1000 g.

PREPARATION Ingredients mixed well.

REARING About 700 flies kept in 12 x 12 x 20" screened cages with water, honey, and mixture of honey and meat. 10-15 flies held for 1-2 h at 95°F to induce oviposition in vial containing lean ground beef. Larger numbers of flies kept in 3½ x 3½ x 6' cages, draped with paper towelling and held in darkness. Trays of meat and incubated meat juices covered by wooden grating placed in cage for oviposition; adults attracted to tray by lamps. After 4 h temperature lowered to 32°F to immobilize flies, and tray taken from cage. Eggs removed with spatula and divided into 6 g lots. Transferred to 26 x 16 x 4" pans with ½" layer of diet (using plasma instead of whole blood) for 3 h at 100°F and 95% RH; then to 4 x 5' vats, with 1½" deep layer of fresh diet using whole blood. 4-6-day-old larvae crawl into large funnels leading to sand pupation trays and moved automatically to separator after 8 h where pupae sifted out. Kept at 80°F and 85% RH.

DEVELOPMENT Eclosion in 16 h; larval period, 5-9 days; adult emergence from 8 days.

Gingrich, R.E. 1964. Nutritional studies on screw worm larvae with chemically defined media. ANN ENTOMOL SOC AM 57: 351-360.

COMPOSITION Basic diet (mg/ml): vitamin-free casein, 156.0; yeast extract, 12.5; cholesterol, 1.88; McCollum-Davis salt mixture No. 185, 1.25; agar, 1% of total volume; and water to 48 ml.

Defined diet (mg/ml): amino acid mixture, 32.8; RNA, 1.0; cholesterol, 1.0; McCollum-Davis salt mixture No. 185, 1.25; agar, to 1% of total volume; NaOH, to pH 6.0-6.2; distilled water to 48 ml; and (µg): vitamin mixture, 277.575.

amino acids (mg/ml): alanine, 2.0; arginine (free base), 1.1; aspartic acid, 3.0; cystine, 0.1; glutamic acid, 4.0; glycine, 0.5; histidine (free base), 0.7; hydroxyproline, 0.5; isoleucine, 3.0; leucine, 2.8; lysine hydrochloride, 1.8; methionine, 1.8; phenylalanine, 2.8; proline, 2.0; serine, 0.8; threonine, 1.9; tryptophane, 0.9; tyrosine, 0.1; valine, 3.0.

vitamins (µg/ml): thiamine, 15.0; sodium riboflavin-5-phosphate, 15.0; niacin, 15.0; calcium pantothenate, 15.0; folic acid, 2.5; biotin, 0.025; B_{12}, 0.05; *p*-aminobenzoic acid, 15.0; *i*-inositol, 100.0; choline chloride, 100.0.

PREPARATION All B-vitamins, choline, inositol and nucleic acids prepared as stock solutions. Agar prepared in stock solution to provide concentration of 1% in final diet. Cholesterol added as aqueous suspension by dissolving 4.8 g powdered cholesterol in 180 ml of hot 95% alcohol. This solution poured slowly into 150 ml cold water and 0.5 ml Tween 80, and heated to evaporate alcohol to 100 ml volume. Suspensions of other sterols prepared similarly. Dry ingredients ground and poured into 50 ml graduated centrifuge tubes. Stock solutions and suspensions added by pipette, and dissolved over water bath. pH to 6.0-6.2 with NaOH. Hot molten agar added and distilled water to 48 ml. Diet dispensed to 2.5 x 9 cm shell vials. Vials plugged with cotton, autoclaved at 121°C for 15 minutes and cooled slanting.

REARING Eggs sterilized by immersion in 1% NaOH for 10 minutes; washed in water and transferred to sterile petri dish containing 0.1% benzaconium chloride for maximum of 30-45 minutes. 4-6 eggs scooped up with square sterile cotton organdy and placed in each culture vial. Mature larvae removed aseptically after migration from diet, washed, 'blotted' dry, weighed and placed in tube with sand to complete pupation.

DEVELOPMENT On basic chemically defined diet; pupal wt, 36.2 mg in 10.9 days; pupation, 75%; emergence, 31.3% of total. Larval growth and development on defined diet nearly equal to basic diet of casein and yeast.

→ Gingrich, R.E.; Graham, A.J.; Hightower, B.G. 1971. Media containing liquified nutrients for mass-rearing larvae of the screw worm. J ECON ENTOMOL 64: 678-683.

Several liquid formulations tested; best 3 diets for various larval stages.

COMPOSITION Starting diet (first 48 h): dried whole chicken egg, 2-3%; dried whole bovine blood, 5-6%; calf suckle, 1.6-4%; dissolved or suspended in water with 0.2% formalin. Intermediate diet (next 24 h): same ingredients except whole egg replaced by 0.5% sucrose. Finishing diet (remainder of development time): as intermediate diet except; sucrose increased to 1%, and 3% dried cottage cheese added.

PREPARATION Mixed in 55-gal cans. Water and soluble ingredients added first, followed by solid protein ingredients requiring considerable agitation to suspend or dissolve them. 0.2% formalin finally added to retard putrefaction. Stored at 1°C until needed.

REARING About 1500 neonate larvae placed in pan on cotton saturated with liquid diet, and held at 37°C and 100% RH for 48 h. Then transferred to 48 x 38 x 3.8 cm rearing vat with 3 cm thick layer of unbleached cotton saturated with liquid diet, and held at 35°C. Liquid from larvae removed by sponges and fresh liquids provided as needed. Fluids replaced every 8 h period during first 36 h then every 4 and 2 h for increased larval feeding activity.

DEVELOPMENT Final larval weights $\gg$ those of larvae on fresh meat-blood diets. Further improvements for mass rearing required.

See also: Gingrich (1972).

→ Graham, A.J.; Dudley, F.H. 1959. Culture methods for mass rearing of screw worm larvae. J ECON ENTOMOL 52: 1006-1008.

COMPOSITION (%): lean ground meat, 54; water, 30; citrated bovine blood, 15; formalin, 0.2.

PREPARATION Ingredients mixed with double propeller agitator in 30-gal soup kettle.

REARING At 95°F and 100% RH in batches of 100000 larvae. Initial larval growth on diet in 18 x 24 x 4" trays; completed in 10' x 4' x 1" deep vats. Each vat formed from top of table with bottom and sides enclosed. Galvanized iron troughs fitted to sides, to catch and directed mature migrating larvae to sand tubs on floor. Diet heated by light underneath, and air circulated to prevent ammonia accumulation. Pupae placed in 24 x 24 x 5" vats full of sand @ 10000/4 l sand. Separated out after 12 h and washed; kept at 80° F and 95% RH. Adults oviposit on diet of 1 part fluid of thawed frozen beef heart incubated for 72 h at 97° F with 9 parts lean ground meat. To obtain eggs, two wide mouth jars containing 50 g diet placed in 22 x 12 x 12" cardboard cage with sleeves at either end. Adults fed meat-honey mixture, honey fountain and water provided by inserting pint jar over paper towelling in petri dish. Eggs collected after 2 h and weighed into 10 g lots. Kept on moistened paper in petri dishes 14-16 at 80° F.

DEVELOPMENT Mature larvae after 5-6 days. Fecundity, 200-300 eggs/♀, from 8 day old flies at 80° F.

→ Melvin, R.; Bushland, R.C. 1941. The nutritional requirements of screw worm larvae. J ECON ENTOMOL 33: 850-852.

Two best of several diets tested.

COMPOSITION Artificial diet (ml): water, 1000; formalin (39% formaldehyde), 6; and (g): finely ground lean beef, 1000; and (ml): citrated beef blood (3 g sodium citrate/l blood), 500. (Amount of formalin important factor).

Aseptic diet (%): egg yolk, 25; whole sweet milk, 25; agar, 1; water, 49.

PREPARATION Artificial diet: ingredients thoroughly mixed in above order in 15" high and 24" diam galvanized tubs to depth of 1".

Aseptic diet: ingredients mixed; 5 ml autoclaved in 250 ml Erlenmeyer flasks.

REARING Artificial diet: at 85-100° F. Neonate larvae fed for 3 days, then transferred to tub of fresh diet for further 3-4 days. On about 6th day, tub tilted over another containing ca. 1" moist sand for pupation. After 2 days, pupae sifted and placed in buckets of moist sand for emergence.

Aseptic diet: At 89-94° F. Diet poured into 50 ml beaker, then beaker placed in wide mouthed jar containing 1" of washed sea sand. 1" diam copper tubing soldered into jar lid and plugged with cotton. Unit autoclaved. Eggs sterilized in alcoholic $HgCl_2$ solution and aseptically dispensed through tubing. On maturation, larvae migrate from beaker and pupate in sand.

DEVELOPMENT Larval period, 6-7 days; pupal period, 2 days. 88 consecutive generations reared.

See also: Melvin & Bushland (1936).

Lucilia cuprina Wiedemann

→ Lennox, F.G. 1939. Studies of the physiology and toxicology of blowflies. 1. The development of a synthetic medium for aseptic cultivation of larvae of *Lucilia cuprina*. AUST COMMONW COUNC SCI IND RES PAM 90: 7-24.

COMPOSITION (g): dry yeast, 6.7; salt mixture, 0.3; and 1½% agar solution, 93 ml.

salts (g): NaCl, 10; $CaCO_3$, 10; K_2HPO_4, 10; K_2SO_4, 5; KI, 1; $FeSO_4.7H_2O$, 0.7.

PREPARATION Diet sterilized at 20 lb pressure for 20 minutes. Shaken while cooling to mix components thoroughly.

REARING At 30 ± 0.5°C in 2-oz bottles or in small petri dishes on top of thin sand layer in larger petri dishes. Egg masses separated by treatment with 1% aqueous NaS solution for 15 minutes; rinsed with sterile distilled water; washed in 3% lysol for 2 minutes and again rinsed with sterile water.

DEVELOPMENT Completed. Larval weight at 3.5 days, 28.9 mg.

Lucilia sericata (Meigen) and *Chrysomya rufifacies* Macquart also reared.

→ Mackerras, M.J.; Freney, M.R. 1933. Observations on the nutrition of maggots of Australian blow-flies. J EXP BIOL 10: 237-246.

COMPOSITION Liver; bakers' yeast; nutrient agar.

PREPARATION Small cube of liver placed in wide mouthed bottle, yeast and agar added, bottle plugged with cotton wool and autoclaved for 20 minutes at 15 lb pressure.

REARING At 31°C. Eggs collected on fresh liver and surface sterilized in 1% NaS, rinsed in distilled water and 0.4% $HgCl_2$ for 10 minutes. Eggs poured into filter funnel with little cotton wool, washed in sterile distilled water and aseptically transferred to diet.

Diet for *Chrysomya rufifacies* Macquart as above, except agar replaced by nutrient broth.

DEVELOPMENT Normal.

→ Williams, K.L.; Nurmi, S.; Birt, L.M. 1974. Developmental and biochemical characteristics of sterile cultures of the blowfly, *Lucilia cuprina*. LAB ANIM 8: 177-187.

COMPOSITION Sheep liver.

PREPARATION Connective tissue removed from liver; liver homogenized; poured into 100 ml glass jars to depth of 20 mm. Metal lids screwed on and autoclaved at 121°C for 20 minutes. Cooled and stored at 0-4°C.

REARING At 30°C with continuous lighting. Eggs incubated at 100% RH for 5 h at 37°C. Separated in excess 0.1 M NaOH or NaS. Washed with sterile distilled water and sterilized in (g): $HgCl_2$, 0.5; NaCl, 6.5; concentrated HCl, 1.25 ml; absolute ethanol, 250 ml; distilled water, to 1-l. 1 ml eggs with 30 ml fluid. Liquid decanted and eggs soaked in 50% ethanol for 10 minutes, then washed in sterile distilled water. Eggs to produce 100-200 larvae in each jar; lids replaced loosely for ventilation. Lids removed after 3 days and sterile 100 ml bottles, with double lids, containing 20 mm layer sand connected to jar. Mature larvae separated from sand by sieving and placed in dry sand in tray with metal lid. Pupae placed in sterile plastic petri dishes. 50 flies placed in 900 ml jar with mixture of sand and silica gel (5:1) to 20 mm depth. 20 ml beaker of 0.5 M sucrose and wick placed in jar to feed adults.

DEVELOPMENT No difference between sterile and non-sterile cultures. Larval period, 5 days; pupal period, 6 days. Adults mature in 4 days.

See also: Roberts & Kitching (1974); Webber (1958).

Lucilia sericata (Meigen) [previously *Phaenicia*], European green blow fly

→ Barlow, J.S.; Kollberg, S. 1971. An improved chemically defined diet for *Lucilia sericata* (Diptera: Calliphoridae). CAN ENTOMOL 103: 1341-1345.

Chemically defined diet originally designed for *Agria housei* by House* (1954a) modified and improved by addition of Na^+ ions.

COMPOSITION (g/l): amino acids, 50.0; lipids, 6.25; salts, 2.2; glucose, 12.5; agar, 30.0; ribonucleic acid, 1.88; NaOH, 0.04; KOH, 0.04; and (mg/l): vitamins, 282.8.

l-amino acids (g/l): alanine, 2.57; arginine, 2.11; aspartic acid, 3.04; cysteine hydrochloride, 0.71; glutamic acid, 5.46; glycine, 0.70; histidine hydrochloride, 0.70; hydroxyproline, 0.96; isoleucine, 3.04; leucine, 5.59; lysine hydrochloride, 3.52; methionine, 1.88; phenylalanine, 2.79; proline, 3.99; serine, 3.51; threonine, 1.88; tryptophan, 0.94; tyrosine, 3.28; valine, 3.28.

lipids and fatty acids (g/l): palmitic acid, 1.10; stearic acid, 0.50; oleic acid, 2.40; linoleic acid, 0.75; linolenic acid, 0.25; and cholesterol, 1.25.

salts (g/l): $FeCl_3.6H_2O$, 0.065; K_2HPO_4, 1.344; $MgSO_4.7H_2O$, 0.472; $MnSO_4.H_2O$, 0.0015; $CoCl_2.6H_2O$, 0.017; $CaCl_2$, 0.110; $CuSO_4.5H_2O$, 0.020; $ZnCl_2$, 0.025; $NaHPO_4.7H_2O$, 0.139.

vitamins (mg/l): biotin, 0.7; calcium pantothenate, 11.3; choline chloride, 52.3; folic acid, 8.9; inositol, 149.0; nicotinic acid, 8.1; *p*-aminobutyric acid, 22.6; pyridoxine hydrochloride, 22.5; riboflavin, 4.1; thiamine hydrochloride, 3.3.

PREPARATION See Barlow (1965) and House & Barlow (1956).

DEVELOPMENT 68% to 3rd instar or pupae after 6 days; 90% survival to 3rd instar; average pupal wt, 28.4 mg. Larval weight nearly doubled if 0.04 g/l NaOH and KOH added.

→ Hobson, R.P. 1935a. On a fat-soluble growth factor required by blowfly larvae. I. Distribution and properties. BIOCHEM J 29: 1292-1296.

COMPOSITION 15% peptone; yeast extract, small amount; NaOH to pH 6; oil under test, 2 mg.

PREPARATION Lean beef digested with pepsin for 4 days at pH 2, filtered and concentrated on water bath. Aqueous yeast extract added; pH to 6 with NaOH and diluted until peptone 15%. Solution refrigerated in chloroform. Samples heated before use to remove chloroform. 0.3 ml peptone solution absorbed on 60 mg cotton wool, pressed into pellet in small test tube. Oil measured on to platinum loop and smeared over cotton wool.

REARING At 27° C in moist incubator; one egg/test tube.

DEVELOPMENT Larval maturity, 70-80 h. Oils, in decreasing potency (mg): wool wax, 0.1 > beef muscle oil, 0.4 > ovalecithin, 2 > wheat germ oil, 3 > egg yolk, 4 > cod liver oil, 6 > butter, 8.

Larval weight on 4th day on 0.1 mg wool wax, 45-49 mg.

See also: Hobson (1933, 1935b).

Kadner, C.G.; LaFleur, F.M. 1951. The vitamin requirements of *Phaenicia sericata* (Meigen) larvae (Diptera: Calliphoridae). Wasmann J BIOL 9: 129-136.

COMPOSITION (g): Labco vitamin-free casein, 2.0; cholesterol, 0.1; Wesson's salt mixture, 0.05; phosphate buffer, 0.1 M to pH 7, 10.0 ml; riboflavin, 24.0 μg; distilled water, 10.0 ml; and (μg): thiamine chloride, 16; choline, 195; nicotinic acid, 72; pyridoxine hydrochloride, 16; calcium pantothenate, 16.

PREPARATION Casein spread evenly over pad of glass wool in flask; remaining constituents except vitamins poured on to casein and autoclaved for 30 minutes at 15 lb pressure. Vitamins prepared in aqueous solution, passed through microbiological filter, and added to autoclaved diet.

REARING Egg masses sterilized in White's solution for 60 minutes, then washed in sterile water. 10-15 eggs transferred to flask and kept at 25 ± 1° C. Larvae transferred to pill boxes prior to pupation. Pupae placed in vials until emergence.

DEVELOPMENT Larval period, 11-13 days.

Michelbacher, A.E.; Hoskins, W.M.; Herms, W.B. 1932. The nutrition of flesh fly larvae, *Lucilia sericata* (Meig.). 1. The adequacy of sterile synthetic diets. J EXP ZOOL 64: 109-131.

COMPOSITION (g): purified casein, 3; brewers' yeast, 0.3 (or Fleischmann's bakers' yeast, 1.5); butter fat or cod liver oil, 0.1; salt mixture, 0.03; cystine, 0.02; and agar solution 1½%, 7 ml.

salts (g): NaCl, 10; $CaCO_3$, 10; K_2HPO_4, 10; K_2SO_4, 5; KI, 0.1; $FeSO_4.7H_2O$, 0.7.

PREPARATION Dry components mixed in tubes; liquid components added and stirred; diet sterilized in autoclave at 17 lb pressure for 30 minutes.

REARING At 25 ± 1° C in 8" test tubes or Erlenmeyer flask. Eggs laid on fish heads, sterilized in 3% lysol for 2 minutes, rinsed in sterile distilled water and placed on diet.

DEVELOPMENT Larval period, 6-7 days; pupation, 95%; pupal period, 8-9 days. Emergence nearly, 100%.

Tenquist, J.D. 1971. Rearing of *Lucilia sericata* (Diptera: Calliphoridae) on a modified *Musca domestica* medium. N Z ENTOMOL 5: 30-31

COMPOSITION (g): whole milk powder, 50; dried yeast, 5; and tap water, 51 ml.

PREPARATION Components mixed thoroughly and poured on to paper flock.

REARING 200-300 1st instar larvae placed on diet in 450 ml waxed cardboard container. Gauze cover attached. Containers kept at 30° C for 4 days then tops removed and containers inverted in tray of sawdust for pupation. After 2 days trays removed and jars discarded. Pupae sieved from sawdust. Adults held at 27° C and 55-60% RH in 23 cm cube stainless steel cages with disposable plastic containers for food and water. Fed mixture of whole milk powder and sugar. Fresh sheep liver given to encourage oviposition. Water provided in petri dish and filter paper kept moist. Cages cleaned twice weekly and food and water renewed.

DEVELOPMENT To pupa in 6 days at 30° C. *Musca domestica* also reared.

→White, G.F. 1937. Rearing maggots for surgical use. pp. 418-427. *in* J.G. Needham *et al.* (Eds) Culture Methods for Invertebrate Animals. Comstock Publishing Co., Ithaca, N.Y. (Dover Edition 1959).

COMPOSITION (g): coarsely ground lung, 150; yeast, 1; hen's egg, ¼-½; chopped liver, 50; and agar, 1.5%.

PREPARATION Lung and liver autoclaved, coarsely ground and placed in 4 x 8.5" diam larval pan. Liquified agar, egg and yeast added. Lid replaced, metal tube inserted and whole autoclaved for few minutes at 15 lb pressure on 2 successive days.

REARING At 26°C or 10°C. Eggs shaken in shell vial with 2% NaOH; 5% NaOH added for ½ h. 700 eggs pipetted on to gauze and washed in 20 ml of 0.02% HCl. Eggs transferred on gauze to metal tube of rearing pan. Metal tube and gauze removed after 1 day. Full grown larvae migrate to wood shavings to pupate. Adults placed in 10 x 15 x 18" cage with water fountain; fed ripe banana and sugar lumps on 1st day, then sliced liver every day until eggs laid on fresh beef provided. Rearing pan, shell vials, rearing cabinet and adult cage illustrated.

DEVELOPMENT Normal. Pupal period, 2 weeks at 10°C. *Phormia regina* Meigen also reared.

See also: White (1931).

Phormia regina (Meigen), black blowfly

→Brust, M.; Fraenkel, G. 1955. The nutritional requirements of the larvae of a blowfly, *Phormia regina* (Meig.). PHYSIOL ZOOL 28: 186-204.

Best 2 of many diets tested.

COMPOSITION Diet 1 (g): casein, 4; dried brewers' yeast, 0.4; McCollum-Davis salt mixture No. 185, 0.04; *l*-cystine, 0.027; cholesterol, 0.04; wheat-germ oil, 0.093; agar, 1.2% of water; methyl *p*-hydroxybenzoate, 2.5% of solids; water (75-77.5%), 14 ml.

Diet 2 (g): casein, 4.0; McCollum-Davis salt mixture No. 185, 0.04; cholesterol, 0.04; agar, 0.16; methyl *p*-hydroxybenzoate, 0.09; and (ml): vitamin solution, 0.48; water, 14.

vitamins (μg/g): thiamine, 60; riboflavin, 30; nicotinic acid, 120; pyridoxine, 30; pantothenic acid, 60; choline chloride, 1200; *i*-inositol, 600; folic acid, 6; biotin, 1.

PREPARATION Solids weighed into 2-oz wide-mouthed glass jars. Vitamins from stock solution, and water added. Jars stoppered with gauze-covered cotton plugs and autoclaved for 20 minutes at 15 lb pressure.

REARING Adults in large wire-screen cages at 29-30°C, and uncontrolled humidity. Fed cane sugar and water. Oviposition on liver. Eggs separated in 1% Na_2SO_3 then placed on paper; sterilized in 0.1% Zephiran chloride for 25 minutes, incubated for 4-6 days, then transferred aseptically to diet bottles for emergence.

DEVELOPMENT Average weight of 6 day old larvae, 48.7 mg; pupation, 75.8%; adult emergence, 89.1%.

Cheldelin, V.H.; Newburgh, R.W. 1959. Nutritional studies on the blowfly. ANN N Y ACAD SCI 77: 373-383.

Two diets tested.

COMPOSITION Basal diet (%): vitamin-free casein, 29; yeast extract - Difco, 3; agar, 1; water, 67; and (mg/g basal diet): cholesterol, 0.67; NaCl, 0.60; KCl, 0.013; $CaCl_2$, 0.013; $NaHCO_3$, 0.013; $NaH_2PO_4.H_2O$, 0.35; K_2HPO_4, 0.30.

Chemically defined diet: *l*-amino acids (mg/ml): *dl*-alanine, 2.58; arginine, 1.06; *dl*-aspartic acid, 3.06; cystine, 0.36; glutamic acid, 5.48; glycine, 0.7; histidine, 0.7; hydroxyproline, 0.48; *dl*-isoleucine, 3.06; leucine, 2.8; lysine, 1.76; *dl*-methionine, 1.88; *dl*-phenylalanine, 2.8; proline, 2.00; *dl*-threonine, 1.88; *dl*-tryptophan, 0.94; tyrosine, 1.64; *dl*-valine, 3.30.

vitamins (μg/ml): thiamine hydrochloride, 5; riboflavin, 15; pyridoxine, 15; inositol, 100; nicotinic acid, 15; choline chloride, 100; calcium pantothenate, 15; *p*-aminobenzoic acid, 15; folic acid, 5; biotin, 0.025.

Others (mg/ml): ribonucleic acid, 1.0; cholesterol, 1.0; inorganic salts, 2.4; agar, 25; water to 2 ml; pH 5.6.

PREPARATION Basal diet; inorganic salts solution mixed with cholesterol suspension, then dry ingredients added and mixed. Dispensed to tubes @ 6 ml/tube. Tubes covered with aluminium caps, autoclaved, slanted and cooled.

Chemically defined diet: amino acids mixed dry, ground in mortar and added with agar to tube (@ 75 or 5 mg/tube). Vitamins, RNA and salts added in solution; cholesterol in suspension. Tubes covered with aluminium caps, autoclaved, slanted and cooled.

REARING At 30°C. Egg masses sterilized in 2% NaOH for 15 minutes, rinsed twice in 70% ethanol, and immersed in 70% ethanol for 15 minutes. Transferred aseptically to tubes.

DEVELOPMENT From inoculation to pupation, 12-14 days. Emergence, 19-20 days. Larval weight at 7 days, 11.5 mg. Development on basal and chemically defined diets comparable, though larval and pupal weights higher on basal diet.

Hill, D.L.; Bell, V.A.; Chadwick, L.E. 1947. Rearing of the blowfly, *Phormia regina* Meigen, on a sterile synthetic diet. ANN ENTOMOL SOC AM 40: 213-216.

COMPOSITION (g): casein, 630; brewers' yeast, 63; powdered agar, 21; and (g/30 g of above): anhydrous lanolin, 3; or hydrous lanolin, 4; Belar solution, 60 ml.

Belar solution (g/2-l): NaCl, 1.8; $CaCl_2$, 0.04; KCl, 0.04; $NaHCO_3$, 0.04; $NaH_2PO_4.H_2O$, 1.35; K_2HPO_4, 1.55.

PREPARATION Casein, yeast and agar mixed in large fibre drum. 30 g portions to ½ pint culture bottles. Lanolin and Belar solution added. Bottles plugged and autoclaved for 15 minutes at 15-17 lb pressure.

REARING At 25°C and 70% RH. Eggs laid on 3" cube of ground horse meat left in cage for 24 h; removed and placed in 2% NaOH for 15-20 minutes; NaOH decanted and eggs rinsed in 70% ethanol for 15 minutes. 500 eggs pipetted from ethanol to each culture bottle. On 8th day plugs removed and autoclaved sawdust added and covered with 40-80 mesh wire; or larvae removed to large shallow pan with ¼" sawdust. 100 or more pupae placed in small cup. Transferred, with sugar and water to 1-qt Mason jars for emergence. Technique described. Optimum adult density 300/cu ft. Fresh meat daily for oviposition.

DEVELOPMENT Eclosion, <24 h; larval development, ca. 10 days; pupal stage, 5-6 days. Oviposition, 7 days from emergence.

See also: Cohen & Stürckow (1971).

→ Hodgson, E.; Rock, G.C. 1971. Insect Nutrition. pp. 105-145. *in* Kerkut, G.A. (Ed.). Experiments in Physiology and Biochemistry Vol. 4. Academic Press, New York and London.

Two diets.

COMPOSITION Chemically defined diet (mg/ml): *dl*-alanine, 2.07; *l*-arginine hydrochloride, 1.03; *dl*-aspartic acid, 2.46; *l*-cystine, 0.29; *l*-glutamic acid, 4.35; glycine, 0.56; *l*-histidine hydrochloride, 0.76; hydroxy-*l*-proline, 0.39; *dl*-isoleucine, 2.46; *l*-leucine, 2.25; *l*-lysine, 1.77; *dl*-methionine, 1.51; *dl*-phenylalanine, 2.25; *l*-proline, 1.61; *dl*-threonine, 1.51; *dl*-tryptophan, 0.76; *l*-tyrosine, 1.32; *dl*-valine, 2.65; casein, 4.00; ribonucleic acid, 1.60; cholesterol, 0.80; and (μg): thiamine hydrochloride, 10.00; riboflavin, 30.0; pyridoxine hydrochloride, 30.0; calcium pantothenate, 30.0; folic acid, 10.0; *meso* inositol, 200.0; nicotinic acid, 30.0; *p*-amino benzoic acid, 30.0; biotin, 0.10; choline chloride, 200.0; NaCl, 496.0; KCl, 10.60; $NaHCO_3$, 10.60; $NaH_2PO_4.H_2O$, 291.0; K_2HPO_4, 248.0; $CaCl_2$, 10.60.

Mass rearing diet (g): casein, 260; agar, 10; dried brewers' yeast, 30; and (ml): cholesterol suspension, 1.0; salt solution, 50; water, 19.

salts (g): NaCl, 60; $NaH_2PO_4.H_2O$, 35; K_2HPO_4, 30; KCl, 1.3; $NaHCO_3$, 1.3.

PREPARATION Chemically defined diet: amino acids mixed, homogenously ground, and stored in desiccator over $CaCl_2$. Salts ground in pestle and mortar and stored over $CaCl_2$; added to distilled water, $CaCl_2$ solution last of all. 600 mg RNA dissolved in 12 ml 2N NaOH. 300 mg cholesterol ground in 6.0 ml ethanol to uniform slurry. Salt, cholesterol and RNA solutions mixed together. Amino acids, agar and casein poured into flasks, solutions added, and flasks stoppered with plug of surgical cotton wrapped in 4 layers of cheesecloth. Sterilized at 15 lb pressure for 15 minutes in autoclave, then allowed to gel.

Mass rearing diet: 1.34 g cholesterol ground in mortar in 4.0 ml ethanol. 16.0 ml water added and stirred. 5.1 g salt mixture dissolved in 1000 ml water and 4.0 ml $CaCl_2$ solution (300 mg/100 ml distilled water). 30 g dry mix, 1.0 ml cholesterol suspension, 50 ml salt solution and 19 ml water all put into 500 ml Erlenmeyer flask. Flask stoppered with cotton plugs wrapped in cheesecloth and autoclaved for 15 minutes at 15 lb pressure.

REARING Eggs separated by washing in 2% NaOH; rinsed three times in distilled water, and surface sterilized in Zephiran chloride for 10 minutes, rinsed and transferred to flasks @ 5 eggs/flask. Incubated at constant temperature between 75-90° F and >70% RH for 7 days, then larvae washed into Mason jar fitted with screen lid; washed several times with water, then jar filled with sawdust and lid replaced till larvae pupate. Pupae, plus water and sugar cubes, placed in cage for adult emergence. Beef or pork liver placed in cage for 2 days. Adult cages illustrated.

DEVELOPMENT Av wt/larva on chemically defined diet, 40.6 mg. Rearing method and diet also used for *Sarcophaga bullata* Parker (Sarcophagidae).

See also: Hodgson & Dauterman (1960, 1964); Hodgson *et al.* (1956); Rock *et al.* (1975).

McGinnis, A.J.; Newburgh, R.W.; Cheldelin, V.H. 1956. Nutritional studies on the blowfly, *Phormia regina* (Meig.). J NUTR 58: 309-324.

Best two of ten diets described.

COMPOSITION Diet 1 (%): vitamin-free casein, 26; yeast extract, 3; agar, 1; water (containing supplements), 70.

supplements (mg/g diet): cholesterol, 0.78; NaCl, 0.7; KCl, 0.015; $CaCl_2$, 0.015; $NaHCO_3$, 0.015; $NaH_3PO_4.H_2O$, 0.41; K_2HPO_4, 0.35.

Diet 2: chemically defined diet (mg/ml): *dl*-alanine, 2.58; *l*-arginine, 1.06; *dl*-aspartic acid, 3.06; *l*-cystine, 0.36; *l*-glutamic acid, 5.48; glycine, 0.7; *l*-histidine, 0.7; hydroxy-*l*-proline, 0.48; *dl*-isoleucine, 3.06; *l*-leucine, 2.8; *l*-lysine, 1.76; *dl*-methionine, 1.88; *dl*-phenylalanine, 2.8; *l*-proline, 2.0; *dl*-threonine, 1.88; *dl*-tryptophan, 0.94; *l*-tyrosine, 1.64; *dl*-valine, 3.3; ribonucleic acid, 1.0; cholesterol, 1.0; inorganic salts, 2.4; agar, 25.0.

vitamins (µg/ml): thiamine, 5; riboflavin, 15; pyridoxine, 15; inositol, 100; nicotinic acid, 15; choline chloride, 100; calcium pantothenate, 15; *p*-aminobenzoic acid, 15; folic acid, 5; biotin, 0.025.

PREPARATION Diet 1: salts dissolved in water; cholesterol suspension prepared as in House* (1949a). Solutions combined and dry ingredients added. 100 g mixed in 500 ml Erlenmeyer flasks and sterilized at 16 lb pressure for 15 minutes; diet dispensed to 6" test tubes and slanted while cooling.

Diet 2: amino acids homogenously ground and 75 mg added with 50 mg agar to culture tube. Vitamin solutions prepared so 0.1 ml solution sufficient for final volume (2 ml) in tube. Salts dissolved in water. Cholesterol suspension prepared as in diet 1 and RNA solution prepared by dissolving 200 mg of yeast RNA in 4 ml of 2N NaOH. Then 1 ml of solution containing 44 parts salt solution, 2 parts cholesterol suspension and 4 parts RNA solution added to diet tube. Contents mixed thoroughly, sterilized at 16 lb pressure for 15 minutes and slanted while cooling. pH 5.6.

REARING Eggs sterilized by separating in 2% NaOH, rinsing twice with 70% ethanol and immersing in 70% ethanol for 15 minutes; then transferred aseptically to culture tubes @ 5/tube. Tubes incubated at 30°C in water bath.

DEVELOPMENT Complete larval development in 8 days; adults emergence after 16 days. Growth from egg to adult almost equal on chemically defined diet and natural food so long as amino acid concentration <60 mg/ml diet.

See also: Hodgson *et al.* (1956).

→ Orr, C.W.M. 1964. The influence of nutritional and hormonal factors on egg development in the blowfly *Phormia regina* (Meig.). J INSECT PHYSIOL 10: 53-64.

COMPOSITION Oviposition medium and larval diet (parts): water, 100; dried bakers' yeast, 6; dried whole milk, 6; agar, 1; methyl *p*-hydroxybenzoate, 1/100.

PREPARATION Mixture brought slowly to boil with constant stirring. Poured into 1-pint cylindrical cardboard cartons until half-full. Stored at 4°C till use.

REARING Adults at 24 ± 1°C with 16 h photoperiod in 12" cube wire cages; supplied with fresh water and dry sucrose crystals changed weekly. Eggs laid on small pieces of above medium, replaced daily. 100 eggs placed on top of diet in cartons, and covered with sawdust. Mature larvae migrate to sawdust to pupate. Pupae separated from sawdust by sieving and transferred to clean Mason jars.

DEVELOPMENT Larval period, ca. 10 days.

→ Rasso, S.C.; Fraenkel, G. 1954. The food requirements of the adult female blow-fly, *Phormia regina* (Meigen), in relation to ovarian development. ANN ENTOMOL SOC AM 47: 636-645.

Best diet tested.

COMPOSITION Adult diet: soluble casein, 6%; sucrose, 1 cube; B-vitamin solution, 1 ml; Wesson's salt mixture, 0.1 g; water, 15 ml.

vitamins (mg/10 ml water): thiamin, 5.0; riboflavin, 2.5; nicotinic acid, 10.0; pantothenic acid, 5.0; pyridoxin, 2.5; choline chloride, 100; inositol, 50.0; folic acid, 0.5; biotin, 0.1.

PREPARATION Dry components weighed into Syracuse staining dish with 15 ml tap water. Paper towelling floated on surface to facilitate feeding.

REARING At 23°C. Larvae fed on fresh beef liver; pupated in sawdust. 250 pupae placed in 7½" cube screen cage. Emerging adults supplied with cube sugar, and water in glass vials with exposed cotton wicks.

DEVELOPMENT Egg development slower than on liver diet.

→ Stoffolano, J.G. Jr 1974. Influence of diapause and diet on the development of the gonads and accessory reproductive glands of the black blowfly, *Phormia regina* (Meigen). CAN J ZOOL 52: 981-988.

COMPOSITION (g): powdered whole milk, 200; brewers' yeast, 200; Tegosept, 8.5; agar, 40; and water, 3-1.

PREPARATION See Dethier & Goldrich (1971). Diet autoclaved for 7 minutes at 121°C, and poured hot into sterilized flasks plugged with cotton.

REARING Eggs placed on diet and incubated at 27°C until pupation. Pupae placed in quart glass jars until emergence. Adults, both diapausing and non diapausing, fed on granulated sucrose, water, and powdered milk. Fresh beef liver provided daily for oviposition. Adults maintained at 26°C and 50% RH with 16 h photoperiod.

See also: Calabrese & Stoffolano (1974); Dethier & Goldrich (1971).

Protophormia terraenovae (Robineau-Desvoidy)

➔ Harlow, P.M. 1956. A study of ovarial development and its relation to adult nutrition in the blowfly, *Protophormia terrae-novae* (R.D.). J EXP BIOL 33: 777-797.

COMPOSITION Whale meat mixture.

PREPARATION Whale meat and water homogenized in Waring Blendor and strained through butter muslin.

REARING At 25°C and 70-80% RH with 24 h photoperiod. Newly laid eggs transferred to fresh meat in small plastic dish covered with muslin. After 6 days meat and larvae transferred to petri dishes on sand for pupation.

DEVELOPMENT Complete. Adult ♀♀ maintained on sucrose and water; ovarial growth not promoted by chemically defined diet of 10 B-vitamins, RNA, mineral salt and cholesterol.

Miscellaneous - Calliphorids

➔ Ashby, G.J. 1972c. Blowflies. pp. 554-555. *in* Universities Federation for Animal Welfare, the UFAW Handbook on the Care and Management of Laboratory Animals. Fourth edition Churchill Livingstone, Edinburgh and London.

COMPOSITION AND PREPARATION. See Hill *et al*,* (1947).

REARING At 25-27°C and 50-60% RH. 600-700 eggs transferred to culture jar containing food. Sawdust, dry peat or Vermiculite sprinkled around food, 3 cm deep, for pupation. Adult cage with pan of sugar and water always available. 40 g liver or lean beef placed in cage daily for oviposition.

DEVELOPMENT Variable between species. Flies breed more prolifically on meat or fish diet. *Calliphora* sp and *Lucilia* sp. pupate in 4-5 days, *Phormia* sp. in 6-7.

➔ Greenberg, B. 1960b. Host-contaminant biology of Muscoid flies. I. Bacterial survival in the pre-adult stages and adults of four species of blow flies. J INSECT PATHOL 2: 44-54.

Four species reared:- *Phormia regina* (Meigen), *Lucilia sericata* (Meigen), *Calliphora vicina* (Robineau-Desvoidy), *Cynomyopsis cadaverina* (Robineau-Desvoidy) [previously *Cynomyia*].

COMPOSITION Fresh fish.

REARING At 23°C and 45% RH. Adults fed on cane sugar and water in 18" cube cages. Eggs laid on fresh fish in jar with moist sawdust. Pupation in sawdust.

DEVELOPMENT Prepupal period of *P. regina* and *L. sericata*, 1½-3 days; *C. vicina*, 2-3 days; and *C. cadaverina*, 3-5 days. Pupal period, 5-5½ days for *P. regina*; 6-7 days for *L. sericata*; 12 days for *C. vicina*; and 9 days for *C. cadaverina*.

→ Murdoch, F.F.; Smart, T.L. 1931. A method of producing sterile blowfly larvae for surgical use. U S NAV MED BULL 20: 406-417.

Four species reared:- *Lucilia sericata* (Meigen), *Cynomyopsis cadaverina* Robineau-Desvoidy [previously *Cynomyia*], *Calliphora erythrocephala* (Meigen), *Phormia regina* (Meigen).

COMPOSITION Fresh meat.

REARING At 26°C and 60-70% RH. Eggs placed in larval pots in exhaust draft in incubator. Fresh meat placed on moist sand, old meat replaced daily. Meat supply withdrawn when larvae matured and sand left to dry out. Muslin pot cover replaced by single layer of gauze at end of pupal stage. When adults started to emerge, pupae separated from sand by sieving and placed in 30 cm cube cage for emergence. Food supplied in 3 dishes replaced daily; water fountain with sprinkle of glucose, 2% glucose solution fountain with orange juice and pieces of orange, pineapple and apple laid around paper, glucose sprinkled over top, and half petri dish of raw meat and pineapple on saturated filter paper. Cages placed near light. Oviposition tubes with moist cotton and fresh meat introduced nightly. Egg surface sterilized in Dakin's solution (1.6% Na_2CO_3 and 0.39% Cl_2) added in 1:2 dilution and tube shaken 1 minute. Eggs washed in sterile distilled water several times and sterilized in 2000 ml of 1/1000 bichloride in 0.5% HCl solution for 30 minutes. Eggs washed in sterile salt solution and transferred to culture bottles at 37°C. Newly hatched larvae placed in 5.5 x 11 cm bottles of 5% agar, 2% glucose and piece of liver. Tubes previously sterilized for 3 successive days at 15 lb for 20 minutes.

DEVELOPMENT Complete life cycle, 18-26 days. Egg hatch, 10-24 h; larval period, 5-7 days; prepupal period, 6-8 days; pupal period, 6-8 days.

Ceratopogonidae

Atrichopogon jacobsoni (de Meijere)

→ Drake, E.F. 1971. Life cycle and laboratory diet for *Atrichopogon jacobsoni* (de Meijere) (Diptera: Ceratopogonidae). PROC HAWAII ENTOMOL SOC 21: 63-66.

COMPOSITION AND PREPARATION See Steffan* (1966) diet developed for sciarids.

REARING At 20°C in darkness. Several gravid ♀♀ placed in each vial. Larvae left for 3 days then mixture of brewers' yeast and pulverized straw sprinkled on surface. Brewers' yeast added daily. Adults fed honey-water solution in 2 x 2 x 4' organdie cage with *Panax* sp shrub (30-40/cage).

DEVELOPMENT Egg stage, 5-6 days; larval stage, 16 days; pupal stage, 4-5 days. Adults, 7-10 days.

Culicoides variipennis (Coquerel)

→ Jones, R.H.; Potter, H.W. Jr; Baker, S.K. 1969. An improved larval medium for colonized *Culicoides variipennis*. J ECON ENTOMOL 62: 1483-1486.

COMPOSITION (g): powdered alfalfa meal, 500; powdered Misco high-protein-supplement livestock feed (Peavey Co., Minneapolis, Minn. 55415), 250; powdered egg albumen, 5; Difco brain-heart infusion, 5; and nutrient broth (Difco) added as 100x fluid concentrate.

REARING Larval pans of unplasticized PVC, 63 cm wide x 6 cm deep, set up with 2 strips of inert dacron fibrous matting (10 x 40 cm). One end of strips fastened at front with masking tape to bottom of dry pan. Dechlorinated water added to 35 mm depth so tops of strip level with water surface. 1 ml microbial starter added and food for microorganisms added daily. 20000 viable eggs placed in each pan on strips. Dry ingredients added at rate of 24 g/larval-pan and 12 g daily thereafter. Nutrient broth concentrate added at rate of 2-4 ml daily. In younger pans, clear diet with little scum indicates adequate nutrition; in older pans, diet thick with suspended material may prove lethal to larvae. Pans discarded at 9-10 days and larvae transferred to holding pan set up previous day with 48 g dry food, 10 ml nutrient broth concentrate, and 1 ml microbial starter, for pupation temperature of fluid maintained at 25-28° C by closed system of pipes carrying warm water under pans. Room temperature ca. 24° C. Slowly turning paddles through surface of diet maintain slight current and reduce scum formation. Light standardized using row of fluorescent bulbs midway above each shelf; 13 h photoperiod. Dacron strips from discarded pans weighed down with ½" galvanized screen in 60-l tub with overflow spout leading to 100 mesh sieve. After 45 minutes, pupae picked off with wide-mouthed rubber bulb and placed for emergence in 1-gal cages; manipulated so that flies are all of known age. 1-2-day-old flies fed blood for egg production.

DEVELOPMENT Pupae 9-10 days from inoculation. Production of pupae/pan, 1735-5613 when inoculated with ca. 20000 eggs.

See also: Jones (1964).

Culicoides nubeculosus (Meigen)

Culicoides riethi (Kieffer)

Culicoides variipennis (Coquerel)

→ Boorman, J. 1974. The maintenance of laboratory colonies of *Culicoides variipennis* (Coq.), *C. nubeculosus* (Mg.) and *C. riethi* Kieff. (Diptera, Ceratopogonidae). BULL ENTOMOL RES 64: 371-377.

COMPOSITION Dry food (parts): grass meal, 4; Bemax wheat germ, 1; Oxoid No. 2 nutrient broth powder, 80 g; and 100 ml water.

REARING Larvae reared in 18 x 14 x 2" white enamelled metal pans, with 2 ½ x 3 x 14" fibreglass pads taped lengthwise on each pan with free space of 1" left around and between strips. Both pans and pads first rinsed with water, then water added to cover pads surface. Air bubbles in pads pressed out. Water stored for 24 h before use. Room temperature 27-28° C, pan temperature 24-26° C, 16 h photoperiod. Filter paper discs with eggs placed on pads, face up so eggs wet but not submerged. 4 ml nutrient broth concentrate and 1 heaped teaspoon dry food added to pan. New pans seeded with 3-5 ml diet culture made by incubating 0.5 g fresh cow manure in 50 ml bacteriological nutrient broth at room temperature for 48 h. Each pan given 2 ml broth and 1 level teaspoon dry food 3 times/week. Water in pans circulated using motorized paddles. When pupation effected, water added so that pads completely submerged; pupae float to surface and collected into conical flasks using suction pumps. Pupae then placed in single layer over towelling in container and water added. Adults in 2½" waxed card pill boxes. Adult fed 10% sugar solution and mice offered for blood meal.

C. riethi kept at 26-27° C room temperature.

C. nubeculosus as for *C. variipennis*.

DEVELOPMENT *C. variipennis:* pupation after 10 days; adult emergence, 2-3 days later. *C. riethi:* pupation after 14-18 days, adult emergence, 2-3 days later. *C. nubeculosus:* details not given. Life cycle of all 3 species, 10-21 days at 23-25° C.

See also: Becker (1958).

→ Kettle, D.S.; Wild, C.H.; Elson, M.M. 1975. A new technique for rearing individual *Culicoides* larvae (Diptera: Ceratopogonidae). J MED ENTOMOL 12: 263-264.

COMPOSITION 1% agar solution to which small nematodes added. (Nematode species not given). Nematodes reared on 'porridge' made from "Cerekite" (mixture of barley, corn and wheat).

PREPARATION Agar dissolved in distilled water in boiling water bath. Dispensed in 5 mm layers to petri dishes. Small nematodes added when cooled, and lid replaced.

REARING At 27° C in darkness. Nematodes burrow into agar. *Culicoides* larvae reared individually or in group on petri dishes. Pupation on agar. Nematodes replenished as numbers decline.

DEVELOPMENT 8 species reared. *Culicoides austropalpalis* Lee & Reye, *C. brevitarsis* (Kieffer), *C. marksi* Lee & Reye, *C. marmoratus* (Skuse), *C. molestus* (Skuse), *C. narrabeenensis* Lee & Reye, *C. victoriae* McFie and *Culicoides* sp. Three other ceratopogonids *Paradasyhelea minuta*, *Stilobezzia pictipes* and *Bezzia* sp. also successfully reared.

Chironomidae

Chironomus sp.

Goeldichironomus holoprasinus (Goeldi)

Tanypus grodhausi Sublette

→ Biever, K.D. 1971. Effect of diet and competition in laboratory rearing of chironomid midges. ANN ENTOMOL SOC AM 64: 1165-1169.

Best 2 of many diets tested.

COMPOSITION "Dog kisses" (commercial dog candy) 0.25 g/unit or/50 larvae; or powdered liver.

REARING At 26.7 ± 0.5° C with 14 h photoperiod. In 475 ml plastic container connected to constant air supply of 20-30 ml air/minute; top of container ventilated through 25 mm hole covered with screen. 1 cm deep sand layer moistened with 180 ml water. 50, 1st instar larvae introduced/cage. Half food added on day 1, rest on day 6. Rearing unit illustrated.

DEVELOPMENT Adult emergence, 88%. Developmental time, 252 h; survival, 95%.

Chloropidae

Hippelates collusor Townsend, eye gnat

Mulla, M.S.; Barnes, M.M. 1957. On laboratory colonization of the eye gnat, *Hippelates collusor* Townsend. J ECON ENTOMOL 50: 813-816.

COMPOSITION Larval diet (%): alfalfa meal, 20; dry Vermiculite No. 4, 80; and brewers' yeast, 150-200 g/6 lb dry mix; and water, 2 gallons.

Adult diet (g): honey, 60; MRT protein hydrolysate, 22.5; partially hydrolyzed yeast, 7.5; Staley's sauce base 2, 15.0; human whole fluid blood, 30.0; sodium propionate, 0.5; cholesterol, 1.0; corn oil, 4.0; and (%): KH_2PO_4, 0.1; K_2SO_4, 0.32; $MgSO_4$, 0.04; NaCl, 0.092; $C_6H_5O_7Fe.5H_2O$, 0.052; $MnSO_4.4H_2O$, 0.036; $(CH_3COO)_2Cu.H_2O$, 0.001; $(CH_3COO)_2Zn.2H_2O$, 0.024; H_3BO_3, 0.0125; $(NH_4)_6Mo_7O_{24}.4H_2O$, 0.0350; $CaCl_2$, 0.092; choline chloride, 0.25; water to 650 ml.

PREPARATION Larval diet: dry ingredients mixed thoroughly and water added. Adult diet: all components except corn oil and cholesterol mixed. 3" dental rolls soaked in diet and sprayed with ether solution of corn oil and cholesterol. Rolls dried at 95-100°F for 8-10 h.

REARING At 25°C. See Mulla* (1962).

Adults provided 2 dental rolls of food/cage, 2000-3000 gnats/cage. Rolls changed after 3-4 days. Kept at 85 ± 2°F and 20-25% RH.

DEVELOPMENT Egg to adult stage, 2-3 weeks. Adult longevity from 6-8 days after oviposition. Two generations reared/mix. Fungus gnat, *Bradysia* sp. (Sciaridae) and scavenger gnat, *Scatopse fuscipes* Meigen (Scatopsidae) also reared.

Mulla, M.S. 1962. Mass rearing of three species of *Hippelates* eye gnats (Diptera: Chloropidae). ANN ENTOMOL SOC AM 55: 253-258.

COMPOSITION Larval diet (kg): Vermiculite No. 4, 3.732; alfalfa meal, 0.932; whole wheat flour, 0.200; brewers' yeast, 0.100; Tedion 0.1% dust, 0.010.

Adult diet (g): honey, 350; human whole fluid blood, 350; MRT protein hydrolysate, 50; sodium benzoate or calcium propionate, 5; corn oil, 20; water to 2700.

PREPARATION Larval diet: dry ingredients mixed and 160 g placed in wide mouth quart fruit jar. 400 ml water added just before eggs introduced.

Adult diet: all ingredients except corn oil dissolved in tap water. Bunches of dental rolls soaked in solution. Spread on foil or wax paper on trays. Wicks atomized with corn oil in acetone or alcohol. Trays placed in forced air oven at 100°F for 15 h. Wicks shredded before offering to adults.

REARING At 82 ± 1°F in darkness. Eggs washed into 50-100 ml water and added to diet. Covered with muslin and kept for 5 days; then 20-50 ml water sprinkled over diet. Jars transferred to 13 x 10 x 11" screened emergence cages. Adults fed on dental rolls soaked in 5% sugar solution. Transferred to 8 x 5 x 5" metal holding cage with food wicks and water. Three 1" holes in roof of cage for vials and 30 ml plastic water bottle. After 3-4 days transferred to oviposition chamber of 11.5

x 0.8 cm diam glass tubing; discs fitted to one end with 3 cm hole covered with 40 mesh plastic screen for aeration. Disc provided with 2 holes for food and water vials. Moist filter paper placed in petri dish and covered with corrugated board in oviposition cage for high humidity. Eggs laid under food wick on corrugated board and recovered by anaesthetizing gnats and washing eggs into beaker. Eggs collected every other day. Oviposition chamber and emergence cage illustrated.

DEVELOPMENT Adult emergence, 12 days from seeding eggs. *Hippelates collusor* Townsend : 76% egg viability, 50 generations reared; *H. robertsoni* Sabrosky: 72% and 25; *H. pusio* Loew: 80% and 25; *H. dorsalis* Loew reared on small scale; mass rearing not successful.

Hippelates pusio Loew, eye gnat

→ Legner, E.F.; Bay, E.C. 1965. Culture of *Hippelates pusio* (Diptera: Chloropidae) in the West Indies for natural enemy exploration and some notes on behaviour and distribution. ANN ENTOMOL SOC AM 58: 436-440.

COMPOSITION Adult diet (g): soluble beef blood powder, 10; powdered blood ablumin, 15; brewers' yeast, 5; sodium benzoate, 3; honey, 40; powdered gelatin, 2 packets; and water, 400 ml.

REARING Fed honey water solution soaked on cotton for 3-4 days after emergence. Later fed split dried prunes soaked in above diet.

DEVELOPMENT Average number of eggs laid 2.85/♀ in 14 days.

→ Schwartz, P.H. Jr; Turner, R.B. 1966. A dietary study of the adult eye gnat, *Hippelates pusio* (Diptera: Chloropidae). ANN ENTOMOL SOC AM 59: 277-280.

COMPOSITION Larval diet (parts by volume): CSMA Standard Fly Larval Medium (Chemical Specialties Manufacturers Association), 1; Vermiculite, 5; water, 2.

Adult diets: 60 combinations of carbohydrates (sucrose, glucose, fructose, honey), acid-hydrolized casein, and yeast extract in liquid and dry form.

PREPARATION Diets prepared as solutions in distilled water containing 0.18% sodium benzoate as mould inhibitor, and frozen until use. Undiluted honey dried at room temperature for 3 days to prevent excessive stickiness. All diets prepared on w/v basis except honey, diluted with water on v/v basis. Liquid diets not replaced during 10 day feeding period unless considerable microbial growth.

REARING At 28 ± 2°C and 60-70% RH under continuous artificial light. Diet immersed in water, pupae collected in 30 mesh screen, transferred to paper towelling and dried. Transferred to ½-gal jars and separated in air stream. Adults collected within 24 h of emergence and transferred to test cages of 1-pint plastic cartons @ 10 pairs/carton. Lids pierced with three, ½" diam holes; two holes covered with cloth, and other plugged with cotton dental roll, 1½" long. Carton inverted over 10-dram glass vial filled with liquid diet or water. Dry diet given then wicks saturated with liquid food, oven-dried under vacuum, and stored in desiccator until used. Dried wicks placed inside 1-pint carton and water used in vial. Adults killed after 10 days, eggs in each carton washed into petri dish, and held in carton in warm water to induce hatching.

DEVELOPMENT In either liquid or dry form carbohydrate diets at concentrations of 2-20% and honey at concentrations of 5-100% led to partial or complete adult development (for carbohydrates, average survival of 63% for liquid and 40% for dry; for honey, 94% for liquid and 64% for dry); eggs mostly viable, production reaching peak with 5% sucrose (dry and liquid, average of 10 and 16 eggs/♀ respectively) and 5% dry, and 10% liquid glucose, average of 9 eggs/♀ each. In liquid form, glucose and fructose produce greatest egg viability at 10% (average range of 2 replicates, 0-70% and 56-70% respectively). In liquid diet, honey dilution of as much as 1:4 gave no appreciable change in number of eggs, viability or survival. In dry form, egg yield on honey can. ½ that on liquid diets. In general, egg production on liquid greater than on dry diets. When acid-hydrolyzed casein at 0.1% and 0.5%, or yeast extract at 0.2%, or both at 0.1% and 0.2%, respectively, added to sucrose or honey, survival, egg production and egg viability usually not substantially improved. Exceptions noted when dry honey diets supplemented with both casein and yeast. Survival and egg production increased by ca. 12 eggs/♀ when 50% liquid honey supplemented with 0.2% of casein. Liquid form of diet generally results in increased survival and greater fecundity. Carbohydrate alone responsible for adult survival and fecundity. Addition of protein and/or yeast extract of very little effect. Two generations reared.

Culicidae

There are numerous papers on the rearing of various species of mosquitoes, especially on diets of yeast and liver with some variations and modifications. These methods are standardized in various laboratories. Here, only a few selected references are given; those which deal with a new species have been repeated. Other work is listed under "see also" at the end of the Culicidae section.

Aedes aegypti Linnaeus, yellow fever mosquito

→ Akov, S. 1962a. A qualitative and quantitative study of the nutritional requirements of *Aedes aegypti* L. larvae. J INSECT PHYSIOL 8: 319-335.

COMPOSITION (μg/ml): KH_2PO_4, 600; K_2HPO_4, 600; $FeSO_4.7H_2O$, 12; $MnSO_4.4H_2O$, 12; NaCl, 12; $CaCl_2$, 12; $MgSO_4.7H_2O$, 200; ribonucleic acid, 1000; cholesterol, 6; pyridoxine hydrochloride, 4.0; riboflavin, 2.0; thiamine hydrochloride, 2.0; calcium pantothenate, 10.0; pteroylglutamic acid, 0.6; niacinamide, 10.0; choline chloride, 100.0; biotin, 0.1. These nutrients in solution added to 50 mg of casein.

PREPARATION Casein ground in mortar, sieved through No. 80 sieve; 50 mg placed into cotton-stoppered test tubes, and sterilized in oven at 160°C for 1½-2 h. Cholesterol emulsion (quantity for 100 tubes):- 3.0 mg cholesterol dissolved in 2 ml alcohol and 18 ml distilled water containing 5 mg Tween 40; 0.2 ml emulsion used/test tube. Inorganic salts, except for phosphates, added to emulsion. Tubes autoclaved and contents added to sterile casein. RNA (quantity for 100 tubes):- 0.5 g RNA dispersed in ca. 20 ml water; dissolved by addition of NaOH solution, pH adjusted to 7.0, and volume brought to 50 ml; autoclaved and 0.5 ml added to each tube. Vitamins and phosphates (and cysteine, sucrose and glutathione when used) made to solution containing each ingredient in 10 times required concentration. Solution filtered through Selas filter and 0.5 mm added to each tube. Volume in each tube made to 5.0 ml. Final pH of diet 6.5.

REARING Aseptically at 28°C. One-month-old eggs sterilized in autoclaved 10% benzalkonium chloride solution for 30-60 minutes, then washed in sterile water and immersed in 80% alcohol for 15-30 minutes. Washed twice more in sterile water, transferred to test tubes with sterile distilled water, and placed in desiccator; pressure reduced by 700 mm for 10-20 minutes to remove air from water and to stimulate hatching. Neonate larvae transferred aseptically to test tubes containing diet.

DEVELOPMENT Larval period on basal casein diet, 7-8 days; shortened by 2 days by addition of brewers' yeast. Survival to adult, 90-100%.

See also: Akov (1962b, 1967).

→ Boorman, J. 1967. Aseptic rearing of *Aedes aegypti* Linn. NATURE LONDON 213: 197-198.

COMPOSITION White bread, 1 g; distilled water, 80 ml; 5% 'Marmite' solution, 20 ml; Penicillin, 100 µg/ml; Streptomycin, 100 µg/ml.

PREPARATION Bread and water placed in 8-oz culture bottles and autoclaved for 10 minutes at 10 lb pressure. Marmite added after sterilizing through Seitz filter, then Penicillin and Streptomycin.

REARING At 28°C. 400-500 neonate larvae transferred to bottles; screw caps replaced with 'Oxoid' metal cap. Larval numbers thinned out to other bottles during growth. Eggs laid on filter paper. Sterilized in 5 ml 10% benzalkonium chloride for 40 minutes, and 80% ethanol for 15 minutes then rinsed twice in sterile distilled water. Eggs and paper transferred to 40 ml tube with 25 ml sterile water and induced to hatch by reducing pressure in vacuum desiccator.

DEVELOPMENT Larval period, 10-12 days; pupae comparable in size with those of laboratory colony. Method in use for 2 years for production of sterile pupae.

→ Hartzell, A.; Burchfield, H.P.; Cotty, V.F. 1958. Improvement in culturing larvae of *Aedes aegypti* (L.) for use in insecticide bioassay. CONTRIB BOYCE THOMSON INST 19: 321-326.

COMPOSITION (g): brewers' yeast, 50; blood albumen, 15; sucrose, 10.

PREPARATION Stock nutrient prepared by blending brewers' yeast, blood albumen, and granular sucrose; stored under refrigeration in screw top bottle. Diet prepared by grinding 2.5 g dry mixture with 10 ml of deionized water in mortar to smooth paste then diluting to 1 litre. Aliquots of agitated suspensions added to flasks containing larvae. Deionized water for rearing test larvae obtained by passing once distilled water through Barnstead "Bantam" Demineralizer. Water tested as passed through resin column and maintained at less than 0.5 ppm solids, calculated as NaCl. Thus rearing diet reproducible to standard ionic strength by adding physiological salt.

REARING Larvae for testing: 2000 larvae reared in 8 l rearing solution in 12-l round bottom flasks. Rearing solution prepared as follows: 800 mg U.S.P. XIV salt mixture added to 500 ml deionized water and heated to 45°C for 5 minutes. Solution cooled to room temperature, adjusted to pH 6 with 1 N HCl and 1.25 N NaOH. Salt solution added to 7.5 l of deionized water containing 4 ml of 1% aqueous Pluronic F-68 solution (a surface active agent supplied by Wyandotte Chemical Co.). Flask containing this mixture placed in water bath held at 31° ± 0.5°C; on equilibrium, 2000 hatched larvae added and flask stoppered with cotton plug. After 18.5 h, 225 ml of freshly prepared nutrient suspension added.

Rearing of culture larvae: 250 larvae placed in 2-l flasks containing 1 l distilled water and 50 ml nutrient suspension. Flasks stoppered with cotton and kept at 31° C. After 3 days larvae separated from culture medium by 60 mesh sieve and placed in flask with 1 l distilled water and 100 ml nutrient suspension. After further 2 days larvae and pupae separated by 60 mesh sieve and transferred to 1-pint Mason jar with 180 ml distilled water and 50 ml nutrient suspension. Small corks floated on surface for adults. Cheesecloth placed over jar and held in place with open ring screw cap. Adult emergence within 3 days. Adults released daily into 30" cube, wooden culture cages with 23 x 11" glass windows and 10 x 13" openings covered by muslin sleeve on front panels. Adults fed 1.25% dextrose solution in 125 ml Erlenmeyer flasks with cheesecloth wick. Two guinea pigs introduced for 1 h daily. 250 ml beaker lined with 1.5" wide paper towelling strips, half full of distilled water placed in cage daily. Eggs deposited on towelling. Strips collected, washed free of debris, and stored between moist filter paper in closed container for 5 days. Eggs air dried and kept as reserve supply. Egg hatching: pyrex test tubes containing 10 ml 0.15% U.S.P. XIV salt mixture and 1% dextrose plugged with cotton and heated in water bath for 5 minutes to remove dissolved oxygen. Tubes cooled to 33° C and strips of paper towelling containing moist 5-10-day-old eggs placed in tubes for 20-30 minutes to hatch. Larvae poured off into open petri dish with 100 ml deionized water.

DEVELOPMENT Larval development, 5-6 days. Adult emergence, 6-8 days from inoculation.

See also: Burchfield *et al.* (1952, 1953).

→ De Meillon, B.; Golberg, L.; Lavoipierre, M. 1945. The nutrition of the larva of *Aedes aegypti* L. I. J EXP BIOL 21: 84-89.

Best diet tested.

COMPOSITION Brewers' yeast, suspended in 0.01 M $CaCl_2$, 0.5 g; 4% liver extract in 0.01 M $CaCl_2$, 10 ml. Liver extract replaced by B-vitamins also successful.

vitamins (μg): thiamin, 100; riboflavin, 100; pyridoxin, 50; calcium pantothenate, 50; nicotinic acid, 100; biotin, 0.025; *p*-aminobenzoic acid, 100; *i*-inositol, 250; choline chloride, 250.

PREPARATION Tubes containing fresh brewers' yeast autoclaved for 30 minutes at 15 lb pressure. Liquid fraction of liver extract or B-vitamins or both adjusted to pH 5.8, and passed through Seitz filter. Sterile filtrate collected in special apparatus to facilitate rapid and reliable dispensing into tubes containing sterile solid fraction.

REARING Gravid ♀♀ confined singly in sterile test tubes with 2-3 ml sterile distilled water at 28° C, and liberated after eggs laid. Water removed with sterile pipette, leaving eggs. Tubes half filled with White's solution and left for 15 minutes. Eggs washed several times with sterile distilled water, then poured into sterile petri dish and covered. Removed 5 or more at once with sterile pipette, introduced to tubes containing agglutinating broth and stored at 37° C before transfer to tubes containing diet.

DEVELOPMENT Egg incubation period, 3 days. Larval development on yeast and liver extract, 4-5 days; on yeast and vitamins and liver extract, 6-7 days. Pupal stage on yeast and liver extract, 2.2 days. 43 out of 60 larvae to adults. Slight increase in developmental period on replacement of liver extract by B-vitamins.

Golberg, L.; De Meillon, B. 1948. The nutrition of the larva of *Aedes aegypti* Linnaeus. 4. Protein and amino-acid requirements. BIOCHEM J 43: 379-387.

COMPOSITION (g): lipid mixture (lecithin, cholesterol, kephalin, 2:1:1), 0.25; finely powdered cellulose, 10; and (g/l): glucose, 10; yeast autolysate, 20; Tatum's (1941) salt mixture, 1.5; and (mg/l): aneurin chloride, 10; riboflavin, 20; pyridoxin hydrochloride, 10; calcium *d*-pantothenate, 10; nicotinic acid, 10; soluble casein or yeast, 1.0%.

Tatum's salt mixture (g): K_2HPO_4, 25; KH_2PO_4, 25; $MgSO_4$, 10; NaCl, 0.5; $MnSO_4$, 0.5; $FeSO_4$, 0.5; $CaCl_2$, 0.5.

PREPARATION Cellulose and lipid mixture mixed and sterilized in autoclave for 30 minutes at 15 lb pressure. Remaining ingredients, except protein, mixed and sterilized by filtration through pad. Protein in solution autoclaved for 45 minutes at 18 lb pressure and added.

REARING See De Meillon *et al.** (1945).

DEVELOPMENT With 1% sodium caseinate in diet, total development time, 10.8 ± 1.09 days.

See also: Golberg & De Meillon (1947); Golberg *et al.* (1947); Tatum (1941).

Lea, A.O.; Dimond, J.B.; DeLong, D.M. 1956a. A chemically defined medium for rearing *Aedes aegypti* larvae. J ECON ENTOMOL 49: 313-315.

COMPOSITION Larval diet (mg): amino acids: *dl*-alanine, 300; *l*-arginine hydrochloride, 500; *l*-aspartic acid, 200; *l*-cysteine monohydrochloride, 800; *l*-glutamic acid, 1250; glycine, 700; *l*-histidine hydrochloride, 200; *dl*-isoleucine, 650; *l*-leucine, 800; *l*-lysine hydrochloride, 800; *dl*-methionine, 200; *dl*-phenylalanine, 1500; *l*-proline, 200; *dl*-threonine, 400; *l*-tryptophane, 400; *l*-tyrosine, 200; *dl*-valine, 1250.

vitamins (mg/50 ml water): biotin, 0.05; calcium pantothenate, 15; choline chloride, 50; glutathione, 25; nicotinamide, 25; nicotinic acid, 5; pteroylglutamic acid, 10; pyridoxine hydrochloride, 10; riboflavin, 10; thiamine hydrochloride, 10.

sugar/lipid/salts (mg): glucose, 2000; cholesterol, 200; Wesson's salt mixture, 500; ammonium acetate, 500; dissolved in 500 ml distilled water.

PREPARATION Vitamin mixture dissolved in 50 ml distilled water. Sugar and salts dissolved in 500 ml distilled water, and cholesterol in 5 ml of 95% alcohol added. 25 mg amino acid mixture dissolved in 0.9 ml of sugar-lipid-salt solution, then 5 mg RNA (dissolved in 0.2 N NaOH) added. 0.1 ml of vitamin stock added and mixture brought to pH 6.5-7.0 by addition of NaOH. Finally 10 mg agar added/ml medium and solution brought to boil. Five ml medium dispensed/rearing bottle which then plugged and autoclaved for 15 minutes at 20 lb pressure. When nutrient agar medium hardened, 5 ml sterile distilled water (pH 6.5-7.0) pipetted into bottles.

REARING Aseptically. Eggs laid on moist paper towelling strips. Eggs surface sterilized by scraping into glass tubing (1 ml section of 5 ml pipette) covered at one end with fine mesh nylon, and submerging for 2 minutes in beaker with 9 ml Clorox, 1 ml Roccal, and 40 ml distilled water. Eggs withdrawn into blunt eyedropper and forced back into tube several times in 2 minutes. Then agitated similarly in 75% alcohol. Alcohol poured off after 5 minutes, eggs drained and beaker and tube laid on side. Few eggs transferred to each rearing bottle using wire spatula. Transfers successful even 15-20 minutes after alcohol bath. 1-oz French Square bottles plugged with cotton used for rearing. Adults emerging from pupae taken from sterile nutrient media, and placed in clean water in non-sterile cage. Given blood meal.

DEVELOPMENT Medium not optimal for larval growth; larval period 11-13 days.

See also: Lea (1957).

→ Dimond, J.B.; Lea, A.O.; Hahnert, W.F. Jr; DeLong, D.M. 1956. The amino acids required for egg production in *Aedes aegypti*. CAN ENTOMOL 88: 57-62.

COMPOSITION Adult diet: 5 g/100 ml each of fructose and glucose; 0.15 g/100 ml salt mixture, and sufficient NaOH to neutralize solutions. Best mixture of amino acids for egg production (g/100 ml diet): *l*-arginine hydrochloride, 0.38; *l*-cystine, 0.15; *l*-glutamic acid, 1.00; *l*-histidine hydrochloride, 0.15; *dl*-isoleucine, 0.50; *l*-leucine, 0.75; *l*-lysine hydrochloride, 0.75; *dl*-methionine, 0.15; *dl*-phenylalanine, 1.20; *dl*-threonine, 0.30; *l*-tryptophane, 0.30; *dl*-valine, 1.00; total, 6.64.

PREPARATION Diets prepared as solutions in water. Adults fed from saturated cotton pads.

REARING Pads replaced in adult cages daily. Oviposition dishes in cages; eggs removed daily and counted. After emergence, adults held 4 days on 10% glucose diet, then inactivated with low temperature and counted into test cages in groups of 400 females.

DEVELOPMENT On control diet, 6961 eggs laid by 400 ♀♀ after 14 days.

See also: Lea *et al.* (1956b).

→ Singh, K.R.P.; Brown, A.W.A. 1957. Nutritional requirements of *Aedes aegypti* L. J INSECT PHYSIOL 1: 199-220.

Two larval diets.

COMPOSITION Diet 1 (g/l): glucose, 10.0; insoluble residue of yeast, 6.25; ribonucleic acid, 0.50; McCollum-Davis No. 185 inorganic salts, 1.50; and (mg/l): thiamine, 10; riboflavin, 20; pyridoxine, 10; calcium pantothenate, 10; nicotinic acid, 10; folic acid, 30; choline chloride, 25; total, 115.

substances substituted for insoluble residue of yeast (g/l): vitamin-free casein, 6.25; and (mg/l): lecithin, 78; cephalin, 39; cholesterol, 39; biotin, 0.5; B_{12}, 0.05; B_t (i.e. carnitine), 4.0; glutathione, 50.0. Lipids, lecithin, cephalin, and cholesterol totalling 0.156 g absorbed on to casein at rate of 1 part to 40.

Diet 2 (g/l): glucose, 5; ribonucleic acid, 1.00; McCollum-Davis No. 185 inorganic salts, 1.00.

l-amino acids (g/l): leucine, 1.00; isoleucine, 1.12; threonine, 0.75; valine, 1.20; phenylalanine, 1.00; lysine hydrochloride, 0.66; arginine hydrochloride, 0.39; histidine hydrochloride, 0.25; tryptophane, 0.36; methionine, 0.70; cystine, 0.13; proline, 0.75; hydroxyproline, 0.18; serine, 1.40; glycine, 0.25; alanine, 1.00; tyrosine, 0.50.

vitamins (mg/l): thiamine, 5.00; riboflavin, 20.00; pyridoxine, 15.00; calcium pantothenate, 15.00; nicotinic acid, 15.00; folic acid, 50.00; choline chloride, 200.00; biotin, 0.50; B_t, 4.00; glutathione, 200.00; B_{12}, 0.05; *p*-aminobenzoic acid, 25.00; *i*-inositol, 100.00. Lipids, lecithin, cephalin, and cholesterol treated as in diet 1.

PREPARATION Solid fraction of lipids absorbed on to insoluble casein by dissolving 0.25 g lipid mixture in redistilled diethyl ether and adding to 10 g vitamin-free casein. Liquid fraction contained all other ingredients. Amino acids, vitamins, glucose, inorganic salts and RNA weighed into 250 ml flask. RNA and tyrosine dissolved separately in minimum quantities of 0.1 N NaOH; cystine in 1 N HCl; folic acid in 1% NH_4OH; and phenylalanine and biotin in hot water. Solution adjusted to pH 6.6 before sterilizing through Seitz filter. Distributed to sterile test tubes using special dispensing flask. Solid fraction dispensed into 18 x 120 mm test tubes, @ 50 mg/tube. Tubes stoppered with cotton wool and autoclaved at 18 lb/sq" for 1 h. Transfer chamber sterilized by UV for 1 h and sprayed with White's solution. 8 ml of liquid fraction transferred to each test tube inside chamber. Insoluble residue of yeast prepared by boiling brewers' yeast powder with 4 volumes water and few drops capryl alcohol for 15 minutes then filtering. Process repeated 3-4 times. Final residue extracted continuously with boiling water in Soxhlet apparatus. Dried in vacuum, ground and passed through 80 mesh sieve. Diet with lipids absorbed on to cellulose at rate of 1 part to 40. Amino acid mixture added to liquid fraction.

REARING At 30°C and 70% RH. One larva inoculated to each tube and cotton wool plug replaced. Adults obtained by rearing larvae on sterile amino acid diet. Emerging adults removed by suction in stream of sterilized air, and 20 ♀♀ and 10 ♂♂ placed in cages of inverted, 7" high lantern globes screened at top and standing on moistened filter paper in petri dish. Adult diet given in 2 ml portions on 2 cm^3 pads of sterilized absorbent cotton wool; fresh pad supplied daily for 6 days. Eggs collected after 7 days and treated with 0.05% MD No. 185 hatching solution.

DEVELOPMENT Diet 1: 8.6 days to adult emergence; 37 pupae from 39 larvae; survival to adults, 95%. Eggs laid in 3-6 days after emergence.

Diet 2: 17.3 days to adult emergence; survival to adults, 72%.

→ Trager, W. 1935a. The culture of mosquito larvae free from living micro-organisms. AM J HYG 22: 18-25.

COMPOSITION Liver extract, 0.5 g; and (ml): distilled water, 100; 1 N NaOH, 0.25-0.30; yeast suspension, 1 ml/14 ml liver extract.

PREPARATION Liver extract added to distilled water, filtered, and pH adjusted to 7.0 with 1 N NaOH. Sterilized through Berkefeld "N" filter, or in autoclave. 2-4-day-old yeast growth in Blake bottles with dextrose agar suspended in sterile tap water; centrifuged, resuspended in sterile distilled water, centrifuged again and finally suspended in sterile distilled water so that 10-11 ml suspension contained yeast from 1 Blake bottle. Yeast suspension killed by heating at 80-85°C for 30 minutes or by autoclaving 30 minutes at 15 lb. 1 ml of suspension added to each 22 x 180 mm test tube containing 12-14 ml liver extract. Liver extract and yeast suspension may be mixed before sterilization and autoclaved together. Simpler medium made by using 1% suspension of Harris' brewers' yeast in liver extract, and autoclaving in tubes.

REARING At 27-28°C in dimly lit incubator. Eggs laid on filter paper partly immersed in distilled water. Eggs sterilized for 5-7 minutes in soap solution. Transferred aseptically to 80% alcohol for 15-17 minutes, then washed in sterile water. 10-40 eggs dropped into tube of culture medium.

DEVELOPMENT Egg stage, 3 days; larval stage, 6-15 days; pupal stage, 2 days. Normal developmental times @ 10 larvae/tube.

See also: Trager (1935b, 1936, 1937a b).

➔ Trager, W. 1948. Biotin and fat-soluble materials with biotin activity in the nutrition of mosquito larvae. J BIOL CHEM 176: 1211-1223.

COMPOSITION (mg/ml): KH_2PO_4, 0.60; K_2HPO_4, 0.60; $FeSO_4.7H_2O$, 0.012; $MnSO_4.4H_2O$, 0.012; NaCl, 0.012; $MgSO_4.7H_2O$, 0.20; $CaCl_2$, 0.012; ribonucleic acid, 1.0; *l*-cystine, 0.2; cholesterol, 0.03; *i*-inositol, 0.04; sucrose, 2.0; and (µg/ml): pyridoxamine dihydrochloride, 0.02; pyridoxine hydrochloride, 4.0; riboflavin, 2.0; thiamine hydrochloride, 2.0; niacin, 2.0; calcium pantothenate, 6.0; pteroylglutamic acid, 0.6; *p*-aminobenzoic acid, 2.0; niacinamide, 10.0; glutathione, 10.0; choline chloride, 20.0. Casein, 30-40 mg and biotin at various concentrations.

PREPARATION Casein supplied in solid form rather than in solution, thus providing bottom feeding larvae with some particulate food. 30 to 40 mg of dry powdered casein dispensed to 18 x 160 mm test tubes and sterilized in dry heat at 160-170°C for 2 h. Sufficient sterile distilled water to give final volume of 6 ml pipetted into each tube, followed by 3 ml of double strength medium and 0.1 ml of cholesterol suspension (300 mg cholesterol dissolved in 20 ml 95% ethyl alcohol). Biotin in solution finally added. 1 ml of this suspension added to 9 ml distilled water, and autoclaved for 20 minutes at 15 lb. Stored in refrigerator. From this 0.1 ml sterile suspension added aseptically to diet tubes containing 6 ml of final medium. Remainder of medium prepared at twice final concentration as follows: cystine dissolved in minimal volume of hot 6 N HCl and added to suspension of RNA in volume of water half that of double strength medium. RNA dissolved in 2 N NaOH to give pH of 6.8-6.9. Sucrose dissolved in water. Salts and growth factors added from concentrated stock solutions kept under toluene in refrigerator, except for $CaCl_2$, nicatinamide, glutathione, choline and inositol which added from freshly prepared concentrated solutions. pH readjusted 6.8-6.9. Medium diluted to correct volume and sterilized by filtration through Selas 03 porcelain filter. Refrigerated and used within 1 week of preparation.

To test biotin and other fat soluble materials: biotin added from autoclaved solution standardized by microbiological assay and stored in refrigerator. Water soluble materials for testing similarly added from autoclaved stock solutions. Materials insoluble in water, such as oleic acid and biotin-active oil from hydrolyzed horse plasma emulsified by ethyl alcohol.

REARING At 28-29°C. Tubes inoculated with axenic 1st instar larvae obtained from eggs sterilized in White's solution.

DEVELOPMENT If biotin absent, then larval growth very slow and no adults produced. Optimum biotin @ 50 µg/ml medium; best larval development 10 days.

See also: Subbarow & Trager (1940).

Aedes albopictus (Skuse)

Moore, C.G.; Fischer, B.R. 1969. Competition in mosquitoes. Density and species ratio effects on growth, mortality, fecundity, and production of growth retardant. ANN ENTOMOL SOC AM 62: 1325-1331.

Also for *Aedes aegypti* (Linnaeus).

COMPOSITION Pelleted sheep food at 0.25 g/day.

PREPARATION Sheep food finely ground and sieved added to 1.5 l tap water.

REARING Oviposition jars (250 ml beakers) lined with paper towelling, removed from cages after 72 h, drained, and conditioned at 25 ± 2° C and 80% RH for periods of up to 29 days. Then eggs placed in 5 ml vials and submerged for 48 h at 27 ± 1° C in diluted (1:10 in tap water) hay infusion medium deoxygenated under vacuum. 1st stage larvae placed in 38 x 23 x 6 cm white enamel pans with 1.5 l tap water. Pelleted sheep food added each day. Pupae collected daily and placed in 1-pint icecream carton cages for emergence.

DEVELOPMENT High egg mortality of *Aedes albopictus; Aedes aegypti* pupated sooner. Larval period prolonged at high larval density levels. Optimum density, 0.1 larva/ml. When reared individually, median pupation time for *A. aegypti* 133 h and for *A. albopictus* 159 h.

See also: Moore & Whitacre (1972).

Aedes taeniorhynchus (Wiedemann)

Nayar, J.K. 1966. A method of rearing salt-marsh mosquito larvae in a defined sterile medium. ANN ENTOMOL SOC AM 59: 1283-1285.

COMPOSITION (mg/100 ml): lactalbumin purified, 2000; sucrose or glucose, 1000; Alphacel, 500; ribonucleic acid, 100; KH_2PO_4, 100; K_2HPO_4, 100; pyridoxine hydrochloride, 2.5; riboflavin, 2.5; thiamine chloride, 2.5; niacin, 10.0; calcium pantothenate, 5.0; folic acid, 5.0; choline chloride, 50.0; biotin, 0.1; ascorbic acid, 100; *i*-inositol, 50; and 100 ml of 10% artificial sea water.

artificial sea water (g/l distilled water): NaCl, 23.991; KCl, 0.742; $CaCl_2.6H_2O$, 2.240; $MgCl_2.6H_2O$, 10.893; $Na_2SO_4.10H_2O$, 9.100; $NaHCO_3$, 0.197; $NaBr.2H_2O$, 0.115; H_3BO_3, 0.027. Diluted 1:9 with distilled water.

PREPARATION Lactalbumin, sucrose or glucose, RNA and Alphacel in amounts for 20 ml of medium weighed into 50 ml Erlenmeyer flasks and dissolved in 10 ml of 10% artificial sea water. Flasks stoppered with non-absorbent cotton and autoclaved at 120° C and 18 lb pressure; cooled rapidly and shaken to prevent clumping. B complex vitamins, inositol, ascorbic acid and phosphates dissolved in 10% artificial sea water so that each present in twice required concentration, then filtered through Seitz bacterial-filter and 10 ml added aseptically to each flask.

REARING At 28° C with 12 h photoperiod. 5-6 first instar larvae, ca. 30 minutes old, inoculated to each flask, which placed in covered clear, plastic box to prevent evaporation of medium. Aseptic techniques used throughout.

DEVELOPMENT Average larval period, 6½ days; compared with 5 days for larvae reared under non sterile conditions on yeast and liver. 80-90% pupation with 100% emergence. Adult appearance normal.

Normal development of *Aedes sollicitans* (Walker) also supported. Normal growth of *Aedes aegypti* when concentration of salt solution reduced to 2% artificial sea water.

Corollary

Nayar, J.K.; Sauerman, D.M, Jr 1971. The effects of diet on life span, fecundity and flight potential of *Aedes taeniorhynchus* adults. J MED ENTOMOL 8: 506-513.

COMPOSITION Larval medium (mg): dry brewers' yeast, 320; liver powder, 40; seawater (diluted/10), 350 ml.

REARING At 27° C with 12 h photoperiod. 200 larvae/24 x 34 x 6 cm enamel tray containing medium. Liver powder added after 24 h; 160 mg dry brewers' yeast added after 48 h and every subsequent 24 h until 50% pupation.

Or reared as described by Nayar & Sauerman (1968, 1969). Within 6-8 h of emergence, 300 pairs transferred to varying nutrient combinations in 22 x 30 x 33 cm cages, and maintained at 27° C and 75% RH with 12 h photoperiod. Sugar solution provided via small cloth covered vial inverted through holes in tops of cages. Chickens presented 3 times weekly for blood feeding. Eggs collected on small, moist pads of folded cheesecloth.

DEVELOPMENT ♀♀ outlived ♂♂. Adult survival best on 10% sucrose solution and chicken blood; average life span, 3 weeks.

See also: Nayar & Sauerman (1968, 1969).

Aedes triseriatus (Say)

Romoser, W.S.; Tucker, D.D, 1975. The effect of various constant temperatures on the growth of the ventral diverticulum in *Aedes triseriatus* (Diptera: Culicidae). ANN ENTOMOL SOC AM 68: 771-776.

COMPOSITION "Purina Laboratory Chow", 10 mg.

PREPARATION Finely ground.

REARING At 27 ± 1° C. Neonate larvae placed in plastic 12½ x 10 x 4" "crisper" water bath with wire screen platform supporting 25 1⅜ x 2" diam cups with 30 ml boiled tap water, 10 mg medium and 20 larvae in each cup. Fed daily. Pupae placed in 2 x 1" plastic vials with 15 ml boiled tap water at 27° C.

DEVELOPMENT Larval period ca. 7 days.

Anopheles albimanus Wiedemann

Rabbani, M.G.; Seawright, J.A.; Leatherwood, L.B. 1976. A method for culturing single families of *Anopheles albimanus*. MOSQ NEWS 36: 100-102.

COMPOSITION Larval diet for first 3 days. 200 ml water (from subterranean well) infused with 2 ml of 2% liver powder:yeast (2:1) suspension. Larval diet on 4th day, 0.5 ml of 2%, suspension of hog supplement (40% protein, Ralston-Purina Co.).

PREPARATION Hog supplement pellets roasted at 120°C for 1.5 h, ground and sieved through 300 µm screen. Best results if suspension prepared then stored at 7°C for 24 h before use.

REARING At 28 ± 0.5°C. Individual gravid ♀♀ isolated in 2.5 x 6 cm plastic tubes lined with filter paper, and containing 5 ml water infused with 0.5 ml of liver powder yeast suspension; plugged with cotton. Eggs laid within 24 h and hatching completed on third day when vials opened and ♀♀ removed. Larvae then transferred to 16-oz clear plastic cups (Sweetheart Plastics) containing diet. Cups covered with plastic lids with cotton plugged 2 cm hole in centre. Larvae not fed again until fourth day when hog suspension added. Hog suspension added daily in decreasing quantities, and stopped when >50% pupation complete. Adults emerge inside cups; aspirated out through hole in lid. Rearing equipment illustrated.

DEVELOPMENT Survival from larvae to adults, 90%.

Anopheles quadrimaculatus Say

→ Knierim, J.A.; Lea, A.O.; Dimond, J.B.; DeLong, D.M. 1955. Feeding adult mosquitoes on preserved blood to maintain egg production. MOSQ NEWS 15: 176-179.

COMPOSITION Successful adult diets: skimmed milk, or 10% solution of egg albumin, or 10% solution of proteose-peptone. Feeding better if 1 part honey added to 9 parts of protein solution.

REARING Feeding apparatus: large 32 x 200 mm Pyrex test tube fitted with 2 x 3 x ⅛" Masonite collar. Tubes wrapped in blood-soaked cotton and maintained at 35-38°C by circulating hot water from tap or reservoir through tubes, or by electrically heating water in tubes by 20 or 40 watt, thermostatically controlled aquarium heater. Tubes suspended by collar from framework at top of stock cages. Culture thus maintained for over nine months. Unit illustrated.

→ Stahler, N.; Terzian, L.A. 1953. Studies in the laboratory mating habits of *Anopheles quadrimaculatus* Say. J EXP ZOOL 124: 317-328.

COMPOSITION Finely powdered "Purina Laboratory Chow" added when necessary to water.

REARING At 80° ± 1°F and 70-75% RH. 200-300 larvae reared in 20 x 17 x 2½" white enamel photographic pans containing 5 l tap water. Food added as required. Adults fed on 4% sugar solution, changed daily.

DEVELOPMENT Pupation in 9-11 days.

See also: Terzian & Stahler (1954).

Anopheles stephensi (Liston)

→ Rosales-Ronquillo, M.C.; Simons, R.W.; Silverman, P.H. 1973. Aseptic rearing of *Anopheles stephensi* (Diptera: Culicidae). ANN ENTOMOL SOC AM 66: 949-954.

COMPOSITION Larval medium (mg/1000 ml): cholesterol, 10; Tween 80, 10; $MgSO_4.7H_2O$, 80; $FeSO_4.7H_2O$, 18.5; $MgSO_4.7H_2O$, 18.5 [sic]; NaCl, 18.5; $CaCl_2$, 18.5; liver extract, 400; sucrose, 10000; brewers' yeast, 4000; RNA, 1000; pyridoxine hydrochloride, 4.59; riboflavin, 2.3; thiamine hydrochloride, 2.3; calcium pantothenate, 11.4; pteroylglutamic acid, 1.37; niacinamide, 11.4; biotin, 0.114; choline chloride, 114; KH_2PO_4, 685; K_2HPO_4, 685; casein, 50 mg/flask.

Adult diet: autoclaved 10% white sugar solution. Blood meals for ♀♀ from rabbits.

PREPARATION Cholesterol emulsion and RNA prepared as in Akov* (1962a). Vitamins dissolved in 500 ml distilled water and KH_2PO_4 and K_2HPO_4 added; then sterilized by passing through 0.22 µm Millipore filter. Liver extract and sucrose dissolved in 200 ml distilled water, autoclaved and introduced to flask containing autoclaved brewers' yeast. Salts dissolved in 200 ml distilled water and filtered (0.22 µm Millipore). All solutions mixed, and brought to final volume of 1000 ml with sterilized distilled water. 15 ml of medium dispensed to 125 ml Bell's rearing flasks containing 50 mg dry casein, sterilized as in Trager* (1935a). Medium made up immediately before use.

REARING Seven neonate larvae pipetted into each flask which then put into surface sterilized plastic cages. Pupae removed with sterilized pipette and transferred singly to rearing unit with distilled water. Each unit (illustrated) of 16.1 x 81.4 mm autoclear, polycarbonate inner tube with screw cap, and 22 x 100 mm outer tube with disposable foam rubber plug. 1 cm wide and 4 cm long 'window' cut from the lower part of inner tube and replaced by nylon netting. Parts of unit autoclaved separately, assembled, and pupa pipetted to inner tube. When adult emerged, inner tube removed and transferred to sterilized plastic cage. Food from 10% white sugar solution on wad of sterilized cotton wool placed on netted screw cap. Aseptic rearing at 27 ± 1°C and 78% RH with 12 h photoperiod. Conventional rearing as in Thompson & Bell (1968) and Brust (1971).

DEVELOPMENT Egg to pupae in 9 days; adult emergence 3 days later.

Culex molestus Forskål

→ Lichtenstein, E.P. 1948. Growth of *Culex molestus* under sterile conditions. NATURE LONDON 162: 227.

Four diets tested.

COMPOSITION Diet 1: dried or fresh autoclaved yeast, 1.2 g; distilled water, 15 ml.

Diet 2: diet 1 with (µg/ml): thiamin, 10; riboflavin, 10; pantothenic acid, 10; niacin, 10; pyridoxin, 10; folic acid, 10; biotin, 0.14.

Diet 3: liver extract and dried autoclaved yeast (Trager's medium for *Aedes aegypti*).

Diet 4 (%): glucose, 0.1; salt mixture (not specified), 0.2; casein hydrolysate, 2; vitamins as diet 2 but (µg/ml): biotin, 7.5; ascorbic acid, 150; and water.

REARING Aseptically at 28°C; 25 eggs added to each medium in tubes.

DEVELOPMENT On diet 1: egg to adult, 10.5-13 days, eggs inviable. Diet 2: egg to adult, 7.5 days; eggs viable. Diet 3: egg to adult, 14 days; eggs viable. Diet 4: egg to adult, 25 days; eggs viable.

Culex pipiens (Linnaeus), northern house mosquito

→ Buddington, A.R. 1941. The nutrition of mosquito larvae. J ECON ENTOMOL 34: 275-281.

Also for *Aedes aegypti* (Linnaeus).

Best diet tested.

COMPOSITION Dried yeast (amount not specified); 70% ethyl alcohol (amount not specified); glucose, 0.1%; Osborne-Mendel salt mixture, 0.1%. pH 6.8-7.0.

PREPARATION Dried yeast covered with 70% ethyl alcohol and shaken periodically for ca. 6 weeks. Alcohol removed by vacuum distillation and sterile water added to make heavy suspension. Added to autoclaved glucose and salt mixture suspension and transferred to rearing tubes (amount of water not specified, perhaps 100 ml?).

REARING Adults of both spp kept in small cages at 28° ± 1°C. Females fed on blood of pigeons, guinea pig, and man. Males fed on soaked raisins. Eggs collected on floating filter paper. *Culex* eggs teased out of rafts. Sterilized by submergence in dilute soap solution for 5 minutes, rinsing in sterile water and immersing in 1:1000 solution of hexylresorcinol for 2 minutes. Sterile eggs or sterile 1-day-old larvae placed in medium.

DEVELOPMENT *A. aegypti:* larvae to 4th instar, 75%; pupation, 71%; adult emergence, 69.5%. *C. pipiens:* larvae to 4th instar, 74%; pupation, 71%; adult emergence, 65%.

→ Dadd, R.H.; Gomez, I.; Nambe, M. 1973. Requirement for ribonucleic acid in a semi-synthetic larval diet for the mosquito *Culex pipiens*. J MED ENTOMOL 10: 47-52.

COMPOSITION Larval diet (mg/100 ml): casein (BDH light white soluble), 1000.0; ribonucleic acid (NBC reagent, yeast), 300.0; soluble starch (NBC reagent), 200.0; thiamine, 0.5; riboflavin, 0.5; nicotinamide, 1.0; pyridoxine, 0.5; calcium pantothenate, 5.0; folic acid, 0.1; biotin, 0.1; choline chloride, 10.0; K_2HPO_4, 60.0; KH_2PO_4, 60.0; $MgSO_4$, 20.0 (pers. comm.); calcium gluconate, 5.0; NaCl, 2.0; sequestrene salts (Geigy) Fe, 2.0; Zn, 2.0; Mn, 2.0; Cu, 0.5; KOH to pH 7.5 and water to 100 ml.

PREPARATION RNA dissolved in ca. 80 ml water. Phosphates and minerals, then casein added whilst stirring. KOH added to maintain pH at neutrality. Stirred for 30 minutes (no heat) until all casein dissolved and solution opalescent and colloidal. Vitamins added, and KOH to pH 7.5. Stirred and water added to 100 ml. Five ml medium transferred to 150 x 25 mm culture tubes; tubes capped and autoclaved at 125°C and 17 lb pressure for 15 minutes.

REARING One larvae inoculated/sterile tube. Tubes then kept in darkness in a BOD incubator at 25 ± 1°C.

DEVELOPMENT With optimum RNA of 0.3%, survival to pupation, 95%; and to adult, 82%.

See also: Dadd (1972); Hosoi (1964).

→ Lang, C.A. 1963. The effect of temperature on the growth and chemical composition of the mosquito. J INSECT PHYSIOL 9: 279-286.

COMPOSITION (%): commercial rabbit feed, 1; dried yeast powder, 0.5.

PREPARATION 10 g infusion of above mixture added to 990 ml distilled water, stored at 4°C; prepared weekly.

REARING Eggs hatched in distilled water, and 200, 18-20-h-old first instar larvae transferred to pans. One 1 medium poured to 3 cm depth giving 1.5 cm^2 surface area/larva. 10 ml infusion pipetted into each pan on alternate days until 4th instar, then given daily. Volume maintained by adding water, surface swept daily with absorbent paper to minimize scum formation. Pans kept at 16-17°, 25-26°, or 30-31°C. Pupae transferred to covered dishes of distilled water and incubated.

DEVELOPMENT *Culex pipiens:* 64, 71, and 82% pupation at 16, 25, and 30°C in average of 24, 11, and 9 days respectively.

Culex molestus: 23 and 44% at 16 and 25°C in average of 42 and 14 days.

Opifex fuscus Hutton, New Zealand mosquito

→ Haeger, J.S.; Provost, M.W. 1964. The laboratory colonization of *Opifex*. BULL W H O 31: 451-452.

COMPOSITION (mg): powdered beef liver, 200; yeast, 100; ground dog food pellets, 201; rabbit food, 100.

PREPARATION Various quantities and combinations of components sprinkled on 75% sea water.

REARING At 25°C. Oviposition on pads of tightly rolled cheese cloth soaked in water; removed after 2 days. Stored for 3 days, then air-dried for 2-3 days. Pledgets (oviposition pads of cheesecloth) placed in porcelain or plastic crisper at 8-10°C in petri dishes. 300-500 eggs placed in 6" bowl with 75% sea water, 50 mg liver, 50 mg yeast, 1 mg dog pellets. Left 4 days and transferred to 16 x 8 x 2" pan with 50 mg yeast, 50 mg liver. On 5th and 6th day, fed 50 mg liver, 100 mg dog food. on 9th day 75, 3rd instar larvae placed in each pan, and fed 50 mg liver, 50 mg rabbit food, 50 mg dog pellets. 10th and 11th days as day 9. 12th-17th day fed 50 mg dog pellets, 50 mg rabbit food. Pupae transferred to 18 x 18 x 3" pan filled to 2" depth with 75% sea water. Pupal pan placed in 2' cube adult cage with ♂ adults. ♀♀ in small 12 x 9 x 8" cage fed either water or blood for oviposition.

DEVELOPMENT Eggs incubation period, 72 h; larval period, 12-20 days; pupal period, 3 days; 12 generations reared.

Theobaldia incidens (Thomson)

→ Frost, F.M.; Herms, W.B.; Hoskins, W.M. 1936. The nutritional requirements of the larva of the mosquito, *Theobaldia incidens* (Thom.). J EXP ZOOL 73: 461-479.

Best of many diets tested.

COMPOSITION Fresh bakers' yeast, 3 g; distilled water, 1 l.

PREPARATION 10 ml portions of mixture placed in test tubes and autoclaved for 1 h at 17 lb pressure. pH at 6.0.

REARING At 21°C. Eggs sterilized in in 5% soap solution, then 80% alcohol. Hatched in sterile water; larvae placed in small flasks in lots of 6.

DEVELOPMENT Average of 25% adult emergence in 16-18 days.

Toxorhynchites rutilus septentrionalis (Dyar & Knab)

➞ Trimble, R.M.; Corbet, P.S. 1975. Laboratory colonization of *Toxorhynchites rutilus septentrionalis* (Diptera: Culicidae). ANN ENTOMOL SOC AM 68: 217-219.

COMPOSITION Larval medium: liver powder (NBC), ¼ tsp; in 35 x 22.5 x 37 cm white enamel pan to depth of 3 cm; eggs of *Aedes aegypti* (Linnaeus), 10000.

PREPARATION Components mixed thoroughly in pan.

REARING At 24 ± 3°C and 75 ± 5% RH with 15 h photoperiod; optimum of 300 adults held in 61 cm^3 aluminium and screen cages, with 1.3 x 122 x 46 cm layer of water-soaked plastic foam resting on 1.6 cm thick layer of glass wool on top of cage. Whole covered with clear polyethylene to retard evaporation. Four, honey-soaked, synthetic sponge wicks (15 x 2.5 x 2.5 cm) set upright in 6 x 4 cm diam glass jars on floor of cage. Several water wicks of folded, water-soaked paper towelling held in 100 ml beakers also provided. Water and foam renewed daily; honey when necessary. Eggs laid in black, glass, 13 x 7.5 cm diam jars containing water to within 5 cm of rim; removed from water surface every 24 h and placed in water-filled jars for 24 h. 50 eggs transferred to enamel pan containing larval medium. Pan left for 24 h for host and prey eggs to hatch. Prey kept smaller than host larvae by using only ¼ tsp liver powder to maintain desired size ratio, adding when necessary. Pan water changed when scum appears. Ratio of prey to host larvae always in excess of 10:1. Prey larvae may need to be added. Pupae removed, washed in clean water and placed in 330 ml styrofoam cups, 50 in each.

DEVELOPMENT Eggs laid 5-7 days after peak emergence. 3 generations reared using induced mating technique involving anaesthetizing ♀♀ and inseminating with ♂♂ prepared by removing legs and head immediately before joining to ♀♀.

➞ Bruen, J.P.; Goldberg, L.J.; Tanabe, A.M.; Watkins, H.M.S.; Ford, I.J. 1976. Quantitative rearing of small numbers of mosquito larvae. MOSQ NEWS 36: 95-97.

COMPOSITION Basal diet (g): Laboratory rabbit chow pellets (Ralston Purina Co.), 0.8; Spur pellets (Albers, Div. of Carnation Co., Van Nuys, Calif. 91412), 0.8; alfalfa foliage, 16.0; and distilled water, 300 ml.

PREPARATION Ingredients blended for 2 minutes, autoclaved at 1.4 kg/cm^2 for 45-60 minutes, cooled, then filtered through coarse glass wool to remove large particulate matter. pH, ca. 7.2. Stored at 4°C for maximum of 7 days. Then 50 ml of previous week's diet, containing mixed microflora necessary for larval development, added to each newly prepared lot of sterilized basic diet, and volume made to 1000 ml with sterilized distilled water. Diet and eggs incubated overnight at 30°C on shaker to enhance growth of aerobic bacteria. After first day of larval growth, diet supplemented with basic diet enriched with heat-killed yeast; dried baker's yeast (Red Star Yeast Co., Oakland, Calif. 94607) added to distilled water @ 25 g/l and autoclaved for 20 minutes at 1.4 kg/cm^2; 100 ml added to 900 ml diluted basic diet.

REARING At 25°C individually or in pairs in disposable clear polystyrene rearing trays (Champion Packages Co., Belvidere, Ill. 61008) with 5 rows, with 5 wells/row, each individual well measuring 2.8 x 4.1 x 1.6 cm with 15 ml capacity. 19 x 26 x 0.6 cm plexiglass sheet provided with 2 plastic dowels positioned at diagonal corners; dowels pass through preformed holes in tray lip; second tray inverted over first to retard evaporation and provide additional air space; inverted tray weighed down by Plexiglass, to stabilize unit and facilitate handling, counting, observation and to minimise space. 10 similar units held at 25°C in slotted clear Plexiglass storage rack. Unit illustrated.

DEVELOPMENT *Culex pipiens pipiens* Linnaeus, *Culex tarsalis* Coquillett and *Aedes taeniorhynchus* (Wiedemann) reared. Larval and pupal mortality <10%; good larval mobility, pupation rate and development period comparable to those mass reared in large containers; >80% adult emergence from all 3 spp. Adult survival and fecundity same as standard diet. % egg eclosion, high.

See also: Abdel-Malek & Adham (1974); Asahina (1964); Atkin & Bacot (1917); Baker (1964); Brust (1971); Coluzzi (1964); Gahan & Smith (1964); Lang & Stephan (1967); Lang *et al.* (1965); Laurence (1964); Martínez-Palacios (1964); Pillai & Madhukar (1969).

Cuterebridae

Dermatobia hominis (Linnaeus Jr), torsalo

→ Banegas, A.D. 1968. Progress report on the rearing of torsalo larvae *Dermatobia hominis* (Diptera: Cuterebridae). pp. 7-18. *in* Radiation, Radioisotopes and Rearing Methods in the Control of Insect Pests. Proceedings of a Panel Organized by the Joint FAO/IAEA Division of Atomic Energy in Food and Agriculture.

Best of many diets tested.

COMPOSITION (%): ground lean beef, 60; distilled water, 15; whole citrated bovine blood, 25; formalin, 0.2; and Mycostatin, 100 μg.

PREPARATION See Melvin & Bushland* (1941).

REARING At 34-35°C. 10 g diet placed in shell vials with 15 neonate larvae. Eggs sterilized in 0.25% NaClO for 20 minutes.

DEVELOPMENT 3 out of 135 larvae survived after 7 days. One larvae survived 25 days to 3rd instar, 13 mm long. Further improvements needed to diet.

Drosophilidae

Drosophila ampelophila Loew

→ Guyénot, E. 1917. Recherches expérimentiles sur la vie aseptique et la développement d'un organisme *(Drosophila ampelophila)* en fonction du milieu. BULL BIOL FR BELG 51: 1-330. (In French)

COMPOSITION Basal diet (g): peptone, 1.0; glucose, 1.5; KCl, 0.2; $Na_3PO_4.12H_2O$, 0.2; $MgCl_2$, 0.4; $CaH_2(PO_4)_2$, 0.1; $FeCl_3.6H_2O$, 0.01; tributyrine [2 drops]; crystallized acetic acid [1 drop]; distilled water to 100 g.

PREPARATION Mineral salts dissolved in water plus acetic acid and added to glucose and peptone. Fat added at last moment and emulsified by prolonged stirring. 10 ml dispensed to rearing tubes with 1 g of cotton wool to give diet substance.

REARING Aseptically; individually in tubes at 24°C.

DEVELOPMENT Initial growth rapid, but most larvae dead by 11th day. Limited survival to 30 days. Lack of essential "vitamin" of yeast recognized.

See also: Baumberger & Glaser (1917); Delcourt & Guyénot (1910, 1912); Guyénot (1913).

Drosophila busckii Coquillett

→ Singh, P. 1976a. Artificial diet for rearing *Drosophila busckii* (Diptera: Drosophilidae). UNPUBLISHED.

COMPOSITION (g): bran, 25; brewers' yeast, 1; and water, 75 ml.

PREPARATION Yeast and bran mixed in 600 ml milk bottles; water added and mixed again. Used immediately or after 24 h.

REARING At 25°C in milk bottles plugged with cotton wadding. Diet in bottles to depth of ca. 5 cm.

DEVELOPMENT Life cycle, 9-10 days; at least 16 generations reared.

Drosophila hydei Sturtevant

→ Doane, W.W. 1970. A quick and easy method for rearing large quantities of 'clean' larvae. DROS INF SERV 45: 189.

COMPOSITION (%): brewers' yeast, 20; cane sugar, 10; water, 70; and acid mix of Lewis (1960).

acid mix (ml): propionic acid, 836; distilled water, 164; added to 85% H_3PO_4, 83; and distilled water, 917. pH 0.8-0.9.

PREPARATION Paste made by mixing sterile aqueous stocks of yeast and sugar. Lewis mixture added, and spread over ¼-½" thick foam plastic pad. Stock paste refrigerated; plastic pads re-usable after cleaning and autoclaving.

REARING Adults in plastic boxes, ventilated at one end, with diet impregnated lining to base. After oviposition adults removed by suction into cotton gauze bag. Larval numbers unregulated but more diet added if necessary. Larvae easily washed off and decanted into collecting vessels. Method also useful for rearing *D. melanogaster*.

Drosophila melanogaster Meigen, vinegar fly

There have been numerous papers published on the dietetics of *D. melanogaster* and its allied species. Diets, from a simple banana, yeast, and agar mixture to a highly complex chemically defined diet, have been described, but no single publication has covered all aspects adequately. Furthermore, since techniques are continually changing, any such publication would be out of date by the time it appeared. Satisfactory coverage of most changes appears in the Technical Notes Section of Drosophila Information Service (DIS).

DIS provides an annual compendium of stock lists, research and technical notes and an inventory of laboratories engaged in various types of *Drosophila* research. Information concerning subscription to this publication can be obtained from the Department of Biology, University of Oregon, Eugene, Oregon, 97403, U.S.A.

In this chapter, only a few selected diets tracing the developments of *Drosophila* dietetics, and some other standard diets, are listed. References relating to diets similar in composition are listed in the bibliography. Diets for species other than *D. melanogaster* have been included, regardless of their similarity to previously abstracted diets.

➞ Begg, M. 1956. Some observations on water-soluble factors which accelerate the growth in aseptic culture of larvae of *Drosophila melanogaster*. J EXP BIOL 33: 142-154.

COMPOSITION Aseptic basal larval diet (g): casein, 6; *dl*-tryptophane, 0.5; ergosterol, 0.5; cholesterol, 0.5; glucose, 10; ribose nucleic acid (from yeast), 0.3; agar (0.35 g/culture tube), 10; and (μg/culture tube): biotin, 0.2; folic acid, 4.0; choline chloride, 1000; aneurin hydrochloride, 4.0; pyridoxin, 12.0; calcium pantothenate, 12.0; riboflavin, 6.0; nicotinic acid, 12.0; thymine, 25.0; and (mg/culture tube): K_2HPO_4, 2.5; KH_2PO_4, 2.5; $MgSO_4$, 1.0; $MNSO_4$, 0.5; NaCl, 0.5; $FeSO_4$, 0.5; $CaCl_2$, 0.5; total diet/culture tube, 5 ml.

PREPARATION RNA extract prepared by autolyzing live brewers' yeast for 4 days at 35° under reflux with ethyl ether, then filtering and centrifuging resulting mass until cell-free solution obtained. Solution dried under reduced pressure at 40-50° C. 1% NH_4OH added at start of autolysis yields adequate extract. Agar gel made up using 0.3 g autolysate to 5 ml gel. Other components added. Culture tubes removed from autoclave, agitated in cold running water than slanted or rotated. pH 6.7-7.0.

REARING Adults kept for 2-3 days at 18-19° C in half pint milk bottles containing diet with smear of bakers' and brewers' yeasts. Then shaken into clean empty bottles where they oviposit for 8 h at 26° C on thin layer of yeast. Fullter's earth paste (to which trace of ethanol and acetic acid added) smeared on agar gel in watch glasses. Eggs separated from medium and sterilized in special apparatus (illustrated) by washing for ½ h in solution of 4% Cetavlon to free from surface debris; then dechorionated by rinsing in 1% filtered solution of calcium hypochlorite for ½ h. All solutions autoclaved and operations carried out under dust free hood previously exposed to UV light. Eggs pipetted aseptically on to carmine-stained sterile agar slabs. Hatch over 24 h at 25° C; neonate larvae or eggs transferred to diet with platinum spoons, @ 25 larvae or 40-50 eggs/tube and reared at 25° C.

DEVELOPMENT Larval development, 9-11 days.

See also: Begg (1949); Begg & Robertson (1948, 1949); Begg & Sang (1950).

➞ Cohen, B. 1968. The influence of larval crowding on the development time of populations of *Drosophila melanogaster* on chemically defined medium. CAN J ZOOL 46: 493-497.

Two diets described.

COMPOSITION Diet 1: standard brewers' yeast-sucrose diet (g): agar, 76; brewers' yeast, dry powder, Anheuser-Busch type, 200; sucrose, 212; and (ml): propionic acid, 18; water (dechlorinated), 4800.

Diet 2: chemically-defined diet (g): agar, 30.0; cholesterol, 0.3; choline chloride, 0.075; lecithin, 4.0; ribonucleic acid (yeast extracted), 4.0; sucrose, 7.5; biotin, 0.00016; B_{12}, 0.00003; calcium pantothenate, 0.016; folic acid, 0.003; niacin, 0.012; pyridoxine, 0.0025; riboflavin, 0.01; thiamine, 0.002; Na_2HPO_4, 1.89; KH_2PO_4, 1.83; $NaHCO_3$, 1.4; $MgSO_4$, 0.3; $CaCl_2$, 0.013; $Fe_2(SO_4)_3$, 0.013; $MnSO_4.4H_2O$, 0.013; NaCl, 0.013.

amino acids (g): *l*-arginine, 0.642; *l*-cysteine, 0.538; *l*-glutamic acid, 4.950; *l*-glycine, 1.960; *l*-histidine, 0.538; *l*-isoleucine, 1.412; *l*-leucine, 2.632; *l*-lysine, 1.500; *l*-methionine, 0.380; *l*-phenylalanine, 1.132; *l*-tryptophan, 1.960; *l*-threonine, 0.850; *l*-valine, 1.524; and water to 1000 ml.

PREPARATION Diet 1: agar dissolved in boiling dechlorinated tap water. Heat turned off and yeast and sucrose stirred in. Propionic acid added to 70° C and 50 ml diet portions dispensed to 225 ml milk bottles.

Diet 2: three times listed quantity of vitamins (except for biotin, B_{12}, and folic acid) dissolved in 2000 ml of sterile distilled water over hot water bath. 666 ml used/1000 ml diet. RNA and $NaHCO_3$ dissolved in 100 ml portion; all minerals plus choline chloride added to remaining 566 ml. Both added to dry agar and sucrose in beaker, made to 900 ml with distilled water and heated to 80° C over water bath. Blended for 5 minutes at high speed with addition of biotin and B_{12}. Lecithin and cholesterol dissolved in 10 ml of 100% ethanol over hot water bath and slowly blended into main mixture. Three times above quantity of folic acid dissolved in 30 ml of 20% ethanol over water bath (until no Schlieren effect apparent) and 10 ml blended into main mixture. Dry-prepared amino acid mixture thoroughly ground with mortar and pestle and blended into main mixture. pH adjusted to 6.5 with concentrated NaOH and total volume brought to 1000 ml. 5 ml sufficient to support complete development of ca. 200 individuals.

REARING At 24 ± 1° C with constant illumination. Stock cultures from brewers' yeast-sucrose diet. 2-day-old larvae transferred to fresh diet to avoid crowding. Adults, when 3-5 days old, placed on fresh brewers' yeast diet one day before use @ 250 pairs/225 ml milk bottle with 30 ml diet for egg collection. Concentrated suspension of autoclaved live yeast placed on centre of diet to stimulate oviposition; ca. 5000 eggs laid in 4 h. Eggs washed from bottles with water into 20 ml test tubes, washed 5 times with sterile distilled water and once with 4% cetyl-trimethyl-ammonium bromide; then decanted into sterile test tubes with 3% NaClO solution; after 15 minutes washed with sterile centrimide followed by sterile distilled water; repeated ten times. Finally, water decanted and excess fluid taken up by sterile absorbent cotton gauze. Eggs transferred with sterile wire to 5 ml chemically defined diet. Surface area in tube after slanting, 1.5 cm^2.

DEVELOPMENT Larval stage, 4.1 days on optimal live yeast diet; 9 days on amino acid diet.

→ David, J.; Clavel, M.F. 1965. Interaction entre le génotype et la milieu d'élevage conséquences sur les caractéristiques du développement de la drosophile. BULL BIOL FR BELG 99: 369-378. (In French with English summary)

Three diets tested on 3 strains of *D. melanogaster*.

COMPOSITION Diet 1: axenic - see David (1962) (g): agar, 15; dried brewers' yeast, 115; cornflour, 110; Nipagin, 5.75; water, 1200.

Diet 2 (g): agar, 20; cornflour, 100; sucrose, 120; dried brewers' yeast, 15; Nipagin, 0.5; water, 1000.

Diet 3, as diet 2 but Nipagin increased to 1.5 g and salts added (g): KH_2PO_4, 1.5; $MgSO_4.7H_2O$, 0.5; $(NH_4)_2$, 2.0. Diets 2 and 3 mixed with suspension of live bakers' yeast just before use.

PREPARATION Agar dissolved in boiling water. Other ingredients added and autoclaved for 15 minutes at 115°C. Dispensed to culture tubes.

REARING In tubes at 24-25°C.

DEVELOPMENT Variable between strains and diets, but in summation, for all strains, development on axenic diet better than on both live yeast diet.

See also: Van Herrewege (1974).

→ Ellis, J.F. 1959. Reversal of an adenine and a cytidine requirement in axenic *Drosophila* culture. PHYSIOL ZOOL 32: 29-39.

Two diets tested.

COMPOSITION Casein diet (mg/ml): Labco-vitamin-free casein, 20-40-50 in different experiments; sucrose, 7.5; cholesterol, 0.15; lecithin, 2.0; agar, 25.0; $MgSO_4.7H_2O$, 0.246; $MnSO_4.4H_2O$, 0.0129; $FeSO_4$, 0.02; Na_2HPO_4, 3.04; KH_2PO_4, 0.71; NaCl, 0.0129; $CaCl_2$, 0.0129; and (μg/ml): biotin, 0.06; calcium pantothenate, 3; choline chloride, 40; niacinamide, 1; pteroylglutamic acid, 0.6; pyridoxine, 0.4; riboflavin, 2; thiamine, 1.

Amino acids diet (mg/ml): arginine, 0.794; cysteine, 0.480; glutamic acid, 4.418; glycine, 1.745; histidine, 0.484; isoleucine, 1.260; leucine, 2.345; lysine, 1.337; *dl*-methionine, 0.339; phenylalanine, 1.008; *dl*-threonine, 0.745; tryptophane, 1.745; *dl*-valine, 1.355; sucrose, 7.5; cholesterol, 0.1; agar, 15.0; $MgSO_4.7H_2O$, 0.246; $MnSO_4.4H_2O$, 0.0129; $FeSO_4$, 0.0129; KH_2PO_4, 0.606; K_2HPO_4, 0.606; NaCl, 0.0129; $CaCl_2$, 0.0129; and (μg/ml): biotin, 0.02; calcium pantothenate, 6; choline chloride, 20; niacinamide, 10; pteroylglutamic acid, 6; pyridoxine, 30; riboflavin, 2.4; thiamine, 1.5.

PREPARATION Ingredients mixed; pH adjusted to 7 with 1 N NaOH prior to autoclaving. 4 ml diets pipetted into 15 x 150 mm test tubes, plugged with non absorbent cotton and autoclaved for 15 minutes at 15 lb pressure. Tubes then shaken vigorously and slanted before cooling.

REARING Collection and sterilization of eggs as in Sang* (1956), except that sterile spoons from cellophane dialyzing tubing easier for counting and picking up larvae from sterile agar plates. 25, 6-h-old larvae inoculated aseptically to each tube.

DEVELOPMENT Larvae of strain Oregon-K unable to grow and develop at normal rate on diet with casein unless RNA included; near normal development with high hydrogen-ion concentration replacing RNA. On 20 mg/ml casein, 58% pupation in 12.8 days. 86.3% pupae to adults. On 50 mg/ml casein at pH 5.1, 47.1% pupation in 10.6 days. 85.8% pupae to adults.

→ Erk, F.C.; Sang, J.H. 1966. The comparative nutritional requirements of two sibling species, *Drosophila simulans* and *D. melanogaster*. J INSECT PHYSIOL 12: 43-51.

Diet modified from that developed for *Drosophila melanogaster* (Meigen) by Sang* (1956).

COMPOSITION (g): casein (Genatosan-low vitamin), 5.50; agar (Oxoid, agar No. 3), 3.00; sucrose, 0.75; lecithin, 0.40; RNA (yeast), 0.40; cholesterol, 0.03; aneurin, 0.0002; riboflavin, 0.0001; nicotinic acid, 0.0012; calcium pantothenate, 0.0016; pyridoxine, 0.00025; biotin, 0.000016; folic acid, 0.0003; $NaHCO_3$, 0.150; KH_2PO_4, 0.183; Na_2HPO_4, 0.189; $MgSO_4.7H_2O$, 0.062; and water to 100 ml.

PREPARATION Ingredients autoclaved for 15 minutes at 20 lb pressure in 1" boiling tubes. Shaken whilst cooling for even dispersal of casein. Cultures not slanted.

REARING At 25°C and 80% RH. Eggs collected and sterilized (see Sang*, 1956). 40 neonate larvae (26 ± 2 h after laying)/tube.

DEVELOPMENT Development similar in both *D. melanogaster* and *D. simulans*. Life cycle from egg to adult completed in <10 days.

→ Fleming, W.E.; Parker, L.B.; Maines, W.W.; Plasket, E.L.; McCabe, P.J. 1962. Bioassay of soil containing residues of chlorinated hydrocarbon insecticides with special reference to control of Japanese beetle grubs. U S DEP AGRIC TECH BULL 1266, 44 pp.

COMPOSITION (g): agar, 22.8; yellow cornmeal, 190.0; and (ml): corn syrup, 300.0; 20% sodium propionate, 45.0; water, 1500.

PREPARATION Agar dissolved in 1 l boiling water; corn syrup added. Cornmeal mixed in 500 ml of water and added to boiling mixture, cooked until fairly thick then sodium propionate added. Diet dispensed to 500 ml Erlenmeyer flasks @ 200 ml/flask and sterilized at 20 lb for 15 minutes.

REARING At 74°F and 40% RH with 24 h photoperiod. Half teaspoon of dry bakers' yeast distributed evenly over diet surface and 100 flies/flask (50 ♀♀, 50 ♂♂) introduced; flies removed on 11th day and an additional ¼ teaspoon of bakers' yeast spread over diet. Emerging flies removed each day and transferred to plastic randomizing cylinder for testing.

DEVELOPMENT Average production/flask over 2 week period, ca. 350 flies/day.

→ Geer, B.W. 1963. A ribonucleic acid-protein relationship in *Drosophila* nutrition. J EXP ZOOL 154: 353-364.

COMPOSITION (g): agar, 1.5; casein, 5.0; fructose, 1.0; cholesterol, 0.03; RNA, 0.1; thiamine, 0.0002; riboflavin, 0.001; nicotinic acid, 0.0012; calcium pantothenate, 0.0016; pyridoxine, 0.00025; biotin, 0.00002; folic acid, 0.001; choline, 0.008; $FeSO_4.7H_2O$, 0.00129; $CaCl_2$, 0.00129; $MgSO_4.7H_2O$, 0.0246; $MnSO_4.H_2O$, 0.00129; NaCl, 0.00129; KH_2PO_4, 0.183; Na_2HPO_4, 0.189; and water to 100 ml.

PREPARATION Casein and agar suspended in phosphate buffer solution; NaCl, nicotinic acid, thiamine and pyridoxine added from stock solutions. Neutralized with dilute NaOH; and remaining B-vitamins, choline and salts added. Cholesterol added in suspension prepared by dissolving in warm ethanol, adding water and removing ethanol by autoclaving for 10 minutes. Final neutralization with NaOH. Mixture autoclaved at 15 lb pressure for 15 minutes and fructose added. Aseptically dispensed in 5 ml amounts to shell vials.

REARING Bottomless half pint milk bottles set in petri dishes with warm cornmeal grape-juice diet. When solidified, several hundred ♀♀ from 4-5-day-old stock cultures introduced to bottles. Eggs collected by removing flies, lifting milk bottles and sweeping eggs from diet surface with camel hair brush; sterilized in successive rinses of *Drosophila* saline, 0.25% NaClO and 0.04% n-alkyl dimethyl benzyl ammonium chloride. Saline and hypochlorite rinses in centrifuge tubes, other rinses in apparatus of Begg* (1956). Aseptic transfer of eggs to shell vials @ 30-80/vial carried out in hood sterilized with germicidal UV lamp. Incubated at 25° C.

DEVELOPMENT Larval development, 9 days; pupation, 60%; survival to adult, 50%.

Geer, B.W.; Vovis, G.F. 1965. The effect of choline and related compounds on the growth and development of *Drosophila melanogaster*. J EXP ZOOL 158: 223-236.

COMPOSITION (mg): *l*-arginine hydrochloride, 80; *l*-cystine, 30; *l*-glutamic acid, 840; *l*-glycine, 40; *l*-histidine hydrochloride, 100; *dl*-isoleucine, 300; *l*-leucine, 200; *l*-lysine hydrochloride, 190; *dl*-methionine, 80; *dl*-phenylalanine, 130; *dl*-threonine, 200; *l*-tryptophan, 50; *dl*-valine, 280; *l*-tyrosine, 80; agar, 1500; sucrose, 1000; yeast ribonucleic acid, 100; cholesterol, 30; thiamine hydrochloride, 0.2; nicotinic acid, 1.2; sodium riboflavin-5-phosphate, 1.0; calcium pantothenate, 1.6; pyridoxine hydrochloride, 0.25; biotin, 0.03; folic acid, 1.0; choline chloride, 8.0; $FeSO_4$, 1.0; $CaCl_2$, 1.29; $MgSO_4.7H_2O$, 24.6; $MnSO_4.4H_2O$, 1.29; $NaHCO_3$, 100; KH_2PO_4, 183; Na_2HPO_4, 189; and water to 100 ml.

PREPARATION RNA dissolved in phosphate buffer solution; KH_2PO_4 and Na_2HPO_4 in amounts equivalent to 0.4% of final volume. All amino acids except cystine dissolved in warm RNA solution. Neutralized with dilute NaOH; B-vitamins, salts and water added. Cholesterol added in suspension prepared by dissolving in warm ethanol, adding water, and removing ethanol by autoclaving for 10 minutes. Cystine added, dissolved in minimal 1 N HCl. Final neutralization with NaOH and agar and sucrose added. Mixture warmed and dispensed in 5 ml amounts to 6-dram shell vials. Vials sterilized by autoclaving at 15 lb pressure for 15 minutes.

REARING Eggs sterilized with successive rinses of sterile distilled water, 0.125% NaClO solution and 8 minutes in peracetic acid plus 0.1% Nacconol. Aseptically transferred to culture diet in room sterilized with UV lamp and microfilter blower system. Cultures kept at 23.8 ± 1.1° C and 50% RH with 10 h photoperiod from standard illumination.

DEVELOPMENT Larval period, 12.3 ± 1.7 days; pupation, 90.6%; pupal wt, 17.6 mg; emergence, 85.3%.

See also: Geer (1966); Geer *et al.* (1970).

Hinton, T.; Dunlap, A. 1958. The genetical basis for nutritional requirements in insects. PROC 10 INT CONGR ENTOMOL MONTREAL (1956) 2: 123-126.

Best of 4 diets tested.

COMPOSITION (mg/ml): *l*-alanine, 1.085; *l*-arginine, 0.559; *l*-aspartic acid, 1.221; *l*-cysteine, 0.484; *l*-glutamic acid, 4.418; *l*-glycine, 1.745; *l*-histidine, 0.484; *l*-hydroxyproline, 0.384; *l*-isoleucine, 1.260; *l*-leucine, 2.345; *l*-lysine, 1.337; *dl*-methionine, 0.339;

l-phenylalanine, 1.008; *l*-proline, 1.682; *l*-serine, 1.260; *dl*-threonine, 0.756; *l*-tryptophane, 1.745; *l*-tyrosine, 1.240; *l*-valine, 1.355; sucrose, 7.5; cholesterol, 0.4; inosine, 0.25; thymine, 0.04; *p*-aminobenzoic acid, 0.002; inositol, 0.042; B_{12}, 0.000028; $MnSO_4.4H_2O$, 0.0129; $MgSO_4.7H_2O$, 0.246; $FeSO_4$, 0.0129; KH_2PO_4, 0.606; K_2HPO_4, 0.0606; NaCl, 0.0129; $CaCl_2$, 0.0129.

Nucleic acids required by some strains.

PREPARATION See Sang* (1956).

REARING Aseptically at 25°C. Eggs collected and washed in repeated rinses of 70% alcohol, then still in alcohol, placed in sterile wire mesh baskets and transferred to sterile petri dishes containing non-nutrient agar. Neonate larvae transferred aseptically to test tubes of autoclaved diet.

DEVELOPMENT Pupation in average of 9.9 days, 90%; emergence, 93.9%.

See also: Hinton (1952, 1955, 1956a b); Hinton *et al.* (1951).

→ Hoog, E.G. van't 1935b. Aseptic culture of insects in vitamin research. Z VITAMINFORSCH 4: 300-324.

Three diets described.

COMPOSITION Diet 1: ripe banana, 1; boiling water, 100 ml; agar, 1½%; brewers' yeast, sprinkled.

Diet 2 (mg): purified casein, 250; rice-starch, 235; cane sugar, 160; butterfat, 80; McCollum's salt mixture, 15; agar, 75; dried brewers' yeast, 375; and water to 10 ml.

Diet 3 (g): sugar, 5; unsaponifiable fraction of arachis oil, 0.003; agar, 1.8; and (ml): casein-peptone, 10; dialysate from aqueous brewers' yeast-extract, 10; Pearl's salt mixture (double strength), 17.5; and water to 50 ml.

Pearl salt mixture, part A (g): cane sugar, 500; $KNaC_4H_4O_8.4H_2O$, 50; $(NH_4)_2SO_4$, 12; $MgSO_4.7H_2O$, 3; $CaCl_2$, 1.5; and water to 3 1. Part B (g): tartaric acid, 30; agar-agar, 135; KH_2PO_4, 6; and water to 3 1.

PREPARATION Diet 1: ripe banana pounded in mortar, mixed with boiling water containing agar, poured into 1-1 preserving jar rotating around sides of jar. When cooled, brewers' yeast sprinkled on surface, repeated every 4 days.

Diet 2: no details.

Diet 3: casein-peptone prepared by diluting 25 ml concentrated HCl with 75 ml water, then adding 4 ml to 1 1 water with 2 g pepsin and 100 g casein. This left for 12-18 h until solid content at 150 mg/1 ml. pH to 3.5-4.0 with 10% NaOH. Salts added as in S101 of Pearl *et al.* (1926) but in double strength and without agar or cane sugar. These added in equal quantities until pH 3.5. Vitamins extracted from yeast with water and 500 ml filtrate placed in collection bag in 12 1 acidified water. Dialysate concentrated by vacuum-distillation until solid content at 185 mg/ml.

REARING At 23°C. 100 flies/jar; food covered by filter paper. Food renewed after 14 days. Eggs collected on filter papers soaked in diet 1. Eggs disinfected within 15 h using mixture of 44 ml 96% alcohol and 6 ml 5% chloramine for 35 minutes, then washing in sterilized distilled water.

DEVELOPMENT Incubation period, 1 day; larval period, 5-6 days; pupal period, 4-5 days; first eggs laid, 2-3 days from emergence. Life cycle complete in ca. 12 days.

See also: Hoog (1935a, 1936).

→ Kouni, M.H.; Nash, D. 1974. Biology of 5-fluoro-2'-deoxyuridine sensitivity of *Drosophila melanogaster* larvae. J INSECT PHYSIOL 20: 1481-1490.

COMPOSITION Diet 1: modified from Sang's* (1956) chemically defined medium (g): agar, 2.60; vitamin-free casein, 5.50; and (mg): sucrose, 750; cholesterol, 30; lecithin, 400; thiamine, 0.2; riboflavin, 0.1; nicotinic acid, 1.2; calcium pantothenate, 1.6; pyridoxine, 0.25; biotin, 0.016; folic acid, 0.3; $NaHCO_3$ (anhydrous), 140; KH_2PO_4 (anhydrous), 183; K_2HPO_4 (anhydrous), 189; $MgSO_4$ (anhydrous), 62; Streptomycin, 20; Penicillin, 25000 I.U.; and water to 100 ml; and, if added, (mg): thymidine, 121.1; uridine, 122.1; deoxyuridine, 114.1; deoxycytidine, 113.6.

Diet 2 (g): brewers' yeast, 12.5; sucrose, 10.0; agar, 2.0; and (ml): water, 90; propionic acid, 1.0; and Streptomycin, 20 mg; Penicillin, 25000 I.U.

Diet 3, egg laying medium (ml): water, 100; propionic acid, 1; and agar, 1.5 g; Streptomycin, 20 mg; Penicillin, 25000 I.U.

PREPARATION Penicillin and propionic acid added after autoclaving. Details of preparation not given.

REARING At 25°C. Neonate larvae to 3 x 1" shell vials with diet; or reared in petri dishes. Later generations maintained using diet 2 without antibiotics or propionic acid. Adults fed on fresh bottles of diet 2 for 48 h before egg laying.

→ Lafon, M. 1939. Récherches sur quelques aspects du besoin qualitatif d'azote. 2e Mémoire Essai sur le besoin qualitatif d'azote chez un insecte: *Drosophila melanogaster* Meig. ANN PHYSIOL PHYSICOCHIM BIOL 15: 215-260. (In French)

COMPOSITION Basal diet (g): agar gelose, 1.0; Nipagin sodium, 0.5; egg lecithin, 0.5; peptone from beef muscle, 2.0; and (ml): saline solution, 10.0; hydroalcoholic extract of yeast (0.5 g dry wt), 10.0; distilled water to 100.0.

REARING At 26°C.

DEVELOPMENT Adults normal. Larval period, 7 days; pupal period, 6 days.

→ Sang, J.H. 1956. The quantitative nutritional requirements of *Drosophila melanogaster*. J EXP BIOL 33: 45-72.

Three diets tested.

COMPOSITION Diet 1 (g): agar, 3.00; casein, 4.00; fructose, 1.00; cholesterol, 0.02; yeast nucleic acid, 0.25; thiamine, 0.0002; riboflavin, 0.0003; nicotinic acid, 0.0005; calcium pantothenate, 0.0004; pyridoxine, 0.0001; biotin, 0.00004; folic acid, 0.0001; choline, 0.006;

thymine, 0.002; $NaHCO_3$, 0.126; $FeSO_4.7H_2O$, 0.002; $CaCl_2$, 0.002; $MgSO_4$, 0.002; $MnSO_4.4H_2O$, 0.020; NaCl, 0.002; KH_2PO_4, 0.183; Na_2HPO_4, 0.189; water to 100 ml.

Diet 2 (g): agar, 3.00; casein, 5.00; fructose, 0.75; cholesterol, 0.015; lecithin, 0.20; yeast nucleic acid, 0.40; thiamine, 0.00001; riboflavin, 0.0002; nicotinic acid, 0.0001; calcium pantothenate, 0.0003; pyridoxine, 0.00004; biotin, 0.000006; folic acid, 0.00006; choline, 0.004; remaining as diet 1.

Diet 3 (g): agar, 3.00; casein, 5.50; fructose, 0.75; cholesterol, 0.03; lecithin, 0.40; yeast nucleic acid, 0.40; thiamine, 0.0002; riboflavin, 0.0010; nicotinic acid, 0.0012; calcium pantothenate, 0.0016; pyridoxine, 0.00025; biotin, 0.000016; folic acid, 0.0003; $NaHCO_3$, 0.140; KH_2PO_4, 0.183; Na_2HPO_4, 0.189; water to 100 ml.

PREPARATION 5 ml of diet autoclaved for 15 minutes at 20 lb pressure, and cooled with shaker to disperse casein evenly in 1" boiling tubes. Cultures not sloped.

REARING Egg sterilization on 2% agar in watch glasses (see Begg & Sang, 1950). Eggs transferred aseptically, using sterile paper spoons to agar plates. Incubated at 25°C for 20-22 h, then transferred @ 40-50 neonate larvae/culture tube.

DEVELOPMENT Pupal period, 4.3 days; adult emergence, 10-12 days from eclosion.

Corollary

→ Royes, W.V.; Robertson, F.W. 1964. The nutritional requirements and growth relations of different species of *Drosophila*. J EXP BIOL 156: 105-135.

COMPOSITION See Sang* (1956): Diet C, abstracted as diet 3.

DEVELOPMENT Species reared:- *Drosophila subobscura* Collin; *D. ambigua* Pomini; *D. obscura* Fallén; *D. funebris* Fabricius; *D. immigrans* Sturtevant.

See also: Bryant & Sang (1969); Burnet & Sang (1963, 1968); Cooke & Sang (1970); Falk & Nash (1972); Sang (1957, 1959, 1962a b, 1964, 1972b); Sang & Burnet (1963); Sang & King (1961); Van Herrewege (1973a b).

→ Singh, P.; Taylor, B.; Rodriguez, J.G. 1968. Bioassay of insecticides for control of *Drosophila melanogaster*. J ECON ENTOMOL 61: 336-337.

COMPOSITION (g): active dry yeast, 12; agar, 6; cornmeal, 60; and (ml): malt, 90; propionic acid, 1; water, 600.

PREPARATION Agar dissolved in boiling water. Cornmeal added slowly until mixture thickened, then yeast and malt added and well mixed. Acid mix added and diet poured into petri dishes or bottles.

REARING At 76-78°F and 50-60% RH. In plastic cage, from 30.5 x 25.5 x 10 cm tray with lid, and 6 equally spaced holes of 5.5 cm diam on bottom to fit plastic petri dishes, 7 cm diam, containing diet. Tops of petri dishes with holes to match cemented onto bottom of cages. Lid with 2 cm diam hole plugged with cotton for ventilation. Tray mounted on 4 plastic specimen tubes 5 cm high. Cage draped with black plastic sheeting during oviposition to keep flies close to diet. Three of 6 petri dishes replaced with new dishes daily and kept in quart plastic jar until emergence. In this way continuous supply of 1-day-old flies obtained. Cage illustrated.

→ Villee, C.A.; Bissell, H.B. 1948. Nucleic acids as growth factors in *Drosophila*. J BIOL CHEM 172: 59-66.

COMPOSITION Solution A (g): agar, 7.5; casein hydrolysate, 10.0; *l*-cystine, 0.2; *l*-tryptophan, 0.1; cholesterol, 0.05; lecithin, 0.05; ribonucleic acid, 0.5; sucrose, 25.0; KH_2PO_4, 0.303; K_2HPO_4, 0.303; $CaCl_2$, 0.00645; $FeSO_4.7H_2O$, 0.00645; $MgSO_4.7H_2O$, 0.123; $MnSO_4.4H_2O$, 0.00645; NaCl, 0.00645; water, 437 ml.

Solution B (mg): choline, 10.0; *p*-aminobenzoic acid, 1.0; inositol, 21.0; niacinamide, 5.0; calcium pantothenate, 3.0; folic acid, 3.0; pyridoxine hydrochloride, 1.5; riboflavin, 1.2; thiamine hydrochloride, 0.75; biotin, 75 µg and water, 63 ml.

PREPARATION Solution A made up in 437 ml water and autoclaved at 15 lb for 20 minutes. Solution B made up in 63 ml water and filtered through medium sintered glass filter. 0.5 ml B and 3.5 ml A transferred aseptically to each tube, mixed thoroughly and slanted while cooling.

REARING At 25° C. Eggs collected from ½-pint culture bottles, washed in insect-Ringers solution, sterilized in 70% ethanol and transferred aseptically to tubes of diet @ 10-20 eggs/tube.

DEVELOPMENT Egg to adult, 14-17 days.

Drosophila obscura Fallén

→ Gershenson, S. 1928. A new sex-ratio abnormality in *Drosophila obscura*. GENETICS 13: 488-507.

COMPOSITION (g): fermented raisins, 100; potatoes, 400; agar, 4; and water, 800 ml.

REARING At 21-25° C, in sterilized flat bottomed glass cylinders. Yeast sprinkled on diet. Cylinders plugged with thick cotton wool stoppers.

DEVELOPMENT Total development, 20 days; emergence, 50.4%.

Drosophila pachea Patterson & Wheeler

→ Goodnight, K.C.; Kircher, H.W. 1971. Metabolism of lathosterol by *Drosophila pachea*. LIPIDS 6: 166-169.

COMPOSITION (g): soy protein, enzymatically hydrolyzed, 40; brewers' yeast, 23.5; flake agar, 10; fructose, 10; carboxymethyl cellulose (Hercules 740 F), 8; vitamin mixture, 2; lathosterol, 0.25; and (ml): distilled water, 1000.

vitamins (mg): *i*-inositol, 225; choline chloride, 210; nicotinic acid, 23.8; thiamine hydrochloride, 8.4; calcium pantothenate, 6.82; riboflavin, 3.25; pyridoxine hydrochloride, 1.71; folic acid, 1.1; biotin, 0.11; mixed with 1.52 g extracted yeast.

REARING Axenic culture obtained by modification of method of Doll *et al*. (1963) using peracetic acid for egg sterilization.

See also: Heed & Kircher (1965).

Miscellaneous - Drosophilids

Komai, T. 1927. The culture medium for *Drosophila*. SCIENCE 65: 42-43.

COMPOSITION (g): peeled banana, 100; Kôji, 10.0; agar-agar, 10; 800 ml water; and powdered magic yeast added to mixture.

PREPARATION Kôji (culture of *Aspergillus oryzae* on rice) soaked overnight before use.

REARING In ½-pint milk bottles.

DEVELOPMENT Species reared: *Drosophila melanogaster* Meigen; *D. immigrans* Sturtevant; *D. virilis* Sturtevant.

Wheeler, M.R. 1972. Fruitflies. pp. 543-548. *in* Universities Federation for Animal Welfare (Ed.), The UFAW Handbook on the Care and Management of Laboratory Animals. Fourth edition. Churchill Livingstone. Edinburgh and London.

Three diets listed.

COMPOSITION Diet 1 (g): agar, 15; brewers' yeast, 30; cornmeal (maize meal), 100; (ml): water, 1000; dark corn syrup ('Karo'), 100; propionic acid, 5; 95% ethanol, 5.

Diet 2 (g): agar, 20; banana (ripe, crushed), 300; brewers' yeast, 50; sucrose, 25; and (ml): water, 1000; propionic acid, 4; 95% ethanol, 4.

Diet 3 (g): agar (fine grain), 13.5; yeast (dried brewers'), 50; Kellogg's 'Special K' cereal, 10; Kellogg's 'concentrate' cereal, 5; Gerber's 'High Protein' cereal, 15; Kretschmer's 'Wheat Germ' cereal, 15; and (ml): water, 1000; ethanol, 5; propionic acid, 5; and 1 jar (4¾-oz) banana baby food.

PREPARATION Diet 1: agar dissolved in 750 ml boiling water. Ethanol added to yeast; syrup, cornmeal and remaining water added to boiling agar solution. Boiled 10-15 minutes. Cooled to 43°C and propionic acid added. Poured into containers, and covered with sterile cloth while cooling. When cooled sprayed with live yeast suspension, plugged with cotton and stored at cool room temperature before use.

Diet 2: agar dissolved in boiling water. Banana added and cooked for few minutes. Yeast moistened with ethanol; sugar added and diet cooked till homogenous. Cooled to 43°C and propionic acid added. Then as diet 1.

Diet 3: dry ingredients blended in Waring Blendor for 3-5 minutes. Agar dissolved in boiling water. Yeast moistened with ethanol, added to dry ingredients, and cooked in water/agar solution for 8 minutes in jars. Cooled and plugged with cotton.

REARING Most species reared on diet 3. Stock changed every 7-10 days. Rearing methods depend on pupation behaviour.

DEVELOPMENT Species reared on diet 1: *Drosophila melanogaster* Meigen; *D. simulans* Sturtevant; *D. obscura* Fallén.

Species reared on diet 2: *D. immigrans* Sturtevant; *D. funebris* Fabricius; *D. virilis* Sturtevant.

Species reared on diet 3: all above species.

See also: Bacot & Harden (1922); Baumberger (1917, 1919); Bridges (1932); Bridges & Darby (1933); David (1959, 1962); David *et al.* (1970); Demal (1961); Demerec & Kaufman (1962); Friedman *et al.* (1974); Hollingsworth & Burcombe (1970); Kerr (1954); King (1965); Lewis (1960); Loeb (1915a b); Loeb & Northrop (1916); Nagasawa (1972); Northrop (1926); Pearl (1926); Pearl *et al.* (1926); Schader (1941); Schultz *et al.* (1946); Scudo & Peschiera (1971); Shankland *et al.* (1968); Spencer (1950); Strickberger (1962); Sturtevant (1937); Tatum (1939); Travaglini & Tartof (1972); Umbachi & Tsuchitani (1955); Wilson & Birch (1944); Winchester (1933).

Muscidae

Atherigona varia soccata Rondani, sorghum shootfly

→ Dang, K.; Doharey, K.L.; Srivastava, B.G.; Jotwani, M.G. 1971. Artificial diet for the mass rearing of *Atherigona varia soccata* Rondani. ENTOMOL NEWSL 1: 65.

COMPOSITION (parts): agar, 2; propionic acid, 0.3; yeast, 5; water, 93.

DEVELOPMENT Larval and pupal periods on diet, 9.7 and 7.2 days respectively. Pupation, 45.6%; adult emergence from pupae, 90%. Further improvements needed to diet.

Fannia canicularis Linnaeus, little house fly

→ Lewallen, L.L. 1954. Biological and toxicological studies of the little house fly. J ECON ENTOMOL 47: 1137-1141.

COMPOSITION (g): dried brewers' yeast, 30; finely ground laboratory chow, 112; milled red wheat bran, 225; and (ml): concentrated dimalt, 12.5; water, 598.

PREPARATION Ingredients blended.

REARING At 50° F and 65% RH in 12" square screened cage with plywood base and front circular opening fitted with muslin sleeve. Cotton pad saturated in 20% sugar solution provided as adult food. Eggs laid on 2-day-old pads impregnated with 2.25 g protein hydrolysate and 50 ml water. Pads with 200-300 eggs removed from petri dish in cage and placed on diet in 1-gal jar covered with muslin. Pupation at top of diet. Emerging adults transferred to breeding cages.

DEVELOPMENT Eclosion in 36-48 h; larval period, 8-10 days; pupal period, 8-12 days. Adult wt: ♂, 6.89 mg; ♀, 7.35 mg. Complete cycle from egg to egg, 24-29 days.

→ Steve, P.C. 1960. Biology and control of the little house fly *Fannia canicularis* in Massachusetts. J ECON ENTOMOL 53: 999-1004.

COMPOSITION 20% molasses solution; sufficient CSMA dry mix; yeast cake, 1/1000 ml water.

PREPARATION Yeast cake and molasses solution mixed and added to dry CSMA mix to near saturation.

REARING At 80° F and 65% RH. Adults fed molasses and canned evaporated milk solutions both diluted 1:1 with water. Provided in 5-oz paper cups with towelling and replaced twice weekly. Oviposition on towelling in household ammonia diluted with equal amounts of water and dry CSMA medium. Eggs placed on 1½" larval diet in 6" diam battery jars; 100 larvae/jar. Pupae collected on tightly rolled cardboard on diet surface or by washing diet through cone-shaped strainer.

DEVELOPMENT Eclosion in 1½-2 days; larval period, 8-10 days; pupal period, 9-10 days. Egg to adult, 18½-22 days; egg to egg, 22-27 days. 50% adult longevity: ♂, 14 days; ♀, 24 days.

Haematobia irritans Linnaeus, horn fly

Eschle, J.L.; Schmidt, C.D,; Matter, J.J. 1973. Horn fly: Preliminary evaluation of media for rearing larvae. ANN ENTOMOL SOC AM 66: 352-354.

Best of 4 diets tested.

COMPOSITION Dry mix (parts by weight): ground sugar cane bagasse ($\frac{3}{16}$" screen in hammer mill), 41; whole wheat flour, 8; fish meal, 6; baking soda, 1. Dry mix, 1.13 kg; fresh bovine faeces, 1.59 kg; water, 3 l.

PREPARATION Faeces from Hereford steer fed 10-15 lb alfalfa hay daily. Dry mix added to slurry of faeces and water.

REARING At 27°C and 60% RH with constant light. 75 g diet placed in 8-oz polyethylene container; seeded with 50 eggs less than 8 h old. Pupae separated by flotation in water; air dried and placed in plastic cups covered with cheesecloth.

DEVELOPMENT Larval period, 6 days; pupal period, 2 days; survival from egg to pupa, 42.3%; pupation, 74.4%; average pupal wt, 4.19 mg; emergence from pupae, 93.9%.

See also: Kunz & Eschle (1971).

Harris, R.L. 1962. Laboratory colonization of the horn fly, *Haematobia irritans* (L.). NATURE LONDON 196: 191-192.

COMPOSITION (parts): blood, 1; beef extract, 1; satndard antibiotic mixture (as below).

PREPARATION 480 ml blood drawn from steer into vacuum transfusion bottle containing 120 ml acid citrate dextrose anticoagulant solution of 1.32 g sodium citrate U.S.P., 0.44 g citric acid U.S.P. and 1.47 g dextrose U.S.P./100 g. Blood mixed immediately with other ingredients and unused portion stored in refrigerator. Antibiotic mixture of 1 mg Streptomycin, 1000 units Penicillin, 250 units Mycostatin/ml diet. Beef extract obtained by centrifuging fresh ground beef muscle tissue for 15 minutes at 3000 rpm.

REARING At 90°F with continuous lighting. 10 pairs adults placed in 3 x 1½" diam cylindrical plastic cages, covered top and bottom with light-weight unbleached muslin, and fitted with sliding sheet-metal divider to trap flies in one part of cage while other cleaned. Flies fed through muslin cover from absorbent cotton pads soaked in diet mixture. Pads covered with plastic dish to retard evaporation, and replaced twice daily; cage cleared daily and new muslin cloth daily. Eggs collected daily, placed on moist paper towelling in vial and held at 80°F for hatching. Cages for rearing colony 6" diam with plastic screened bases placed over cow manure for oviposition.

DEVELOPMENT Preoviposition period, 4 days; hatch, 40%. Ten generations reared.

→ Harris, R.L.; Frazer, E.D.; Grossman, P.D. 1967. Artificial media for rearing larvae of hornflies. J ECON ENTOMOL 60: 891-892.

Best of several diets tested.

COMPOSITION (g): ground sugar cane pulp, 264; whole wheat flour, 48; dehydrated bovine plasma, 12; $NaHCO_3$, 6; and 1100 ml of water.

PREPARATION Sugar cane pulp originally prepared for chicken litter ground in hammer mill equipped with pulverizing screen to reduce particle size. Wheat flour prepared for baking. Ingredients mixed together.

REARING In 75 x 115 mm diam cages. Adults fed daily.

DEVELOPMENT Pupal weights lower but more pupae produced on artificial diet than on bovine manure. Percent hatch of eggs from adults reared on diet slightly lower, but equal numbers of eggs laid in both cases. Ten generations reared.

See also: Depner (1962); Wright & Oehler (1971):- except bovine plasma replaced by fish meal.

Musca autumnalis DeGeer, face fly

→ Fales, J.H.; Bodenstein, O.F.; Keller, J.C. 1961. Face fly laboratory rearing. First successful rearing in the laboratory of *Musca autumnalis*, a serious pest of livestock, is major step forward in developing controls. SOAP CHEM SPEC 37: 81-83.

COMPOSITION Adult diet, supplied separately, (g): fresh cow manure, 175; ground raw stew beef, 2; skimmed milk and granulation sugar solution, 2:1 v/v ratio; Diamalt mixture, 0.35 g.

Diamalt: to 50 ml Diamalt add 2.5 ml 5% w/v water solution of dried mucous membrane of animal intestines, 2.5 ml of 7.5% w/v water solution of brain-heart infusion, 2.5 ml filtrate of water solution of equal parts mustard and hedge pollen, made by soaking 18.5 g of each pollen in total volume of warm water equal to 200 ml. After 2 h solution filtered through 2 layers cheesecloth.

REARING At 80°F and 50-55% RH with 24 h photoperiod in 10" cube cages. Food provided weekly. Oviposition in manure in sponge cups. Dishes of manure removed daily and placed in another cage for 3 days for larval development; then contents transferred to 15 mm crystallizing dish partly covered with fine sand. Dishes covered with cheesecloth. In 4-5 days, pupae removed and placed in shell vials stoppered with cotton plugs. Vials placed in cages for emergence.

DEVELOPMENT Eclosion, ca. 12 h; larval period, 8-9 days; pupal period, 8 days. Eight generations reared.

→ Turner, E.C. Jr; Hair, J.A. 1967. Effect of diet on longevity and fecundity of laboratory-reared face flies. J ECON ENTOMOL 60: 857-860.

COMPOSITION Adult diet (% approximate): egg solids, 3.7; powdered milk, 22; sugar, 22; and equal weight of water. Various combinations of sugar and proteinaceous materials tried, e.g. powdered milk, egg solids, whole beef blood, yeast hydrolysate, peptone and blood albumin. Most successful protein diets with 5 or 10% blood, and 5% milk added to 5% sugar solution.

PREPARATION Diets calculated in g/ml distilled water placed in 1-oz disposable paper cups.

REARING Newly emerged flies anaesthetized with CO_2 and 25-60 placed in clean 9" cube cages. Diet changed daily. On 6th day Fischer aluminium weighing dishes containing cattle faeces placed in cages for oviposition. Dishes supplied every other day for 21-30 days; removed after 1 day and incubated at 32° C.

DEVELOPMENT Average longevity on 5% blood and 5% milk in 5% sugar solution, 6.7 days; and average number of larvae/♀, 26.0. On 10% blood and 5% milk in 5% sugar solution, average longevity, 8.9 days; and average of 23.1 larvae/♀ produced.

→ Valder, S.M.; Hopkins, T.L. 1968. Effects of nutrition on egg development and survival in the face fly, *Musca autumnalis*. ANN ENTOMOL SOC AM 61: 827-834.

COMPOSITION Adult diet (parts): sucrose, 47; protein source, 47; sodium oleate, 2; Wesson's salt mixture, 4; ribonucleic acid, 0.1.

vitamins: see Monore* (1960).

amino acid mixture substituted for protein (%): *l*-arginine, 4.42; *l*-histidine, 2.56; *l*-lysine hydrochloride, 6.54; *l*-tyrosine, 4.42; *dl*-tryptophan, 1.06; *dl*-phenylalanine, 5.66; *l*-cystine, 1.41; *dl*-methionine, 3.89; *dl*-threonine, 3.98; *dl*-serine, 6.46; *l*-leucine, 8.49; *dl*-isoleucine, 5.84; *dl*-valine, 7.07; *l*-glutamic acid, 17.25; *l*-aspartic acid, 7.25; *dl*-glycine, 2.56; *dl*-alanine, 4.33; *l*-proline, 6.72.

PREPARATION Ingredients weighed and ball-milled for 4 h. One ml of B-vitamin mixture added to 10 g diet, then mixed for 10 minutes with kitchen electric mixer. Diet then air-dried with fan and processed through 0.012" hammer-mill screen. 0.1% solution of cholesterol in dichloromethane then added @ 1 ml/g diet; excess solvent added to ensure thorough wetting and even mixing of cholesterol with dry diet. Solvent evaporated under nitrogen with occasional stirring, and diet stored at -20° C until needed. Alternatively, all ingredients but protein formulated as diet base, and protein added later as needed. Gives advantage of large-scale preparation of protein-free diet base from which further diets rapidly formulated.

REARING At 80 ± 4° F and 70 ± 5% RH with 16 h photoperiod. Adults maintained in 12 x 10 x 8" cages, top and sides wire screen stapled to wooden floor and end boards. Entrance by 18" nylon fabric sleeve stapled to screen and floor and closed with rubber band. New colonies started with 800-2000 pupae. Adults given dry food and water in small waxed cups with styrofoam float in cup to minimize drowning. 8-10 days after emergence fresh bovine manure in 8 x 8 x 2" aluminium pans placed in cage for oviposition. After eggs laid pans removed to larger containers and placed on layer of sand for pupation. Pupae sifted from sand after 2 days and placed in cages.

DEVELOPMENT 77% survival after 10 days on diet with both *l*- and *dl*-amino acids. Development on amino acid diet less than with sodium caseinate and egg albumin but equal to or better than most of other proteins tested. Egg development and adult survival best when fed diets containing undefined components such as egg solids; dried blood and sucrose (1:1); dried milk, dried blood and sucrose (1:1:4); and laboratory diet (dried soluble blood, milk, and sucrose used as control). Average adult survival and of egg development (%) respectively, for egg solids, 71, 82; for dried blood sucrose, 88, 76; for dried milk dried blood sucrose, 95, 78; and for undefined laboratory diet, 93, 77. Egg development and adult survival on diets containing different proteins showed best results with gastric mucin, sodium caseinate, and egg albumin. Adult survival and of egg development (%), respectively: gastric mucin, 39, 90; sodium caseinate, 86, 82; egg albumin, 87, 81. Large scale tests demonstrate that egg albumin and sodium caseinate diets about equal to undefined laboratory diet in promoting reproduction and far superior to other protein or amino acid mixtures tested.

→ Wang, C.M, 1964. Laboratory observations on the life history and habits of the face fly, *Musca autumnalis* (Diptera: Muscidae). ANN ENTOMOL SOC AM 57: 563-569.

COMPOSITION (ratio): powdered skim milk, 2; granulated sugar, 1.

REARING At 25-30°C and 50-70% RH in 12" cube screen cages. Diet, cup of clean water and 10 ml citrated bovine blood provided daily. Fresh cow dung in 10-oz cardboard containers for oviposition also given every day. Cartons transferred to flat pan containing sand. Eggs extracted from dung with forceps and transferred to water. Larvae of various ages similarly picked up from manure. Reared in 15½ x 4½ x 2½' wooden and clear plastic cabinet with 3 doors on each side. Heater, air circulator and light source all provided.

DEVELOPMENT Egg to adult, 11½-12 days at 25-30°C; 9-9½ days at 30-35°C; and 7½-8½ days at 35-40°C.

Musca domestica (Linnaeus), housefly

Musca domestica is one of the most commonly reared insects for entomological research. Numerous papers on the housefly have been published, including a bibliography of over 5700 references (West & Peters, 1973). Here only selected works on housefly dietetics, from simple to complex diets, are listed.

→ Ashby, G.J. 1972d. The housefly. pp. 549-553. *in* Universities Federation for Animal Welfare (Ed.) The UFAW Handbook on the Care and Management of Laboratory Animals. Fourth edition Churchill Livingstone. Edinburgh and London.

Three larval diets listed.

COMPOSITION Diet 1 (g): soft wheat bran, 1200; alfalfa meal, 600; bakers' yeast, 30; and (ml): malt extract, 50; water, 2700.

Diet 2 (g): dried yeast powder, 3; dried malt extract, 4; grass meal, 60; toppings, 120; water, 390.

Diet 3 (parts by volume): fish meal, 2; ground rabbit pellets, 2; dried yeast, 2; sugar, 1; bran, 6; and water to dampen.

PREPARATION Dry ingredients mixed thoroughly with water.

REARING At 27.5 ± 1°C and 50-55% RH; various types of cage described. 50 cm^3 space needed/individual. Fresh food and water as adults emerge. Water from cotton wick in lid of water jar. Cotton wad pads soaked in equal parts of milk and water for oviposition. Eggs collected by gentle shaking in water, (0.1 ml contains 700 eggs) and transferring to diet 10 cm deep in jars. Pupation in upper, crusty part of diet. Pupae removed by sieving, rolling, winnowing or flotation in warm water. Mating and oviposition cages described. Wooden, 45 cm cube cages with 6 mesh brass wire gauze sides. Clear glass for upper front panel resting on plywood lower panel with 2, 13 cm diam holes fitted with metal flange and muslin sleeves.

Similar cage from 18-gauge aluminium with sliding glass door in front also described, as is cage of circular-knitted muslin cloth drawn over cylindrical 23 x 15 cm wire frame and knotted at either end for holding 100-150 flies.

DEVELOPMENT Egg to pupa in 8 days. Eggs laid 4-5 days after emergence.

→ Basden, E.B. 1947. Breeding the housefly (*Musca domestica* L.) in the laboratory. I. Introduction and technique. BULL ENTOMOL RES 37: 381-387.

COMPOSITION Larval diet 1 (g): wheat bran, 1474; alfalfa meal, 794; bakers' yeast, 56; and (ml): water, 5250; Diamalt, 25.

Diet 2 (g): coarse wheat bran, 1800; alfalfa meal, 900; compressed yeast, 45; and (ml): water, 4050; malt extract, 75.

Diet 3 (g): middlings, 1814.5; grass meal, 907.25; dry malt extract, 60; dried yeast, 45; and water, 6000 ml.

PREPARATION Middlings and grass meal sterilized in water-jacketted oven, thoroughly mixed and stored dry. Suspension of yeast and dry malt extract prepared in warm water and mixed with middlings and grass meal.

REARING At 27.5 ± 0.5° C and 60-70% RH with 16 h photoperiod. 160 one-day-old adults introduced to 18 x 18 x 19" oviposition cages with zinc gauze sides and glass upper front panel resting on plywood lower half with 2, 5" diam holes with muslin sleeves. Supplied with water fountain, dish of granulated sugar, and pad of cotton wool soaked with milk diluted with equal volume water, or shallow layer of dried skimmed milk powder in open dish. 5½ x 3½" diam jar with freshly mixed larval food to within ½" of top, covered with milk pad and put into cage for oviposition; removed after 24 h, fitted with muslin top, and incubated for 24 h. Contents of oviposition jar mixed with freshly prepared warm food, and transferred to 12, 8 x 6" diam culture jars to depth of only 4". Jars fitted with muslin cover. 7000 larvae/jar thus obtained. Pupae collected in various ways; best by flotation.

DEVELOPMENT Egg incubation period, 12 h; larval period, 5-6 days; total developmental period, 17 days.

→ Bridges, R.G.; Ricketts, J.; Cox, J.T. 1965. The replacement of lipid-bound choline by other bases in the phospholipids of the housefly, *Musca domestica*, J INSECT PHYSIOL 11: 225-236.

COMPOSITION (g): salt mixture, 0.1; casein, 7.5; ribonucleic acid, 0.25; agar, 0.75; methyl *p*-hydroxybenzoate, 0.23; and (ml): vitamin solution, 10; cholesterol suspension, 10 (0.1 g cholesterol).

vitamins (mg/400 ml water): biotin, 1.0; folic acid, 2.5; nicotinic acid, 50; *d*-pantothenic acid (calcium salt), 25.0; pyridoxine hydrochloride, 12.5; riboflavin, 12.5; aneurine hydrochloride, 25.0.

salt mixture: see House & Barlow* (1958).

Solutions of choline chloride, acetyl-β-methylcholine chloride, *dl*-carnitine hydrochloride, N-dimethyl amino ethanol, and N-monomethyl amino-ethanol added as required in amounts equal to molar equivalent of 12.5 mg choline/250 ml flask. Distilled water added to give total volume of water of 37.5 ml.

Amino acid diet as above except that casein replaced by 0.703 g mixture of *l*-amino acids as in House & Barlow* (1958).

PREPARATION Dry constituents mixed in pestle and mortar poured into 250 ml wide-mouthed conical flasks. Cholesterol suspension, vitamin solution, solutions of choline or choline substitutes and water added. Whole stirred with glass rod, flasks plugged with cotton wool and autoclaved at 15 lb/in^2 and 120° C for 15 minutes.

REARING Eggs collected on milk pads and washed in several changes of distilled water; then surface sterilized in 2.5% formalin solution and rinsed in sterile water. 100-150 eggs brushed on to small pieces of sterile black paper and transferred to diet flasks. Flasks kept at 26-27° C. Larvae removed after 6 days for pupation in covered dishes if adult flies to be examined. Otherwise, pupae transferred to muslin-topped jars containing glucose and distilled water. Adults fed on 1% dilutions of test solutions; those of glucose and control made with water saturated with methyl *p*-hydroxybenzoate and glucose.

DEVELOPMENT Best mean larval wt, 16.9 mg from casein diet and acetyl-β-methylcholine solution.

→ Brookes, V.J.; Fraenkel, G. 1958. The nutrition of the larva of the housefly, *Musca domestica* L. PHYSIOL ZOOL 31: 208-223.

Best of 7 diets tested.

COMPOSITION (g/tube): casein, 2.0; cholesterol, 0.02; McCollum's salt mixture No. 185, 0.02; ribosenucleic acid, 0.05; lecithin, 0.07; adenine, 0.0012; guanine, 0.0012; uracil, 0.0012; agar, 0.15; methyl *p*-hydroxybenzoate, 0.046; water to 7.5.

vitamins (μg/tube): biotin, 5.0; choline, 2500.0; folic acid, 12.5; inositol, 1250.0; nicotinic acid, 250.0; pantothenic acid, 125.0; pyridoxine, 62.5; riboflavin, 62.5; and thiamine, 125.0.

PREPARATION Cholesterol, salts, RNA, agar and fungicides weighed in amounts for 50-200 tubes, ground in mortar, and mixed for 4-8 h in ball mill. Then ground with casein. Vitamins prepared in aqueous solution. Adenine and uracil dissolved in water, and guanine in 1 N KOH. Two solutions combined, adding water to bring concentration to 1 mg/ml. Lecithin emulsified in water, vitamin and purine-pyrmidine solutions combined with lecithin, mixed with dry mix and water added to 7.5 ml. Tubes stoppered with cotton wad covered with gauze wrapper, autoclaved for 15 minutes at 15 lb pressure and 120° C, then shaken to suspend insoluble particles. Tubes slanted and cooled.

REARING Adults fed condensed milk diluted to 50% with water, and cane or brown sugar. Milk removed before oviposition. Eggs laid on folded paper towelling immersed in bottles of diluted condensed milk and collected within 3 h. Sterilized 8 h later by removing from towelling and placing in 12 or 15 ml tubes with 10 ml distilled water; egg clumps then broken up and washed 4-6 times in water, and twice with Zepherin chloride. Poured on to strips of construction paper floating in Zepherin chloride in petri dishes for 20-30 minutes. Strips transferred to tubes with sterile forceps.

DEVELOPMENT Larval development comparable on both synthetic and natural diets except that rate of growth ca. 10% slower and maximal average weight 30-40% lower on synthetic diet. Best results when 15-30 larvae/tube. Fertile adults produced.

→ Greenberg, B.; Burkman, A.M. 1963. Effect of B-vitamins and a mixed flora on the longevity of germ-free adult houseflies, *Musca domestica* L. J CELL COMP PHYSIOL 62: 17-22.

COMPOSITION Adult diet (g): sucrose, 5; Wesson's salt mixture, 0.2; and (ml): 5% casein hydrolysate, 80; distilled water to 100 and B-vitamin solution, 20 ml.

B-vitamins (mg): thiamine hydrochloride, 6; riboflavin phosphate, 9; pyridoxine hydrochloride, 8; nicotinamide, 50; calcium *d*-pantothenate, 20; *d*-biotin, 1; cyanocobalamin, 0.2; choline chloride, 1000; folic acid, 5; and distilled water to 1000 ml.

PREPARATION Ingredients, except vitamins, mixed and autoclaved in serum vials @ 25 ml/vial and stored at 4°C. Vitamin solution passed through Seitz filter into 10 ml serum vials and stored at 4°C. 5 ml B-vitamin solution added to 25 ml sterile basal solution.

REARING At 24°C and 45% RH with 16 h photoperiod. Newly emerged adults isolated in 25 x 200 mm cotton-plugged, sterile, pyrex tubes with 1 cm^2 piece of filter paper on to which nutrient dispensed with hypodermic needle. Fed daily.

DEVELOPMENT Longevity of ♀♀, 62.6 days; of ♂♂, 43.7 days. Longevity of adults not affected by lack of vitamins in diet or presence of bacteria in the diet.

See also: Greenberg (1959, 1960a).

→ House, H.L.; Barlow, J.S. 1958. Vitamin requirements of the housefly, *Musca domestica* L. (Diptera: Muscidae). ANN ENTOMOL SOC AM 51: 299-302.

COMPOSITION (mg/ml distilled water): *l*-amino acids: alanine, 1.00; arginine hydrochloride, 1.22; aspartic acid, 1.28; cysteine hydrochloride, 0.26; glutamic acid, 2.10; glycine, 1.56; histidine hydrochloride, 0.48; hydroxyproline, 0.34; isoleucine, 1.04; leucine, 1.54; lysine, 1.16; methionine, 0.60; phenylalanine, 1.34; proline, 1.50; serine, 1.34; threonine, 1.08; tryptophane, 0.28; tyrosine, 0.70; valine, 1.18; total, 20.00.

salts No. 2 U.S.P. XII (mg/ml): NaCl, 0.0287; $MgSO_4.7H_2O$, 0.0904; NaH_2PO_4, 0.0576; K_2HPO_4, 0.1583; $CaH_4(PO_4)_2$, 0.0896; $C_6H_5O_7Fe.5H_2O$, (17.5% Fe), 0.0196; $Ca[CH_3CH(OH)COO]_2$, 0.2157; total, 0.66.

lipids and fatty acids (mg/ml): cholesterol, 1.00; linoleic acid, 0.60; linolenic acid, 0.20; oleic acid, 1.92; palmitic acid, 0.88; stearic acid, 0.40; Tween 80, 2.70; total, 7.70.

vitamins (mg/ml): biotin, 9×10^{-7}; calcium pantothenate, 0.0044; choline chloride, 0.0200; folic acid, 0.0034; inositol, 0.0596; nicotinic acid, 0.0030; *p*-aminobenzoic acid, 0.0090; pyridoxine hydrochloride, 0.0090; riboflavin, 0.0016; thiamine hydrochloride, 0.0012.

miscellaneous (mg/ml): dextrose, 5.0; ribonucleic acid, 0.75; KCH, 2.47; agar, 15.0.

PREPARATION See House & Barlow (1956) diet for *Pseudosarcophaga affinis*. In brief, diet compounded in 2 parts. Part 1 contained dextrose and water soluble vitamins and sterilized via microbiological filter; part 2 made up of amino acid mixture, salt mixture, lipid and fatty acids emulsion, RNA, agar, and water. Autoclaved. Two parts combined aseptically. pH, 6.8-7.2.

REARING At 22.5°C laboratory colony maintained on Richardson (1932) formula; adults fed milk, and females oviposit on sour milk. Egg mass shaken and stirred in soapy water to separate eggs, transferred to small crucibles and rinsed several times in sterile distilled water. Eggs covered with 2.5% solution of formaldehyde for 3-4 minutes, rinsed again in sterile distilled water, and transferred, to large sterile test tube containing strip of moist filter paper; allowed to hatch at room temperature. Single larvae transferred aseptically to $\frac{3}{8}$ x 3" test tubes containing 1 ml diet. After 7 days larvae transferred to containers for pupation and emergence.

DEVELOPMENT For every 100 newly hatched larvae put on diet, 64 reached maturity, 56 pupated, and 27 adults emerged. Puparia measured 5.5 ± 0.4 mm long and 2.4 ± 0.6 mm wide. Several second generation larvae obtained.

→ Monroe, R.E. 1960. Effect of dietary cholesterol on housefly reproduction. ANN ENTOMOL SOC AM 53: 821-824.

COMPOSITION Adult diet (parts): vitamin-free casein, 47.0; sucrose, 47.0; sodium oleate, 2.0; Wesson's salt mixture plus 0.02% $ZnCl_2$, 4.0; yeast ribosenucleic acid, 0.1; cholesterol (final concentration 0.1%); and vitamin mixture.

vitamins (mg/100 g diet): thiamine hydrochloride, 50.0; riboflavin, 25.0; nicotinic acid, 100.0; pantothenic acid, 50.0; pyridoxine hydrochloride, 25.0; choline chloride, 1000.0; inositol, 500.0; folic acid, 5.0; biotin, 1.0.

PREPARATION Casein, sucrose, sodium oleate, salts and RNA ball-milled for 4-5 h. Cholesterol added in dichloromethane and mixture stirred while solvent evaporating. B-vitamins dissolved in distilled water and 2-4 drops NH_4OH. 10 ml of vitamin, and $ZnCl_2$ solution added to 100 g diet and mixture ground in mortar. Diet dried at room temperature and held at -30°C until use.

REARING At 28 ± 2°C and 50% RH. 600 pupae in beakers placed in screen cages for emergence. Diets placed in cages and changed every third day. Distilled water provided in waxed cups with thin piece of styrofoam on surface. Dead flies removed twice daily. Muslin bags containing absorbent cotton saturated with ammonium carbonate solution placed in petri dishes inside cages for oviposition. Eggs collected on 6th day and thereafter on alternate days. Eggs transferred to centrifuge tube with water and thoroughly washed. 200 eggs placed in petri dish for hatching.

DEVELOPMENT Adult survival at 22 days: ♂, 4-7%; ♀, 82-85%; ca. 92% adult emergence from pupae.

→ Monroe, R.E. 1962. A method for rearing house fly larvae aseptically on a synthetic medium. ANN ENTOMOL SOC AM 55: 140.

COMPOSITION Larval diet (parts): casein (micro-pulverized), 70.0; celluflour (Chicago Dietetic Supply House, Inc., Chicago, Ill.), 3.0; sodium oleate, 2.0; ribosenucleic acid, 1.0; Wesson's salt mixture, 4.0; agar, 20.0; cholesterol, 0.2; and vitamins, 0.5 ml/5.0 g of diet. For vitamin mixture see Monroe* (1960).

PREPARATION Components except cholesterol and B-vitamins ball-milled for 4-5 h. Cholesterol and vitamins prepared as in Monroe* (1960). 15 g dry ingredients placed in beaker with 100 ml distilled water.

1.5 ml vitamin solution added and mixture stirred constantly over Bunsen burner to near boiling. Suspension divided equally into 4-oz pyrex jars and autoclaved at 15 lb and 121°C for 20 minutes. Bottles removed, swirled and immersed in iced water to solidify diet.

REARING Eggs washed in distilled water and surface sterilized with 0.1% NaClO solution for 20 minutes. 50-70 eggs transferred to 4-oz jars with sterilized calibrated pipette. Kept at 34°C for 2 days then 26 ± 2°C for 3 days. Larvae and pupae washed out of diet with warm water through No. 20 standard sieve. Larvae held on filter paper for pupation within 24 h.

DEVELOPMENT Weight of final instar larvae, 24-25 mg; adult wt: ♀, 20-22 mg; ♂, 16-17 mg. 3 generations reared.

See also: Bauer & Monroe (1969); Dutky *et al.* (1967); Monroe & Lamb (1968); Monroe & Leyrer (1973); Monroe *et al.* (1967).

→ Morrison, P.E.; Davies, D.M. 1964a. Feeding of dry chemically defined diets, and egg production in the adult house-fly. NATURE LONDON 201: 104-105.

Two adult diets tested.

COMPOSITION Diet 1 (g): *l*-amino acids: arginine, 0.38; histidine, 0.15; isoleucine, 0.25; leucine, 0.75; lysine, 0.75; methionine, 0.15; phenylalanine, 1.20; threonine, 0.15; tryptophan, 0.30; valine, 0.50; glutamic acid, 1.00; and Wesson's salt mixture, 0.15; sucrose, 3.42.

Diet 2: as above, with 0.4 g ribonucleic acid and vitamins (mg): folic acid, 0.34; niacin, 0.30; calcium pantothenate, 0.44; pyridoxal, 0.90; riboflavin, 0.16; thiamine, 0.12; biotin, 0.05; inositol, 5.95; choline chloride, 2.00.

PREPARATION Both diets finely powdered and stored at 5°C. Sucrose supplied either simultaneously as 0.1 M solution on absorbent cotton wick or as dry diet to encourage aggregation and feeding.

REARING 12 ♀♀ and 3 ♂♂ separated after pupation into clean, 180 x 150 nm diam glass cylinders covered with nylon screening; placed on aluminium trays covered with filter paper. Test diets and sugar-water renewed daily. Fresh milk as control renewed daily.

DEVELOPMENT Both diets acceptable. Fecundity low, but no significant difference in ovarian development. Mean eggs/♀ on diet 1, 20; on diet 2, 16; on fresh milk control, 101.

See also: Bodnaryk (1968); Davies & Anderson (1964); Davies *et al.* (1964); Morrison & Davies (1964b).

→ Perry, A.S.; Miller, G.; Buckner, A.J. 1965. The essential role of folic acid and the effect of antimetabolites on growth and metamorphosis of housefly larvae *Musca domestica* L. J INSECT PHYSIOL 11: 1277-1287.

COMPOSITION (g): vitamin-free casein, 75; agar, 15; Wesson's salt mixture, 4; cellulose powder, 3; sodium oleate, 2; cholesterol, 0.2; RNA, 1; and (mg): thiamine, 50; riboflavin, 25; nicotinic acid, 100; pantothenic acid, 50; pyridoxine hydrochloride, 25; choline chloride, 1000; inositol, 500; folic acid, 5; biotin, 1.

PREPARATION Casein, celluflour, salt and sodium oleate thoroughly mixed. Cholesterol dissolved in CH_2Cl_2 and ground into mixture. Vitamins, except folic acid, dissolved in 10 ml of distilled water and solution made to pH 6.2 with NH_4OH. Folic acid dissolved in mixture of 1 ml of 5% Na_2CO_3, 1 ml of ethanol, and 8 ml of distilled water and added separately. RNA dissolved in 0.1 N NaOH and added. Solids, vitamin-concentration, and 1 ml of 10% solution of Tween 80 added to 750 ml of distilled water and mixture homogenized for 2 minutes; boiled, poured into half-pint milk bottles (50 g/bottle) and autoclaved at 15 lb and 120°C for 15 minutes. Bottles removed, shaken to suspend solids, and then solidified.

REARING Eggs surface sterilized with 10% neutral formalin with 0.05% Tween 80 for 20 minutes and thoroughly rinsed with distilled water. Diet in bottles chopped up with sterile glass rod and 200 ± 20 eggs introduced/bottle. Bottles kept at 34 ± 1°C for 3 days and then at 26 ± 2°C in darkness for 1 day. Larvae removed to sand for pupation. Adult emergence in cages.

DEVELOPMENT Dietary folic acid essential for normal growth and metamorphosis of larvae reared under aseptic conditions.

→ Singh, P.; Jerram, E.M. 1976. Rearing housefly larvae in polythene bags. N Z J ZOOL 3: 57-58.

COMPOSITION Larval diet (g): whole milk powder, 100; dried brewers' yeast, 25; agar, 30; methyl *p*-hydroxybenzoate, 0.3; streptomycin sulphate, 0.1; water to 1 l.

Adult diet (g): sugar, 900; skim milk powder, 900; autolyzed yeast powder, 200; egg yolk powder, 150; cholesterol, 10.

PREPARATION Agar dissolved in 500 ml water and heated to 80-90°C. Other dry ingredients mixed in second beaker with remaining water. Two solutions mixed and heated to 70°C; poured into 5 plastic bags and left to set at room temperature. Stored at 4°C overnight.

REARING At 25°C and 50% RH. 800 eggs seeded with pipette into each bag. Bags kept open with organdie covering. Perlite spread to depth of 50-75 mm on top of diet after 4 days; after 8 days bag inverted on to sieve and rinsed under water, leaving pupae on sieve. Pupae remaining in bag removed by flotation. Pupae rinsed and dried on paper towelling. 400 pupae removed to 30 x 15 cm metal framed, muslin covered, cages for emergence; adults fed from two plastic pottles of 2.5% sugar solution and 1 cup of adult food/cage. Food and sugar renewed every 3 days. Oviposition cup placed in cage every 3 days; 60 ml plastic pottle with 7 dental rolls soaked in:- yeast, 5 g; milk powder, 5 g; ammonium carbonate solution, 5 ml and water, 320 ml. Solution sufficient for 84 rolls. Pottle left in cage for 4 h, stored overnight at 18°C in tray of water. Eggs washed from rolls and rinsed in tap water before seeding.

DEVELOPMENT Larval period, 6-7 days; pupal wt, 20.6 ± 0.3 mg. Total life cycle, egg to adult, ca. 16 days; several consecutive generations reared and colony strong after 2 years.

→ Singh, P. 1976b. Improved chemically defined diet for rearing housefly larvae. UNPUBLISHED.

Diet modified from House & Barlow* (1958) for *M. domestica*.

COMPOSITION *l*-amino acids (mg/100 g): alanine, 100; arginine, 122; aspartic acid, 128; cysteine hydrochloride, 26; glutamic acid, 210; glycine, 156; histidine, 48; hydroxyproline, 34; isoleucine, 104; leucine, 154; lysine, 116; methionine, 60; phenylalanine, 134; proline, 150; serine, 134; threonine, 108; tryptophan, 28; tyrosine, 70; valine, 118.

cholesterol/lipids (mg/100 g): cholesterol, 50; stearic acid, 20; palmitic acid, 44; (ml/100 g): oleic acid, 0.1125; linoleic acid, 0.0330; linolenic acid, 0.0110; Tween 80, 0.1250.

salts (mg/100 g): K_2HPO_4, 48; Na_2HPO_4, 17.5; $MgSO_4$, 27.4; NaCl, 8.7; $CaHPO_4$, 27.1; $C_6H_5O_7Fe.5H_2O$, 5.9; $Ca[CH_3CH(OH)COO]_2$, 65.4.

sugar/vitamins (mg/100 g): folic acid, 0.51; inositol, 8.94; *p*-amino benzoic acid, 1.35; pyridoxine hydrochloride, 1.35; riboflavin, 0.24; thiamine hydrochloride, 0.18; nicotinic acid, 0.45; calcium pantothenate, 0.60; choline chloride, 3.00; biotin, 0.0038; B_{12}, 0.00015; glucose, 750.00.

others (mg/100 g): agar, 1500; cellulose powder, 500; ribonucleic acid, 400. 4 M KOH, 24 drops; and distilled water to 100 g.

PREPARATION Dietary ingredients prepared in bulk. Amino acids ground in mortar and refrigerated in air-tight bottles. Fatty acids and Tween 80 added to hot water; cholesterol (dissolved in 95% ethanol) added to hot fatty acid-Tween mixture; cooled, emulsified, volume adjusted with water, and stored till use. Salt mixture prepared by grinding salts, dissolving in water, oven drying and regrinding. Sugars and vitamins dissolved in water and stored until use.

Dry dietary mixture prepared by grinding in ball mill (g): amino acids, 20; salts, 2; RNA, 4; agar, 15; cellulose powder, 5.

To prepare 100 g diet: 4.60 g dry dietary mixture dissolved in 20 ml cholesterol/fatty acid solution. 59.0 ml distilled water added, and pH adjusted to 6.5 with 4 M KOH. Stirred well, autoclaved at 103.35 kPa for 20 minutes and cooled to 60°C. 16 ml of sugar and vitamin solution, previously sterilized through microbiological filter, added and mixed well. Diet dispensed aseptically @ 1 g/test tube; slanted and left for 24 h before inoculation.

REARING Aseptically, at 26 ± 1°C in dark, individually in 10 x 12 mm test tubes, or in flasks. Eggs washed several times in distilled water and then sterilized in 5% (v/v) formalin for 5 minutes, again washed in sterile distilled water, then treated with White's solution for 1 h with periodic shaking. Finally washed 5 times with sterile distilled water and transferred to sterile filter paper in petri dish. Neonate larvae transferred individually with sterile brush to test tubes. Or, 200 eggs dispensed to Erlenmeyer flasks for rearing.

DEVELOPMENT Larval period, ca. 8 days; pupal period, 5 days; pupation, 82%; average pupal wt, ca. 11 mg; adult emergence, 74.0%; 2 generations reared.

→ Spiller, D. 1966. House flies. pp. 203-225. *in* C.N. Smith (Ed.), Insect Colonization and Mass Production. Academic Press, New York and London.

COMPOSITION Larval diet: equal parts of whole cream dried milk and flocked paper; dried yeast, 5% by weight; water, 2.1 ml/g mix.

PREPARATION Ingredients mixed. 63 g diet and 135 ml tap water placed in each rearing jar.

REARING At 25°C. Adults in cylindrical wire cages; ends covered with cake tin lids, suspended from roof. 20 g cleaned dry pupae placed in 6-oz polyethylene bottle. Emerging adults fed on 2.5% sugar solution and 1:1 dry mix of whole cream dried milk; cane sugar supplemented with 10% autolyzed yeast and 0.1% cholesterol. Eggs collected on muslin dipped into solution of thick milk and saturated ammonium carbonate. Left overnight, then muslin shaken in water to free eggs; rinsed and 5 ml volume placed in each larval rearing jar. Pupae separated by flotation after 8-9 days. Washed, excess water blotted off and pupae dried under air blower.

DEVELOPMENT Larval period, 5-6 days. Adult longevity: ♂, 17.4 days; ♀, 29.4 days.

See also: Spiller (1963, 1964a b).

→ Wilkes, A.; Bucher, G.E.; Cameron, J.W.M.; West, A.S. Jr 1948. Studies on the housefly (*Musca domestica* L.). 1. The biology and large scale production of laboratory population. CAN J RES D26: 8-25.

COMPOSITION Modification of Richardson (1932) formula: soft wheat bran, (approx. 15000 g); alfalfa meal, (approx. 7500 g); malt extract, 50 ml; bakers' yeast cake, 94 g; water, (approx. 13600 ml).

PREPARATION Ingredients mixed.

REARING Diet placed in 9 x 6" diam cylindrical battery jars to within 2" of top. 2500 eggs stirred lightly beneath diet surface and jar covered with damp cloth; dry cloth after 3 days. After 9 days, top inch of diet removed, spread on trays, and placed in front of fan until sufficiently dry for clean separation of pupae by inclining tray in direct path of air stream. Pupae from different jars mixed and kept at 26.7°C for emergence. Adults supplied with mixture of equal parts water and milk in small stender dishes, 2/cage, in which paper towelling placed. Eggs deposited on towelling or moist edges of dish. Eggs removed daily and agitated in water. Adults kept at 27°C and 50% RH in 12" cube cages.

DEVELOPMENT Larval survival, 80.8%; pupal wt, 19.41 mg; adult emergence from puparia, 93.0%; eggs/♀/day, 23.0; several generations reared.

See also: Richardson (1932).

Musca domestica nebulo Fabricius, Indian house fly

→ Deoras, P.J. 1954. Breeding the Indian house-fly (*Musca domestica nebulo* Fabr.) for experimental studies. PARASITOLOGY 44: 304-309.

COMPOSITION Adult diet (g): milk, 5; cane sugar, 10-15; banana, 25; agar, 1; and (ml): water, 250; formalin, 2 drops.

Larval diet (parts): wheat bran, 10; Quaker oats, 1; and (ml): yeast water (4 tablets of 4 grain [sic] (perhaps gram) - each in 200 ml of water), 10; water, 50.

PREPARATION Adult diet: fresh ripe banana crushed in mortar to fine paste and sugar ground in. Milk powder gradually added, then agar, previously soaked overnight in water. Formalin and water added. Cotton wads, dipped in this mixture, and hung over strings in rearing and stock cages.

Larval diet: wheat bran and oats mixed and yeast solution gradually added. Sufficient water added to give consistency of soft dough.

REARING At 30°C and 65% RH. In 41 x 17 x 35.5 cm wooden rearing cages with complete sliding door at front. Fine wire gauze on 3 sides of cage, one with 8 cm diam hole covered by sliding door. String suspended from cage top for fly support. Food provided on cotton wads, changed daily. Petri dish with larval diet for oviposition on bottom of cage. Or, in 27 x 15 x 21.5 cm galvanized sheet iron breeding cages with 16 mesh wire gauze covering. Central strip of sheet metal with hooks inside from which food suspended. 9.5 cm diam hole on one side with tin cover. 0.5 cm layer of loose dry earth on bottom of cage. Or, in 46 x 36 x 27 cm galvanized sheet iron stock cages, with 9 cm diam holes in roof and 2 sides for light bulb and natural lighting. All cage types illustrated.

Eggs laid in crevices of moist food. Larval diet added each day over dried crust of previous day's food for 4 days. Flies emerge and cage discarded after 10th day. Successful emergence from early pupae kept at 4°C for 20 days.

DEVELOPMENT Egg hatch, 6-12 h; larval period, 4-5 days; pupal period, 4 days; adult emergence on ca. 10th day from seeding eggs. Oviposition from 3rd day after emergence.

See also: Panda (1966).

Musca domestica vicina Macquart, common housefly

→ Chang, J.T.; Wang, M.Y. 1958. Nutritional requirements of the common housefly, *Musca domestica vicina* Macq. NATURE LONDON 181: 566.

COMPOSITION (mg/ml): agar, 20; ribonucleic acid, 1.0; cholesterol, 0.1; dextrose, 15.0; U.S.P. XIII, No. 2, mineral mixture, 2.0; casein, 11.0.

vitamins (mg/100 ml): thiamine, 2.0; riboflavin, 2.5; ascorbic acid, 30.0; niacinamide, 20.0; pyridoxine hydrochloride, 0.2; calcium pantothenate, 1.0; and vitamin A, 5000 U.S.P. units; rachitosterol D, 1000 U.S.P. units.

PREPARATION Casein and RNA dissolved separately in distilled water, with addition of 2 N NaOH to casein. Mineral mixture dissolved in hot water. Solutions mixed and agar added; dissolved by heating; whole mixture autoclaved at 15 lb for 2 minutes. Cholesterol dissolved in 95% ethyl alcohol, emulsified with 5% gum arabic solution, and added, with dextrose solution and vitamins, to diet. pH to 6.4 with 2 N NaOH. Diet dispensed @ 2 ml/vial; stoppered with cotton, and autoclaved at 15 lb for 15 minutes.

REARING Eggs collected from oviposition dishes and washed and sterilized with Vanderzant & Reiser (1956a) solution modified by omitting sodium glycocholate and reducing mercuric chloride to 0.15 g. Eggs submerged twice in solution, each time for 10 minutes, with 1 h interval. Rinsed in sterile water and transferred singly to cotton stoppered vials with sterile diet.

DEVELOPMENT First instar, 1 day; 2nd, 1 day; 3rd, 3-4 days; pupal period, 3-5 days; pupal wt, 15-18 mg.

→ Levinson, Z.H. 1960. Food of housefly larvae. NATURE LONDON 188: 427-428.

Two diets tested.

COMPOSITION Diet 1 (%): wheat bran, 40; milk powder, 2.5; and water, 57.5; incubated for varied periods (22, 46, 72, and 96 hours) at 30°C with 4 x 10^4 of *Escherichia coli*.

Diet 2 (g): vitamin-free casein, 2.5; cholesterol, 0.025; ribonucleic acid, 0.025; salt mixture No. 2, 0.016; Nipagin M, 0.07; powdered filter paper, 8.0; water, 17.5.

vitamins (μg/g dry diet): thiamine, 47; riboflavin, 47; pyridoxine, 38; pantothenic acid, 95; nicotinic acid, 238; folic acid, 19; choline chloride, 2377; biotin, 0.4; inositol, 1902; *p*-aminobenzoic acid, 95; and vitamin B_{12}, 0.2.

REARING See Levinson & Bergman (1959). Newly laid eggs from axenic laboratory strain grown on mixture of 40% (w/v) wheat bran and water. Eggs washed, immersed for 30 minutes in 5% formalin and 1% KOH, and rinsed with sterile water. Larvae hatched in sterile water and transferred immediately to experimental diets.

DEVELOPMENT Diet 1: larval growth optimal when bacterial concentration increased to 8 x 10^8/g of diet (about 72 h incubation); rates of pupation and adult emergence then comparable to values obtained on stock culture or natural housefly, e.g.: out of 35 larvae, 31 pupated and 29 adults emerged.

Diet 2: inferior. Out of 35 larvae, 23 pupated and 21 adults emerged.

See also: Arevad (1963, 1974); Ascher & Levinson (1956); Benschoter (1967); Bergman *et al.* (1958); Fisher & Jursic (1958); Frings (1948); Gerberich (1948); Glaser (1924, 1938); Greenberg (1954, 1959, 1960a); Hafez (1948); Hammen (1956, 1957); Harvey & Howell (1965); Hays & Amerson (1966); Henry & Cotty (1957); Ignoffo & Gard (1970); Keiding & Arevad (1964); Levinson & Bergman (1957, 1959); Levinson & Silverman (1954); Moreland & McLeod (1957); Richardson (1932); Rockstein (1957b); Silverman & Levinson (1954); Silverman & Silverman (1953); Sinha (1974).

Musca sorbens Wiedemann

→ Greenberg, B. 1969. Sterile culture of *Musca sorbens*. ANN ENTOMOL SOC AM 62: 450.

COMPOSITION CSMA larval diet (Raulston Purina Co.).

REARING At 30°C and 50% RH. CSMA larval diet in 500 ml Erlenmeyer flasks to depth of 2"; moistened with tap water and heated at 37°C for 3 days. Mixed thoroughly and 2-3" layer of sawdust added. Flasks plugged with cotton and autoclaved at 121°C for 30 minutes. Eggs, laid on moistened sheets of filter paper on top of mixture of monkey and pig faeces, collected, washed in distilled water and sterilized in 1% NaC10 for 10 minutes. Rinsed 3 times and 150 eggs seeded to each flask.

DEVELOPMENT Pupal period, 4-5 days; adult emergence, 8-9 days under axenic conditions. On human and animal faeces under non-sterile conditions, adult emergence, 7 days.

→ Hafez, M.; Attia, M.A. 1958. Rearing and culturing *Musca sorbens* Wied. in the laboratory. BULL ENTOMOL RES 49: 633-635.

Best of several diets tested.

COMPOSITION (g): coarse wheat bran, 120; diluted milk, 240.

PREPARATION Mixed in small jam jar.

REARING At 27-28°C. Eggs laid in 9 cm petri dish on cotton wool soaked in water and milk (1:2 v/v). Transferred to loosely packed diet in jam jar tightly covered with muslin. Dry layer of sand or wheat bran placed over diet for pupation. After 5-6 days pupae separated by flotation, dried and transferred to cage for emergence. Adults supplied with milk pad for oviposition, water fountain and 2 sugar cubes.

DEVELOPMENT Average pupal wt, 11.8 mg. Average duration from egg to adult, 8.4 days. Comparable with flies reared on human excrement.

Muscina stabulans (Fallén), false stable fly

→ Grieve, E.G. 1937. The culture of *Muscina stabulans*. pp. 432-434. *in* Needham, J.G. *et al.* (Eds). Culture Methods for Invertebrate Animals. Comstock Publishing Company, Ithaca, N.Y. (Dover Edition 1959).

COMPOSITION (parts): alfalfa leaf meal, 6; soybean meal, 1; water, 11; dried skim milk, 1.

PREPARATION Water and meals boiled for 10 minutes. Dried skim milk added and mixed.

REARING At 76-80°F with natural lighting. Adults in 18 x 18 x 72" cages supplied with water, sugar, and cake of moist Fleischmann's yeast. Chopped boiled egg added twice weekly. Pan with 3-4 cups of larval food provided after 12 days for oviposition. Removed after 2 days, and kept moist at 72-76°F. Mature larvae removed if required by placing culture in 8 mesh basket and washing in running water; if not, then pans placed in cage for emergence.

DEVELOPMENT Larval period, 8-10 days; pupal period, 7-12 days. Many generations reared.

Stomoxys calcitrans Linnaeus, stable fly

→ Ashrafi, S.H. 1964. The cultivation and nutritional requirements of *Stomoxys calcitrans*. BULL W H O 31: 519-520.

COMPOSITION (g): alfalfa-bran mixture, 375.0; dry yeast, 3.5; cholesterol, 1.35; and (ml): Diamalt, 10.0; water, 500.0.

PREPARATION Ingredients mixed.

REARING At 28 ± 2°C and 50 ± 2% RH with 14 h photoperiod in 4.5-1 plastic jar, 17.5 x 20 cm diam, half filled with diet. 0.5 ml eggs buried 0.6 cm deep, in furrows in diet; ½" layer sand placed on diet after 1 day. Water sprinkled over sand, repeated on 6th day. On 9th day, 75 ml water added, pupae removed by flotation and air dried. Placed in 90 cm cube cage for emergence. Adults fed on fresh blood with 250 ml of 5% sodium citrate/gallon, stored in deep freeze. Thawed for 24 h at 3°C, poured into pie dish and covered with thin layer of cellulose cotton. Blood changed daily. Eggs laid in 20 x 8.7 x 10.6 cm plexiglass box on

7.5 x 10 x 2.5 cm cellulose sponge soaked in water and covered with paper towel. Eggs collected daily by spraying them off with water.

DEVELOPMENT Larval period, 9 days; pupal period, 2-3 days; pre-oviposition period, 5 days.

Bailey, D.L.; Whitfield, T.L.; LaBrecque, G.C. 1975. Laboratory biology and techniques for mass producing of the stable fly, *Stomoxys calcitrans* (L.) (Diptera: Muscidae). J MED ENTOMOL 12: 189-193.

COMPOSITION (parts): wheat bran, 3; bagasse, 1; water, 5.

PREPARATION 2 x 1.25 cm diam bagasse pellets and other ingredients mixed together.

REARING Adults at 24°C and 70% RH with 12 h photoperiod in 46 x 38 x 38 cm aluminium framed cages with solid aluminium floor, and cloth sleeve on front panel. 5000 pupae placed in cages and citrated bovine blood (3 g sodium citrate and 1 ml formalin/1 whole blood) provided on 6 g absorbent cotton with 160 ml blood in 450 ml waxed paper cup. Food renewed daily. Eggs laid on saturated cotton. Special apparatus to remove eggs from blood and cotton using funnel and wire screen. Larvae reared at 21°C and 70% RH with 12 h photoperiod in 51 x 38 x 8 cm trays. 2 ml eggs placed on diet and tray placed in black polyester bag. Held for 14 days. Upper portion of diet removed and pupae separated by flotation.

DEVELOPMENT At 25°C, eclosion in 2 days; larval period, 10 days; pupal period, 7 days; av. pupal wt, 10.6 mg; pre-oviposition period, 5 days; total life cycle egg to egg, 24 days; adult emergence from eggs, 67%; 1 million flies produced/week at cost of $150.

Christmas, P.E. 1970. Laboratory rearing of the biting fly *Stomoxys calcitrans* (Diptera: Muscidae). N Z ENTOMOL 4: 45-49.

COMPOSITION Larval diet (g): ground sugar cane pulp, 264; wheat flour, 100; fish meal, 24; and $NaHCO_3$, 6.

PREPARATION 200 g dry mix mixed with 500 ml water and placed in 10 x 15" glass jar.

REARING 250 adults in 20 x 20 x 15" wood frame cage with 12 mesh, 33 gauge wire gauze panels. 6 x 6" door on top right rear corner. Held at 18-24°C. Adults fed on citrated bovine blood (90 ml of 2% sterilized sodium citrate solution in 480 ml of fresh blood). Blood stored at 4.4°C for 7 days maximum. Fed from 4 x 2 x 0.5" pieces of absorbent cellulose dipped in blood suspended from roof @ 3/cage. Black cloth placed over cage. Oviposition on finely cut, moistened, compressed, fermenting ryegrass placed in 3" diam evaporating dish in cage. Eggs to diet in jar @ 300-500/jar; covered with organdie and incubated at 25°C and 45-55% RH.

DEVELOPMENT Eclosion in 24-72 h; larval period, 10-14 days; pupal period, ca. 6 days; 2% mortality from egg to adult.

Gingrich, R.E. 1960. Development of a synthetic medium for aseptic rearing of larvae of *Stomoxys calcitrans* (L.). J ECON ENTOMOL 53: 408-411.

COMPOSITION Basal diet (mg/ml): agar, 20; RNA, 2; guanine, 0.03; uracil, 0.004; cholesterol, 0.1; dextrose, 15.0; Wesson's salt mixture, 2.0; amino acid mixture, 28.08; yeast extract, 0.075; and (μg): vitamin mixture, 322.09; and (ml): water, 1000; NaOH to pH 6.5.

l-amino acids (mg/ml): *dl*-alanine, 1.09; arginine, 1.67; aspartic acid, 0.49; cysteine hydrochloride, 0.48; glutamic acid, 4.42; glycine, 1.75; histidine, 0.48; cystine, 1.00; *dl*-isoleucine, 1.26; leucine, 2.35; lysine hydrochloride, 1.34; methionine, 0.34; *dl*-phenylalanine, 2.00; proline, 1.68; *dl*-serine, 0.88; threonine, 1.00; tryptophan, 1.23; tyrosine, 0.62; *dl*-valine, 3.00; hydroxyproline, 1.00.

vitamins (μg/ml): biotin, 0.05; B_{12}, 0.04; calcium pantothenate, 30.0; choline chloride, 100.0; folic acid, 20.0; nicotinic acid, 20.0; pyridoxine hydrochloride, 20.0; riboflavin, 20.0; thiamine hydrochloride, 20.0; inositol, 40.0; *p*-aminobenzoic acid, 2.0; niacinamide, 50.0.

PREPARATION All amino acids except glutamic acid, aspartic acid, tyrosine and tryptophan weighed in quantities for 1 l of diet and ground in mortar. Tyrosine dissolved in 100 ml hot water and 3 ml concentrated HCl. Tryptophan dissolved in 100 ml hot water. Aspartic and glutamic acids each dissolved in 65 ml warm water. 4 amino acid solutions combined and others added. 30 mg guanine and 4 mg uracil mixed with 75 ml hot water and 0.25 ml concentrated HCl and heated over steam bath for 1 h. 2 g RNA dissolved in 70 ml water and 20 ml 2 N NaOH. 2 g Wesson's salt mixture dissolved in 100 ml hot water. 15 g dextrose dissolved in 25 ml hot water. Stock suspension of cholesterol made by dissolving 500 mg in 20 ml of hot 95% ethyl alcohol, and pouring slowly into 100 ml cold water and 0.3 ml Tween 80. Milky white suspension warmed until volume 60 ml. Vitamins, choline, inositol, and *p*-aminobenzoic acid prepared separately; riboflavin dissolved in 0.02 N warm acetic acid and folic acid in 20% ethyl alcohol. Others dissolved in distilled water. Yeast extract prepared by slowly adding 75 g dried bakers' yeast to 100 ml warm distilled water and heating to 100° C; allowed to stand until insoluble portion settled out. Supernatant decanted and stored. Vitamin solutions poured into 1-l beaker, then salt and amino acid solutions. RNA, guanine and uracil solutions added. pH to 6.5 with 2 N NaOH. Agar added and mixture autoclaved for 2 minutes at 15 lb pressure. Cholesterol suspension, dextrose solution and yeast extract added, and finally distilled water to bring final volume to 1 l. Flask placed in oven at 80° C. 2 ml amounts dispensed to 1-dram shell vials; plugged with cotton, autoclaved at 15 lb pressure for 30 minutes. Slanted while cooling.

REARING Eggs laid on moist sponge, washed into plastic container and stored in water at 4° C. Eggs sterilized in 60 x 10 mm diam glass tubing covered at both ends with organdie by washing in 2.5% lysol solution for 2½ minutes, then rinsing in sterile distilled water for 5 minutes. Eggs removed to sterile petri dish and aseptically transferred, individually to vials. Vials kept in dark at 28° C.

DEVELOPMENT 2nd instar in 2.1 days; 3rd instar in 11.0 days; pupation after, 21.3 days; adult emergence, 27.6 days after inoculation. Under natural conditions, total development time, ca. 14 days.

→ Stones, L.C. 1972. The Stablefly. pp. 556-564. *in* University Federation of Animal Welfare, (Ed.). The UFAW Handbook on the Care and Management of Laboratory Animals. Fourth edition, Churchill Livingstone, Edinburgh and London.

COMPOSITION Larval diet (g): coarse bran, 1200; dried grass meal or alfalfa, 900; crushed whole oats, 600; dried yeast, 45; and (ml): extract of malt, 75; water, 4500.

PREPARATION Bran, grass meal and oats mixed, then yeast and malt dissolved in water added. Mixed well, packed lightly into closed container and put in cold store. pH, 5.8.

REARING At 26 ± 1°C and 50 ± 3% RH with 8 h photoperiod. Larval cultures set up in 150 ml beakers, half full of diet maintained at insectary temperature for 24 h. 1000 eggs brushed on to diet, moistened with water, and covered with cotton. After 5th day transferred to larger dish of 16.5 cm diam and 6 cm deep. 2.5 cm layer of fresh diet added and 30 ml water sprinkled on top. After further 4 days fresh diet and 30 ml water added. Pupae removed after 9 or 10 days and put in single pupal container of wire gauze in adult fly cage. Adults given two beef blood-tubes with 0.40% sodium citrate in morning and afternoon for first 2 days. For next 2 days, given 8 tubes morning and afternoon. After 7-9 days, ♀♀ ready to oviposit. On 10th day transferred to 2 smaller oviposition cages. Cages placed under black cloth under fluorescent lights, and 4 feeding tubes supplied twice daily. Flies oviposit on black cloth. Eggs collected twice daily and counted to 1000/batch. Both adult and oviposition cages described. Adult cage 45 cm cube of plywood.

DEVELOPMENT Larval period, 9-10 days; first adult emergence on 14th day.

Oestridae

Hypoderma lineatum (de Villers), common cattle grub

→ Chamberlain, W.F. 1970. *In vitro* culture of the common cattle grub for six weeks after hatching. ANN ENTOMOL SOC AM 63: 1465-1466.

For development from egg to early oesophageal stage.

COMPOSITION See Morgan *et al.* (1950) and Earle (1943). Morgan's diet 199; 90% salt base plus 10% gamma calf serum (Hyland Laboratories, 4501 Colorado Blvd, Los Angeles, Calif. 90039). In addition, 3 mg streptomycin sulphate and 1000 units potassium penicillin G/10 ml diet added.

REARING Sterilized eggs hatched in sterile Ringer's solution at 32°C, and within 1 h placed in 15 ml screw-capped vials (3-8/vial) with 10 ml of culture diet. After 3 weeks, diet changed, and number of larvae/vial reduced to 1-4. Three weeks later diet changed again, and larvae reduced to 1/vial.

DEVELOPMENT Activity and gross morphology of larvae reared *in vitro* similar to those from oesophagi. From 6 weeks, decrease in larval activity; at 9 weeks only limited survival.

Phoridae

Megaselia scalaris Loew

→ Aboul-Nasr, A.E.; El-Miniawi, S.F. 1974. Successful mass-rearing of *Megaselia scalaris* Loew, on a low cost pabulum in Egypt. *in* FAO/IAEA Information Circular on Radiation Techniques and their Application to Insect Pests No. 18.

COMPOSITION (g): dried sugar cane baggasse (Komombo), 240; dried ground corn hulls, 60; proteilan, 30; granulated sugar, 90; dried chicken's yeast, 60; sodium benzoate, 2; and (ml): benzoic acid, 1; 2 N HCl, 22-25; tap water, 1250-1500.

DEVELOPMENT Egg-adult survival, 86%; and egg-pupa survival, 89.8%; 160000 adults within 5 weeks from original 600 gravid females.

→ El-Miniawi, S.F.; Moustafa, M.A. 1965. On the biology of *Megaselia scalaris* Loew (Diptera: Phoridae). BULL SOC ENTOMOL EGYPTE 49: 81-91.

COMPOSITION (g): ground, dried carrot (28 mesh), 300; dried yeast, 70; sodium benzoate, 2; and (ml): 2 N HCl, 14; distilled water, 1200.

PREPARATION Ingredients thoroughly mixed; solutions of benzoate and HCl in water added. Poured into thin 1½" waxed cardboard cubes to depth of 1 cm below rim.

REARING At 27°C. Diet cubes arranged in circle in fly cage. Adults fed on solution of (ml): distilled water, 70; cow's milk (3% fat), 30; honey, 5; protein hydrolysate, 1.5 g. Cubes left in cage for 2 h for oviposition. Larvae reared in cubes. Mature larvae collected in 0.7 x 5 cm tubes. Adult ♀♀ kept in tubes with piece of wet cotton wool covered with filter paper at bottom. Male introduced into tube for mating.

DEVELOPMENT (Average): eclosion in 17 h; hatch, 90%; larval period: ♂, 4.8 days; ♀, 5.8 days; pupal period, 9 days; emergence: ♂, 79%; ♀, 86%. Adult longevity: ♂, 7.3 days; ♀, 17.3 days. Pre-oviposition period, 38 h; fecundity, 391/eggs/♀. Life cycle, ca. 18 days from egg to adult.

Piophilidae

Piophila casei (Linnaeus), cheese skipper

→ Anonymous. 1971. Diet for *Piophila casei*: rearing method. INSECT TOXICOL INF SERV 14: 47-48.

COMPOSITION (g): agar, 20; powdered milk, 80; dried yeast, 50; water, 1000; Nipagin, 1; and ethanol, 10 ml.

Nipagin can be substituted with aqueous solution of propionic acid, or 10% ethanol solution of o-diphenol.

PREPARATION Agar, milk powder, and yeast stirred in water, and suspension boiled for at least 30 minutes. Nipagin dissolved in alcohol added and diet kept for several days at 5°C.

REARING At 27-30° C. Adults in muslin or wire net cages, fed on 3 cm Pablum cubes or diet, on which eggs also laid. Eggs transferred to 6 x 12 cm jars with diet. Pupation in cotton provided.

DEVELOPMENT Larval period, ca. 9 days; pupal period, 5 days. Mating immediately after emergence; oviposition from 3rd day. Adult longevity, ca. 20 days.

→ Jones, W.R.; Cochran, D.G.; Graham, P.P.; Kelly, R.F. 1973. Growth and development of the cheese skipper under sterile, unsterile and sterile-inoculated conditions. J ECON ENTOMOL 66: 1252-1254.

COMPOSITION & PREPARATION Diet used that of Greenberg (1954) for *Musca domestica*. CSMA. Six Erlenmeyer flasks (3 for sterile eggs and 3 for unsterile eggs), each containing 2.5 cm of sawdust moistened with 50 ml tap water, plugged with cotton stoppers, covered with aluminium foil, autoclaved for 2 h at 121° C, and cooled. Shells of 18-day-old embryonated chick eggs disinfected with $HgCl_2$, caps cut away with sterile scissors and chick aseptically added to each flask. Incubated at 32° C.

REARING 12-h-old eggs washed in 1% sodium sulphite solution until separated. Eggs transferred aseptically to Millipore filter and sterilized in sequence:- 30 minutes in 10% formalin; 2 rinses in sterile distilled water; 20 minutes in 2% NaClO; 2 more water rinses; 30 minutes in White's solution + 5 ml of 0.2% $HgCl_2$/100 ml of White's solution; and 7 water rinses. Sterile eggs dispensed into flasks containing incubation medium.

Adults reared in 100 x 42 x 42 cm cages enclosed with 30-mesh screen. Flies fed fresh water and dog biscuits daily. Larvae reared on dog food until pupae. Transferred to 0.5-1 cups and held at 5° C until more flies needed.

DEVELOPMENT On sterile and unsterile medium respectively:- pupation, 51 ± 7.0 and 63 ± 3.9%; mean larvae period, 5.03 ± 0.1 and 5.61 ± 0.15 days; mean pupal wt: 5.06 ± 0.13 and 4.34 ± 0.21 mg; number of adults emerging from pupae, 310 and 379. Normal flora of *P. casei* eggs has no beneficial influence on growth and development, and little adverse influence. *Proteus vulgaris* (bacterium) has no significant effect on growth or development.

→ Zuska, J. 1975a. Concentration of dietary glucose or sucrose, determining longevity of adult *Piophila casei* (Diptera, Piophilidae). ACTA ENTOMOL BOHEMOSLOV 72: 80-86.

COMPOSITION Adult diet. 50-80% sucrose for ♀, 70-80% sucrose for ♂. Or 50% glucose for both sexes.

REARING At 25 ± 1° C and 85% RH in darkness. Pupae in 50 x 10 mm glass vials plugged with cotton. Adults transferred to 250 ml flasks with cellulose cotton plug containing feeding device, by which liquid food in glass bell soaks through cotton. Feeding device illustrated.

DEVELOPMENT Longevity: of ♀♀ 35 days, of ♂♂ 38 days on glucose or sucrose.

See also: Zuska (1975b).

Psilidae

Psila rosae (Fabricius), carrot rust fly

McClanahan, R.J.; Niemczyk, H.D. 1963. Continuous rearing of the carrot rust fly, *Psila rosae* (Fab.). CAN ENTOMOL 95: 827-830.

COMPOSITION Adult diet (%): soybean flour, 70; brewers' yeast, 30; crystalized honey and water provided separately.

REARING At 64 ± 2°F and 40% RH, with 16 h photoperiod. Adults kept in plastic screen covered cages with cotton sleeve at each end. Fed honey, above diet, and water supplied in 4-oz jar inverted over filter paper. Potted carrot plants for oviposition sites. Half inch layer of 1:1 mixture moist peat moss and muck soil placed around carrot crown, removed with eggs every 3-4 days. Eggs transferred to gallon jars containing several carrots layered over muck soil autoclaved for 1 h at 28 lb pressure. Water added weekly. Pupae sifted out, placed in small waxed cartons of muck soil or left in jar for emergence.

DEVELOPMENT Mating 24 h from emergence. 70-95% eggs viable; eclosion from 6 days on moist filter paper at room temperature. Larval period, 39 days; pupal period, 29 days. Total developmental time, 80 days.

Städler, E. 1971. An improved mass-rearing method of the carrot rust fly, *Psila rosae* (Diptera: Psilidae). CAN ENTOMOL 103: 1033-1038.

Diet used that of Hagen & Finney (1950) for tephritid flies.

COMPOSITION Adult diet (part): cane sugar, 4; enzymatic yeast hydrolysate, 1; water, 1.5. Also 1.5% water-agar mixture in separate container.

PREPARATION Diet coated on to 15 cm x 15 cm x 4 mm foam rubber on roof of cage.

REARING At 21 ± 1°C and 70 ± 5% RH with 18 h photoperiod in 30 cm^3 metal-frame cages. Front of cage made from plexiglass with 3 openings. Bottomless plastic beaker glued to large opening to act as entrance. Two small holes in bottom half of frame used for removing dead flies. Ceiling and sides of wire mesh. Floor of galvanized sheeting with 4 holes of ca. 10 cm diam covered with black nylon net. Water-agar mixture fed through yoghurt beaker inverted on top of cage; adult diet in foam rubber laid on ceiling. Food and water changed 3 times/week. Carrot leafage in bottle used as oviposition stimulant. Eggs deposited through nylon net on 9.5 cm petri dishes containing 2.5 cm thick wetted foam rubber lining covered with black cloth. Adult emergence in special containers made from bottomless yoghurt beakers glued to ca. 7 cm plastic petri dish with 200 µm nylon gauze forming bottom of beaker where pupae are laid. Beaker covered and 100% RH maintained. Larvae reared on carrots bedded in damp soil.

DEVELOPMENT Oviposition enhanced by feeding protein to adults; addition of yeast hydrolysate to sugar diet gives 50% increase in oviposition, to average of 75 eggs/♀. Larval period, 42 days; emergence, 80%.

Psychodidae

Lutzomyia trapidoi (Fairchild & Hertig), Panamanian phlebotomine sand fly

→ Chaniotis, B.N. 1975. A new method for rearing *Lutzomyia trapidoi* (Diptera: Psychodidae), with observations on its development and behaviour in the laboratory. J MED ENTOMOL 12: 183-188.

COMPOSITION Adult diet: 30-50% sucrose solution. Blood meal of suckling hamsters offered.

REARING Combination cage for oviposition and larval rearing described and illustrated. 5 mm diam hole cut in base of 0.24-1 styrofoam cup and 50 g plaster of Paris mixed with 25 ml water poured in 30 seconds later, cup rotated to coat bottom and 60-80% of lower interior wall. Plastic tumbler with 30 mm diam hole cut in base and covered with fine nylon mesh inverted over cup. 1 or 2 cork-stoppered small holes cut in side of tumbler to provide flexibility in transferring sand flies. Interior of styrofoam cup kept wet by placing directly in water or on wet sponges. Bottom covered with thin layer of decomposed leaf litter on which eggs sprinkled for immature larval rearing. Alternative vessel (also illustrated) used: 500 ml Nalgene plastic jar containing bottom layer of fine sand covered with 100 ml layer peat moss or decomposed leaf litter. 30 mm hole cut in plastic lid and stoppered with screened cork. 35 x 30 x 25 cm plywood framed, nylon screened cages with sleeve attached to 15 cm diam hole in back panel provided. Cage supplied with 1 or 2, 500 ml water flasks with paper towel wick. Kept at 23-27° C and 80% RH. Sugar solution supplied. Blood meal offered using styrofoam cups as holding vessels. Females aspirated from feeding cage and blown lightly into cup through hole in tumbler, @ 50 ♀♀/cup. Eggs laid on wet plaster coating and washed into small beaker with water and fine camel hair brush. 1 drop detergent/50 ml water added. Eggs washed 2-3 times and transferred with Pasteur pipette on to organic substrate in rearing vessel. Liver powder sprinkled on surface after 2nd instar; 200-300 larvae/jar.

DEVELOPMENT Adult emergence from 30 days at 29° C; 45 days at 23-25° C. Mean number eggs laid/♀, 27.4.

Phlebotomus longipes Parrot & Martin

→ Gemetchu, T. 1971. Liver and yeast as larval diets in colonization of a sandfly (*Phlebotomus longipes*). TRANS R SOC TROP MED HYG 65: 682-683.

COMPOSITION Various diets from liver powder, yeast, rabbit faeces, and beef blood.

PREPARATION Liver powder and yeast autoclaved before use.

REARING At 25° C; 25 eggs placed in centre of petri dish with small quantities of food placed equidistant from each other for preference tests. Yeast or liver preferred by 1st instar larvae; rabbit faeces and beef blood by 4th instars. Migration by 4th stage larvae for pupation away from yeast and liver to rabbit faeces, beef blood or no food.

DEVELOPMENT Adults from: yeast, 70.1%; liver, 82.0%; rabbit faeces, 61.4%; beef blood, 7%. Reared for 4 generations on either liver or yeast. Life cycle from 1st instar larvae to pupa: 24.6 days (liver); 30.1 days (yeast); 37.7 days (rabbit faeces); and 52.8 days (beef blood).

Phlebotomus papatasii (Scopoli)

→ Hafez, M.; Zein-el-Dine, K. 1964. Culturing *Phlebotomus papatasii* (Scopoli) in the laboratory. BULL ENTOMOL RES 54: 657-659.

COMPOSITION (parts): ground sieved guinea pig faeces (60 mesh), 1; sieved dry earth (60 mesh), 2; water suspension of *Alcaligenes viscosus* bacterium or Salbrol M solution (Hoechst laboratories).

PREPARATION Ingredients mixed until damp and friable.

REARING 18 mesh moist sand poured to depth of 5 cm in 6-8 cm diam unglazed, earthenware breeding pots. Granules of diet placed over sand in 1 cm thick layer. Pot kept at 30° C for 3 days at 60-70% RH then placed in small cage of muslin. Dish of water placed in cage. 30 adult pairs introduced; ♀♀ fed on white mice; ♂♂ fed sterile 50% glucose solution. Oviposition in pots.

DEVELOPMENT Culture maintained for several years.

See also: Hertig (1964); Safyanova (1964); Schmidt (1964).

Sarcophagidae

Agria housei Shewell [previously *Agria affinis* (Fallén) or *Pseudosarcophaga affinis* (Fallén)]

→ Coppel, H.C.; House, H.L.; Maw, M.G. 1959. Studies on dipterous parasites of the spruce budworm, *Choristoneura fumiferana* (Clem.) (Lepidoptera: Tortricidae). VII. *Agria affinis* (Fall.) (Diptera: Sarcophagidae). CAN J ZOOL 37: 817-830.

COMPOSITION Fresh pork liver.

REARING At 21 ± 1° C and 60% RH. 1000 adults kept in screened 12 x 12 x 18" cage. Fed 10% aqueous honey solution from cotton wad, or water and sugar lumps. Fresh pork liver placed on wax mat in cage for larviposition. Larvae collected daily. >1000 larvae reared on fresh liver in covered ventilated dish; transferred to box of dry sawdust for pupation. Pupae collected by sieving. >1000 pupae put in 4 x 4 x 2 " wooden container, black on inside, rigged so that adults pass through sleeve into breeding cage.

DEVELOPMENT Larval period, 5-8 days; pupal period, 10-14 days. Total development, 28 days. Also suitable for rearing *Sarcophaga aldrichi* Parker.

See also: Arthur & Coppel (1953); House (1958a b); House & Traer (1948).

→ Hegdekar, B.M. 1970. Amino acid analogues as inhibitors of insect reproduction. J ECON ENTOMOL 63: 1950-1956.

COMPOSITION Basal adult diet (g): sucrose, 10.00; Wesson's salt mixture, 0.15; $ZnCl_2$, 0.02; cholesterol, 0.01; sodium benzoate, 0.012.

l-amino acids (g/100 ml water): arginine, 0.38; histidine, 0.15; isoleucine, 0.25; leucine, 0.75; lysine, 0.75; methionine, 0.15; phenylalanine, 1.20; threonine, 0.15; tryptophan, 0.30; valine, 0.50; glutamic acid, 1.00.

vitamins (mg/100 ml water): inositol, 5.00; riboflavin, 0.16; folic acid, 0.34; biotin, 0.05; choline chloride, 2.00; calcium pantothenate, 0.44; thiamine hydrochloride, 0.12; pyridoxine hydrochloride, 0.90; nicotinic acid, 0.30.

PREPARATION Nutrient mixture dissolved in distilled water; vitamins added, and blended at high speed for 1 h. Solution stirred slowly for further 1 h, while being neutralized by addition of 2 N NaOH. Stored at -10° C. Experimental diets prepared by adding or deleting amino acid analogues.

REARING At 22 ± 1° C and 50-55% RH in 46 x 30 x 30" screen cages. Adults fed on 20% solution of whole powdered milk in tap water through dental cotton wick; sugar cubes, and fresh pork liver. Diet renewed daily.

→ House, H.L. 1954a. Nutritional studies with *Pseudosarcophaga affinis* (Fall.), a dipterous parasite of the spruce budworm, *Choristoneura fumiferana* (Clem.). I. A chemically defined medium and aseptic-culture technique. CAN J ZOOL 32: 331-341.

COMPOSITION (mg/ml water): amino acid mixture, 20.0; ribonucleic acid, 1.0; cholesterol, 1.0; dextrose, 20.0; salt mixture (U.S.P. XII No. 2), 2.0; agar, 7.5.

vitamins (μg/ml water): thiamine, 5.0; riboflavin, 15.0; pyridoxine, 15.0; inositol, 100.0; nicotinic acid, 15.0; choline chloride, 100.0; calcium pantothenate, 15.0; *p*-aminobenzoic acid, 15.0; folic acid, 5.0; biotin, 0.025.

amino acids (mg/ml water): *dl*-alanine, 1.29; *l*-arginine monohydrochloride, 0.53; *dl*-aspartic acid, 1.53; *l*-cysteine, 0.18; *l*-glutamic acid, 2.74; glycine, 0.35; *l*-histidine hydrochloride, 0.35; *l*-hydroxyproline, 0.24; *dl*-isoleucine, 1.53; *l*-leucine, 1.40; *l*-lysine hydrochloride, 0.88; *dl*-methionine, 0.94; *dl*-phenylalanine, 1.40; *l*-proline, 1.00; *dl*-serine, 1.76; *dl*-threonine, 0.94; *dl*-tryptophane, 0.47; *l*-tyrosine, 0.82; *dl*-valine, 1.65.

PREPARATION Amino acids and salts heated gently in 250 ml flask with 30 ml distilled water. Suspension of cholesterol made by mixing 5 g cholesterol in minimal 95% ethyl alcohol, and adding to 0.25 g Tween 80 in water to 100 ml; added to amino-acids and salts solution. 0.1 g RNA dissolved in 2 ml 2 N NaOH and 7 ml distilled water and added to other ingredients. pH to 5.8. 0.75 g agar dissolved in 39 ml distilled water. Aqueous solution of vitamins and dextrose prepared, sterilized and filtered through Seitz filter. Other solutions autoclaved for 20 minutes at 15 lb pressure. Before agar gels 20 ml filtered solutions of dextrose and vitamins added and flask shaken. Diet dispensed into small test tubes @ 1 ml/tube, plugged with non-absorbent cotton and slanted. All equipment autoclaved and aseptic techniques used.

REARING Stock cultures reared on liver and fish. Larvae dissected from gravid ♀♀ submerged in 2.5% formalin in aseptic chamber. Larvae transferred individually with sterile brush to diet tube. Kept at 23° C and 60% RH in darkness, except during examination times. Mature larvae transferred to moist moss or sawdust for pupation. Pupae collected and stored at 1° C until diapause broken. Incubated again at 23° C until emergence.

DEVELOPMENT Survival to 3rd instar, 83.9%; pupation, 59.9%. Mean pupal length and width, 6.6 mm ± 0.7 mm and 2.9 ± 0.4 mm respectively. Larval period, 9.2 days. 66% of adults females.

See also: House (1954b c d); House & Barlow (1956).

House, H.L.; Barlow, J.S. 1960. Effects of oleic and other fatty acids on the growth rate of *Agria affinis* (Fall.) (Diptera: Sarcophagidae). J NUTR 72: 409-414.

COMPOSITION (mg/100 ml distilled water): *l*-amino acids: alanine, 64.5; arginine hydrochloride, 52.9; aspartic acid, 76.5; cysteine hydrochloride, 18.0; glutamic acid, 137.1; glycine, 17.6; histidine hydrochloride, 17.6; hydroxyproline, 24.0; isoleucine, 76.5; leucine, 140.3; lysine hydrochloride, 88.2; methionine, 47.3; proline, 100.2; serine, 88.2; phenylalanine, 70.1; threonine, 47.3; tryptophan, 23.6; tyrosine, 82.2; valine, 82.6.

salts: $CaH_4(PO_4)_2$, 8.96; $Ca[CH_3CH(OH)COO]_2$, 21.58; $C_6H_5O_7Fe.5H_2O$, (17.5% Fe), 1.96; $MgSO_4.7H_2O$, 9.04; K_2HPO_4, 15.83; $NaH_2PO_4.2H_2O$, 5.76; NaCl, 2.87.

lipids: cholesterol, 100; linoleic acid, 60; linolenic acid, 20; oleic acid, 192; palmitic acid, 88; stearic acid, 40; Tween 80, 270.

vitamins: biotin, 0.00009; calcium pantothenate, 0.44; choline chloride, 2.00; folic acid, 0.34; inositol, 5.96; nicotinic acid, 0.30; *p*-aminobenzoic acid, 0.9; pyridoxine hydrochloride, 0.9; riboflavin, 0.16; thiamine hydrochloride, 0.12.

others: glucose, 500; RNA, 75; 2 N KOH, 247; agar, 750.

PREPARATION Fatty acids dissolved with 250 mg cholesterol in 1 ml hot ethanol and emulsified in 100 ml hot water with 0.5 ml Tween 80. 40 ml used/100 ml diet. For details see House* (1954a).

REARING At 23°C individually in test tubes. See House & Barlow (1957). Pupation in sawdust in 2.5 x 2.5 x 3.5" cage (50 pupae) with access to 7 x 5.5" breeding cage for emerging adults.

DEVELOPMENT Third instar after 6 days.

Corollary 1

House, H.L. 1967a. The nutritional status and larvicidal activities of C_6 to C_{14} - saturated fatty acids in *Pseudosarcophaga affinis* (Diptera: Sarcophagidae). CAN ENTOMOL 99: 384-392.

COMPOSITION (mg/100 ml distilled water): *l*-amino acid mixture, 2250.00; ribonucleic acid, 140.60; 2 N KOH, 62.32; cholesterol, 187.50; inorganic salt mixture, 165.00; B-vitamin mixture, 21.50; glucose, 937.50; and agar, 750.00.

l-amino acids (mg/100 ml distilled water): alanine, 112.50; arginine hydrochloride, 137.25; aspartic acid, 144.00; cysteine hydrochloride, 29.25; glutamic acid, 236.25; glycine, 175.50; histidine hydrochloride, 54.00; hydroxyproline, 38.25; isoleucine, 117.00; leucine, 173.25; lysine hydrochloride, 130.50; methionine, 67.50; phenylalanine, 150.75; proline, 168.75; serine, 150.75; threonine, 121.50; tryptophane, 31.50; tyrosine, 78.75; and valine, 132.75.

inorganic salts (mg/100 ml distilled water): $FeCl_3.6H_2O$, 4.7483; K_2HPO_4, 99.0286; $Na_2HPO_4.12H_2O$, 13.6842; $MgSO_4.7H_2O$, 34.7277; $MnSO_4.H_2O$, 0.0105; $CoCl_2$, 1.2755; $CaCl_2$, 8.0705; $CuSO_4.5H_2O$, 1.4787; and $ZnCl_2$, 1.8812.

B-vitamins (mg/100 ml distilled water): biotin, 0.00017; calcium pantothenate, 0.8250; choline chloride, 3.7500; folic acid, 0.6375;

inositol, 11.8125; nicotinic acid, 0.5625; *p*-aminobenzoic acid, 1.6875; pyridoxine hydrochloride, 1.6875; riboflavin, 0.3000; and thiamine, 0.2250.

REARING At 23 ± 0.1°C. Larvae reared individually in test tubes for 7 days.

DEVELOPMENT No saturated C_6-C_{14} fatty acids fed free as dietary components improved rate of growth or development; several toxic to larvae. On control, 90% pupation, with 90% emergence from pupae.

Corollary 2

House, H.L. 1967b. The role of nutritional factors in food selection and preference as related to larval nutrition of an insect, *Pseudosarcophaga affinis* (Diptera, Sarcophagidae), on synthetic diets. CAN ENTOMOL 99: 1310-1321.

Two best of 8 diets tested.

COMPOSITION Diet 1 (mg/100 ml distilled water): *l*-amino acid mixture, 225.00; glucose, 750.00; agar, 750.00; ribonucleic acid, 112.50; 2 N KOH, 78.54; lipid mixture, 750.00; salt mixture, 66.00; and B-vitamin mixture, 17.20.

l-amino acids (mg): alanine, 500; arginine hydrochloride, 610; aspartic acid, 640; cysteine hydrochloride, 130; glutamic acid, 1050; glycine, 780; histidine hydrochloride, 240; hydroxyproline, 170; isoleucine, 520; leucine, 770; lysine hydrochloride, 580; methionine, 300; phenylalanine, 670; proline, 750; serine, 670; threonine, 540; tryptophan, 140; tyrosine, 350; valine, 590; (totals, 10000 mg).

lipids (mg): cholesterol, 150.00; palmitic acid, 132.00; stearic acid, 60.00; oleic acid, 292.62; linoleic acid, 87.14; linolenic acid, 28.24; the level of each was 1.5 times that used previously, House & Barlow (1956).

salts (mg/g): $FeCl_3.6H_2O$, 28.7; K_2HPO_4, 600.0; $Na_2HPO_4.12H_2O$, 83.8; $MgSO_4.7H_2O$, 210.0; $MnSO_4.H_2O$, 0.6; $CoCl_2.6H_2O$, 7.7; $CaCl_2$, 48.8; $CuSO_4.5H_2O$, 9.0; and $ZnCl_2$, 11.4.

B-vitamins (mg): biotin, 0.000136; calcium pantothenate, 0.66; choline chloride, 3.00; folic acid, 0.51; inositol, 9.45; nicotinic acid, 0.45; *p*-aminobenzoic acid, 1.35; pyridoxine hydrochloride, 1.35; riboflavin, 0.24; and thiamine, 0.18.

Diet 2 as for diet 1, except that *l*-amino acid mixture reduced to 1125.00 mg; and glucose doubled to 1500.00 mg.

PREPARATION See House* (1954a). Except that salts, agar and aqueous emulsion of lipids (House & Barlow*, 1960) autoclaved, and all other components dissolved in water and filtered; appropriate amounts of both fractions then combined aseptically to give complete diet.

DEVELOPMENT Larval maturity in 6 and 7 days on diets 1 and 2 respectively.

Corollary 3

House, H.L. 1972b. Inversion in the order of food superiority between temperatures effected by nutrient balance in the fly larva *Agria housei* (Diptera: Sarcophagidae). CAN ENTOMOL 104: 1559-1564.

Three larval diets tested.

COMPOSITION Diet 1 (g): amino acid mixture, 1.25; glucose, 2.5; oleic acid, 0.5; palmitic acid, 0.05; stearic acid, 0.05; salts, 0.066; RNA, 0.0625; cholesterol, 0.075; Tween 80, 0.15; agar, 0.75; and (mg): calcium pantothenate, 0.062; thiamine hydrochloride, 0.003; and (ml): vitamin solution, 15; water, 85.

Diet 2, similar to 1, except (g): amino acid mixture, 0.75; glucose, 0.1; oleic acid, 0.1; palmitic acid, 0.25; stearic acid, 0.25; and (mg): calcium pantothenate, 0.05; thiamine hydrochloride, 0.0005.

Diet 3, similar to 2, except: amino acid mixture, 1.5 g; glucose, 1.5 g; calcium pantothenate, 0.33 mg; thiamine hydrochloride, 0.09 mg.

amino acids (g): alanine, 6.02; arginine hydrochloride, 7.34; cysteine hydrochloride, 1.56; glycine, 9.39; histidine, 2.89; hydroxy-proline, 2.05; isoleucine, 6.26; leucine, 9.27; lysine hydrochloride, 6.98; methionine, 3.61; phenylalanine, 8.06; proline, 9.02; serine, 8.06; threonine, 6.50; tryptophan, 1.68; tyrosine, 4.21; valine, 7.10; (total, 100.00 g).

salts: as in House & Barlow (1965).

vitamins (mg/100 ml of water): riboflavin, 0.8; nicotinic acid, 1.5; choline chloride, 10.0; *p*-aminobenzoic acid, 4.5; inositol, 29.8; pyridoxine, 4.5; folic acid, 1.7; biotin, 0.01075; B_{12}, 0.000435.

Corollary 4

House, H.L. 1972a. Insect Nutrition. pp. 513-573. *in* Fiennes, R.N. (Ed.). Biology of Nutrition I.E.F.N. Vol. 18. Pergamon Press, Oxford and New York.

COMPOSITION Chemically defined diet (mg/100 ml water): *l*-amino acids: alanine, 112.0; arginine hydrochloride, 137.25; aspartic acid, 144.0; cysteine hydrochloride, 29.25; glutamic acid, 236.25; glycine, 176.0; histidine hydrochloride, 54.0; hydroxyproline, 38.25; isoleucine, 117.0; leucine, 173.25; lysine hydrochloride, 130.5; methionine, 67.25; phenylalanine, 150.75; proline, 168.75; serine, 150.75; threonine, 121.5; tryptophan, 31.5; tyrosine, 78.75; valine, 132.75; total, 2250.

inorganic salts: $FeCl_3.6H_2O$, 3.5875; K_2HPO_4, 75.0; $Na_2HPO_4.12H_2O$, 10.475; $MgSO_4.7H_2O$, 26.25; $MnSO_4.H_2O$, 0.075; $CoCl_2.6H_2O$, 0.9625; $CaCl_2$, 6.1; $CuSO_4.5H_2O$, 1.125; $ZnCl_2$, 1.425; total, 125.

lipids: cholesterol, 187.5; linoleic acid, 112.4; linolenic acid, 37.6; oleic acid, 377.2; palmitic acid, 165.0; stearic acid, 75.0; polysorbate, 506.2; total, 1460.9.

vitamins: biotin, 1.69×10^{-4}; calcium pantothenate, 0.825; choline chloride, 3.75; cyanocobalamin, 1.63×10^{-5}; *meso*-inositol, 11.175; nicotinic acid, 0.5625; *p*-aminobenzoic acid, 1.6875; pteroylmonoglutamic acid, 0.6375; pyridoxal hydrochloride, 1.6875; riboflavin, 0.3; thiamine hydrochloride, 0.225; α-tocopherol, 1.875; total, 22.7.

miscellaneous: glucose, 937.5; 2 N KOH, 247.0; agar, 750.0; total, 1934.5.

2 N KOH solution to adjust pH to ca. 6.5.

See also: Barlow (1965); Barlow & House (1956, 1960); House (1964a, 1965c, 1966a, 1969a b, 1970, 1971a b); House & Barlow (1957a b, 1964, 1965).

Sarcophaga aldrichi Parker

→ Arthur, A.P.; Coppel, H.C. 1953. Studies on dipterous parasites of the spruce budworm, *Choristoneura fumiferana* (Clem.) (Lepidoptera: Tortricidae). I. *Sarcophaga aldrichi* Park. (Diptera: Sarcophagidae). CAN J ZOOL 31: 374-391.

COMPOSITION (parts): salmon, 1; pork liver, 3; diet developed by House & Traer (1948) to rear *Pseudosarcophaga affinis* (Fall.). Suitable for rearing *Sarcophaga aldrichi*.

Sarcophaga bullata Parker, fleshfly

→ Barnes, F.J.; Goodfellow, R.D. 1971. Mevalonate kinase: localization and variation in activity during the development of *Sarcophaga bullata*. J INSECT PHYSIOL 17: 1415-1427.

Diet modified from that of Brust & Fraenkel* (1955).

COMPOSITION (g): salt mixture MD No. 185, 5; methyl *p*-hydroxybenzoate, 11.25; cholesterol, 5.0; brewers' yeast, 50.0; bacto-agar, 20.0; vitamin-free casein, 500.0; and (ml): water, 1300.

PREPARATION Ingredients mixed in above order and autoclaved. Dispensed to jars or dishes when cool.

REARING At 28°C and 50% RH with 15 h photoperiod. Larviposition on calf liver. Larvae transferred to jars or dishes; after 7 days migrate to autoclaved sand to pupate. Pupae transferred to petri dishes and placed in cages with water and sugar supply.

DEVELOPMENT Larval period, ca. 7 days; pupal period, ca. 14 days. Adults emergence, ca. 21 days from larviposition.

Corollary

→ Goodfellow, R.D.; Barnes, F.J.; Graham, W.J. Jr 1971. Dietary sterol effects on mevalonate kinase activity in axenic and non-axenic *Sarcophaga bullata* (Diptera). J INSECT PHYSIOL 17: 1625-1635.

COMPOSITION (g/60 g diet): bacto-agar, 2.00; salt mixture MD No. 185, 0.50; sterol (to 1% concentration), 0.60; vitamin-free casein, 55.70; methyl *p*-hydroxybenzoate, 1.20; and vitamin mixture, 0.72 ml; and distilled water, 2.50 ml/g dry weight.

vitamins (mg/1.5 ml): thiamine hydrochloride, 1.875; riboflavin, 1.250; niacin, 6.250; calcium pantothenate, 1.875; pyridoxine hydrochloride, 0.375; folic acid, 0.250; biotin, 0.025; choline, 50.00; inositol, 62.500; ascorbate, 12.500.

Frings, H.; Frings, M. 1953. Dog biscuit as a larval medium for *Sarcophaga bullata*. J ECON ENTOMOL 46: 183.

COMPOSITION Moistened dog biscuit (amount of moisture critical).

REARING Adults kept in cages with water and sugar. Larviposition on small pieces of meat (skinned albino mice) or moist dog biscuit. Larvae placed in 2-quart jars half full of diet, covered with fine screening (60 mesh/inch). Layer of sawdust or dry sand added for pupation. Pupae sieved and placed in cages for emergence.

Shannon, R.C. 1923. Rearing dipterous larvae on nutrient agar. PROC ENTOMOL SOC WASH 25: 103-104.

COMPOSITION Beef infusion from 1 lb hamburger steak; 3% agar; water to 1 l.

REARING In petri dish.

DEVELOPMENT Life cycle of 3 species completed, *Sarcophaga cimibicis* Townsend, *Sarcophaga sarracenioides* Aldrich, *Limosina* sp. Sphaeroceridae [previously Borboridae].

Sciaridae

Bradysia impatiens (Johannsen), fungus gnat

Kennedy, M.K. 1973. A culture method for *Bradysia impatiens* (Diptera: Sciaridae). ANN ENTOMOL SOC AM 66: 1163-1164.

COMPOSITION Brewers' yeast.

REARING At 25°C, with 16 h photoperiod. One ♀ and 2 ♂♂ introduced to 25 x 95 mm shell vials with sterile 2% bacto-agar and sterile, blended dried grass clippings. Mating within 24 h, and oviposition 1-2 days later on agar surface. Neonate larvae fed on 1:4 yeast:sterile blended grass added every 2 days or as required. Discontinued at onset of pupation, ca. 9 days from eclosion. 100-150 adults reared from each vial. About 30 vials kept to avoid difficulties of asynchronous emergence, ♀♀ being monogenic.

DEVELOPMENT Egg to adult, 15.4 days; survival, 70-80%; 20 generations reared with no significant decline in culture viability.

Kennedy, M.K. 1974. Survival and development of *Bradysia impatiens* (Diptera: Sciaridae) on fungal and non-fungal food sources. ANN ENTOMOL SOC AM 67: 745-749.

Fungal diets: best of 5 tested.

COMPOSITION Diet 1: *Alternaria tenuis* C.G. Nees, 39 g; 2% potato dextrose agar, 1 l.

Diet 2: brewers' yeast (*Saccharomyces cerevisiae* Hansen).

PREPARATION Diet 1: *A. tenuis* cultures at 20°C on 2% potato dextrose agar, slanted in 18 x 150 mm culture vials capped with plastic bacti-capalls.

Diet 2: brewers' yeast cultured on potato dextrose agar in 25 x 95 mm glass vials stoppered with foam plugs.

REARING As in Kennedy* (1973). Eggs brushed from filter paper to spot plates for surface sterilization in 0.1% Amphyl solution for 5 minutes, then transferred with sterile pipette to glass vials with diet.

DEVELOPMENT Survival on *A. tenuis*, 91.6%; on brewers' yeast, 87.9%. Developmental time: on *A. tenuis* 25.1 ± 0.2 days; on brewers' yeast, 24.6 ± 0.4 days. Development 3 times longer on non-fungal diets such as lima bean agar.

Rhynchosciara sp.

Mattingly, E.; Ogle, S. 1969. A new culture method for *Rhynchosciara*. ANN ENTOMOL SOC AM 62: 94-96.

COMPOSITION Larval diet: yellow corn meal, 30 g; 5% non-nutrient agar, 1 l.

PREPARATION Yellow corn meal suspended in agar and dispensed to petri dishes.

REARING Adults in 11.5 x 7.5" mating cage with wet coconut fibre and sugar water placed on damp cotton floor cover. Eggs laid in coconut fibres and removed with forceps after adults dead; placed on sterile bacto-agar containing 100 mg/ml streptomycin sulphate and 30 mg/ml tetracycline hydrochloride; transferred to corn meal agar when black larval head capsules visible. Fermented plant material (corn ensilage or peat moss) given as extra food. Pupation on peat moss.

DEVELOPMENT Adult life span, 5 days; egg development, 14 days; larval period, 40 days; pupal period, 5 days.

Steffan, W.A. 1966. A generic revision of the family Sciaridae (Diptera) of America North of Mexico. UNIV CALIF BERKELEY PUBL ENTOMOL 44, 77 pp.

COMPOSITION (g): corn meal agar, 4.2; bacto-agar, 2.0; and distilled water, 200 ml.

PREPARATION Corn meal and agar dissolved in water in boiling water bath; dispensed to 25 x 95 mm glass vials to depth of 1": vials plugged with cotton and autoclaved for 15 minutes; slanted and stored up to 3 weeks.

REARING In diet vials. Light layer of sterile, chopped straw sprinkled over agar surface 15 minutes before introduction of adults to remove excess moisture and provide suitable environment for oviposition and larval feeding. When lower layers of mycelial growth consumed by larvae, remaining mycelial growth removed and larvae fed mixture of finely chopped grass compost and brewers' yeast (3-4:1) every 2-3 days.

DEVELOPMENT Eggs hatch, 6-8 days. Under optimum conditions 3 generations reared in one culture tube. Species of 13 genera reared: *Phytosciara, Bradysia, Scatopsciara, Metangela, Zygoneura, Eugnoriste, Trichosia, Sciara, Epidapus, Pnyxia, Plastosciara, Cornyoptera, Lycoriella*; possibly including *Lycoriella agraria* (Felt), *Bradysia coprophila* (Lintner), *Lycoriella fenestralis* (Zetterstedt), *Bradysia impatiens* (Johannsen), *Scatopsciara nacta* (Johannsen), *Lycoriella mali* (Fitch), *Lycoriella multiseta* (Felt), *Bradysia prolifica* (Felt), *Bradysia reynoldsi* (Metz), *Lycoriella similans* (Johannsen) and *Bradysia tricti* (Coquillett).

Simuliidae

Wood, D.M.; Davies, D.M. 1965. The rearing of simuliids (Diptera). PROC 12 INT CONGR ENTOMOL LONDON (1964): 821-823.

COMPOSITION Dilute bakers' yeast suspension.

PREPARATION Suspended with blender; kept in suspension by stirrer.

REARING Eggs in darkened boxes at 60°F with dry sucrose and water. After 10 days, blades of grass floated on water surface in dim light. Eggs laid on edge of leaf. Larvae reared in apparatus consisting of activated granular carbon filter, sand and gravel; head tank; troughs of larvae; and device for dispensing food. Troughs ½" x 1" x 4' long with reservoir to prevent escape. Water flow and food flow continuously maintained. Apparatus illustrated.

DEVELOPMENT 12 species reared from eggs or young larvae. *Simulium aureum* Fries: 50-60% adults from eggs. *S. pictipes* Hagen: some larvae reared. *Boophthora erythrocephala* (de Geer) reared from pupae through 2 generations of mating and blood feeding (Wenk, 1963).

See also: Sutcliffe & McIver (1975).

Sphaeroceridae

Leptocera caenosa (Rondani)

Fredeen, F.J.H.; Glen, G.S. 1970. The survival and development of *Leptocera caenosa* (Diptera: Sphaeroceridae) in laboratory cultures. CAN ENTOMOL 102: 164-171.

Best of 10 organically rich diets tested.

COMPOSITION Green fescue grass meal and Vermiculite in equal volume; and water.

PREPARATION Grass meal milled from green leaves, mixed with Vermiculite. Water added and diet poured into 500 or 1000 ml Erlenmeyer flasks to depth of ca. 1". Flasks plugged with cotton, autoclaved at 122°C and 16 psi for 30 minutes and stored till required.

REARING At 17°C. Moist compressed air circulated through flask to flush out gaseous products of fermentation. 10 or more ♀♀/flask. Water added when necessary to maintain level at ca. ½" below diet surface.

DEVELOPMENT Life cycle, 31 days. Egg development, 2 days; larval and pupal periods, 13 days each. 60 generations reared.

Syrphidae

Metasyrphus corollae (Fabricius)

Ružička, Z. 1974. A note on substitutes of plant pollen in the adult diet of *Metasyrphus corollae* (Fabr.) (Diptera). VESTN CESK SPOL ZOOL 38: 148-151.

3 substitutes for pollen tested.

COMPOSITION Most suitable (ratio): torula dried yeast, 1; casein, 5; fructose, 5; choline chloride, ascorbic acid and water also provided.

REARING At 20°C and 55-70% RH, with 16 h photoperiod. Fluon painted on inside edge of 500 ml rearing jars to prevent escape of adults with amputated wings. Water, 20% fructose solution and pollen substitute changed daily. After 8 days oviposition stimulated by *Acyrthosiphon pisum* in small nylon bags. 2nd and 3rd oviposition periods at 4 day intervals. Larvae reared on green form of *A. pisum* in 150 ml plastic boxes with moist cotton wool on bottom @ 10 larvae/box. Aphids changed daily and provided in excess.

DEVELOPMENT Fecundity on pollen (*Corylus avellana*), 72.6%; on diet, 61.3%. Larval period, 8-9 days; pupal wt, 27-30 mg.

Tachinidae

Phryxe caudata Rondani

Grenier, S.; Bonnot, G.; Delobel, A. 1975. Définition et mise au point de millieux artificiels pour l'élevage *in vitro* de *Phryxe caudata* Rond. (Diptera, Tachinidae). II. Croissance et mues larvaires du parasitoide en millieux définis. ANN ZOOL ECOL ANIM 7: 13-25. (In French, with English summary)

COMPOSITION Basal diet (mg/ca. 100 ml): malic acid, 33.0; α-cetoglutaric acid, 41; succinic acid, 7; fumaric acid, 7; citric acid, 33; pyruvic acid, 33, maltose, 1690; glucose, 113; levulose, 45; NaCl, 69; KCl, 90; H_2PO_4, 90; $MgSO_4.7H_2O$, 417; $CaCl_2$, 160; KI, 0.02; $Co(CH_3CO_2)_2.4H_2O$, 0.02; $Mn(CH_3CO_2)_2.4H_2O$, 5.5; $Cu(CH_3CO_2)_2.H_2O$, 0.2; $Fe(NO_3)_3.9H_2O$, 8.9; $Zn(NO_3)_2.6H_2O$, 2.3, adenine, 6; thymine, 6; guanine, 6; cytosine, 3; uracyl, 6; ATP, 65; RNA, 100; glutathion, 11; choline hydrochloride, 100; glucosamine, 40; asparagine, 170; threonine, 70; serine, 68; glutamine, 220; proline, 96; glycine, 80; alanine, 100; valine, 102; methionine, 30; cysteine, 4; isoleucine, 80; leucine, 120; tyrosine, 70; phenylalanine, 70; β-alanine, 42; lysine, 152; histidine, 76; arginine, 138; tryptophane, 40; phosphoethanolamine, 2; thiamine, 0.15; riboflavine, 1.0; pyridoxine, 0.3; cyanocobalamine, 0.015; biotin, 0.04; ascorbic acid, 8.2; nicotinic acid, 4.4; calcium pantothenate, 2.2; folic acid, 0.11; inositol, 13.1; plus NaOH-KOH equimolar to pH 6.6; distilled water to 390 mOsm.

diet improved by addition of (mg/100 ml): cholesterol, 150; vitamin A, 2; vitamin E, 2; Tween 80, 100.

PREPARATION See Grenier *et al.* (1974). Ingredients dissolved in above order in minimum of distilled water, except phosphorylate products dissolved separately and added at end. Oligoelements and vitamins added in concentrated solution form. pH adjusted to 6.6 with 1:2 2 N NaOH: 2 N KOH solution and osmotic pressure brought to 390 mOsm with distilled water. Preparation then sterilized by filtration through 0.45 μm Gelman membranes. Liposoluble agents added with or without emulsifier. Diet dispensed to 6 cm diam siliconated petri dishes @ 5 or 6 drops/dish.

REARING Aseptically at 22 ± 0.5°C and 100% RH in darkness. Fertile ♀♀ killed and sterilized by immersion in alcoholic solution of Cetavlon (1/1000) and dissected. Uterus immersed in 70% ethanol, dried on filter paper and rinsed 3 times in physiological fluid (Grenier *et al.* 1974) on microscope slides. Incision made after final rinse and eggs expulsed by light pressure of mounted needle; taken up in sharp haematological pipette and placed in several ml of physiological solution to hatch. Neonate larvae transferred by micropipette and rinsed in 203 ml of culture medium; then ready for transfer to definitive diet. Saturated humidity maintained by salivary plug soaked in physiological liquid (not distilled water) to limit dilution of diet by gaseous diffusion.

DEVELOPMENT With addition of cholesterol, growth obtained from egg to beginning of final instar.

See also: Grenier *et al.* (1974).

Tephritidae

Anastrepha fraterculus Wiedemann, South American fruit fly

Ceratitis capitata Wiedemann, Mediterranean fruit fly

→ Simon, J.E. 1969. Present stage of research into the eradication of the Mediterranean and South American fruit flies and the cotton stainer in Peru by the sterile male technique. pp. 115-121. *in* Sterile Male Technique for Eradication or Control of Harmful Insects. Proceedings of a Panel Organized by the Joint FAO/IAEA Division of Atomic Energy in Food and Agriculture.

Best of many diets tested for *A. fraterculus*.

COMPOSITION (%): water, 79.20; sodium benzoate, 0.14; HCl, 0.70; fermented molasses, 5.66; powdered carrot, 14.15; Nipagin, 0.125; pH 3.8.

REARING At 27°C and 50% RH with continuous light. Adults in 0.45 cm cube cages. Water supplied by Pelegs system of 1% agar-agar. Eggs laid on waxed cloth domes, washed on to muslin in petri dishes, and incubated for 2 days. Larvae placed on diet in trays and covered with polyethylene or cloth for 4 days. On 7th day larvae sieved and washed from diet into trays of sawdust. Pupae maintained for 12 days at 27°C and 75% RH then sieved from sawdust and transferred to adult cages for emergence.

DEVELOPMENT Egg stage, 3 days; larval period, 7 days; pupal period, 12 days; preoviposition period, 8 days; adult longevity, 45 days; oviposition for 30 days.

Best of many diets tested for *C. capitata*.

COMPOSITION (g): yeast, 29.27; sodium benzoate, 0.41; Nipagin, 0.41; and (ml): water, 200; HCl, 2.04; sugar, 41.91; bagasse, 25.09.

REARING Eggs laid in perforated polyethylene bottles. 1 ml eggs placed in tray with 4.5 kg diet.

DEVELOPMENT Life cycle, 29 days; egg stage, 2 days; larval period, 7 days; pupal period, 10 days.

Anastrepha ludens (Loew), Mexican fruit fly

→ Rhode, R.H.; Spishakoff, L.M. 1965. Tecnicas usadas en la cultivo de *Anastrepha ludens* (Loew). II. Memorias del dia Parasitologia, Departo de Parasitologica. ESC NAC AGR CHAPINGO (1964): 23-28. (In Spanish)

COMPOSITION Larval diet (%): water, 80.92; HCl, 0.6; brewers' yeast, 4.0; sodium benzoate, 0.08; methyl *p*-hydroxybenzoate, 0.1; enzymatic yeast hydrolysate, 0.3; granular dehydrated carrot, 7.0; powdered dehydrated carrot, 7.0.

PREPARATION Diet dispensed to plastic-sheet lined, 33 x 75 x 5 cm iron trays @ 7.5 l/tray.

REARING Eggs sterilized on organdie disc in 0.07% sodium benzoate, and incubated at 27.8° C at 100% RH. Neonate larvae transferred to diet at optimum of 1 larva/cm^3 diet, and maintained at 26.7° C and 80% RH. Mature larvae sieved from diet for pupation in sand/vermiculite mixture @ 3-4 larvae/ml at 27.8° C and 80% RH. Mature pupae transferred to 60 x 90 x 120 cm adult cages for emergence and oviposition @ 20000/cage at 23-25° C and 50-60% RH. Adults fed on 3:1 sugar: orange juice crystals impregnated on plywood or plastic strips. Water from wick in flask. Oviposition on interior of dome; washed with water on to organdie discs.

DEVELOPMENT Egg incubation period, 4 days; larval period, ca. 10 days; pupal period, ca. 16 days; eclosion, ca. 80%; survival to pupa, ca. 50%. Production cost 2500 pesos (Mexican)/10^6 larvae. Effects of various nutrients discussed.

See also: Benschoter & Paniagua (1966); Rhode (1957).

Anastrepha ludens (Loew), Mexican fruit fly

Anastrepha suspensa (Loew), Caribbean fruit fly

→ Lopez, D.F. 1970. Sterile male technique for eradication of the Mexican and Caribbean fruit flies. Review of current status. pp. 111-117. *in* Sterile Male Technique for Control of Fruit Flies. Proceedings of a Panel Organized by the Joint FAO/IAEA Division of Atomic Energy in Food and Agriculture.

Two diets tested.

COMPOSITION Diet 1 (%): water, 80.92; HCl, 0.60; sodium benzoate, 0.08; Tegosept, 0.10; brewers' yeast, 4.00; yeast hydrolysate, 0.30; granulated dehydrated carrot, 7.00; powdered dehydrated carrot, 7.00.

Diet 2 (%): water, 70; HCl, 0.6; sodium benzoate, 0.08; Tegosept, 0.1; torula yeast, 5; sugar, 6.22; corn cob grits, 13; powdered carrots, 5.

DEVELOPMENT For *A. ludens:* survival to mature larvae, 50%. For *A. suspensa:* survival to mature larvae on carrot diet, 15%. 7 generations reared.

Anastrepha suspensa (Loew), Caribbean fruit fly

→ Burditt, A.K. Jr; Lopez D. F.; Steiner, L.F.; Windeguth, D.L. von; Baranowski, R.; Anwar, M. 1974. Application of sterilization techniques to *Anastrepha suspensa* Loew in Florida, United States of America. pp. 93-101. *in* Sterility Principle for Insect Control 1974. Proceedings of a Panel Organized by the Joint FAO/IAEA Division of Atomic Energy in Food and Agriculture.

Two diets described.

COMPOSITION Diet 1 (%): water, 64.5; Tego, 0.15; sodium benzoate, 0.10; sugar, 6.0; wheatgerm, 3.0; torula yeast, 4.0; HCl, 0.7; ground corn cob (20-40 mesh), 21.55.

Diet 2 (%): water, 80.95; agar, 0.3; Tego, 0.1; sodium benzoate, 0.1; cholesterol, 0.05; sugar, 6.0; Embo (defatted wheat germ), 6.0; torula yeast, 6.0; HCl, 0.5.

PREPARATION Diet 1: all ingredients except corn cobs mixed for 20 minutes, then corn cob added, and mixing continued for further 15 minutes. Three litres diet poured into 23 x 64 cm tray.

For diet 2: agar dissolved in about 50% of boiling water; other half used for mixing remaining ingredients. Agar solution added between 42-45° C, mixing continued for 2-3 minutes, then diet poured into trays.

REARING Adults at 25-27° C and 70-80% RH with 12 h photoperiod in 0.2 m^3 cages holding approximately 28000 flies. Water provided from 4 blocks of 1% agar; food supply of 454 g of cubed sugar. Thick mixture of hydrolyzed yeast and water applied to 20 x 60 cm 16-mesh screen panel hung lengthwise in cage. Fresh yeast placed on top of cage at weekly intervals. Oviposition through panels of cheesecloth impregnated with hot mixture of paraffin/petrolatum (2:1 by weight) containing red candle dye. Two panels, arranged to form one side of cage, covered externally with moistened cellulose sponges and wood to prevent desiccation. Eggs collected from panels by fine mist spray of water and sterilized in 0.025% NaClO for 20-30 seconds, rinsed several times in water, and then mixed with 0.03% solution of sodium benzoate in water. Eggs spread uniformly on paper towelling @ 1 ml/100 cm^2 of towelling (22500 eggs/ml).

On diet 1, larvae crawl off trays of diet into trays of Vermiculite for pupation. On diet 2, larvae washed from diet, then excess water drained off and larvae placed on Vermiculite for pupation. Pupae held at 26° C and separated from Vermiculite with 10-mesh screen.

DEVELOPMENT Mass produced successfully. Adult emergence from 13th day after seeding.

→ Kamasaki, H.; Sutton, R.; Lopez, D.F.; Selhime, A. 1970. Laboratory culture of the Caribbean fruit fly, *Anastrepha suspensa,* in Florida. ANN ENTOMOL SOC AM 63: 639-642.

PREPARATION Diet prepared in 1.5 l lots. Details in Rhode & Spishakoff* (1965).

REARING At 27° C with 9 h photoperiod. Adults in 91.4 x 61.0 x 38.1 cm screened cages. 2:1 mixture of granulated sugar and orange juice crystals plus separate containers of type M hydrolyzed protein and water given for food. Moist oviposition shells in cage for 6 h daily. Eggs removed by rinsing with 0.07% aqueous sodium benzoate and measured in 0.1 ml aliquots (ca. 1840 eggs) on to discs of batiste; placed on surface of 300 ml diet in 10 cm cube plastic containers. Maintained for 8 days then larvae washed through 14-mesh sieve. Mature larvae transferred to moist sand and Vermiculite, and 3 days before emergence, pupae sifted and placed in stock cages.

DEVELOPMENT Egg incubation period, 3 days; larval period, 8 days; pupal period, 12 days; emergence, 57%; three generations reared.

Ceratitis capitata (Wiedemann), Mediterranean fruit fly

→ Awadallah, A.; Fares, F. 1973. A low-cost artificial medium for the mass rearing of the Mediterranean fruit fly, *Ceratitis capitata* (Wied.). AGRIC RES REV 51: 63-67.

COMPOSITION (g): dried ground carrot, 900; dried yeast, 250; sodium benzoate, 15; and (ml): 2 N HCl, 30; water, 4000.

PREPARATION Ingredients mixed to homogenous paste. 5-6 kg dispensed to 58 x 29 x 5 cm plastic trays to depth of 3.5 cm.

REARING At 26°C and 65-75% RH. Oviposition on diet surface in rearing cages. 37000 eggs/tray. Trays transferred to 75 x 50 x 40 cm screened cages with sand in removable lower drawer for pupation. Sand removed and sieved daily to separate pupae.

DEVELOPMENT Equal if 400 g carrot powder or 500 g of either wheat or rice bran used: larval period, 8.3 days; pupal period, 10.5 days; emergence, 89.7%. Mixture of carrot powder and wheat bran gives larval period, 6.1 days; pupal period, 10.4 days; emergence, 94.0%. Mixture of carrot powder and rice bran gives larval period, 6.1 days; pupal period, 10.4 days; emergence, 89.6%.

→ Cavalloro, R.; Girolami, V. 1969b. Miglioramenti nell'allevamento in massa di *Ceratitis capitata* Wiedemann (Diptera, Trypetidae). REDIA 51: 315-327. (In Italian with English summary)

COMPOSITION Adult diet (parts): sugar, 10; debittered brewers' yeast, 2; enzymatic brewers' yeast hydrolysate, 1.

Larval diets. Diet 1 (g): ground cornstalks, 245; sugar, 120; torula yeast flakes, 80; and (ml): HCl, 40; benzoic acid, 5; water, 510.

Diet 2 (g): alfalfa meal, 300; sugar, 110; torula yeast flakes, 75; and (ml): HCl, 40; benzoic acid, 5; water, 470.

PREPARATION & REARING Optimum of 30000 eggs/kg diet.

DEVELOPMENT Satisfactory over 72 generations. Larval maturation earlier on cornstalk than on alfalfa diet.

→ Hafez, M.; Shoukry, A. 1972. Effect of irradiation on adult fecundity and longevity of the Mediterranean fruit fly, *Ceratitis capitata* Wied., in Egypt (Diptera: Tephritidae). Z ANGEW ENTOMOL 72: 59-66.

COMPOSITION (g): dried powdered carrot, 300; brewers' yeast, 85; sodium benzoate, 5; and (ml): 1 N HCl, 15; water, 1250-1350. pH 4-4.5.

REARING At 25°C and 60-65% RH. Adults reared in 60 x 40 x 40 cm cages and fed on mixture of 800 g sucrose and 100 g enzymatic yeast hydrolysate. Water provided. Oviposition through muslin side of cage. Eggs collected daily with brush. 1.5-2.0 kg larval diet dispensed to 40 x 25 x 7 cm trays. 1 ml eggs sprinkled with little water from salt shaker on to 1 kg diet. Trays covered for 3 days with polyethylene cover punctured with fine needle. Replaced by muslin cover on 4th day. On 8th day larvae separated from diet by 'Berlese funnel' method into dry sand or wood wool for pupation. From emerging adults 25 ♂♂ and 25 ♀♀ caged together to check fecundity and viability of eggs.

DEVELOPMENT Larval period, ca. 8 days.

See also: Scherney & Haisch (1967).

Monro, J. 1968. Improvement in mass rearing the Mediterranean fruit fly, *Ceratitis capitata* Wied. pp. 91-104. *in* Radiation, Radioisotopes and Rearing Methods in the Control of Insect Pests. Proceedings of a Panel Organized by the Joint FAO/IAEA Division of Atomic Energy in Food and Agriculture.

A. Nine semi-solid larval diets used for mass rearing.

COMPOSITION Water, 1500 ml; and (g): peanut paste, 50; sucrose, 300; Nipasol (propyl *p*-hydroxybenzoate), 3; Nipagin, 3. Remaining ingredients varied in each diet as follows (g):

diet SYP - brewers' yeast, 200; and sawdust, 360.

diet CeYP - brewers' yeast, 200; and dried cellulose pulp, 200.

diet BaYP - brewers' yeast, 200.

diet MYP - brewers' yeast, 200; and dried maize stalk, 200.

diet BrYP - brewers' yeast, 200; and wheat bran, 600.

diet BaCoP - solvent extracted cotton seed, 500; and dried extracted sugarcane (bagasse), 175.

diet MCoP - solvent extracted cotton seed, 500; and dried maize stalk, 175.

diet BrGYP - wheat germ, 150; brewers' yeast, 200; and wheat bran, 500.

diet BeYP - brewers' yeast, 200; and dried extracted sugar beet, 350.

PREPARATION Diets adjusted to pH of 4.0-4.5 with HCl. Sawdust, dried cellulose pulp, dried extracted sugarcane (bagasse), dried maize stalks, wheat bran, and dried extracted sugar beets chiefly as carriers. Quantities varied, when necessary, to achieve firm consistency.

REARING At 25°C and 70-75% RH with 14-16 h photoperiod. 30000 pupae sealed in 100 x 50 x 30 cm rearing cages of aluminium or PVC sheeting; roof of plastic fly screen; long wall facing light source of terylene gauze. Adults fed paste of 4 parts sucrose and 1 part enzymatic yeast hydrolysate, smeared on paper towels placed on roof. Water dispensed automatically for 1 minute in every 30 to strip of filter paper on roof. Food and watering pads changed weekly. Eggs laid through wall of terylene gauze, fed into water trough and filtered daily. Larvae reared in open trays @ 20 eggs/g diet. IR lamps used to drive mature larvae through screen into pupation tray of sawdust. Pupae stored in layers 2 cm deep.

DEVELOPMENT All diets gave >50% pupal yield. 5 generations reared on BaCoP diet.

B. Liquid diet for mass rearing.

COMPOSITION (g): cholesterol, 0.3; sucrose, 100.0; brewers' yeast, 50; Nipagin, 0.8; Nipasol, 0.8; and (ml): water, 500; and HCl to pH 4.0; and one vitamin pill.

vitamin pill: vitamin A, 5000 I.U.; vitamin D_3, 500 I.U.; and (mg): vitamin E, 0.5; vitamin K, 0.05; vitamin B_1, 5; vitamin B_2, 5; vitamin B_4, 5; vitamin B_6, 0.5; folic acid, 0.5; nicotinamide, 50.0; vitamin C, 75.0; cobalt, 0.5; copper, 1.0; fluorine, 0.2; iron, 20.0; magnesium, 25.0; manganese, 2.0; potassium, 0.2; sodium, 1.0; and zinc, 1.0; and vitamin B_{12}, 0.5 µg.

REARING 600 larvae/5 ml diet as thin film in small petri dishes. After 6 days reduced to 25 larvae/5 ml.

DEVELOPMENT Survival rate only 50% after 3 days in liquid diet compared with 80-90% in standard carrot-yeast diet and diet BrGYP. Thereafter, survival rates equal. Early mortality due to drowning; avoidable by keeping young larvae on thin sheets of paper dampened with diet. Addition of 0.19% β-carotene or 5% concentration of dried, water-soluble residue of fresh carrots resulted in considerable increase in growth and survival.

→ Mourad, A. 1975. Preliminary experiments in reducing the cost of Med. fly larval diets. pp. 237-240. *in* Sterility Principle for Insect Control 1974. Proceedings of a Panel Organized by the Joint FAO/IAEA Division of Atomic Energy in Food and Agriculture.

Best of several diets tested.

COMPOSITION (%): water, 52; wheat bran, 25; sucrose, 13; brewers' yeast, 9; 37% HCl, 0.8; sodium benzoate, 0.2.

PREPARATION Diet prepared by blending HCl, aqueous sodium benzoate, yeast, and sucrose successively with water; then mixing in wheat bran. pH adjusted to 4-4.5. Diet placed in 1.5 x 15 cm diam petri dishes @ 150 g/dish.

REARING At 25 ± 1°C and 70-90% RH; 3000 eggs seeded/dish and covered to maintain humidity. After 4 days, covers slightly opened for ventilation. One day before pupation, covers removed and dishes placed in 25 x 20 x 10 cm paper boxes; wheat bran added to boxes to encourage pupation. Bran and pupae collected daily for 3-4 days.

DEVELOPMENT Period to pupation, 10.3 days; mean pupal wt, 8.9 mg; pupal recovery from eggs, 51%; adult emergence, 90%.

→ Nadel, D.J. 1970. Current mass-rearing techniques for the Mediterranean fruit fly. pp. 13-19. *in* Sterile Male Technique for Control of Fruit Flies. Proceedings of a Panel Organized by the Joint FAO/IAEA Division of Atomic Energy in Food and Agriculture.

Many diets tested.

COMPOSITION 'Washing diet' (USDA) (%): sodium benzoate, 0.1; Nipagin, 0.1; granulated sugar, 12.4; dry brewers' yeast, 3.1; concentrated HCl, 0.9; tap water, 59.8; wheat shorts, 5.0; Gelgard M, 0.8; middlings, 17.8.

'Popping diet' (USDA) (%): as diet 1, except tap water, 64.8; wheat shorts, 17.8; and no middlings.

'Popping diet' (ORISA) (%): sodium benzoate, 0.1; Nipagin, 0.1; granulated sugar, 13.0; dry brewers' yeast, 7.8; concentrated HCl, 0.7; tap water, 63.5; wheat germ, 7.1; bagasse, 7.7.

'Popping diet' (IAEA) (%): sodium benzoate, 0.1; Nipagin, 0.1; granulated sugar, 13.0; dry brewers' yeast, 8.5; concentrated HCl, 2.3; tap water, 50; wheat bran, 26.

'Popping diet' (Palemo) (%): sodium benzoate, 0.2; granulated sugar, 8.0; dry brewers' yeast, 5.6; concentrated HCl, 2.0; tap water, 66.3; lucerne grass powder, 9.4; crushed straw, 8.5.

REARING At 25-27°C and 80% RH with 16 h photoperiod or continuous lighting. Oviposition cages of various sizes. Water provided from 0.75-1% agar. Adults fed mixture 1 part yeast hydrolysate and 3-5 parts sugar. Eggs collected daily from perforated plastic or synthetic cloth screen and seeded on to larval diet. Larvae kept at 27°C for 5 days, then 20°C for 2 days; mature larvae collected by sieving into moist Vermiculite or in boxes under larval trays. Pupation in dry sawdust or bran. After 4 days pupae separated by sifting and 1 cm layer placed in screen bottomed trays.

DEVELOPMENT Highly satisfactory for mass rearing on all diets.

→ Peleg, B.A.; Rhode, R.H. 1970. New larval medium and improved pupal recovery method for the Mediterranean fruit fly in Costa Rica. J ECON ENTOMOL 63: 1319-1321.

COMPOSITION Bagasse diet (g): water, 5500.00; methyl *p*-hydroxybenate, 11.25; sodium benzoate, 11.25; 1 N HCl, 66.0; sucrose, 1135.00; torula yeast type B, 680.00; wheat germ, 620.00; bagasse, 675.00.

Carrot diet (g): water, 5400; methyl *p*-hydroxybenzoate, 9.25; sodium benzoate, 9.25; 1 N HCl, 5.5; brewers' yeast 33-B, 360.00; carrot powder, 1260.0.

PREPARATION Ingredients mixed in above sequence. Wheat germ essential for larval development and for providing suitable diet texture (see Monro 1965). Slight variation in particle size and moisture content of bagasse requires adjusting quantity so as to obtain very loose mixture. pH from 4.3-4.5.

REARING 153000 newly collected eggs placed on diet in half-filled 81 x 36 x 5 cm plastic lined trays. Reared for 9 days at 27°C in screened cabinets holding 26 trays. Mature larvae induced to leave diet (method and construction of cabinets described) and pupate in sawdust at base of cabinet.

DEVELOPMENT On bagasse and carrot diets, averages of 35.1% and 37.1% pupal recovery respectively. Cost of producing 1 million pupae on bagasse diet, US$14.10; on carrot diet, US$97.90.

See also: Nadel & Peleg (1968); Peleg *et al.* (1968).

→ Tanaka, N.; Okamoto, R.; Chambers, D.L. 1970. Methods of mass rearing the Mediterranean fruit fly currently used by the U.S. Department of Agriculture. pp. 19-23. *in* Sterile Male Technique for Control of Fruit Flies. Proceedings of a Panel Organized by the Joint FAO/IAEA Division of Atomic Energy in Food and Agriculture.

Two diets given.

COMPOSITION Diet 1 (% by weight): sodium benzoate, 0.1; Nipagin, 0.1; granulated sugar, 12.4; dried torula yeast, 3.1; concentrated HCl, 0.9; wheat shorts, 5.0; Gelgard M (moisture control agent), 0.8; middlings, 17.8; tap water, 59.8.

Diet 2: as diet 1, except no middlings; wheat shorts increased to 17.8% and water adjusted to 64.8%.

PREPARATION Water, preservatives, granulated sugar, and torula yeast mixed in 127-l capacity concrete mixer for 5 minutes. HCl added and mixed. Then for diet 1: Gelgard M and shorts blended, added to mixer and mixed till even. Middlings added and whole mixed until smooth; for diet 2: Gelgard M blended with ½ shorts and added to mixer. Remaining shorts added and mixed until smooth.

REARING Mass rearing technique described in detail. Briefly, 30000 adults in 30.5 x 61.0 x 121 cm screened cage for oviposition. Fed on 453 g granulated sugar and 226 g enzymatic yeast hydrolysate, with water from 0.75% agar gel on roof of cage. Kept for 3 weeks at 27.0 ± 2°C and 50-70% RH with egg collections 3 times/week. Eggs laid on water saturated 25 x 2.5 x 5 cm cellulose sponge over 18-20 h period in cage, then washed out.

DEVELOPMENT Larval period, 6 days; pupal period, 4 days; recovery, 90%. Cost $10.00/million.

Ceratitis rosa (Karsch) [previously *Pterandrus*], Natal fruit fly

→ Barnes, B.N. 1976. Mass rearing the Natal fruit fly *Pterandrus rosa* (Ksh.) (Diptera: Trypetidae). J ENTOMOL SOC S AFR 39: 121-124.

Adapted from Nadel* (1970) for rearing *Ceratitis capitata* and Monty (1973) for rearing *Pterandrus rosa*.

COMPOSITION Larval diet (g): torula yeast powder, 55; white sugar, 156; digestive bran, 300; Nipagin, 2; fresh carrot pulp (with all liquid retained), 840; sodium benzoate, 3; and (ml): acetone, 100; 1 N HCl, 172; water, ca. 600.

Adult diet: 5:1 mixture of white sugar and enzymatic yeast hydrolysate.

PREPARATION Yeast, sugar and digestive bran thoroughly mixed. Nipagin dissolved in acetone and added. Entire mixture spread out to dry. Carrot, sodium benzoate (dissolved in little water), and water added and mixed well. pH of diet, 4.0 to 4.5.

REARING Larval rearing at 25°C and 60-70% RH with 9 h photoperiod in 24 x 30 x 3 cm plastic trays covered with cheesecloth (illustrated) to prevent contamination of diet by *Drosophila* spp. Trays filled to 3 cm depth with ca. 2 kg of larval diet; inoculated with 25000 eggs on damp filter paper. Trays of mature larvae inverted on to large wire-gauze screens above sand in larval collecting boxes (illustrated) at 12°C. Pupae sifted from sand.

Adults reared at 20.5-27.5°C and 35-55% RH in net sleeve-cages 1.2 m long suspended between two 30 cm diam plastic basins (illustrated). Base of upper basin replaced by plastic gauze; water from wet sponge on gauze. Food from hose-pipe in troughs attached to inside of basins. 15000-20000 pupae placed in bottom of each cage amongst loosely crumpled paper. Oviposition in device of 8 cm diam plastic funnel perforated with ca. 0.3 cm diam holes lined with single layer of Whatman's No. 1 or 2 filter paper dampened with guava juice (illustrated). If left in cage for >12 h plug of cotton wood wetted with guava juice added to prevent desiccation. Most eggs laid through filter paper; soaked in beaker of water, filtered off and kept in saturated atmosphere until just before hatching when placed on diet.

DEVELOPMENT Egg hatch, 75-85%; pupal period, 10-11 days; pupal emergence, 75-95%; diet costs, 18.2 cents/kg. Production cost/100000 pupae, ca. R2.40.

See also: Monty (1973).

→ Étienne, J. 1973. Conditions artificielles nécessaires à l'élevage massif de *Ceratitis rosa* (Diptera: Trypetidae). ENTOMOL EXP APPL 16: 380-388. (In French with English summary).

Two diets required for mass-rearing each generation.

COMPOSITION Diet 1 (g): dehydrated carrot powder, 100; yeast, 50; dehydrated cooked potato, 75; methyl *p*-hydroxybenzoate, 2; sodium benzoate, 2; and (ml): water, 1000.

Diet 2 (g): dehydrated carrot powder, 50; dehydrated cooked potato, 50; torula yeast, 125; wheatgerm, 100; sugar, 200; methyl *p*-hydroxybenzoate, 4; sodium benzoate, 4; bagasse (sugar cane waste), 220; and (ml): 16.5% HCl, 60; water, 2000. To raise one million insects, 5 kg of diet 1 and 25 kg of diet 2 required.

PREPARATION All ingredients stirred to form paste.

REARING Sterilized eggs placed directly on to diet 1, in 10 x 15 cm boxes and kept at 26 ± 1° C for 2 days with solid lid, and further 1 day with ventilated lid. Larvae then transferred to diet 2, on which development completed. Contents of each box upturned on to surface of 25 x 30 cm dish with 2 kg of diet 2. These dishes placed on shelves, separated by 15 cm in laterally enclosed apparatus. Top open for ventilation and base made from funnel leading to plastic receptacle into which larvae fall when ready for pupation. Larvae reared at 21-25° C with 12 h photoperiod, collected from boxes each day and allowed to pupate in 20% moist Vermiculite at 21-25° C and 75-84% RH with 12 h photoperiod.

DEVELOPMENT (Days): total developmental time, 33-45; incubation, 2-3; larva, 8-12; prepupa, 1-3; pupa, 12-15; preoviposition, 10-12. Fertility of eggs, 85-90%; pupation, 75-80%; emergence, 85-90%.

Dacus cucumis French, cucumber fly

→ Bailey, P.; Ship, E. 1970. Corpus allatum size and ovarian environment in irradiated cucumber fly, *Dacus cucumis*. J INSECT PHYSIOL 16: 1293-1299.

COMPOSITION See Monro & Osborn* (1967).

REARING At 27° C and 80% RH with normal lighting. Eggs collected in thin slices of cucumber. Adults kept in 10 x 8 x 8" terylene covered cages with water, sugar, and Fleishman's yeast. For larval rearing see Monro & Osborn* (1967).

DEVELOPMENT Eggs mature 14 days after adult emergence; mating not necessary for egg development.

Corollary

Bailey, P. 1973. Respiration and nutrition of radiation - sterilized female *Dacus cucumis* (Diptera: Tephritidae). ENTOMOL EXP APPL 16: 433-444.

Feeding apparatus: 50 x 25 mm diam length of glass specimen tube sealed at one end with plastic stopper into which gauze-covered air hole and 2 holes for capillary tubes cut. Capillary tubes 100 µl capacity, 1 mm bore and open at both ends; one filled with 10% solution Fleishman's yeast and one with 50% sucrose solution. Each unit to hold 5 flies of same age.

Dacus cucurbitae Coquillet, melon fly

Chawla, S.S. 1966c. Studies on the nutritional requirements of the fruit fly, *Dacus cucurbitae* Coquillet (Diptera: Trypetidae). III. Development of a synthetic medium for *D. cucurbitae* larvae. RES BULL PANJAB UNIV SCI 17: 243-250.

COMPOSITION (g): vitamin-free casein, 1.5; cholesterol, 0.05; McCollum's salt mixture No. 185, 0.15; lecithin, 0.01; *l*-tryptophan, 0.02; *l*-cystine, 0.04; ribosenucleic acid, 0.07; adenine, 0.014; uracil, 0.014; glutathione, 0.010; yeast, 0.5; agar, 2.0; glucose, 5.0; and water, 100 ml; and (µg): thymine, 400.

vitamins (µg): thiamin, 250; riboflavin, 250; nicotinic acid, 250; pantothenic acid, 250; pyridoxin, 250; *p*-aminobenzoic acid, 250; choline chloride, 30000; inositol, 25000; biotin, 10; folic acid, 40; B_t, 400; B_{12}, 5.

PREPARATION Casein mixed with cholesterol dissolved in chloroform; solvent evaporated. Salt mixture, tryptophan, cystine, lecithin, RNA, glutathione and yeast added and ground in mortar. Adenine and uracil dissolved in water, and guanine in 1% NH_4OH, added. Then thymine and agar dissolved separately in boiling water; solutions added and mixed thoroughly. Transferred to test tubes, plugged with non-absorbent cotton and autoclaved at 10 lb pressure for 10 minutes. Aqueous solution of vitamins prepared and added to glucose in flask. Solution diluted so 1 ml sufficient/test tube, added while diet hot, stirred, tubes slanted; maintained at 35° C for 72 h.

REARING At 27° C. Eggs sterilized and inoculated aseptically to diet.

DEVELOPMENT Hatch within 24 h of inoculation. Larval period, 15-17 days. Larval period on control of yeast, 6-8 days. Development incomplete on diet but adequate for studying role of individual nutrients.

See also: Chawla (1966a b d).

Pant, N.C.; Ghai, S.; Chawla, S.S. 1959. A simple unfortified rice medium for mass-rearing of the melon fruit-fly, *Dacus cucurbitae* Coquillett under laboratory conditions. CURR SCI 28: 288-289.

COMPOSITION Finely powdered rice, 25 g and 100 ml water.

PREPARATION Mixture boiled with frequent stirring till thick paste. Dispensed to 8.5 x 6.5 x 2" enamel trays to depth of ¾".

REARING Large numbers of eggs or neonate larvae placed directly on food and covered with thin layer of moist cotton. Trays covered with fine muslin and kept at 26-29°C. Mature larvae transferred to jar of sand for pupation; sand sifted after 3 days, and pupae to oviposition chambers. Adult flies fed from hydrolyzed protein, sugar and water.

DEVELOPMENT Pupation, 80%; egg to adult emergence, 12-15 days.

→ Schroeder, W.J.; Miyabara, R.Y.; Chambers, D.L. 1970. A reusable diet support for rearing the melon fly on fluid larval medium. J ECON ENTOMOL 63: 2002.

COMPOSITION (g/100 g): torula yeast type 200, 4.0; granulated sugar, 6.0; extracted cottonseed embryo (Proflo), 1.0; Gelgard M, 0.84; methyl *p*-hydroxybenzoate, 0.1; sodium benzoate, 0.1; concentrated HCl, 0.2; water, 87.7.

PREPARATION Inhibitors added to water; other ingredients mixed dry and then added. Mixture poured over nylon grass mat (Astro Turf, Monsanto Co.) in rearing pan to depth of 2 cm.

REARING At 27 ± 2°C and 50-80% RH. Automated mass rearing; 13000 eggs/l of diet placed on diet surface on strips of toilet tissue. Vermiculite placed below each tray to catch larvae leaving diet before washing on 6th day.

DEVELOPMENT Mean weight/pupa, 13.4 mg. Pupal yield/l, 7317 ± 259 (60% return). Average adult emergence, 91%.

Corollary

→ Schroeder, W.J.; Miyabara, R.Y.; Tanaka, N.; Chambers, D.L. 1971. A fluid larval medium for rearing the melon fly. J ECON ENTOMOL 64: 1221-1223.

COMPOSITION As Schroeder *et al.** (1970) except Gelgard M increased from 0.84 to 1.30 g, and amount of water adjusted accordingly.

DEVELOPMENT Three generations reared. Pupal yield, 72%; adult emergence, 96%. Fecundity in successive generations increased with addition of 0.05 g cholesterol.

Dacus dorsalis Hendel, oriental fruit fly

→ Rejesus, R.S. 1975. Control of fruit flies by the sterile-insect technique. Studies on sterilization, population dynamics, fruit infestation sequence, and dispersal of *Dacus dorsalis* Hendel. pp. 124-125. *in* Controlling Fruit Flies by the Sterile-insect Technique. Proceedings of a Panel Organized by the Joint FAO/IAEA Division of Atomic Energy in Food and Agriculture.

COMPOSITION (g): cooked yellow sweet potato, 50; rice bran, 100; brewers' yeast, 33; sugar, 10; sodium benzoate, 1.5; water, 250; HCl to pH 4.5.

REARING Diet dispensed to trays within muslin cloth cabinets. Eggs seeded on to strips of paper towelling laid on diet surface, and covered with newspaper. 8-9-day-old larvae washed out of diet and put in moist sawdust for pupation. Stock cages contained cotton balls saturated with mixture of sugar, yeast hydrolysate and water. Oviposition on perforated plastic cups from which eggs brushed or rinsed.

DEVELOPMENT Larval development reduced to 6.5-7.0 days and preoviposition to 6.2-7.7 days when yeast added. Cost of 1 million pupae, US$25-30.

→ Taguchi, T. 1963. Evaluation of the dosage of various ingredients and the possible substitutes for carrot in the carrot-yeast medium in the larval culture of the oriental fruit fly, *Dacus dorsalis* Hendel. RES BULL PLANT PROT SERV JPN 38: 17-27.

COMPOSITION (g): fresh carrot, 100; dried brewers' yeast, 4; butyl *p*-hydroxybenzoate, 0.14; and (ml): water, 50; concentrated HCl, 0.4.

PREPARATION All ingredients except carrot briefly boiled. Fresh carrot finely sliced and added to mixture, then homogenized in blender.

REARING At 25° C and 80-90% RH. Adults in boxes covered with silken net fed on artificial diet of protein hydrolysate of yeast, soy bean and sucrose given separately from petri dishes. Water supplied from cotton ball soaked in water. Oviposition through pinholes into hollow plastic lemon with cotton ball soaked in orange juice inside. 100 eggs inoculated on to 50 g diet in 9 cm diam petri dish. Mature larvae placed in petri dish of sand with 10% moisture.

DEVELOPMENT Larval period, 5-6 days; larval wt, 21.6 mg. 236 out of 261 larvae to adult.

→ Taguchi, T. 1966. Influence of different diets upon the longevity of the adult of oriental fruit fly, *Dacus dorsalis* Hendel. RES BULL PLANT PROT SERV JPN 20: 16-19. (In Janapese with English summary)

COMPOSITION Best adult diet (g): soy hydrolysate and yeast hydrolysate, 4; sucrose, 6; McCollum's salt mixture, 400 mg; and sufficient water.

PREPARATION Hydrolysates mixed. Salts, yeast, sugar and water provided in separate dishes.

REARING At 25° C and 70-75% RH in Biotran with automatic light cycle of 380-9530-380 lux over 12 h period. 50 pairs of newly emerged adults in 30 cm cube cages.

DEVELOPMENT Longevity: ♀, 91.6 days; ♂, 82.9 days. Total of 13366 eggs laid with 71.1% hatch. Adult survival for ca. 3 months. Longevity and fecundity affected by protein content of diet.

Dacus oleae Gmelin, olive fruit fly

→ Cantarino, M.L.; Rey, J.M. 1970. Improvements in the artificial rearing of *Dacus oleae*. BOLL R SOC ESP HIST NAT SECC BIOL 68: 181-188.

Modification of Rey (1970) diet for *D. oleae*.

COMPOSITION (g): corncob powder, 165; chickpea sprouts, 200; brewers' yeast (Barcia), 75; Nipagin, 2.5; potassium sorbate, 0.5; and (ml): water, 505; olive oil, 20; 2 N HCl, 35. pH 4.4-4.5.

PREPARATION Chickpea sprouts blended for 2 minutes. ⅓rd of water, olive oil, most of HCl solution, Nipagin and potassium sorbate added. Mixed for 2-3 minutes then remaining ingredients mixed in. 100 g dispensed to 5 x 9 cm diam glass jars.

REARING At 25 ± 1°C. Cages for adult maintenance as described by Tzanakakis *et al.* (1967). Oviposition device (illustrated) from roughened petri dish covers with 7 holes each of 15 mm diam. Small domes of soft cerosine dyed black fixed over holes. Bottom of dish filled with water to maintain high humidity. Eggs collected by flushing water from cerosine domes and kept on black filter paper. 80 neonate larvae transferred to roughened surface of diet with fine brush. Pupae collected in sieve under running water and placed on moist cellulose in petri dish to hatch.

DEVELOPMENT Larval yield, 54%; pupation, 60%; mean pupal wt, 4.94 mg. Pupation completed in 18 days.

See also: Rey (1970).

→ Cavalloro, R. 1967. Orientamenti sull'allevamento permanente di *Dacus oleae* Gmelin (Diptera, Trypetidae) in laboratorio. REDIA 50: 337-344. (In Italian with English summary)

COMPOSITION Adult diet (parts): sucrose, 4; debittered brewers' yeast, 2; enzymatic brewers' yeast hydrolysate, 1.

Larval diet (g): ground cornstalks, 200; debittered brewers' yeast, 75; shelled "arachidi", 60; enzymatic brewers' yeast hydrolysate, 30; sucrose, 20; cholesterol, 1; Nipagin, 2.5; sorbic acid, 0.5; and (ml): water, 550; olive oil (Virgin), 20; 2 N HCl, 33; Tween 80, 8.

Mass rearing technique described.

See also: Cavalloro & Girolami (1969a).

→ Hagen, K.S.; Santas, L.; Tsecouras, A. 1963. A technique of culturing the olive fly, *Dacus oleae* Gmel., on synthetic media under xenic conditions. pp. 333-356. *in* Radiation and Radioisotopes Applied to Insects of Agricultural Importance. Proceedings of a Panel Organized by the Joint FAO/IAEA Division of Atomic Energy in Food and Agriculture.

COMPOSITION Adult diet (g): enzymatic protein hydrolysate of yeast, 20.0; sucrose, 80.0; choline chloride, 0.1; streptomycin sulphate, 0.5; and 100.0 ml water.

Larval diet (g): dehydrated carrot (Puccinelli Packing Co., Turlock, Calif.), 25.0; brewers' yeast, 15.0; agar, 2.00; enzymatic protein hydrolysate of soya, 6.0; choline chloride, 0.05; sodium benzoate, 0.3; Butaben (*n*-butyl parahydroxybenzoate), 0.05; and (ml): olive oil (Virgin), 15.0; Tween 80, 5.0; 2 N HCl, 7.0; and water, 125.0. pH about 4.1.

PREPARATION Larval diet. All ingredients except HCl and dehydrated carrot mixed in water and boiled. Cooled to 70°C and HCl or citric acid added. Homogenized at high speed for 1 minute then powdered dehydrated carrot mixed in. 50 ml dispensed to 9 cm diam dish to depth of 4 cm.

REARING At 25 ± 1° C and 75 ± 20% RH with natural lighting. 200 adults in 30 cm cube cages. Liquid adult diet brushed twice daily in streaks on to wax paper sealed to glass cage front facing light. Sucrose and water on cotton-wool pads provided in dishes on floor of cage. Eggs laid on paraffin wax 'domes' (construction described) attached to glass cage front in morning and removed in afternoon. Eggs removed by rinsing in distilled water. Incubated on moist filter paper at 25° C for 2.5 days, then mature eggs or neonate larvae transferred to grooves in diet with fine brush. Covered tightly and kept at 25° C. Pupae collected daily and placed in petri dishes with soaked cotton wool to hatch.

DEVELOPMENT Larval period, 11 days. Average pupal wt, 6.2 mg; pupal recovery, 32%; adult emergence, 100%. Preoviposition period, 13 days.

See also: Hagen (1966); LaBrecque & Keller (1965).

→ Manoukas, A.G. 1975. Low-cost larval diets for mass production of the olive fruit fly. J ECON ENTOMOL 68: 22-24.

COMPOSITION Basal diet (g): brewers' yeast, 7.5; soy hydrolysate, 3.0; sucrose, 2.0; potassium sorbate, 0.05; Nipagin, 0.20; and (ml): water, 55.0; olive oil, 2.0; Tween 80, 0.75; 2 N HCl, 3.00.

control diet 1: basal diet with SS cellulose No. 123, 30.0 g.

2: basal diet with SF cellulose, Grade BW-40, 27.5 g.

experimental diets 1: basal diet with MN cellulose No. 402a, 27.5 g.

2: basal diet with corn cob grits No. 4060, 25.0 g.

3: basal diet with corn cob meal 12-40 mesh, 25.0 g.

PREPARATION Nipagin, potassium sorbate, olive oil, Tween, and yeast homogenized in half amount of water. Sugar, soy hydrolysate, remaining water and 2 N HCl added in mixer and blended. Inert ingredient added and mixed until diet of granular appearance to give higher pupal yield. 60 g diet dispensed to round plastic containers to depth of 2-3 cm.

REARING At 25° C. 5 eggs/g diet placed on diet and mixed gently with top layer. Pupae removed from 12th day onwards.

DEVELOPMENT 90% pupation in 13-15 days; pupal yield, 30-76%; pupal wt, 4.3-5.8 mg; adult emergence from 67-83%, depending on diet.

See also: Manoukas (1974).

→ Mittler, T.E.; Tsitsipis, J.A. 1973. Economical rearing of larvae of the olive fruit fly, *Dacus oleae*, on a liquid diet offered on cotton towelling. ENTOMOL EXP APPL 16: 292-293.

Diet derived from that of Tzanakakis *et al.** (1970).

COMPOSITION (g): brewers' yeast, 12.0; sucrose, 2.75; soy hydrolysate, 4.0; and (ml): olive oil, 2.75; Tween 80, 1.0; preservative, 100.0.

preservative (%): methyl *p*-hydroxybenzoate, 0.2; potassium sorbate, 0.05; 2 N HCl, 4.0.

PREPARATION Ingredients blended for 1 minute and poured on to cotton towelling at rate of 2-4 g diet/g towel (approx. 20 cm^2).

REARING Eggs seeded on to towel directly, thus avoiding necessity of incubation on damp filter paper when using dry diet. Larvae burrow between layers of towels, feeding on readily accessible diet. Pupae collected for hatching. Practical advantages of use of cotton towelling discussed.

DEVELOPMENT Mean pupal weights of 4-6 mg comparable with that of diet of Tzanakakis *et al.**(1970).

See also: Mourad (1975).

Moore, I. 1962. Further investigations on the artificial breeding of the olive fly, *Dacus oleae* Gmel. under aseptic conditions. ENTOMOPHAGA 7: 53-57.

COMPOSITION Larval diet (g): vitamin-free casein, 3.8; brewers' yeast, 5.6; powdered cellulose, 5.0; agar, 3.0; Wesson's salt mixture, 2.8; sucrose, 4.0; and (ml): β-sitosterol emulsion (0.18 g of sitosterol and 0.16 ml of Tween 40), 20.0; choline chloride solution (0.084 g of choline chloride), 0.25; and distilled water to 120 ml.

Adult diet (parts): brewers' yeast, 1; yeast hydrolysate, 1; sucrose, 4; water, 5.

PREPARATION Stock emulsion prepared by dissolving sterol in boiling 95% ethanol, transferring to Tween 40/water solution, evaporating alcohol in autoclave and adding water to required volume. Diet sterilized by autoclaving at 15 lb for 20 minutes if in 100 ml flasks; or at 5 lb to dissolve agar then dispensed at 5 ml/tube and tubes autoclaved.

REARING Adults at room temperature and RH, in 1-pint muslin-covered jars containing single 12.0 x 2.5 cm glass slab on to which adult food atomized. Tap water on cotton wool in small plastic containers added. Eggs collected on paraffin dome sealed directly on to food slab. Eggs surface sterilized with mercuric chloride (1/1000) then incubated in sterile petri dishes with filter paper at 27 ± 1°C for 24-48 h. Neonate larvae aseptically transferred to test tubes at 27 ± 1°C and 90% RH.

DEVELOPMENT Total development, 24 days. 76.6% survival of neonate larvae to adults. Larval period, 14-16 days; pupal period, 9.4 days.

See also: Moore (1959).

Silva, G.M. de 1973. The use of parafilm as an oviposition device. AGRON LUSIT 35: 85-90.

REARING Two 45 x 120 cm sides of 30 x 45 x 120 cm cage covered by Parafilm stretched to half-thickness; 40 W lamps positioned 15 cm from either side. Eggs laid through parafilm and collected by manual flushing with water twice daily across parafilm. Technique illustrated.

DEVELOPMENT Fertility, 90%. Fecundity, 100-300 eggs/♀.

→Tischlinger, A. 1975. Rearing olive fly at Seibersdorf. pp. 137-139. *in* Controlling Fruit Flies by the Sterile-insect Technique. Proceedings of a Panel Organized by the Joint FAO/IAEA Division of Atomic Energy in Food and Agriculture.

COMPOSITION Larval diet (g/kg): potassium sorbate, 0.5; Nipagin, 1.0; sugar, 40; dried brewers' yeast, 100; soy hydrolysate, 30; yeast hydrolysate, 10; Vanderzant's vitamins, 10; cholesterol, 1; Wesson's salt mixture, 5; cellulose, 227.5; and (mg): Aureomycin, 50.0; and (ml): water, 480; olive oil, 80; linseed oil, 3; 1 N HCl, to adjust pH to 4.

Adult diet: 4:1 mixture of carbohydrate (sucrose: 9 parts; fructose: 1 part) and yeast hydrolysate. Water provided through wicks.

PREPARATION Larval diet: potassium sorbate, Nipagin, cholesterol and Aureomycin blended with some water for 2 minutes then added to premixture of soy hydrolysate, yeast hydrolysate, Vanderzant's vitamins and Wesson's salts. Whole mixed until smooth. Then following ingredients mixed in order: more water and sugar, olive and linseed oils, yeast and more water, cellulose and remaining water containing HCl. Diet then dispensed to paper cups @ 100 g/cup.

REARING At 25°C. Eggs sterilized in 1% formaldehyde for 2 minutes. Each cup inoculated with ca. 435 eggs and cup sealed. After 8 days, small holes punched in bottom to allow pupation in bran placed in trays below cups. Adult flies maintained in rotating 190 x 36 cm diam sleeve cages with 500 ♀♀ and 100 ♂♂/cage. Each cage with 6 wax cones for oviposition. Cages discarded after 6 weeks.

DEVELOPMENT Larval period, 10-16 days; larval mortality, 70-75%; pupal wt, 2.9-5.6 mg; adult emergence, 10-11 days from pupation at rate of 20-70%.

→Tsitsipis, J.A. 1975. Mass rearing of the olive fruit fly, *Dacus oleae* (Gmelin), at "Democritos". pp. 93-100. *in* Controlling Fruit Flies by the Sterile-insect Technique. Proceedings of a Panel Organized by the Joint FAO/IAEA Division of Atomic Energy in Food and Agriculture.

Best of 4 diets tested.

COMPOSITION (g): cellulose powder, 30.0; brewers' yeast, 7.5; enzymatic soy hydrolysate, 3.0; sucrose, 2.0; potassium sorbate, 0.05; Nipagin, 0.2; and (ml): tap water, 55; olive oil, 2.0; Tween 80, 0.75; 2 N HCl, 3.0.

REARING Eggs transferred to filter paper impregnated with 0.3% propionic acid in petri dishes for 24 h at 25°C, then surface sterilized with 2% Clorox (5-6% NaClO) for 10 minutes, rinsed in tap water for 10 minutes, suspended in 0.3% propionic acid and sprinkled on to diet. Mature larvae either crawl out of diet to pupate in sawdust or pupate on or inside diet. Pupae separated from diet by flotation in water.

DEVELOPMENT Average pupal yield/g diet, 2.54 (range 1.0-3.96). Pupal wt, 5.5-6.5 mg. Cost/1000 pupae, US$1.00. Quality of pupae comparable to that on previous diets.

→ Tzanakakis, M.E.; Economopoulos, A.P. 1967. Two efficient larval diets for continuous rearing of the olive fruit fly. J ECON ENTOMOL 60: 660-663.

COMPOSITION Diet 1 (g): agar, fine powder, 2.0; brewers' yeast powder, 9.0; enzymatic soya hydrolysate, 3.5; roasted peanuts, 6.0; sucrose, 2.4; olive oil, 2.4; Nipagin, 0.2; potassium sorbate, 0.05; and (ml): tap water, 66.0; Tween 80, 0.8; 2 N HCl, 3.5.

Diet 2 (g): cellulose powder (No. 123 for column chromatography), 26.0; brewers' yeast powder, 7.5; soy hydrolysate, enzymatic, 3.0; roasted peanuts, 6.0; sucrose, 2.0; olive oil, 2.0; Nipagin, 0.2; potassium sorbate, 0.05; and (ml): tap water, 55.0; Tween 80, 0.75; and 2 N HCl, 3.0.

PREPARATION Diet 1: ca. 45 ml tap water blended with roasted peanuts for 2 minutes. Mixed with Nipagin and potassium sorbate, then soya hydrolysate, sugar and agar. Mixture transferred to large Erlenmeyer flask where olive oil, Tween 80, and remaining water added. Boiled, with continuous stirring, and yeast added. Stirred for 1 minute. HCl added when cooled to below 70°C; finally stirred at high speed for 4 minutes. Dispensed while warm to 19 x 11 x 7.5 cm plastic boxes or 4.0 x 8.5 cm diam yoghurt cups.

Diet 2: unheated; as diet 1, except yeast added with soya hydrolysate and sugar. Stirred for 2 minutes and transferred to 'egg-beater' cup of mixer where remaining water, olive oil, Tween 80 and HCl added; stirred for 1 minute, then ca. 70% of cellulose powder added; after 2-3 minutes of stirring at high speed remaining cellulose added gradually, followed by another minute of stirring. Diet 2 considerably thicker than diet 1, needing spatula for transfer to rearing containers.

pH of both diets 3.8-4.0.

REARING Adults given water, solid sucrose, and liquid food of 1:4:5 mixture of enzymatic hydrolysate of yeast; sucrose; water with 0.05% choline chloride. If no choline, then enriched with 2% egg yolk and 6% olive fruit juice, and 50% of sucrose replaced by honey. Eggs incubated for 1 day on moist filter paper at 25 ± 1°C, then 600-800 eggs placed in grooves on diet in boxes; 50-100 eggs in cups.

DEVELOPMENT Adult emergence on diet 1 in one case, 75.6% and in another, 58.1%; on diet 2, 79%. Fecundity on both diets satisfactory. Egg viability on diet 1, 91.7-99.2%; on diet 2, 55.2-80.0%; on control (olive-reared), 63.8-78.2%. Adult longevity on both diets comparable to that of flies produced on olives. 7 generations reared on diet 1; 3 reared on diet 2.

Yield not affected if 'Minerva' olive oil (brand well-known on Greek market) replaced by other olive oils; but agar (fine powder by E. Merck A.G., or agar shreds by Fluka A.C.) replaced by powder supplied by Biolyon pupal yield reduced to almost zero.

See also: Economopoulos & Tzanakakis (1967); Santas (1965); Tzanakakis *et al.* (1968).

Corollary

→ Tzanakakis, M.E.; Economopoulos, A.P.; Tsitsipis, J.A. 1970. Rearing and nutrition of the olive fruit fly. I. Improved larval diet and simple containers. J ECON ENTOMOL 63: 317-318.

COMPOSITION (g): agar, fine powder, 0.5; cellulose powder, 20.0; brewers' yeast powder, 7.5; soy hydrolysate enzymatic, 3.0; roasted peanuts, 6.0; sucrose (granular), 2.0; and (ml): olive oil (Minerva), 2.00; Tween 80, 0.75; and (g): potassium sorbate, 0.05; Nipagin, 0.20; and (ml): 2 N HCl, 3.0; tap water, 55 ml.

DEVELOPMENT Pupal yield dependent on egg density in diet; 30-40% yield from two eggs/g diet on 32 cm round trays; 80% in 9 cm diam cups. Pupal weights from 4.5-5.8 mg; increased with yeast tablets instead of yeast powder. Ten generations reared.

See also: Cavalloro & Delrio (1973); Tzanakakis (1971); Tzanakakis *et al.* (1966, 1967, 1968); Wehrstein (1975).

Dacus tryoni (Froggatt) [previously *Strumeta*], Queensland fruit-fly

→ Barton-Browne, L.L. 1956. The effect of light on the fecundity of the Queensland fruit-fly, *Strumeta tryoni* (Frogg.). AUST J ZOOL 4: 125-145.

COMPOSITION Adult diet (parts): pawpaw paste (*Carica papaya*), 60; banana paste, 10; orange paste, 6; honey, 1.

Larvae reared on apples.

PREPARATION Mixed together and placed in 2" watch-glasses.

REARING At 26.6 ± 0.5°C and 75 ± 2% RH with natural dusk. Adults in wire-framed organdie cages with dry sucrose, diet and cotton wool saturated with water. Eggs laid on pawpaws. Apples artificially infested with eggs; 1 egg to 4 g apple. Intensity of light important factor in mating.

DEVELOPMENT Egg hatch, 92%.

→ Monro, J.; Osborn, A.W. 1967. The use of sterile males to control populations of Queensland fruit-fly, *Dacus tryoni* (Frogg.) (Diptera: Tephritidae). I. Methods of mass-rearing, transporting, irradiating, and releasing sterile flies. AUST J ZOOL 15: 461-473.

COMPOSITION Larval diet (g): Nipagin M, 8; dried powdered carrot, 917; dried brewers' yeast, 917; and (ml): concentrated HCl, 50; water to 9 l.

PREPARATION Ingredients mixed in stainless steel trays.

REARING Two ml eggs and neonate larvae spread over surface of 2.25 l of diet. 7-8 days later, trays emptied into clothes washing machine, water added, agitated 1 minute and pumped into terylene net with 1 mm^2 holes. Excess water removed by spinning, and bag emptied on to tray above sawdust in cage; pupation in sawdust within 24 h at 25°C. Pupae kept at 25°C and 70-80% RH for 8 days when sawdust sieved off, and 4000-5000 pupae/cup transferred for emergence to adult cages. Adults kept at 25-27°C and 70-80% RH with natural dusk needed for mating. Adults provided with sugar, enzymatic hydrolysate of yeast and water, changed weekly. Apple-hollows (domes of apple skin waxed on to watchglass), placed in cages after 3, 4, 5 and 6 weeks, for 7 h for oviposition. Eggs washed into beaker of water, and 2 ml ejected on to Buchner funnel; (12000 eggs/ml). Stored on damp filter paper in covered petri dishes at 28 ± 1°C until hatching after 3 days.

DEVELOPMENT Oviposition period, 2 weeks; larval period, 7-8 days; pupal period, >8 days.

See also: LaBrecque & Keller (1965).

Dacus zonatus (Saunders)

→ Qureshi, Z.A.; Ashraf, M.; Bughio, A.R.; Hussain, S. 1974. Rearing, reproductive behaviour and gamma sterilization of fruit fly, *Dacus zonatus* (Diptera: Tephritidae). ENTOMOL EXP APPL 17: 504-510.

COMPOSITION Larval diet (g): wheat shorts, 100; brewers' yeast, 17.0; granulated sugar, 33.0; agar, 3.5; Nipagin, 0.5; and (ml): 1 N HCl, 20.0; water, 400.

Adult diet: see Anwar *et al.* (1966).

PREPARATION Wheat shorts, yeast, sugar and Nipagin blended. Agar dissolved in boiling water; cooled to 50°C and main mix added, followed by HCl; blended for 2-3 minutes at pH 4.5 and poured on to 22.6 x 17.6 x 2.5 cm enamel trays. Tissue paper placed on diet for egg seeding.

REARING At 27°C and 60-70% RH with 9 h photoperiod. Eggs collected from cylindrical plastic domes and seeded on to diet @ 5000 (0.5 ml)/ tray.

DEVELOPMENT Pupal recovery, 69.3%; adult emergence, 93.5%; fertility, 77.9%; egg and larval period, 9.3 days; preoviposition period, 14.2 days; pupal wt, 12.9 mg; fecundity, 212.1 eggs/♀. Four consecutive generations reared.

Rhagoletis cerasi Linnaeus, European cherry fruit fly

→ Boller, E.F.; Remund, U. 1975. Application of sit on the European cherry fruit fly, *Rhagoletis cerasi* L. in north west Switzerland. pp. 77-82. *in* Controlling Fruit Flies by the Sterile-insect Technique. Proceedings of a Panel Organized by the Joint FAO/IAEA Division of Atomic Energy in Food and Agriculture.

COMPOSITION (%): crude sugar, 10.0; brewers' yeast, 10.0; wheat germ, 7.5; water, 55.9; paper pulp (dry), 13.0; Gelgard M, 1.0; cholesterol, 0.35; choline chloride, 0.07; Wesson's salt mixture, 0.18; ascorbic acid, 0.8; propionic acid, 0.4; citric acid, 0.8. pH 4.0-4.1.

REARING At 24°C and 60-70% RH with low light intensity.

DEVELOPMENT Survival from egg to adult, 20%; due to pupal diapause, only 2 generations/year, but possible to stock pile pupae throughout the year.

→ Haisch, A. 1968. Preliminary results in rearing the cherry fruit-fly (*Rhagoletis cerasi* L.) on a semi-synthetic medium. pp. 69-78. *in* Radiation, Radioisotopes and Rearing Methods in the Control of Insect Pests. Proceedings of a Panel Organized by the Joint FAO/IAEA Division of Atomic Energy in Food and Agriculture.

COMPOSITION (%): water, 61.10; carrot powder, 13.55; benzoic acid, 0.05; *dl*-malic acid, 1.20; citric acid, 0.80; Nipagin, 0.25; sugar, 5.00; torula yeast, 10.00; *Pseudomonas* sp., 5.00; choline chloride, 0.05; *myo*-inositol, 3.00.

PREPARATION Solid ingredients mixed, added to water and stirred well. Dispensed to 3 cm diam petri dishes. Addition of 5% dead *Pseudomonas* sp. beneficial. Bacteria isolated from cherries infested by *R. cerasi* larvae.

REARING At 25-27° C with 16 h light. 20-30 pairs of adults in 28 cm high plastic gauze cages. Water provided on cotton pad; water, sugar and brewers' yeast in 10:8:4 ratio provided on sponge renewed twice weekly. Oviposition on artificial cherry made of 2 cm diam foam ball, covered by thin stretched parafilm and coated with cherry extract; eggs also fall through gauze bottom of cage on to wet filter paper, collected daily. Optimum incubation temperature, 25° C; 40-80 neonate larvae placed on 5.0 g diet in petri dish.

DEVELOPMENT Larval period, 11-13 days; pupae smaller than on natural food.

→ Haish, A. 1970. Rearing experiments with the European cherry fruit fly. pp. 107-109. *in* Sterile-male Technique for Control of Fruit Flies. Proceedings of a Panel Organized by the Joint FAO/IAEA Division of Atomic Energy in Food and Agriculture.

COMPOSITION (%): wheat germ diet, 10.4; cellulose, 13.4; powdered carrots, 4; dried brewers' yeast, 4; sugar, 4; vitamin fortification mixture (NBC), 2.4; water, 44; formalin, 0.04; benzoic acid, 0.05; Nipagin M, 0.10; pH 4.5.

wheat germ diet (%): casein, 28; sucrose, 28; wheat germ, 24; Wesson's salt mixture, 8; Alphacel, 12.

REARING Oviposition on black wax domes.

DEVELOPMENT 3 days to 2nd instar larvae; 9 days to 3rd instar; pupae in 16 days; survival to adult, 10%.

→ Haisch, A.; Boller, E.F. 1971. Genetic control of the European cherry fruit fly *Rhagoletis cerasi* L. Progress report on rearing and sterilization. pp. 67-76. *in* Sterility Principle for Insect Control or Eradication. Proceedings of a Panel Organized by the Joint FAO/IAEA Division of Atomic Energy in Food and Agriculture.

Best of 5 diets tested.

COMPOSITION (%): wheat germ, 5; paper pulp, 20; Gelgard, 1.2; brewers' yeast, 5; sugar, 5; Wesson's salt mixture, 0.18; choline chloride, 0.07; cholesterol, 0.35; ascorbic acid, 0.1; citric acid, 1.0; Nipagin M, 0.12; water, 62.0; pH, 4.0.

PREPARATION Solids mixed dry, or if soluble, dissolved in boiling tap water, cooled and added to dry portion. Gelgard and citric acid added in two lots.

REARING Eggs laid on wax domes in oviposition cages. Units of 64 cm diam drum shaped cages with 500 wax domes on removable floor. Eggs removed by flushing inside of domes with water jet and filtering. 100 larvae placed on 18-20 g diet.

DEVELOPMENT Larval period, 12.7 ± 0.7 days; pupal yield, 22%.

See also: Haisch (1974).

Corollary

Haisch, A. 1975. Raising and multiplying of the European cherry fruit fly, *Rhagoletis cerasi* L. pp. 120-121. *in* Controlling Fruit Flies by the Sterile-insect Technique. Proceedings of a Panel Organized by the Joint FAO/IAEA Division of Atomic Energy in Food and Agriculture.

COMPOSITION (g): torula yeast, 5.6; wheat germ, 4.0; sucrose, 4.0; Vanderzant's vitamin mixture, 1.0; agar-agar, 4.2; and (ml): propionic acid, 0.4; 4 N HCl, 0.4; water, 80.4. pH 4.0.

DEVELOPMENT Pupation, 9-14 days; pupal yield, 46%; mean pupal wt, 3.9 mg.

Prokopy, R.J.; Boller, E.F. 1970. Artificial egging system for the European cherry fruit fly. J ECON ENTOMOL 63: 1413-1417.

COMPOSITION Adult diet (g): enzymatic yeast hydrolysate, 1; brown sugar, 4.

PREPARATION Made to thick slurry on filter paper strips and drip dried.

REARING Adults kept in 22 cm^3 wood frame cages of aluminium hardware cloth, plexiglass and Saran wrap. Mercury vapour lamps placed 30 cm above cages for 18 h photoperiod. Kept at 25 ± 1° C (light phase), 21 ± 1° C (dark phase) and 75 ± 5% RH. Strips of food hung on screened walls of cages, sugar cubes and dishes of cotton soaked in tap water provided on floor. Oviposition on artificial domes. Mass production of ceresin wax domes described in detail and procedure illustrated. Eggs removed from domes by flushing inside domes with water. Eggs incubated on moist filter paper at 25° C.

DEVELOPMENT Preoviposition period, 4 days; fecundity, 258 eggs/♀ yielding 234 larvae. Adult longevity, 30-32 days.

See also: Boller (1968).

Rhagoletis pomonella (Walsh), apple maggot

Boush, G.M.; Baerwald, R.J.; Miyazaki, S. 1969. Development of a chemically defined diet for adults of the apple maggot based on amino acid analysis of honeydew. ANN ENTOMOL SOC AM 62: 19-21.

Three adult diets tested.

COMPOSITION Diet 1 (mg): *dl*-alanine, 11.50; *l*-arginine-mono-hydrochloride, 35.80; *l*-asparagine, 117.60; *l*-aspartic acid, 6.20; γ-amino-butyric acid, 14.15; *l*-glutamic acid, 20.80; *l*-glutamine, 117.60; glycine, 15.90; *l*-histidine hydrochloride, 5.30; *dl*-isoleucine, 6.60; *l*-leucine, 2.20; *l*-lysine, 5.30; *l*-methionine, 0.90; *l*-ornithine, 10.60; *l*-phenyl-alanine, 4.00; *dl*-serine, 53.05; *l*-threonine, 53.05; *l*-tyrosine, 3.55; *l*-valine, 15.90; ascorbic acid, 0.250; biotin, 0.00025; *dl*-calcium pantothenate, 0.125; choline chloride, 1.250; folic acid, 0.125; *meso*-inositol, 1.250; nicotinic acid, 0.125; pyridoxine hydrochloride, 0.075; riboflavin, 0.125; thiamine hydrochloride, 0.075; fructose, 955.70; and salt mixture No. 2 USP, 45.00. Total weight 1500.00 mg.

Diet 2: as diet 1 with addition of: 10 mg *l*-tryptophan; 10 mg *l*-cysteine; and 10 mg *l*-cystine.

Diet 3 (mg): fructose, 2000; soy hydrolysate, 2000; brewers' yeast, 1000; and salt mixture No. 2 U.S.P., 500.

Adults also supplied with distilled water and ripe Winesap apple for oviposition.

PREPARATION Not described. To achieve semiliquid consistency vitamins added to 0.25 ml of water and then blended finally into diet.

REARING Pupae placed in open petri dishes within inverted battery jars and held at 23 ± 4° C and 60-80% RH with continuous lighting. Newly emerged adults removed daily and placed on test diets in growth chambers at 21 ± 1° C and 80 ± 6% RH with 16 h photoperiod. Diets changed every other day.

DEVELOPMENT Diet 1 unsatisfactory. Diet 2: ♀ maintenance for >1 month; resulted in 5 observed matings and production of 33 pupae, and appaeared comparable to control diet 3. Diet 3 tested with only 3 mated pairs; 3 matings observed and 21 pupae produced.

→Neilson, W.T.A. 1973. Improved method for rearing apple maggot larvae on artificial media. J ECON ENTOMOL 66: 555-556.

Based on Neilson (1969) diet for *R. pomonella*.

COMPOSITION (g): agar, 1.0; sucrose, 4.0; brewers' yeast, 2.5; cholesterol, 0.35; choline chloride, 0.7; salt mixture No. 2 U.S.P. XIII, 0.35; shredded tissue paper, 2.0; and (ml): distilled water, 100; formalin (36% formaldehyde), 0.05. Diet also prepared without formalin.

PREPARATION Water, sucrose, and agar placed in beaker, heated to 93.3° C, and then stirred at low speed in Waring Blendor for 3 minutes. Cooled for 10 minutes and cholesterol added, again blending for 1 minute. Remaining ingredients added at 1 minute intervals. Autoclaved at 107° C in 5 lb/in^2 for 20 minutes. Dispensed to 85 x 25 mm plates.

REARING Eggs surface sterilized and 10-30 placed on diet plates. Covered, sealed and incubated at 26.7° C with 16 h photoperiod. Pupae removed and placed on moist Vermiculite in pill boxes; held at 22° C and 75-90% RH with 16 h photoperiod for adult emergence. 100-200 adults placed in 18 x 12 x 12" Saran-screened cages and held at 22 ± 1° C and 80 ± 5% RH with 16 h photoperiod. Fed sucrose from cubes in petri dish; and water from cotton wool in petri dish; and amino acid mixture (g): (vitamin- and salt-free casein acid hydrolysate, 93; tryptophane, 2; arginine, 2; histidine, 2; cystine, 1) mixed with minerals (salt mixture No. 2 U.S.P. XIII) in 4:1 ratio and sprinkled on filter papers. Vitamins supplied separately on filter paper.

DEVELOPMENT Fecundity, 2-6 times higher than adults raised on formalin-free diet than on apple or diet with formalin.

See also: Nielson (1965, 1969); Neilson & McAllen (1964a b, 1965).

Prokopy, R.J. 1967. Artificial diet for apple maggot larvae. J ECON ENTOMOL 60: 1161-1162.

COMPOSITION (g/1 diet): methyl *p*-hydroxybenzoate, 0.96; dehydrated carrot powder, 114.80; brewers' yeast powder, 53.60; and (ml/1 diet): 12 N HCl, 4.0; and water, 928.0.

PREPARATION Water, HCl and methyl *p*-hydroxybenzoate blended for 1 minute; brewers' yeast and dehydrated carrot added and blended for 5 minutes.

REARING At 22 ± 1° C and 33% RH for 1 day prior to egg seeding and at 75% RH thereafter.

DEVELOPMENT Ca. 90% egg hatch; 32% survival to pupae in average of 15 days. Adult longevity and fecundity equal to field population.

Miscellaneous - Tephritids

Maeda, S.; Hagen, K.S.; Finney, G.L. 1953. Artificial media and the control of microorganisms in the culture of tephritid larvae (Diptera: Tephritidae). PROC HAWAII ENTOMOL SOC 15: 177-185.

Three species reared:- *Dacus dorsalis* Hendel, oriental fruit fly; *Dacus cucurbitae* Coquillett, melon fly; *Ceratitis capitata* (Wiedemann), Mediterranean fruit fly.

Best 2 of many diets tested.

COMPOSITION Diet 1 (g/100 ml distilled water): flaked agar-agar, 1.1; sucrose, 4.0; cholesterol, 0.35; (U.S.P. XIII) salt mixture, 0.35; brewers' yeast, 3.25; choline chloride, 0.07.

Diet 2 (g/100 ml water): agar, 1.3; Butaben, 0.12; glucose, 4.9; wheat germ oil, 0.175; cholesterol, 0.175; (U.S.P. XIII) salt mixture, 0.35; brewers' yeast, 3.0; choline chloride, 0.07.

PREPARATION Diet 1: agar dissolved in water, and boiled for few minutes with other ingredients. pH ca. 5.5. 5 ml dispensed to 12 x 125 mm test tubes; plugged, autoclaved at 15 lb for 15 minutes, slanted and shaken just before agar set to ensure uniform distribution of nutrients in diet.

Diet 2: ingredients dissolved in boiling water. pH to 4.5 with 0.1 N HCl. When agar set, diet blended at high speed till pulpy consistency, then refrigerated.

REARING Eggs of *D. dorsalis* and *C. capitata* laid in orange peel sections sealed to glass slide with paraffin; washed into water, and pipetted into 20 ml centrifuge tube with 1:1000 $HgCl_2$ solution. Shaken periodically; tube then rinsed 3 times in sterile distilled water, and eggs pipetted on to damp, thin, white cloth in sterilized petri dish. Incubated for 24 h; 6 neonate larvae placed on diet and kept at 80 ± 8° F and 62% average RH. Larvae reared on diet 2 in trays.

DEVELOPMENT Average larval wt, 23.34 mg; larval recovery, 100%. *C. capitata* reared successfully on diet 1 and pumpkin pulp based diet.

D. cucurbitae reared successfully on diet 1.

See also: Hagen (1958).

Schroeder, W.J.; Miyabara, R.Y.; Chambers, D.L. 1972. Protein products for rearing three species of larval tephritidae. J ECON ENTOMOL 65: 969-972.

Three species mass-reared: *D. cucurbitae*, *D. dorsalis*, and *C. capitata*.

COMPOSITION Basal diet (g/100 g): torula yeast type 200, 3.0; granulated sugar, 6.0; wheat shorts (milling by-product containing 14-16% minimum protein, 3.5% minimum lipids, 7.0% maximum fibre), 15.0; sodium benzoate, 0.1; concentrated HCl, 0.2; Gelgard M, 0.7; water, 75.0.

Test diets: torula yeast type 200 omitted and alternative protein sources substituted: (i) solvent-extracted cottonseed products (50-55% protein; Proflo or Pharamedia, Traders Protein Division); (ii) fatted or lecithinated soy flour (52% protein); (iii) soy flour (53% protein); (iv) special soy protein (60% protein, Textrol, Central Soya, Chemurgy Division); (v) torula yeast type B (50% protein); (vi) marine protein concentrate (70% protein); (vii) *Saccharomyces fragilis* Jörgenson, combined with whey protein (total protein 58-60%).

PREPARATION Inhibitors, sodium benzoate, and HCl added to water. Other ingredients mixed dry and then added. Dispensed to plastic-lined pans.

REARING See Schroeder *et al.** (1970).

DEVELOPMENT Diets with torula yeast type 200 (standard), cottonseed protein, soybean protein, *Saccharomyces fragilis* combined with whey protein, torula yeast type B, and fluid larval diet (cottonseed + torula yeast), all nutritionally and physically adequate, as determined by regression analysis, for rearing *D. cucurbitae* and *D. dorsalis*. Only diets formulated with protein from yeast products, i.e. torula yeast and *S. fragilis*, adequate for rearing *C. capitata*. Diet with marine protein concentrate inadequate for all species.

Steiner, L.F.; Mitchell, S. 1966. Tephritid fruit flies. pp. 555-583. *in* C.N. Smith (Ed.). Insect Colonization and Mass Production, Academic Press, New York and London.
Three species reared: *C. capitata*; *D. dorsalis*; *D. cucurbitae*.

COMPOSITION For *C. capitata* and *D. dorsalis* (kg): sodium benzoate, 0.1; concentrated HCl, 0.5; brewers' yeast powder, 3.2; dehydrated carrot powder, 12.0; and water, 90.7 l; total, 106.5 kg; to make 100 litre diet.

For *D. cucurbitae* (kg): sodium benzoate, 0.1; concentrated HCl, 0.5; brewers' yeast, 4.25; dehydrated squash powder, 4.0; dehydrated carrot powder, 6.4; and water, 91.21; total, 106.5 kg; to make 100 l of diet.

PREPARATION Diet for mass production prepared in batches of 180 l, in large vats and electric mixers. pH 4.5.

REARING At 25-27°C and 60-80% RH in subdued light with simulated dusk periods to stimulate mating. Eggs incubated on wet paper in moist chamber. Pupation in moist sand, sawdust or fine Vermiculite. Mass production technique described, utilizing cages holding 25000 adult *D.dorsalis* or *D. cucurbitae*, or 30000 *C. capitata*. Egg collection from 4-9 days after emergence.

DEVELOPMENT Average yield of flies/g dry ingredients: *D. dorsalis*, 28; *D. cucurbitae*, 29; and *C. capitata*, 35.5. Pupae larger than wild ones. Mature larvae of *C. capitata* and *D. dorsalis* in 6 days; of *D. cucurbitae* in 5 days. Pupation in 2 days.

See also: Mitchell *et al.* (1965).

→ Tanaka, N.; Steiner, L.F.; Ohinata, K.; Okamoto, R. 1969. Low-cost larval rearing medium for mass production of oriental and Mediterranean fruit flies. J ECON ENTOMOL 62: 967-968.

Two species reared: *C. capitata*; *D. dorsalis*.

COMPOSITION Larval diet (g): methyl *p*-hydroxybenzoate, 102.2; sodium benzoate, 102.2; granulated sugar, 13333.0; dried torula yeast, type 200, 3333.0; concentrated HCl, 991.2 (833.0 ml); wheat shorts, 5333.0; wheat middlings, 19111.0; Gelgard M, 833.0; and tap water, 64.4 l. Total, 107538.6 g (amount for 100 l).

PREPARATION Water, preservatives, sugar and yeast mixed in 127-l capacity concrete mixer for ca. 5 minutes, HCl added and mixed, followed by Gelgard M. Middlings added last and mixed till even consistency with pH of 4.5. Dispensed to 3.5 x 30.5 x 87.3 cm polyethylene sheet lined aluminium pans to depth of 2.0 cm (4.2 l by volume).

REARING At 28 ± 1°C. 45000 mature eggs on 10.8 x 66 cm strips of toilet tissue seeded on to diet by placing across top of each pan.

DEVELOPMENT Pupal recovery, average pupal size, and adult eclosion of both species on this and carrot diet comparable. Larval periods, 5 and 6 days on present and carrot powder diet respectively. Cost reduced from US$40 to $20/million flies.

See also: Castillón *et al.* (1974); Cavicchi (1971); Christenson *et al.* (1956); Féron *et al.* (1958); Finney (1956); Hagen & Finney (1950); Madariaga *et al.* (1972a b, 1974); Marlowe (1934); Marucci & Clancy (1950); Municio *et al.* (1973); Tanaka *et al.* (1972).

Hemiptera

A. Heteroptera

Anthocoridae

Anthocorus nemoralis Fabricius

Anthocorus nemorum Linnaeus

→ Bronnimann, H. 1964. Rearing anthocorids on an artificial medium. COMMONW INST BIOL CONTROL TECH BULL 4: 147-150.

COMPOSITION (parts): water, 1; sweetened condensed milk (Nestlé "La Laitère"), 2.

PREPARATION Mixed.

REARING In 40 mm lengths of 10 mm diam glass tubing. Each end corked, with pierced ventilation hole covered with fine gauze. Strip of paper in tube touching both corks. Three methods of feeding (a) From soaked cotton wool in ventilation hole (no gauze) changed every 2 days. (b) Food inside plastic drinking straw glued to paper strip. Diet introduced to straw by drawing soaked cotton wool through with fine wire and forceps. Straw frozen and cut into 1 cm length for gluing. Useful method for rearing large numbers individually. Cage illustrated. (c) Cloth-gauze sided cage, and paper or glass plate sprayed with diet and dried before insects introduced. Water sprayed on surface daily.

DEVELOPMENT Larval development: 26.75 days on artificial diet; 28.77 days on control of natural prey - psyllids. Eggs laid if host-plant given. Viable eggs laid and reared to 3rd and 5th instar on artificial diet.

Lyctocoris beneficus (Hiura)

Xylocoris galactinus (Fieber)

→ Chu, Y.I. 1969. On the bionomics of *Lyctocoris beneficus* (Hiura) and *Xylocoris galactinus* (Fieber) (Anthocoridae, Heteroptera). J FAC AGRIC KYUSHU UNIV 15: 1-136.

Four diets tested.

COMPOSITION Diet A: as in Bronnimann* (1964).

Diet B: modified from Hagen & Tassan* (1965). (g): enzymatic protein hydrolysate of yeast, 10.0; enzymatic protein hydrolysate of casein, 10.0; fructose, 17.5; ascorbic acid, 2.0; and (mg): choline chloride, 25.0; and water, 100.0 ml.

Diet C: modified from Beck *et al.** (1958). (g): soluble starch, 2.0; dextrose, 2.0; sodium caseinate, 2.0; beef extract, 2.0; brewers' yeast powder, 4.0; cholesterol acetate, 0.1; Wesson's salt mixture, 0.1; and water, 100.0 ml.

Diet D: modified from Scheel *et al.* (1957). (g): dextrose, 2.0; soluble starch, 2.0; sodium caseinate, 2.0; brewers' yeast powder, 2.0; beef extract, 0.5; Wesson's salt mixture, 0.2; and (mg): cholesterol, 80.0; and water, 100.0 ml.

PREPARATION See respective references.

REARING Best of 4 methods evaluated. 1 cm cube of Uretane Foam soaked with diet and placed in flat-bottomed 2.5 x 8 cm glass tube. Tube corked and diet replaced daily.

DEVELOPMENT Diet C inferior for nymphal survival. No significant superiority observed among 3 remaining tested formulations. Diet D best for longevity and developmental rate of both species but egg production <33% of that on natural food. Oviposition period much shorter (1:7) in individual groups of both species on artificial diets compared with natural food.

Xylocoris flavipes (Reuter)

→ Mukherjee, A.B.; Som Choudhury, A.K. 1971. Artificial diets for rearing of *Xylocoris flavipes* (Reuter), a predator on some pests of stored cereals. INDIAN J ENTOMOL 33: 356-358.

Reared on two artificial diets.

COMPOSITION Diet 1: modified from Bronnimann* (1964). Sweetened condensed milk + water + honey, 2:1:1 by volume.

Diet 2: modified from Chu* (1969). (g): dextrose, 2.0; soluble starch, 2.0; sodium caseinate, 2.0; commercial yeast powder, 2.0; fish extract, 0.5; salt mixture, 0.2; cholesterol, 0.08; water, 100 ml.

salts (parts): steamed bone meal, 40; fine ground limestone, 40; iodized salt, 19; $MnSO_4$, 1; 30 g $ZnSO_4.7H_2O$ added to salt mixture.

PREPARATION For both diets, ingredients mixed thoroughly.

REARING Diet on soaked cube of foam in conical flask with nymphs.

DEVELOPMENT On diet 1; mean of 26.1 days to adults; survival, 15%; 5.8 eggs laid/♀. On diet 2: mean of 22.8 days to adults; survival, 45%; 20.2 eggs laid/♀; figures lower than when *Tribolium castaneum* larvae offered as food.

Lygaeidae

Oncopeltus fasciatus (Dallas), large milkweed bug

Scheel, C.A.; Beck, S.D.; Medler, J.T. 1958. Feeding and nutrition of certain Hemiptera. PROC 10 INT CONGR ENTOMOL MONTREAL (1956) 2: 303-308.

Best of several diets listed.

COMPOSITION (g): glucose, 6.25; starch, 6.25; sodium caseinate, 6.25; cholesterol, 0.25; Wesson's salt mixture, 0.50; brewers' yeast, 5.00; and (ml): corn oil, 1.5; and distilled water, 15.

PREPARATION See Scheel *et al.* (1957). Ingredients mixed, moistened with water and rolled into small pellets.

REARING At 70-80° F and 60% RH. In 1-pint icecream carton covered by petri dish bottom. Water provided from jar directly beneath carton via glass-sleeved dental cotton wick. Edge of carton sand papered for uniform contact with dish to prevent escape of small nymphs. Feeding chamber illustrated. Diet renewed daily.

DEVELOPMENT Growth from second instar to adulthood satisfactory. 40% and 80% survival on diet and milkweed seed respectively. Rearing from first instar unsuccessful.

Euschistus variolarius Palisot de Beauvois, one-spot stink bug, (Pentatomidae), (with non feeding first instar) reared from second instar to adult.

See also: Feir & Beck (1963); Scheel *et al.* (1957).

Corollary

Beck, S.D.; Edwards, C.A.; Medler, J.T. 1958. Feeding and nutrition of the milkweed bug, *Oncopeltus fasciatus*. ANN ENTOMOL SOC AM 51: 283-288.

Two diets tested.

COMPOSITION Diet A: granular purified diet (g): dextrose, 5; soluble starch, 5; sodium caseinate, 5; corn oil, 1; cholesterol acetate, 0.2; Wesson's salt mixture, 0.5; brewers' yeast powder, 4; and (ml): water, 21.

Diet B: liquid purified diet (g): dextrose, 2; soluble starch, 2; sodium caseinate, 2; corn oil, 0.8; cholesterol acetate, 0.1; mineral salts, 0.1; yeast hydrolysate, 1; and (ml): B-vitamin mixture, 2; water, 98.

B-vitamins: as in Beck (1953). (μg/g dry diet): choline chloride, 1000; thiamine hydrochloride, 12; riboflavin, 18; nicotinic acid, 100; calcium pantothenate, 40; pyridoxine hydrochloride, 16; inositol, 2000; *p*-aminobenzoic acid, 50; folic acid, 5; and biotin, 1.

FEEDING Insects confined in stender dishes ½ filled with paraffin wax. Hollow cut from centre and replaced by absorbent cotton wick, moistened either with water or liquid diet; or shallow holes cut in wax surface to take moistened granular purified diets. Lid of stender dish replaced by flat transparent plastic with small hole cut immediately above moist wick. Feeding chamber and other, photographic, equipment illustrated.

Srivastava, P.N.; Auclair, J.L. 1970. Responses of the large milkweed bug, *Oncopeltus fasciatus* (Hemiptera: Lygaeidae) to chemically defined diets. ENTOMOL EXP APPL 13: 208-216.

Three diets tested.

COMPOSITION Diet 1: modified after Auclair* (1965) for *Acyrthosiphon pisum*. *l*-amino acids (mg): alanine, 100; arginine, 400; asparagine, 300; aspartic acid, 100; cysteine, 50; cystine, 5; γ-amino-butyric acid, 20; glutamic acid, 200; glutamine, 600; glycine, 20; histidine, 200; *dl*-homoserine, 800; isoleucine, 200; leucine, 200; lysine monohydrochloride, 200; methionine, 100; phenylalanine, 100; proline, 100; serine, 100; threonine, 200; tryptophan, 100; tyrosine, 20; valine, 200.

vitamins (mg): ascorbic acid, 10.0; biotin, 0.1; calcium pantothenate, 5.0; choline chloride, 50.0; folic acid, 1.0; *i*-inositol, 50.0; nicotinic acid, 10.0; *p*-aminobenzoic acid, 10.0; pyridoxine hydrochloride, 2.5; riboflavin, 5.0; thiamine hydrochloride, 2.5.

other ingredients (mg): cholesterol benzoate, 2.5; K_3PO_4, 500; $MgCl_2.6H_2O$, 200; salt mixture No. 2 USP XIII, 5.0; and (g): sucrose, 5.0; and total water to 100 ml. pH 7.0.

Diet 2: modified after Dadd & Mittler* (1966) for *Myzus persicae*. *l*-amino acids (mg): alanine 100; arginine, 270; asparagine, 550; aspartic acid, 140; cysteine, 40; glutamic acid, 140; glutamine, 150; glycine, 80; histidine, 80; isoleucine, 80; leucine, 80; lysine monohydrochloride, 120; methionine, 40; phenylalanine, 40; proline, 80; serine, 80; threonine, 140; tryptophan, 80; tyrosine, 40; valine, 80.

vitamins: as for diet 1 except: folic acid and riboflavin reduced to 0.5 mg; *p*-aminobenzoic acid omitted.

other ingredients (mg): KH_2PO_4, 500; $MgCl_2.6H_2O$, 200; sequestrene salts-Cu, 0.4; Fe, 1.5; Mn, 0.8; Zn, 0.8; and sucrose, 5.0 g; and total water to 100 ml. pH 7.0.

Diet 3: *l*-amino acids (mg): alanine, 100; arginine, 250; asparagine, 500; aspartic acid, 100; cysteine, 50; cystine, 5; γ-aminobutyric acid, 10; glutamic acid, 150; glutamine, 150; glycine, 80; histidine, 80; *dl*-homoserine, 100; isoleucine 80; leucine, 80; lysine monohydrochloride, 120; methionine, 50; phenylalanine, 50; proline, 80; serine, 80; threonine, 140; tryptophan, 80; tyrosine, 40; valine, 80.

vitamins: as for diet 1 except: folic acid and *p*-aminobenzoic acid reduced to 0.5 and 5.0 mg respectively.

other ingredients (mg): KH_2PO_4, 500; $MgCl_2.6H_2O$, 200; sequestrene salts-Cu, 0.6; Fe, 1.5; Mn, 1.0; Zn, 1.0; and sucrose, 5.0 g; and water to 100 ml. pH 7.0.

PREPARATION See Auclair* (1965); Dadd & Mittler* (1966).

REARING At 20 ± 2°C and 64 ± 10% RH with 16 h photoperiod. For details of cage and rearing technique see Auclair & Raulston (1966).

DEVELOPMENT Adults not obtained but 10 fold nymphal weight gain during first or second moult on different diet solutions. Fourth instar nymphs reached adult stage and survived for 2-5 months. Better developed on diet with 5% sucrose, with weight gain equal to that on milkweed seed.

See also: Auclair *et al.* (1973); Gordon (1974).

Miridae

Lygocoris pabulinus (Linnaeus)

Wightman, J.A. 1968. Rearing and feeding *Lygocoris pabulinus* (Heteroptera: Miridae). ANNU REP AGRIC HORT RES STN UNIV BRISTOL 157-160.

COMPOSITION (g): potato dextrose agar, 39; casein extract, 1; yeast extract, 1; sucrose, 20; and water, 1000 ml.

PREPARATION Mixture autoclaved for 20 minutes at 15 lb/in^2 and stored in sterile flasks at 0° C.

REARING Fed from 1 x 1½" diet filled paper 'boat' glued to inside lid of disposable cup; or by placing petri dish with diet in 4 x 4 x 5½" plastic box; or by pouring diet into plastic cup to depth of 5 mm; or by pouring diet into 2 x 1" glass vial. Diet changed every 2-3 days. Adults fed on 10% sucrose solution from wick of paper tissue rolled in filter paper and fixed to bottom of plastic cup. Paper cone with apex removed put over paper tube to prevent drowning in solution in cup. Cages illustrated.

DEVELOPMENT Slower on diet than on natural food; no ovarian development.

Lygus disponsi Linnavurori

Hori, K. 1972. Utilization of sucrose and starch by the bug, *Lygus disponsi* Linnavuori (Hemiptera: Miridae). APPL ENTOMOL ZOOL 7: 79-82.

COMPOSITION (mg/100 ml water): casein hydrolysate, 2000; CH_3COOK, 100; CH_3COONa, 50; separate mixture of sucrose or soluble starch or neither (for sucrose, starch, or control diets respectively), 4000; dihydrostreptomycin sulphate, 13.5; penicillin G potassium (100000 units), 7; Chloromycetin (1000 mg/1 g), 32; inositol, 25; cholesterol, 50; sodium ascorbate, 100; $NaH_2PO_4.H_2O$, 80; KH_2PO_4, 160; NaCl, 20.

salts: $CuSO_4.5H_2O$, 0.167; $FeSO_4$, 2.000; $(NH_4)_2SO_4$, 2.180; $MgSO_4$, 22.000; $MnSO_4.H_2O$, 3.400; KI, 0.0065; $NaHCO_3$, 0.240; $Na_2MnO_4.2H_2O$, 0.050; $NaSeO_3$, 0.0218; $ZnSO_4.7H_2O$, 2.200; fructose-1, 6-diphosphate monocalcium, 200.

vitamins: ascorbic acid, 0.625; biotin, 0.005; B_{12}, 0.00025; *d*-calcium pantothenate, 0.25; choline bitartrate, 20.00; folic acid, 0.050; menadion sodium bisulphite, 0.025; niacin, 0.625; pyridoxine hydrochloride, 0.0625; riboflavin phosphate, 0.15; thiamine hydrochloride, 0.25; *d*-α-tocopherol acetate, 0.50; vitamin A, 0.200; vitamin D_3, 0.0005; linoleic acid sodium salt, 50.00; Tween 85, 150.00.

PREPARATION Casein hydrolysate solution, salts, vitamins and acetate mixture stored separately at 2-5°C; diet prepared by bringing ingredients to room temperature before mixing. Sucrose or starch added last.

REARING In glass tubing: 75 x 50 mm diam, 40 x 20 mm diam and 25 x 10 mm diam for 5, 3, and 1 bugs/tube respectively. Details of cages given in Landes & Strong* (1965).

DEVELOPMENT From 3rd instar to adult if sucrose and starch present in diet. Starch diet gave higher 3rd and 4th instar mortality than did sucrose diet, which resulted in higher mortality in 5th instar and during emergence. Rates of development on both diets similar, with ca. 60% survival to adults.

Lygus hesperus Knight

Raulston, J.R.; Auclair, J.L. 1968. Responses of *Lygus hesperus* to chemically defined diets. ANN ENTOMOL SOC AM 61: 1495-1500.

Diet modified from Auclair* (1965) developed for *Myzus persicae*.

COMPOSITION As in Auclair* (1967a): except, *l*-glutamine and *dl*-homoserine omitted; total amino acid concentration reduced from 4.315% to 2.915%; 35.0 g of sucrose reduced to 3.0-5.0 g; and pH of diet reduced from 7.6 to 5.0-6.0.

PREPARATION Amino acids and amides mixed, powdered in mechanical mortar for 1.5 h and stored at 0°C. Stock solution of (i) vitamins and (ii) cholesterol benzoate and salts (other than $MgCl_2.6H_2O$) prepared, each at 10 times desired concentration. Cholesterol benzoate partially dissolved by boiling for 10 minutes in 30-40 ml of water. 3rd solution of sucrose and $MgCl_2.6H_2O$ prepared at twice desired concentration. All solutions frozen and stored in darkness. For 100 ml of diet, 2.915 g of amino acids mixed with 10 ml of vitamin solution and 50 ml of sucrose-$MgCl_2.6H_2O$ solution in 100 ml volumetric flask. Water added to 100 ml. Components of diet usually dissolved within 2 h, except for traces of cholesterol benzoate and small amount of fine precipitate. pH of freshly prepared diet adjusted by addition of microlitre amounts of concentrated H_3PO_4 or saturated NaOH.

REARING At 24 ± 1.2°C and 60 ± 2% RH with 14 h photoperiod. On food sachet between 2 parafilm membranes as in Auclair* (1965). Cage developed because of nymphal cannibalism: diet sachet mounted on 35 x 35 mm glass ring. Aluminium disc, 35 x 3 mm, with four 12 mm holes into which small cages inserted, placed on upper membrane of sachet. Small cages made from size S-1 hollow polyethylene stoppers as in Auclair & Raulston (1966). Four nymphs fed on each diet mount. Improved feeding cage illustrated.

DEVELOPMENT Maximum survival, 49 days. Nymphs moulted 3 times to adult stage with up to 800% increase in weight. Only 3rd instar nymphs used. 100% 1st or 2nd instar mortality. Growth on diet ca. $\frac{2}{3}$ that on beans; 1st adults 2 days earlier on beans than on diet.

See also: Auclair & Raulston (1966).

Strong, F.E.; Kruitwagen, E. 1970. Gustatory discrimination between meridic diets by the bug, *Lygus hesperus*. J INSECT PHYSIOL 16: 521-530.

Diet modified from Strong & Kruitwagen* (1969).

COMPOSITION (mg/100 g diet): *l*-amino acids: alanine, 150; arginine hydrochloride, 200; asparagine, 150; diglutamic acid mono-calcium hydrate, 600; glutamic acid, 150; glutamine, 150; glycine, 180; histidine hydrochloride, 75; isoleucine, 120; leucine, 180; lysine, hydrochloride, 188; *dl*-methionine, 122; phenylalanine, 188; proline, 150; serine, 75; sodium glutamate mono hydrate, 300; threonine, 98; tryptophan, 33; valine, 150; total, 3249 mg.

salts (mg): $CaCO_3$, 37.8; $Ca_3(PO_4)_2$, 26.1; $CuSO_4.5H_2O$, 0.24; $FePO_4.4H_2O$ 2.6; $FeSO_4(NH_4)SO_4.6H_2O$, 4.1; $MgSO_4$, 37.8; $MnSO_4$, 3.4; $K_2Al_2(SO_4)_4.24H_2O$, 0.02; KCl, 21.0; CH_3COOK, 100.0; KI, 0.02; KH_2PO_4, 54.2; NaCl, 18.4; $CH_3COONa.3H_2O$, 50.0; $NaHCO_3$, 0.2; NaF, 0.1; $Na_2MoO_4.2H_2O$, 0.05; Na_2SeO_3, 0.02; $ZnSO_4.7H_2O$, 2.2; fructose-1,6-diphosphate monocalcium, 300.0; total, 658.25.

vitamins: ascorbic acid, 0.625; biotin, 0.005; B_{12}, 0.00025; *d*-calcium pantothenate, 0.25; choline chloride, 20.0; folic acid, 0.05; inositol, 25.0; menadione sodium bisulphite, 0.025; niacin, 0.625; pyridoxine hydrochloride, 0.0625; riboflavin phosphate, 0.150; sodium ascorbate, 100.0; thiamine hydrochloride, 0.25; *d*-α-tocopherol acetate (1360 I.U./g), 0.50; vitamin A (1000000 I.U./g), 0.20; viatmin D (40000 I.C.U./mg), 0.0005; total, 147.7432.

lipids: cholesterol, 50.0; ethyl linoleate, 50.0; total, 100.0.

miscellaneous: Tween 80, 150.0; sucrose, 4000.0; agar, 1000.0; Penicillin G (1435 I.U./mg), 7.0; Streptomycin (740 I.U./mg), 13.5; total, 5170.5.

Total ingredients, 9.3255 g. Water to 100 ml.

PREPARATION See Strong & Kruitwagen* (1969). All ingredients, except sucrose and agar, mixed at room temperature. Homogenized with an ultrasonic probe for 30 seconds. Sucrose and agar added to predetermined amount of boiling water in second container, dissolved, cooled to ca. 80° C and added to first mix. In experiments involving lipids, final solution rehomogenized with ultrasonic probe for an additional 30 seconds. Final solution poured into small glass tubes at ca. 40° C and immediately stored at 2° C.

REARING Test chambers from Plexiglass rings 10 cm in diam and 1 cm high. Covered with stretched nylon hose; bottoms fitted with removable Plexiglass disc. Diets on top of nylon screen, forcing bugs to feed in inverted position. Chambers positioned on glass plate inside box. Details of viewing box in Strong (1967). Test diets in small plastic dishes 19 mm wide and 3 mm deep. Diets covered with partially stretched parafilm. Two dishes on nylon screen of each test chamber.

DEVELOPMENT Amino acids phagostimulatory; but feeding not influenced by either sucrose, salts or lipids.

Corollary 1

→ Landes, D.A.; Strong, F.E, 1965. Feeding and nutrition of *Lygus hesperus* (Hemiptera: Miridae). I. Survival of bugs fed on artificial diets. ANN ENTOMOL SOC AM 58: 306-309.

REARING In 2 containers, one above other. Lower feeding cage from 11 mm sections of cellulose nitrate centrifuge tubing (16 mm ID). Piece of nylon stocking fastened to 1 end. Other end plugged with cork covered with cigarette paper after bugs introduced. Upper, food chamber, prepared by covering 1 end of small glass tube with membrane, filling tube with diet and covering other end with parafilm. Tube then placed on feeding cage with membrane in contact with nylon screen of feeding cage. Insects feed by piercing membrane. Parafilm best out of many membranes tested. Cages illustrated. Third-instar nymphs caged individually.

Corollary 2

→ Strong, F.E.; Landes, D.A. 1965. Feeding and nutrition of *Lygus hesperus* (Hemiptera: Miridae). II. An estimation of normal feeding rates. ANN ENTOMOL SOC AM 58: 309-314.

Best of 4 rearing methods.

REARING Hydraulic tube method used to obtain direct estimation of nett volume of juice ingested by feeding bugs. System of 3 tubes: A, 15 x 7 mm diam glass tubing; B, 20 mm x 3/16" diam Tygon tubing; and C, 16 cm of ¼ x ½ mm capillary tubing. Tubes assembled by connecting A to B and wrapping parafilm over free end of A. A filled completely with bean juice. C coupled to B, forcing juice into capillary tube. Calibrated with mercury; 10 mm tubing = 1.24 μl volume. Bugs housed individually in 12 mm holes in plexiglass block. Hole corked after bugs introduced. Tubes filled and left to stand 24 h before testing. 3-7-day-old adult used and conditioned to feeding on bean juice in food vials covered with parafilm. Vials removed after 24 h and blocks positioned so parafilm of tube A in contact with nylon hose fabric cover of cages. 24 h later bugs removed and tubes left to stand 24 h to check rate of evaporation. Readings of height of the liquid taken at intervals and converted to μl. Cage illustrated.

Corollary 3

→ Strong, F.E.; Kruitwagen, E. 1969. Feeding and nutrition of *Lygus hesperus*. III. Limited growth and development on a meridic diet. ANN ENTOMOL SOC AM 62: 148-155.

L. hesperus on diet containing high amino acid concentration produced 3 times more eggs as bugs fed on green beans. Growth greater and maturation faster when reared on diets containing 2-6% sucrose than when bean-fed; survival significantly higher on beans than on any diet. Eggs from diet-reared bugs infertile.

See also: Strong (1967).

Lygus lineolaris (Palisot de Beauvois), tarnished plant bug

→ Vanderzant, E.S. 1967b. Rearing *Lygus* bugs on artificial diets. J ECON ENTOMOL 60: 813-816.

Also for *Lygus hesperus* Knight.

Best 2 of 8 diets.

COMPOSITION Diet A (g): alfalfa meal (water extract), 3.0; casein, enzymatic hydrolysate, 2.0; sucrose, 3.0; and (mg): $NaH_2PO_4 \cdot H_2O$, 80; K_2HPO_4, 160; NaCl, 20; $MgSO_4$, 50; choline chloride, 50; inositol, 20; cholesterol, 50; linseed oil, 200; ascorbic acid, 100; and (ml): B-vitamins, 1.0; water to 100. pH to 6.5 with 10% KOH.

B-vitamins as in Vanderzant* (1959).

Diet B: as diet A, except: linseed oil omitted and replaced by 500 mg of soybean lecithin (with oil).

PREPARATION Alfalfa extract prepared by stirring alfalfa meal with ca. 25 times its weight of distilled water at 70°C for 1 h. Refrigerated overnight and centrifuged. Extract divided into portions equivalent to 3 g of dry meal, and frozen until needed. All ingredients, except ascorbic acid, blended in order listed (except that B-vitamins blended between inositol and cholesterol); pH adjusted to 7, and mixture autoclaved. pH adjusted to about 6.5 when ascorbic acid added. Diets refrigerated or frozen. Refrigerated diets containing ascorbic acid used within 7 days.

REARING At 27 ± 1° C with 13.5 h photoperiod. 6 neonate nymphs in 2-oz transparent plastic cup fitted with two holed polyethylene lid. Piece of nylon stocking placed between lid and cup to retain insects. 5-6 drops liquid diet in 12 x 35 mm shell vial, covered with parafilm and inverted through one hole in lid so that film touches nylon. Food changed daily. Adults removed to quart jars with screened covers. Food vial inserted through hole in screen. Water filled sponge placed in dish on bottom of jar. ♀♀ removed after preoviposition period and placed 1/larval-type cup, with screen glued over 1 hole and paper roll for oviposition in the other. Roll from 2 strips of thick filter paper, 4 cm long and 5 cm square of parafilm rolled around 4 mm glass rod with parafilm on outside. Rod removed and roll cut into 1 cm lengths, saturated with distilled water and dipped in diet. Eggs laid on exposed edges of papers.

DEVELOPMENT Average rate of development of *L. lineolaris* nymphs on diet A, 14.6 days; of *L. hesperus* nymphs, 14.9 days. Mean fecundity of *L. lineolaris* on diets A and B, 66 and 72 eggs/♀; average longevity, 21 and 22 days. Results with *L. hesperus* on same diets gave average of 19-24 and 21-32 eggs/♀ and average of 15-19 and 17 days longevity respectively. Developmental period on diet equal to that on beans; but adult longevity on diet shorter.

Three generations of *L. lineolaris* and 2 of *L. hesperus* reared.

Reduviidae

Rhodnius prolixus Stål.

→ Lake, P.; Friend, W.G. 1968. The use of artificial diets to determine some of the effects of *Nocardia rhodnii* on the development of *Rhodnius prolixus*. J INSECT PHYSIOL 14: 543-562.

Seven diets (basic, total vitamin diet, FNPPRT diet, NPPT diet, pantothenic diet, pyridoxine diet and thiamine diet) tested on normal and aposymbiotic 4th instar nymphs.

COMPOSITION Basic diet: fresh rabbit blood, 100 ml; sodium citrate, 3.8 g; and 0.061 g ATP (= 10^{-3} M ATP). Best diet for aposymbiotic larvae (FNPPRT - diet) (ml): basic diet, 100; and (mg): folic acid, 3.0; nicotinic acid, 10.0; calcium pantothenate, 10.0; pyridoxine hydrochloride, 10.0; riboflavin-5-phosphate, 2.0; and thiamine hydrochloride, 10.0.

PREPARATION Vitamin solution prepared day before use; all dissolved in 0.15 M NaCl except folic acid dissolved in N $NaHCO_3$. All vitamins except thiamine, riboflavin and folic acid autoclaved at 15 lb pressure for 15 minutes. Others sterilized by filtration through sterile Swinny Hypodermic adapter and 0.45 µm Millipore filters. Vitamins added as solutions to citrated rabbit blood in 10-130 µl quantities. Adenosine triphosphate 10^{-3} M added to blood to induce gorging.

FEEDING At 37° C in glass funnels. See Friend & Cartwright (1963). Apparatus (illustrated) of ½" diam funnel-shaped glass vessel with two ⅛" ID glass tubes projecting one from the side and one from the base; 1" and 12" lengths of tygon tubing connected to side and base tube respectively. Mouth of funnel capped with membrane of thin Durex rubber held in place with rubber band. Funnel assembly slotted into drilled 1¼" square brass block. Block warmed to blood temperature. All housed in lighted, ventilated 30 x 12 x 24" plywood box. Temperature differential between box and membrane (27° and 37° C) to aid feeding response. Insects in 2½ x ¾" diam long glass vials capped with No. 60 mesh brass strainer cloth held by rubber bands. Membrane soaked in distilled water for 24 h, blotted dry and secured to top of feeding funnel. Funnel filled and side tube blocked with short piece glass rod. Bottom tubing raised or lowered to alter pressure on diet. Close contact needed between brass mesh cloth and membrane. Up to 100 insects/day fed on different diets. Apparatus easy to clean and sterilize.

DEVELOPMENT No difference in average ecdysial times for larvae with *N. rhodnii* (normal) fed on various artificial diets; high percentage moulted.

See also: Friend (1965); Friend & Cartwright (1963).

Corollary

→ Smith, J.J.B.; Friend, W.G. 1970. Feeding in *Rhodnius prolixus* : responses to artificial diets as revealed by changes in electrical resistance. J INSECT PHYSIOL 16: 1709-1720.

Feeding apparatus: modification of Friend & Cartwright (1963). Insects singly in feeding cages of 2" length of 1½" plexiglass tubing. One end covered with No. 60 mesh brass cloth. Other end open, 5th instar larvae being unable to climb walls. Cage placed on diet chamber with brass screen resting on rubber membrane for feeding. Diet chamber of 1¾" diam glass tubing, ¼" plexiglass base and membrane cover. Glass tube cemented to hole drilled in wall to allow filling and electrode insertion. Apparatus illustrated.

See also: Friend & Smith (1971, 1975).

B. Homoptera

Aphididae

Acyrthosiphon pisum (Harris), pea aphid

→ Akey, D.H.; Beck, S.D. 1971. Continuous rearing of the pea aphid, *Acyrthosiphon pisum* , on a holidic diet. ANN ENTOMOL SOC AM 64: 353-356.

COMPOSITION (mg/100 ml distilled deionized water): *l*-amino acids: alanine, 100; arginine hydrochloride, 400; asparagine (anhydrous), 300; aspartic acid, 100; cysteine hydrochloride monohydrate, 50; cystine, 5; γ-amino butyric acid, 20; glutamic acid, 200; glutamine (commercial), 600; glycine, 20; histidine (free base), 200; *dl*-homoserine, 800; isoleucine (allo free), 200; leucine, 200; lysine hydrochloride, 200; methionine, 100; phenylalanine, 100; proline, 100; serine, 100; threonine (allo free), 200; tryptophan, 100; tyrosine, 20; valine, 200.

vitamins: *p*-aminobenzoic acid, 10.0; *l*-ascorbic acid, 100.0; biotin, 0.1; *d*-calcium pantothenate, 5.0; choline chloride, 50.0; folic acid, 1.0; *i*-inositol (meso) dihydrate, 50.0; nicotinic acid, 10.0; pyridoxine hydrochloride, 2.5; riboflavin, 5.0; thiamine hydrochloride, 2.5.

trace metals (cation μg): cupric chloride, 120; ferric chloride, 460; manganese chloride, 220; sodium chloride, 500; zinc chloride, 200.

others (mg): $Ca_3(C_6H_5O_7)_2$, 10.0; cholesterol benzoate, 2.5; $MgCl_2.6H_2O$, 200.0; KH_2PO_4, 250.0; and sucrose, 35.0 g.

PREPARATION Amino acids, vitamins, and salts except trace metals added to sucrose solution of about $\frac{2}{3}$ of total volume and stirred with counter rotating mixer for 1-2 h under N_2 stream. Trace metals - ferric, cupric, zinc, and manganese - individually added at end of mixing. Adjusted with 1.75 M KOH to pH 7.5. Solution then made to final volume and passed through Millipore filter (0.45 μm) into sterile flasks. Diet stored in small quantities at -20°C for up to 1 month.

REARING Feeding cage illustrated. Aphid chamber made of 2 glass sections; lower (85 mm diam x 30 mm) had 10 mm flange. Access hole midway on side for manipulation of aphids and hole plugged with cork. Upper section of aphid chamber was collared (35 mm diam x 30 mm) with 10 mm flanges on top and bottom edges and a side access hole. Diet chamber 35 x 100 mm diam made from glass tubing. Apparatus assembled by stacking diet chamber on top of aphid chamber with ring of filter paper between serving as gasket. All glass parts had opposing ears on both sides so could be held together by rubber bands. Parafilm membrane base to diet chamber forming a cavity holding ca. 15 ml diet. Apparatus sterilized in autoclave or 70% ethanol. Diet chamber assembled and filled in aseptic isolation chamber. Feeding at 23 ± 1°C with 16 h photoperiod.

DEVELOPMENT Eighteen generations reared. Although apparently satisfactory for continuous propagation, diet neither gave optimum growth nor reproduction achieved. Mean adult wt, 1.6 mg or 68% of that of plant-reared aphids. Mean number of progeny/♀/day; 0.54 on diet and 6 on plant. Mean generation period, 13.1 days for first 10 generations.

Corollary 1

→ Akey, D.H.; Beck, S.D. 1972. Nutrition of the pea aphid, *Acyrthosiphon pisum:* requirements for trace metals, sulphur and cholesterol. J INSECT PHYSIOL 18: 1901-1914.

Diet as above but improved by: 1. Trace metals as chloride salts (cation μg/100 ml of diet) Cu^{2+}, 120; Fe^{3+}, 920; Mn^{2+}, 220; Na^{+}, 1000; Zn^{+}, 400. 2. Others: (mg/100 ml diet): $MgCl_2.6H_2O$, 2.5; $MgSO_4.7H_2O$, 242.

DEVELOPMENT Mean adult weight on improved diet over 4 generations, 2.06 ± 0.025 mg; nearly 29% higher than on basal diet and 88% of plant-raised aphid. Average number of progeny/aphid/day, 2.20 over two generations or 40% of that of plant-raised aphids.

Corollary 2

→ Akey, D.H.; Beck, S.D. 1975. Programmable automated system for continuous rearing of the pea aphid, *Acyrthosiphon pisum,* on holidic diets. ENTOMOL EXP APPL 18: 1-16.

REARING Automated rearing consisting of freezers, diet bottles, distribution networks, pumps and diet chambers, all controlled by programmable sequence timer. Diet chambers: double spiral semicircular grooves cut in Teflon discs. Membrane-covered channel formed by bonding over double spiral. Diet pumped through channels of diet chamber 3 times/week. Once every 2 weeks, empty bottles of diet replaced and new diet chambers exchnnged for old ones. Aphids easily inspected with diet chambers connected to system or disconnected. Useful technique for mass production. Detailed instructions described and illustrated.

DEVELOPMENT Average generation time, 13.5 days; over 3 generations. Adult weights and reproduction comparable with manual system.

→ Auclair, J.L. 1965. Feeding and nutrition of the pea aphid, *Acyrthosiphon pisum* (Homoptera: Aphididae), on chemically defined diets of various pH and nutrient levels. ANN ENTOMOL SOC AM 58: 855-875.

COMPOSITION Chemically defined medium. *l*-amino acids (mg): alanine, 100; arginine, 400; asparagine, 300; aspartic acid, 100; cysteine, 50; cystine, 5; γ-aminobutyric acid, 20; glutamic acid, 200; glutamine, 600; glycine, 20; histidine, 200; *dl*-homoserine, 800; isoleucine, 200; leucine, 200; lysine monohydrochloride, 200; methionine, 100; phenylalanine, 100; proline, 100; serine, 100; threonine, 200; tryptophan, 100; tyrosine, 20; valine, 200.

vitamins (mg): ascorbic acid, 10.0; biotin, 0.1; calcium pantothenate, 5.0; choline chloride, 50.0; folic acid, 1.0; *i*-inositol, 50.0; nicotinic acid, 10.0; *p*-aminobenzoic acid, 10.0; pyridoxine hydrochloride, 2.5; riboflavin, 5.0; thiamine hydrochloride, 2.5.

others: cholesterol benzoate, 2.5; K_3PO_4, 500.0; $MgCl_2.6H_2O$, 2.00; salt mixture (No. 2 U.S.P. XIII), 5.0; sucrose, 35.0 g; water to 100 ml; pH of diet 7.6 ± 0.05.

PREPARATION Stock diets prepared in parts. Amino acids mixed and finely powdered. Stored dry in dark at 0° C. Stock solution of 70% sucrose and 0.4% $MgCl_2.6H_2O$ prepared and kept at 0° C. Water soluble vitamins at 10 times final concentration prepared and stored frozen in dark. Then 4.315 g amino acid mixture mixed in 100 ml flask with 25 ml distilled water and 10 ml vitamin solution. 2.5 mg powdered cholesterol benzoate dissolved in few ml boiling distilled water for 10 minutes and added to flask. 50 ml of sucrose and $MgCl_2.6H_2O$ solution, and salts added; then finally distilled water to 100 ml. All dissolved within 2 h to pH 7.6.

REARING At 21 ± 1° C and 50 ± 10% RH with 16 h photoperiod equivalent to ASA Diffuse Density of 2.00. 0.3 ml diet sealed in parafilm sachet attached to end of 25 mm length of 18 mm diam glass tubing. Sachet from 2 squares; one depressed into end of glass tubing, depression filled with diet and another square laid on top to exclude all air. Fused and cut with hot wire. Rubber ring placed over sealed edges.

Lower part of apparatus ½ or ¾-oz glass bottle, top of which covered by rubber ring of feeding tube. Ventilation provided by boring two 3-4 mm holes in sides of bottle. Piece cotton wad soaked with 1 ml water, placed once weekly in bottle to keep humidity high.

DEVELOPMENT Adult wt, 1.8-2.6 mg; feeding rate, 0.38 µl/aphid/day; longevity, 22-35 days; and development, 0.2 days longer than on peas.

See also: Auclair & Cartier (1963, 1964); Cartier & Auclair (1964, 1965); Maltais (1952).

Corollary 1

→ Retnakaran, A.; Beck, S.D. 1967. Aspects of mineral nutrition of the pea aphid, *Acyrthosiphon pisum* (Harris). J NUTR 92: 43-52.

Modified from Auclair & Cartier (1963).

COMPOSITION See Auclair & Cartier (1963) but aspartic acid increased from 100 to 300 mg and phenylalanine from 100 to 200 mg.

PREPARATION See Auclair & Cartier (1963). Dispensed to 25 ml Erlenmeyer flasks and stored at -20° C before use. Diet changed after 3 days under sterile conditions.

REARING At 20° C and 85% RH with 16 h photoperiod. Rearing cage (illustrated) of upper, diet chamber with cotton plugged, ventilated stopper and lower, aphid chamber; stretched parafilm over base of diet chamber. Aphid chamber with screen-covered ventilation hole in side. Filter paper gasket between chambers to prevent adhesion of membrane to aphid chamber. Membrane and screen sterilized in 70% ethanol, and filter paper exposed to germicidal lamp; remainder cage autoclaved at 120° for 1 h before assembly. 4 ml diet dispensed into each cage with aseptic syringe. 10 first instar nymphs, ca. 24 h old, weighed collectively and placed in each cage. Aphids assigned to various diets.

DEVELOPMENT Diet improved by replacing K_3PO_4 with 0.25% of KH_2PO_4 and supplementing diet with 0.010% $Ca_3(C_6H_5O_7)_2$, when growth and reproduction used as criteria.

See also: Retnakaran & Beck (1968).

Corollary 2

→ McLean, D.L. 1971. Probing behaviour of the pea aphid, *Acyrthosiphon pisum*. V. Comparison of *Vicia faba*, *Pisum sativum*, and a chemically defined diet as food sources. ANN ENTOMOL SOC AM 64: 499-503.

COMPOSITION As Auclair* (1965) except: aspartic acid increased to 300 mg; *dl*-homoserine replaced by 600 mg of *l*-homoserine; thiamine hydrochloride replaced by 3.6 mg of thiamine pyrophosphate chloride; cholesterol benzoate replaced by 2 mg of cholesterol; and K_3PO_4 replaced by 367 mg of dibasic K_2HPO_4, and 265 mg of $CaHPO_4$.

REARING Aseptically at 25° C day, 21° C night and 79-84% RH with 16 h photoperiod. Feeding apparatus modified with electrical measuring system (illustrated) to record feeding and salivation behaviour. 20 x 70 mm glass vial nearly filled with distilled water and covered with stretched parafilm. Sterilized under UV for 30 minutes when 3 drops diet placed on parafilm and covered by second stretched parafilm square. Small, cylindrical Plexiglass cages placed over vials, resting on internal rims. Cages contain aphids. Apparatus inverted on to Plexiglass tray, and kept in environmental growth chamber.

DEVELOPMENT Four geographic populations reared on diet with survival of 20.8, 58.3, 68.3, and 75.8%. These values lower than on host plants - *Vicia faba* and *Pisum sativum*.

Corollary 3

→ Srivastava, P.N.; Auclair, J.L. 1971a. An improved chemically defined diet for the pea aphid, *Acyrthosiphon pisum*, ANN ENTOMOL SOC AM 64: 474-478.

Best of 7 chemically defined diets tested; modified after Dadd & Mittler* (1966).

COMPOSITION (mg/100 ml): *l*-amino acids; see Dadd & Krieger* (1967).

vitamins: ascorbic acid, 100; biotin, 0.1; calcium pantothenate, 5.0; choline chloride, 50.0; folic acid, 0.5; *i*-inositol, 50.0; nicotinic acid, 10.0; pyridoxine hydrochloride, 2.5; riboflavin, 0.5; thiamine hydrochloride, 2.5.

salts: KH_2PO_4, 500; $MgCl_2.6H_2O$, 200; sequestrenes: -Cu, 0.4; -Fe, 1.5; -Mn, 0.8; -Zn, 0.8.

others: sucrose, 35.0 g; water to 100 ml. pH to 7.6.

DEVELOPMENT Reared for >7 successive generations. First generation nymphs to adults in 9 days; maximum weight, 1.55 mg; reproduction by 12th day; survival <30 days. Decline in number of progeny with successive generations, and none reached adult stage in 8th generation.

See also: Auclair & Srivastava (1972); Markkula & Laurema (1967); Srivastava & Auclair (1971b, 1974, 1975).

Aphis fabae Scopoli, bean aphid

→ Dadd, R.H.; Krieger, D.L. 1967. Continuous rearing of aphids of the *Aphis fabae* complex on sterile synthetic diet. J ECON ENTOMOL 60: 1512-1514.

COMPOSITION (mg): alanine, 100; arginine, 270; asparagine, 550; aspartic acid, 140; cysteine hydrochloride, 40; glutamic acid, 140; glutamine, 150; glycine, 80; histidine, 80; isoleucine, 80; leucine, 80; lysine hydrochloride, 120; methionine, 40; phenylalanine, 40; proline, 80; serine, 80; threonine, 140; tryptophan, 80; tyrosine, 40; valine, 80; sucrose, 15000; ascorbic acid, 100; thiamin hydrochloride, 2.5; riboflavin, 0.5; nicotinic acid, 10.0; pyridoxin, free base, 2.5; folic acid, 0.5; calcium pantothenate, 5.0; inositol, 50.0; choline chloride, 50.0; biotin, 0.1; KH_2PO_4, 500; $MgCl_2.6H_2O$, 200; sequestrenes: -Fe, 1.5; -Zn, 0.8; -Mn, 0.8; -Cu, 0.4; water to 100 ml; pH to 7.0 with KOH.

PREPARATION & REARING Both described in Mittler & Dadd (1964), Dadd & Mittler (1965), Dadd *et al.* (1967).

REARING At 20°C and 85% RH with 16 h low-intensity photoperiod.

DEVELOPMENT Dependent on levels of trace metals. Mean weight range from 317-425 μg over 20 generations; 25 generations reared.

➡ Ehrhardt, P. 1965. Speicherung anorganischer Substanzen in den Mitteldarmzellen von *Aphis fabae* Scop. und ihre Bedeutung für die Ernährung. Z VGL PHYSIOL 50: 293-312. (In German with English summary)

COMPOSITION & PREPARATION See Mittler & Dadd (1962) and Auclair & Cartier (1963).

REARING At 18-22°C with natural lighting. Diets offered in Parafilm sachets on top of small container or cage. 2-2.5 ml of diet/sachet, changed every other day.

DEVELOPMENT Maximum of 2 generations. For first 7 days moulting normal. Mortality ca. 5% rising to ca. 20% by 14 days. 70% of surviving larvae to adult from 12-19 days; all alatae dead by day 5; apterae survival up to 25 days on diet. Development twice as long than on plants.

See also: Danneel (1969); Ehrhardt & Schmutterer (1966); Leckstein (1974); Leckstein & Llewellyn (1973, 1974); van Emden (1967).

Aphis gossypii (Glover), cotton aphid

➡ Auclair, J.L. 1967a. Effects of pH and sucrose on rearing the cotton aphid, *Aphis gossypii*, on a germ-free and holidic diet. J INSECT PHYSIOL 13: 431-446.

COMPOSITION, PREPARATION & REARING See Auclair* (1965).

DEVELOPMENT Two generations reared. Optimum diet pH 7.4-7.8 and sucrose concentration of 20-30%.

See also: Auclair (1967b); Turner (1971): (35.0 g sucrose replaced by 20.0 g sucrose and 10 g maltose).

Aphis rumicis Linnaeus

➡ Herger, P. 1975. Einfluss von Farbe und Nahrungszusammensetzung auf das Saugverhalten der kunstlich ernahrten Ampferblattlaus, *Aphis rumicis* (Homoptera: Aphididae). ENTOMOL GERM 2: 149-166. (In German with English summary)

COMPOSITION, PREPARATION & REARING As in Herger & Lampel* (1975).

DEVELOPMENT 14-15 generations successfully reared.

Brevicoryne brassicae (Linnaeus), cabbage aphid

➡ White, D. 1972. Effect of varying dietary amino acid and sucrose concentrations on production of apterous cabbage aphids. J INSECT PHYSIOL 18: 1241-1249.

COMPOSITION & PREPARATION See Dadd & Mittler* (1966) for pea aphid, *Myzus persicae*.

REARING At 15-20°C and 70% RH with 14 h photoperiod in 1.0 or 2.6 cm diam perspex cages.

DEVELOPMENT Form of aphid influenced by sucrose and amino acid levels in diet. Highest production of apterae on diet with 0.5% amino acid and 30% sucrose; lowest production with 2.5% amino acids and 20% sucrose.

Chaetosiphon fragaefolii (Cockerell), strawberry aphid

→ Shanks, C.H.Jr; Finnigan, B.F. 1969. Behaviour, survival, and reproduction of the strawberry aphid on holidic diets and host and non-host plants. ANN ENTOMOL SOC AM 62: 1490-1492.

COMPOSITION See Auclair* (1965); Dadd & Mittler* (1966).

PREPARATION As above, except pH to 7.5 with saturated NaOH or KOH.

REARING At 21° C with 20 h photoperiod. Aphid chamber from 18 mm diam glass tubing. Cork, with cloth covered ventilation hole, pushed into tubing to within 1-2 mm of other end. This open end set on Parafilm membrane where aphid confined. Parafilm sachet prepared as described earlier. Sachet frozen at -18° C until used. Parafilm exposed to UV in sterile glove box and diet sterilized by Seitz filter.

DEVELOPMENT Ca. 63% survival after 9 days on all diets; ca. 37% survival after 23 days on those with higher ascorbic acid and 4% on those with lower levels. Poor reproduction on all diets.

Dysaphis plantaginea (Passerini)

→ Arnold, M.L. 1975. Zur sexuales- Produktion von *Dysaphis plantaginea* bei Haltung auf Künstlicher Diät (Homoptera: Aphididae). ENTOMOL GERM 1: 253-257. (In German with English summary)

COMPOSITION & PREPARATION See Dadd & Mittler* (1966); Mittler *et al.* (1970).

REARING At 18° C and 75% RH with 8 h photoperiod.

DEVELOPMENT Several generations reared.

Macrosiphum avenae (Fabricus), English grain aphid

→ Sutter, G.R.; Kieckhefer, R.W. 1969. Performance of the English grain aphid on sterilized and unsterilized artificial diet. J ECON ENTOMOL 62: 254-255.

COMPOSITION As in Kieckhefer & Derr* (1967). amino acids (mg/100 ml diet): alanine, 50; arginine, 200; asparagine, 150; aspartic acid, 50; cysteine, 25; cystine, 2.5; glutamic acid, 100; glutamine, 300; glycine, 10; histidine, 100; *dl*-homoserine, 500; γ-amino butyric acid, 10; isoleucine, 100; leucine, 100; lysine hydrochloride, 100; methionine, 50; phenylalanine, 50; proline, 50; serine, 50; threonine, 100; tryptophan, 50; tyrosine, 10; valine, 100.

(g/100 ml diet): sucrose, 20; cholesterol, saturated water.

vitamins (mg/100 ml diet): ascorbic acid, 10.0; biotin, 0.2; calcium pantothenate, 2.5; choline chloride, 20.0; folic acid, 0.5; iso-inositol, 25.0; nicotinic acid, 5.0; *p*-aminobenzoic acid, 5.0; pyridoxine hydrochloride, 1.5; riboflavin, 2.5; thiamine hydrochloride, 1.5; B_{12}, 0.2.

miscellaneous (mg/100 ml diet): *dl*-carnitine, 10.0; lipoic acid, 2.0; RNA, 50.0; standard salt mix U.S.P. No. 2, 2.5; K_2PO_4, 62.0; $MgCl_2.6H_2O$, 25.0.

PREPARATION See Kieckhefer & Derr* (1967). Dietary constituents in cholesterol saturated water. RNA, lipoic acid, tyrosine, cysteine and folic acid added. Water added to 100 ml; pH, 7.15-7.20.

Diet divided in to 5 ml portions, and stored at -20° C until needed. Liquid diet filtered through 0.20 μm Millipore filter assembly, then into feeding sachets, and irradiated for 1 h with UV lamp at distance of 25 cm.

REARING At 22° C. Aphid cages from 2 cm lengths of 2 cm diam cellulose nitrate tubing, lower end of which covered with No. 88 silk bolting. Top end covered with parafilm sachet containing 0.05 ml diet. Sachets of 2 parafilm membranes stretched to $\frac{1}{10}$ original thickness.

DEVELOPMENT Survival after 48 h, on sterile diets significantly higher than on unsterile diet. After 144 h, only treatments replaced with sterile sachet every 24 h, were significantly different from other treatments. Sterilization of sachet enhanced reproduction.

Macrosiphum euphorbiae (Thomas), potato aphid

→ Chawla, S.S.; Perron, J.M.; Cloutier, M. 1974. Effects of different growth factors on the potato aphid, *Macrosiphum euphorbiae* (Aphididae: Homoptera), fed on an artificial diet. CAN ENTOMOL 106: 273-280.

COMPOSITION & PREPARATION See Srivastava & Auclair* (1971a), basic diet No. 5 for *Acyrthosiphon pisum*.

REARING At 20 ± 0.5° C and 65-75% RH with 18 h photoperiod provided by fluorescent and incandescent light. Sachets covered by yellow paper.

DEVELOPMENT Ten generations reared; significant decline after 8 generations of successful rearing; number of adults declined from 84% in 8th generation to 42 and 14% in 9th and 10th generations respectively; corresponding increase in nymphal developmental period, and decrease in adult weight. Addition of 30 mg chlorogenic acid/100 ml basic diet gave progressively improved adaptation over 17 generations reared successfully.

See also: Cartier (1965); Cartier & Morin (1965).

Myzus persicae (Sulzer), green peach aphid

Since the first publication by Mittler & Dadd (1962) on an artificial diet for rearing the green peach aphid, *Myzus persicae* (Sulzer), diets for rearing this aphid have been progressively improved in numerous publications. Their diet formulated in 1966, which has also been used as a basal diet for rearing many other species of Hemiptera, is given below: (other diets are referred to when applicable).

→ Dadd, R.H.; Mittler, T.E. 1966. Permanent culture of an aphid on a totally synthetic diet. EXPERIENTIA 22: 832-833.

COMPOSITION (mg/100 ml): *l*-amino acids: alanine, 100; arginine, 270; asparagine, 550; aspartic acid, 140; cysteine hydrochloride, 40; glutamic acid, 140; glutamine, 150; glycine, 80; histidine, 80; isoleucine, 80; leucine, 80; lysine monohydrochloride, 120; methionine, 40; phenylalanine, 40; proline, 80; serine, 80; threonine, 140; tryptophan, 80; tyrosine, 40; valine, 80.

vitamins: ascorbic acid, 100; biotin, 0.1; calcium pantothenate, 5.0; choline chloride, 50.0; folic acid, 0.5; *i*-inositol, 50.0; nicotinic acid, 10.0; pyridoxine hydrochloride, 2.5; riboflavin, 0.5; thiamine, 2.5.

salts: KH_2PO_4, 500; $MgCl_2.6H_2O$, 200; sequestrenes: Cu, Fe, Mn, Zn, 0.4, 1.5, 0.8, and 0.8 respectively.

others: sucrose, 15.0 g; water to 100 ml. pH adjusted to 7.0.

PREPARATION See Mittler & Dadd (1963a b). Ingredients dissolved in ca. 80 ml of cholesterol saturated water in following order:- amino acids, sugar and salts. Sequence observed to avoid precipitation of salts. 10 ml vitamin solution then added; pH adjusted with K_2HPO_4 to 7.0, and diet made to 100 ml.

REARING As in Mittler & Dadd (1963a b). Feeding cages from 2 chambers cut from 32 mm diam "lusteroid" (cellulose nitrate) test tubes, one each for aphids and diet. Small square of Parafilm stretched almost to breaking point when 10-20 µm thick, stable and impermeable to water. Lower chamber made by cutting off round end of test-tube few mm above point where sides of tube become parallel. Base sliced off to leave slightly incurving edge which is then levelled and smoothed off. Other end of incurved annullus thus formed covered by silk gauze. Upper chamber made from top 2 cm of lipped test tube and sealing lipped end with parafilm square. Diet poured into upper chamber and enclosed by another parafilm square. Lower chamber pinned, open end upwards, by drawing pin through its gauze floor to cork cemented on glass microscope slide. Aphids placed in lower chamber, covered by upper diet chamber with membrane resting on incurved rim, forming complete seal. Feeding cage illustrated.

DEVELOPMENT Adult weight of 2nd generation considerably lower than that of 1st generation:- 330 µg compared to ca. 500 µg; thereafter average weight of 300-350 µg in each generation, larvae taking ca. 10-20 days to become adults. Diet-reared aphids smaller than plant-reared aphids. 20 consecutive generations reared.

Corollary 1

→ Mittler, T.E.; Koski, P. 1974. Meridic artificial diets for rearing aphids. ENTOMOL EXP APPL 17: 524-525.

COMPOSITION As in Dadd & Mittler* (1966) except:- arginine and methionine increased to 275 and 80 mg respectively; riboflavin and pyridoxine deleted; 123 mg $MgSO_4.7H_2O$ substituted for $MgCl_2.6H_2O$. Sequestrene salts changed to Cu, 0.2; Fe, 11.0; Mn, 0.8; and Zn, 0.8. pH adjusted to 6.8 with 1.5 g $K_2HPO_4.3H_2O$.

DEVELOPMENT On diets containing casein hydrolysate in range 1.5-2.5%, 2% yeast extract and the non-amino acid constituents, the following mean weights (µg ± s.d., n and morph) recorded for aphids of first 8 successive generations: I, 518.3 ± 60.1, 30 alatiform 4th instar larvae; II, 446.7 ± 79.3, 29 apterous adults; III, 547.6 ± 91.7, 23 alatiform 4th instar; IV, 418.2 ± 74.3, 18 apterous adults; V, 430.6 ± 88.8, 28 apterous adults; .....VIII, 513.2 ± 74.4, 18 apterous adults. With all constituents of holidic diet supplemented with 0.5% yeast extract: I, 602.1 ± 62.0, 15 alatiform 4th instar; II, 640.4 ± 87.6, 14 apterous adults; III, 504.0 ± 63.1, 10 alatiform 4th instar; .....V, 446.9 ± 54.5, 24 alatiform 4th instar; .....X, 562.8 ± 55.8, 10 apterous adults; XI, 458.0 ± 48.4, 10 alatiform 4th instar; .....XIII, 577.6 ± 55.2, 10 alatiform 4th instar; .....XV, 538.8 ± 79.0, 10 alatiform 4th instar. With 2.5% yeast extract as only nitrogen, B-vitamin and trace mineral source, 14 generations reared: I, 534.5 ± 34.7, 10 alatiform 4th instar; II, 239.3 ± 27.6, 10 apterous adult; III, 324.2 ± 50.6, 10 alatiform 4th instar; IV, 208.8 ± 18.5, 10 apterous adult; V, 303.2 ± 37.2, 17 alatiform 4th instar; .....XIV, 177.6 ± 19.8, 10 apterous adult.

Corollary 2

Hertel, R. 1974. Einfluss von ATP in einer holidischen Diät auf *Myzus persicae* (Sulz.) (Aphidina). EXPERENTIA 30: 775-776. (In German with English summary)

COMPOSITION As in Mittler & Kleinjan (1970) except:- (mM): ATP.Na_2, 1.11 added.

DEVELOPMENT About 20% more larvae born on diets containing 1.11 mM of ATP than on diets without ATP. Parents of larvae reared on diet containing 1.11 mM ATP, 9.4% heavier than those on diets without ATP.

Species used: *Myzus persicae, Brevicoryne brassicae.*

See also: Dadd (1967, 1968a b); Dadd & Krieger (1968); Dadd & Mittler (1965); Dadd *et al.* (1967); Griffiths *et al.* (1975); Harrewijan (1973); Harrewijan & Noordink (1971); Kunkel & Mittler (1971); Legge & Palmer (1968); Massonié (1971b, 1973, 1974); Mittler (1967a b, 1970a b, 1971a b); Mittler & Dadd (1962, 1963a b, 1964, 1965); Mittler & Kleinjan (1970); Mittler & Kunkel (1971); Mittler *et al.* (1970); Paik (1956); Sutherland & Mittler (1971); Tsitsipis & Mittler (1970); van Emden & Adam (1972); Wearing (1968).

Rhopalosiphum maidis (Fitch), corn leaf aphid

Rhopalosiphum padi (Linnaeus), oat aphid

Macrosiphum avenae (Fabricius), English grain aphid

Kieckhefer, R.W.; Derr, D.F. 1967. Rearing three species of cereal aphids on artificial diets. J ECON ENTOMOL 60: 663-665.

Two diets modified from Auclair & Cartier (1963).

COMPOSITION Diet 1: see Sutter & Kieckhefer* (1969).

Diet 2: as diet 1 except:- 4.0 mg of glutathione added to amino acids. Ribose, 10.0 mg; deoxyribose, 20.0 mg; glucosamine hydrochloride, 200.0 mg added to carbohydrates. Vitamins: ascorbic acid increased to 50 mg; nicotinic acid reduced to 3.0 mg; pyridoxine hydrochloride reduced to 1.0 mg. Niacinamide, 2.0 mg; pyridoxal hydrochloride, 1.0 mg; vitamin A acetate, 0.01 mg; calciferol, 0.01 mg; α-tocopherol, 0.001 mg; menadione, 0.001 mg added. Miscellaneous components modified: *dl*-carnitine reduced to 4.0 mg; $MgCl_2.6H_2O$ (25 mg) omitted. Tween 80, 2.0 mg; thymine, 10.0 mg; ATP-2Na, 10.0 mg; egg lecithin, 2.0 mg; calcium gluconate, 50.0 mg; $MgSO_4.7H_2O$, 25.0 mg; and Fe $(NO_3)_2.9H_2O$, 4.0 mg added.

PREPARATION Cholesterol added to diets by dissolving dietary constituents in water saturated with cholesterol. RNA, lipoic acid, tyrosine, cysteine, and folic acid added as potassium salts. Enough ethanol added to calciferol (19 mg), menadione (1 mg), α-tocopherol phosphate (1 mg), vitamin A acetate (10 mg), and Tween 80 (2 g) to make 10 ml; 0.05 ml of this solution used in diet 2. Water added to both diets to 100 ml after which pH adjusted to 7.15-7.20. Diets divided into 5 ml portions and stored at -20°C until needed.

REARING At 22°C. Aphid cages from 2 cm lengths of 2 cm diam cellulose nitrate tubing, one end of which sealed with silk. Tops of cages covered with parafilm sachet containing 0.05 ml diet. Sachets of 2 parafilm membranes stretched to original thickness.

DEVELOPMENT Diet 1, developed empirically but specifically for *R. maidis*, less useful for other cereal aphids. *R. maidis* survived and reproduced for 3 weeks but nymphs failed to develop beyond 4th instar. *M. avenae* and *R. padi* declined more rapidly than *R. maidis*. Diet 2 theoretically more complete than Diet 1, but did not sustain aphids as well.

Neomyzus circumflexus (Buckton)

Ehrhardt, P. 1968d. Die Wirkung verschiedener Spurenelemente auf Wachstum, Reproduction und Symbioten von *Neomyzus circumflexus* Buckt. (Aphidae, Homoptera, Insecta) bei kunstlicher Ernährung. Z VGL PHYSIOL 58: 47-75. (In German with English summary)

COMPOSITION (mg/100 ml double distilled water): *l*-amino acids: α-alanine, 100; arginine monochloride, 400; asparagine, 300; aspartic acid, 100; cysteine, 50; cystine, 5; γ-aminobutyric acid, 20; glycine, 20; glutamine, 600; glutamic acid, 200; histidine, 200; *dl*-homoserine, 800; isoleucine, 200; leucine, 200; lysine monochloride, 200; methionine, 100; phenylalanine, 100; proline, 100; serine, 100; threonine, 200; tryptophan, 100; tyrosine, 20; valine, 200.

vitamins: ascorbic acid, 100.0; thiamine dichloride, 2.5; riboflavin, 5.0; nicotinic acid, 10.0; pyridoxine chloride, 2.5; folic acid, 1.0; biotin, 0.1; sodium-pantothenate, 5.0; *i*-inositol (meso), 50.0; choline chloride, 50.0.

salts: KH_2PO_4, 500.0; $MgCl_2.6H_2O$, 200.0.

other ingredients: saccharose, 20 g; trace elements (μg/100 ml diet): Fe, 460; Mn, 220; Zn, 190; Ca, 850; Cu, 100.

PREPARATION Ingredients dissolved in double distilled water, trace elements added and volume brought to 100 ml with double distilled water. pH adjusted to exactly 7.1 with KOH. Diet sterilized by filtration and dispensed aseptically in aliquots of 0.15-0.2 ml to sterile parafilm membranes on glass ring. Then immediately covered with membrane to form sachet. Stored at -25° C until required.

REARING At 20° C with 16 h photoperiod in 20 x 19 mm diam glass rings, with diet sachets at one end. 15 insects/cage; food replaced every 2 days by exchanging cages.

DEVELOPMENT Very similar to that of naturally fed aphids.

See also: Ehrhardt (1968a b, 1969).

Schizaphis graminum (Rondani), greenbug

Cress, D.C.; Chada, H.L. 1971a. Development of a synthetic diet for the greenbug, *Schizaphis graminum*. 1. Greenbug development on two synthetic diets. ANN ENTOMOL SOC AM 64: 1237-1240.

COMPOSITION Diet 1: see Auclair* (1965).

Diet 2: *l*-amino acids (mg): alanine, 100; arginine, 400; asparagine, 300; aspartic acid, 100; cysteine hydrochloride, 50; cystine, 5; γ-aminobutyric acid, 20; glutamic acid, 200; glutamine, 600; glycine, 20; histidine, 200; *dl*-homoserine, 800; isoleucine, 200; leucine, 200; lysine monohydrochloride, 200; methionine, 100; phenylalanine, 100; proline, 100; serine, 100; threonine, 200; tryptophan, 100; tyrosine, 20; valine, 200.

vitamins (mg): ascorbic acid, 1.00; biotin, 0.01; calcium pantothenate, 0.50; choline chloride, 5.0; folic acid, 0.10; *i*-inositol (meso), 5.00; nicotinic acid, 1.0; *p*-aminobenzoic acid, 1.00; pyridoxine hydrochloride, 0.25; riboflavin, 0.50; thiamine hydrochloride, 0.25.

other ingredients (mg): cholesterol benzoate, 2.5; K_3PO_4, 500; $MgCl_2.6H_2O$, 200; salt mixture, No. 2, 5; and sucrose, 35 g; and water to 100 ml.

PREPARATION See Auclair* (1965).

REARING At 22°C and 30-50% RH with 16 h photoperiod. Cages and diet sachet as in Auclair* (1965). Also, feeding cages supported in growth chamber in 5 holes of ca. 1 inch diam evenly spaced in circle on lid of half-gallon icecream cartons. Wooden legs glues on outside of lid so that protruding ends of glass rings would not rest on floor of chamber. A 3" square hole cut in bottom of inverted icecream carton and replaced by yellowish-green 11 x 1 Kodak Wratten filter. Light intensity at surface of sachet set at 60 ft-c. Cage and diet sachet illustrated.

DEVELOPMENT Reproduction from 11th day; maximum mean rate of 3.5 nymphs on 18th day. Maximum mean weight gained, 0.304 mg on 20th day; 10-13 fold increase in overall weight. One generation reared.

Corollary

→ Cress, D.C.; Chada, H.L. 1971b. Development of a synthetic diet for greenbug, *Schizophis graminum*. 2. Greenbug development as affected by zinc, iron, manganese and copper. ANN ENTOMOL SOC AM 64: 1240-1244.

COMPOSITION Diet 2 of Cress & Chada* (1971a) plus EDTA salts at various concentrations.

DEVELOPMENT Rates of growth and development on diet with addition of 200, 325, and 200 μg of Zn, Fe and Mn respectively comparable with those on susceptible Roger's barley.

See also: Cress & Chada (1971c).

Toxoptera aurantii (Boyer), black citrus aphid

→ Tahori, A.S.; Hazan, A. 1970. Rearing of the black citrus aphid *Toxoptera aurantii* on chemically defined diets. J INSECT PHYSIOL 16: 1975-1981.

Two diets tested.

COMPOSITION Diet 1: as in Auclair & Cartier (1964).

Diet 2: as in Dadd & Mittler* (1966).

PREPARATION Mittler & Dadd (1964).

DEVELOPMENT Neither diet supported more than 2 generations. Sucrose concentration optimal at 55%. Greater nymphal survival with higher amino acid concentrations. Omission of all vitamins affected neither survival nor production of second generation.

Miscellaneous - Aphididae

→ Herger, P.; Lampel, G. 1975. Zuchtversuche mit mono-bis polyphagen Röhrenblattlaus-Arten auf künstlichem Nährmedium (Homoptera: Aphidina: Aphididae). ENTOMOL GERM 1: 258-278. (In German with English summary)

Modification of Ehrhardt* (1968d) diet for *Neomyzus circumflexus*.

COMPOSITION (mg/100 ml): *l*-amino acids: α-alanine, 100; β-amino butyric acid, 20; arginine, 400; asparagine, 400; aspartic acid, 100; cysteine hydrochloride, 50; cystine, 5; glutamine, 500; glutamic acid, 200; glycine, 50; histidine hydrochloride, 150; *dl*-homoserine, 800; isoleucine, 150; leucine, 150; lysine hydrochloride, 150; methionine, 100; phenylalanine, 50; proline, 100; serine, 100; threonine, 150; tryptophan, 100; tyrosine, 30; valine, 100.

vitamins: *l*-ascorbic acid, 100; thiamine-dichloride, 2.5; riboflavin, 0.1; nicotinic acid, 10; pyridoxin hydrochloride, 2.5; folic acid, 1.0; *d*-biotin, 0.1; sodium-*d*-pantothenate, 5; *meso*-inositol, 50; choline chloride, 50.

salts: $K_2HPO_4.3H$ 0, 700; $MgCl_2.6H_2O$, 200; $CaCl_2.2H_2O$, 3.115; sequestrene salts (Geigy): Cu, 0.268; Mn, 0.793; Zn, 0.396; Fe, 2.228.

sugars: saccharose, 20000.

PREPARATION Stock solution of vitamins deep frozen at -25° C, and of salt solution refrigerated for up to 6 months. Amino acids, sugar and $K_2HPO_4.3H_2O$ mixed with 70-80 ml of deionized double distilled water at 48° C for 2 h. Then 10 ml vitamin and 1 ml trace element solutions added and stirred for ½ h; pH adjusted to 7.1; $MgCl_2$ added and stirred for 10 minutes. Finally adjusted to 100 ml, sterilized through microbiological filter and sachet prepared. Diet storable at -25° C for maximum of 2 months.

REARING In greenhouse where temperature and relative humidity recorded (not given in text) with 18 h photoperiod. Details of sachet in Herger (1973); 20 aphids/sachet kept under yellow filter. Diet renewed every 4 days.

DEVELOPMENT 12 species reared:-

Aphis fabae Scopoli: reproduction, 95%; half generation time, 9.8 days; 18 generations reared.

Aulacorthum circumflexum Buckton: reproduction >84%; half generation time, 11.0 days; 8 generations reared.

Macrosiphum euphorbiae Thomas: reproduction <50%; half generation time, 11.3 days; 17 generations reared.

Myzus persicae (Sulzer): reproduction, 80%; half generation time, 10.6 days; 21 generations reared.

Following species reared from 1-3 generations:-

Aphis newtoni Theobald; *Aphis pomi* De Geer; *Aphis sambuci* Linné; *Brachycaudus cardui* (Linné); *Brevicoryne brassicae* (Linné); *Macrosiphum rosae* (Linné); *Nasonovia ribisnigri* (Mosley); *Rhopalosiphum nymphaeae* (Linné).

See also: Herger (1973).

→ Krieger, D.L. 1971. Rearing several aphid species on synthetic diet. ANN ENTOMOL SOC AM 64: 1176-1177.

COMPOSITION See Dadd & Mittler* (1966).

Species reared with 30% sucrose in diet: *Macrosiphum fragariae* (Walker), *Lipaphis erysimi* (Kaltenbach).

Species reared with 15% sucrose in diet: cotton aphid, *Aphis gossypii* Glover; *Cavariella aegopodii* (Scopoli); the crescent-marked lily aphid, *Neomyzus circumflexus* (Buckton); *Aulacorthum solani* (Kaltenbach); *Myzus ornatus* Laing; and cowpea aphid, *Aphis craccivora* Koch.

Diet probably acceptable for many other species.

→ Cloutier, C.; MacKauer, M. 1975. A modified technique for rearing of the pea aphid on artificial diets. ENTOMOL EXP APPL 18: 395-396.

REARING Mittler & Dadd (1962) sachet modified so as to reduce aphid handling. First parafilm membrane stretched across upper opening of aphid chamber. Second membrane surface-sterilized with UV stretched over first one. Sterile diet placed on to 2nd membrane and sealed with 3rd membrane under aseptic conditions. Diet renewed by peeling sachet formed by upper most membranes gently off base membrane, without disturbing aphids. Cage illustrated.

Used for rearing *Acyrthosiphon pisum*, *Macrosiphum euphorbiae*, and *Macrosiphum albifrons* Essig for 3 generations.

Cercopidae

Aeneolamia varia saccharina (Distant), froghopper

→ Hagley, E.A.C. 1967. Artificial diet for the adult froghopper. NATURE LONDON 213: 414-415.

COMPOSITION Adult diet. (mg): *l*-amino acids: alanine, 15; amino butyric acid, 15; arginine, 30; asparagine, 50; aspartic acid, 15; cystine, 2; glutamic acid, 35; glutamine, 30; glycine, 5; histidine, 5; isoleucine, 2; leucine, 5; lysine, 15; methionine, 1; phenylalanine, 1; proline, 5; serine, 5; threonine, 10; tyrosine, 15; tryptophan, 5; valine, 5; biotin, 0.01; calcium pantothenate, 0.5; choline chloride, 5.0; folic acid, 0.05; *i*-inositol, 5.0; nicotinic acid, 1.0; pyridoxine, 0.5; riboflavin, 0.25; thiamine, 0.25; ascorbic acid, 100; B_{12}, 0.015% (v/v); Wesson's salt mixture, 100; $MgCl_2.6H_2O$, 50; KH_2PO_4, 50; cholesterol, 2; linoleic acid, 0.03%; sucrose, 4 g; and water to 100 ml.

PREPARATION Cholesterol dissolved in 1.5 ml acetone, added to 300 ml distilled water and boiled to remove solvent. Suspension remade to original volume and remaining nutrients added with one drop of Tween 80 and stirred for 20 minutes; boiled for further 5 minutes, and pH adjusted to 7 with 0.8 N KOH. Refrigerated at 4-6° C.

REARING Pairs of newly emerged adults held in 1.5 x 1.2" diam glass cages. Diet impregnated on circular pieces of Whatman No. 1 filter paper between two pieces of parafilm stretched over one end of cage. Diet renewed every 24 h.

DEVELOPMENT Longevity of ♂♂ and ♀♀ on diet averaged 5.8 and 6.7 days; on control (sugar cane leaf) 4.0 and 4.9 days. Average fecundity on diet 90 eggs/♀; on control, 52 eggs/♀.

Cicadellidae

Empoasca fabae (Harris), potato leafhopper

→ Dahlman, D.L. 1963. Survival of *Empoasca fabae* (Harris) (Cicadellidae; Homoptera) on synthetic media. PROC IOWA ACAD SCI 70: 498-504.

Best diet tested.

COMPOSITION (mg): arginine, 27; histidine, 4; isoleucine, 3; leucine, 4; lysine, 12; methionine, 1; phenylalanine, 1; threonine, 14; tryptophan, 4; valine, 4. Added to 100 ml of 1 % sucrose - 2% agar solution.

REARING At 85°F in darkened incubator; in clear plastic 1 x 1 x ½" boxes. 2.5 ml hot diet poured into each half of box. 1" squares of filter paper placed over agar after gelling. 4th-5th instar larvae placed individually in cages and adults transferred to fresh cages every other day.

DEVELOPMENT Survival twice as long on diet with 2% agar plus 1% sugar than on control of 2% agar. On full diet described above, 50% survival time tripled. 7% sucrose optimal for survival.

→ Kirollos, J.Y.; Hibbs, E.T. 1971. Viability of *Empoasca fabae* (Homoptera: Cicadellidae) eggs in sterile media with added monosaccharides, amino acids, or phorate. ANN ENTOMOL SOC AM 64: 32-36.

COMPOSITION Hatching media - various compounds tested in 2% agar at different concentrations.

1. *d* (+) glucose at (0.03, 0.06, 0.12, 0.30) M.

2. *d* (-) fructose at (0.03, 0.06, 0.12, 0.30) M.

3. *l*-valine at (0.0003, 0.0006, 0.0012) M.

4. *l*-isoleucine at (0.0003, 0.0006, 0.0012) M.

PREPARATION 2% agar medium prepared in Pringle's saline solution in flask at 100°C. Test compound added to solution, autoclaved for 20 minutes at 100°C, and poured into sterile petri dishes.

REARING Dishes inoculated aseptically with 72 ± 12 h old eggs. Eggs previously surface sterilized for 3 minutes in 0.1% Hyamine - 2389 solution. 0.1% Hyamine - 2389 prepared by dilution from 50% aqueous solution with Pringle's saline solution (g): NaCl, 9.0; KCl, 0.2; $CaCl_2$, 0.2; glucose, 4.0; and water to 1 l. pH 6.5 with 0.05 M phosphate buffer.

DEVELOPMENT Both glucose and fructose at 0.12 and 0.30 M slowed development. Sugar concentrations >1.08% not tolerated. *L*-valine at 0.0003 and 0.0006 M shortened development time of visible eye spots by 16-18 h but total developmental time unchanged. Visible eye spots used as criterion of egg viability as were eclosion or attempted eclosion. Developmental time on medium and control 120-144 h and 168-192 h.

Inazuma dorsalis (de Motschulsky)

Nephotettix cincticeps (Uhler)

→ Koyama, K. 1973b. Rearing of *Inazuma dorsalis* and *Nephotettix cincticeps* on a synthetic diet. JPN J APPL ENTOMOL ZOOL 17: 163-166. (In Japanese)

Four diets tested.

COMPOSITION Diet 1 (mg): *l*-amino acids: alanine, 100; γ-aminobutyric acid, 20; arginine hydrochloride, 400; asparagine, 300; aspartic acid, 100; cysteine, 50; cystine hydrochloride, 5; glutamic acid, 200; glutamine, 600; glycine, 20; histidine, 200; *dl*-homoserine, 800; isoleucine, 200; leucine, 200; lysine hydrochloride, 200; methionine, 100; phenylalanine, 100; proline, 100; *dl*-serine, 100; threonine, 200; tryptophan, 100; tyrosine, 20; valine, 200.

vitamins: thiamine hydrochloride, 2.5; riboflavin, 5.0; nicotinic acid, 10.0; pyridoxine hydrochloride, 2.5; folic acid, 1.0; calcium pantothenate, 5.0; inositol, 50.0; choline chloride, 50.0; biotin, 0.1; sodium *l*-ascorbate, 100.0; sucrose, 5000.

minerals: $MgCl_2.6H_2O$, 200; K_2HPO_4, 500; $FeCl_3.6H_2O$, 2.228; $CuCl_2.2H_2O$, 0.268; $MnCl_2$, 0.793; $ZnCl_2$, 0.396; $CaCl_2$, 3.115. pH to 6.5 with KOH.

Diet 2: diet 1 and cholesterol, suspended and filtered.

Diet 3: diet 1 and distilled water fed separately.

Diet 4: diet 1 and cholesterol suspension fed separately.

PREPARATION See Ehrhardt* (1968). All substances added to 100 ml distilled water, dissolved and pH adjusted. Liquid diet filtered through microbiological filters, and stored.

REARING From laboratory populations; *I. dorsalis* eggs collected by method of Mitsuhashi (1970). *N. cincticeps* eggs collected on rice seedlings; hatched on wet filter paper and transferred immediately to rearing apparatus (illustrated). Diet obtained through parafilm membrane.

DEVELOPMENT *I. dorsalis:* survival rate on diet 4 almost equal to control on rice seedlings for first 12 days, then reduced to lower level. Normal adults obtained but later than on control. On diets 1, 2, and 3, nearly 100% larval mortality within 15 days. No adults obtained: diet 1, to 4th instar; diet 2, to 3rd instar; diet 3 to 5th instar.

N. cincticeps: survival rate on diet 4 lower than on control; adults obtained but later than on control. Diet 1: nearly 100% larval mortality within 10 days; development only to 5th instar. Difference from diet requirements of *Laodelphax striatellus* (see Mitsuhashi & Koyama (1971)) is cholesterol, but not effective unless given separately from basic diet.

See also: Koyama (1973a).

Macrosteles fascifrons (Stål), aster leafhopper

→ Hou, R.F.; Brooks, M.A. 1975. Continuous rearing of the aster leafhopper, *Macrosteles fascifrons*, on a chemically defined diet. J INSECT PHYSIOL 21: 1481-1483.

COMPOSITION *l*-amino acids (mg): alanine, 20; arginine hydrochloride, 80; asparagine, 60; aspartic acid, 20; cysteine, 10; cystine, 1; aminobutyric acid, 4; glutamic acid, 40; glutamine, 120; glycine, 4; histidine, 16; *dl*-homoserine, 160; isoleucine, 40; leucine, 40; lysine hydrochloride, 40; methionine, 20; phenylalanine, 20; proline, 20; *dl*-serine, 20; threonine, 40; tryptophan, 20; tyrosine, 4; valine, 40; biotin, 0.05; calcium pantothenate, 2.50; choline chloride, 25; folic acid, 0.25; *i*-inositol, 25; nicotinic acid, 5.00; pyridoxine hydrochloride, 1.25; riboflavin, 0.25; thiamine hydrochloride, 1.25; $CuCl_2.2H_2O$, 0.051; $FeCl_3.6H_2O$, 0.534; $MnCl_2.4H_2O$, 0.100; $CaCl_2$, 0.444; Zn-sequestrene, 0.160; $Na_3C_6H_5O_7.2H_2O$, 10.0; sucrose, 5000; $MgSO_4.7H_2O$, 242; KH_2PO_4, 250; ascorbic acid, 100; cholesterol, 5.0; lecithin, 5.0; and water, 100 ml.

PREPARATION Amino acid solution (20 ml) and B-vitamin solution (5 ml) combined in beaker. Trace metals dissolved in water in concentrated stock solution, and requisite aliquots of each added. 2 ml sodium citrate stock solution in 2nd beaker. Sucrose, $MgSO_4.7H_2O$ and KH_2PO_4 added; pH to 6.5 with 1.75 M KOH. Ascorbic acid added and pH to 6.8. Diet made to 90 ml with water and sterilized by filtration through Millipore disposable filter (0.45 μm). 10 ml autoclaved cholesterol dispersion incorporated into filtered solution; total volume 100 ml. Diet stored in 5-8 ml aliquots at -21° C.

Cholesterol dispersion prepared by dissolving 25 mg cholesterol in 2.5 ml hot 95% ethanol, and adding to 1 ml 5% Tween 80. Then 25 mg lecithin dispersed in 10 ml distilled water and poured into cholesterol solution. Made to 45 ml with distilled water, boiled to reduce volume to 20 ml, brought to 50 ml with distilled water and autoclaved at 121° C for 20 minutes.

REARING In feeding chambers similar to those of Mittler & Dadd (1964) at 70-75% RH and 22-25° C with 18 h photoperiod. Diet changed every 2nd day during 1st-3rd instars, and daily from 3rd instar. Eggs laid in sachet, removed and washed in double distilled water, rinsed in 70% ethanol for 1 minute, rinsed twice in double distilled water, and placed on moist filter paper.

DEVELOPMENT Egg incubation period, 9-11 days; larval period, 18 days - equal to control on host plants. Survival and longevity poorer on diet than those reared on plants but fecundity higher on diet with 79% eclosion. Nine generations reared.

→ Mitsuhashi, J.; Maramorosch, K. 1963. Aseptic cultivation of four virus transmitting species of leafhoppers (Cicadellidae). CONTRIB BOYCE THOMPSON INST 22: 165-173.

Four species reared.

COMPOSITION (mg/1000 ml water): *l*-amino acids: arginine, 105; cystine, 24; histidine, 31; isoleucine, 52; leucine, 52; lysine, 58; methionine, 15; phenylalanine, 32; tryptophan, 10; threonine, 48; tyrosine, 36; valine, 46; glutamine, 292; choline, 2; folic acid, 2; inositol, 4; nicotinamide, 2; calcium pantothenate, 2; pyridoxal, 2; riboflavin, 0.2; thiamine, 2; and (g): sucrose, 2; K_3PO_4, 0.5; $MgCl_2.6H_2O$, 0.2; cholesterol, 0.02.

PREPARATION See Mittler & Dadd (1962).

REARING 3rd instar larvae fed from gauze wick in liquid diet. Eggs laid on leaves and removed using fine steel needles. Fastened to strips of waxed paper with egg albumin and sterilized in 0.1% Hymamine solution for 3 minutes. Eggs placed in culture tubes over natural host plant raised axenically.

DEVELOPMENT 3rd instar larvae moulted once but died after 15 days. No reproduction. On axenic plants following species reared:- *Macrosteles fascifrons* easily reared on rye plants at 30° C; generation time 2-3 weeks; adult ♀ longevity, 1 month. Also reared on carrot tissue culture.

Dalbulus maidis reared on corn plants at 30° C; larval period, 14 days. Adult ♀ longevity, 50 days.

Agallia constricta reared on alfalfa or crimson clover; larval period, 36 days at 30° C.

Agalliopsis novella reared on alfalfa and crimson clover; larval period, 38 days.

Coccidae

Kilifia acuminatum (Green) [previously *Lecanium acuminatum* Signoret], soft scale

→ Hafez, M.; Salama, H.S.; Saleh, M.R. 1971. Survival and development of *Lecanium acuminatum* Sign. (Coccoidea) on a host plant and artificial diets. Z ANGEW ENTOMOL 69: 182-186.

Artificial diet developed for *Myzus persicae* (Sulzer), by Mittler & Dadd (1962, 1963, 1964; see Dadd & Mittler* (1966)); and rearing technique for coccids, Salama & Saleh (1968).

PREPARATION Ingredients dissolved in ca. 80 ml of cholesterol-saturated water in following order: amino acids, sugar, and salts; 10 ml of vitamin solution then added. pH to 6.8-7.0, and volume adjusted to 100 ml.

REARING As in Salama & Saleh (1968). Feeding unit from 3 x 2.5 cm plastic cylinder; stretched parafilm membrane over one end with 2 drops diet, then second membrane sealed over first to encapsulate diet. Insect introduced through open end of feeding unit which then covered by glass slide.

DEVELOPMENT Mean adult survival, 6.8 days. Diet incomplete.

Delphacidae

Laodelphax striatellus Fallén, smaller brown planthopper

→ Koyama, K.; Mitsuhashi, J. 1969. Artificial feeding of smaller brown planthopper, *Laodelphax striatellus* Fallén (Hemiptera: Delphacidae). JPN J APPL ENTOMOL ZOOL 13: 89-90.

COMPOSITION & PREPARATION See Mittler & Dadd (1962); except sucrose content altered to 5%, and pH to 6.5.

REARING At 25° C with 16 h photoperiod. On parafilm membrane sachet, as in Mittler & Dadd (1964). Diet renewed every 2 days; and changed from initial rice seedling extract to artificial diet during either 1st or 3rd instar.

DEVELOPMENT Mean period in days for development on diet and rice seedling extract respectively. 2nd instar, 4.8, 3.1; 5th instar, 6.6, 3.9. High larval mortality on artificial diet from 1st instar; some development to 5th instar; none to adult. Larval mortality lower when transferred to artificial diet in 3rd instar, though still higher than control, and adults obtained.

See also: Mitsuhashi (1970).

→ Mitsuhashi, J.; Koyama, K. 1972. Artificial rearing of the smaller brown planthopper, *Laodelphax striatellus* Fallén, with special reference to rearing conditions for the first instar nymphs. JPN J APPL ENTOMOL ZOOL 16: 8-17. (In Japanese with English summary)

COMPOSITION & PREPARATION See Mitsuhashi & Koyama (1971); Mitsuhashi* (1974). Stored at -20° C.

REARING For each generation, 50 larvae reared individually; remainder reared 20-30/container (illustrated). At 25° C with 16 h photoperiod in 100 mm length of 20 mm diam glass tube; diet portion partitioned from rearing chamber by Parafilm membrane. Diet changed every 2 days; 5% sucrose solution provided for egg laying (separated from chamber by parafilm) as alternative from, or simultaneously to diet; eggs stored in distilled water until mature when transferred to wet filter paper for hatching.

DEVELOPMENT Ten generations reared; nymphal growth rate slower on diet than on natural host plant, but size comparable. Reproduction low on diet. Higher humidity and temperatures between 15-23° C gave improved nymphal survival.

See also: Koyama & Mitsuhashi (1973, 1975).

→ Mitsuhashi, J. 1974. Methods for rearing leafhoppers and planthoppers on artificial diets. REV PLANT PROT RES 7: 57-67.

Three diets tested.

COMPOSITION Diet 1 (mg/l): *l*-amino acids: alanine, 1000; γ-aminobutyric acid, 200; arginine hydrochloride, 4000; asparagine, 3000; aspartic acid, 1000; cysteine, 500; cystine hydrochloride, 50; glutamic acid, 2000; glutamine, 6000; glycine, 200; histidine, 2000; *dl*-homoserine, 8000; isoleucine, 2000; leucine, 2000; lysine hydrochloride, 2000; methionine, 1000; phenylalanine, 1000; proline, 1000; *dl*-serine, 1000; threonine, 2000; tryptophan, 1000; tyrosine, 200; valine, 2000.

vitamins: thiamine hydrochloride, 25; riboflavin, 50; nicotinic acid, 100; pyridoxine hydrochloride, 25; folic acid, 10; calcium pantothenate, 50; inositol, 500; choline chloride, 500; biotin, 1; sodium ascorbate, 1000.

minerals: $MgCl_2.6H_2O$, 2000; KH_2PO_4, 5000; sucrose, 50000; $FeCl_3.6H_2O$, 22.28; $CuCl_2.2H_2O$, 2.68; $MnCl_2.4H_2O$, 7.93; $ZnCl_2$, 3.96; $CaCl_2.2H_2O$, 31.15; pH adjusted to 6.5 with KOH.

Diet 2 (mg/l): as diet 1, except *l*-amino acids: alanine, 1500; asparagine, 4500; aspartic acid, 1500; cysteine, 800; glutamic acid, 3000; glutamine, 9000; histidine, 3000; isoleucine, 3000; leucine, 3000; lysine hydrochloride, 3000; methionine, 1500; *dl*-serine, 1500; threonine, 3000.

minerals: $FeCl_3.6H_2O$, 20.00; $CuCl_2.2H_2O$, 3.00; $MnCl_2.4H_2O$, 8.00; $ZnCl_2$, 4.00; $CaCl_2.2H_2O$, 30.00.

Diet 3 (mg/l): as diet 1 except *l*-amino acids: arginine hydrochloride, 2700; asparagine, 5500; aspartic acid, 1400; cysteine, 400; glutamic acid, 1400; glutamine, 1500; glycine, 800; histidine, 800; isoleucine, 800; leucine, 800; lysine hydrochloride, 1200; methionine, 800; phenylalanine, 400; proline, 800; *dl*-serine, 800; threonine, 1400; tryptophan, 800; tyrosine, 400; valine, 800; and γ-amino butyric acid, cystine hydrochloride and *dl*-homoserine deleted.

minerals: $ZnCl_2$, 11.88; $MgSO_4.7H_2O$, 1230; K_2HPO_4, 7500.

PREPARATION See Ehrhardt* (1968d) and Mittler & Dadd (1962).

REARING At 25°C with 16 h photoperiod; 3 different types of glass cages used (illustrated). All covered at one end with parafilm diet sachet. One with various solutions of dilute H_2SO_4 to control humidity, another with sachet at both ends to supply two different diets. Eggs laid on parafilm over 5% sucrose solution in special oviposition cage with sucrose and diet at opposite ends. Eggs always laid on sucrose end; removed from parafilm by pushing into solution with needle and transferring to water with pipette, and then to moist filter paper 24 h before hatching. Best survival rates at 15-20°C for 10 days after hatching, then at 25°C. Optimal humidity >80%.

DEVELOPMENT Ten generations reared on diet 1. Growth of nymphs retarded by ca. 5 days. Size of adult not significantly different from controls fed on rice seedlings.

Other species: *Nilaparvata lugens* Stål: not reared to adult; *Sogatella furcifera* Horvath and *Sogatella elegantissima* Ishihara: reared from 1st instar to adult on diet 1 but fecundity low; *Macrosteles orientalis* Vilbaste: several generations reared; *Balclutha viridis* Matsumura: reared from 1st instar to adult, but no eggs laid; *Inazuma dorsalis* (Motschulsky): reared on diet 1 and cholesterol saturated water; *Nephotettix cincticeps* (Uhler): required cholesterol for development, but no eggs laid.

See also: Mitsuhashi & Koyama (1971, 1972, 1974, 1975).

Pseudococcidae

Planococcus citri (Risso), citrus mealy bug

→ Gothilf, S.; Beck, S.D. 1966. Rearing the citrus mealy bug, *Planococcus citri* (Risso), on a defined diet. J ECON ENTOMOL 59: 489-490.

Best diet tested.

COMPOSITION *l*-amino acids (mg): glutamine, 300; asparagine, 250; glutamic acid, 200; aspartic acid, 200; γ-amino butyric acid, 140; arginine, 140; valine, 100; threonine, 100; lysine, 80; serine, 50; alanine, 50; phenylalanine, 50; methionine, 50; histidine, 50; proline, 40; leucine, 40; isoleucine, 40; tyrosine, 40; tryptophan, 40; glycine, 20; cysteine, 10; cystine, 10.

vitamins (mg): ascorbic acid, 150; choline chloride, 20; inositol, 20; nicotinic acid, 6; calcium pantothenate, 1.5; thiamine hydrochloride, 0.6; riboflavin, 0.3; pyridoxine, 0.3; folic acid, 0.1; biotin, 0.01.

other nutrients (g): sucrose, 30; and (mg): cholesterol acetate, 20; K_3PO_4, 500; $MgCl_2$, 200; USP salt No. 2, 5; and water to 100 ml.

PREPARATION Ingredients stirred for 3 h, and pH adjusted to 6.9-7.0 with KOH. Diet then sterilized through bacterial filters and introduced to rearing units.

REARING At 23.5 ± 0.1°C and 50% RH in total darkness. Unit from 2 pieces of glass tubing; diet chamber, 30 x 9 mm diam, and larval chamber, 25 x 12 mm diam. Feeding end of diet chamber covered by stretched, creased parafilm membrane and inserted into larval chamber to depth of 3 mm. Two chambers sealed with double thickness of parafilm. Bug introduced to larval chamber which then corked. Unit assembled aseptically and kept horizontally.

DEVELOPMENT Average duration of ♂ larval period, 35 days. Diet-reared ♂♂ similar to those reared on potatoes. Average ♀ developmental time on diet, 80 days; on potatoes, 50 days; diet-reared ♀♀ smaller and laid fewer eggs.

Hymenoptera

Aphelinidae

Encarsia formosa Gahan

→ Balboni, E.R. 1964. Influence of chloramphenicol on some aspects of the metabolism of the parasitoid *Encarsia formosa* Gahan. J INSECT PHYSIOL 10: 887-896.

Modified from Vanderzant and Reiser (1956b) diet for *Pectinophora gossypiella.*

COMPOSITION Adult diet (g): vitamin-free casein, 5.00; *l*-cystine, 0.10; glycine, 0.15; cholesterol, 0.15; corn oil, 1.00; Wesson's salt mixture, 0.70; sucrose, 300.00; and water to 200 g.

vitamins (mg): thiamine hydrochloride, 2.50; riboflavin, 1.25; nicotinic acid, 5.00; pyridoxine hydrochloride, 1.25; calcium pantothenate, 2.50; choline chloride, 50.00; biotin, 25.00×10^{-3}; and folic acid, 250.00×10^{-3}.

PREPARATION See Vanderzant & Reiser (1956b).

REARING At 20 ± 1° C and 60-80% RH with 24 h photoperiod in 4-oz jars with lawn cloth covers. Diet in small shell vial inverted over centre of cover such that drop of diet hung underneath. Vials and covers replaced every 24 h. Feeding time 72-96 h except in fecundity studies, when 24-48 h to prevent resorption of eggs.

DEVELOPMENT Mean of 391 eggs laid/♀.

Apidae

Apis mellifera Linnaeus, honey bee

→ Dietz, A. 1973. Longevity and survival of honey bee larvae on artificial diets. J GA ENTOMOL SOC 8: 59-63.

Best of many diets tested.

COMPOSITION (g): egg yolk, 0.3; egg white, 0.4; glucose, 0.1; dried skim milk, 0.6; brewers' yeast, 0.6; and calcium pantothenate, 100 μg; distilled water, 8.0 ml. pH, 6.0.

PREPARATION Mixed, boiled in steam bath, homogenized and refrigerated in small vials.

REARING See Dietz & Haydak (1971). Larvae in desiccator in small glass dishes, at 34.5 ± 0.5°C and 96% RH; fresh food every 12 or 24 h. When ready to pupate, transferred to gelatin capsule maintained at 75% RH for emergence.

DEVELOPMENT Diet incomplete. No adults from neonate larvae; maximum larval survival, 120 h.

→ Gilliam, M.; Argauer, R.J. 1975a. How long is Terramycin stable in diets fed to honey bee colonies for disease control? AM BEE J 115: 230-234.

COMPOSITION (g): mixed bee collected pollen, 167; sucrose, 152; Drivert, 136; Alphacel, 0.76; animal soluble powder Terramycin, 0.5; Fumidil B, 0.5; and water, 33 ml.

PREPARATION See Argauer & Gilliam (1974). Made into 1 lb patty by mixing thoroughly.

REARING Patties of diet placed on top of frames; hives held in 12 x 12 x 9 ft Saran mesh cage. Water provided.

See also: Gilliam & Argauer (1975b).

→ Robinson, F.A.; Nation, J.L. 1966. Artificial diets for honey bees, *Apis mellifera*. FLA ENTOMOL 49: 175-184.

A. Basic solid artificial diet.

COMPOSITION (% by weight): gelatin, 5.00; vitamin-free casein, 5.00; zein, 5.00; egg albumin, 5.00; Wesson's salt mixture, 5.00; Mazola corn oil, 5.00; cholesterol, 0.25; RNA, 1.00; sucrose, 45.00; cellulose, 23.75.

vitamins (mg/20 g diet): riboflavin, 1.2; pyridoxine, 1.0; niacinamide, 8.0; thiamine, 1.0; ascorbic acid, 60.0; pentathenol, 3.0; choline, 12.0; inositol, 12.0; carnitine, 0.46; *p*-aminobenzoic acid, 0.48; glutathione, 1.18; biotin, 0.0298; vitamin B_{12}, 0.00064.

PREPARATION Ingredients mixed, moistened with small amount of filtered honey and formed into cakes of ca. 20 g weight.

DEVELOPMENT Consumption poor. Development ceased at older larval stages. Addition of whole pollen to basic diet gave complete larval development.

B. Liquid-dry diet modified from Weaver* (1964).

COMPOSITION Liquid portion: water, 400 ml; sucrose, 600 g; and (μg): carnitine, 204; inositol, 6000; choline, 6000; glutathione, 600; *p*-aminobenzoic acid, 204; and Deca-Vi-Sol vitamin mixture, 0.6 ml.

vitamins in 0.6 ml Deca-Vi-Sol: vitamin A, 3000 U.S.P. units; vitamin D, 400 U.S.P. units; and (mg): vitamin C, 60; B_1, 1; B_2, 1.2; B_6, 1; niacinamide, 8; panthenol, 3; and (μg): B_{12}, 1; biotin, 30.

Dry portion (g): vitamin-free casein, 200; corn oil, 10; RNA, 10; tryptophan, 2.4; histidine, 5.4; cystine, 1.2; glycine, 3.8; cysteine, 1.0; cholesterol, 10.0; Wesson's salt mixture, 4.0.

liquid portion fed *ad libitum*.

DEVELOPMENT Bees failed to accept this diet.

C. Liquid diet.

COMPOSITION Water, 400 ml; and (g): gelatin, 10; casein, 10; corn oil, 1; cholesterol, 1; Wesson's salt mixture, 0.1; RNA, 0.1; sucrose, 600; Deca-Vi-Sol mixture, 0.6 ml.

PREPARATION A few drops of Triton X-100 emulsifier added; ingredients mixed in Waring Blendor.

DEVELOPMENT Diet readily acceptable.

D. Solid diet.

COMPOSITION (%): gelatin, 10; casein, 10; corn oil, 5; Wesson's salt mixture, 5; RNA, 1; cholesterol, 1; sucrose, 68.

PREPARATION 10 g of diet moistened with filtered honey and 0.6 ml of Deca-Vi-Sol, formed into cake.

REARING Liquid diets placed in pans outside colonies. Solid diets moistened with small amount filtered honey and formed into small cakes. Fresh cake with 20 g diet placed on top of frames every day.

DEVELOPMENT Consumption of controlled diets poor compared with consumption of control diets of honey and pollen.

→ Robinson, F.A.; Nation, J.L. 1968. Substances that attract caged honey bee colonies to consume pollen supplements and substitutes. J APICULT RES 7: 83-88.

Basic artificial diet.

COMPOSITION (%): as basic solid diet (A) of Robinson & Nation* (1966) except, corn oil deleted; cholesterol decreased from 0.25 to 0.1%; cellulose powder adjusted to 23.9%; and vitamins (mg/20 g dry wt): riboflavin, 2.4; choline chloride, 2.0; inositol, 24.0; carnitine, 0.92; *p*-aminobenzoic acid, 0.96; biotin, 0.0596; thiamine, 2.0; B_{12}, 0.0012.

PREPARATION 5 g of various lipid and volatile substance extracts added to 100 g of basic diet. Ingredients mixed with pressure-filtered honey and formed to small flat cakes.

REARING In 2 m^3 bioclimatic chamber at 46.1-46.6°C and 50-60% RH with 14 h photoperiod. One 5-frame hive, placed in corner of chamber containing bees. Dish in centre of chamber with 30% sucrose syrup. Colony supplied with 4 x 10 g cakes changed daily, or 120 g cake every second day.

DEVELOPMENT Various additives greatly affected consumption of foods.

See also: Nation & Robinson (1966, 1968).

→ Standifer, L.N.; Poole, H.K.; Doull, K.M. 1971. Egg production, oviposition and survival of isolated queen honey bees fed experimental diets. ANN ENTOMOL SOC AM 64: 228-232.

Best of many adult diets tested, Nulomoline (Diamond Match Co. Apiary Supplies, Chico, California) - a commercial syrup.

COMPOSITION (%): dextrose-levulose (1:1), 76; water, 24; iron and copper-trace.

REARING At 29.5° C and 75% RH. Cages illustrated; of 1-oz clear plastic cups covered with 5 mesh cloth. Two 9 mm holes punched in sides for food wicks and water containers, and 5-7, 4 mm holes in bottom for ventilation. Cages on two 6 mm strips of wood placed in top or bottom halves of petri dishes lined with wax coated, black filter paper. Queens also held in 15 x 15 x 7.5 cm cage with attendants but isolated in screen-wire push-in cages on beeswax blocks at 22-24° C.

DEVELOPMENT Adult longevity, 4-68 days (average 34 days) in isolation. Maximum of 31 eggs deposited in 48 h and none after.

→ Standifer, L.N.; Haydak, M.H.; Mills, J.P.; Levin, M.D. 1973. Influence of pollen in artificial diets on food consumption and brood production in honey bee colonies. AM BEE J 113: 94-95.

COMPOSITION (parts): soybean flour, 3; dried brewers' yeast, 1; Wheast, ½; meat scraps, ½. 20% pollen added to 4 parts of diet.

PREPARATION Dry ingredients mixed with 2:1 sucrose-water (w/w) solution into patties with doughlike consistency.

REARING Patties placed on bars of frames in hive. All colonies provided with 2:1 sucrose-water solution in Boardman entrance feeders for 3 weeks. Penicillin @ 4.8 g/l and Fumidil B @ 5.0 g/l added to solution for first 3 weeks.

DEVELOPMENT More sealed brood produced than other diet-fed colonies. Highest brood production and protein consumption with 20% pollen in diet.

→ Weaver, N. 1964. A pollen substitute for honey bee colonies. GLEAN BEE CULT 92: 550-553.

COMPOSITION Liquid portion (g): water, 400; sucrose, 600; enzymatic casein hydrolysate, 10; and vitamin mixture, 12 ml.

vitamins (μg/ml): thiamine hydrochloride, 50; pyridoxine hydrochloride, 50; sodium riboflavin-5-phosphate, 50; nicotinic acid amide, 500; calcium pantothenate, 500; ascorbic acid, 500; *m*-inositol, 500; choline chloride, 500; biotin, 33; glutathione, 50; folic acid, 17; carnitine, 17; *p*-aminobenzoic acid, 17; vitamin B_{12}, 3; and acetylcholine chloride, 2 mg.

Solid portion (g): cottonseed flour, 10; corn oil, 0.5; cholesterol, 0.5; Wesson's salt mixture, 0.2.

PREPARATION Liquid portion: casein dissolved in 60% sugar syrup and vitamin solution added.

Solid portion: Cholesterol and corn oil dissolved in chloroform and mixed with cottonseed flour. Chloroform evaporated and salt mixture mixed in.

REARING Colony in 6 x 6 x 36 ft cages. 250 g liquid portion in division board feeder or inverted friction top can with holes in lid. 10 g solid portion given in dish. Supplied 3 times/week.

DEVELOPMENT Poorer than on pollen diet; low survival to adult.

Cephidae

Cephus cinctus Norton, wheat stem sawfly

→ Kasting, R.; McGinnis, A.J. 1958. Note on a method of artificially sustaining larvae of the wheat stem sawfly, *Cephus cinctus* Nort. (Hymenoptera: Cephidae). CAN ENTOMOL 90: 63-64.

Best two of four larval diets tested.

COMPOSITION Diet 1 (%): vitamin-free casein, 29.4; yeast extract, 2.9; cellulose powder, 1.0; water, 66.7; and (mg/g): anhydrous lanolin, 1.0; NaCl, 0.6; $CaCl_2$, 0.013; KCl, 0.013; $NaHCO_3$, 0.013; $NaH_2PO_4.H_2O$, 0.44; K_2HPO_4, 0.51.

Diet 2 (%): vitamin-free casein, 5.5; yeast extract, 0.5; cellulose powder, 27.3; water, 66.7; and (mg/g): anhydrous lanolin, 0.3; other salts as in diet 1.

PREPARATION See McGinnis *et al.** (1956). Liquid portion mixed with dry portions and sterilized at 10 lb pressure for 15 minutes.

REARING Diets packed into drinking straws with glass rod; ends of straws plugged with damp cotton. Hole cut in centre of straw, 1 larva inserted and opening sealed with cellulose tape. Kept at room temperature and 100% RH. Larvae transferred to new straws every 1-2 days.

DEVELOPMENT On both diets larval survival ca. 2 weeks; body weights increased.

See also: McGinnis & Kasting (1962).

→ Villacorta, A.; Bell, R.A.; Callenbach, J.A. 1971. An artificial plant stem as an oviposition site for the wheat stem sawfly. J ECON ENTOMOL 64: 752-753.

COMPOSITION (g): agar, 3; sucrose, 1; methyl paraben, 0.3; sorbic acid, 0.3; and distilled water, 100.0 ml.

PREPARATION Mixture boiled and immediately poured into polyethylene plastic aqueeze bottle with tapered spout for filling 150 x 4 mm diam straws with diet. Both ends of straws then sealed with melted paraffin; 1.5 ml of green food colouring incorporated into 100 ml of diet.

OVIPOSITION After 24 h at 23° C with 15 h photoperiod, 54 eggs deposited compared with control straw of 2-7 eggs.

See also: Villacorta (1970).

Formicidae

Bhatkar, A.; Whitcomb, W.H. 1970. Artificial diet for rearing various species of ants. FLA ENTOMOL 53: 229-232.

COMPOSITION Agar, 5 g; water, 500 ml; 1 whole hen's egg; honey, 62 ml; and 1 vitamin-mineral capsule (McKesson Bexel).

vitamin-mineral capsule: Vitamin A, 25000 U.S.P. Units; Vitamin D, 1250 U.S.P. Units; and (mg): thiamine mononitrate (B_1), 7.5; riboflavin (B_2), 7.5; pyridoxine hydrochloride (B_6), 1.0; cyanocobalamin (B_{12}), 5.0; ascorbic acid (C), 100.0; niacinamide, 100.0; calcium pantothenate, 2.0; and vitamin E (*d*-α-tocopherol acetate concentrate), 3.0 IU, and (mg): iron (dried $FeSO_4$, 84 mg), 25.0; iodine (KI), 0.10; copper ($CuSO_4.5H_2O$), 1.0; manganese ($MnSO_4$), 1.0; magnesium ($MgSO_4$), 1.0; zinc ($ZnSO_4$), 0.5; potassium (K_2SO_4), 5.0; calcium ($Ca(PO_4)_2$), 50.0; phosphorus ($Ca(PO_4)_2$), 39.0; and (%): methyl paraben, 0.064; propyl paraben, 0.016; and excipients, artificial colour and flavouring.

PREPARATION Agar boiled in 250 ml water, and cooled to room temperature. Vitamin-mineral capsule, honey, 250 ml of water, and egg blended in Waring Blendor at 2000 rpm for 3 minutes, and folded into agar; then poured into petri dishes to solidify. Refrigerated and cut into 18 mm squares for feeding.

REARING & DEVELOPMENT Following 28 species of ants representing 4 subfamilies of Formicidae reared successfully. *Odontomachus ruginodis* Wheeler; *Pogonomyrmex badius* (Latreille); *Pheidole bicarinata vinelandica* Forel; *Pheidole dentata* Mayr; *Pheidole morrisi* Forel; *Crematogaster minutissima* Mayr; *Crematogaster clara* Mayr; *Monomorium minimum* (Buckley); *Monomorium pharaonis* (Linnaeus); *Solenopsis geminata* (Fabricius); *Solenopsis saevissima richteri* Forel; *Solenopsis molesta* (Say); *Cyphomyrmex rimosus minutus* Mayr; *Iridomyrmex pruinosum* (Roger); *Conomyrma pyramicus flavus* (McCook); *Camponotus pylartes fraxinicola* M.R. Smith; *Formica schaufussi dolosa* Wheeler; *Pseudomyrmex brunneus* F. Smith; *Pseudomyrmex pallidus* (F. Smith); *Aphaenogaster rudis* Emery; *Crematogaster ashmeadi* Mayr; *Crematogaster atkinsoni* Wheeler; *Leptothorax curvispinosus* Mayr; *Camponotus abdominalis floridanus* (Buckley); *Paratrechina (Nylanderia)* sp.

Following 5 species reared on this diet with difficulty, but with care, sexual forms obtained consistently: *Hypoponera opaciceps* Mayr; *Neivamyrmex opacithorax* Emery; *Pheidole metallescens* Emery; *Trachymyrmex septentrionalis* (McCook); *Conomyrma pyramicus pyramicus* Roger.

Following 2 species failed to mature to sexual forms: *Prenolepis imparis* (Say); *Lasius neoniger* Emery.

Ichneumonidae

Exeristes comstockii (Cresson)

Bracken, G.K. 1965. Effects of dietary components on fecundity of the parasitoid *Exeristes comstockii* (Cress.) (Hymenoptera: Ichneumonidae). CAN ENTOMOL 97: 1037-1041.

COMPOSITION Adult diet (g): *l*-amino acid mixture, 3.000; Wesson's salt mixture, 0.075; cholesterol, 0.100; ribonucleic acid, 0.075;

sucrose, 14.000; and (ml): fatty acid mixture, 40.0; vitamin mixture 20.0; 2 N KOH, 2.5; Tween 80, 0.6; water, 39.0.

l-amino acids (g): glycine, 6.4; alanine, 7.0; valine, 4.5; leucine, 7.7; isoleucine, 5.2; serine, 6.5; threonine, 5.5; cysteine, 1.3; methionine, 3.0; aspartic acid, 6.5; glutamic acid, 10.5; lysine, 5.3; arginine, 5.0; histidine, 4.0; phenylalanine, 5.5; tyrosine, 4.0; tryptophan, 2.0; proline, 8.2; hydroxyproline, 1.9; total 100.0.

fatty acids (mg): oleic acid, 192.0; palmitic acid, 88.0; stearic acid, 40.0; linoleic acid, 60.0; linolenic acid, 20.0.

vitamins (mg): biotin, 0.00009; calcium pantothenate, 0.44; choline chloride, 2.00; folic acid, 0.34; inositol, 5.96; nicotinic acid, 0.30; *p*-aminobenzoic acid, 0.9; pyridoxine hydrochloride, 0.9; riboflavin, 0.16; thiamine hydrochloride, 0.12.

PREPARATION Details in House & Barlow* (1960). Fatty acids and cholesterol dissolved in hot ethanol and emulsified in water with Tween 80 in Waring Blendor. pH to 6.5 with 2 N KOH.

REARING At 23 ± 1°C and 65 ± 1% RH. 5-6 ♀♀/15 x 15 x 12.5 cm cage. For oviposition, coddled *Galleria mellonella* larva put into 5 mm diam hole drilled through 10 x 10 x 30 mm wooden block with 1 mm wide slot cut to expose hole throughout length. Larvae positioned and ends plugged with modelling clay. Slot to prohibit feeding but enable oviposition. 3 larvae provided/cage. Liquid diets offered in bent glass tubes, 1/cage, and one tube of distilled water. All diets and larvae renewed daily.

FECUNDITY Synthetic diet as good as insect tissues of *Galleria mellonella*. Fatty acids in diet decrease fecundity. Eggs/♀, 35 ± 6.4 on diet with no fatty acids. Fecundity and eclosion lowered with lack of vitamins.

Corollary

→ Bracken, G.K. 1969. Effect of dietary amino acids, salts, and protein starvation on fecundity of the parasitoid *Exeristes comstockii* (Hymenoptera: Ichneumonidae). CAN ENTOMOL 101: 91-96.

COMPOSITION Basic adult diet (g/100 ml diet): sucrose, 10.000; *l*-amino acids, 3.000; ribonucleic acid, 0.075; Wesson's salt mixture, 0.075; B-vitamins, 0.011; cholesterol, 0.010; α-tocopherol, 0.008.

FECUNDITY Fecundity on diet with 10 essential amino acids equal to that on control diet with 19 amino acids. Fecundity reduced by 80% if Na^+ substituted for K^+, and by 33% if K^+ only mineral added. Fecundity unaltered by omission of Mn^{2+} or Ca^{2+} from, or addition of Zn^{2+} to, control salt mixture (Wesson's). Females fed control (chemically defined) diet or 10% dilution (sucrose content maintained) for 30 days in absence of host larvae laid more eggs initially when hosts supplied than did females fed 1% control diet, honeydew, or 10% sucrose for similar period.

See also: Bracken (1966).

Exeristes roborator (Fabricius)

Thompson, S.N. 1975. Defined meridic and holidic diets and aseptic feeding procedures for artificially rearing the ectoparasitoid *Exeristes roborator* (Fabricius). ANN ENTOMOL SOC AM 68: 220-226.

Best of 2 diets tested.

COMPOSITION (mg/100 ml): *l*-amino acids: alanine, 300; arginine hydrochloride, 240; aspartic acid, 600; cysteine hydrochloride, H_2O, 60; glutamic acid, 600; glycine, 300; histidine dihydrochloride, 480; hydroxyproline, 240; isoleucine, 240; leucine, 420; lysine dihydrochloride, 420; methionine, 120; phenylalanine, 240; proline, 240; serine, 240; threonine, 240; tryptophan, 60; tyrosine, 240; valine, 720; total, 6000.

vitamins: *d*-biotin, 0.06; choline chloride, 200; cyanocobalamine, 0.02; folic acid, 0.10; *i*-inositol, 15.00; nicotinamide, 5.00; *d*-pantothenic acid (hemi calcium), 2.00; pyridoxine hydrochloride, 0.30; riboflavin, 1.00; thiamine hydrochloride, 0.20; total, 223.68.

lipids: cholesterol, 50; linoleic acid, 50; linolenic acid, 50; palmitic acid, 123; palmitoleic acid, 25; oleic acid, 123; stearic acid, 25; total, 496.

carbohydrate: glucose, 2000.

salts: $CaCl_2$, 30; $CoCl_2.6H_2O$, 5; $FeCl_3.6H_2O$, 20; $ZnCl_2$, 5; K_2HPO_4, 370; $Na_2HPO_4.7H_2O$, 50; $MgSO_4.7H_2O$, 120; $CuSO_4.5H_2O$, 5; $MnSO_4.H_2O$, 1; KOH, 390; total, 606.

PREPARATION Amino acids dissolved in hot water. Riboflavin dissolved in water at 50° C and added to other vitamins dissolved in water. Lipids dissolved in acetone, acetone evaporated with N_2 and lipids emulsified with 0.25 ml Tween 80 and 1 N KOH to pH 7. 50% glucose solution sterilized through Millipore filter. Salts dissolved in water and similarly sterilized. Then lipids added to hot amino acid solution and autoclaved at 121° C for 20 minutes. Cooled aseptically and salt, glucose and vitamin solutions added. 5 g Sephadex LH-20 added and left for 12 h. Diets stirred before use and applied to 1 cm dental rolls in small shell vials.

REARING At 23° C and 60% RH. Reared to pupae in vials and transferred to quart plastic containers; dental rolls removed. Pairs of adults isolated in small plastic containers with honey droplets on side. ♂♂ removed after 24 h and coddled host spread on 4" diam disc and covered with tissue paper, introduced. Eggs removed and sterilized with $HgCl_2$ (see Vanderzant & Davich*, 1958) and individually placed in vials on diet.

DEVELOPMENT Total development time from neonate larva to adult emergence, ca. 22 days, compared with 15 and 17 days on hosts *Pectinophora gossypiella* and *Gnorimoschema operculella* respectively. 80% and 60% survival on diet and hosts respectively. Adult size, fecundity and fertility normal.

Itoplectis conquisitor (Say)

Arthur, A.P.; Hegdekar, B.M.; Batsch, W.W. 1972. A chemically defined, synthetic medium that induced oviposition in the parasite *Itoplectis conquisitor* (Hymenoptera: Ichneumonidae). CAN ENTOMOL 104: 1251-1258.

COMPOSITION Best oviposition medium (M): serine, 0.5; arginine, 0.05; leucine, 0.065; magnesium chloride, 0.025.

PREPARATION In distilled water; pH adjusted to 7.0.

REARING Diet in 35 x 7 mm parafilm tubes.

DEVELOPMENT Mean of 95.6 eggs/♀, compared with 50.6 on wax moth haemolymph and water (diluted 1:2).

→ Yazgan, S. 1972. A chemically defined synthetic diet and larval nutritional requirements of the endoparasitoid *Itoplectis conquisitor* (Hymenoptera). J INSECT PHYSIOL 18: 2123-2141.

COMPOSITION (mg/100 ml diet): *l*-amino acid mixture: alanine, 312.50; arginine hydrochloride, 381.25; aspartic acid, 400.00; cysteine hydrochloride, 81.25; glutamic acid, 656.25; glycine, 487.50; histidine hydrochloride, 150.00; hydroxyproline, 106.25; isoleucine, 325.00; leucine, 481.25; lysine, 362.50; methionine, 187.50; phenylalanine, 418.75; proline, 468.75; serine, 418.75; threonine, 337.50; tryptophan, 87.50; tyrosine, 218.75; valine, 368.75; total, 6250.00.

inorganic salts: $CaCl_2$, 30.5702; $ZnCl_2$, 7.1259; $CoCl_2.6H_2O$, 4.8314; $CuSO_45H_2O$, 5.6010; $FeCl_3.6H_2O$, 17.9858; K_2HPO_4, 375.1083; $MgSO_4.7H_2O$, 131.5444; $MnSO_4.H_2O$, 0.3991; $Na_2HPO_4.12H_2O$, 51.8337; total, 625.00.

lipids and fat-soluble vitamins: cholesterol, 65.6255; linoleic acid, 44.6323; linolenic acid, 20.9932; oleic acid, 112.8729; palmitic acid, 81.3756; palmitoleic acid, 2.6250; Tween 80, 168.9453; vitamin A acetate, 0.2734; vitamin E acetate, 1.5625; total, 498.91.

water-soluble vitamins: ascorbic acid, 8.1619; biotin, 0.0364; calcium pantothenate, 2.1862; choline chloride, 189.4737; folic acid, 0.1093; inositol, 13.1174; nicotinic acid, 4.3725; pyridoxine hydrochloride, 0.2915; riboflavin, 1.0202; thiamine hydrochloride, 0.1458; vitamin B_{12}, 0.0146; total, 218.93; distilled water to 100 ml.

miscellaneous: agar, 468.75; glucose, 2000.00; ribonucleic acid, 234.38; 2 N KOH, 340.00; 2 N KH_2PO_4, 27.00.

PREPARATION Amino acids, lipids, inorganic salts, and water-soluble vitamins solution prepared in bulk. Amino acid mixture, RNA, salt mixture, and agar dissolved in distilled water in 250 ml Virtis flask; lipids added, and pH adjusted to ca. 6.5 by addition of 2 N KOH. Stirred magnetically for 10 minutes at room temperature, autoclaved for 15 minutes at 121° C and 103 kPa and cooled to ca. 50° C. Glucose and water-soluble vitamins adjusted to pH 6.5 and filtered through Seitz filter, then combined aseptically with autoclaved portion in sterile chamber. Diet ground to viscous slurry and dispensed to 10 x 75 mm test tubes @ 0.25 ml/tube.

REARING Axenically at 23° C and 65% RH in total darkness. Eggs of *I. conquisitor* sterilized in $HgCl_2$ solution. Parasitized *Galleria mellonella* pupae held for 24 h before dissection in 0.8% saline to allow some embryonic development of eggs to proceed (less likelihood of damage during sterilization). For sterilization, eggs dipped in $HgCl_2$ solution; large number of eggs treated at one time without injury by using tool made by covering one side of small black loop with black cheesecloth. Sterilized eggs rinsed several times in sterile solution. Finally, transferred to small glass container with sterile saline solution and covered, except for periodic ventilation; eggs left to hatch. Neonate larvae individually transferred with sterile hair brush to tubes with diet.

DEVELOPMENT 73% survival to adults, compared with 36% on *Galleria mellonella* pupae. Mean of 25.5 days to adult on diet; 15.3 days on pupa. Average weights of ♂♂ and ♀♀ from diet 7.2 and 10.4 mg respectively, compared with 24.8 and 33.5 mg from *Galleria* pupae. Weight and size of adults from diet about that from *Galleria* but produced normal second generation on *Galleria* pupae, and lived as long as those obtained from insect host.

See also: Yazgan & House (1970).

Pimpla turionellae (Linnaeus)

→ Bronskill, J.F.; House, H.L. 1957. Notes on rearing a pupal endoparasite, *Pimpla turionellae* (L.) (Hymenoptera: Ichneumonidae), on unnatural food. CAN ENTOMOL 89: 483.

COMPOSITION Pork liver and 0.8% aqueous solution of NaCl.

PREPARATION Equal parts by weight of liver and saline solution homogenized in Waring Blendor and autoclaved at 15 lb pressure for 15 minutes. When cooled, coagulated homogenate transformed into thick watery mixture without breaking asepsis. About 2 ml aliquots pipetted into 12 mm diam sterile test tubes and plugged with cotton.

REARING At 23° C and 72% RH in total darkness individually in test tubes. Mature larvae transferred to small, ventilated gelatin capsules, lined with blotting paper.

DEVELOPMENT 147 out of 152 embryos hatched; 10 (7%) became adults.

Megachilidae

Megachile rotundata Fabricius, alfalfa leaf-cutter bee

→ Nelson, E.V.; Roberts, R.B.; Stephen, W.P. 1972. Rearing larvae of the leaf-cutter bee *Megachile rotundata* on artificial diets. J APIC RES 11: 153-156.

Best of 5 diets tested.

COMPOSITION Natural pollen diet (g): mixed pollen from honeybees, 75; cellulose powder, 25; and sugar solution.

sugar solution (g): sucrose, 500; methyl *p*-hydroxybenzoate, 0.5; and (ml): water, 500; honey, 50.

PREPARATION Dry ingredients blended and sugar solution added to form thin paste; poured into 9.5 cm petri dishes to depth of 3.5 mm.

REARING 25 eggs or larvae transferred with bent spatula to diet and kept at 30° C. Cocoons spun in 5 x 7 mm paraffin block capped with half No. 2 gelatin capsule with food supply.

DEVELOPMENT Normal adults from 2-3-day-old larvae. After 5-6 days mature larval wt, 63-70 mg.

Pteromalidae

Pteromalus puparum (Linnaeus)

→ Hoffman, J.D.; Ignoffo, C.M.; Long, S.H. 1973. *In vitro* cultivation of an endoparasitic wasp, *Pteromalus puparum*. ANN ENTOMOL SOC AM 66: 633-634.

COMPOSITION Haemolymph-plasma from a mature 5th instar *Heliothis zea* (Lepidoptera: Noctuidae) larvae.

PREPARATION *H. zea* larvae immersed in water at 60°C for 6 minutes prior to bleeding to inhibit melanization of haemolymph. Blood cells etc. removed by centrifuging (23000 G, 30 min, 5°C) and filtration (0.3 μm membrane filter). Haemolymph dilutions prepared with sterile distilled water.

DEVELOPMENT Haemolymph above 50% concentration supported development from neonate larva to adult, with no significant variation in degree. Data for 50, 75, and 100% haemolymph were: % survival to (i) prepupa: 65, 49, 49; (ii) pupa: 61, 45, 45; (iii) adult: 47, 40, 39. Average number of days to (i) prepupa: 7.2, 6.6, 6.5; (ii) pupa: 7.7, 7.8, 7.8; (iii) adult: 12.5, 12.4, 12.3.

→ Hoffman, J.D.; Ignoffo, C.M. 1974. Growth of *Pteromalus puparum* in a semisynthetic medium. ANN ENTOMOL SOC AM 67: 524-525.

COMPOSITION Medium (YEG) formulated by equal folumes of: 10% yeastolate (Difco Labs, Detroit, Michigan) in distilled water, foetal bovine serum (Microbiological Assoc. Inc., Bethesda), and Grace's tissue culture medium (Grand Island Biological Co., New York).

PREPARATION Blended then sterilized through 0.22 μm Millipore membrane filter.

REARING Sterile eggs of *P. puparum* dissected from *Pieris rapae* (Lepidoptera: Pieridae) floated on surface of 100 μl of medium in tissue culture plate cells. Mature larvae transferred to moist filter paper to pupate. Incubation in dark at 27°C and 100% RH.

DEVELOPMENT Developmental data, with those of controls reared on *Heliothis zea* haemolymph plasma (Hoffman *et al.** 1973) in brackets: survival to pupa, 27 (66) %; emergence, 44 (89) %; average time of neonate larva to adult, 15, (13) days.

See also: Bouletreau (1968).

Trichogrammatidae

Trichogramma pretiosum Riley

→ Hoffman, J.D.; Ignoffo, C.M.; Dickerson, W.A. 1975. *In vitro* rearing of the endoparasitic wasp, *Trichogramma pretiosum*. ANN ENTOMOL SOC AM 68: 335-336.

COMPOSITION Processed haemolymph plasma from 5th stage larvae of *Heliothis zea* (Boddie). See Hoffman *et al.* (1973).

PREPARATION See Hoffman *et al.** (1973).

REARING Eggs from parasitized *Trichoplusia ni* surface sterilized in 0.125% NaClO for 15 minutes, neutralized in 10% sodium thiosulphate for 15 seconds and rinsed for 30 seconds in sterile distilled water. Parasite eggs dissected from host eggs on to sterile 6 mm diam filter paper discs; water withdrawn and replaced with 10 µl of culture medium. Incubated at 27° C and 100% RH under constant fluorescent light. Excess medium withdrawn after 24 or 96 h.

DEVELOPMENT Larval development and adult survival increased by 30% and 15% respectively. Wing deformities in 66%; but eggs viable. Development to adult in 225 ± 4 h if liquid withdrawn after 24 h; 199 ± 4 h on *T. ni* eggs.

Isoptera

Kalotermitidae

Zootermopsis angusticollis (Hagen), Pacific dampwood termite

→ Cook, S.F.; Scott, K.G. 1933. The nutritional requirements of *Zootermopsis (Termopsis) angusticollis*. J CELL COMP PHYSIOL 4: 95-110.

Best 2 diets of many diets tested.

COMPOSITION Diet 1 : agar; salt mixture 0.4%; casein; glucose; Crisco; cod liver oil; vitamins B and G.

Diet 2: as above except no 'Crisco' (hydrogenated, fat-free oil).

Quantities of ingredients not specified for either diet.

PREPARATION Ingredients combined in 2% agar gel.

REARING At 18-21°C; food and containers changed daily.

DEVELOPMENT On diet 1, weight gain of 14.8% after 37 days; survival, 100%. Diet 2, weight gain of 16.5% after 25 days; survival 89%.

Rhinotermitidae

Reticulitermes flavipes (Kollar), eastern subterranean termite

→ Bready, J.K.; Friedman, S. 1963b. The nutritional requirements of termites in axenic cultures. 2. Studies on the effectiveness of antibiotics in the sterilization of workers of *Reticulitermes flavipes*. ANN ENTOMOL SOC AM 56: 706-708.

COMPOSITION (g): agar, 0.25; glucose, 1.25; vitamin-free casein, 1.25; or bacto peptone, 1.25; and distilled water, 12.5 ml. Various antibiotics also added.

PREPARATION Ingredients placed in 2-oz widemouth glass jar. Methyl *p*-hydroxybenzoate added to concentration of 0.02 M. Jars plugged with gauze-covered cotton, autoclaved 20 minutes at 15 lb pressure and shaken while cooling. Antibiotics added to diet in 1 ml water.

REARING At 22 ± 0.5° C and 100% RH. 3rd or 5th instar workers placed individually in shell vials or together in jars.

DEVELOPMENT Maintenance diet. Antibiotics at 1.2% concentration not effective in killing termite-associated bacteria.

See also: Bready & Friedman (1963a).

Reticulitermes flavipes (Kollar), eastern subterranean termite

Coptotermes formosanus Shiraki, Formosan subterranean termite

→ Mauldin, J.K.; Rich, N.M. 1975. Rearing two subterranean termites, *Reticulitermes flavipes* and *Coptotermes formosanus*, on artificial diets. ANN ENTOMOL SOC AM 68: 454-456.

Best of 16 diets. Maintenance diet for laboratory colony.

COMPOSITION (% wet weight): β-sitosterol or ergosterol, 0.06; distilled water, 65.87; α-cellulose, 34.06; methyl *p*-hydroxybenzoate, 0.01.

REARING At 26.5 ± 1° C.

DEVELOPMENT Survival after 1 year for *R. flavipes*: 28% workers, 3% soldiers, and 4% nymphs. For *C. formosanus* after 3 months: 73.3% workers, 1.2% soldiers.

See also: Becker (1965).

Lepidoptera

Arctiidae

Arctia caja Linnaeus

→Moreau, J.P. 1965a. A propos de la biologie d'*Arctia caja* L. (Lépidoptères Arctiidae). PROC 12 INT CONGR ENTOMOL LONDON (1964) 539. (In French)

COMPOSITION (g): powdered vine (*Vitis vinifera*) leaf, 1.0; cellulose powder, 1.0; agar, 0.2; yeast extract, 0.12; benzoic acid, trace; distilled water, 9.5.

REARING At 22-24°C with constant lighting.

DEVELOPMENT Complete.

Estigmene acrea (Drury), salt-marsh caterpillar

→Earle, N.W.; MacFarlane, J. 1968. A unisexual strain of the salt-marsh caterpillar, *Estigmene acrea*. ANN ENTOMOL SOC AM 61: 949-953.

Modified diet of Vanderzant* (1967a) developed for *E. acrea*.

COMPOSITION (g/100 g diet): wheat germ, 3; vitamin-free casein, 3.5; sucrose, 3.5; Wesson's salt mixture, 1.0; cholesterol, 0.05; choline chloride, 0.10; inositol, 0.04; B-vitamin mixture, 0.0033; ascorbic acid, 0.4; raw linseed oil, 0.25; methyl *p*-hydroxybenzoate, 0.10; 20% acetic acid, 2.1; 40% formaldehyde, 0.05; agar, 2.5; distilled water, 85.

vitamins (mg): niacinamide, 1.0; calcium pantothenate, 1.0; riboflavin, 0.5; thiamine hydrochloride, 0.25; pyridoxine hydrochloride, 0.25; folic acid, 0.25; biotin, 0.02; vitamin B_{12}, 0.002.

PREPARATION See Vanderzant* (1967a).

REARING At 27 ± 1°C with 14 h photoperiod, individually in 1-oz polystyrene cups with cardboard lids. Final instar larvae transferred to fresh diet in plastic petri dishes. Adult pairs held in 1-pint cartons lined with paper towelling and covered with polyethylene sheeting. Moisture provided; lighting of low intensity.

DEVELOPMENT Survival from neonate larvae to adult, 90%; seven generations reared.

→ Vail, P.V.; Henneberry, T.J.; Pengalden, R. 1967. An artificial diet for rearing the salt marsh caterpillar, *Estigmene acrea* (Lepidoptera: Arctiidae), with notes on the biology of the species. ANN ENTOMOL SOC AM 60: 134-138.

Modification of diet by Ignoffo* (1963) for *Trichoplusia ni*.

COMPOSITION (ml): distilled water, 3100; methyl *p*-hydroxybenzoate (15% w/v in 95% ethyl alcohol), 36; choline chloride (0.1 g/ml of water), 36; 4M KOH, 18; formaldehyde (0.1 g/ml of water), 15; vitamin stock, 6; and (g): sucrose, 126; casein, 126; wheat germ, 108; ascorbic acid, 15; Aureomycin, 0.5.

vitamins (mg/100 ml water): niacin, 1200; calcium pantothenate, 1200; riboflavin, 600; thiamine, 300; folic acid, 300; biotin, 24; vitamin B_{12}, 2.4; pyridoxine, 300.

PREPARATION Agar dissolved in 2220 ml hot water. 880 ml water blended, in order, with casein, KOH, all solids, (except ascorbic acid and Aureomycin), methyl *p*-hydroxybenzoate, choline chloride, formaldehyde, vitamin solution, agar solution, and finally ascorbic acid and Aureomycin until homogenous. Then dispensed to cups or rearing trays.

REARING At 25.5° C. Two rearing methods: (1) 75 ml diet cooled in 6-oz cups, and seeded with 5 neonate larvae; transferred to fresh diet after 2 weeks for pupation. Sexed pupae placed in 1-gallon icecream cartons with cloth lids for emergence. Five pairs of adults placed in 1-gallon cartons with Saran-wrap lined with graph paper. (2) In 50.8 x 33 x 6.3 cm porcelain trays @ 125 larvae/tray on diet resting on chicken wire frame 1.27 cm above bottom of tray. Diet replaced weekly. Glass lid, resting on rubber or felt on rim of tray.

DEVELOPMENT Similar to that on host plants. 36.7 days to pupation; pupation, 77%; adult emergence, after 12.8 days. Twelve generations reared.

Hyphantria cunea (Drury), fall webworm

→ Injac, M.; Veyrunes, J-C.; Kuhl, G. 1973. Gajenje dudovca (*Hyphantria cunea* Drury) veštačkom hranom u aseptičnim uslovima. ZAST BILJA 24: 111-117. (In Yugoslav with French summary)

Modification of diet for *Lymantria dispar* by Magnoler* (1970).

COMPOSITION (g): agar, 22.8; casein, 28; Wesson's salt mixture, 10; glucose, 19.8; wheat germ, 57; cellulose powder, 14; ascorbic acid, 5; choline chloride, 1; and (ml): distilled water, 825; 4M KOH, 4.7; linolenic acid, 2.5; cholesterol, 0.2; vitamin mixture, 10.

vitamins (mg/100 ml): *i*-inositol, 2000; niacin, 100; calcium pantothenate, 100; riboflavin, 50; thiamine hydrochloride, 25; pyridoxine hydrochloride, 25; folic acid, 25; biotin, 2; vitamin B_{12}, 0.2.

PREPARATION Agar dissolved in boiling distilled water; remaining ingredients, less vitamins, added and blended in above order. Autoclaved at 100° C for 1 minute, cooled, vitamins added after sterilizing through 0.45 µm Millipore filter. Well mixed and 30 cm^3 dispensed to sterile 150 mm petri dishes. Stored at 4° C.

REARING At 24° C with continuous light. Air sterilized under UV lamps. 20 neonate larvae/dish; diet orientated towards light. Adults in 20 x 10 cm diam plastic cylinders with wire mesh covered ends and thick paper lining for oviposition.

DEVELOPMENT Egg incubation period, 7 days; web spinning from 24 h to 4th moult. From 24-26 days, no feeding, but cocoon spinning. Pupal weight (mg) on mulberry leaves and diet; ♂, 9.88 and 10.5; ♀, 14.25 and 16.4 respectively. Pupal period, 13-22 days. Total developmental time, 48 days; sex ratio 1:1.35 ♀:♂; mean fecundity, 618 ± 123 eggs/♀. Five generations reared.

See also: Beratlief & Boguleau (1974).

Kunimi, Y.; Aruga, H. 1974. Susceptibility to infection with nuclear- and cytoplasmic-polyhedrosis viruses of the fall webworm, *Hyphantria cunea* Drury, reared on several artificial diets. JPN J APPL ENTOMOL ZOOL 18: 1-4. (In Japanese with English summary)

Three diets used.

COMPOSITION Diet 1 (g): potato starch, 10.00; sucrose, 10.00; soybean meal, 40; β-sitosterol, 0.50; soybean oil, 3.00; Wesson's salt mixture, 3.50; agar, 15.00; cellulose, 34.00; Morin, 0.30.

Diet 2 (g): potato starch, 7.42; sucrose, 7.42; soybean meal, 29.68; β-sitosterol, 0.37; soybean oil, 2.23; Wesson's salt mixture, 2.60; agar, 11.13; cellulose, 25.23; Morin, 0.22; mulberry leaf powder, 30.00.

Diet 3 (g): potato starch, 4.84; sucrose, 4.84; soybean meal, 19.36; β-sitosterol, 0.24; soybean oil, 1.45; Wesson's salt mixture, 1.69; agar, 7.26; cellulose, 16.46; Morin, 0.15; mulberry leaf powder, 60.00.

and for all diets (g): KH_2PO_4, 1.00; ascorbic acid, 2.00; citric acid, 0.40; sorbic acid, 0.30; vitamin B mixture (undescribed), 1 ml/10 g; propionic acid, 1 ml/10 g; Spiramycin, 2 ml/100 g; distilled water, 3 ml/1 g.

REARING At 25 ± 1° C. Larvae from 1 egg mass reared to second instar in 9 x 15 cm diam plastic containers. From 3rd instar, reduced to 20 larvae/container. Diet supplied daily.

DEVELOPMENT Growth on semisynthetic diet inferior to that on diets with mulberry leaf powder. Average weight of 4th instar larvae (mg): semisynthetic diet, 5.62; 25% mulberry leaf powder diet, 7.57; 50% mulberry leaf powder diet, 10.07. Similar differences between pupation rates and pupal weights.

Yearian, W.C.; Gilbert, K.L.; Warren, L.O. 1966. Rearing the fall webworm, *Hyphantria cunea* (Drury) (Lepidoptera: Arctiidae) on a wheat germ medium. J KANS ENTOMOL SOC 39: 495-499.

COMPOSITION (g): vitamin free casein, 120; wheat germ, 150; sucrose, 120; Wesson's salt mixture, 36; choline chloride, 4; methyl *p*-hydroxybenzoate, 6.0; ascorbic acid, 15.0; Tetracycline, 1.0; powdered cellulose, 18; agar, 55 and (ml): sodium ascorbate ($NaHCO_3$, 42 g; and ascorbic acid, 94 g/l water), 75; 20% acetic acid, 75; vitamin mixture, 36; water, 3100.

vitamins (mg/ml): nicotinic acid, 1.0; calcium pantothenate, 1.0; riboflavin, 0.5; thiamine hydrochloride, 0.25; pyridoxal hydrochloride, 0.25; folic acid, 0.25; biotin, 0.02; vitamin B_{12}, 0.002.

PREPARATION Agar dissolved in hot water and blended with combined dry ingredients except ascorbic acid. Blended again at 80° C with remaining ingredients and dispensed to 8-dram vials at 15 ml/vial. Vials plugged with cotton and stored at 7° C.

REARING At 26° C with 14 h photoperiod. 5-10 adult pairs in ½-gallon icecream container oviposition cage. Screen top and side sections for ventilation. Eggs laid on 2-4 cheesecloth strips hung in cage; removed and incubated in 20 x 90 mm petri dishes lined with moist filter paper. 4-6 neonate larvae transferred to vials with brush; vials upright for 48 h, then inverted. Pupae transferred to 1-oz plastic cups on moist vermiculite. Adults fed on 50% honey solution.

DEVELOPMENT Incubation of eggs, 6-11 days; pupation, 25 days from eclosion; pupal period, 12 days. Adult longevity: ♂, 12.1 days; ♀, 9.8 days. Average fecundity, 650 eggs/♀. Total life cycle, 52 days.

Bombycidae

Bombyx mori Linnaeus, silkworm

There is a large volume of literature, mostly from Japan, on the development and formulation of artificial diets for the silkworm, *B. mori*. Ito and his associates have published numerous papers in both Japanese and English on the subject. Their preliminary work started in the 1950's; and the first basal artificial diet was published in 1961. Since then more than 50 papers have appeared giving modifications of the diet and refining the rearing technique for normal production of silkworm. Some important papers of Ito's group and also other workers are listed in the bibliography. Here, only selected works are abstracted, and listed in chronological order for each group.

→ Ito, T. 1961a. Nutrition of the silkworm, *Bombyx mori*. I. Nutritive effects of soybean oil for the larva. BULL SERIC EXP STN JPN 16: 311-348. (In Japanese with English summary)

Three basic diets tested.

COMPOSITION Diet 1 (g): mulberry leaf powder, 1.2; potato starch (Wako Pure Chemical Industries Ltd, Tokyo), 0.9; sucrose, 1.5; defatted soybean casein, 1.5; cellulose powder, 0.84; Wesson's salt mixture, 0.06; total, 6.0.

Diet 2 (g): as diet 1 except: mulberry leaf powder reduced from 1.2 to 0.6 g; sucrose and soybean casein both increased from 1.5 to 1.8 g.

Diet 3 (g): as diet 2 except: sucrose and soybean casein both reduced from 1.80 to 1.32 g; cellulose powder increased from 0.84 to 1.80 g; and 2.37 mg extra vitamins, added.

All diets contained 9 ml water/6 g diet. Some diets also contained butyl *p*-hydroxybenzoate or sodium sorbinate or sodium dehydroacetate as mould inhibitors at 0.5, 5, 5 mg/6 g diet respectively. Amounts of soybean oil varied.

vitamins (µg/6 g dry diet): biotin, 10; calcium pantothenate, 100; choline chloride, 1000; folic acid, 10; inositol, 1000; niacin, 100; pyridoxine hydrochloride, 50; riboflavin, 50; thiamine hydrochloride, 50; total, 2370.

PREPARATION Mulberry leaves dried at room temperature, pulverized and sifted to 150 mesh when required. Soybean powder prepared by grinding raw soybean and sieving to 80 mesh. When soybean oil, casein or other ingredients used, other ingredients adjusted to maintain total dry diet 6 g.

Diet constituents mixed thoroughly, ground in mortar, and transferred to petri dish. Water added alone, or with vitamin solution. Stirred, autoclaved for 15 minutes, cooled and refrigerated until use.

REARING 20 neonate larvae in 9 cm petri dishes at 25°C on 1 mm thick slices of diet on filter paper. Diet replaced twice daily.

DEVELOPMENT Larval development reduced from 24 to 15 days with diet improvement. Highest efficacy of soybean oil at 6%; larval growth, development and weight accelerated and larval mortality reduced with addition of 6 or 12 mg soybean sterol to diet 3.

→ Ito, T.; Tanaka, M. 1962. Rearing of the silkworm by means of aseptic technique. J SERIC SCI JPN 31: 7-10. (In Japanese with English summary)

First report of rearing *B. mori* aseptically.

COMPOSITION (% dry diet): mulberry leaf powder, 20; potato starch, 10; glucose, 5; sucrose, 5; soybean casein, 25; soybean oil, 2; β-sitosterol, 0.2; Wesson's salt mixture, 1; ascorbic acid, 2; cellulose powder, 29.8; and vitamin mixture (Ito*, 1961a), 4.74 mg; distilled water, 1.5 ml/g dry diet.

PREPARATION Mixed, autoclaved for 15 minutes at 100°C, sliced into 500 ml Erhlenmeyer flasks. Flasks autoclaved daily for 2 days each for 30 minutes at 10 lb pressure.

REARING Sterilized silkworm eggs seeded @ 10/flask and reared at 25°C.

DEVELOPMENT Growth and development in aseptic culture inferior to that in septic culture. Only 1 larva spun cocoon, 41 days after hatching; cocoon wt, 0.106 g.

Corollary

→ Ito, T.; Arai, N. 1965a. Nutrition of the silkworm, *Bombyx mori*. VIII. Amino acid requirements and nutritive effects of various proteins. BULL SERIC EXP STN JPN 19: 345-373. (In Japanese with English summary)

Two diets tested.

COMPOSITION Amino acid diet (% dry diet): agar, 4.0; potato starch, 13.5; sucrose, 7.5; amino acid mixture, 10.0; soybean oil, 3.0; β-sitosterol, 0.3; Wesson's salt mixture, 4.0; ascorbic acid, 2.0; cellulose powder, 55.4; morin, 0.3; total, 100.0.

Protein diet (% dry diet): agar, 5.0; potato starch, 15.0; sucrose 15.0; protein, 15.0; other ingredients similar to amino acid diet except cellulose powder reduced from 55.4 to 44.4.

Both diets included B-vitamin mixture @ 4.74 mg/6 g dry diet (see Ito & Horie, 1962); mulberry leaf fraction @ 50 mg/6 g dry diet (see extraction procedure Horie, 1962 and Horie & Ito, 1963); and distilled water, 1.5 ml/g dry diet.

l-amino acids (%): alanine, 7.0; arginine hydrochloride, 6.0; sodium aspartate, 11.0; cystine, 2.5; sodium glutamate, 13.0; glycine, 4.5; histidine hydrochloride, 2.5; hydroxyproline, 1.0; isoleucine, 5.5; leucine, 8.5; lysine hydrochloride, 6.0; methionine, 2.5; phenylalanine, 5.5; proline, 4.0; serine, 3.5; threonine, 5.0; tryptophan, 4.0; tyrosine, 2.0; valine, 6.0; total, 100.0.

proteins tested: soybean protein, α-protein, vitamin-free casein, Hammarsten casein, edestin, gluten, zein, bovine albumin, egg albumin, lactalbumin, gelatin, peptone and soybean meal.

DEVELOPMENT On amino acid diet, average weight of surviving larvae after 20 days, 27.5 mg; 55% survival to 4th instar in 30 days. Best single protein for growth, development and survival, soybean meal @ 15% in diet.

→Ito, T.; Arai, N. 1966b. An amino-acid diet for the silkworm, *Bombyx mori* L. J AGRIC CHEM SOC JPN 40: 110-112. (In Japanese with English summary)

COMPOSITION (g): potato starch, 5.0; sucrose, 15.0; amino acid mixture, 20.0; refined soybean oil, 3.0; β-sitosterol, 0.5; Wesson's salt mixture, 4.0; ascorbic acid, 2.0; cellulose powder, 35.2; morin, 0.3; agar, 15.0; and (mg/g): mulberry leaf fraction, 10; and (μg/g): biotin, 4; choline chloride, 1500; folic acid, 4; inositol, 400; nicotinic acid, 300; calcium pantothenate, 150; pyridoxine hydrochloride, 30; riboflavin, 20; thiamine hydrochloride, 20.

l-amino acids (g): alanine, 7.0; arginine hydrochloride, 6.0; sodium aspartate, 12.0; cystine, 2.5; sodium glutamate, 12.0; glycine, 4.5; histidine hydrochloride, 2.5; hydroxyproline, 1.0; isoleucine, 5.5; leucine, 8.5; lysine hydrochloride, 6.0; methionine, 2.5; phenylalanine, 5.5; proline, 4.0; serine, 3.5; threonine, 5.0; tryptophan, 4.0; tyrosine, 2.0; valine, 6.0.

PREPARATION Mulberry leaf fraction extracted with 80% methanol, defatted with ether, precipitated with lead acetate; lead removed with H_2S. Amino acids and vitamins added in aqueous solution; all ingredients combined, heated and cooled to gel.

REARING At 25°C.

DEVELOPMENT Survival of 33 out of 40 neonate larvae to maturity in 31 days. Weight of 7-day-old 5th instar larvae (g): ♂, 1.81; ♀, 1.91. Several larvae to pupae without spinning; no emergence. First report on rearing entirely on amino acid diet.

→Ito, T. 1969. Rearing of the silkworm under aseptic condition. JPN AGRIC RES Q 4: 33-35.

COMPOSITION (g): dried mulberry leaf powder, 50; potato starch, 20; soybean meal, defatted, 20; mineral, 1; ascorbic acid, 2; citric acid, 0.5; cellulose powder, 8; agar, 12; total, 113.5; and distilled water, 300 ml. Also antiseptic and B-vitamin mixtures.

vitamins (μg/g): biotin, 2; choline chloride, 1500; folic acid, 2; inositol, 2000; niacin, 100; calcium pantothenate, 150; pyridoxine hydrochloride, 30; riboflavin, 20; thiamine hydrochloride, 20.

Composition of minerals and antiseptics not given.

PREPARATION Ingredients ground, mixed, and transferred to petri dish with vitamins and antiseptic in solution. Dish covered, heated to 95°C in water bath for 10 minutes, then cooled to gel. Diet sliced and placed in presterilized flasks of test tubes; then autoclaved for 20 minutes at 69 kPa.

REARING At 25°C. Eggs sterilized by immersion in 2% formalin for 15 minutes. Rinsed in sterile water and transferred to test tube or flasks. Rearing conducted in vinyl isolator (illustrated). Main chamber floor area 120 x 55 cm^2; height 55 cm. Air intake filter, outlet filter and outlet trap provided. Internal sterilization with 2% peracetic acid. For further details see Ito *et al.* (1967).

DEVELOPMENT Normal larval development, giving aseptic pupae and adults.

→ Horie, Y.; Inokuchi, T.; Watanabe, K. 1970. Effects of dietary amino acid balances on growth and cocoon quality in the silkworm, *Bombyx mori* L. BULL SERIC EXP STN JPN 24: 345-365. (In Japanese with English summary)

Two diets tested.

COMPOSITION Diet 1 (mg): agar, 150; potato starch, 75; soybean oil (refined), 30; sucrose, 125; β-sitosterol, 5; Wesson's salt mixture, 40; ascorbic acid, 20; citric acid, 5; cellulose powder, 342; essential amino acids, 112; morin, 2; total,906. Also B-vitamin mixture and antiseptic. Mulberry leaf fraction and distilled water added at 5 mg/g and 3 ml/g dry weight of diet respectively.

Diet 2, as diet 1 except: non-essential amino acids and essential amino acids added at 48 and 50 mg; mulberry leaf fraction omitted. Total weight, 1004 mg.

essential amino acid mixture (mg): arginine hydrochloride, 12; histidine hydrochloride, 6; isoleucine, 10; leucine, 16; lysine hydrochloride, 14; methionine, 6; phenylalanine, 12; threonine, 10; tryptophan, 6; valine, 12; proline, 8; total, 112.

non-essential and acidic amino acid mixture (%): alanine, 29.2; cystine, 8.3; glycine, 20.8; serine, 29.2; tyrosine, 12.5; total, 100.0; acidic amino acids (%): potassium glutamate, 48.0; potassium aspartate, 52.0.

vitamins (mg/g dry diet): choline chloride, 1.5; inositol, 2.0; niacin, 0.3; calcium pantothenate, 0.15; and (μg/g dry diet): biotin, 4; folic acid, 4; pyridoxine hydrochloride, 30; riboflavin, 20; thiamine hydrochloride, 20.

PREPARATION & REARING See Ito* (1961a).

DEVELOPMENT Maximum larval weight when dietary mixture of 49% non-essential and 51% essential amino acids. Effects of individual amino acids on cocoon spinning described.

Ito, T.; Horie, Y.; Watanabe, K.; Takamiya, K.; Furuyama, M.; Miyabayashi, M.; Yamamoto, K.; Nagashima, M. 1974. Rearing of larvae of the silkworm, *Bombyx mori*, entirely on semi-synthetic diets. J AGRIC CHEM SOC JPN 48: 403-407. (In Japanese with English summary)

Three different semi-synthetic diets: "starter" (for 1st instar), "grower" (for 2nd to 4th instar) and "producer" (for 5th instar) formulated.

COMPOSITION "Starter" (g): potato starch, 10; sucrose, 10; soybean oil (refined), 3.0; soybean meal (defatted), 30.0; β-sitosterol, 0.5; Wesson's salt mixture, 3.5; K_2HPO_4, 1.0; ascorbic acid, 2.0; cellulose powder, 34.0; agar, 15.0; citric acid, 0.5; sorbic acid, 0.2; morin, 0.2; total, 109.9; and vitamin B mixture (μg/g dry diet): biotin, 2; choline chloride, 1500; folic acid, 2; inositol, 200; nicotinic acid, 100; calcium pantothenate, 150; pyridoxine hydrochloride, 30; riboflavin, 20; thiamine, 20; total, 2024. Antiseptic (not specified) and distilled water, 3.0 ml/g diet.

"Grower" : as "starter" except: soybean meal increased from 30.0 to 40.0 g; and morin decreased from 0.2 to 0.1 g; total, 119.9 g.

"Producer" : as "starter" except: potato starch increased from 10.0 to 20.0 g; sucrose replaced by 12 g glucose; soybean meal increased from 30 to 60 g; K_2HPO_4, cellulose powder and morin deleted; total, 105.2. Also 2.6 ml instead of 3.0 ml distilled water/g dry diet used.

COMPOSITION & REARING Ito* (1961a).

DEVELOPMENT Development to maturity, 21-25 days. Larval weight of artificially reared silkworm higher than that of mulberry reared; cocoon weight heavier, but cocoon: shell ratio lower.

Ito, T.; Mizuta, Y.; Takamiya, K.; Ueda, S.; Kimura, R.; Higuchi, T.; Takahashi, S. 1975. Growth, development, and cocoon production in artificial-diet-rearing of the silkworm, *Bombyx mori*, with or without application of a synthetic juvenile hormone analog. J AGRIC CHEM SOC JPN 49: 39-48. (In Japanese with English summary)

COMPOSITION (g): mulberry leaf powder, 25; corn starch, 7.5; sucrose, 8.0; soybean meal (defatted), 36.0; soybean oil (refined), 1.5; soybean sterol, 0.2; salt mixture, 3.0; cellulose powder, 15.0; agar, 7.5; ascorbic acid, 1.0; citric acid, 4.0; total, 108.7; also vitamin B mixture (Ito*, 1961a); antiseptic; and distilled water, 2.57 ml/g dry diet.

antiseptic: sorbic acid; propionic acid, chloramphenical.

REARING At 26°C and 75% RH in total darkness.

DEVELOPMENT On mulberry leaf: larval period, ca. 24 days; 84.0% to cocoon; wt, 1.59 g. On artificial diet: larval period, ca. 28 days; 95.3% to cocoon; wt, 1.75 g. Quality of cocoon on artificial diet superior to that on mulberry-leaf.

Inokuchi, T. 1969a. Nutritional studies of amino acids in the silkworm, *Bombyx mori*. I. Effect of dietary cystine on methionine requirement and its metabolism. BULL SERIC EXP STN JPN 23: 371-387. (In Japanese with English summary)

COMPOSITION Basal dry diet (%): agar, 15.0; potato starch, 7.5; sucrose, 13.0; amino acid mixture, 20.0; soybean oil (refined), 3.0; β-sitosterol, 0.5; salt mixture, 4.2; ascorbic acid, 2.0; cellulose powder, 34.0; citric acid, 0.5; morin, 0.3; total, 100.0. Also vitamin mixture, distilled water (3 ml/g dry diet), and antiseptic.

l-amino acids (A) (mg/g dry diet): arginine hydrochloride, 6; histidine hydrochloride, 3; isoleucine, 5; leucine, 8; lysine hydrochloride, 6; phenylalanine, 6; threonine, 5; tryptophan, 3; valine, 6; proline, 4; potassium aspartate, 13; potassium glutamate, 12; alanine, 7; glycine, 4; serine, 5; tyrosine, 2.

or *l*-amino acids (B): as (A), with methionine and cystine @ 3 and 2 mg/g dry diet respectively.

vitamin-B mixture according to Horie *et al.* (1966) (μg/g of dry diet): biotin, 2; choline chloride, 1500; folic acid, 2; inositol, 2000; nicotinic acid, 100; calcium pantothenate, 150; pyridoxine hydrochloride, 30; riboflavin, 20; thiamine, 20; total, 3824.

Salt mixture and antiseptic not specified.

PREPARATION Details in Ito & Arai (1966b). Morin added to stimulate larval feeding, sorbic acid added to retard fungal growth. Ingredients heated to 90°C for 10 minutes, and, when cooled, sliced and fed to larvae.

DEVELOPMENT 2mg/g cystine markedly increased larval growth. No larval growth in absence of methionine.

See also: Akhtar & Ali (1972); Arai & Ito (1964, 1967); Hamamura *et al.* (1966); Horie (1962); Horie & Ito (1963, 1965); Horie & Nakasone (1968, 1971); Horie *et al.* (1966); Inokuchi (1969b, 1970, 1971); Inokuchi *et al.* (1967); Ito (1960, 1961b c d, 1962); Ito & Arai (1965b, 1966a, 1967, 1973); Ito & Horie (1962, 1966); Ito & Mukaiyama (1964); Ito & Nakasone (1966, 1967); Ito & Nimura (1966); Ito & Tanaka (1960); Ito *et al.* (1961, 1964, 1967a b, 1972, 1975a); Kamioka *et al.* (1971); Kato & Suminoto (1968); Kato & Yamada (1966a b); Kuwata *et al.* (1970); Matsubara (1965); Mukaiyama & Ito (1962); Yamada & Kato (1967).

Fukuda, T.; Suto, M.; Higuchi, Y. 1960a. Silkworm raising on the artificial food. J SERIC SCI JPN 29: 1-3. (In Japanese)

COMPOSITION (g): powdered mulberry leaves, 5; starch, 1.5; soybean meal, 1; defatted frozen bean curd, 1; and (ml): 20% sucrose solution, 5; antimicrobial solution, 10.

antimicrobials (g): vitamin K_3, 0.1; dehydro sodium acetate, 0.2; sodium sorbitate, 0.2; dissolved in 100 ml hot water.

PREPARATION Powdered mulberry leaves freeze-dried and finely ground; frozen bean curd defatted with ether and finely ground; all powdered ingredients sieved through 80 mesh. Mixture autoclaved for 15 minutes; cut into 0.1-0.3 mm thick segments and autoclaved for another 15 minutes; refrigerated for 12 h before use. Maximum storage, 7 days.

REARING At 24-25°C and 92-96% RH in petri dishes with wet filter paper, waxed paper and artificial diet; held in semi-sterile cabinet. Diet given at 12 h intervals.

DEVELOPMENT Successfully reared through 1 generation. 45 days from hatching to completion of cocoons (twice time taken on mulberry leaf); pupal stage, 14 days (near normal). 36 out of 110 larvae cocooned and 2♀ and 22♂ pupated and emerged. Weight (g): mature larvae, 1.13; pupae (6 days before emergence), 0.62 (♂); adults, 0.33 (♂).

Average cocoon silk production 86 mg 20% of that on mulberry leaf diet. 1♀ laid 227 eggs. Further improvement required. *Philosamia cynthia ricini* (Lep., Saturniidae) reared to 4th moult with substitution of powdered castor-oil leaf (*Ricinus communis*).

See also: Fukuda *et al.* (1960b).

→ Fukuda, T.; Suto, M.; Kameyama, T.; Kawasugi, S. 1962. Synthetic diet for silkworm raising. NATURE LONDON 196: 53-54.

Best of three diets to rear newly-moulted fifth-instar larvae to maturity.

COMPOSITION (g): potato starch, 1.5; sucrose, 1.0; amino acid mixture, 1.0; mineral mixture, 0.3; cellulose powder, 4.8; alcohol extract of mulberry leaves, 0.4; and (mg): vitamin mixture, 16.0; β-sitosterol, 50.0; and 11.0 ml of water.

amino acids (g): alanine, 1.0; arginine, 1.0; aspartic acid, 2.1; cystine, 1.0; glutamic acid, 2.3; glycine, 1.2; histidine hydrochloride, 0.7; isoleucine, 1.2; leucine, 1.8; lysine hydrochloride, 1.5; methionine, 0.3; phenylalanine, 1.1; serine, 1.2; proline, 1.1; threonine, 1.1; tryptophan, 0.3; tyrosine, 0.7; valine, 1.4; total, 21.0.

minerals (g): $NaH_2PO_4.H_2O$, 1.067; $MgSO_4$, 2.360; $CaCO_3$, 4.590; K_2SO_4, 4.518; NaCl, 0.813; total, 13.348.

vitamins (mg): *p*-aminobenzoic acid, 2.0; ascorbic acid, 400.0; biotin, 0.1; β-carotene, 20.0; choline chloride, 550.0; ergosterol, 400.0; folic acid, 0.4; inositol, 100.0; nicotinic acid, 40.0; calcium pantothenate, 30.0; pyridoxin hydrochloride, 4.0; riboflavin, 15.0; thiamin, 4.0; total, 1565.5.

PREPARATION Dry mulberry leaves extracted with hot absolute CH_3OH for 5 h. Sucrose, amino acids, vitamin mixture (except for ergosterol and β-carotene) and water heated; hot solution poured into beaker containing potato starch, cellulose powder (with β-sitosterol, ergosterol, β-carotene and alcohol extract of mulberry leaves when necessary) and mineral mixture. Contents well stirred, steamed for 3 minutes and refrigerated.

REARING See Fukuda *et al.**(1960a). At 25°C and 73% RH. Diet fed as 0.1 mm slices in boxes; renewed twice daily.

DEVELOPMENT Fourfold increase in larval weight. 100% cocoon formation containing average 83 mg silk, or ca. 70% of production on mulberry leaf powder diet.

See also: Suto *et al.* (1966).

→ Legay, J.M.; Pascal, M. 1963-1964. Essais préliminaires à la mise au point d'une alimentation préparée à partir de feuilles de mûrier cuites pour de ver à soie *Bombyx mori* L. REV VER SOIE 15-16: 101-104. (In French)

COMPOSITION (g): whole, preserved, (bottled) mulberry leaves, 15.0; lactic casein, 1.0; whole, powdered soya bean, 1.5; potato starch, 1.5; sucrose, 1.0; and (ml): distilled water, 10-12; and (μg): vitamin B_1 + vitamin B_2, 150 each.

PREPARATION Ingredients homogenised and autoclaved for 10 minutes.

REARING In glass boxes or tubes. Eggs preferably disinfected.

DEVELOPMENT Silk worm development below normal. 2 successive generations of *Mamestra brassicae* Linnaeus (Lep., Noctuidae) reared on diet. *Arctia caja* Linnaeus (Lep., Arctiidae) developed to final instar before virus ended experiment.

→Matsubara, F.; Kato, M.; Hayashiya, K.; Kodama, R.; Hamamura, Y. 1967. Aseptic rearing of silkworm with prepared food. J SERIC SCI JPN 36: 39-45. (In Japanese with English summary)

COMPOSITION (g): powdered, dried mulberry leaves, 2.0; cellulose powder, 3.5; defatted soybean, 1.5; sucrose, 1.0; starch, 1.5; Wesson's salt mixture, 0.15; vitamin mixture, 0.04; β-sitosterol, 0.05; vitamin C, 0.05; inositol, 0.05; K_2HPO_4, 0.05; acetylcholine, 0.01; dehydro-streptomycin, 0.01; and (ml): distilled water: for 1st-3rd instar, 16; for 4th instar, 15; and for 5th instar, 14.

vitamins (mg): biotin, 0.02; calcium pantothenate, 0.2; folic acid, 0.02; nicotinamide, 0.2; VB_6-HCl, 0.1; VB_2 (FMN-Na), 0.1; VB_1-HCl, 0.1; VB_T, 0.1; VC, 39.16.

PREPARATION Dehydro-streptomycin, choline hydrochloride, vitamin C, and K_2HPO_4, dissolved and mixed with other ingredients; spread 0.5-1 cm thick on parchment paper, cut into narrow strips and placed in 20 x 3 cm diam test tubes; stoppered with cotton-wool and autoclaved at 120°C for 30-40 minutes.

REARING At 25°C. 1-3 days before eclosion, eggs disinfected for 15-20 minutes in 70% ethanol and then in 0.1% $HgCl_2$ or 2-4% formalin. Eggs incubated in sterile test tubes. 40-50 larvae transferred aseptically to each culture tube. Number/tube reduced with growth. Diet renewed when necessary.

DEVELOPMENT Slightly slower than control on mulberry leaves, but body weight and mortality similar. Cocoon quality inferior: cocoon wt, 0.22-0.33 g and ratio cocoon wt/total wt, 17.5-19.5%, compared with 0.43-0.45 g and 20.3-24.0% on control.

→Matsuda, M.; Matsuura, Y. 1967. The germ free rearing of the silkworm (*Bombyx mori* L.) in the flexible plastic isolator. J SERIC SCI JPN 36: 403-408. (In Japanese with English summary)

COMPOSITION (g): mulberry leaf powder, 4.85; soybean powder, 2.50; starch, 1.0; glucose, 0.50; ascorbic acid, 0.15; agar, 1.0; and water, 30 ml.

PREPARATION Agar dissolved in water, mixed with other ingredients, cooled in 100 ml beakers, covered with polyester film and sealed with polyester tape. Autoclaved for 30 minutes at 120°C and 1 kg/cm^2, either in separate sterilizing cylinders or in autoclave directly connected with isolator from sterile lock side.

REARING At 25 ± 1°C, axenically in plastic isolator (described). Diet renewed every 5 days up to 3rd instar, and every 2 days for 4th and 5th instars.

DEVELOPMENT Similar to non-axenic rearing. Larval maturity, 88%; cocoons in 29-35 days; average pupal wt: ♀, 1.57 g; ♂, 1.52 g. Average fecundity, 692 eggs/♀.

Nihmura, M. 1972. Rearing of the silkworm, *Bombyx mori*, on the artificial feed by adapting the changes in feeding palatability according to larval growth. J SERIC SCI JPN 41: 375-382. (In Japanese with English summary)

Two diets tested.

COMPOSITION Diet 1 (%): defatted soybean meal, 50.0; potato starch, 10.0; sucrose, 10.0; cellulose powder, 9.6; agar, 10.0; salt mixture, 2.0; soybean oil, 3.2; soybean sterol, 0.3; citric acid, 1.0; ascorbic acid, 2.0; morin, 0.3; propionic acid, 1.0; and water (3 ml/g dry feed) and vitamin B mixture.

Diet 2 (%): defatted soybean meal, 45.0; corn starch, 40.9; agar, 5.0; salt mixture, 1.0; soybean oil, 2.7; soybean sterol, 0.3; citric acid, 1.0; ascorbic acid, 2.0; propionic acid, 1.5; sorbic acid, 0.2; chloramphenicol, 0.01; and water (3 ml/g dry feed) and vitamin B mixture.

salts (g): as in Ito & Nihmura (1966). K_2HPO_4, 4.6; $CaCO_3$, 2.6; KCl, 1.5; $MgSO_4$, 1.1; $FePO_4.4H_2O$, 0.2.

vitamins (diet 1) (mg) as in Horie & Ito (1963): choline chloride, 3.33; inositol, 3.33; nicotinic acid, 0.33; calcium pantothenate, 0.33; pyridoxine hydrochloride, 0.17; riboflavin, 0.17; thiamine hydrochloride, 0.17; biotin, 0.03; folic acid, 0.03.

vitamins (diet 2) (mg) as in Horie *et al.* (1966): choline chloride, 15; inositol, 20; nicotinic acid, 1; calcium pantothenate, 1.5; pyridoxine hydrochloride, 0.3; riboflavin, 0.2; thiamine hydrochloride, 0.2; biotin, 0.02; folic acid, 0.02.

PREPARATION Ingredients mixed and autoclaved at 115°C for 20 minutes. Cooled, cut up, and refrigerated.

REARING At 28°C and 65% RH.

DEVELOPMENT Both diets acceptable to 4th and 5th instar larvae giving 90% survival to maturity. Development of neonate larvae to maturity on both diets with addition of 10% mulberry leaf powder giving 100% survival to 4th instar. Larval period, 22-26 days. Cocoon wt, 1.79-1.95 g.

See also: Nihmura (1973a b c); Nihmura & Kirimura (1974); Nihmura *et al.* (1972).

Yoshida, T.; Matsuoka, M.; Kimura, K. 1960. On the rearing of silkworm larvae with an artificial diet containing dried mulberry leaf powder as its basic material. BULL SERIC EXP STN JPN 15: 543-586. (In Japanese with English summary)

COMPOSITION (g): mulberry leaf powder (freeze dried), 5.0; bean-curd powder (freeze dried and fat removed), 1.0; soybean powder, 1.0; starch, 1.5; cane sugar, 1.0; and (ml): antiseptic solution, 10.0; distilled water, 5.0.

antiseptic solution (g): vitamin K_3, 0.1; sodium-dehydroacetate, 0.2; sodium sorbate, 0.2; dissolved in 100 ml water.

PREPARATION Ingredients mixed, autoclaved at 15 lb for 15 minutes in petri dishes, cooled and sliced. Sprayed with fine mulberry leaf powder for improved results.

REARING Younger larvae at 24-25°C and 80-90% RH; older larvae at 24°C and 75-80% RH. In petri dishes in chamber equipped with UV lamp and formalin fumigation apparatus. Diet renewed and faecel pellets removed twice daily.

DEVELOPMENT Complete development to pupal stage.

See also: Hamamura (1970); Hayashiya *et al.* (1965); Kuwata *et al.* (1970); Morisaki *et al.* (1974).

Cossidae

Prionoxystus robiniae (Peck), carpenterworms

→ Leppla, N.C.; Solomon, J.D.; Doolittle, R.E. 1975. Culturing the carpenterworm. ANN ENTOMOL SOC AM 68: 683-685.

COMPOSITION (g): pinto beans (ground, 275; wheat germ, 200; soybean protein, 125; casein, 100; brewers' yeast, 125; Gelcarin, 46; Alphacel, 40; vitamin mixture (400 g Hoffman-LaRoche/1000 ml H_2O), 18; cholesterol, 1.5; choline chloride, 1.5; methyl *p*-hydroxybenzoate, 8; sorbic acid, 4; Tetracycline, 250 mg; and (ml): water, 3000; soybean oil, 3; 40% formalin, 10.

PREPARATION Dry ingredients except ascorbic acid, sufficient to make up 1-gallon, poured into 1.5-gallon pan containing warm demineralized water. Sorbic acid and methyl *p*-hydroxybenzoate added in 100 ml 95% ethanol; whole mixture heated for 10 minutes at 78°C; formalin added after first 5 minutes. Vitamins, Tetracycline and ascorbic acid added with diet to high speed blender, mixed for 30 seconds and poured into plastic cups. "Cakes" of gelled diet removed from cups and baked at 60°C for 5 h.

REARING Fertile moths held for oviposition at 25 ± 2°C and 60 ± 15% RH with 16 h photoperiod. Eggs incubated in darkness at 27 ± 2°C and 95 ± 5% RH. Emerging larvae placed in 16-oz cups. After 90 days of development, transferred to dark at 5 ± 2°C and 85 ± 5% RH, and weekly after 30 days, 15 cups of larvae removed and held at 27 ± 2°C for pupation.

DEVELOPMENT Larval maturation, 57%; pupation and emergence, 13%; larval period, ca. 192 days; minimum life cycle, 220 days.

→ Rivas, A.M.; Buchanan, W.D. 1958. A new technique for rearing carpenterworms. J ECON ENTOMOL 51: 406-407.

COMPOSITION (g): sawdust (host plant), 290; agar, 20; and water, 1000 ml.

PREPARATION Sawdust thoroughly mixed with hot agar solution and placed in glass or transparent plastic cage, 10 x 6 x 0.7"; packed to exclude air and left to cool.

REARING 0.7" square hole cut in one edge ca. 1.5" from bottom of cage for introduction of larvae from wire basket fastened over hole, and for ejection of frass. Cage covered with screen wire, transparent tape and aluminium foil to prevent larval escape and conserve moisture.

DEVELOPMENT Minimum larval survival, 41 days.

Solomon, J.D. 1966. Artificial rearing of the carpenterworm, *Prionoxystus robiniae* (Lepidoptera: Cossidae), and observations of its development. ANN ENTOMOL SOC AM 59: 1197-1200.

COMPOSITION (g): sawdust (*Quercus nuttallii*), 30-50; agar, 10; soybean protein, 10; sucrose, 10; Wesson's salt mixture, 1.5; cholesterol, 0.25; choline chloride, 0.5; and (ml): B-vitamin stock solution, 5; sorbic acid and methyl parasept stock solution, 3; distilled water, 300.

B-vitamin solution of Vanderzant* (1957).

antimicrobial solution (g): sorbic acid, 10; methyl parasept, 7.5; and 95% ethanol, 100 ml.

PREPARATION Agar dissolved in boiling water and removed from heat. Soybean protein, sucrose, Wesson's salt mixture, cholesterol and choline chloride added and mixed in Waring Blendor. Sawdust stirred in slowly until thick consistency obtained. B-vitamins and antimicrobial agents added after cooling, and blended. Dispensed to petri dishes to depth of 3-5 mm.

REARING At ca. 78°F and 45-75% RH; 25 neonate larvae/petri dish. Transferred to fresh food fortnightly, reducing numbers/ dish each transfer. Pupae in uncovered dishes placed in screen cage.

DEVELOPMENT Average pupal period, 21 days. Pupation, 59%; emergence to adult, 88%; 9-23 months after egg hatch. ♀♀ slightly smaller than field-reared moths, and fecundity lower. Adult longevity, 5 days.

See also: Solomon (1967).

Zeuzera pyrina (Linnaeus), leopard moth

Moore, I.; Navon, A. 1966. The rearing and some bionomcs of the leopard moth, *Zeuzera pyrina* L., on an artificial medium. ENTOMOPHAGA 11: 285-296.

COMPOSITION (g): full fat soybean, 36.0; sucrose, 48.0; brewers' yeast, 24.0; agar-agar flakes, 24.0; Nipagin, 1.5; and (ml): acetic acid (20% v/v), 30.0; distilled water, 570.0.

Two modifications of above diet described: first, with 30.0 ml sodium ascorbate (10% v/v) and only 540.0 ml distilled water; second with 70.0 ml pear bark homogenate (10 g/70 ml of water) and 500.0 ml distilled water.

PREPARATION Agar flakes dissolved in 200 ml distilled water in 1-litre Erlenmeyer flask. Remaining dry ingredients homogenised with remaining water in Waring Blendor. Homogenate transferred to flask with agar and mixed. Autoclaved at 15 psi for 20 minutes; cooled to 45°C, and acetic acid solution added and thoroughly mixed. Pear bark homogenate or sodium ascorbate then added if required. Diet dispensed to plastic, 3.5 x 3.0 cm diam boxes, and kept overnight at 20-25°C, inverted over filter paper.

Pear bark homogenate prepared by blending fresh bark with distilled water, and filtering through 80 mesh nylon gauze.

REARING At 26 ± 1°C. Neonate larvae introduced, and boxes left inverted for 24 h. After 1st month boxes covered with tight fitting plastic lid; later lids replaced with copper grid of 100 mesh. Larvae transferred to fresh food every fortnight after 6 weeks. Pupae kept in wire-mesh cage covered with cheese cloth for emergence. Adults kept at 22 ± 2°C in paper bags around wire frame, 1 pair in each. Oviposition in folds of bag. Egg masses cut out and incubated at 22 ± 2°C.

DEVELOPMENT Diet with pear bark homogenate best, giving larval survival of 80%, larval period, 100 days; pupal period, 18.6 days. Pupation, 62%; and adult emergence, 37%. Fecundity, 400-900 eggs/♀. Incubation period, 14 days.

Gelechiidae

Anarsia lineatella Zeller, peach twig borer

→ Anthon, E.; Smith, L.O.; Garrett, S.D. 1971. Artificial diet and pheromone studies with peach twig borer. J ECON ENTOMOL 64: 259-262.

Diet modified from Ignoffo* (1963) for *Trichoplusia ni*.

COMPOSITION (g): wheat germ, 30; casein, 35; sucrose, 35; Wesson's salt mixture, 10; cholesterol, 3; agar, 20; sorbic acid, 1.2; Alphacel, 25; methyl *p*-hydroxybenzoate, 0.2; Wheast (yeast cultured on cottage cheese whey), 35.13; Vanderzant's fortification mixture, 12; and (ml): formaldehyde (37% actual), 3; distilled water, 1000.

PREPARATION Ingredients, except Vanderzant's fortification mixture, mixed in 3 l flask. Autoclaved for 20 minutes at 103 kPa, cooled to ca. 55°C, homogenised in Waring Blendor until cooled to 45-49°C, then vitamin mixture stirred in by hand. 15 ml plastic jelly cups ½-¾ filled with diet. Cups placed below two 15-watt germicidal lamps, and allowed to cool and dry for 4-6 h. Cups fitted with tight lids and refrigerated for maximum of 7 days.

REARING At 16.1°C for 8 h and 22.8°C for 16 h with 60% RH. Larvae removed from infested twigs and placed in diet-filled cups @ 1-4 larvae/cup. Cups inverted on to sterilised glass plates. Various types of oviposition cages: plastic cylinders lined with wax paper, battery jars and large Mason fruit jars and glass lantern chimney-type cages. Eggs laid on cotton organdie, removed each day and incubated at 21.1-26.7°C and 70-80% RH with 17 h photoperiod. Neonate larvae transferred to cups with ventilated lids in darkened room. Cups removed after 15 days and placed on glass plate covered with 3 layers paper towelling. Cone of dark grey construction paper placed over cups and kept at 25.6-26.7°C and 30% RH with 18 h photoperiod. Lighter coloured cone after 2 days. Pupation in container. Adults fed honey solution diluted 20:1 with distilled water on sponge. Adults held at 25.6-26.7°C and 40% RH with 14½ h photoperiod.

DEVELOPMENT Complete developmental periods (days): egg, 5; larva, 28.2; pupa, 6; and adult, 6.9. Two generations reared.

Pectinophora gossypiella (Saunders), pink bollworm

→ Adkisson, P.L.; Vanderzant, E.S.; Bull, D.L.; Allison, W.E. 1960. A wheat germ medium for rearing the pink bollworm. J ECON ENTOMOL 53: 759-762.

Three diets described.

COMPOSITION Diet 1 (g): vitamin-free casein, 3.5; wheat germ, 3.0; sucrose, 3.5; Wesson's salt mixture, 1.0; choline chloride, 0.1; agar, 2.5; sodium alginate, 0.5; and (ml): vitamin mixture, 1.0; and water, 85.0.

Diet 2 (g): casein, 3.0; wheat germ, 3.0; sucrose, 5.0; Wesson's salt mixture, 1.0; choline chloride, 0.1; agar, 2.0; and (ml): vitamin mixture, 1.0; water, 80.

Diet 3 (g): vitamin-free casein, 5.00; cystine hydrochloride, 0.10; glycine, 0.15; sucrose, 5.00; Wesson's salt mixture, 1.20; cholesterol, 0.05; corn oil, 0.25; α-tocopherol, 0.01; choline chloride, 0.10; cellulose, 4.00; agar, 3.00; sodium alginate, 0.50; and (ml): vitamin mixture, 1.0; and water, 80.0.

vitamins (mg/100 g): niacinamide, 1.00; calcium pantothenate, 1.00; thiamine hydrochloride, 0.25; riboflavin, 0.50; pyridoxine hydrochloride, 0.25; folic acid, 0.25; biotin, 0.02; vitamin B_{12}, 0.002; and inositol, 20.00 (inositol omitted in diet 3).

mould inhibitors: sorbic acid, methyl *p*-hydroxybenzoate and butyl *p*-hydroxybenzoate dissolved in 95% ethyl alcohol, and added at concentration equal to 0.2% by weight of each.

PREPARATION Agar dissolved in boiling water and added to dry ingredients in Waring Blendor. Liquids added, blended, and dispensed so as to ⅓ fill sterilised 2 dram shell vials.

REARING See Vanderzant & Reiser* (1956a). At 85° F and 75-85% RH with 12 h photoperiod. Vials plugged with non-absorbent cotton and autoclaved. When cool, 1 neonate larva from sterilised egg placed in each vial.

DEVELOPMENT On diet 1, larval period, 16.3 days; prepupal period, 2.2; pupal period, 8.8. On diet 2, larval period, 10.1 days; prepupal period, 3.8; pupal period, 8.1. On diet 3, larval period, 16.5 days; prepupal period, 3.3; pupal period, 8.6. Pupae slightly smaller than those collected from cotton bolls, but females more fecund. 81.5% survival from larvae to adult on diet.

→ Adkisson, P.L.; Bull, D.L.; Allison, W.E. 1960. A comparison of certain artificial diets for laboratory cultures of the pink bollworm. J ECON ENTOMOL 53: 791-793.

COMPOSITION Diet 1, see Vanderzant & Reiser* (1956a) (g): casein, vitamin-free, 5.0; *l*-cystine, 0.1; glycine, 0.15; sucrose, 8.0; corn oil, 1.0; cholesterol, 0.3; Wesson's salt mixture, 0.7; choline chloride, 0.1; Alphacel, 4.0; agar, 3.0; sodium alginate, 0.5; water to 100 g and (ml): vitamin mixture, 1; antimicrobial mixture, 10.

Diet 2 (g): cottonseed meal, 16.25; sucrose, 2.0; brewers' yeast, 1.5; Alphacel, 1.5; albumin, 1.0; agar, 1.5; wheat germ oil, 0.5; cholesterol, 0.15; and (ml): water, 75; antimicrobial mixture, 10.

vitamins (mg): niacinamide, 1; calcium pantothenate, 1; thiamine hydrochloride, 0.25; riboflavin, 0.50; pyridoxine hydrochloride, 0.25; folic acid, 0.25; biotin, 0.02; vitamin B_{12}, 0.0002; inositol, 20.

antimicrobial mixture (g): butyl paraben, 0.2; methyl paraben, 0.2; sorbic acid, 0.2; 95% ethyl alcohol to 10 ml.

PREPARATION Agar dissolved in boiling water; remaining dry ingredients added and boiled for 2-3 minutes; poured into Waring Blendor, antimicrobial and vitamin mixtures in diet 1 added and blended. Poured into plastic squeeze bottles, and dispensed to ⅓ full sterilised 2-dram shell vials. Vials loosely wrapped in heavy paper and allowed to cool.

REARING At 85 ± 2° F and 90% RH with 24 h photoperiod. Eggs collected daily from cotton leaves, washed in distilled water and transferred to small vial on piece of filter paper. Vial plugged with cotton.

Neonate larvae transferred individually to diet vials; brush dipped in 3000 ppm solution of "Roccal" during procedure. Vials plugged with non-absorbent cotton.

DEVELOPMENT On diet 1, larval period, 18-37 days; prepupal period, 2.3; pupal period, 7.6. Pupal wt: ♀, 14.5 mg; ♂, 12.8 mg. Fecundity, 10.6 eggs/♀. On diet 2, larval period, 10-20 days; prepupal period, 2.3; pupal period, 7.9. Pupal wt: ♀, 21.0 mg; ♂, 18.6 mg. Fecundity, 45.4 eggs/♀.

Corollary

→ Adkisson, P.L. 1961b. Fecundity and longevity of the adult female pink bollworm reared on natural and synthetic diets. J ECON ENTOMOL 54: 1224-1227.

REARING Adults at 80 ± 2°F in darkness. Pairs placed in 1-pint carton oviposition cages. Cotton leaf placed daily over screened hole in top of cage for oviposition. Inside, cartons coated with paraffin, and provided with small cotton-stoppered vial of sugar-water solution.

DEVELOPMENT Average female life span, 15.4-17.2 days. Fecundity (eggs/♀) on 1% cotton meal diet, 336.7; wheat germ diet, 312.2; sucrose, 98.1; and cotton bolls, 204.3.

See also: Adkisson (1961a); Bull & Adkisson (1962).

→ Beckman, H.F.; Bruckart, S.M.; Reiser, R. 1953. Laboratory culture of the pink bollworm on chemically defined media. J ECON ENTOMOL 46: 627-630.

COMPOSITION (%): dextrose, 49; egg albumin, 25; cottonseed oil, 20; salts, 5; cholesterol, 1; B_{12} concentrate (12.5 mg of B_{12}/lb), 0.6; choline chloride, 0.25; inositol, 0.125.

vitamins (mg/kg): niacin, 70; calcium pantothenate, 25; pyridoxine hydrochloride, 8; riboflavin, 8; thiamin hydrochloride, 6; *p*-aminobenzoic acid, 3; folic acid, 2.5; biotin, 0.25; menadione, 15; carotene, 330; vitamin D_3, 0.08; mixed tocopherols (170 mg D-tocopherol/g), 10 g/kg.

salts (g): $CaCO_3$, 1200.0; K_2HPO_4, 1290.0; $CaHPO_4.2H_2O$, 300.0; $MgSO_4.7H_2O$, 408.0; NaCl, 670.0; $FeC_6H_5O_7.6H_2O$, 110.0; KI, 3.2; $MnSO_4.4H_2O$, 20.0; $ZnCl_2$, 1.0; $CuSO_4.5H_2O$, 1.2.

mould inhibitors: 1 g *n*-butyl *p*-hydroxybenzoate dissolved in 10 ml ethyl alcohol.

PREPARATION 100 g dry ingredients mixed well with 150 ml distilled water; 2 ml mould inhibitor added. Ca. 150 mg purified cotton linters (Buckeye Cotton Oil Co., Memphis, Tenn.) placed in 15 x 45 mm vials and diet added until linters just saturated. Moisture content of diet kept constant. If desiccation evident, then few drops of water added.

REARING Eggs surface sterilised in 0.1% $HgCl_2$ then brushed on to wall of vial just above diet @ 5 eggs/vial. Kept at 26-29°C and high humidity.

DEVELOPMENT 97 eggs hatched from 158; 6 pupae survived; 6 infertile adults emerged.

→ Clark, E.W.; Richmond, C.A.; McGough, J.M. 1961. Artificial media and rearing techniques for the pink bollworm. J ECON ENTOMOL 54: 4-9.

Two best of many diets tested.

COMPOSITION Diet 1 (g): peanut flour, 14; agar, 2; brewers' yeast, 2; carboxymethyl cellulose, 1; cottonseed oil, 4; dextrose, 2; and distilled water, 75 ml.

Diet 2 (g): cottonseed meal, 10; agar, 2; albumin, 1; brewers' yeast, 2; sodium alginate, 2; cholesterol, 0.2; dextrose, 2; and distilled water, 57 ml.

PREPARATION Agar dissolved in hot distilled water added to dry constituents except carboxymethyl cellulose and blended until homogenous. Then carboxymethyl cellulose added. Diets placed in 15 x 150 mm test tubes or 125-250 ml Erlenmeyer flasks, plugged with non-absorbent cotton, autoclaved at 20 lb pressure for 15 minutes and held for 2-3 days at 28° C.

REARING Aseptically at 28° C and 65-75% RH with 10 h photoperiod. Diets inoculated with 4 to 5-day-old eggs or 1-day-old 1st instar larvae from sterile egg storage vials with sterile brush @ 1-5 eggs or larvae/ culture tube, or 10-30/flask. Antimycotic agents necessary for large scale rearing; 22 compounds tested at 0.01-2% levels by blending with cottonseed meal diet. Occasional bacterial contamination of little importance. Effective agents: Butoben, Captan, formaldehyde, methylparaben, potassium sorbate, sodium benzoate, sodium propionate and sorbic acid. In 0.8-2% levels, 80% fungus contamination controlled. For mass rearing, Butoben (0.2%), sodium benzoate (0.2%) and sorbic acid (0.3%) required. 8-10 non-sterilised eggs placed in flask or petri dishes, ca. ¼ filled with diet, remaining space filled with absorbent cotton and two halves sealed by pushing top and bottom together. 20-200 adult moths placed in 1 cubic foot wire gauze oviposition box. Food of aqueous fructose-glucose solution in cotton stoppered vial. Oviposition on washed cotton leaves, filter paper, or black paper placed on wire gauze. Eggs removed by washing in 50 ppm solution of Tween 80 for 2-5 minutes, collected in Buchner filter, and either sterilized or placed directly on diets.

DEVELOPMENT Complete on both diets; larvae reared on peanut-meal diet of pale and greasy-grey appearance.

Development of boll weevil, *Anthonomus grandis* Boheman (Col., Curculionidae) and humpbacked fly, *Megaselia* sp. (Dipt., Phoridae) also completed on cottonseed-meal diet.

→ Martin, D.F. 1966. Pink bollworm. pp. 355-366. *in* Smith, C.N. (Ed.), Insect Colonization and Mass Production. Academic Press, New York and London.

COMPOSITION (g): casein, 126; sucrose, 126; wheat germ, 108; Wesson's salt mixture, 36; α-cellulose, 18; agar, 90; and (ml): methyl *p*-hydroxybenzoate (15% w/v in 95% ethyl alcohol), 36; choline chloride, (0.1 g/ml water), 36; 4M KOH, 18; formaldehyde (0.1 g/ml of water), 13; vitamin stock, 6; and distilled water, 3100.

vitamins (mg/100 ml water): riboflavin, 300; pyridoxine, 150; thiamine, 150; folic acid, 150; niacin, 600; calcium pantothenate, 600; biotin, 12; B_{12}, 1.2.

PREPARATION Agar dissolved in ca. 2200 ml boiling water. Casein blended in 900 ml water and 18 ml KOH in 1-gal Waring Blendor. All solids then added and blended, while choline chloride, formaldehyde, methyl *p*-hydroxybenzoate and vitamin solution successively added. Dissolved agar added and entire mixture blended until homogenous. Poured into trays, and when solid, cut into ¼" cubes.

REARING In one-quart cylindrical icecream containers coated inside with melted paraffin; with paper lids replaced by wire screen covers for oviposition cages. Screen wire 2-3" diam disc placed on screen cover with egg-laying pad. Pads placed in flask at 80-85° F. Adults fed 5% sugar solution. Egg hatch in 4 days. Larvae migrate to lip of flask under illumination and removed with brush to cups or vials. Kept at 80° F Larvae reared in 2-dram shell vials with 3 ml diet at 80-85° F in darkness or in 9-oz wax paper cups @ 25 neonate larvae/cup. Thin layer of sterile cotton on bottom, then layer of 25, ¼" diet cubes; larvae on diet then covered with cotton. 4-5 layers diet and cotton in each cup. Plastic snap-on lids; kept in dark at 80-85° F. After several days, cups placed in emergence cages (described).

DEVELOPMENT Eggs laid 5-7 days from emergence; incubation period, 4 days.

See also: Richmond & Martin (1966).

→ Ouye, M.T. 1962. Effect of antimicrobial agents on micro-organisms and pink bollworm development. J ECON ENTOMOL 55: 854-857.

COMPOSITION (g/100 ml): agar, 2.5; Alphacel, 0.5; casein, vitamin-free, 4.0; Wesson's salt mixture, 1.0; sucrose, 4.0; wheat germ, 3.0; choline chloride, 0.1; and (ml): vitamin solution, 1.0; 4M KOH, 0.5; water to 100 ml; and (%): formaldehyde, 0.05; methyl paraben, 0.15; pH 6.0 to 6.5.

vitamins (mg): niacin, 1; calcium pantothenate, 1; riboflavin, 0.5; thiamin hydrochloride, 0.25; pyridoxine hydrochloride, 0.25; folic acid, 0.25; biotin, 0.02; B_{12}, 0.002.

PREPARATION Casein blended with little water; agar dissolved in remainder. KOH and remaining ingredients added to casein and all blended. Dissolved agar added and blended again. Diet squeezed into 2-dram vials @ 3 ml/vial from plastic squeeze bottle. Vials stoppered with absorbent cotton and held at room temperature for 24 h before use.

REARING At 78-85° F, individually in vials from neonate larvae to adult.

DEVELOPMENT Larval period, 17.4 days; pupal period, 8.5 days. Pupal wt: ♀, 26.0 mg; ♂, 20.6 mg. Best antimicrobial agent, 0.05% formaldehyde plus 0.15% methyl paraben.

Corollary 1

→ Ouye, M.T.; Vanderzant, E.S. 1964. B-vitamin requirements of the pink bollworm. J ECON ENTOMOL 57: 427-430.

Vitamin solution (mg/100 g diet): nicotinamide, 4; calcium pantothenate, 2; riboflavin, 1; thiamine, 0.5; pyridoxine, 0.5; folic acid, 0.5; biotin, 0.02; B_{12}, 0.002.

Diet also modified so that final concentration of inhibitors (%): methyl *p*-hydroxybenzoate, 0.15; sorbic acid, 0.1.

DEVELOPMENT Average larval period, 20 days. Pupation, 80%; emergence, 80%.

Corollary 2

Mangum, C.L.; Ridgway, W.O.; Brazzel, J.R. 1969. Large-scale laboratory production of the pink bollworm for sterilization programmes. U S DEP AGRIC PRESENTATION PAP ARS 81-35, 7 pp.

REARING At 29°C and 65 to 75% RH with constant dim lighting. Ca. 2000 adults (1♂:1♀) placed in 3-gal fibre cans for oviposition. Cans fitted with copper mesh screen lids over which plastic screen rings placed. Oviposition rings, corresponding in size to plastic rings, cut from paper towel. Six 2" diam iron discs weighing 70 g, distributed over paper towel rings (moths prefer to oviposit on weighted paper). Adults fed through absorbent cotton moistened with 5% sucrose solution in centre of lid.

Larvae reared in 1-gal glass-lined fibre cans with ca. 750 g diet/can. Egg pads treated with 25% formaldehyde solution for 10 minutes and then heat-treated for 30 minutes at 42°C before implanting. Rearing containers held at dark room for 8-10 days. When larvae at late 4th instar, cans emptied on to hardware cloth trays fitted over fibreboard shelves in 8 x 8' rack; each rack with 3 shelves. To collect pupae, wax paper placed over fibreboard, then 1" layer vermiculite spread over paper. Alternative method for collecting pupae also described and illustrated. More than 95% larvae pupated within 3 days. Pupae placed in 26 x 18 x 4" boxes; transferred to adult collecting device (illustrated). Moths collected daily.

Raulston, J.R. 1971. A practical diet containing cottonseed for rearing the pink bollworm. J ECON ENTOMOL 64: 1021-1023.

Modified diet from Adkisson et al.* (1960).

COMPOSITION (g): casein, 3.71; wheat germ, 3.17; sucrose, 3.71; Alphacel, 0.53; methyl *p*-hydroxybenzoate, 0.159; Wesson's salt mixture, 1.06; agar, 2.0; and (ml): 4M KOH, 0.53; vitamin mixture, 0.35; 15% choline chloride, 0.75; 10% formaldehyde, 0.44; 25% acetic acid, 1.18; water, 88.0.

vitamins (mg/ml): calcium pantothenate, 12; niacin, 6; riboflavin, 3; folic acid, 3; thiamine hydrochloride, 1.5; pyridoxine hydrochloride, 1.5; biotin, 0.12; B_{12}, 0.012.

PREPARATION Basically as in Adkisson *et al.** (1960). Test diets formulated from stock mixture of control diet without casein, or, on occasions, with no other dietary components. Instead, cottonseed test product placed in Waring Blendor, and stock added to correct volume.

Mixture dispensed to vials. Cottonseed products (glanded meats and meal, glandless meats and meal, and blended meals with oil extracted by direct solvent or expellor processes) obtained from R.J. Miravalle, National Cottonseed Products Association Inc., USA. Rinsed with 1% solution of sodium propionate prior to use to reduce fungal contamination.

REARING Individual 1st instar larvae in 1-dram vial; held at 26.7-29.4°C and ca. 50% RH.

DEVELOPMENT Pupal wt, 22.7 mg on basal diet; time to pupation, 18.1 days. Diet containing 5% glandless cottonseed meats and 2% brewers' yeast nutritionally adequate, and superior to casein-wheat germ diet with regard to fecundity. Cost of cottonseed diet containing no mineral supplement and ½ normal vitamin supplement ca. 25% less than standard casein-wheat germ base diet.

→ Vanderzant, E.S.; Reiser, R. 1956a. Aseptic rearing of the pink bollworm on synthetic media. J ECON ENTOMOL 49: 7-10.

Best of 10 diets tested.

COMPOSITION (g): egg albumen, 11.5; glucose, 8.0; corn oil (Mazola), 3.5; cholesterol, 0.3; Wesson's salt mixture, 1.0; choline chloride, 0.1; inositol, 0.1; alginic acid, 2.0; dried forage juice, 1.0; and (ml): vitamin mixture, 1; water, 73.

vitamins (mg/100 g diet): thiamin, 0.6; riboflavin, 0.9; calcium pantothenate, 2.00; niacin, 6.25; pyridoxin, 0.8; folic acid, 0.125; biotin, 0.018; *p*-aminobenzoic acid, 1.5; vitamin B_{12}, 0.005.

PREPARATION Dry ingredients and fat homogenised with choline chloride solution, vitamin solution and water in Waring Blendor. Alginic acid added and blended; pH to 6 with 2M KOH. 3 ml diet placed in 17 x 60 mm vials, plugged with non-absorbent cotton or Bakelite screw caps. Autoclaved at 15 lb pressure for 25 minutes and kept at room temperature for 1-2 days before use.

REARING At 86 ± 1°F and 80-90% RH. Moths placed in 8" cube internally waxed, screen oviposition cages. Sugar-water provided in cotton wad. Cotton leaf covered with cellophane placed on screen for oviposition. Eggs transferred for 20 minutes to solution of (g): $HgCl_2$, 0.25; NaCl, 6.5; $CH_2(OH).COONa$, 0.5; and (ml): conc. HCl, 1.25; ethyl alcohol, 250; distilled water, 750; then dried filter paper in sterile vial. Neonate larvae transferred individually to diet vials with small rolls of dry, sterile filter paper inside. Caps loosened at 4th instar.

DEVELOPMENT Larval period, 24 days; pupal wt, 13.3 mg. Three generations reared.

→ Vanderzant, E.S. 1957. Growth and reproduction of the pink bollworm on an amino acid medium. J ECON ENTOMOL 50: 219-221.

COMPOSITION (mg/100 ml diet): arginine, 311; histidine hydrochloride, 75; isoleucine, 113; leucine, 175; lysine hydrochloride, 113; methionine, 75; phenylalanine, 150; threonine, 100; tryptophan, 75; valine, 138; alanine, 150; aspartic acid, 100; glutamic acid, 125; glycine, 150; proline, 50; serine, 50; cystine, 50; tyrosine, 50; choline chloride, 50; niacinamide, 1.0; calcium pantothenate, 1.0; thiamin hydrochloride, 0.25; riboflavin, 0.50; pyridoxin hydrochloride, 0.25; folic acid, 0.25; biotin, 0.02; vitamin B_{12}, 0.002; K_2HPO_4, 160; NaH_2PO_4, 80; NaCl, 20; $MgSO_4.7H_2O$, 100; $FeSO_4.7H_2O$, 20; $CuSO_4.5H_2O$, 0.25; $MnSO_4.H_2O$, 2.50; $ZnCl_2$, 0.12; glucose, 2.5 g.; corn oil (Mazola), 300; Tween 80, 150; cholesterol, 50; α-tocopherol, 10; KI, 0.4; $CaCO_3$, 120; agar, 3 g.

PREPARATION Filtered aliquots of amino acid, vitamin, choline, NaCl, sodium and potassium phosphates, and magnesium sulphate solutions mixed. Glucose added and solution diluted to 48 ml; pH 6.1-6.2. Mixture filtered through Seitz filter. Iron, copper, manganese and zinc salts solutions filtered at end, and final volume adjusted to 50 ml. Heat sterilized portion of diet prepared by weighing corn oil, Tween 80 and cholesterol in 250 ml Erlenmeyer flask; α-tocopherol added as 10 mg in 1 ml ethanol. Several ml ethyl ether added until all dissolved. Ether evaporated and 25 ml water added. KI solution, $CaCO_3$, agar and water added, flask stoppered and autoclaved for 15 minutes at 15 lb. Sterile filtered solution added and rotated gently. Diet poured aseptically into vials and plugged with cotton.

REARING See Vanderzant & Reiser* (1956a).

DEVELOPMENT Of 86 pink bollworms reared in three experiments, 73 became adults, 3 failed to emerge from pupae, and 10 larvae remained in resting stage. Larval period, 16 to 24 days; pupae normal. Average weight of 26 ♀ pupae, 18.4 mg, and of 41 ♂ pupae, 14.7 mg. Eggs fertile, and second generation reared.

→ Vanderzant, E.S.; Kerur, D.; Reiser, R. 1957. The role of dietary fatty acids in the development of the pink bollworm. J ECON ENTOMOL 50: 606-608.

COMPOSITION (g/100 g diet): casein (vitamin free), 5.0; *l*-cystine, 0.1; glycine, 0.15; sucrose, 5.0; corn oil (Mazola), 0.25; cholesterol, 0.05; Wesson's salt mixture, 1.2; choline chloride, 0.1; α-tocopherol, 0.01; cellulose, Whatman ashless powder, 4.0; bacto-agar, granules, 3.0; sodium alginate (Kelco Company), 0.5; 2M KOH to pH 6.5 and water to 100 g; and (mg): niacin, 4.0; calcium pantothenate, 2.0; riboflavin, 1.0; thiamin hydrochloride, 0.5; pyridoxine hydrochloride, 0.5; folic acid, 0.5; biotin, 0.02; B_{12}, 0.002.

PREPARATION See Vanderzant & Reiser* (1956a). Choline chloride added in solution of 0.1 g/ml water. Cholesterol ground with sucrose in mortar. Casein, cystine and glycine mixed with $\frac{1}{3}$ water and 2M KOH to dissolve casein; pH to 7. Agar dissolved in $\frac{1}{3}$ water over boiling water bath. Remaining ingredients mixed and blended with casein mixture in Waring Blendor. pH to 6.5 with 2M KOH. Agar solution added and diet heated in boiling water bath until fluid. When cooled, 1.5-2 g plugs placed in screw capped vials and autoclaved for 20 minutes at 15 lb.

REARING See Vanderzant & Reiser* (1956a).

DEVELOPMENT Results with linoleic acid similar to those with corn oil. Near-normal emergence obtained with 0.125 g linoleic acid/ 100 g diet. Linolenic acid more effective than linoleic acid in promoting emergence. Linseed oil at 0.05 g level very effective, containing 0.025 g linolenic acid and 0.01 g linoleic acid.

See also: Graham & Mangum (1971); Vanderzant (1958); Vanderzant & Reiser (1956b); Vanderzant & Richardson (1963).

Phthorimaea operculella (Zeller), potato tuber moth

→ Singh, P.; Charles, J.G. 1976. An artificial diet for the potato tuber moth (*Phthorimaea operculella* (Zeller)) (Lepidoptera: Gelechiidae). UNPUBLISHED.

COMPOSITION Diet as in Singh* (1974c); except that grade of cellulose powder changed from CF11 to CF1; lactic casein (N.Z. Co-op Dairy Co. Ltd) substituted for fat and vitamin-free casein, and water increased to 75 ml.

PREPARATION Group 1 dry ingredients ground and mixed. Cholesterol dissolved in dichloromethane, mixed with group 1 and solvent evaporated. Group 3 mixed in hot water and emulsified. Group 4 dissolved in water; aqueous solutions of biotin and B_{12} prepared separately. Then group 1 mixed with water and pH adjusted to ca. 6.5 with 4N KOH. Lipids in solution added; mixture autoclaved at 120° C and 15 lb/in^2 for 20 minutes. When cooled to 60° C, group 4 added, then mould inhibitor. Diet finally mixed, solidified and dispensed in 2 g plugs to 75 x 12 mm diam plastic tubes; kept at 30° C for 18 h before use.

REARING At 30 ± 2° C with 16 h photoperiod individually in tubes. Adults removed to 90 x 150 mm diam gauze covered dishes for mating. Oviposition on underside of filter paper soaked in 10% sugar solution placed on gauze for food. Paper renewed every 2 days. Incubation in plastic pottles.

DEVELOPMENT Incubation period, ca. 2.5 days; larval period, 12-17 days; pupal period, 5-6 days; mean development period from neonate larva to adult, 20.39 ± 0.17 days, compared with 14-17 days on potato; survival to adult, 70.7%. Increased particle size of cellulose powder important factor in larval establishment and development.

Sitotroga cerealella (Olivier), Angoumois grain moth

→ Chippendale, G.M. 1970a. Development of artificial diets for rearing the Angoumois grain moth. J ECON ENTOMOL 63: 844-848.

Two refined diets described.

COMPOSITION Diet 1 (g/100 g): corn starch, 56.39; vitamin-free casein, 10.53; wheat germ, 1.65; wheat bran, 1.65; wheat germ oil, 1.50; salt mixture, 1.50; cholesterol, 0.30; glycerol, 18.96; brewers' yeast, 7.52.

Diet 2, as above except: casein decreased to 9.03 g; wheat germ and wheat germ oil increased to 2.65 and 2.00 g, respectively.

salt mixture as in Medici & Taylor* (1966) (%): KH_2PO_4, 67.40; $MgSO_4$, 16.00; $Ca(H_2PO_4)_2.H_2O$, 8.00; NaCl, 5.00; $FeSO_4$, 1.1; $Zn(C_2H_3O_2)_2$, 0.5; $MnSO_4.H_2O$, 0.5; $CuSO_4.5H_2O$, 0.25.

PREPARATION Ingredients ground and mixed; left overnight to improve consistency. Ca. 400 mg diet packed into No. 0 gelatine capsules (Parke Davis & Co., Detroit, Mich.).

REARING At 30° C and 55 ± 5% RH in darkness. Stock culture reared in gallon jars with whole field corn. Adults transferred to quart mason jars. Folded wax paper strips for oviposition collected daily and incubated at 30° C in small glass jars for 5 days. 2 neonate larvae/capsule.

DEVELOPMENT Mean weight of 26-day old larvae; 3.2 mg on diet 1 and 7.2 mg on ground corn control. 25% survival to adults on diet 1; 70% on ground corn. Diet 2 slightly better than first; maximum larval wt, 5.5 mg; and adult emergence, 30%.

→ Chippendale, G.M. 1971a. Growth and development of the Angoumois grain moth, *Sitotroga cerealella* in an artificial diet. J INSECT PHYSIOL 17: 109-115.

COMPOSITION (g): whole wheat flour, 38.0; corn meal, 29.3; honey, 13.7; glycerol, 13.1; brewers' yeast, 5.9.

PREPARATION Wheat grains and white degermed corn meal powdered and mixed with other components. Packed to 5 mm diam pellets of ca. 440 mg.

REARING One pellet and 2 neonate larvae in No. 0 transparent gelatine capsule (Parke Davis & Co., Detroit). Capsules held vertically in holes of peg board; incubated at 30°C and ca. 70% RH in darkness.

DEVELOPMENT Four larval instars. Maximum wt: ♂, 7.2 mg at 22 days; ♀, 12.3 mg at 24 days. 50% ♂ pupation 5 days before ♀♀. Mean weight of newly emerged adult ♀♀ 77% higher than that of ♂♂.

→ Chippendale, G.M. 1971b. Observations on the physical and chemical composition of diets for the Angoumois grain moth. J INSECT PHYSIOL 17: 1257-1266.

Five artificial diets tested.

COMPOSITION Diet 1 (g/100 g): glycerol, 11-31; vitamin-free casein, 9.04; brewers' yeast, 7.52; wheat germ, 2.64; wheat germ oil, 2.00; wheat bran, 1.64; cholesterol, 1.20; salt mixture, 0.00-1.50.

Diet 2, as diet 1; except brewers' yeast from 0-15 g/100 g diet, and wheat germ oil omitted.

Diet 3, as diet 1; except wheat germ from 0-7.9 g/100 g diet, and wheat germ oil omitted.

Diet 4, as diet 1; except cholesteral from 0-2.4 g/100 g diet.

Diet 5, as diet 1; except salt mixture from 0-1.5 g/100 g diet.

In all 5 diets, corn starch adjusted to 100 g; diets 2-5 with 19.0 g glycerol.

salt mixture: See Chippendale*(1970a).

PREPARATION & REARING See Chippendale*(1971a). At 30°C and 75% RH.

DEVELOPMENT Development complete on diets in pellet form when corn starch from 49-64%, glycerol 15-19%, casein 9%, yeast 7.5%, and wheat germ 2.6%. Larvae reared on natural food ca. 60% heavier than on diets at 22 days.

See also: Chippendale (1971c); Chippendale & Mann (1972); Peters *et al.* (1972).

→ Chippendale, G.M. 1973. Metabolic reserves of larvae and pupae of the Angoumois grain moth, *Sitotroga cerealella*. INSECT BIOCHEM 3: 1-10.

COMPOSITION (g/100 g): corn starch, Argo, 59.8; glycerol, 19.0; wheat germ, whole (Nutritional Biochemicals Corporation, contains ca. 10% oil), 10.6; casein, vitamin-free (NBC), 9.0; vitamin mixture (NBC), 1.0; cholesterol, 0.6.

PREPARATION & REARING Components mixed with glycerol in mortar and pestle, and packed to 5 mm diam pellets of ca. 440 mg. Pellets placed singly in gelatin capsules. See Chippendale* (1971a).

DEVELOPMENT Mean mature larval weights of ♂♂ and ♀♀ on natural diet and refined diet: 6.6 ± 0.2, 7.7 ± 0.5 and 11.7 ± 0.2, 11.2 ± 0.4 mg respectively. Similarly, mean newly-ecdysed pupal weights, 5.1 ± 0.2, 5.6 ± 0.1 and 8.3 ± 0.3, 8.8 ± 0.7 mg. Also, 50% completion of larval-pupal ecdysis in 22, 24, and 27, 29 days respectively.

Geometridae

Alsophila pometaria (Harris), fall cankerworm

→ Lyon, R.L.; Brown, S.J. 1970. Contact toxicity of insecticides applied to fall cankerworm reared on artificial diet. J ECON ENTOMOL 63: 1970-1971.

Larval diet modified from Addy* (1969) for *Malacosoma disstria*.

COMPOSITION (g): vitamin-free casein, 98.00; dextrose, 98.00; Wesson's salt mixture, 28.0; cholesterol, 8.40; potassium sorbate, 3.36; choline chloride, 2.80; wheat embryo, 140.00; Alphacel, 84.00; wheat germ oil, 7.50; agar, 70.00; sodium alginate, 14.00; potassium sorbate, 14.00; Vanderzant's fortification mixture, 40.00; and distilled water, 2100 ml.

PREPARATION Agar dissolved in hot distilled water and autoclaved. All dry ingredients except sodium alginate, ascorbic acid and Vanderzant's fortification mixture, added to agar solution and mixed in 1-gal stainless steel blender. Cooled to 65°C with periodic stirring; remaining ingredients added, and mixed for 3 minutes. Immediately poured into cheesecloth sheets and allowed to solidify.

REARING 1st-4th instar larvae reared in plastic sterile 100 x 20 mm petri dishes @ 10-20/dish at 22-26°C and 85% RH. Eggs disinfected in 6% formaldehyde solution for 15 minutes followed by 3 rinses in distilled water; incubated at 5°C.

DEVELOPMENT Ca. 95% survival to 4th instar in ca. 3 weeks with no fresh diet renewed.

Operophtera brumata (Linnaeus), winter moth

→ Feeny, P.P. 1968. Effect of oak leaf tannins on larval growth of the winter moth *Operophtera brumata*. J INSECT PHYSIOL 14: 805-817.

Modified from David & Gardiner* (1965) for *Pieris brassicae*.

COMPOSITION Group (a): distilled water, 110 ml; casein (light, white, soluble), 21.9 g.

group (b) (g): sucrose, 12.6; wheat embryo (Bemax), 1.5; Wesson's salt mixture, 3.6; cellulose powder, 7.2.

group (c): choline chloride, 0.3 g; methyl *p*-hydroxybenzoate, 0.4 g; vitamin solution (McMorran*, 1965), 3.0 ml.

group (d): distilled water, 200 ml; agar (fine Japanese powder), 9.0 g.

group (e): *l*-ascorbic acid, 1.5 g; Aureomycin, 0.8 g; gibberellic acid, 0.01 g.

PREPARATION See David & Gardiner* (1965). Ingredients of groups a and b blended. Agar solution added at 95°C; groups c and e blended in and diet dispensed to sterile 1 x 1.5" diam plastic pots @ 0.5 g/pot and further sterilised under UV light for 1 h.

REARING At room temperature in darkness. Disinfected larvae reared individually; pots inverted for frass to fall on to filter paper lined lid.

DEVELOPMENT Mean pupal wt on artificial diet, 30.3 ± 2.1 mg; on host plant, 30.8 ± 1.2 mg.

Hepialidae

Wiseana cervinata (Walker), porina caterpillar

→ Wood, J. 1970. Rearing *Wiseana* species in the laboratory. N Z ENTOMOL 4: 3-7.

COMPOSITION (parts): sterilized earth, 2; old pine sawdust, 1; fine sand, 1-1.5.

PREPARATION Mixed and placed in 5" plastic flower pots. Fresh clover leaves placed on surface.

REARING At 8-23°C with natural lighting. Eggs collected and kept in 9 x 12" plastic bags; apparatus for separating neonate larvae from eggs described. 200 larvae placed in each diet pot. After 3 months transferred to 5" squat pots divided into 4 by metal spacers. 1 larva in each. Pots sprayed occasionally.

DEVELOPMENT Incubation period, 17-45 days. Egg to adult, 243-363 days; emergence, 37%.

Wiseana spp.

→ Dodgshun, T.J. 1970. A semi-synthetic maintenance diet suitable for hepialid and melolonthid larvae. N Z ENTOMOL 4: 78-79.

Based on Shorey & Hale* (1965) diet for nine noctuid species.

COMPOSITION (g): fresh cut grass (for hepialids), 533 in 260 ml of distilled water; fresh grated carrots (for melolonthids), 533 in 260 ml of distilled water; brewers' yeast, 80; ascorbic acid, 8; methyl *p*-hydroxybenzoate, 5; sorbic acid, 2.5; agar, 32 in 1600 ml of distilled water; and (ml): 40% formaldehyde, 5.

PREPARATION For porina larvae, grass homogenised in blender with little water. All dry ingredients mixed with 300 ml water, formaldehyde added; and mixed with grass in blender for ca. 1 minute. Agar dissolved by heating in 1600 ml of water, cooled to about 70°C, mixed with other ingredients, and dispensed to petri dishes. Same procedure for melolonthid larvae, substituting carrot for grass.

DEVELOPMENT Third-instar larvae of porina or 2nd- or 3rd-instar larvae of melolonthid placed in holes cut in diet. Larvae sustained for ca. 4 months.

Lasiocampidae

Malacosoma disstria Hübner, forest tent caterpillar

→ Addy, N.D. 1969. Rearing the forest tent caterpillar on an artificial diet. J ECON ENTOMOL 62: 270-271.

COMPOSITION (g): vitamin-free casein, 98.00; dextrose, 98.00; Wesson's salt mixture, 28.00; cholesterol, 8.40; sorbic acid, 3.36; choline chloride, 2.80; whole wheat germ, 140.00; Alphacel, 84.00; wheat germ oil, 7.50; agar, 70.00; sodium alginate, 14.00; ascorbic acid, 14.00; vitamin diet fortification mixture, 40.00; and 2100 ml water.

PREPARATION Agar dissolved in 2100 ml of hot distilled water. All dry ingredients except sodium alginate, ascorbic acid, and vitamin diet mixture, added to agar solution, mixed in 1-gal stainless steel blender, and autoclaved for 15 minutes. Mixture cooled to 65°C with periodic stirring, then remaining ingredients added and complete diet mixed for 3 minutes, dispensed immediately to cheesecloth sheets and allowed to solidify.

REARING At 25°C and 70% RH with continuous lighting. 100 1st instar larvae placed in cylindrical quart icecream carton rearing containers. Bottoms of cartons replaced by fine mesh organdie and tops by clear plastic (illustrated). Diet applied to 4" wide strip of cheesecloth and draped down sides. Fluted filter paper sheets for pupation placed in cage when larvae 5th instar. Pupae transferred to 12 x 24 x 24" wire mesh cages, and kept at 22°C and 70% RH, with 12 h photoperiod for emergence. Eggs laid on ½" diam twigs, removed, incubated at 25°C and 75% RH for 3 weeks, then in cold storage for 16 weeks.

DEVELOPMENT Average of 80% survival from egg to adult.

→ Grisdale, D.G. 1976. Laboratory methods for rearing the forest tent caterpillar. BIMON RES NOTES CAN FOR SERV 32: 1.

COMPOSITION & PREPARATION See Grisdale* (1973); except raw linseed oil added @ 0.015 ml/100 ml diet to correct crippled wing condition.

REARING Field collected eggs stored at -2°C for 4 months then incubated for 5 days to hatch. Eggs surface sterilized in NaClO solution. Larvae reared at 23 ± 1°C and 60% RH with 16 h photoperiod in inverted ¾-oz plastic creamer cups containing ca. 15 ml diet. Neonate larvae climb sides of cup to establish on diet. After 12 days, 3rd instars thinned to 10/cup. Final instars transferred to cups with ca. 6 ml diet. Cups placed upright; pupation on underside of lid ca. 30 days from eclosion. Pupae separated from cocoons and lids by soaking in NaClO solution and sexed. ♂♂ initially held 4°C lower than ♀♀ to synchronise adult emergence at 2 weeks from pupation. Adults mated in 1-pint cylindrical cardboard containers containing aspen branch tip for oviposition. Held at 20 ± 1°C and 90-95% RH; 2 pairs of adults/chamber for maximum egg yield.

DEVELOPMENT Egg to adult, ca. 45 days. Eclosion from 11.6-94.7% with average of 77.9% after 7 months storage. Mating success, 63.7% from 13 matings with 3 pairs/container.

Lymantriidae

Hemerocampa pseudotsugata McDunnough, Douglas-fir tussock moth

→ Chauthani, A.R.; Claussen, D. 1968. Rearing Douglas-fir tussock moth larvae on synthetic media for the production of nuclear-polyhedrosis virus. J ECON ENTOMOL 61: 101-103.

Modified from diets of Lyon & Flake* (1966).

COMPOSITION Stock tussock diet (g): vitamin-free casein, 160.0; sucrose, 150.0; wheat germ, 100.0; Alphacel, 18.0; ascorbic acid, 20.0; Terramycin, 2.5; Wesson's salt mixture, 30.0; sorbic acid, 3.0; Tegosept, 5.0; Dithane, 1.0; (premixed).

vitamins (mg/100 ml water): niacin, 600.0; calcium pantothenate, 600.0; riboflavin, 300.0; thiamine, 150.0; pyridoxine, 150.0; folic acid, 150.0; biotin, 12.0; vitamin B_{12}, 1.2.

PREPARATION Agar dissolved in 1800 ml hot water. 1700 ml distilled water poured into 1-gal Waring Blendor, and following added in order, stirring after each addition: 2.5 ml 37% formaldehyde; 5 lb tussock moth medium; 21.0 ml 4M KOH; 40.0 ml 10% choline chloride; 10.0 ml vitamin solution, and agar solution. Diet poured while still warm and fluid into 50 x 50 x 5 cm cardboard trays in sterile transfer room to depth of ca. 2.5 cm. Cut up when set. Storage possible without contamination for 2-3 days at 50°F.

REARING At 25 ± 1°C with 14 h photoperiod. Eggs sterilized in 10% formaldehyde and agitated in 1% Clorox solution to separate egg masses; held at 4°C until use. Rearing @ 4 larvae/1-oz plastic containers in trays. Diet supplied weekly.

DEVELOPMENT Reared from egg to 4th and 5th instar before infection, reaching average weight of 310 mg. Mortality during larval stage prior to infection very low.

See also: Thompson (1965).

→ Lyon, R.L.; Flake, H.W. Jr 1966. Rearing Douglas-fir tussock moth larvae on synthetic media. J ECON ENTOMOL 59: 696-698.

Modification of medium by McMorran* (1965).

COMPOSITION Water (distilled), 640 ml; vitamin-free casein, 105 g; 4M KOH, 15 ml; and (g): Alphacel, 15; Wesson's salt mixture, 30; sucrose, 78; wheat embryo, 90. Antimicrobial agents: formalin (36% formaldehyde), 1.5 ml; methyl *p*-hydroxybenzoate, 4.5 g; potassium sorbate, 3.0 g; Aureomycin, 0.9 g; and vitamin preparation: water (distilled), 200 ml; vitamin diet fortification mixture (NBC), 31 g; and ascorbic acid, 10.6 g; and agar solution: water (distilled), 1800 ml; and agar, 75 g.

PREPARATION In batches of 3100 g in 1-gal Waring Blendor. Ingredients added in order given above, blending for 30 seconds after each addition, and 5 minutes at end.

REARING At 22-25°C. 30 1st instar larvae transferred to 100 x 20 mm sterile, plastic petri dishes containing 35 x 35 x 18 mm block of diet. Lids taped on, and diet renewed when larvae 4th or 5th instar.

DEVELOPMENT Average weight of mature, 7th instar larvae, 300 mg (in 30 days). Normal pupation and adults.

Porthetria dispar (Linnaeus) [previously *Lymantria*], gypsy moth

Leonard, D.E.; Doane, C.C. 1966. An artificial diet for the gypsy moth, *Porthetria dispar* (Lepidoptera: Lymantriidae). ANN ENTOMOL SOC AM 59: 462-464.

COMPOSITION (g): vitamin-free casein, 35; Wesson's salt mixture, 10; sucrose, 12; fructose, 23; wheat germ meal (type A >25% crude protein, 4.0% crude fibre and 9.1% crude fat), 60; choline chloride, 1; methyl *p*-hydroxybenzoate, 1.5; agar (dissolved in 620 ml distilled water), 25; Aureomycin, 0.3; ascorbic acid, 4; and (ml): distilled water, 240; 4M KOH, 5; formaldehyde, 0.5; 55% linolenic acid, 4.3; vitamin solution, 10.

vitamins (mg/100 ml water): niacin, 100; calcium pantothenate, 100; riboflavin, 50; thiamine hydrochloride, 25; pyridoxal hydrochloride, 25; folic acid, 25; biotin, 2; vitamin B_{12}, 0.2.

PREPARATION Wheat germ finely ground. All ingredients blended and dispensed while hot to sterilised ¾-oz cups, or petri dishes; refrigerated until use.

REARING Egg masses dehaired by rubbing over cheesecloth and fastened to 11 mm discs of adhesive tape. Sterilised in 0.05% NaClO for 45 minutes then transferred to cups with ½" diet. Reared at 23.3 ± 1.3° C and 35-55% RH with 16 h photoperiod. Pupae and adults at 21-24.5° C. Late instar larvae placed in sterilized 5 x 7 x 2⅜" ventilated plastic boxes and fed twice weekly.

DEVELOPMENT ♂ pupal wt, 430 mg; 215 normal adults reared from 319 larvae. ♀ pupal wt, 1602 mg; 99 normal adults reared from 319 larvae; fecundity, 750 eggs/♀.

Magnoler, A. 1970. A wheat germ medium for rearing of the gypsy moth, *Lymantria dispar* L. ENTOMOPHAGA 15: 401-406.

COMPOSITION Distilled water, 20 ml; vitamin-free casein, 2.8 g; 4M KOH, 0.47 ml; and (g): Wesson's salt mixture, 1; glucose, 1.98; cholesterol, 0.02; wheat germ, 5.7; sodium alginate, 0.5; Alphacel, 1.4; choline chloride, 0.1; sorbic acid, 0.08; methyl *p*-hydroxybenzoate, 0.08; and distilled water, 62 ml; and (g): agar, 2.28; ascorbic acid, 0.5; and linseed oil, 0.25 ml; Aureomycin, 0.01 g; vitamin solution, 1 ml.

vitamins (mg/100 ml): *i*-inositol, 2000; niacin, 100; calcium pantothenate, 100; riboflavin, 50; thiamine hydrochloride, 25; pyridoxine hydrochloride, 25; folic acid, 25; biotin, 2; vitamin B_{12}, 0.2.

PREPARATION First 12 ingredients blended in above order in Waring Blendor. Agar then dissolved in hot water and blended with ingredients. Cooled to 60° C, ascorbic acid, linseed oil, and Aureomycin added and completed diet thoroughly stirred for 3 minutes. Poured into sterile containers and sectioned when set.

REARING At 23 ± 1° C. Egg masses dehaired and surface sterilized with 10% formalin solution for 1 h. Larvae reared in petri dishes: 15 x 55 mm for 1st, 15 x 80 mm for 2nd to 4th, and 100 mm diam for 5th to 6th instars. Pupae placed in plastic containers. Adults paired and placed in small screen cages for mating and oviposition as described in Odell & Rollinson* (1966).

DEVELOPMENT Mean larval period, 37.23 ± 0.11 days; 96.8% survival from neonate larvae to adult. Average adult emergence, 51.38 days.

Corollary

→ Vasiljevic, L.J.; Injac, M. 1971. Artificial diet for the gypsy moth (*Lymantria dispar* L.). ZAST BILJA 22: 389-396.

REARING At 22.5°C and 55% RH. Hairs removed by brush and eggs rinsed; sterilized for 3 minutes in 50 ml NaClO, and rinsed in sterile distilled water aseptically for 1-2 h. Sterilized eggs transferred to vacuum flask plugged with sterile gauze and rinsed at 26°C. Egg cluster then kept for 5 months in fridge at 0-5°C. 50 neonate larvae transferred with sterile loop to each of 50 mm petri dishes. Diet cut into 6 short strips and glued to upper surfaces of dishes. 2nd instars placed in 150 mm dishes @ 10/dish; 4th instar, 5/dish; 5th and 6th instar, 3/dish. Pupae placed in glass cylinders lined with paper.

DEVELOPMENT Incubation period, 6 days; after 3 and 5 months storage, hatch 53% and 95-100%; larval period, 41-56 days; pupal wt: ♂, 450 mg; ♀, 1.35 g; pupal period, 8-14 days; average fecundity, 511 ± 93 eggs/♀.

→ Odell, T.M.; Rollinson, W.D. 1966. A technique for rearing the gypsy moth, *Porthetria dispar* (L.), on an artificial diet. J ECON ENTOMOL 59: 741-742.

COMPOSITION (g): vitamin-free casein, 2.5; fructose, 1.8; Alphacel (purified cellulose), 3.0; Wesson's salt mixture, 1.0; cholesterol, 0.3; sorbic acid, 0.12; sodium alginate, 0.5; choline chloride, 0.1; wheat germ, 5.0; agar (granular), 2.5; ascorbic acid, 0.5; 55% linolenic acid, 0.05; 75% methyl linoleate, 0.05; Aureomycin, 0.03; and 1 ml vitamins.

vitamins (mg/100 ml water): *i*-inositol, 2000; calcium pantothenate, 100; niacin, 100; riboflavin, 50; pyridoxine hydrochloride, 25; thiamine hydrochloride, 25; folic acid, 25; biotin, 2; vitamin B_{12}, 0.2.

PREPARATION First 9 listed ingredients blended with 20 ml water. Agar dissolved in 57 ml hot water, added to first mix; stirred, and cooled to 55-60°C. Remaining ingredients added, with ascorbic acid in 5 ml water. Diet dispensed to petri dishes.

REARING Eggs separated from hairy egg mass and placed in plastic screen packets @ 700 eggs/packet. Immersed in 0.5% Clorox bath for 20 minutes, then 20 minutes rinse in running water. Each packet put in 15 x 100 mm sterile plastic petri dish containing moistened filter paper. Dish taped around edges to maintain high humidity and kept at 25-27°C, with 16 h photoperiod. Neonate larvae brushed on to diet in petri dishes. Dishes taped and inverted to force larvae to feed at the top of dishes and allow frass to drop away from food (larvae prefer to feed upside down position). Dishes not reverted for 6th instar larvae. After each moult larvae transferred to new sterile dishes. 1st: 75-100 larvae, 4 cubes diet @ 0.5 cm^3; 2nd: 15-20 larvae, 4 cubes @ 1 cm^3; 3rd: 10 larvae, 4 cubes @ 1 cm^3; 4th: 5 larvae, 3 cubes @ 2 cm^3; 5th: 2-3 larvae, 3 cubes @ 2 cm^3; 6th: 1 larva, 1 cube @ 4 cm^3. Pupae in 8-cm diam dish with screen cover.

DEVELOPMENT 4-5 days in each instar, except 6th instar took 6-7 days. Hatching to adult emergence, 40-45 days. Pupal wt (g): ♂, 0.62; ♀, 2.00; (field collected: ♂, 0.500, ♀, 1.400).

See also: Rollinson (1964).

→ Ridet, J.M. 1972. Étude des conditions optimales d'élevage et d'alimentation de *Lymantria dispar* L. ANN SOC ENTOMOL FR 8: 653-668. (In French with English summary)

Best of 3 diets tested.

COMPOSITION (g): casein, 2.50; fructose, 1.80; brewers' yeast, 2.00; cellulose powder, 3.00; Wesson's salt mixture, 1.00; wheat germ, 5.00; cholesterol, 0.30; agar, 2.50; sorbic acid, 0.1; and (ml): Nipagin M (20 g in 96% alcohol), 2.00; safflower oil, 0.5; distilled water, 78; and (g?): Vanderzant's fortification mixture, 0.3; choline chloride, 0.1; ascorbic acid, 0.5.

REARING At 25° C with 16 h photoperiod.

DEVELOPMENT Normal adults; 49 days to adult (40 days on apple); weight of ♀ pupae, 1.258 g (2.422 g on apple). Diet nutritionally deficient.

Megalopygidae

Megalopyge opercularis J.E. Smith, puss caterpillar

→ Khalaf, K.T. 1974. Non aseptic wheat germ diet for *Megalopyge opercularis* (Lepidoptera: Megalopygidae). FLA ENTOMOL 57: 377-381.

COMPOSITION Part 1 (g): wheat germ premix (Vanderzant's NBC), 14; sorbic acid, 0.05; Methyl paraben, 0.1; and (ml): formaldehyde, 0.05; 4M KOH, 0.4; raw linseed oil, 0.2; cholesterol, 0.03 g; distilled water, 25 ml.

Part 2: agar (granulated), 1.5 g; distilled water, 35 ml; vitamin premix (Vanderzant's NBC), 1.5 g; Aureomycin, 0.015 g; kanamycin sulphate (powder), 0.01 g.

PREPARATION Ingredients continuously mixed in order listed, with maximum of 30 ml distilled water, with electric or magnetic stirrer. Agar dissolved in hot water, cooled to 52° C, stirred in with first mix; then vitamins, Aureomycin and kanamycin added quickly and stirred. Diet poured into 16-oz rearing carton, and left for 1 h.

REARING 16-oz icecream cartons adapted for rearing; 3-gal icecream tub adapted for emergence chamber and mating/oviposition cage (illustrated). Adults non-feeding.

DEVELOPMENT Incubation period, ≤ 6 days; larval period, 3½-5 months. Two generations reared.

Other species reared: saddleback caterpillar, *Sibine stimulea* (Clemens) (Lep., Limacodidae), and salt-marsh caterpillar, *Estigmene acrea* (Drury) (Lep., Arctiidae).

Megathymidae

Megathymus streckeri (Skinner)

→ Petterson, M.A.; Wielgus, R.S. 1973. Acceptance of artificial diet by *Megathymus streckeri* (Skinner) (Megathymidae). J RES LEPID 12: 197-198.

COMPOSITION (g): agar, 90; casein, 129; alfalfa meal, 54; sugar, 129; wheat germ, 108; Wesson's salt mixture, 36; Alphacel, 18; ascorbic acid, 15; Aureomycin, 0.5; and (ml): water, 2500; 4M KOH solution, 18; 10% aqueous choline chloride, 36; 27% formalin (10% formaldehyde gas in water), 15; 38% solution sorbic acid in 95% ethanol, 20.6; 38% solution methyl *p*-hydroxybenzoate in 95% ethanol, 20.6; vitamin solution, 12.

vitamins (mg/400 ml water): niacin, 2400; calcium pantothenate, 2400; riboflavin, 1200; thiamin, 600; pyridoxin, 600; folic acid, 600; biotin, 48; 0.1% vitamin B_{12}, 4800.

PREPARATION See Henneberry & Kishaba* (1966a). Hot liquid diet poured into mould of 7.62 cm diam plexiglass tube at least 12.7 cm long and allowed to solidify at 23°C.

REARING At 25.6°C, and 40% RH. Section of diet 12.7 cm long removed from mould and placed in polyethylene sandwich bag. Bag then sealed to slow loss of moisture, inverted and supported upright in pint glass jar; 10 mm hole cut in bag. Paper "tent" set ca. 6.5 mm into diet and larva placed inside tent or larvae placed around slice of *Yucca caudex* plug, 25.4 mm long with triangular top 3.2 mm side and slice inserted into diet so broad end flush with diet. Larvae established in both methods.

DEVELOPMENT Complete.

Corollary

→ Wielgus, R.S. 1974. Artificial diet: the key to the mass rearing of *Megathymus* larvae. J RES LEPID 13: 271-277.

REARING Improved rearing techniques described and illustrated. For tent-building species, *Megathymus ursus ursus*, PVC irrigation pipe 80 x 24 mm inside diam filled with diet. One end of pipe capped with 27 mm square of 0.5 mm sheet aluminium. Other end of pipe capped with square of clear plastic sheet of same dimension; small hole cut in centre of each plastic square for insertion of artificial tent of rolled paper. Tent pushed into diet surrounding *Yucca* plug inserted earlier. Egg or larvae placed in paper tent; container then placed upright in small boxes or bowls. Maintained at room temperature. Larval transfer easy; and on maturity ends of pipe wrapped with plastic to reduce moisture loss of diet. Pupation in pipes after several months of feeding.

Non-tent builders (all other species listed below) kept in drinking straw filled with diet to within 20 mm of one end. Small *Yucca* plug pushed into diet from other end. Egg attached to *Yucca* leaf introduced into straw at first end, then sealed. Straws, in clusters of ≯ 20, stood upright in glass jelly jars. Larvae feed on diet and moult; 5th instars transferred to another container consisting of semi-transparent plastic tube 30.5 cm x 9 mm ID with wall 2 mm thick. Diet filled to within 4 cm of one end. Both ends of tube capped with 10 mm square of 0.5 mm sheet aluminium. Third container type as above, but tubes of 12 mm diam capped with aluminium over bottom end only; upper ends plugged with wads of cotton. Diet changes necessary, larvae removed by uncapping ends and blowing gently to force contents of tube out. Larva collected and placed in new container.

DEVELOPMENT Larval period reduced from 10 months to ca. 4 months using this method. Species reared: *Megathymus streckeri* (Skinner), *M. texanus texanus* B.T. McD., *M. ursus ursus* (Poling), *M. cofaqui* (Strecker), *M. harrisi* Freeman, *M. texanus leussleri* Holland.

Noctuidae

Achaea catella Guerin-Meneville

Achaea finita Guenée

Anua tirrhaca Cramer

→ Bot, J. 1967b. Rearing three species of fruitsucking moths on artificial diets. S AFR J AGRIC SCI 10: 1009-1014.

COMPOSITION Diet for *A. catella* and *A. finita* (g): dehydrated carrot, finely ground, 20.0; castor oil leaf powder, 16.0; wheat germ, 28.0; vitamin-free casein, 5.0; brewers' yeast (ground live *Saccharomyces utywala africana*), 20.0; powdered agar, 5.0; ascorbic acid, 3.0; inositol, 0.2; cholesterol, 0.2; choline chloride, 0.4; cysteine hydrochloride, 0.06; methyl *p*-hydroxybenzoate, 2.0; and water, 400.0.

Diet for *A. tirrhaca* (g): dehydrated carrot, 10.0; *Rhus lancea* leaf powder, 30.0; wheat germ, 24.0; vitamin-free casein, 3.0; brewers' yeast (as above), 9.0; powdered agar, 6.0; ascorbic acid, 3.0; choline chloride, 0.2; inositol, 0.1; white sugar, 4.0; methyl *p*-hydroxybenzoate, 1.2; and water, 360.0.

PREPARATION Host plant powder: castor oil (*Ricinus communis*):- young, fully developed, fresh, dry leaves submerged for 10-15 minutes in re-distilled acetone and fan dried. Powdered and stored in deep freeze. Karee (*Rhus lancea*):- treated as above, or oven dried at 95°C. Carrot, wheat germ, leaf powder, casein, and yeast thoroughly mixed; methyl *p*-hydroxybenzoate and cholesterol (where applicable) dissolved in ca. 5.0 ml mixture of 25% ethyl ether and 75% acetone and stirred in. Mixture then spread to thin layer on glass plate for solvent evaporation. Agar dissolved in ca. 300 ml water, added to diet and boiled. When applicable, ascorbic acid, choline chloride, inositol, cysteine hydrochloride, and sugar dissolved in remaining portion of water and added, together with carrot, wheat germ, leaf powder, and casein-yeast mixture, to boiling agar solution.

Diet dispensed to 4" or 6" by 1" heat sterilized tubes to depth of 1½". Tubes slanted till set then placed upright till cool. After several hours, stoppered with heat sterilized non-absorbent cottonwool plugs enclosed in organdie. Strip of sterile filter paper added to each tube of *A. catella* and *A. finita*, before stoppering to supply foothold for larvae.

REARING At 22-28°C and 70% RH. Moths kept in 3 x 4 x 4', 18 mesh mosquito gauze cages and fed daily on canned guava juice from ½ x ¾" diam dishes. Oviposition on gauze. Larvae removed from gauze with camel hair brush, and placed individually in each tube. After 8-10 days, tubes inverted to remove faeces. Pupae removed from tubes to damp vermiculite in hatching tray. Larger (4 x 4 x 4') cage needed for *Anua tirrhaca* adults.

DEVELOPMENT Larval period in days on natural host plant and on artificial diet respectively: *A. catella*, 14-16 and 12-15; *A. finita*, ca. 19 and 18-24; and *A. tirrhaca*, 45-64 and 37-50. Average wt (g) of two-day old pupae reared on natural host plant and artificial diet respectively: *A. catella*, 0.71 and 0.75; *A. finita*, 0.83 and 0.70; and *A. tirrhaca*, 1.21 and 1.26. Oviposition of *A. catella*, *A. finita*, and *A. tirrhaca* starts 5-6 days, 7-8 days, and 10-12 days from emergence respectively, continuing for 2-3 weeks. Adult longevity > 4 weeks. All 3 species reared for 18 months continuously.

Achaea janata Linnaeus, castor semi-looper

➞ Pant, J.C.; Anand, M. 1975. An artificial diet for rearing castor semi-looper, *Achaea janata* Linnaeus. ENTOMOL NEWSL 5: 49-50.

Modified from Siddiqui & Chatterji* (1972) for *Chilo zonellus*.

COMPOSITION (g): agar, 1.25; castor leaf powder, 5.00; rajmah (*Phaseolus vulgaris*), 13.7; wheat flour, 5.00; yeast, 1.00; ascorbic acid, 0.42; sorbic acid, 0.10; methyl *p*-hydroxybenzoate, 0.20; vitamin E, 0.025; and (ml): formaldehyde, 0.25; water, 63.00.

PREPARATION See Siddiqui & Chatterji* (1972). Castor leaves dried at 50° C for 2 h and ground in mortar. Diet poured into 15 x 10 cm diam glass jars to depth of ca. 3 cm.

REARING At 27 ± 2° C @ 25 larvae/glass jar.

DEVELOPMENT Larval period, 12-15 days; pupal period, 5-8 days; larvae to adult, 18-22 days; 60% survival to adults; viable eggs laid. Larval period on natural host, 8-12 days; pupal period same as on diet.

Agrotis ipsilon (Hufnagel), black cutworm

➞ Mangat, B.S. 1970. Rearing the black cutworm in the laboratory. J ECON ENTOMOL 63: 1325-1326.

COMPOSITION (g): whole wheat germ, 24.0; vitamin-free casein, 28.0; sucrose, 28.0; bacto-agar, 20.0; Wesson's salt mixture, 8.0; sorbic acid, 1.6; methyl *p*-hydroxybenzoate, 1.6; and (ml): 10% KOH, 16; 95% ethanol , 15; vitamin solution, 26.4; water, 650.

vitamins (g/1): niacinamide, 1.00; calcium pantothenate, 1.00; thiamine hydrochloride, 0.25; riboflavin, 0.50; pyridoxine hydrochloride, 0.25; vitamin B_{12} in mannitol, 2.00; inositol, 20.0; folic acid, 0.25; biotin, 0.02; choline chloride, 50.0; α-tocopherol powder, 8.00; ascorbic acid, 270.00.

PREPARATION Liquid ingredients (except vitamins) and agar mixed in 1200 ml stainless steel beaker and dissolved by constant stirring and heating in water bath at 90-95° C; or autoclaved. Inhibitor solution (prepared by adding 1.6 g of each antimicrobial agent to 15 ml of 95% ethanol) added; solution allowed to cool to 50° C, blended with vitamin solution and other dry ingredients. Poured into 5 x 9 cm sterile crystallising dishes to depth of 2.5 cm, and lid placed over each dish. Stored in refrigerator at 4° C until needed (no more than 7 days). Plugs of diet ca. 2.5 x 2.0 cm diam cut and transferred to 95 x 25 mm cotton plugged, sterile shell vials.

REARING At 24 ± 0.5° C; 15 neonate larvae/vial. 3rd instar larvae separated and reared individually. Fresh food every 4 days. Pupae placed in groups of 20 in 500 ml glass beakers with peat moss, and covered with cheesecloth. Adults transferred to 28 x 28 x 38 cm oviposition cages. Fed 10% honey solution and kept at 27 ± 1° C and 85-90% RH in darkness. Eggs laid on muslin strip and surface sterilized for 5 minutes in 100 ml 0.15% NaClO, washed with sterile distilled water and incubated in sterile test tubes at 24 ± 0.5° C in darkness.

DEVELOPMENT Average of 65.1% survival to adult. Life cycle from eggs to adult emergence completed in 42 ± 2 days. Three consecutive generations reared.

See also: Mangat (1971).

Reese, J.C.; English, L.M.; Yonke, T.R.; Fairchild, M.L. 1972. A method for rearing black cutworms. J ECON ENTOMOL 65: 1047-1050.

COMPOSITION Larval diet (g): finely ground pinto beans, 560; Vanderzant-Adkisson wheat germ diet (NBC), 200; torula yeast, 130; agar, 120; ascorbic acid, 40; wheat germ oil, 20; methyl *p*-hydroxybenzoate, 8; Vanderzant's vitamin mixture, 8; cellucotton, 5; sorbic acid, 4; Cerophyl, 4; and (ml): 40% formaldehyde, 10; distilled water, 3200.

Adult diet (ml): distilled water, 500; beer, 300; honey, 200; and (g): ascorbic acid, 10; *l*-cysteine hydrochloride, 2.

PREPARATION Pinto beans, torula yeast, wheat germ diet, wheat germ oil, methyl *p*-hydroxybenzoate, sorbic acid, vitamin mixture, and 1500 ml water blended in stainless steel mixing bowl. Diet more easily handled when ingredients allowed to stand for at least 30 minutes. Agar, cellucotton, Cerophyl and remaining water placed in double boiler and cooked for at least 20 minutes until very stiff, then added to ingredients in mixing bowl and blended. Formaldehyde and ascorbic acid added after mixture cooled. Poured into 1-oz plastic cups or into flat-bottomed 28 cm diam dishes to depth of 25 mm. Dishes covered with paper towelling, allowed to cool, and refrigerated.

REARING At 20-25° C and 75-85% RH with 12 h photoperiod. 1st instar larvae placed on diet with brush @ 1 larva/cup and cups placed in egg cartons. Pupation chambers of 2-oz plastic cups with plastic lids punched with holes and ⅓ filled with moist fine vermiculite. Pupae placed in wire emergence cages 11-14 x 8" diam with 1-oz cups of adult diet via wick. 15 adults in each oviposition cage of 1-gal paper carton. Sheet of paper towel folded over edge for oviposition. Two diet cups given every 2-3 days. Egg covered towel placed in icecream carton for hatching.

DEVELOPMENT Average larval period, 23.9 days; average pupal wt, 0.49 g; larval survival rate, > 80%.

Agrotis orthogonia Morrison, pale western cutworm

McGinnis, A.J.; Kasting, R. 1959. Nutrition of the pale western cutworm, *Agrotis orthogonia* Morr. (Lepidoptera: Noctuidae). I. Effects of underfeeding and artificial diets on growth and development, and a comparison of wheat sprouts of Thatcher, *Triticum aestivum* L., and Golden Ball, *T. durum* Desf., as food. CAN J ZOOL 37: 259-266.

Best two of four diets described.

COMPOSITION Diet 1 (%): cellulose powder, 33.3; plant extract, 66.7.

Diet 2 (%): vitamin-free casein, 6.1; yeast extract, 0.6; glucose, 6.1; cellulose powder, 26.3; lanolin, 0.1; corn oil, 0.2; amino acid mixture, 1.6; vitamin solution, 37.0; salt solution, 22.0; plant extract, 47.0.

vitamin solution, salt solution and plant extract reduced to 55% of original volume at 45° C *in vacuo* to keep final moisture content at 59%.

amino acids (2 parts by weight): *dl*-alanine, *l*-arginine, *l*-glutamic acid, glycine, *l*-leucine, *l*-proline, *dl*-serine, *dl*-tryptophan, and *dl*-valine, and 1 part by weight of each of the following: *l*-aspartic acid, *l*-cystine, *l*-histidine, hydroxy-*l*-proline, *dl*-isoleucine, *l*-lysine, *dl*-methionine, *l*-phenylalanine, *dl*-threonine, and *l*-tyrosine.

vitamins (μg/ml): thiamin, 5; riboflavin, 15; pyridoxine, 15; inositol, 100; nicotinic acid, 15; choline chloride, 100; calcium pantothenate, 15; folic acid, 5; and biotin, 0.025.

salts (g/l): NaCl, 1.8; $CaCl_2$, 0.04; KCl, 0.04; $NaHCO_3$, 0.04; $NaH_2PO_4.H_2O$, 1.35; and K_2HPO_4, 1.55.

PREPARATION Plant extract: juice pressed from Golden Ball wheat leaves collected when plants in milk stage and kept frozen until required. Centrifuged when thawed and supernatant added to diet.

REARING Individually or in groups of 10 in 60 x 15 mm petri dishes; transferred daily to clean dishes until emergence.

DEVELOPMENT On diet 1 all 9 larvae moulted at least three times and showed greatest weight increases and longest survival periods, but seven died during the prepupal period; 2 pupated, and 1 emerged. On diet 2, 6 instars matured in 4 weeks; increased weight from 0.25 mg-700 mg. Wheat leaf juice necessary for weight increases.

→ Kasting, R.; McGinnis, A.J. 1967. An artificial diet and some growth factor requirements for the pale western cutworm. CAN J ZOOL 45: 787-796.

Best of 7 diets.

COMPOSITION (g): vitamin-free casein, 3.5; sucrose, 3.5; Wesson's salt mixture, 1.5; choline chloride, 0.1; ascorbic acid, 0.4; agar, 2.5; water, 77.0; methyl *p*-hydroxybenzoate, 0.2; wheat germ, 3.0; and B-vitamins, 8.1 ml.

B-vitamins (mg/8.1 ml solution): niacinamide, 1.00; calcium pantothenate, 1.00; thiamine hydrochloride, 0.25; riboflavin, 0.50; pyridoxine hydrochloride, 0.25; folic acid, 0.25; biotin, 0.10; inositol, 20.00; and 2 μg of B_{12}.

PREPARATION All dry components ground in Wiley Mill to pass 40-mesh screen, and tumble-mixed for ca. 24 h. Water, B-vitamins, and choline chloride blended with dry mixture. Oil or wheat germ added to diet during blending. Diet steam sterilized for 17 minutes at 15 psi. Ascorbic acid, previously dissolved in water and exposed to ultraviolet light for 2 h, finally added to diet.

REARING At ambient temperature in dark larvae reared individually in 60 x 15 mm petri dishes. Eggs from 4th generation laboratory reared moths placed on diet 1 or 24 h after hatching. Larvae transferred daily to clean petri dishes lined with moist blotting paper and fresh food.

DEVELOPMENT Survival to pupation, 60%; days to pupation, 70 ± 11; pupal wt, 270 ± 70 mg; adult survival, 50%.

Sutter, G.R.; Miller, E.; Calkins, C.O. 1972. Rearing the pale western cutworm on artificial diet. J ECON ENTOMOL 65: 1470-1471.

COMPOSITION Larval diet. See Sutter & Miller* (1972) diet for *Euxoa auxiliaris*.

Adult diet: beer, 12-oz; sucrose, 25 g; and ascorbic acid, 12.5 g.

PREPARATION See Sutter & Miller* (1972).

REARING At 26.5°C and 80% RH in darkness. 5-8 neonate larvae placed on diet in cups, covered with lids and kept for 14 days. Then each larva transferred to 1-oz cup with fresh diet, and kept at 24°C and 65% RH for 24-26 days until pupation. Pupae removed and placed in screened cages at 24 ± 1°C and 70 ± 5% RH with 12 h photoperiod. Adults fed; oviposition in petri dish with 10 mm soil. Eggs removed by sieving through 60 mesh sieve, surface sterilized for 30 minutes in 10% formalin, rinsed in sterile, distilled water, transferred to filter pad in Buchner funnel and air dried. Eggs placed in 1-oz jelly cups, capped and incubated at 26.5°C and 80% RH.

DEVELOPMENT Eight larval instars completed in 34-45 days with 96% survival. Pupal period, 21-31 days; adult emergence, 87%; 11 generations reared.

Agrotis segetum (Schiffermiller)

Uspenskaya, N.V.; Kozhaeva, K.I. 1974. A new semi-synthetic nutrient medium for rearing larvae of *Agrotis segetum* (Lepidoptera: Noctuidae). ZOOL ZH 53: 782-784. (In Russian with English summary)

COMPOSITION (g): lucerne meal, 150; cottonseed meal, 50; agar, 10; autolysate brewers' yeast, 40; ascorbic acid, 4; methyl *p*-hydroxybenzoate, 2; sorbic acid, 1; 40% formaldehyde, 1.1 ml; made to 1 kg with distilled water.

Anagrapha falcifera (Kirby), celery looper

Autographa californica (Speyer), alfalfa looper

Treat, T.L.; Halfhill, J.E. 1973. Rearing alfalfa loopers and celery loopers on an artificial diet. J ECON ENTOMOL 66: 569-570.

COMPOSITION (g): agar, 36; alfalfa leaf meal powder, 150; brewers' yeast powder, 105; wheat germ, 195; ascorbic acid, 12; methyl *p*-hydroxybenzoate, 12; vitamins ABCDE (Park Davis & Co.), 1 capsule; and (ml): formaldehyde, 9; water, 2400.

PREPARATION Agar dissolved in 800 ml water and cooled to 70°C. Remaining ingredients mixed with 1600 ml water in 1-gal blender. Agar mixture added to blender, mixed for 3-5 minutes and diet poured into 30 x 50 x 6 cm stainless steel pan. Pan covered with Saran wrap, cooled for 1 h and refrigerated.

REARING At 21-23°C and 50-60% RH with 14 h photoperiod. Eggs on blotting paper collected from 20 pairs moths in 30 x 35 x 54 cm Saran screen covered units for 2-3 days. Adults fed on sugar from vial. Eggs sterilised in 0.1% NaClO, dried, and placed, with 100 g diet @ 40 eggs/8-oz Lilly cup. 30-35 pupae/cup sterilized in 0.25% NaClO solution, and 150 transferred to plastic shoe boxes on moist vermiculite for emergence.

DEVELOPMENT *A. californica*: eggs incubation period, 3-5 days; larval period, 20-25 days; pupal period, 7-10 days; with 95% pupation. Development of *A. falcifera* slightly longer.

Ceramica picta (Harris), zebra caterpillar

→ Tamaki, G.; Turner, J.E.; Wallis, R.L. 1972. Life tables for evaluating the rearing of the zebra caterpillar. J ECON ENTOMOL 65: 1024-1027.

COMPOSITION & PREPARATION See Shorey* (1963) diet developed for *Trichoplusia ni*.

REARING At average temperature of 25°C (19-28) with 16 h photoperiod:- 3 methods used. (1) 100 larvae placed singly in jelly cups with diet; (2) 100 larvae placed in cardboard tray with diet; see Yamamato* (1969) for detail; and (3) 100 larvae placed in tray with sugarbeet leaves. Fresh diet added daily to trays; but cups of (1) not opened. Pupae placed in moist vermiculite in separate, ventilated jelly cups. Moths kept in 473 ml icecream cartons lined with paper towelling. Wet vermiculite on bottom and cotton wick saturated with 10% sucrose solution provided.

DEVELOPMENT Larval survival, 23%; adult emergence, 10-19%. Females laid viable eggs.

Diparopsis castanea Hampson, red bollworm

→ McKinley, D.J. 1974. An attempt to develop a completely artificial diet for rearing the red bollworm, *Diparopsis castanea*. CENT OVERSEAS PEST RES MISC REP 16: 1-4.

Two diets tested.

COMPOSITION Diet 1, casein wheat germ diet (g): agar, 1.7; casein (light, white, soluble), 11.0; wheat germ, 3.0; sucrose, 0.5; choline chloride, 0.25; cellulose (Whatman CF 11), 13.6; mineral salts (Glaxo DL6), 0.5; vitamin C, 0.8; Mycostatin powder, 0.05; and (ml): distilled water, 68.7; maize oil, 2.0; vitamin B mixture (NBC), 0.8; and linolenic acid, 4 drops.

Diet 2, cotton meat diet (g): agar, 1.5; cotton meats, 16.5; cellulose (CF 11), 5.2; choline chloride, 2.6; cholesterol, 0.7; vitamin C, 2.6; xylose, 2.6; Mycostatin, 0.05; and distilled water, 66.0 ml.

PREPARATION Diet 1: agar dissolved in hot water; other ingredients except vitamin C, vitamin B mixture and Mycostatin stirred in. Mixture kept hot during mixing and for further 5 minutes. Remaining ingredients added, cooled and diet allowed to set.

Diet 2: agar dissolved in hot water; remaining ingredients except choline chloride, vitamin C, xylose and Mycostatin added, but mixed and cooled quickly to avoid meat turning brown. Other ingredients added when cool.

DEVELOPMENT Larval establishment, 70%; growth rapid compared with that on cotton buds until 6 day, then retarded. Three underweight pupae produced; adults died during eclosion.

Earias fabia Stoll, spotted bollworm

Pant, J.C.; Anand, M. 1972. An artificial diet for the spotted bollworm, *Earias fabia* Stoll. ENTOMOL NEWSLET 2: 45-46.

COMPOSITION (g): vitamin-free casein, 2.5; sucrose, 1.0; fructose, 0.8; cholesterol, 0.3; U.S.P. Salt Mixture No. 2, 1.0; choline chloride, 0.1; wheat germ, 5.0; agar, 2.0; ascorbic acid, 0.5; linoleic acid, 0.1; methyl *p*-hydroxybenzoate, 0.3; and (ml): formaldehyde, 0.5; vitamin solution, 5.0; and water to 100 ml.

vitamins (mg): inositol, 200; pantothenic acid, 100; nicotinic acid, 100; riboflavin, 50; pyridoxine, 25; thiamine, 25; folic acid, 25; biotin, 2; B_{12}, 0.2; and water to 100 ml.

PREPARATION See Leonard & Doane* (1966) for *Porthetria dispar*. Refrigeration up to 6 weeks before use.

REARING At 34 ± 1° C and 70-80% RH.

DEVELOPMENT Neonate larvae to adult in 20-22 days; survival to adult, 60-70%.

Earias insulana (Boisduval)

Poitout, S.; Bues, R.; Le Rumer, C. 1972. Élevage sur milieu artificiel simple de deux noctuelles parasites du coton *Earias insulana* et *Spodoptera littoralis*. ENTOMOL EXP APPL 15: 341-350. (In French with English summary)

Also for *Spodoptera littoralis* (Boisduval), cotton leafworm.

COMPOSITION (g): agar, 5.0; corn semolina, 28.0; wheat germ, 7.0; brewers' yeast, 7.5; ascorbic acid, 1.0; benzoic acid, 0.3; Nipagin, 0.25; aureomycin chlorohydrate, 0.03; and (ml): water, 170.0.

PREPARATION See Poitout & Bues* (1970).

REARING Larvae of both spp at 25 ± 1° C and 70% RH with 16 h photoperiod. Adults at 20 ± 2° C with natural lighting. Larvae of *E. insulana* reared individually or in groups; individually in 8 cm^3 boxes with 0.5 cm cube of diet. New diet plug added after 6-7 days; when reared in groups, larvae transferred to new diet in clean cage. Adult yield less when reared singly. 5-8 pairs of adults in 80 x 105 mm diam boxes, with gauze lid and filter paper on bottom. Adults fed 30% honey solution from dental wick. Oviposition on gauze strip held between lid and base; renewed every 2 days; eggs incubated at 25° C.

For *S. littoralis*, see Poitout & Bues* (1970). Larvae reared individually when in 5th and 6th instars during nutrition studies. Pupation in sterile sand. Adults reared singly or in groups: singly in 80 mm diam box, with 20 x 20 x 10 mm container for 30% honey solution. Eggs laid directly on sides of cage. In groups same principle, but cages 250 x 200 mm diam; 25 pairs/cage. Eggs laid on filter paper.

DEVELOPMENT *E. insulana:* mean development periods from larvae to adult, 27.1 and 26.5 days for ♂♂ and ♀♀ respectively. Yield, 40-50% increasing by 10% on addition of 2 g Wesson's salt mixture.

S. littoralis: mean development periods from larvae to adult, 28.0 and 26.7 days for ♂♂ and ♀♀ respectively. Replacement of 10 g of corn semolina by 10 g dried cabbage leaf gave slight reduction in larval and pupal period.

Ramanankasina-Razafialisoa, E. 1970. Alimentation artificielle de la larvae d'*Earias insulana* Boisd. Noctuidae-Westermaniinae. C R ACAD SCI PARIS SER D 271: 408-410. (In French)

Diet modified from that of Guennelon (1968).

COMPOSITION Basal diet (%): water, 80; agar, 2.0; cellulose, 2.0; brewers' yeast, 2.0; casein, 2.0; sucrose, 2.0; ascorbic acid, 0.5; cholesterol, 0.4; cotton-seed oil, 0.4; Wesson's salt mixture, 1.0; vitamin mix, 1.003; choline, 0.1; benzoic acid, 0.6; To this, cotton flour meal added to give final diet (concentration not given).

PREPARATION Agar dissolved in boiling water. Ascorbic acid and vitamin mix dissolved in cold water. Other ingredients homogenised. Cholesterol and cotton-seed oil dissolved in ethyl alcohol. Dried. Cooled agar and vitamin solutions added, and homogenised again. Dispensed to containers to 1.5 cm depth; 1 mm diam holes bored to facilitate larval establishment.

REARING At 27-30° C and 80% RH.

DEVELOPMENT Complete. Pupation after about 16 days.

Euxoa auxiliaris (Grote), army cutworm

Sutter, G.R.; Miller, E. 1972. Rearing the army cutworm on an artificial diet. J ECON ENTOMOL 65: 717-718.

COMPOSITION Group A: distilled water, 13650 ml; agar, 270 g.

group B (g): wheat germ, 550; casein, 645; sucrose, 645; Alphacel, 275; Wesson's salt mixture, 185; methyl paraben, 20; sorbic acid, 12.5; cholesterol, 1.25; raw linseed oil, 5.

group C: ice cubes, 3150 g.

group D (ml): Vanderzant's fortification mixture, 300; 10% KOH, 200; formalin, 30.

PREPARATION Agar dissolved in water by heating in autoclave for 20 minutes, then blended with group B ingredients for 1 minute. Ice cubes added to cool mixture to 50° C, then group D ingredients added, and entire diet blended for 1 minute. Dispensed to 1-oz jelly cups.

REARING Larvae reared at 26.5° C and 80% RH. See Sutter *et al.** (1972). Eggs surface sterilized and stored at 4° C for up to 30 days with no ill effect. Possible to maintain adults for up to 3 months at 4° C and then to incubate at 24° C with 12 h photoperiod for egg production.

DEVELOPMENT Larval stages completed in 29-33 days with 97% survival. Pupal stage, 19-24 days; adult emergence, 88%. Viable eggs laid; hatch in 5 days. Reared for 10 generations.

Euxoa subalba Walker

Theron, P.P.A. 1945. Rearing *Euxoa* on artificial media. J ENTOMOL SOC SOUTH AFR 8: 165-166.

COMPOSITION (ml): maize meal, 100; water, 60.

PREPARATION Mixed in 1 lb honey jars and inoculated with nutritive fungus (*Mucor* or *Rhizopus*) after sterilization at 20 lb pressure for 30 minutes.

REARING At 80° F. Oviposition on muslin or waxed paper. Eggs sterilized in 4% formaldehyde and transferred aseptically to jars. Maximum of 11 larvae/jar. Mature larvae and young pupae placed on slightly damp soil.

DEVELOPMENT Larval stage, 30 days. Larval size same as when reared on cabbage.

Feltia subterranea (Fabricius), granulate cutworm

→ Lee, B.L.; Bass, M.H. 1969. Rearing technique for the granulate cutworm and some effects of temperature on its life cycle. ANN ENTOMOL SOC AM 62: 1216-1217.

Modification of diet of Shorey* (1963) for *Trichoplusia ni*.

COMPOSITION (g): soaked lima beans, 1200; brewers' yeast, 120; agar, 30; ascorbic acid, 12; methyl *p*-hydroxybenzoate, 9; sorbic acid, 3; and (mg): Aureomycin, 300-900; and 2200 ml of water.

PREPARATION Lima bean seeds (Henderson Bush variety) soaked in water at 22-24° C for 24 h. Then seeds, yeast, methyl *p*-hydroxybenzoate, and ascorbic acid blended with half of total water. Agar dissolved in remaining water, then cooled to below 70° C. Agar solution added to ingredients in blender and mixture briefly blended again. Diet poured into wide-mouth Mason jars.

REARING At 27° C in darkness. Adults fed 10% sugar solution in large mouth 1-gal jars containing 10 pairs. Top of jar covered with cheesecloth strip for oviposition. Eggs surface sterilized for 3 minutes in 3% NaClO solution, placed in Buchner funnel and solution drawn off. Then egg cloth to pint jar. Neonate larvae placed individually in 1-oz plastic cups with 15 ml diet. Pupae placed in gallon jars for emergence.

DEVELOPMENT Incubation period, 4.8 days; larval period, 24.6 days; pupal period, 16.1 days, and preoviposition period, 5 days. Average of 50.5 days to complete life cycle. Adult longevity, 13.3 days; adult emergence, 27%.

Heliothis armigera (Hübner), cotton bollworm

→ Bot, J. 1966b. Rearing *Heliothis armigera* Hubn. and *Prodenia litura* F. on an artificial diet. S AFR J AGRIC SCI 9: 535-538.

COMPOSITION (g): wheat germ, 24.0; vitamin-free casein, 4.0; finely ground brewers' yeast (dry, live *Saccharomyces utywala africana*), 22.0; powdered agar, 6.0; ascorbic acid, 2.5; inositol, 0.1; cholesterol, 0.1; choline chloride, 0.2; cysteine hydrochloride, 0.03; Nipagin M, 1.2; and water, 225.

PREPARATION Wheat germ, casein and yeast thoroughly mixed, then Nipagin M and cholesterol (dissolved in ca. 20 ml of mixture of 75% acetone and 25% ether) evenly mixed in. Solvents then allowed to evaporate. Agar powder added, followed by inositol, choline, cysteine and ascorbic acid dissolved in 225 ml of water. Boiled while stirring, then boiling mixture dispensed to sterile 4 x 1" or 6 x 1" tubes to depth of ca. ¾". Tubes slanted, allowed to cool, and stoppered with cotton. Free excess moisture removed by inverting tubes for 12-24 h, or with filter paper strip or vermiculite in tubes.

REARING At 28°C, in tubes. When prepupal, tubes unstoppered and stacked over large tray filled to depth of 4" with vermiculite. Adults kept in 12" cube, 18 mesh mosquito-gauze cage at 25°C and 75% RH, and fed 10% honey solution. *H. armigera* eggs laid on gauze; neonate larvae transferred individually to tubes. *P. litura* eggs laid on filter paper discs; paper placed in bottles stoppered with cotton wool, larvae congregate on cotton wool and transferred @ 2-4 larvae/tube.

DEVELOPMENT *H. armigera:* larval period, 12-15 days; pupal period, 12-16 days. Average pupal weight, 0.35 g.

P. litura: larval period, 10-12 days; pupal period, 12-14 days, average pupal wt, 0.44 g. Adults of both species larger and more robust than field collected specimens. Oviposition commenced 3-4 days after emergence and fertility almost 100%. Many generations of both species reared.

Corollary

→ Whitlock, V.H. 1974. A substitute for casein in an artificial rearing medium for the American bollworm, *Heliothis armigera* (Hübn). PHYTOPHYLACTICA 6: 133-134.

COMPOSITION See Bot* (1966); except casein substituted by equal amount of dried milk powder (both skimmed milk and full cream milk powder).

DEVELOPMENT No difference in development on skimmed and full cream milk powder. On skimmed milk powder and casein diet respectively, % larval mortality: 20.2, 25.6; larval period, 14.5, 14.3 days; pupal wt: 0.325, 0.336 g; % ♀♀: 55, 52; adult longevity: 6.6, 6.5 days; % diet contamination: 0, 42; price Rand 0.50/kg, Rand 29.00/kg. No difference in oviposition or fecundity. 11 generations reared.

See also: Whitlock (1973).

→ Mourikis, P.A.; Vassilaina-Alexopoulou, P. 1969. Notes on the rearing on semisynthetic diet of the cotton bollworm, *Heliothis armigera* (Hbn). ANN INST PHYTOPATH BENAKI 9: 46-53.

COMPOSITION (g): flour alfalfa, 70; carrot powder, 30; cornflour, 20; brewers' yeast, 20; milk powder, 20; sugar, 10; vitamin C, 2; sodium benzoate, 0.12 and (mg): vitamin E, 100; Kantrexil (Kanamycin sulphate), 25; and (ml): Ferbelan, 5; water, 250.

Ferbelan (BDH) (mg/5 ml): vitamin B_1, 2; nicotinamide, 5; riboflavin, 0.5; green iron and ammonium citrate, 200; and vitamin B_{12}, 2.5 μg.

PREPARATION Water, sodium benzoate, Ferbelan, vitamin E and Kanamycin stirred well for 1-2 minutes. Remaining ingredients mixed and whole diet homogenised and dispensed to boxes or cups.

REARING At 25 ± 1°C and ≯70% RH with natural lighting. Five pairs of adults in 35 x 30 x 30 cm wooden cages. Oviposition on paper changed daily. Fed on 10% sugar and honey solutions. Humidity maintained by placing piece of water soaked hydrophile cotton in cages. Eggs incubated at 25 ± 1°C and high % RH in boxes containing diet; reared individually from 3rd instar in 1-oz plastic cups to prevent cannibalism.

DEVELOPMENT Incubation period, 70 h; larval period, 12-25 days; pupal period, 10-18 days. Total generation time, 35-45 days. Seven successive generations reared.

Nachiappan, R.M.; Subramaniam, T.R. 1973. Studies on mass-culturing of *Heliothis armigera* Hbn. (Noctuidae: Lepidoptera) on semi-synthetic diets. I. Testing with various type of semi-synthetic diets. MADRAS AGRIC J 60: 561-564.

Best of 6 diets tested.

COMPOSITION (g): Bengal gram flour, 5.0; dried brewers' yeast, 0.5; ascorbic acid, 0.1; agar, 0.2; and distilled water, 30 ml.

PREPARATION See Shorey & Hale* (1965). Agar strips dissolved in 15 ml boiling water. Cooled for 5 minutes and then other components added with remaining water. 0.5 ml of 40% formaldehyde and traces of streptomycin sulphate added as antimicrobial agents. Diet mixed and poured into clear, sterile petri dishes. Stored at 5°C until use.

DEVELOPMENT Larval survival to adult, 76%. Adult emergence comparable to that from host plants - tomato or lab-lab (*Dolichos lablab*).

Patel, R.C.; Patel, J.K.; Patel, P.B.; Singh, R. 1968. Mass breeding of *Heliothis armigera* (Hbn.). INDIAN J ENTOMOL 30: 272-280.

COMPOSITION (g): powdered lucerne, 100; agar, 3; and (ml): inhibitor stock solution, 5; acetic acid, 5; multivitamin syrup, 0.5; water, 300; and formaldehyde, 4 drops.

inhibitors: sorbic acid, 4.0 g; methyl *p*-hydroxybenzoate, 3.0 g; ethyl alcohol, 34.0 ml.

Diet improved by addition of 1% HCl, 0.25% acetic acid, and 0.25% ascorbic acid.

PREPARATION Agar dissolved in ca. 150 ml boiling water then mixed with lucerne powder. Inhibitor stock solution then added in ca. 75 ml water. Cooked for 30 minutes at 15 psi in pressure cooker, and allowed to cool. Formaldehyde, acetic acid, and vitamin syrup added to remaining water and mixed into cooled diet. Diet dispensed to ½" depth in 3 x 1" plastic specimen tubes.

To prepare lucerne leaf powder, green lucerne fodder air-dried, then shaken vigorously to separate leaves and tender plant parts, ground in mill, sieved through 20-mesh screen and stored in plastic-lined gunny bags.

REARING At 80 ± 2°F, individually in tubes. Pupae collected and transferred to 12" cube cages containing moist soil; three sides of cage fitted with plastic sheets and 4th with muslin cloth with sleeve for introduction of pupae and removal of adults. 15 ♂ and 5 ♀ adults/cage, fed with 5-10% honey solution on cotton wool plugs. On 3rd day ♀♀ taken out of cage and confined to 4" diam and 2" high plastic container lined with blotting paper. Eggs laid on blotting paper.

DEVELOPMENT Average larval period, 20.71 days; average pupal period, 12.24 days; emergence from pupae, 89.57%. Fecundity, 1142.3 ± 360.6 eggs/♀.

Twine, B.H. 1971. Cannibalistic behaviour of *Heliothis armigera* (Hübn.). QUEENSL J AGRIC ANIMAL SCI 28: 153-157.

COMPOSITION Semi-artificial diet prepared in 2 parts. Part 1 (g): soaked navy beans (dry beans soaked in water over night), 235; methyl *p*-hydroxybenzoate, 2.2; yeast, 35; ascorbic acid, 3.5; sorbic acid, 1.1;

and (ml): 10% formaldehyde, 8.8; water, 350. Part 2: agar, 14 g; and water, 350 ml.

PREPARATION Agar dissolved in boiling water, cooled to about 70° C, then mixed with part 1 ingredients previously blended. Diet dispensed to vials, or 4 x 9 cm diam plastic containers and refrigerated until use.

REARING At 26.6 ± 0.6° C in darkness.

→ Vandamme, P.; Angelini, A. 1968. Comparaison de trois milieux nutritifs artificiels pour l'élevage d'*Heliothis armigera* Hubn. COTON FIBRES TROP 23: 417-422. (In French with English summary)

Best of 3 diets tested.

COMPOSITION (g): wheat germ, 11; brewers' yeast, 10; Alphacel, 1.8; Wesson's salt mixture, 3.6; ascorbic acid, 1.5; agar, 9.0; and (ml): choline chloride (0.1 g/ml water), 3.6; methyl *p*-hydroxybenzoate (15% in 95% ethanol), 3.6; chlortetracycline, 50; water, 310.

PREPARATION Solids, except ascorbic acid and chlortetracycline dissolved in 110 ml distilled water, and blended for 2 minutes. Methyl *p*-hydroxybenzoate plus choline chloride added and blended for further 2 minutes. Agar dissolved in 200 ml boiling water, and then ascorbic acid and chlortetracycline added. Whole blended for 3-4 minutes. Dispensed to sterile polystyrene tubes @ 15 ml/tube.

REARING At 27° C and 85% RH. Eggs sterilised in 0.2% NaClO for 10 minutes. Rinsed in sterile water and incubated until eclosion. Neonate larvae transferred to diet tubes with sterile forceps @ 2 larvae/tube.

DEVELOPMENT Pupation, 71%; 60% adult emergence after 26.8 days. Results comparable with rearing on cotton squares and bolls.

See also: Khlistovskiy (1965).

Heliothis virescens (Fabricius), tobacco budworm

Heliothis zea (Boddie), bollworm, corn earworm, tomato fruitworm

→ Berger, R.S. 1963. Laboratory techniques for rearing *Heliothis* species on artificial medium. U S DEP AGRIC PRESENTATION PAP ARS 33-84, 4 pp.

COMPOSITION Water, 880 ml; 22.5% KOH, 18 ml; casein, vitamin-free, 126 g Wesson's salt mixture, 36 g; sucrose, 126 g; 10% formaldehyde, 13 ml; 15% methyl *p*-hydroxybenzoate in 95% ethyl alcohol, 36 ml; 10% choline chloride, 36 ml; wheat germ, 108 g; Alphacel, 18 g; vitamin suspension, 6 ml; agar, hot, dissolved in 2200 ml of boiling water, 90 g; ascorbic acid, 14.4 g; and Aureomycin, 500 mg.

vitamins (mg/ml water): niacin, 6.0; calcium pantothenate, 6.0; riboflavin, 3.0; thiamine hydrochloride, 1.5; pyridoxine hydrochloride, 1.5; folic acid, 1.5; biotin, 0.12; B_{12}, 0.12.

PREPARATION Ingredients blended in above order. Dispensed while hot @ 15-20 g/8-dram shell vials.

REARING At 80° F with constant lighting, individually in vials. Pupae washed in NaClO and placed in 1-gal wide-mouthed jars on moist vermiculite for emergence. 10 pairs adults transferred to lantern jar for oviposition on cheesecloth cover to jar. Fed 5% sugar solution from cotton wad in petri dish. High humidity essential. Eggs surface sterilized on cheesecloth for 5 minutes in 0.15% NaClO, then sodium thiosulphate added and cloth washed in distilled water. Incubated in jar covered with paper towelling at 75°F for 3-4 days. Larvae also reared in 20 x 150 mm petri dishes with 75-100 g diet in pieces.

DEVELOPMENT Complete.

Corollary 1

Raulston, J.R.; Lingren, P.D. 1969. A technique for rearing larvae of the bollworm and tobacco budworm in large numbers. J ECON ENTOMOL 62: 959-961.

COMPOSITION & PREPARATION Modified diet of Berger* (1963) for *Heliothis* spp. Modifications include addition of methyl *p*-hydroxybenzoate in solid form instead of in alcohol solution and incorporation of 40 ml 25% glacial acetic acid/300 ml of diet. Diet poured into 61 x 30.5 x 4 cm aluminium tray with 1.5 cm lip. Cardboard hexacel matching tray size with 1.9 cm diam cells (ca. 625 cells) coated previously with fibreglass resin inserted into diet before setting.

REARING Ca. 1000 1st instar larvae scattered over hexacel and capped with wooden-frame lid 61.0 x 30.5 x 1.27 cm covered with plastic sheet. For 5-6 days lid placed so that sheet uppermost for larvae to disperse. Then frame inverted to isolate larvae in hexacel to avoid cannibalism. Pupae shaken from hexacel. Unit then cleaned and sterilized with 5% formalin solution. Unit illustrated.

DEVELOPMENT Yield of *H. zea* pupae 397.6/unit with range of 330-506/unit. Yield of *H. virescens* pupae 438.2/unit with range of 328-546/unit.

Corollary 2

Raulston, J.R.; Shaver, T.N. 1970. A low agar casein-wheat germ diet for rearing tobacco budworms. J ECON ENTOMOL 63: 1743-1744.

Modified from Berger* (1963) for rearing *Heliothis* spp. Modifications: (1) methyl *p*-hydroxybenzoate solid instead of in ethanol solution; (2) 40 ml 25% glacial acetic acid added; (3) 12 ml vitamin suspension used; (4) total water reduced from 3080 to 3000 ml; (5) agar reduced from 90 to 75 g; and (6) Alphacel omitted, and corn cob grits (Grit O'Cobs, -60 grade, Anderson Cob Mills, Box 119, Maumee, Ohio) substituted (350 g/3.4 l of diet). Later agar reduced to 20 g/3.4 l of diet. Ingredients mixed in Waring Blendor, corn cob grits added after blending.

REARING At 27-29°C and 50% RH. Diet poured into 61 x 31.5 x 4 cm aluminium trays (see Raulston & Lingren*, 1969). Method modified. Hexacel replaced by polystyrene light diffusion grid with 800, 1.2 cm^3 cells glued together to form rearing units. Unit inoculated with 1 ml (7000) 1-day-old eggs and capped with layer of polypropylene cloth, layer polyurethane foam and another layer of cloth using wooden frame and clamped to trays. Larvae kept separate. On 17th day, pupae harvested and fecundity tested in 1-pint oviposition chambers. Adults fed 5% sucrose solution. Larvae also reared in ¾-oz plastic cups.

DEVELOPMENT Compared well with high agar diet of Berger* (1963); 25% cheaper. Five generations reared. Average of 584 eggs laid/♀; average longevity: ♀♀, 17.3 days; ♂♂, 17.8 days.

See also: Brewer & Martin (1976); Sell & Bodznick (1971).

Corollary 3

Shaver, T.N.; Raulston, J.R. 1971. A soybean-wheat germ diet for rearing the tobacco budworm. ANN ENTOMOL SOC AM 64: 1077-1079.

COMPOSITION (ml): water, 3000; 4M KOH, 18; choline chloride (15%), 25; formaldehyde (10%), 15; acetic acid (25%), 40; and (g): Nutrisoy flour, 242; wheat germ, 108; Wesson's salt mixture, 36; sucrose, 44; methyl *p*-hydroxybenzoate, 5.4; sorbic acid, 3.25; agar, 15; corncob grits, 250; and (mg): Aureomycin, 500; vitamin solution, 12; total, 3.4 l.

vitamin solution: see Raulston* (1971).

PREPARATION See Raulston & Shaver* (1970).

DEVELOPMENT Reared for 2 generations with no deleterious effects on growth or development. Cost of diet reduced from US$0.41 to $0.24/l by using soybean flour instead of casein, and substituting corncob grits for agar. Mass rearing method described.

See also: Raulston (1975); Shaver & Lukefahr (1969); Vanderzant (1975); Vinson (1967).

Brazzel, J.R. 1970. Second conference on test methods for resistance in insects of agricultural importance. Standard method for detection of insecticide resistance in *Heliothis zea* (Boddie) and *H. virescens* (F). Tentative methods for detection in *Diabrotica* and *Hypera*. BULL ENTOMOL SOC AM 16: 147-153.

COMPOSITION (g): casein, 35; Wesson's salt mixture, 10; wheat germ, 30; methyl paraben, 1.5; agar, 25; Alphacel, 5; sugar, 35; ascorbic acid, 4; sorbic acid (dissolved in ethanol), 1; choline chloride, 1; and (ml): formaldehyde (40%), 1; Vanderzant's fortification mixture, 15; water, 850.

PREPARATION Water boiled, dry ingredients added and blended for 2 minutes; then liquid ingredients added and blended for 2 minutes. Dispensed to plastic cups.

See also: Brazzel *et al.* (1961).

Burton, R.L. 1969. Mass rearing the corn earworm in the laboratory. U S DEP AGRIC PRESENTATION PAP ARS 33-134, 8 pp.

COMPOSITION (g): pinto beans, 6900 (3360 g dry beans soaked overnight and drained); dried torula yeast, 1024; wheat germ, 1600; ascorbic acid, 104; methyl *p*-hydroxybenzoate, 64; sorbic acid, 32; 10% formaldehyde, 256; water in blender, 10400; agar, 408; water for dissolving agar, 10080.

PREPARATION Beans preferably soaked overnight; but if short of time, coarsely fragmented and added dry. Same amount of water added so that total volume reduced. Diet prepared in 10-gal blender and processed using machine, cups, and trays described by Burton & Cox (1966), except that larval dispenser and capper omitted. Good alternative dispensing technique for smaller amounts described by Patana (1967).

REARING At 29° C and 80% RH in 1-oz plastic cups for 8-10 days, then 29° C and 40% RH until pupation. Pupae collected using machine of Harrell *et al.* (1973), and held in shallow cardboard trays at 75° C in wrapping paper lined room. Adults collected at 55° F using modified vacuum cleaner. 10 pairs adults placed in 1-gal icecream containers with wrapping-paper liners. Oviposition on cotton sheeting cover. Fed on beer soaked on cotton in disposable cup. Eggs surface sterilized with 0.157% NaClO and 10% sodium thiosulphate. Dried and dispensed to diet cups @ 2-6 eggs/cup.

DEVELOPMENT Complete. Cost; $16/1000 containers.

→ Burton, R.L. 1970. A low-cost artificial diet for the corn earworm. J ECON ENTOMOL 63: 1969-1970.

COMPOSITION (g): water in blender, 930; 10% formaldehyde, 4; CSM, 374; torula yeast, 20; ascorbic acid, 7; methyl *p*-hydroxybenzoate, 4; sorbic acid, 2; agar, 26; and water to dissolve agar, 630.

CSM, Blended Food Product, Child Food Supplement, Formula No. 2 (%): corn meal, gelatinised, 63.8; soy flour, defatted, toasted, 24.2; non fat dry milk, 5.0; soy oil, refined, stabilised, 5.0; vitamin-mineral premix, 2.0.

PREPARATION Ingredients, except agar and water for agar, mixed in heavy duty blender. Agar and remaining water boiled, added and mixture blended again.

REARING See Burton* (1969); Burton & Harrell (1966).

DEVELOPMENT Oviposition and % egg hatch comparable to those on control bean diet; 11 generations reared. 100 adults reared/1 diet at cost of 13 cents/1.

See also: Burton & Snow (1970).

→ Burton, R.L.; Perkins, W.D. 1972. WSB, a new laboratory diet for the corn earworm and the fall armyworm. J ECON ENTOMOL 65: 385-386.

Also for fall armyworm, *Spodoptera frugiperda* (J.E. Smith) (Lep., Noctuidae).

COMPOSITION (g): water, 930; WSB, 375; torula yeast, 20; ascorbic acid, 7; sorbic acid, 2; methyl *p*-hydroxybenzoate, 4; 10% formaldehyde, 4; agar, 25; and water to dissolve agar, 630.

WSB (%): wheat fractions, 73.35; soy flour, 200.00; soybean oil, stabilised, 4.00; mineral premix, 2.55; vitamin premix, 0.10.

PREPARATION Ca. 12 g diet dispensed to 1-oz plastic cups.

REARING At 29° C and high humidity for first 8 days to prevent desiccation of diet. Larvae reared individually in cups; pupae placed in cages for emergence; 5 pairs/cage mated.

DEVELOPMENT *H. zea:* larval period, 13.43 days; pupation, 92%; emergence, 81.3%; average eggs/♀, 1328.

S. frugiperda: larval period, 13.45 days; pupation, 89.2%; emergence, 91.9%; average eggs/♀, 1901.

Eight generations reared. Cost of 12 cents/1. In both species, artificial diet WSB, superior to previous bean diet.

Other lepidopteran spp. reared on WSB medium include sugarcane borer, *Diatraea saccharalis* (Fabricius) (Pyralidae); European corn borer, *Ostrinia nubilalis* (Hübner) (Pyralidae); cabbage looper, *Trichoplusia ni* (Hübner) (Noctuidae); and tobacco budworm, *Heliothis virescens* (Fabricius) (Noctuidae).

See also: Hare *et al.* (1973).

→ Hardwick, D.F. 1965. The corn earworm complex. MEM ENTOMOL SOC CAN 40: 247 pp.

COMPOSITION Adult diet (oz): yellow sugar, 16; Barbados molasses, 40; clover honey, 16; masticated preserved peaches, 28; water, 40; and dried bread yeast, 1-1½ tablespoons.

PREPARATION Ingredients thoroughly mixed and allowed to ferment for 24 h. Strained, boiled for 3 minutes and stored in sterilized containers.

REARING At 25 ± 0.5°C. Two pairs adults to oviposition jars. Fed via wick from vials replaced every 2nd day. Eggs laid on blotting paper collected every 2nd day and put into film cans with pea pods. Cans sealed with tape and food added every 2nd day. 3rd and 4th instar larvae reared in shell vials with chopped peas. Mature larvae placed in vials with moistened peat moss. Pupae kept on damp moss in candy jar until emergence.

DEVELOPMENT Species reared: *Helicoverpa armigera* (Hübner), *H. assulta* (Guenée); *H. confusa* (Hardwick); *H. gelotopoeon* (Dyar); *H. hawaiiensis* (Quaintance & Brues); *H. punctigera* (Wallengren); *H. zea* (Boddie); and other species of *Helicoverpa* group.

See also: Lukefahr & Martin (1964).

→ Herzog, G.A.; Phillips, J.R. 1974. Selection for a nondiapause strain of the bollworm, *Heliothis zea* (Lepidoptera: Noctuidae). ENVIRON ENTOMOL 3: 525-527.

COMPOSITION (g): casein, 120; Alphacel, 18; Wesson's salt mixture, 36; sugar, 126; wheat germ, 150; choline chloride, 4; methyl parasept, 6; ascorbic acid, 15; Aureomycin, 18; and (mg): Sulphamethazine, 500; and (ml): water, 3032; 10 M acetic acid, 75; vitamin B mixture, 100.

vitamins (mg/1000 ml): nicotinic acid, 1000; calcium pantothenate, 1000; riboflavin, 500; thiamine hydrochloride, 250; pyridoxal hydrochloride, 250; folic acid, 250; and (ml): biotin (50 mg/100 ml water), 40; vitamin B_{12} (100 μg/ml), 2; and choline chloride, 5 g.

REARING At 20° ± 2°C with 10 h photoperiod, in cups.

DEVELOPMENT Strain developed that reduced incidence of diapause from 64.2% to 5.65% after 14 generations.

→ Ignoffo, C.M. 1966. Insect viruses. pp. 501-530. *in* C.N. Smith (Ed.) Insect Colonization and Mass Production. Academic Press, New York and London.

Also for cabbage looper, *Trichoplusia ni* (Hübner) (Noctuidae).

COMPOSITION (ml): distilled water, 3100; methyl *p*-hydroxybenzoate (15% w/v in 95% ethyl alcohol), 36; choline chloride (0.1 g/ml of water), 36; 4M KOH, 18; formalin (0.1 g/ml), 15; vitamin stock, 6; and (g): casein, 126; sucrose, 126; wheat germ, 108; leafmeal (only for *Trichoplusia* diet), 54; agar, 90; Wesson's salt mixture, 36; Alphacel, 18; ascorbic acid, 15; antibiotic, 0.5.

vitamins (mg/100 ml water): niacin, 600; calcium pantothenate, 600; riboflavin, 300; thiamine, 150; pyridoxine, 150; folic acid, 150; biotin, 12; B_{12}, 1.2.

PREPARATION Agar dissolved in 2200 ml of boiling water; casein mixed with 880 ml water and 18 ml KOH in 1-gal blender. All solids, except ascorbic acid and antibiotics, added and mixture blended for 2-3 minutes. Dissolved agar added, followed by ascorbic acid and antibiotic. Entire mixture blended for 2 minutes. Vitamins added and mixture blended again. Hot liquid diet dispensed to rearing containers and allowed to solidify prior to seeding with either larvae or eggs.

REARING *Heliothis virescens* and *H. zea*. At 30°C; 2-3 noenate larvae brushed on to 10-25 ml diet in cups; cups capped and stacked. After 14 days pupae removed, washed in 0.2% NaClO for 15 minutes and placed on vermiculite in shallow cardboard trays. Trays placed in emergence cages at 24-26°C. Adults kept in large lantern globe with base of pie pan covered with 8 x 8" cheesecloth. Adults fed 5 or 10% sterile sugar solution; cage topped with 8 x 8" oviposition pad kept moist with filter paper. Adults at 23-25°C and 30-90% RH. Oviposition pads collected daily and soaked for 5-7 minutes in 0.2% NaClO; 10% sodium thiosulfate added and free liquid removed. Eggs transferred to waxed paper cups and incubated at 30°C.

Trichoplusia ni. At 30°C; 10 neonate larvae transferred to rearing container covered with cheesecloth. Pupae placed in emergence cages after washing with 1 part 1% NaClO to 1 part 5% Na_2CO_3 for 30 minutes. Adults immobilized with CO_2 and placed in cages. Oviposition cage cylindrical, wrapped in paper towelling, and kept at 25°C and 30% RH. Gravity feeding device drops, 5 or 10% sugar solution on absorbent paper pad on top of cage. Eggs removed from towelling as for *Heliothis*.

DEVELOPMENT *Heliothis virescens* and *H. zea*: Incubation period, 3-4 days; larval period, 14 days. Weight of adults: ♀♀, 249 mg; ♂♂, 208 mg. Cost/larva, US $0.07.

T. ni: Incubation period, 2-3 days; larval period, 12 days; pupal period, 7 days. Adult longevity, 7-10 days. Cost/larva, US $0.01.

See also: Ignoffo (1965); Ignoffo & Boening (1970).

Oliver, B.F.; Maxwell, F.G.; Jenkins, J.N. 1970. Utilization of glanded and glandless cotton diets by the bollworm. J ECON ENTOMOL 63: 1965-1966.

Modified from Daum *et al.* (1969) for *Anthonomus grandis*.

COMPOSITION (g): agar, 9; lyophilised cotton square powder, 30; and (ml): mould inhibitor, 7 (sorbic acid, 20 ml, methyl parasept, 30 ml, ethanol, 170 ml); water, 3000; and Calco Oil Red N-1700 dye, 1 g.

PREPARATION Agar, dye, mould inhibitor, and water mixed in Waring Blendor. Cotton square powder added and blended until dye dispersed homogenously throughout mixture. Ca. 7 ml poured into each plastic container and covered with perforated plastic caps.

Patana, R.; McAda, W.C. 1973. Tobacco budworms: use of dry diet flakes in rearing. J ECON ENTOMOL 66: 817-818.

COMPOSITION See Patana* (1969).

PREPARATION Diet flakes made by pouring hot diet on to 182.9 x 45.7 cm sheets of waxed paper resting on 57.8 x 182.9 x 51.1 cm wood-framed rack covered with 0.6 cm mesh hardware cloth, and spreading it to depth of ca. 0.3 cm with 22.9 cm smooth paint roller. Papers sandwiched 5 deep. Racks then placed in drying chamber. Temperature in drying chamber 48.9-54.5° C.

REARING At 28° C. See Patana (1967). 50-150 neonate larvae/box.

DEVELOPMENT From 50 larvae/box: larval period, 14 days; pupal wt : ♀, 284 ± 38 mg; ♂, 283 ± 35 mg. Fecundity, 1247 ± 536 eggs/♀. Method in use for ca. 3 years.

Rock, G.C.; Hodgson, E. 1971. Dietary amino requirements for *Heliothis zea* determined by dietary deletion and radiometric techniques. J INSECT PHYSIOL 17: 1087-1097.

Two diets tested.

COMPOSITION Diet 1: *l*-amino acids (g/100 g): arginine (free base), 0.135; histidine (free base), 0.108; isoleucine (allo free), 0.156; leucine (methionine free), 0.243; lysine hydrochloride, 0.186; methionine, 0.120; phenylalanine, 0.114; threonine, 0.126; tryptophan, 0.135; valine, 0.204; alanine, 0.246; aspartic acid, 0.276; cystine, 0.045; glutamic acid, 0.285; proline, 0.153; serine, 0.126; tyrosine, 0.090; glycine, 0.252.

others: vitamin-free casein, 0.8; sucrose, 3.0; Wesson's salt mixture, 1.0; safflower oil, 0.8; cholesterol, 0.2; Vanderzant's vitamin mixture, 1.0; ascorbic acid, 0.5; cellulose, 5.0; agar, 3.0; anti-microbial agents, 1.2 ml; and water to 100 g.

antimicrobials: sorbic acid, 20 g; methyl *p*-hydroxybenzoate, 15.0 g; dissolved in 170.0 ml ethanol.

Diet 2: (g): vitamin-free casein, 3.0; wheat germ, 4.0; sucrose, 0.5; Wesson's salt mixture, 1.0; Alphacel, 1.0; Vanderzant's vitamin mixture, 0.5; agar, 3.0; and distilled water, 50.0; total, 63.0 g.

PREPARATION No details; prepared in 150 g batches and dispensed @ 6-8 g/vial.

REARING At 27° C with 16 h photoperiod individually in vials. 10-12 day old larvae transferred to new vials with 5 g food.

DEVELOPMENT Diet 1: pupation, 80%; adult emergence, 62.5%; larval period: ♂, 22.2; ♀, 22.8 days; pupal wt: ♂, 442.0; ♀, 462.5 mg.

Diet 2: pupation, 81.5%; adult emergence, 61.7%; larval period: ♂, 20.7; ♀, 20.8 days; pupal wt: ♂, 391.4; ♀, 406.7 mg.

Although larval growth rates, pupal weights, and survival to pupal and adult stages on diet 1 comparable to those on diet 2, most adults reared on diet 1 showed wing and/or body deformities.

See also: Chou *et al.* (1973).

→ Vanderzant, E.S.; Richardson, C.D.; Fort, S.W. Jr 1962. Rearing of the bollworm on artificial diet. J ECON ENTOMOL 55: 140.

COMPOSITION (g): cold-rolled wheat germ, 3.0; vitamin-free casein, 3.5; sucrose, 3.5; Wesson's salt mixture, 1; choline chloride, 0.1; ascorbic acid, 0.4; agar, 2.5; and (ml): B-vitamins (with inositol), 1; water, 85.

vitamins (mg/100 g diet): niacinamide, 1.00; calcium pantothenate, 1.00; thiamine hydrochloride, 0.25; riboflavin, 0.50; pyridoxine hydrochloride, 0.25; folic acid, 0.25; biotin, 0.02; vitamin B_{12}, 0.002; inositol, 20.00.

For non-aseptic rearing, 0.2 g potassium sorbate and 0.2 g methyl *p*-hydroxybenzoate included as antimicrobial agents.

PREPARATION Diet mixed, sterilized and cooled to 40°C before ascorbic acid added; dispensed to sterile rearing vials.

REARING Eggs at brown ring stage surface sterilized and 1 egg placed in each vial (Vanderzant *et al.** 1962). After 1 week larvae transferred to larger vials with fresh diet. Pupae moved to cages for emergence. Adults kept in screened cages at 85°F with 13½ h dim light, and fed combinations of sugars, honey, ascorbic acid with Cotton flowers.

DEVELOPMENT Adult yield, 67-88%. Larval period, 12 days; pupal period, 9 days; pupal wt, 0.41 g, or, if antimicrobial agents added, 0.39 g. Total developmental time, 25 days.

See also: Brewer & Tidwall (1975).

→ Vanderzant, E.S. 1968. Dietary requirements of the bollworm, *Heliothis zea* (Lepidoptera: Noctuidae), for lipids, choline, and inositol and the effect of fats and fatty acids on the composition of the body fat. ANN ENTOMOL SOC AM 61: 120-125.

COMPOSITION (g): casein, vitamin-free, 4.0; cysteine hydrochloride, 0.1; sucrose, 4.0; Wesson's salt mixture, 1.0; corn oil, 0.5; cholesterol, 0.05; choline chloride, 0.10; inositol, 0.04; β-carotene, water soluble, 0.025; ascorbic acid, 0.4; cellulose powder, 5.0; agar, 3.0; and (ml): cobalt, molybdenum and zinc salts (Vanderzant* 1959), 0.5; fat-soluble vitamins, 1.0; B-vitamins, 2.0; water, 80.0. pH adjusted to 6.2 with 10% KOH.

fat-soluble vitamins (mg): α-tocopherol acetate, 10.0; vitamin D_3, 0.5; and menadione in ethanol, 0.5.

B-vitamins: see Vanderzant* (1959).

PREPARATION All ingredients, except ascorbic acid, mixed thoroughly and heat sterilized. Filter-sterilized aqueous solution of ascorbic acid then added, and diet dispensed to small sterile vials.

Concentrations of fat, cholesterol, choline, and inositol changed as required.

REARING At 28 ± 1°C with 13½ h photoperiod. Eggs at brown-ring or black-head stage surface sterilized with 0.15% NaClO for 3 minutes, rinsed in sterile distilled water and placed individually in vials. After 1 week, larvae transferred aseptically to 20 x 70 mm vials with fresh diet. Pupae moved to gallon jars covered with screen or cloth. Moths fed diluted honey or sucrose solution with ascorbic acid and eggs collected on cloth placed in jars.

DEVELOPMENT 2.5 mg cholesterol and 25 mg choline chloride/100 g diet required for normal development. Linseed oil added to promote growth and emergence; yield of adults from larvae, 95%. Average pupal wt: ♀, 0.420 g; ♂, 0.421 g.

See also: Barnett & Berger (1970a b).

Hydroecia micacea (Esper), potato stem borer

→ Chawla, S.S.; Perron, J.M.; Cloutier, M. 1975. Preliminary observations on an artificial diet for *Hydroecia micacea* (Esp.) (Lepidoptera: Noctuidae). ANN ENTOMOL SOC QUE 20: 8-10.

COMPOSITION (g): dry corn leaf powder, 10.0; casein, 14.0; soybean flour, 40.0; brewers' yeast, 13.3; Wesson's salt mixture, 1.0; sucrose, 10.5; sorbic acid, 0.83; wheat germ, 12.0; Alphacel, 6.0; vitamin diet fortification mixture (NBC), 4.0; ascorbic acid, 2.8; agar, 13.5; and (ml): water, 285; 15% methyl *p*-hydroxybenzoate, 8.0; 22.5% KOH, 4.0; 36.7% formaldehyde, 1.4; and water (with agar), 284.0.

PREPARATION See Chawla & Harwood* (1968).

REARING In thick cardboard disposable salad dishes at 23°C for 14 h light then 16°C for 10 h dark. Final instar larvae transferred to fresh diets to avoid stress and starvation. Diapause eggs incubated for 10 days; neonate or 7-day-old larvae transferred to diet.

DEVELOPMENT Larval period, 39-88 days; pupal wt: ♂, 560 mg; ♀, 695 mg; pupal period, 3-4 weeks; 40% pupation of 7-day-old larvae compared with 10% of neonate larvae. Adults fertile and fecund. Dietary reared pupae heavier than those reared on corn plants.

Mamestra brassicae Linnaeus [previously *Barathra*], cabbage army worm

→ Agui, N.; Ogura, N.; Okawara, M. 1975. Rearing of the cabbage armyworm *Mamestra brassicae* L. (Lepidoptera: Noctuidae) and some lepidopterous larvae on artificial diets. JPN J APPL ENTOMOL ZOOL 19: 91-96. (In Japanese with English summary)

Two diets tested.

COMPOSITION Diet 1 (g): agar, 7.5; potato starch, 5.0; sucrose, 5.0; soy bean flour, 20.0; Wesson's salt mixture, 1.75; K_3PO_4, 0.5; citric acid, 0.25; dried brewers' yeast, 7.5; cholesterol, 0.25; soy bean oil, 1.5; powdered filter paper, 11.25; *l*-ascorbic acid, 2.0; finely chopped cabbage leaves, 65.0; and (ml): propionic acid, 0.35; water, 100.

Diet 2: as diet 1, except 11.25 g powdered filter paper replaced by 20 g wheat germ; and cabbage leaves increased from 65.0 to 75.0 g.

PREPARATION For non-aseptic rearing, solid ingredients mixed; cabbage leaves pulped with water in blender and, with soy bean oil and propionic acid, mixed uniformly with solid ingredients. Autoclaved at 0.5-0.6 kg/cm^2 for 20-25 minutes, stirred and water-cooled. For aseptic rearing, 100 g diet as prepared above placed in 200 ml conical flask, stoppered with cotton wool and wax-paper, and then again autoclaved.

REARING At 25 ± 2°C with 16 h photoperiod. Aseptically from 1st to 5th instar. Eggs sterilized in 70% ethanol and 0.1% $HgCl_2$ and placed in diet flasks on sterilized filter paper. 1-2 days before moulting

5th instar larvae transferred to sterile plastic dish and given 3.3 g of diet daily/10 larvae in early 6th instar, and 7 g in late instar. Pupation between damp paper towels.

DEVELOPMENT *Mamestra brassicae* reared for 2 generations on diets 1 and 2; mean larval and pupal weights greater on diet 1, and pupal weights and pupal period on both artificial diets superior to those on cabbage leaf diets. *Plutella xylostella* Linnaeus (Lep., Yponomeutiidae) (on diets 1 and 2) and *Mamestra illoba* Butler (Lep., Noctuidae) (on diet 1) reared for 2-3 generations. Other species reared on diet 1 for one generation only:- *Pieris rapae crucivora* Boisduval (Lep., Pieridae), *Pieris melete* Menetries (Lep., Pieridae), *Ancylolomia japonica* Zeller (Lep., Pyralidae), and *Spilarctia seriatopunctata* Motschulsky (Lep., Arctiidae).

→ David, W.A.L.; Gardiner, B.O.C. 1966a. Rearing *Mamestra brassicae* (L.) on semi-synthetic diets. BULL ENTOMOL RES 57: 137-142.

COMPOSITION See David & Gardiner* (1965, 1966b), diet for rearing *Pieris brassicae*.

REARING At 20 ± 1° C and 60 ± 5% RH with 18 h photoperiod in 30 x 75 mm tubes or 1-lb jam jars with diet renewed weekly. 5.5 cm diam foil cups half full of peat in 2-lb jam jars for pupation. Adults kept in ventilated plastic boxes or small cages with paper wad soaked in 10% sugar or honey solution. Walls lined with paper for oviposition.

DEVELOPMENT Larval development as good as that on fresh cabbage leaves. Pupal weights and percent pupation higher on diet. Diet in which dried cabbage replaced by linseed oil as satisfactory as diet with dried cabbage. Average days to pupation, 28; pupation, 95-100%. Average pupal wt, 385 mg; and adult emergence, 92-98%.

→ Kamano, S. 1964a. Rearing the cabbage armyworm, *Barathra brassicae* Linné, on artificial diets. JPN J APPL ENTOMOL ZOOL 8: 340-342. (In Japanese)

COMPOSITION Water, 100.0 ml; and (g): agar, 3.0; powdered filter paper, 5.0; sucrose, 2.0; glucose, 1.0; casein, 2.0; dried yeast, 2.0; Wesson's salt mixture, 0.3; cholesterol, 0.1; choline chloride, 0.2; sodium ascorbate, 0.3; Na-Cu-chlorophene? (translator not sure), 0.01; dehydro-sodium-acetate, 0.05; sodium sorbate, 0.15; finely chopped cabbage leaves, 20.0. Diet also without cabbage leaves.

PREPARATION Ingredients except sodium-ascorbate mixed in 200 ml beaker, covered with wax paper and sterilized for 30 minutes in Koch's steam sterilizer. Cooled to below 50° C and sodium ascorbate, dissolved in sterilized water, added; cooled until set and refrigerated until use.

REARING At 23-24° C with long-day photoperiod. Diet renewed twice in first 10 days, then at 2-3 day intervals. 1st-5th days, 100-200 larvae, and 6th-10th days, 40-50 larvae/9 cm petri dish; 11th-13th days, 25-30 larvae and thereafter 15 larvae/12 cm petri dish. Pupation in sand under natural temperature conditions for 2 months to break diapause, then returned to 23-24° C for emergence. One pair placed in petri dish on 'semi-rice-paper' with cotton-wool soaked in 1% sucrose, for egg-laying.

DEVELOPMENT i) On artificial diet with cabbage-leaf: length of larval period, and larval and pupal weights, almost equal to control on cabbage leaf diet; fecundity higher; but mortality, especially in pre-pupal stage, 3 times that of control.

ii) On artificial diet without cabbage-leaf: development to pupal stage, but high mortality, and adult emergence incomplete.

→ Maleki-Milani, H. 1971. Élevage de la noctuelle du chou, *Mamestra (Barathra) brassicae* L. sur milieu artificiel. ANN SOC ENTOMOL FR 7: 39-56. (In French with English summary)

COMPOSITION (g): casein, (lactic soluble), 17.00; sucrose, 15.00; powdered cabbage, 7.5; Wesson's salt mixture, 5.2; cellulose powder (Whatman CF 11), 3.0; brewers' yeast (dry powdered), 18.0; wheat germ, 15.0; cholesterol, 0.7; Nipagin M, 2.3; ascorbic acid, 3.0; agar, 10.0; and (ml): choline chloride (10% aqueous soln), 5.0; distilled water, 400.

PREPARATION First 7 ingredients homogenised (Mix A). Nipagin and cholesterol ground and dissolved in 25 ml 95% ethanol, stirred for 20 minutes (Mix B). A and B mixed thoroughly and alcohol evaporated. Agar dissolved in 230 ml boiling water; 140 ml water mixed with choline chloride; AB progressively added till mixture homogenous. Hot agar then added and stirred. When cooled to 50°C, ascorbic acid in 30 ml water added. Whole diet homogenised for 3-4 minutes and dispensed to rearing dishes at pH of 5.5.

REARING At 20 ± 1°C and 70% RH or 25 ± 1°C and 75% RH both with 17 h photoperiod. Eggs sterilised in 4% NaClO and incubated in petri dishes. Neonate larvae reared individually in 80 ml volume dishes. Aeration through gauze-covered hole in lid. Rearing dishes stored in larger sealed boxes to maintain even humidity until 5th instar when transferred to similar but well aerated boxes. Diet replaced either once or several times. On foliage, first three instars reared on *Plantago lanceolata*, last three on *Brassicae oleracea*. At end of larval development, large boxes half-filled with sawdust for pupation. Adults reared in pairs, each in plastic cylinders 8 x 9 cm diam. Roll of paper of same dimensions, covered with stretched muslin held by rubber band inserted inside cylinder. Eggs laid on muslin or paper surface. Adults fed on 10% sugar solution from piece of sponge, renewed daily. Eggs collected daily.

DEVELOPMENT Following data for rearing on foliage, and artificial diet at 20°C and 25°C respectively: larval period (days): 47, 51; 33.5, 35. Pupal wt (mg): 438 ± 12, 455 ± 12; 431 ± 13, 438 ± 15. Emergence (%): 97, 100; 94, 96. Average fecundity (eggs/♀): 1378, 1718; 1700, 1566. No relation between ♀ pupal weight and number of eggs laid. Sterility after 4th generation.

Peridroma saucia (Hübner), variegated cutworm

→ Harper, J.D. 1970. Laboratory production of *Peridroma saucia* and its nuclear polyhedrosis virus. J ECON ENTOMOL 63: 1633-1634.

COMPOSITION As in Shorey & Hale* (1965), except pinto beans replaced by lima beans.

REARING At 25°C. Adult ♀♀ placed in wide mouth 1-gal jars containing petri dish with cotton saturated in 8% sugar solution, and 1 or 2 folded paper towels in pleats for oviposition. Eggs rubbed from towels and surface sterilized in 0.8% Clorax solution for 30 minutes; rinsed, air dried and incubated at 25°C or 4°C for storage. 20 larvae transferred to 12 x 15 x 50 mm diet plugs in petri dishes at 90-100% RH until 4th instar, when diet replaced. 6th instars placed in 1-gal jars @ 30/jar, with 2.5 cm sand wetted with 100 ml water and covered with 5 cm soil. Several leaves of curly dock (*Rumex crispus*) added to jar daily. After 4-5 days foliage removed and emerging adults transferred to oviposition jars.

DEVELOPMENT Egg incubation period, 3-5 days; pupal period, 14 days. Three generations reared; in 4th generation fecundity dropped to 0. When fed diet and dock at 6th instar stage, 11 generations reared.

Plusia acuta Walker, tomato looper

→ Bot, J. 1966a. An artificial diet for the plusia looper, *Plusia acuta* Walker. S AFR J AGRIC SCI 9: 67-70.

COMPOSITION (g): wheat germ, 20.0; vitamin-free casein, 4.0; finely ground dried brewers' yeast (dry live *Saccharomyces utywala africana*), 24.0; powdered agar, 5.0; ascorbic acid, 2.0; inositol, 0.08; choline chloride, 0.08; cholesterol, 0.08; methyl *p*-hydroxybenzoate, 0.8; and 150.0 ml water.

PREPARATION Wheat germ, casein, and yeast thoroughly mixed. Methyl *p*-hydroxybenzoate and cholesterol dissolved in 20 ml 95% ethanol and added; alcohol evaporated. Agar added, followed by inositol, choline, and ascorbic acid dissolved in water. Dispensed to 4 x 1" specimen tubes to depth of ca. ¾"; tubes slanted, stoppered and placed in oven at 90°C for 30 minutes until soft, light brown crust formed on diet. Tubes then cooled in upright position. Tubes with condensation set aside for 24 h to allow equilibrium.

REARING At 25°C, @ 2-3 larvae/tube. Pupae removed to damp vermiculite and kept at 25°C and 65% RH until emergence. Adults held in 18-mesh mosquito gauze cages at 75% RH with normal daylight and fed 10% aqueous honey solution. In addition, two 10 cm petri dishes filled with water-soaked vermiculite placed in cage for high humidity. Eggs laid on gauze; neonate larvae brushed into vials.

DEVELOPMENT Duration of larval stage favourably compared with that on natural food. On diet; larval period, 14-18 days; pupal period, 7-9 days. 16 generations reared.

Prodenia eridania (Cramer), southern armyworm

→ Redfern, R.E.; Raulston, J.R. 1970. Improved rearing techniques for the southern armyworm. J ECON ENTOMOL 63: 296-297.

Modification of diet of Berger* (1963) for *Heliothis* spp.

COMPOSITION (ml): water, 666; 4M KOH, 18; and (g): vitamin-free casein, 126; and (ml): water, 66; and (g): Wesson's salt mixture, 36.0; sucrose, 126.0; Alphacel, 18.0; methyl parasept, 10.8; and (ml): water, 66; vitamin solution, 12; 15% choline chloride, 21; 10% formaldehyde, 15; 25% acetic acid, 40; raw linseed oil, 8.5; and agar, 70-75 g in 2300 ml water; ascorbic acid, 14.5 g; Aureomycin, 500 mg.

vitamins (g/1000 ml water): calcium pantothenate, 12.0; niacin, 6.0; riboflavin, 3.0; folic acid, 3.0; thiamine hydrochloride, 1.5; pyridoxine hydrochloride, 1.5; biotin, 0.12; B_{12}, 6.0.

PREPARATION Ingredients blended in order listed above in 1-gal Waring Blendor at low speed. Agar boiled in water first.

REARING At 27°C with 12 h photoperiod. Ten pairs adults in 15 x 20 cm battery jars lined with waxed paper and covered with unbleached muslin. Fed on 10% sugar solution. Oviposition on fresh lima bean foliage. Eggs collected daily, dipped in 3-10% solution of Captan, dried on blotting paper and incubated in 10 x 15 cm petri dishes with filter paper and wad of moist cotton. At black head stage, 500-1000 transferred to 15 x 15 cm petri dish with artificial diet, covered with 2, 16.5 cm rapid flow milk filter discs faced with single gauze layer and placed smooth side down. After 5-6 days bottom filter disc transferred to fresh diet. 4th instar larvae transferred to large aluminium pans with diet. Bottom lined with paper towelling and sides painted with petroleum jelly. Mature larvae transferred to pan with moist vermiculite and small amount of diet, covered with unbleached muslin. After 7-9 days pupae separated by sieving and placed in battery jars with moist vermiculite for emergence. N-shaped tongue depressed over cheesecloth for adults to expand wings.

DEVELOPMENT Sufficient eggs produced by 10-20 pairs to maintain culture of 1000-2500 larvae/week. Complete generation reared in 32-36 days, with no diapause.

See also: Redfern (1967).

→ Rehr, S.S.; Feeny, P.P.; Janzen, D.H. 1973. Chemical defence in central American non-ant-acacias. J ANIM ECOL 42: 405-416.

Diet modified from Feeny* (1968) developed for *Operophtera brumata*.

COMPOSITION (g/300 g): cellulose, 3.0; cholesterol, 0.3; linolenic acid, 0.15; sucrose, 3.75; Wesson's salt mixture, 0.3; wheat germ, 0.75; choline chloride, 0.15; methyl *p*-hydroxybenzoate, 0.3; ascorbic acid, 0.45; vitamin-free casein, 3.0; agar, 3.75; leaf powder, 15.0; and (ml): vitamin solution, 1.5; water, 267.0.

vitamin solution: see McMorran* (1965).

leaf powder: leaves of test plant freeze-dried and pulverized in Waring Blendor. Leaves of *Phaseolus vulgaris* used in control.

PREPARATION Cellulose, cholesterol and linolenic acid mixed with little ethyl ether and allowed to dry. Cellulose mixture, sucrose, Wesson's salt mixture and wheat germ combined with half of water in blender, and agar, dissolved in remaining water, added at 65°C. Diet blended, dispensed and stored at 4°C until use.

REARING 4th instar larvae reared individually in plastic pots containing excess of diet at 21°C day/19°C night and 30-40% RH with 16 h photoperiod. Pots half filled with damp vermiculite immediately prior to pupation.

See also: Elliot (1955).

→ Soo Hoo, C.F.; Fraenkel, G. 1964a. The resistance of ferns to the feeding of *Prodenia eridania* larvae. ANN ENTOMOL SOC AM 57: 788-790.

COMPOSITION (g): vitamin-free casein, 1.5; bactoagar, 0.55; Wesson's salt mixture, 0.15; brewers' yeast, 0.1; cholesterol, 0.04; choline chloride, 0.01; Solka floc, 40-mesh cellulose, 0.25; powdered cellulose, pure vegetable fibre, 0.75; water, 20.

PREPARATION Dry ingredients mixed together, water added, and heated at 80°C for 5 minutes in water bath. Hot slurry then poured into 15 cm petri dish and cooled. 1 cm discs cut from hardened diet and laid singly on wax paper in 5 cm petri dish bottom lined with moist filter paper.

REARING One newly moulted starved 5th instar larva/petri dish.

DEVELOPMENT Food eaten within 2 h. Frass count, 12.1 in 12 h.

See also: Soo Hoo & Fraenkel (1964b, 1966).

Prodenia litura Fabricius, Egyptian cotton leaf worm, tobacco cutworm

→ Levinson, Z.H.; Navon, A. 1969. Ascorbic acid and unsaturated fatty acids in the nutrition of the Egyptian cotton leafworm, *Prodenia litura*. J INSECT PHYSIOL 15: 591-595.

Two diets tested.

COMPOSITION Diet 1 (%): ground soy germs (aged 24 h), 5.0; vitamin-free casein, 3.5; sucrose, 3.5; salt mixture, 1.0; choline chloride, 0.1; B-vitamins, 0.06; *l*-ascorbic acid, 0.5; cholesterol, 0.3; wheat germ oil, 0.3; agar, 2.5; methyl *p*-hydroxybenzoate, 0.2; propyl *p*-hydroxybenzoate, 0.05; and distilled water, 83.0.

Diet 2 (%): α-soy protein, 1.5; vitamin-free casein, 3.5; sucrose, 5.0; salt mixture, 2.0; choline chloride, 0.1; B-vitamins, 0.09; *l*-ascorbic acid, 0.5; cholesterol, 0.3; 99% methyl linoleate, 0.56; 99% methyl linolenate, 0.1; *dl*-α-tocopherol, 0.02; polysorbate 80, 0.16; potassium sorbate, 0.14; agar, 2.5; cellulose powder, 4.8; methyl *p*-hydroxybenzoate, 0.2; propyl *p*-hydroxybenzoate, 0.05; and distilled water, 79.0.

salt mixture according to Beck *et al.** (1968).

B-vitamins: for diet 1 (mg/g diet): *i*-inositol, 0.4600; calcium pantothenate, 0.0850; niacin, 0.0330; *p*-aminobenzoic acid, 0.0155; riboflavin, 0.0048; pyridoxine, 0.0037; thiamine, 0.0028; folic acid, 0.0011; biotin, 0.0008; and vitamin B_{12}, 0.0002. Levels increased by one-third for diet 2.

REARING Neonate larvae reared to adults aseptically on both diets at 25 ± 0.5°C and 60-70% RH. Larvae given 1 cm^3 diet to 4th instar and 2 cm^3 cubes after 4th instar. Diet prepared every 4th day; food cubes renewed daily and vermiculite provided for pupation. Single adult pairs kept in plastic cages with wrapping paper for oviposition; fed 10% sucrose solution.

DEVELOPMENT On both diets almost all larvae developed normally. Average pupal wt: ♀, 377 mg; ♂, 338 mg; adult yield, 85%. Omission of ascorbic acid resulted in poor larval growth and death within 3 weeks. Omission of wheat germ oil resulted in slow growth, and moths with crippled and scaleless wings; addition of linoleic and linolenic acid largely compensated for lack of oil.

Sato, Y. 1965. The artificial diet for mass rearing of the tobacco cutworm, *Prodenia litura* Fabricius and the common armyworm, *Leucania separata* Walker. JPN J APPL ENTOMOL ZOOL 9: 99-105.

COMPOSITION (g): wheat bran, 100; chemical substances, 10.

chemical substances (g): cellulose powder, 50; sucrose, 25; soluble starch, 25; glucose, 25; casein, 50; Wesson's salt mixture, 5; cholesterol, 1; sodium *l*-ascorbate, 1; and (mg): choline chloride, 120; inositol, 120; thiamine, 10; riboflavin, 10; nicotinic acid, 20; pyridoxine, 10; folic acid, 10; calcium pantothenate, 20; *p*-aminobenzoic acid, 10; biotin, 0.5.

PREPARATION Ingredients mixed, 20 ml water added to 10 g dry dietary mixture and again thoroughly mixed.

REARING At 25 ± 1°C with 24 h photoperiod. Larvae reared for first 3 days on host plant in 30 x 26 x 6 cm plastic box. Thence individually in 3 x 6.5 cm diam plastic dish with holes in lid. Sterilized soil for pupation placed in container when larvae mature. Adults transferred to 30 cm cube wooden boxes and fed 10% sugar solution. Oviposition of *P. litura* on paper, and of *L. separata* on corn plant leaf.

DEVELOPMENT *P. litura:* pupation, 32%; emergence, 21%; fecundity, 2.5 egg masses/♀.

L. separata: pupation, 14%; emergence, 14%; fecundity, 1.5 egg masses laid/♀.

See also: Neelgund & Mathad (1974).

Tsukuda, R.; Kiyoku, M. 1966. Laboratory cultures of the tobacco-cutworms *Prodenia litura* Fabricius on the synthetic media. REP FAC AGRIC OKAYAMA UNIV 28: 13-18. (In Japanese with English summary)

COMPOSITION (g): bacto agar, 0.4; cellulose powder, 1.0; casein, 1.0; glucose, 0.5; sucrose, 0.5; yeast extract (Ebios), 1.5; choline chloride, 0.15; Wesson's salt mixture, 0.3; cholesterol acetate, 0.05; sodium carboxy-methyl-cellulose, 0.5; *l*-ascorbic acid, 0.2; sorbic acid, 0.08; sodium propionate, 0.08; and water, 55.0 ml. Sometimes peanut oil or vitamin E included.

PREPARATION Sodium sorbate dissolved in ether and absorbed on to cellulose powder; ether evaporated. Vitamins and sodium propionate dissolved in 10 ml of water. Agar dissolved in 45.0 ml water over water bath; mixture of dry ingredients except choline chloride stirred in for 2-3 minutes, then choline chloride added and stirring continued. Cooled to below 70°C, and solution of vitamins, etc. added. Stored at 5°C. (Diet not autoclaved).

REARING At 26°C, from second instar larvae. Neonate to second instar larvae reared on sweet-potato leaves in 12 cm petri dishes. 2nd instar transferred to 8 x 9 cm diam petri dishes @ 30 larvae/dish. Covered with paper; diet changed every 4th day for early stages, every 3rd day for mid-stages, and every second day or daily for mature larvae.

DEVELOPMENT Pupation, 50-70%; complete life cycle in ca. 5 weeks; infertile eggs laid by dietary reared adults.

Corollary

Kiyoku, M.; Tsukuda, R. 1966. Successive laboratory cultures of the tobacco-cutworms *Prodenia litura* Fabricius on some simple semi-synthetic media. REP FAC AGRIC OKAYAMA UNIV 28: 1-11. (In Japanese with English summary)

DEVELOPMENT Diet 1, soybean diet: pupation, 46.7%; emergence, 56.6%; adult yield, 23.0%; larval period, 21.9 days; pupal period, 10.8 days; duration of egg stage, 3.9 days; pupal wt, 327.30 mg. No. of eggs/♀, 1171.0; emergence from eggs, 84.6%.

Diet 2, kidney bean diet: pupation, 36.9%; emergence, 36.9%; adult yield, 22.5%; larval period, 26.4 days; pupal period, 11.9 days; duration of egg stage, 4.0 days; pupal wt, 343.20 mg. No. of eggs/♀, 1057.6; emergence from eggs, 81.6%.

Diet 3, control (sweet potato): pupation, 83.4%; emergence, 96.0%; adult yield, 80.0%; larval period, 14.1 days; pupal period, 10.6 days; duration of egg stage, 4.0 days; pupal wt, 325.0 mg. No. of eggs/♀, 968.0; emergence, 94.2%.

Seven successive generations reared.

Pseudaletia unipunctata (Haworth), armyworm

Goodwin, R.H. 1966. Axenic culture of the armyworm *Pseudaletia unipunctata* (Haworth) on a meridic diet. NATURE LONDON 212: 799-800.

COMPOSITION Fraction A (g): agar, 2; soluble casein, 3.5; and folic acid, 0.5 mg; and water, 72 ml. Fraction B (g): linseed oil, 0.5; polysorbate 80, 0.3; and α-tocopherol, 25 mg. Fraction C (g): wheat germ flour, 3.0; Wesson's salt mixture, 1.3; Alphacel, 1.8; sucrose, 3.5; β-sitosterol, 0.2; cholesterol, 0.2. Fraction D: 4M KOH, 1.5 ml (sterile). Fraction E (ml): vitamin solution, 2.0; water, 8.0; and ascorbic acid, 2.0 g.

vitamin solution (g): choline chloride, 10; inositol, 2.0; and (mg): nicotinic acid, 100; calcium pantothenate, 100; riboflavin-5-phosphate, 50; thiamine hydrochloride, 25; pyridoxine hydrochloride, 25; biotin, 2.0; cyanocobalamin, 0.2; water, 100 ml.

PREPARATION Fraction A warmed and shaken, and blended with B for 1 minute. C components mixed dry then AB, and C and D autoclaved for 15 minutes at 118°C and mixed in sterile room. Cooled to 40°C and filtered E added and all mixed in blender. Dispensed from sterile flask in 10 ml lots to autoclaved rearing dishes and bottles.

REARING At 27°C with 16 h photoperiod. Neonate larvae hatched in 125ml cotton stoppered bottles and fed towelling strips coated with diet until 3rd instar, when reared in 100 x 20 mm deep glass petri dishes floored with towelling circles; 0.125" thick cardboard discs taped on to top; ½-oz wax paper cups with diet in dishes. For axenic rearing eggs surface sterilixed in 10% formalin for 20 minutes, 0.1% 'Triton X-100' for 30 minutes, and 0.5% NaClO for 10 minutes and rinsed 3 times in sterile distilled water for 5 minutes each. Food changed when 6th instar larvae. Rearing chambers incubated in sterilised plastic boxes ventilated through filter. Axenic adult pairs reared in pint cardboard cartons with vermiculite and small waxed paper cups filled with cotton and filter sterilized feeding solution of (ml): honey, 5; larval vitamin solution, 2; 4M KOH, 1.5; water, 100; ascorbic acid, 2 g. Oviposition in accordion shaped wax paper folds.

DEVELOPMENT Life cycle completed in 40 days. Four generations reared axenically.

Spodoptera exigua (Hübner) (Lep., Noctuidae), *Prodenia ornithogalli* Guenée (Lep., Noctuidae) and *Peridroma saucia* (Hübner) (Lep., Noctuidae) also reared axenically on diet. For *P. saucia* diet soybean oil used instead of linseed oil. *Colias eurythema* Boisduval (Lep., Pieridae) also reared if 1.5% royal jelly added to diet with soybean oil.

Pseudoplusia includens (Walker), soybean looper

→ Jensen, R.L.; Newsom, L.D.; Gibbens, J. 1974. The soybean looper: effects of adult nutrition on oviposition, mating frequency and longevity. J ECON ENTOMOL 67: 467-470.

COMPOSITION Larval diet - as in Burton (1967), except all ingredients reduced ⅛ weight or volume; torula yeast replaced by brewers' yeast and 200 ml addition water added to enhance pouring quality.

REARING At 26 ± 5°C. Oviposition cages from open ended 0.57 l cylindrical cartons. Cages placed on paper towels on moist vermiculite in 45 x 65 x 3 cm trays, and covered with "Kimwipe" held by petri dish cover. 8 x 23 cm strip of paper towel placed in cage for oviposition; fed on 10% honey and sugar solution via cotton wicks. Fresh cotton blossom and soybean leaves also provided. Food replaced every 5 days. Kimwipes replaced daily and inner liners every 2 days.

DEVELOPMENT Reared from egg to adult. Fecundity, 510 eggs/♀. Adult longevity: ♀, 8.6 days; ♂, 9.2 days.

See also: Kogan & Cope (1974).

Scotia segetum Dennis & Schiffermüller [*Scotia* synonymous *Agrotis*]

→ Weismann, L.; Podmanická, D. 1970. Einfluss der Nahrung und der Zuchttemperatur auf die Länge der Larvalentwicklung von *Scotia segetum* Den. et Schiff. BIOLOGIA BRATISLAVA 25: 537-545. (In German with English summary)

Best of 5 diets tested.

COMPOSITION (g): casein, 3.5; pangamin, 3.5; saccharose, 3.5; cellulose powder, 3.0; flour extracted from crushed soybean, 5.0; vitamin mix, 2.0; ascorbic acid, 0.1; choline chloride, 0.1; cholesterol, 0.3; Wesson's salt mixture, 2.0; agar-agar, 1.0; preservative solution, 2.0 ml; distilled water, 100.0 ml.

PREPARATION Diet components blended; agar solution cooled to 55°C and incorporated into diet. Poured into 16-cm diam petri dishes; stored at 2-5°C.

REARING Complete larval development in 3-6 cm diam petri dishes at three temperatures; 18, 25 and 30°C. 1 x 3 x 3 cm diet plugs replenished as necessary.

DEVELOPMENT Shortest larval period, 21.7 days at 30°C.

→ Podmanická, D.; Weismann, L. 1974. Effects of gradually increased doses of B-group vitamins on the development and vitality of *Scotia segetum* Den. & Schiff. BIOLOGIA BRATISLAVA 29: 641-648. (In Slovak with English summary)

Best of many diets tested.

COMPOSITION (g): boiled beans, 200; glucose, 5; mixture of vitamins A, C and D in glucose, 10; sodium benzoate, 1.5; B-vitamins in glucose, 10; agar, 20; and (ml): preserving solution, 22; distilled water, 735; B-vitamins/100 g diet, 29.88.

preserving solution (g): sorbic acid, 20; Nipagin, 15; and ethyl alcohol, 170 ml.

B-vitamins (mg): *p*-amino-benzoic acid, 300; niacin, 300; riboflavin, 150; pyridoxin, 150; thiamine, 90; calcium pantothenate, 210; biotin, 13; folic acid, 60; inositol, 660; glucose, 248063 g (sic).

PREPARATION Agar boiled and cooled to 50° C. Vitamins and glucose dissolved in some water and added to agar, followed by remaining ingredients. Mixed well. Poured into sterilized 12 cm diam petri dishes, cooled and kept at 2-5° C until use.

REARING At 25 C° in 3-6 cm petri dishes until pupation. Adults in 5-1 jars lined with corrugated paper; fed 20% glucose solution.

DEVELOPMENT Average pupal wt, 132.7 ± 4.1 mg.

Sesamia cretica Ledeber, sugarcane borer

→ Salama, H.S.; Tolba, R.A. 1971. Development of the sugarcane borer, *Sesamia cretica* Led. on a semi-artificial diet. Z ANGEW ENTOMOL 68: 74-75.

COMPOSITION (g): corn plant powder, 13; agar, 2; cholesterol, 0.25; ascorbic acid, 0.5; yeast, 4.0; sucrose, 1.5; glucose, 1.5; mashed potatoes, 10.0; and water, 100.0.

PREPARATION Potatoes boiled, mashed, and blended with other ingredients, except agar, in water. Agar separately dissolved in boiling water, cooled to 60° C and then added. 2.5 ml sorbic acid in ethanol (22.5%) used as mould inhibitor. Two-oz aliquots of diet poured into glass tubes and plugged with cotton.

REARING Eggs isolated from *Nerium oleander* plants. Sections of leaves glued to internal surface of glass tubes with 2-oz diet and plugged with cotton.

DEVELOPMENT Mean larval period on diet, 37 days; on corn, 17 days. Mean pupal period, pupal wt, and adult longevity equal on both. Complete development in ca. 50 days on diet, and ca. 29 days on corn. Adults reared on diet laid infertile eggs.

Sesamia inferens Walker, pink stem borer, pink rice borer

→ Chatterji, S.M.; Sharma, G.C.; Siddiqui, K.H.; Panwar, V.P.S.; Young, W.R. 1969. Laboratory rearing of the pink stem borer, *Sesamia inferens* Walker, on artificial diet. INDIAN J ENTOMOL 31: 75-77.

COMPOSITION & PREPARATION See Chatterji *et al.** (1968) diet for maize stem borer, *Chilo zonellus*.

DEVELOPMENT Survival from larva to adult, 60-70%; complete life cycle in ca. 115 days. Developmental data comparable to that when reared on maize stem. Three generations reared.

Qureshi, Z.A.; Anwar, M.; Ashraf, M.; Chata, N.U.; Arif, M.D. 1972. Rearing, biology and sterilization of the pink rice borer, *Sesamia inferens* Walker. JPN PESTIC INF 10: 129.

COMPOSITION (g): rice stem powder, 10.0; agar, 5.0; casein, 7.0; sucrose, 7.0; brewers' yeast, 6.0; Vanderzant's fortification mixture, 4.0; ascorbic acid, 2.0; cholesterol, 0.6; Wesson's salt mixture, 2.0; Nipagin, 0.2; and (ml): linseed oil, 0.6; distilled water, 170.0.

REARING At 27°C and 80% RH.

DEVELOPMENT Life cycle completed in 42-67 days; 194-400 eggs laid/♀.

Sesamia nonagrioides (Lefebure)

Cavalloro, R.; Prota, R. 1972. Metodo di allevamento di *Sesamia nonagrioide* (Lef.) (Lepidoptera, Noctuidae) con diete artificiali. REDIA 53: 395-411. (In Italian with English summary)

Best of seven diets tested.

COMPOSITION (g): ground cornstalks' meal, 195; sugar, 50; debittered brewers' yeast, 50; enzymatic brewers' yeast hydrolysate, 18; vitamin mixture, 3; mineral salt mixture, 1; 2N HCl, 60; Nipagin, 2.5; potassium sorbate, 0.5; and (ml): water, 620.

DEVELOPMENT Larval development in ca. 5 weeks; pupal development in ca. 10 days; 85% survival from larva to pupa. Adults normal. Fecundity and fertility similar to that of wild female.

Spodoptera exempta (Walker), African armyworm, nut grass armyworm, day feeding armyworm

David, W.A.L.; Ellaby, S.; Taylor, G. 1975. Rearing *Spodoptera exempta* on semi-synthetic diets and on growing maize. ENTOMOL EXP APPL 18: 226-237.

Two diets tested.

COMPOSITION Diet 1 (g): casein (light white soluble), 18.9; wheat germ, 16.2; sucrose, 18.9; agar (fine), 13.5; dried maize or grass leaf powder, 8.1; salt mixture, 5.4; cellulose (Whatman CF11 grade), 2.7; choline chloride, 0.5; *l*-ascorbic acid, 2.25; methyl *p*-hydroxybenzoate, 0.8; Aureomycin powder, 1.2; and (ml): 4M KOH, 2.7; vitamin stock, 1.0; *dl*-α-tocopherol acetate, 0.08; linseed oil, 1.5; corn oil, 1.5; formalin (10% w/v soln H.CHO), 2.25; water, 470.

Diet 2: see Bot* (1967).

PREPARATION Leaf powders prepared by drying and grinding. Linseed oil, corn oil, vitamin E added. See David & Gardiner* (1966a) for details. Thoroughly blended diets dispensed to sterilized 75 x 30 mm diam glass tubes to 1 cm depth. When cooled, tubes stored at -20°C in plastic bags.

REARING At 25 ± 1°C and 60 ± 5% RH with 16 h photoperiod. Neonate larvae transferred to diet @ 50-100/tube. Sealed with filter paper held by rubber band and laid on tray with base towards light. Diet changed weekly. 3rd instar larvae reared @ 3/tube. Mature larvae transferred to jars with moist, sterilized soil and small block of diet. Adults kept in 16 x 11 cm diam hurricane lamp glass cages with plastic petri dish base. Food from 2.4 cm tube in hole in base consisting of 5 ml honey, 2 ml vitamin stock, 2 g ascorbic acid, and 0.25 ml of 4% vitamin E in ethanol, made up to 100 ml with water. Eggs laid on paper slips, and removed daily to 1-lb glass jar covered with filter paper. Jar laid on side with base towards light to induce larvae to move to bottom and thus prevent escape.

DEVELOPMENT Eclosion in 74 h, 50%. 6 instars in larval stage taking 17.2 ± 0.3 days. Some larvae to 5th instar only 15.2 ± 0.2 days. Pupal period, 10.4 ± 0.3 days. Pupal yield, ca. 90%; adult yield, 77-91%. Average fecundity, 602 eggs/♀. Six generations reared.

Corollary

→ David, W.A.L.; Ellaby, S. 1975. The viability of the eggs of the African armyworm, *Spodoptera exempta* in laboratory cultures. ENTOMOL EXP APPL 18: 269-280.

Two diets tested.

COMPOSITION Diet 1: see David *et al.** (1975).

Diet 2 (g): casein (light white soluble), 18.9; wheat germ, 16.2; sucrose, 10.8; brewers' yeast, 8.9; agar, 13.5; dried maize leaf powder, 8.1; salt mixture, 5.4; cellulose powder, 2.7; cholesterol, 0.5; inositol, 0.3; *l*-ascorbic acid, 5.0; Aureomycin powder, 1.2; and (ml): choline chloride (10% soln), 5.4; 4M KOH, 2.7; vitamin stock, 2.0; *dl*-α-tocopherol acetate, 0.08; linseed oil, 1.5; corn oil, 1.5; methyl *p*-hydroxybenzoate (15% w/v soln), 5.4; formalin (10% w/v soln H.CHO), 1.5; water, 463.

DEVELOPMENT After 9 generations egg viability reduced to 25% of original due to inbreeding and not diet. No difference between two diets used.

Spodoptera exigua (Hübner), beet armyworm

→ Cobb, P.P.; Bass, M.H. 1975. Beet armyworm (Lepidoptera: Noctuidae): a laboratory rearing technique. J GA ENTOMOL SOC 10: 190-191.

COMPOSITION (g): wheat germ, 210; brewers' yeast, 200; vitamin-free casein, 32; cellulose, 20; ascorbic acid, 15; methyl paraben, 10.4; sorbic acid, 4.0; choline chloride, 1.6; Aureomycin, 1.0; cholesterol, 0.8; inositol, 0.8; linolenic acid, 15.0; agar, 38; distilled water, 2300 ml.

PREPARATION Wheat germ boiled in 1300 ml water. Agar boiled in 1000 ml water and added while hot to all other ingredients; blended and immediately dispensed to larval containers. Ready for use after 2 h.

REARING At 78-85°F with 10 h photoperiod. Neonate larvae placed individually in 1-oz plastic cups with 1 cm^3 diet plug. Cups closed with paper lids, plastic coated on inside surface. Pupation in cups. 25 pupae placed in wide mouthed gallon jars with mouth covered by fine net cloth. Waxed paper strip suspended from rim for oviposition. Aluminium foil cup with cotton moistened with 15% honey-water mixture placed in jar. Cup and paper strip changed every 2 days; eggs placed in pint jars covered with thin porous paper.

DEVELOPMENT Complete. 2.3% larval deformity after 5-10 generations.

→ Loaharanu, S.; Chiravatanapong, S.; Sutantawong, M.; Kaochnong, P. 1974. Mating competitiveness of the radiosterilized male armyworm, *Spodoptera exigua* Hb., and the male mosquito, *Aedes aegypti* L., in field cages. pp. 317-323. *in* Sterility Principle for Insect Control 1974. Proceedings of a Panel Organised by the Joint FAO/IAEA Division of Atomic Energy in Food and Agriculture, Vienna.

Modified from Shorey & Hale* (1965) for noctuid species.

COMPOSITION (g): soaked mungbean, 106; brewers' yeast, 16; ascorbic acid, 1.6; methyl *p*-hydroxybenzoate, 1.0; agar, 6.4; and (ml): 40% formaldehyde, 1.0; water, 320.

DEVELOPMENT Survival from egg to pupa, 66.9 ± 3.9%; from pupa to adult, 88.1 ± 1.48%; 6 generations reared.

→ Petterson, M.A.; Debolt, J.W. 1975. Rearing the beet armyworm in paraffined paper bags. J GA ENTOMOL SOC 10: 165-168.

COMPOSITION & PREPARATION See Henneberry & Kishaba* (1966a) diet for cabbage loopers, *Trichoplusia ni*.

REARING At 26°C and 40% RH with 12 h photoperiod. 27.9 x 20.3 cm diam oviposition cage of 0.6 cm mesh hardware cloth with one closed end. Round cake pan 3.8 x 21.9 cm diam as base. White cellucotton and 10% sucrose solution in 60 x 15 mm plastic petri dish replenished daily. 35, 1-2-day old virgin ♀♀ and 30 ♂♂ of same age placed in cage. Eggs deposited through mesh on to paper towelling, collected 2, 3 and 4 days after introduction. Eggs surface sterilized by soaking towelling in 27% formalin (10% formaldehyde gas in water) for 30 minutes, rinsed with running tap water for 30 minutes and air dried at 24°C. Larvae reared in 16 x 19.5 x 32 cm paraffined paper bags. Same bag (unparaffined) with top 19 cm cut off used as diet mould. 200 g artificial diet poured into mould and allowed to solidify. Diet slab sealed and stored for 24 h at 24°C and transferred to fresh paper bag with 365 g autoclaved white silica sand. Bags infested either by brushing neonate larvae off towelling or placing about 200 eggs on diet. Bags stored at 26.7°C and 40% RH. Pupae collected by cutting sides of bag 3.8 cm above bottom and emptying on to 0.6 cm mesh hardware cloth screen; treated in 1.83% NaClO solution for 1-3 minutes and rinsed under running tap water several times. Dried on paper towelling at 24°C. Rearing cage illustrated.

DEVELOPMENT Larval development, 14-15 days. Normal adult yield, 110 from 200 eggs if no sand used. Average pupal wt, 122 mg. In diet alone normal adults, 80%; in diet and sand 60% normal.

See also: Stimmann *et al.* (1972).

Spodoptera exempta (Walker), nut grass armyworm, lesser armyworm

S. exigua (Hübner), beet armyworm

S. cilium (Guenée), lawn caterpillar

→ Bot, J. 1967a. An artificial rearing medium for three noctuids of economic importance belonging to the genus *Spodoptera* (Lepidoptera). J ENTOMOL SOC SOUTH AFR 29: 157-160.

COMPOSITION (g): wheat germ, 26.0; vitamin-free casein, 4.0; finely ground dried brewers' yeast (dry live *Sacharomyces utywala africana*), 25.0; powdered agar, 6.0; ascorbic acid, 2.5; inositol, 0.1; cholesterol, 0.1; choline chloride, 0.2; methyl *p*-hydroxybenzoate, 1.3; water, 250.

PREPARATION Wheat germ, yeast, and casein thoroughly mixed; methyl *p*-hydroxybenzoate and cholesterol, dissolved in ca. 25 ml of mixture of 75% acetone and 25% ethyl ether, stirred into wheat germ mixture, and spread out to allow evaporation of solvent. Agar dissolved in 200 ml water. Ascorbic acid, choline and inositol dissolved in remaining 50 ml water and added, together with wheat germ mixture to boiling agar solution. Diet boiled again while stirring and dispensed as described by Bot* (1966).

REARING *S. exempta* adults in 18 mesh mosquito gauze 12" cube cages; fed 5% aqueous honey solution. Moisture from petri dishes filled with vermiculite soaked in water. Oviposition on 10-12 cm diam filter paper discs hung inside cage. Eggs cut out and stored in cotton wool stoppered bottles. Larvae transferred to rearing tubes @ 3-5/tube.

S. exigua adults in large organdie covered glass jar, containing 1" thick layer vermiculite, and cotton wool soaked in 5% honey solution. Oviposition on crumbled pieces of filter paper.

S. cilium adults in large glass jars with small tuft of lawn grass for oviposition.

All adults and eggs kept at 22-28°C and 70% RH with normal daylight. Larval tubes kept at 27°C. When larvae mature, stoppers removed from tubes which stacked horizontally over 3-4" of moist vermiculite for pupation.

DEVELOPMENT Adults comparable with field specimens; mostly larger and more robust. Oviposition and fertility normal. Many generations of all species reared.

Spodoptera frugiperda (J.E. Smith), fall armyworm

→ Bowling, C.C. 1967. Rearing of two lepidopterous pests of rice on a common artificial diet. ANN ENTOMOL SOC AM 60: 1215-1216.

Also for *Diatraea saccharalis* (Fabricius), sugar cane borer (Pyralidae).

COMPOSITION (g): dried pinto beans, 100.0; brewers' yeast, 15.0; ascorbic acid, 1.5; methyl *p*-hydroxybenzoate, 1.0; sorbic acid, 0.5; formaldehyde, 1.0; agar, 6.0; and water, 375.0.

PREPARATION Dried pinto beans soaked in water overnight. Other ingredients, except agar, blended with 250 ml water for 10 seconds. Soaked beans boiled in water, drained, added to mixture and blended for 45-60 seconds. Agar dissolved in remaining 125 ml of water then added immediately to mixture and blended for another 30 seconds. Hot diet then poured into plastic squeeze bottle and dispensed to rearing containers.

REARING At 80 ± 3°F and 85-90% RH in darkness (*S. frugiperda*) or with 14 h photoperiod (*D. saccharalis*). Egg masses placed in 100 ml jars. Neonate larvae transferred to ½-oz glass cups with 5 ml diet. Cups covered with paper disc caps. Pupae removed and 5 pairs placed in 1-gal glass jars with 2" moist soil, paper towelling for oviposition and 5% sugar solution on dental roll for adult food.

DEVELOPMENT *S. frugiperda:* larval and pupal development in 31.2 days; mean pupal wt of both sexes, 432 mg; mean fecundity, 696 eggs/♀.

D. saccharalis: larval and pupal development, 38 days; mean pupal wt of both sexes, 95 mg; mean fecundity, 268 eggs/♀.

Spodoptera mauritia acronyctoides (Guenée), lawn armyworm

→ Takei, G.H.; Tamashiro, M. 1974. Axenic rearing of the lawn armyworm, *Spodoptera mauritia acronyctoides* (Guenée), and synxenic inoculation techniques. PROC HAWAII ENTOMOL SOC 21: 475-483.

Diet modified from Tanada & Chang (1968) for *Colias eurytheme*.

COMPOSITION (g): fresh leaves of napier grass (*Pennisetum purpureum*), 74.0 (wet mass); distilled water, 175.0; dried pinto beans, 24.0; dried brewers' yeast, 7.0; Wesson's salt mixture, 2.0; soybean oil, 0.9; Tween 80, 0.6; alfalfa leaf meal, 3.0; agar, 4.0; ascorbic acid, 3.0; 10% formaldehyde, 0.05; methyl *p*-hydroxybenzoate (15% w/v in 95% ethyl alcohol), 0.15; sorbic acid (7.5% in 95% ethyl alcohol), 0.075; α-tocopherol (2.5% in 95% ethyl alcohol), 0.025; Aureomycin, 0.03.

PREPARATION Freshly cut napier grass blended with distilled water. Brewers' yeast, Wesson's salt mixture, dried pinto beans, alfalfa leaf meal, soybean oil and Tween 80 then blended with napier grass. Agar with 60 ml distilled water added and entire mixture blended rapidly; autoclaved, cooled to 48°C, and vitamins and antimicrobial agents added. Dispensed into sterile 8-dram vials to depth of 1.25-2.5 cm and slanted. Vials capped with sterile cotton plugs, covered with Saran wrap and refrigerated.

REARING Axenically at 26.7 ± 2.2°C and 60 ± 10% RH. 1-2 day old egg masses surface sterilized in 30 ml of 10% formaldehyde solution with 0.05 ml of glacial acetic acid for 50 minutes. Rinsed in sterile distilled water three times and incubated in sterile 8-dram vials (1 mass/vial) stoppered with sterile cotton. Neonate larvae placed on axenic diet, and covered, sterilized, 2-10 mesh vermiculite. Larval development to pupation in vial. Oviposition in autoclaved gallon jars with covers cut open and lined with very fine mesh screen. Jars lined with sterilized paper towels and 12 pupae placed for emergence. Adults fed sterilized honey solution through dental wick. Crumpled, sterilized paper towels added in chamber for oviposition.

DEVELOPMENT Survival significantly improved by addition of formaldehyde to diet. Larval growth on diet without formaldehyde faster than that on diet with formaldehyde. Larval and pupal periods and adult emergence similar with or without formaldehyde. Several generations reared.

Spodoptera littoralis (Boisduval), cotton leaf worm

→ Dimetry, N.Z. 1970. A successful technique for mass rearing of the cotton leaf worm, *Spodoptera littoralis* Boisd. on an artificial diet. Z ANGEW ENTOMOL 66: 408-411.

Diet modified from that of Bowling* (1967).

COMPOSITION (g): dried horse beans, 120; brewers' yeast, 15; ascorbic acid, 1.5; sodium benzoate, 1.0; sorbic acid, 0.5; agar, 6.0; and (ml): 40% formaldehyde, 1.0; water, 325.

PREPARATION Beans soaked in water overnight then boiled. Other ingredients blended with ½ water. Agar dissolved in remaining water, cooled to 60° C, added to mixture with beans, and whole diet blended again; poured into cups and used as needed.

REARING At 25° C. 25 neonate larvae/9 x 7 cm diam cup with 1" diet. Cups covered with muslin and kept on sides for 1 week. Then larvae transferred to fresh diet @ 5/cup. Two ♂♂, and 1 ♀ adult placed in 10 cm diam glass jar with muslin cover, and 10% honey solution for food. Eggs laid on blotting paper lining.

DEVELOPMENT Average larval period (days): ♂, 18; ♀, 17; average pupal period (days): ♂, 10; ♀, 9. About 90% adult emergence from 68% pupation. Adult longevity (days): ♂, 8.5; ♀, 11.5. Larval and pupal periods and adult fertility equal to those on natural host of castor oil leaves. Three generations reared.

→ El-Minshawy, A.M.; Zeid, M. 1972. Rearing the larvae of the cotton leaf worm *Spodoptera littoralis* (Boisd.) on semi-artificial diet. Z ANGEW ENTOMOL 70: 101-104.

COMPOSITION (g): ground beans, 293.0; ground peas, 293.0; bakers' yeast, 44.0; ascorbic acid, 4.4; methyl *p*-hydroxybenzoate, 5.4; sorbic acid, 2.6; formaldehyde, 5.4; sawdust, 20.0; agar, 17.4; and water, 633 ml. Viterra (multi-vitamins and minerals), added at 2.45 g/kg of diet.

PREPARATION Ingredients, except agar, blended with half total water. Agar dissolved in remaining water, cooled to 70° C and mixed with other components. Ascorbic acid and Viterra added immediately and mixed. Diet refrigerated until use.

REARING Larvae reared in 5 x 10 cm diam clear plastic containers covered with muslin, @ 10 larvae/cup. Diet given daily. Sawdust added to cups for pupation. Pupae placed in small tube covered with muslin. Each adult pair placed in 30 cm^3 cage and fed 50% sugar solution from small piece of cotton; eggs laid on leaves of *Nerium oleander*.

DEVELOPMENT Average developmental period, 30 days. Pupal weight on castor oil leaves higher than that on diet. Eggs viable; 1000 eggs laid/♀.

→ Khalifa, A.; Salama, H.S.; El-Sharaby, A.F. 1973. Rearing the cotton leafworm *Spodoptera littoralis* Boisd. on a semiartificial diet. Z ANGEW ENTOMOL 73: 129-132.

COMPOSITION (g): dry kidney beans (*Phaseolus* sp.), 150.0; agar, 6.5; brewers' yeast, 16.0; ascorbic acid, 1.5; 20% sorbic acid in 95% ethanol, 2.0; benzoic acid, 2.0; HCl, 10%, 1.0; formaldehyde, 40%, 1.0; vitamin mix, 2.0.

vitamins (mg/100 g diet): niacinamide, 1.0; calcium pantothenate, 1.0; thiamin hydrochloride, 0.25; riboflavin, 0.50; pyridoxine hydrochloride, 0.25; folic acid, 0.25; biotin, 0.02; vitamin B_{12}, 0.002.

PREPARATION Beans soaked in water for 12 h and then boiled. Ingredients, except agar, blended with water. Agar dissolved in 50 ml water, cooled to 60° C and then mixed with other ingredients.

REARING At 30° C in 250 ml glass jars with 30 g diet; jars with glass lids. 3rd instar larvae transferred to 1-l jars with 60 g diet. Pupation within diet; pupae transferred to jars containing moist sawdust for emergence.

DEVELOPMENT Larval period, 11.9 ± 0.3 days; pupal period, 6.9 ± 0.9 days; wt of ♂ and ♀ pupae, 241.0 ± 13.30 and 267.3 ± 13.20 mg respectively. Eggs laid/♀, 472.4 ± 19.10; egg hatch, 99.42%. Data comparable with those of larvae reared on castor oil leaves.

→ McKinley, D.J. 1970. The laboratory culture and biology of *Spodoptera littoralis* Boisduval. REP ANTI-LOCUST RES CENT 18: 12 pp.

COMPOSITION (g): wheat germ, 24; dried bakers' yeast, 22; casein (light, white, soluble), 4; sucrose, 4; Nipagin M, 1.2; cholesterol, 0.1; agar, 4.25; ascorbic acid, 2.5; choline chloride, 0.2; Glaxo salts DL6, 2.0; and (ml): diethyl ether, 8; acetone, 17; distilled water, 225; Vanderzant vitamin mixture, 3.5; and linoleic acid, 2 drops; methyl linolenate, 2 drops.

PREPARATION Wheat germ, yeast, casein and sucrose mixed by hand, then placed on aluminium foil in fume cupboard. Nipagin M and cholesterol dissolved in diethyl ether and acetone, stirred into dry mixture, mixed well, and spread out on foil to dry. Agar dissolved in 200 ml water, cooled slightly; dry mixture poured into hot agar and blended for 3 minutes, boiled for 5 minutes, and blended again until mixture cooled. Vitamins, salts and lipids, mixed with 25 ml distilled water, blended with mixture; diet then cooled rapidly and dispensed to petri dishes while still liquid.

Vitamin mixture prepared by autoclaving 353 g dry Vanderzant mixture for 15 minutes at 10 lb/in^2 1000 ml distilled water. Refrigerated.

REARING At 25 ± 1° C with 12 h photoperiod. Adults in 35 x 20 cm diam perspex cages covered with muslin, with strip suspended for oviposition. Fed from small pad of cotton wool soaked in 5% sucrose and honey solution, renewed daily. Egg batches on muslin strip surface sterilised with 0.5% NaClO, and transferred to 9 cm diam sterile petri dishes with cabbage leaf washed in 0.5% bleach. (Leaf to provide suitable humidity and food if transfer to diet delayed.) Diet surface scored with sterile needle, and 60-100 neonate larvae transferred to each dish. Diet renewed after 6-8 days. Pupation in cellulose wool; pupae removed when hard, washed in 2% bleach for 10 minutes, dried on paper towel, and placed in sterile petri dishes with slightly moistened cotton wool. 5-6 days later transferred to perspex cage on damp vermiculite for adult emergence.

DEVELOPMENT Larval period, 18-28 days; pupal period, 8-14 days; varying pupal wt between strains: ♂, 0.22-0.30 g; ♀, 0.24-0.44 g. Adult longevity, 6-14 days.

Moore, I.; Navon, A. 1969. Calcium alginate: a new approach in the artificial culturing of insects, applied to *Spodoptera littoralis* (Boisduval). EXPERIENTIA 25: 221-222.

COMPOSITION Fraction A (ml): distilled water, 200; and (g): triethanolamine, 0.7; vitamin-free casein, 10.0; full fat soy, finely ground, 45.0; torula yeast, 35.0; methyl *p*-hydroxybenzoate, 1.3; sorbic acid, 0.7; and dibasic calcium phosphate, 2.7.

fraction B: 300 ml distilled water and 11.0 g sodium alginate.

fraction C: 50 ml distilled water, 5.4 g glucono-δ-lactone, and 3.32 g ascorbic acid.

PREPARATION Ingredients of A blended in order given. B blended separately then mixed with A. Finally, fraction C mixed rapidly with AB and mixture allowed to set. Triethanolamine used instead of KOH so as to dissolve casein and soy protein and to emulsify lipids.

REARING At 25 ± 1° C; food renewed daily.

DEVELOPMENT Adult yield, 64.0% when individually reared. Larval period, 21 ± 0.2 days; pupal wt, 283 ± 12 mg. Sex ratio unity in all of 4 generations reared; fecundity and fertility normal.

Medium also satisfactory for codling moth, *Laspeyresia pomonella* (Linnaeus) (Olethreutidae).

Moore, I.; Navon, A. 1971. Rearing of the cotton worm, *Spodoptera littoralis* (Boisduval), on a calcium alginate medium. pp. 81-86. *in* Application of Induced Sterility for Control of Lepidopterous Populations. Proceedings of a Panel Organised by the Joint FAO/IAEA Division of Atomic Energy in Food and Agriculture.

COMPOSITION (g): triethanolamine, 28.0; vitamin-free casein, 200.0; full-fat soy meal, 450.0; torula yeast, 350.0; methyl *p*-hydroxybenzoate, 13.0; calcium carbonate, 6.0; sodium hexametaphosphate mixture (Calgon), 7.0; sodium alginate, 110.0; ascorbic acid, 35.5; and (ml): glacial acetic acid, 67.0; distilled water, 5500.0.

PREPARATION Casein dispersed with triethanolamine in 4800 ml water until pH of about 7. Calgon dissolved in 200 ml water and added, followed by sodium alginate. Soy, Nipagin, and calcium carbonate mixed in (calcium carbonate added last to delay gelling). Ascorbic and acetic acids in 500 ml water then very rapidly mixed in (in ca. 15 seconds) and diet allowed to set for 15 minutes. Stored at 10° C on grid to allow draining of any synergetic fluid.

REARING At 22-27° C and 60% RH in 25 x 33 x 7.5 cm plastic boxes with flat rims. Empty boxes as lids with base replaced by gauze. Rearing box divided into 40 cells, each cell with 5 larvae and 1-2 g diet, renewed every 3 days. Pupation in sawdust; pupae removed to 46 x 26 cm paper bags @ 10 pairs/bag for adult emergence. Fed solution of 10% sucrose, 0.5% ascorbic acid and 0.1% yeast extract from inverted vial in neck of bag. Egg masses collected and adults transferred to new bags every 2 days.

DEVELOPMENT Larval period, ca. 30 days; with 60-67% survival to adult. Adults normal, laying viable eggs. Three generations for individually reared larvae, and 10 generations for larvae reared together.

See also: Ishaaya *et al.* (1971).

→ Moore, I.; Navon, A. 1973. A new gelling technique and improved calcium-alginate medium for rearing *Spodoptera littoralis*. J ECON ENTOMOL 66: 565-566.

COMPOSITION (g): soybeans, 450; triethanolamine, 20; vitamin-free casein, 200; sodium alginate, 130; torula yeast, 350; $CaCO_3$ (precipitated), 10; cholesterol, 6; methyl *p*-hydroxybenzoate, 13; ascorbic acid, 32.5; powdered cellulose, 60; and (ml): distilled water, 5400; glacial acetic acid, 65.

PREPARATION Soybeans autoclaved with equal volume of water at 15 lb/in^2 for 20 minutes, then ground when cool. Casein dissolved in triethanolamine in some water. Sodium alginate dissolved in more water. Yeast, $CaCO_3$, cholesterol, methyl *p*-hydroxybenzoate blended with remaining water. Then all ingredients mixed together into smooth paste. Glacial acetic acid homogenously absorbed on cellulose by stirring. Mixture deep frozen. When needed ground in mortar and poured on to surface of paste together with ascorbic acid. Mixed at slow speed for 45 seconds, and allowed to set.

REARING See Moore & Navon* (1971).

DEVELOPMENT Mean pupal yield and wt/100 neonate larvae, 25.0 ± 4.5 and 327 ± 8.7 mg for ♂♂, and 22.3 ± 1.2 and 377.0 ± 20.3 mg for ♀♀. Thirty generations reared successfully.

→ Sundaramurthy, V.T.; Subramaniam, T.R. 1969. A laboratory technique for mass rearing of the tobacco caterpillar *Spodoptera littoralis* (Boisd.), *Prodenia littura* (F.) (Noctuidae: Lepidoptera) on synthetic diet. CURR SCI 23: 569-570.

COMPOSITION (g): soaked peas (*Pisum sativum*), 50; yeast tablet (Alembic) 2; and (mg): agar, 300; ascorbic acid, 500; calcium pantothenate, 150; nicotinic acid, 100; folic acid, 50; pyridoxin hydrochloride, 50; β-sitosterol, 50; methyl paraben, 50; and 40% formalin, 10 drops; and water, 80 ml.

PREPARATION Agar boiled with 50 ml water. Seedcoat removed from peas, and peas blended in 30 ml water with other ingredients. Agar stirred into food mixture; diet dispensed in small quantities to sterile 7½ x 2½ cm test tubes.

REARING At 85 ± 2°F @ 10 neonate larvae/tube, inverted for first few days. Larvae separated after 5 days and kept @ 2/tube until 10th day when reared individually. Fresh food supplied every 5 days.

DEVELOPMENT *S. littoralis:* larval survival, 92%; larval period, 14.59 days; pupal period, 12.13 days; pupal wt, 518 mg; adult wt: ♂, 227 mg; ♀, 483 mg; fecundity, 452.6 eggs/♀. Data comparable to control on castor leaves except larval survival higher on diet.

No data on *Prodenia litura*.

Achaea janata Linnaeus, castor semi-looper, also reared.

Spodoptera litura (Fabricius), tropical armyworm

→ Miyashita, K. 1971. Effects of constant and alternating temperatures on the development of *Spodoptera litura* F. (Lepidoptera: Noctuidae). APPL ENTOMOL ZOOL 6: 105-111.

COMPOSITION (g): kidney bean (*Phaseolus vulgaris*), 400; beet leaf (*Beta vulgaris*), 100; dry yeast, powder, 80; cellulose powder, 40; agar-agar, 20; sodium *l*-ascorbate, 7; sodium sorbate, 2; sodium dehydroacetate, 2; and (ml): water, 600; formalin, 4; 4N HCl, 5.

PREPARATION Not given, but kidney beans soaked in water for one day before use.

REARING At 16, 19, 25, 28 or 31°C with 16 h photoperiod. Humidity maintained by wet cotton in dishes. Egg to 2nd instar larvae @ 20-30 in 12 cm diam glass dishes. 3rd instar to prepupa @ 20-30 in 15 x 23 x 6 cm polyester containers; food renewed every second day.

DEVELOPMENT At 25°C, larval period on beet leaves, 14.3 days; on artificial diet, 19.0 days. Development during larval and pupal stages accelerated under alternating temperature conditions.

Mochida, O.; Miyahara, Y. 1974. Effect of gamma radiation on male pupae and adults of *Spodoptera litura* (F.) (Lepidoptera: Noctuidae) following pupal exposure. APPL ENTOMOL ZOOL 9: 41-47.

COMPOSITION (g): kidney bean (*Phaseolus vulgaris*), 100; wheat germ, 100; dry yeast, 40; agar, 12; sodium ascorbate, 4; methyl *p*-hydroxybenzoate, 2.8; sorbic acid, 1.4; and water, 800 ml.

REARING At 25°C with 16 h photoperiod, in 9 and 15 cm diam petri dishes and 40 x 30 x 6 cm plastic trays. Mature larvae transferred to boxes with diet, and sawdust for pupation. Pupae transferred to 3100 ml polyethylene boxes for emergence. Adults paired within 24 h of emergence, each in 25 x 20 cm paper bag. 10% sugar-water solution supplied; bags renewed daily and egg masses transferred to test tubes.

DEVELOPMENT Incubation period, 4 days; larval period, 22 days; pupal period, 10 days; adult longevity, 11 days. Number of eggs/♀, 1357-1554; emergence, 46-54%; 56-60% of ♀♀ fertile.

Patel, R.C.; Patel, J.C.; Patel, J.K. 1973. Biology and mass breeding of the tobacco caterpillar, *Spodoptera litura* (F.). ISR J ENTOMOL 8: 131-142.

COMPOSITION (g): lucerne leaf powder, 600; agar, 6; ascorbic acid, 12; and (ml): water, 1800; acetic acid, 12; formaldehyde, 1; antimicrobial solution, 30; vitamin solution, 30.

antimicrobial solution (g): sorbic acid, 4.0; methyl *p*-hydroxybenzoate, 3.0; dissolved in 34 ml ethanol.

vitamins (mg/100 ml water): biotin, 12.0; nicotinic acid, 600; calcium pantothenate, 600; riboflavin, 300; thiamine hydrochloride, 150; folic acid, 150; pyridoxine hydrochloride, 150; vitamin B_{12}, 1.2.

PREPARATION Agar dissolved in 1350 ml water and mixed with lucerne powder. Antimicrobial solution added and autoclaved at 15 psi for 25 minutes. Remaining ingredients added when diet cooled.

REARING At average of 28.6°C. Eggs laid on 25-30 x 3-5 cm plastic strips, scraped off and separated. Neonate larvae reared in glass jars on castor bean (*Ricinus communis*) leaves until 3rd instar, when transferred to artificial diet in 10 x 35 cm diam glavanised iron cage at optimum rate of 200 larvae/cage. Adults fed on 5% honey solution.

DEVELOPMENT Eclosion within 4 days at 26.7 ± 1° C; 6 larval instars for each sex. Mean larval and pupal developmental periods for ♂♂ and ♀♀ at 28.6° C: 12.8, 13.1 and 7.3, 6.1 days respectively. Average fecundity, 2518 eggs/♀; 90% eggs laid between 2nd and 5th day from emergence.

Trichoplusia ni (Hübner), cabbage looper

→ Chippendale, G.M.; Beck, S.D. 1965. A method for rearing the cabbage looper, *Trichoplusia ni*, on a meridic diet. J ECON ENTOMOL 58: 377-378.

COMPOSITION (g): whole wheat germ, 30.0; vitamin-free casein, 21.0; dextrose, 21.0; cellulose powder, 18.0; bacto-agar, 15.0; Wesson's salt mixture, 6.0; sodium alginate, 3.0; ascorbic acid, 3.0; cholesterol, 1.8; wheat germ oil, 1.6; sorbic acid, 0.8; choline chloride, 0.55; and (ml): vitamin solution, 60.0; distilled water, 450.0.

choline chloride added in solution: 6 ml of squeous solution containing 10 g choline/100 ml.

vitamins (mg/l water): *myo*-inositol, 2000; calcium pantothenate, 100; niacin, 100; riboflavin, 50; pyridoxine hydrochloride, 25; thiamine hydrochloride, 25; folic acid, 25; biotin, 2; and vitamin B_{12}, 0.2.

PREPARATION All ingredients except ascorbic acid blended in 1-l stainless steel beaker and autoclaved at 17 psi and 121° C for 20 minutes. Diet cooled to below 60° C, and ascorbic acid added as solution in sterile distilled water; thoroughly stirred and dispensed to 5 x 9 cm diam crystallizing dishes to depth of 2.5 cm. Each dish covered with filter-paper lined lid of petri dish (10 cm diam) and stored at 4° C.

REARING At 30° C and 60% RH in darkness in sterile cotton-plugged 85 x 23 mm shell vials with 5 g diet in each. Larvae transferred from agar slant with nichrome wire and seeded @ 2 larvae/vial. 50 pupae placed on vermiculite for adult emergence in 30 x 20 cm diam wax paper lined, cylindrical, wire mesh cages, placed on moist paper towelling. Adults fed 8% sucrose solution from cotton wick and kept at 24° C. Eggs collected daily after 4-5 days, brushed from waxed paper into 0.038% Clorox solution with 4-5 drops Triton-X100 for 30 minutes, washed in 70% ethanol and transferred to sterile agar slants in cotton plugged 28 x 200 m test tubes. After 2 days at 30° C, 2 pieces opaque paper wrapped around each end of tube; larvae congregate by light slit between pieces of paper.

DEVELOPMENT 50% pupation in 10.7 ± 0.34 days. Several generations reared.

Corollary 1

→ Chippendale, G.M.; Beck, S.D.; Strong, F.M. 1965. Nutrition of the cabbage looper, *Trichoplusia ni* (Hübn.). I. Some requirements for larval growth and wing development. J INSECT PHYSIOL 11: 211-223.

COMPOSITION (g): vitamin-free casein, 3.5; dextrose, 3.5; bacto-agar, 2.5; Wesson's salt mixture, 1.0; sodium alginate, 0.5; cholesterol, 0.3; sorbic acid, 0.12; choline chloride, 0.10; ascorbic acid, 0.50; and (ml): vitamin solution, 10.0; distilled water, 75.0.

vitamins: as in Chippendale & Beck* (1965). To this preparation added 5.0 g whole wheat germ, 3.0 g powdered cellulose, and 250 mg wheat germ oil.

PREPARATION All constituents except ascorbic acid blended, autoclaved for 20 minutes at 17 psi, and allowed to cool to below 55°C. Ascorbic acid added as solution in sterile distilled water. Diet thoroughly stirred and dispensed to 5 x 9 cm diam crystallizing dishes to depth of 2 cm and stored at 4°C prior to use. 5 g diet placed in sterile shell vials for rearing.

See also: Chippendale *et al.* (1964).

Corollary 2

→ Gothilf, S.; Beck, S.D. 1967. Larval feeding behaviour of the cabbage looper, *Trichoplusia ni.* J INSECT PHYSIOL 13: 1039-1053.

Two diets.

COMPOSITION Diet 1 (g): agar, 2.36; powdered cellulose, 8.95; vitamin-free casein, 3.30; dextrose, 3.30; Wesson's salt mixture, 0.94; cholesterol, 0.28; wheat germ oil, 0.24; ascorbic acid, 0.47; choline chloride, 0.09; and (ml): B-vitamin solution, 9.40; distilled water, 70.65.

vitamins: as in Chippendale & Beck* (1965).

Diet 2 slightly modified from that of Shorey* (1963) (% fresh weight): soaked lima beans, 25.0; wheat germ, 6.75; dried brewers' yeast, 3.20; ascorbic acid, 0.60; sorbic acid, 0.15; agar, 1.30; and water, 63.0.

→ Getzin, L.W. 1962. Mass rearing of virus-free cabbage loopers on an artificial diet. J INVERT PATHOL 4: 486-488.

Modified diet of Vanderzant *et al.** (1962) for *Heliothis zea*.

COMPOSITION (g): wheat germ, 3; vitamin-free casein, 3.5; sucrose, 3.5; sodium alginate, 0.5; Wesson's salt mixture, 1; cholesterol, 0.3; choline chloride, 0.1; ascorbic acid, 0.5; sorbic acid, 0.12; agar, 2.5; macerated cabbage tissue, 18; and (ml): B-vitamins plus inositol, 1; water, 67.

vitamins: see Adkisson *et al.** (1960).

PREPARATION Components mixed, sterilized and cooled to 60°C. Ascorbic acid added and diet dispensed to 2 x 10 cm diam petri dishes.

REARING At 25°C. Eggs sterilized in 0.05% NaClO and 0.02% Triton X-100 for 1 h. Washed on Buchner funnel and placed on moist filter paper in petri dish. 10-12 neonate larvae placed on diet and dishes inverted until 3rd or 4th instar, when righted again. Pupae removed to moist vermiculite in 16 x 16 x 18" rearing cage. Adults fed on 2% sugar solution on absorbent cotton. Oviposition on broccoli leaf in flask of water. Constant light for adults from 20 watt bulb.

DEVELOPMENT Incubation period, 3 days; larval period, 12 days; pupal period, 8 days. Generation completed in 4 weeks. Pupal wt, 0.22 ± 0.02 g. Fecundity, 300-600 eggs/♀. 17 generations reared.

→ Henneberry, T.J.; Kishaba, A.N. 1966a. Cabbage loopers. pp. 461-478. *in* Smith, C.N. (Ed.). Insect Colonization and Mass Production. Academic Press, New York and London.

Four diets developed by various workers compared. Modified diet of Ignoffo* (1963) for mass rearing *T. ni* described.

COMPOSITION Part A: agar, 900 g; and water, 22200 ml.
part B: casein, 1260 g; alfalfa meal, 540 g; 4M KOH solution, 180 ml; and water, 8500 ml.
part C (g): sucrose, 1260; wheat germ, 1080; Wesson's salt mixture, 360; and Alphacel, 180; and (ml): choline chloride (10% aqueous), 360; formaldehyde (10% aqueous), 150; vitamin stock, 120; and 38% solutions of sorbic acid, 180; and of methyl *p*-hydroxybenzoate in 95% ethanol, 180.
part D (g): Aureomycin, 5; ascorbic acid, 150; and water, 300 ml.

vitamins (mg): niacin, 600; calcium pantothenate, 600; riboflavin, 300; thiamine, 150; pyridoxine, 150; folic acid, 150; biotin, 12; B_{12}, 1.2; in 100 ml of water.

PREPARATION Ten gallon diet prepared in 60-gal, steam-jacketed, stainless steel kettle (Legion Utensil Co., New York, N.Y.). Agar dissolved in 222 l water. Remaining ingredients added, continuously mixing with ¼ H.P. Lightning Mixer (Mixing Equipment Inc., Rochester, New York); casein, alfalfa meal and 4M KOH mixed in 8500 ml of water and added; followed by mixture of sucrose, wheat germ, Wesson's salt mixture and Alphacel, choline chloride (aqueous solution), formaldehyde (aqueous solution), and vitamin stock solution; and then sorbic acid and methyl *p*-hydroxybenzoate solutions added individually. Finally ascorbic acid and Aureomycin added, suspended in 300 ml water, when temperature of kettle contents cooled to below 70°C. Hot diet poured into 6-oz paraffin coated Lily cups (Lily-Tulip Corporation, New York, N.Y.) @ 70-80 ml/cup and allowed to cool.

REARING At 27°C; air dried towelling with formaldehyde-sterilized eggs cut into strips each with ca. 30 eggs and 1 strip stapled to lid of rearing carton with diet. Pupae removed to container with 1-1.5% NaClO for 10-15 minutes, rinsed in tap water and dried on paper towelling. Emergence cages of inverted 1-gal cylindrical cardboard containers, each with lid stapled to corner of 14" square of ⅜" plywood. Nylon cloth top. Adults kept at 24°C, anaesthetized with CO_2; oviposition cage of Ignoffo* (1963) using paper towelling as oviposition site.

DEVELOPMENT Pupal yield from eggs, 75-80%; from larvae >90%; pupal mortality <5%; pupal wt: ♂, 266 mg; ♀, 251 mg. Rearing cost ca. 0.36 cents/pupa.

See also: Kishaba & Henneberry (1966); Kishaba *et al.* (1968); Leppla *et al.* (1974).

→ Henneberry, T.J.; Kishaba, A.N. 1966b. Pupal size and mortality, longevity, and reproduction of caggabe loopers reared at several densities. J ECON ENTOMOL 59: 1490-1493.

Modification of diet of Ignoffo* (1963) for mass-rearing cabbage loopers.

COMPOSITION (g/100 ml diet): agar, 2.370; alfalfa leaf meal, 1.430; water, 81.000; casein, 3.320; sucrose, 3.320; wheat germ, 2.850; Wesson's salt mixture, 0.980; Alphacel, 0.430; 4M KOH, 0.100; ascorbic acid, 0.390; choline chloride, 0.100; methyl *p*-hydroxybenzoate, 1.800; sorbic acid, 1.800; formaldehyde (0.1 g/ml water), 0.040; Aureomycin, 0.013.

vitamins (mg/100 g diet): niacin, 0.95; calcium pantothenate, 0.95; riboflavin, 0.48; thiamine hydrochloride, 0.24; pyridoxine, 0.24; folic acid, 0.24; biotin, 0.02; vitamin B_{12}, 0.002.

PREPARATION Agar dissolved in boiling water; casein blended in part of water, and KOH added. All solids except ascorbic acid and Aureomycin blended, then methyl *p*-hydroxybenzoate, choline chloride, formaldehyde, vitamin solution, and dissolved agar successively added and blended; ascorbic acid and Aureomycin added last and entire mixture then blended for 2 minutes. Hot diet poured into rearing jars, cooled, capped and stored.

REARING At 80 ± 2° F and 40-50% RH in 6- or 8-oz cups with 70 ml diet and 20-50 eggs seeded on to lid. Details in Henneberry & Kishaba* (1966a).

DEVELOPMENT Larval period, 14 days. Pupal wt: ♂, 257 mg; ♀, 238 mg. Percent pupation and pupal weights reduced at high larval densities; but mating, adult mortality, oviposition, and egg viability unaffected by high or low larval densities. High adult density reduced oviposition.

See also: Henneberry & Kishaba (1967); Vail *et al.* (1968).

Corollary

Smilowitz, Z.; Smith, C.L. 1970. Distributions and frequencies of weight of cabbage looper larvae reared on artificial diet. J ECON ENTOMOL 63: 1106-1107.

COMPOSITION Modified from Henneberry & Kishaba (1966a) diet for *T. ni*; except 10% vitamin stock (Vanderzant, NBC) used in place of previous vitamin solution; and 5.5% 1:1 mixture of raw linseed oil (Archer-Daniels-Midland Co.) and corn oil (Mazola) containing 1.0% *dl*-α-tocopherol (NBC) incorporated into diet.

REARING At 29 ± 1° C, and ca. 40% RH with constant lighting. About 30 larvae reared in 8-oz Mira Glaze Dixie cups.

DEVELOPMENT Complete larval development in 12 days.

See also: Toba & Kishaba (1972).

Ignoffo, C.M. 1963. A successful technique for mass-rearing cabbage loopers on a semisynthetic diet. ANN ENTOMOL SOC AM 56: 178-182.

COMPOSITION (ml): distilled water, 3100; methyl *p*-hydroxybenzoate (15% w/v in 95% ethyl alcohol), 36; choline chloride (0.1 g/ml water), 36; 4M KOH, 18; formaldehyde (0.1 g/ml water), 15; vitamin stock, 6; and (g): sucrose, 126; casein, 126; wheat germ, 108; cotton-leaf stock, 54; agar, 90; Wesson's salt mixture, 36; Alphacel, 18; ascorbic acid, 15; Aureomycin, 0.5.

cotton-leaf stock prepared by drying cotton leaves at 120° C for 1 to 2 h and then finely grinding.

vitamins (mg/100 ml water): niacin, 600; calcium pantothenate, 600; riboflavin, 300; thiamin, 150; pyridoxin, 150; folic acid, 150; biotin, 12; vitamin B_{12}, 1.2.

PREPARATION Agar dissolved in 2220 ml water. Casein blended with 880 ml water, and KOH, and all solids except ascorbic acid and Aureomycin added and blended. Methyl *p*-hydroxybenzoate, choline chloride, formaldehyde and vitamin solution added and blended. Dissolved agar, followed by ascorbic acid and Aureomycin added, then blended for 2 minutes; poured into 5 x 2" sterile rearing jars, filled with 35 ml diet; capped with metal lids, tilted and cooled.

REARING At 29 ± 0.1°C and 60-80% RH, @ 13 neonate larvae/jar, or individually in test tubes with 15 ml diet. After 14 days pupae removed to 6½ x 2¾" diam lantern globes, with petri dish bottom and screen-lid with 3 sheets of paper towelling, for emergence. Ten pairs of adults placed in oviposition cage; cylinder enclosed at both ends with pie pans with 1" wet sand on bottom; fed 50% honey solution from feeding vial. Eggs laid on damp white paper towelling wrapped tightly around outside of screened cage. Eggs soaked for 5 minutes in 0.3% NaClO, then 10% sodium thiosulphate, and washed in distilled water. Incubated on filter paper in petri dish. Oviposition, emergence and rearing jars illustrated.

DEVELOPMENT Average duration of stages (days): egg, 2.3 ± 0.2; larval, 8.1 ± 0.5; prepupal, 1.1 ± 0.1; pupal, 7.6 ± 0.1; adult, 13.1 ± 0.5; generation, 23.0 ± 0.6. On broccoli, generation completed in 25-27 days.

→ Shorey, H.H. 1963. A simple artificial rearing medium for the cabbage looper. J ECON ENTOMOL 56: 536-537.

COMPOSITION (g): soaked lima beans, 100; water, 200; dried brewers' yeast, 10; ascorbic acid, 1; methyl *p*-hydroxybenzoate, 1; and agar, 4.

PREPARATION Lima bean seeds soaked in water at 22-24°C for 24 h, then blended with yeast, methyl *p*-hydroxybenzoate and ascorbic acid with ½ of water. Agar dissolved in remaining water, cooled to 70°C, and blended briefly with other ingredients. Poured to depth of 1" in wide mouth Mason jars; sufficient to rear 60 larvae.

REARING At 27°C and 20-40% RH @ 10 larvae/jar, covered by disc of blotting paper. 3 pairs adults/1-gal paper lined cylindrical carton, with 14 mesh screen lid. Fed 8% sucrose solution; oviposition on paper lining. Eggs washed, dried and removed to rearing jars.

DEVELOPMENT Larval period, 14.9 ± 1.5 days; pupal period, 7.2 days. Adult longevity (days): ♂, 19.3 ± 2.9; ♀, 15.0 ± 2.6. Twelve consecutive generations reared. Development favourable compared with that on natural food.

See also: Shorey (1966).

Corollary

→ Chauthani, A.R.; Snideman, M.; Rehnborg, C.S. 1971. Comparison of commercially produced *Bacillus thuringiensis* var. *thuringiensis* with two bioassay techniques based on toxicity units. J ECON ENTOMOL 64: 1291-1293.

Modified from Shorey* (1963) artificial medium for *T. ni*.

COMPOSITION (g): brewers' yeast, 120; pinto bean flour, 16 mesh, 400; Tegosept, 7.5; sorbic acid, 5.0; Dithane, 2.0; Aureomycin (Chlortetracycline bisulphate, feed grade (American Cyanamid Co., Princeton, N.J.)), 2.5; ascorbic acid, 12.0; agar, 60.0; and (ml): formalin, 40%,

10; distilled water, 3700.

REARING Larvae reared for 6 days from eclosion at 28 ± 2°C and 50-60% RH.

→ Splittstoesser, C.M.; McEwen, F.L. 1961. A bioassay technique for determining the insecticidal activity of preparations containing *Bacillus thuringiensis* Berliner. J INSECT PATHOL 3: 391-398.

COMPOSITION (%): sucrose, 3; Wesson's salt mixture, 0.1; Tegosept, 0.2; agar, 2.0; wheat germ, 3.0; dried leaf (broccoli most satisfactory), 2.0; alginate, 0.1; streptomycin sulphate, 5 μg/ml.

PREPARATION Freshly picked leaves steamed for 5 minutes, dried at 55-60°C, and pulverised to 20 mesh. Sucrose, agar, salts, and Tegosept added to water and heated to 85-90°C; leaf powder, wheat germ and alginate mixed dry and stirred into hot solution. Diet dispensed to bottles and cooled to 45°C. Streptomycin added. pH 6.4.

REARING At 25°C. 3rd instar larvae placed in vials with 2 g diet, or in petri dish on diet.

Miscellaneous - Noctuids

→ Poitout, S.; Bues, R. 1972. Nutrition des Insectes - Mise en évidence de besoins différents en acide linolénique entre Lépidoptères Noctuidae-Trifinae, appartenant à différentes sous-familles, et Lépidoptères Noctuidae-Quadrifinae de la sous-famille des Plusiinae. C R ACAD SCI PARIS 274: 3113-3115. (In French)

Based on diet of Poitout & Bues* (1970).

Three species reared: *Chrysodeixis chalcites* Esper, *Autographa gamma* Linnaeus, *Macdunnoughia confusa* Stephens.

COMPOSITION Basal diet (g): agar, 5.0; corn semolina, 28.0; wheat germ, 7.0; brewers' yeast, 7.5; ascorbic acid, 1.0; benzoic acid, 0.3; Nipagin, 0.25; aureomycin chlorohydrate, 0.03; and (ml): water, 170.0.
best four of nine additives (% by wt): linolenic acid, 0.11; powdered cabbage leaves, 4.6; powdered mallow (*Malva* sp.) leaves, 4.6; powdered goosefoot (*Chenopodium* sp.) leaves, 4.6.

PREPARATION & REARING As in Poitout & Bues* (1970).

DEVELOPMENT Mean larval development (days) and percent adults of *C. chalcites* deformed on basal diet alone and basal diet plus each of 4 additives respectively:- 14.5, 95.8; 14.1, 7.5; 13.1, 6.5; 13.1, 1.6; 13.3, 0. Satisfactory development of both *A. gamma* and *M. confusa* on addition of powdered cabbage leaf to basal diet.

→ Shorey, H.H.; Hale, R.L. 1965. Mass-rearing of the larvae of nine noctuid species on a simple artificial medium. J ECON ENTOMOL 58: 522-524.

COMPOSITION (g): soaked pinto beans, 2133; dried brewers' yeast, 320; ascorbic acid, 32; methyl *p*-hydroxybenzoate, 20; sorbic acid, 10; 40% formaldehyde, 20; agar, 128; water, 6400.

PREPARATION Ingredients except agar blended with ½ total water. Agar dissolved in remaining water, cooled to 70°C then mixed with first batch.

REARING At 27 ± 1°C. All species reared in 6-oz waxed paper cups with 2-oz diet in each. Ten adult pairs placed in 1-gal, cylindrical screen-topped, cardboard breeding cartons with 8% sucrose solution. Walls of carton and feeding bottle covered with blotting paper for oviposition. Paper with *Trichoplusia ni* and *Autographa californica* eggs sectioned and glued on to cardboard lids of 6-oz diet cups @ 25 eggs/section. Lids removed after 21 days. *Prodenia ornithogalli, Spodoptera exigua* eggs glued to lids of 16-oz paper cups with small amount diet for emergence, then 1st to 3rd instar larvae transferred to 6-oz cup @ 5 *P. ornithogalli* and 15-20 *S. exigua* larvae/cup. *Heliothis zea* and *H. virescens* larvae held initially in 16-oz cups then transferred to 6-oz cups stored above moist vermiculite.

DEVELOPMENT 9 species reared satisfactorily. *Trichoplusia ni* (Hübner): larval survival, 83%; pupal survival, 92%; survival from egg to adults, 52%; egg to adult, 26 days. *Autographa californica* (Speyer): eggs laid, but no larvae reared. *Spodoptera exigua* (Hübner): larval survival, 89%; pupal survival, 94%; survival from egg to adult, 44%; egg to adult, 26 days. *Prodenia ornithogalli* Guenée: larval survival, 79%; pupal survival, 84%; survival from egg to adult, 17%; egg to adult, 37 days. *Heliothis zea* (Boddie), larval survival, 71%; pupal survival, 91%; survival from egg to adult, 51%; egg to adult, 30 days. *Heliothis virescens* (Fabricius): larval survival, 80%; pupal survival, 74%; survival from egg to adult, 86%; egg to adult in 35 days.

Heliothis phloxiphaga Grote & Robinson, *Pseudaletia unipuncta* (Howorth), *Peridroma saucia* (Hübner) successfully reared for more than 3 generations; *Euxoa* sp. reared for more than 4 generations; and *Autoplusia egena* (Guenée) reared for 1 generation, adults laying infertile eggs.

→ Terriere, L.C.; Grau, P.A. 1972. Dietary requirements and tissue levels of fatty acids in three Noctuidae. J INSECT PHYSIOL 18: 633-647.

Three species reared on meridic diet; cabbage looper, *Trichoplusia ni* (Hübner), alfalfa looper, *Autographa californica* (Speyer), and corn earworm, *Heliothis zea* (Boddie).

COMPOSITION (%): wheat-germ residue, lipid-free, 4.50; vitamin-free casein, 3.50; dextrose, 3.50; Alphacel, 3.00; Wesson's salt mixture, 1.00; cholesterol, 0.30; choline chloride solution (10 g choline chloride/100 ml water), 1.00; vitamin solution, 10.00; ascorbic acid, 0.50; sodium alginate, 0.50; sorbic acid, 0.15; bacto-agar, 2.50; fatty acid source (when present), 0.75; and water, 68.70.

vitamins: as in Chippendale & Beck* (1965).

PREPARATION See Chippendale & Beck* (1965). Diet rendered virtually lipid-free by extracting whole wheat-germ with 4 times its weight of chloroform-methanol (2:1 v/v) in Waring Blendor for 2 minutes, then removing solvent by filtration under vacuum; this step repeated with ¼ less solvent. Residue then refluxed for 1 h with 5 times its weight of solvent, filtered free of solvent, washed 3 additional times with excess solvent, and oven dried.

After autoclaving and reblending, 100 g portions of diet fortified with appropriate levels of lipid under study. Lipid emulsion prepared by shaking 1 ml of ethanol solution of lipid with 5 ml sterile water (with 1% Tween 80 and 33 mg α-tocopherol). Emulsion incorporated into diet by mechanical stirring at 65°C. Diet dispensed to sterilized 23 x 85 mm shell vials stoppered with sterile cotton, or to 6-oz waxed paper cups.

REARING At 30 C and 95% RH. Neonate larvae transferred to diet in cups for *T. ni* and *A. californica* and in vials for *H. zea*. Pupae placed in pint jars with vermiculite and paper strips for emergence. Adults kept in 1-gal cylindrical cartons with wire mesh lids with 16 h photoperiod, and fed 8% sucrose solution from bottle with cellucotton wick. Oviposition of *T. ni* and *A. californica* on waxed paper lining; of *H. zea* on moist cheesecloth. Eggs surface sterilized in 0.03% NaClO solution and Triton X-100 for 30-40 minutes, then washed in 70% ethanol and sterile distilled water. Incubated in sterile petri dish on moist filter paper.

DEVELOPMENT 90-100% pupation of *T. ni* in 13-14 days resulting in normal adults if adequate dietary lipid present. Similar results obtained with *A. californica*. For *H. zea*, higher lipid levels required, and pupation rate only 50%.

→ Vail, P.V.; Anderson, S.J.; Jay, D.L. 1973. New procedures for rearing cabbage loopers and other lepidopterous larvae for propagation of nuclear polyhedrosis viruses. ENVIRON ENTOMOL 2: 341-344.

Three species reared: *Autographa californica* (Speyer), alfalfa looper; *Trichoplusia ni* (Hübner), cabbage looper; *Spodoptera exigua* (Hübner), beet armyworm.

Best of 7 diets tested.

COMPOSITION (g): agar, 90; wheat germ, 200; pinto beans, 275; torula yeast, 128; Aureomycin, 0.5; ascorbic acid, 15; and (ml): water, 3560; formalin (100 ml/170 ml H_2O), 15; methyl *p*-hydroxybenzoate (23.9 g/38 ml 95% ethanol), 7.9; sorbic acid, (23.9 g/38 ml 95% ethanol), 7.9; vitamin mixture (see Henneberry & Kishaba*, 1966a), 12.0.

PREPARATION Pinto beans soaked overnight with several changes of water. Beans and other ingredients blended for 5 minutes. Heated to 63° C and dispensed to bags.

REARING 16 x 19.5 x 32 cm paper bags coated with 15-20 g of paraffin by dipping at 97° C. Paraffin hardened on cooling and bag filled with liquid diet and cooled. Eggs laid on paper towelling stapled inside bag and bag closed by folding and stapling. Pupae collected after 16-20 days.

DEVELOPMENT 115 pupae out of 250 neonate larvae recovered from each bag. 96% normal adults; adult longevity: ♀, 15 days; ♂, 14 days. Fecundity, 788 eggs/♀; egg hatch, 63%. Reared for 6 generations; in 6th generation, 75.3% larval survival to adults.

Olethreutidae

Argyroploce leucotreta Meyrick, false codling moth

→ Bot, J. 1965. An aseptic artificial rearing medium for the false codling moth, *Argyroploce leucotreta* Meyr. S AFR J AGRIC SCI 8: 1075-1078.

COMPOSITION (g): maize meal, 200.0; wheat germ, 40.0; finely ground dried brewers' yeast (live *Saccharomyces utywala africana*), 60.0; vitamin-free casein, 10.0; ascorbic acid, 1.5; choline chloride, 0.3; methyl *p*-hydroxybenzoate, 3.0; water, 250.0.

PREPARATION Maize meal, wheat germ, casein and yeast thoroughly mixed. Methyl *p*-hydroxybenzoate dissolved in 40 ml of 95% ethyl alcohol, then added; alcohol evaporated. Ascorbic acid and choline dissolved in water and stirred into mixture. Diet dispensed to jars.

REARING At 26°C and 60% RH. Oviposition on paper strips in wire gauze cage; at black-head stage eggs and diet jars sterilized by formalin vapour; seeded @ 2-300 eggs/jar.

DEVELOPMENT Larval period, 12-18 days; pupal period, 10-12 days; pupal production/100 g diet, 252. Thirteen generations reared.

Ripley, L.B.; Hepburn, G.A.; Dick, J. 1939. Mass breeding of false codling-moth, *Argyroploce leucotreta* Meyr., in artificial media. SCI BULL DEP AGRIC FOR UNION S AFR 207, 18 pp.

COMPOSITION (volume): oat meal, medium grade, 5; water, 3; also moulds - best, *Mucor hiemalis*.

PREPARATION Ingredients mixed in jars or tubes and autoclaved at 126°C. Spores of *M. hiemalis* introduced to infected egg papers.

REARING At 26-30°C and 80-95% RH. Eggs in red and black stage sterilized for 2 minutes in 8% formalin and rinsed in sterile water; spores introduced with last wash. Then eggs seeded aseptically to diet. Adults removed to screen cages, @ >20/cage. Olfactory stimulus required for oviposition provided by slices of orange near cage.

DEVELOPMENT Larval period, 12 days; oviposition, 5-13 days from emergence; fecundity, 105 eggs/♀.

Citrus snout-beetle, *Sciobius granosus* Faoharaeus (Col., Curculionidae) also reared.

Bactra verutana Zeller, nutsedge moth

Garcia, C. Jr.; Frick, K.E. 1975. *Bactra verutana*, a possible biological control agent of purple and yellow nutsedge: large scale rearing on artificial diet. ANN ENTOMOL SOC AM 68: 15-18.

COMPOSITION Diet developed by Adkisson *et al.** (1960), 150; plus (g): vitamin-free casein, 98; sucrose, 98; wheat germ, 84; Wesson's salt mixture, 28; agar, 95; ascorbic acid, 13; vitamin diet fortification mixture (NBC), 24; milled nutsedge powder (*Cyperus rotundus*), 15; chlortetracycline hydrochloride, 0.5; sulphathiazole, 0.5; methyl paraben (dissolved in 43 ml 95% ethanol), 3.75; sorbic acid (dissolved in 43 ml 95% ethanol), 5.0; and (ml): distilled water, 3200; formaldehyde (37%), 5; 4M KOH, 18.

PREPARATION Nutsedge washed, and dried at 135-140°C for 1-1.5 h. Ground in Wiley Mill to 60-mesh and autoclaved at 121°C for 20 minutes. Agar suspended in 800 ml cold water, added to 1500 ml boiling water, then heated until dissolved. Vitamin-free casein and KOH blended in 900 ml water for 30 seconds. Then sucrose, wheat germ, Adkisson diet and mineral salts added and blended for further 30 seconds; agar solution added and blended for 2 minutes, then mould inhibitors and remaining ingredients added. Entire mixture blended for 2 minutes, then dispensed @ 12-14 ml/30 ml plastic cup and cooled. Diet surface scored, lids replaced, and diet stored at room temperature until used.

REARING Neonate larvae transferred to diet and held at 30 ± 2° C until adult emergence. Adults collected daily, and paired in oviposition cages @ 10 pairs/cage at 31-32° C and 80% RH with 12 h light, then 26-27° C and 80% RH for 12 h darkness. Paper oviposition cages 10.8 x 13.5 cm diam with circular hole 3.3 cm from bottom for food vial with mixture of 10% sugar and honey solutions and cotton wick. Oviposition on Saran wrap pleated foil, and purple nutsedge.

DEVELOPMENT Incubation period, 2.5-3.0 days; adult emergence, 18-20 days from oviposition; survival to adult, 60-61%; pupal wt: ♂, 12.86 mg; ♀, 19.14 mg; fecundity, 135 eggs/♀; adult longevity up to 21 days. 35 generations reared.

→ Sieckert, E.E.; Ritenour, G.L.; Davis, H.G. 1974. An artificial medium and rearing techniques for the colonization of *Bactra verutana*. ENVIRON ENTOMOL 3: 196-197.

Modification of wheat germ diet for *Laspeyresia pomonella* by Howell* (1971).

COMPOSITION (g): agar, 64.0; vitamin-free casein, 126.0; sucrose, 96.0; pinto beans, 1000.0; wheat germ, 108.0; vitamin-mineral mixture (Vita Vurv, William Cooper and Nephews, Inc., Chicago, Il., USA), 25.0; ascorbic acid, 20.0; Wesson's salt mixture, 36.0; ground autoclaved leaves of nutsedge (*Cyperus esculentus*), 20.0; Alphacel, 20.0; methyl *p*-hydroxybenzoate (in 10 ml 95% ethanol), 7.0; sorbic acid (in 10 ml 95% ethanol), 8.0; and (ml): distilled water, 3600.0; 4M KOH, 18.0; propylene glycol, 30.0; formaldehyde (37%), 15.0.

PREPARATION *C. esculentus* leaves dried at 70° C for one day, then ground and autoclaved at 100° C for 15 minutes. Water added to pinto beans, boiled, then cooled and soaked for 2 h at room temperature. Reboiled and simmered for 2 h till soft, when blended in 200 ml batches each with 100 ml distilled water. Agar autoclaved for 15 minutes at 120 C with 2000 ml water; leaves, beans and agar solution mixed. 1000 ml sterile cold water added and whole mixture blended for 30 seconds. Sorbic acid, methyl *p*-hydroxybenzoate, formaldehyde, KOH, propylene glycol, and vitamin mixture in safflower oil added and blended; finally ascorbic acid mixed in 500 ml water, added to diet and blended slowly. Dispensed in ca. 450 ml lots to 3.7-l glass jars, laid horizontally to cool and solidify. 0.75-1.0 ml beeswax layer spread over diet surface to help exclude microbial contaminants and reduce desiccation. 5 ml of 1.0% sorbic acid solution applied to beeswax surface to retard surface contamination.

REARING At 26° C ± 3° C and 50% RH with 16 h photoperiod. Oviposition cages from 8 x 10 cm wax paper lined cardboard cylinders with cheesecloth covered ends. About 100 adults/cage, fed on 10% sugar solution dispensed from glass-vial with cotton-plug in hole in cage. Moths chilled at 3° C for 5 minutes every 2 days and transferred to clean cage; wax sheets with eggs removed, cut into 6 cm^2 sections, washed in 4.0% formaldehyde solution for 30 seconds then rinsed in sterile water, dried and placed on diet. Neonate larvae burrow into diet; grow and pupate within diet. Emerging adults collected in vacuum aspirator, chilled at 3° C for 5 minutes and transferred to oviposition chambers.

DEVELOPMENT Incubation period, 3-4 days; adult emergence, 35-40 days; (20-25 days in wild); survival from eggs to adults, 40%. Five generations reared.

Barbara colfaxiana (Kearfott), Douglas-fir cone moth

→ Ruth, D.S.; Hedlin, A.F. 1969. Rearing the Douglas-fir cone moth, *Barbara colfaxiana* (Kearfott), on an artificial diet in the laboratory. J ENTOMOL SOC B C 66: 22-25.

COMPOSITION As in McMorran* (1965) diet for *Choristoneura fumiferana*.

REARING At 80° F. Two 1st instar larvae/1-dram vial with diet plug, renewed every 2-3 days. Reared individually from 2nd instar. Pupae removed to gelatin capsules on absorbent cotton. Adults to drum shaped cages; oviposition on wax paper lining; eggs removed at black head stage to 1-gal jar with high humidity for emergence.

DEVELOPMENT Survival from egg to pupa, 30%.

Corollary

→ Hedlin, A.F. 1965. Rearing cone insects in the laboratory on an artificial diet. BI-MON RES NOTES CAN FOR SERV 21: 4.

DEVELOPMENT *Barbara colfaxiana* reared to pupae in 6 weeks.

Laspeyresia youngana (Kearfott) and *L. piperana* Kearfott (Olethreutidae) reared from late stages of larval development.

Eucosma monitorana Heinrich

→ Barras, S.J.; Norris, D.M. Jr 1965. *In vitro* establishment and development of *Eucosma* sp. larvae from cones of *Pinus resinosa* on an artificial nutrient medium. J ECON ENTOMOL 58: 1033-1034.

COMPOSITION & PREPARATION See Vanderzant *et al.** (1962) diet developed for *Pectinophora gossypiella*.

REARING Larvae reared individually on diet plugs in sterile 23 x 90 mm vials. Diet added as needed. Vial plug removed when larvae mature, and vials inverted over 1½" sand for pupation.

DEVELOPMENT To 5th instar in 7-14 days; 116 pupae formed from 150 larvae placed on diet.

Grapholitha molesta (Busck), oriental fruit moth

→ Ivaldi-Sender, C. 1974. Techniques simples pour un élevage permanent de la tordeuse orientale, *Grapholitha molesta* (Lépidoptèra: Tortricidae) sur milieu artificiel. ANN ZOOL ECOL ANIM 6: 337-343. (In French with English summary)

Modified from Sender* (1969) for *Laspeyresia pomonella*.

COMPOSITION (g/1000 g): agar, 18; corn semolina, 50; wheat germ, 50; powdered apple (or further corn semolina), 50; brewers' yeast, 50; ascorbic acid, 4.5; Nipagin, 1.5; benzoic acid, 1.5; and (ml): water, 780.

PREPARATION Agar dissolved in boiling water and previously homogenised mixture of dry powders added and stirred in. Diet dispensed to 22 x 22 x 9 cm high transparent plastic rearing boxes to depth of 1 cm.

When cooled, aluminium foil sheet pressed on to diet surface to protect against microbial contamination and desiccation. 0.5 mm diam holes punched in foil @ 10/cm^2, for larval penetration. Stored at 4-11°C for maximum of 14 days.

REARING At 25°C and 70% RH with 18 h photoperiod. Diet boxes illuminated from underneath, as aluminium sheet obscures light which may thus induce larval diapause. Eggs placed on foil and box covered with fine gauze for 6 days while eggs hatch; then gauze replaced by solid lid. Condensation avoided. Mature larvae leave diet and pupate in sterilized strips of corrugated paper placed at sides of boxes. Strips collected every 2 days. 200-300 adults collected, utilising their very pronounced phototropism, and maintained in 1.5-l plastic bottles with bases replaced by nylon gauze at 21 ± 1°C and 75% RH with natural lighting. Cotton wick in water trough through hole cut in side of bottle. Eggs laid at dusk after 2-3 days, collected optimally after 8 days, and disinfected by submersion in NaClO solution for 2 minutes.

DEVELOPMENT Mean of 434 pupae/350 ml diet. Larval mortality low. Generation completed in about 4 weeks. Four generations reared.

→ Laing, D.R.; Hagen, K.S. 1970. A xenic, partially synthetic diet for the oriental fruit moth, *Grapholitha molesta* (Lepidoptera: Olethreutidae). CAN ENTOMOL 102: 250-252.

COMPOSITION (g): Food Wheast (Knudsen Creamery Co., Los Angeles, Calif.), 20.0; dehydrated carrot, 40.0; sodium benzoate, 0.6; methyl *p*-hydroxybenzoate, 0.2; ascorbic acid, 0.64; agar, 4.0; and 250.0 ml distilled water.

PREPARATION Agar dissolved in half of water; other dry ingredients except ascorbic acid mixed with remaining water, then both solutions blended together with addition of ascorbic acid. Dispensed to 8-oz wax coated cups, covered and held for 20 h before use.

REARING At 80°F and 50% RH with constant lighting. Neonate larvae transferred to cups @ 10-20/cup. Pupation in corrugated paper taped to snap-on lid of cup. Adults collected in sleeve cage; oviposition cage as George & Howard (1965).

DEVELOPMENT Average adult yield from larvae reared on apples and diet 61 and 67% respectively. Pupal weights on apples greater than those on diet. Development from neonate larvae to adult (days): 24 on apple, 26 on diet. Average cost/adult, 0.16 cent on apple and 0.08 cent on diet.

→ Matsumoto, Y.; Yasue, Y. 1952. Artificial sterile breeding of the 'Princess' pear core borer, *Laspeyresia* (= *Grapholitha*) *molesta* Busck, in an artificial medium. J AGR RES 40: 205-208. (In Japanese)

Best of 4 diets tested.

COMPOSITION Water, 90 ml; (g): agar, 3.0; cellulose (filter paper 60 mesh powder), 1.0; sodium carboxymethyl cellulose (CMC), 1.5; sucrose, 4.0; fructose, 4.0; casein, 1.0; cholesterol, 0.2; olive-oil 0.25 ml; dried yeast, 3.0 g; choline chloride, 0.1; McCollum & Simmond's salt mixture, 0.5; peach leaves (white peach, dried under reduced pressure and powdered), 1.8; organic acids (malic acid, 0.95 g; citric acid, 0.035 g; tartaric acid, 0.125 g), 0.1.

PREPARATION Agar, powdered filter paper, CMC and 50 ml distilled water stirred for several minutes over water bath; remaining water and ingredients added and stirred for 20 minutes, then cooled in running water. Diet cut, placed in 1.5 x 10.5 cm test-tubes, plugged with cotton and sterilized in Koch's steam sterilizer. Final pH 5.8. (Addition of CMC to prevent precipitation at high temperature of sterilization by keeping diet homogenous).

REARING At 26-27°C and 70-75% RH in darkness; individually in tubes. Eggs on waxed paper sterilized at black-head stage; washed with 0.1% $HgCl_2$, then soaked in 0.1% $HgCl_2$ for 4 minutes, washed several times in 70% alcohol or sterilized water, and transferred to test-tube walls.

DEVELOPMENT Larval emergence, 66%; 25% survival to adult in 25-28 days.

See also: Matsumoto (1954).

→ Tzanakakis, M.E.; Phillips, J.H.H. 1969. Artificial diets for larvae of the oriental fruit moth. J ECON ENTOMOL 62: 879-882.

Best 3 of many diets tested.

COMPOSITION Diet 1 (g): tap water, 75.0; potassium sorbate, 0.1; methyl *p*-hydroxybenzoate, 0.15; agar, granular, 3.0; alfalfa leaf meal, 5.0; carrot baby food (H.J. Heinz of Canada Ltd), 10.0; apple juice, vitamized, 10.0; casein, 0.5; roasted peanuts, ground, 2.0; soy hydrolyzate, enzymatic, 0.5; brewers' yeast powder, 6.0; bacto-yeast extract, 2.0; vitamin solution (Adkisson *et al.** (1960)), 0.5; ascorbic acid, 0.5; Wesson's salt mixture, 0.5.

Diet 2: as diet 1, except brewers' yeast decreased to 2.9 g, bacto-yeast extract decreased to 1.5 g, and ascorbic acid increased to 1.0 g.

Diet 3: as diet 1, except agar reduced to 2.5 g; ascorbic acid and bacto-yeast extract omitted, and 2.0 g creamy peanut butter substituted for roasted ground peanuts.

PREPARATION Agar added to water; after 10-20 minutes casein added and flask half immersed in boiling water for 30 minutes. Hot agar with solution poured into mixer bowl and remaining ingredients added and mixed in following sequence: preservatives, salts, proteins, hydrolyzed proteins, peanuts, carbohydrates, vegetable and fruit juices, vitamin sources, yeast, cellulose, and alfalfa leaf meal. Dispensed to petri-dishes or cups while still warm, then, when cooled and solidified, surface scored or punctured and sprayed or smeared with hot paraffin wax.

REARING At 25-27°C and 90-95% RH with 16 h photoperiod. Neonate larvae transferred to diet and strips waxed paper placed on diet; dish immediately covered with unbleached cotton.

DEVELOPMENT Time required for 50% of larvae to spin cocoons on diets 1, 2 and 3; 13, 13, and 12 days respectively; mean pupal wt (mg); ♂, 10.9, 9.1, and 9.8; ♀, 11.9, 11.8 and 10.8 respectively. Three generations completed.

Gretchena bolliana (Slingerland), pecan bud moth

Schroeder, W.J. 1970. Rearing the pecan bud moth on artificial diet. J ECON ENTOMOL 63: 650-651.

Diet based on pinto bean diet of Shorey & Hale* (1965) developed for noctuids.

COMPOSITION (g): soaked pinto beans, 250; torula yeast, type 361 (St Regis Paper Co., Rhinelander, Wisc. 54501), 32; wheat germ, 42; ascorbic acid, 3.5; vitamin stock, 4.5; sorbic acid, 1.5; methyl *p*-hydroxybenzoate, 2.0; tetracycline hydrochloride, 0.055; agar, 13; and 640 ml distilled water.

PREPARATION Pinto beans soaked overnight, ground in blender with half volume of water, and poured into food mixer. Other ingredients then added; except agar, which dissolved in remaining volume of water, cooled to 70°C and then added to diet. When cold, 20 g lots dispensed to 8-oz waxed paper cups.

REARING At 27 ± 2°C and 5-30% RH with 13 h photoperiod. Wax paper strip placed on diet and 20-30 1st and 2nd instar larvae placed on paper. Cups capped with plastic-coated lids. Pupae collected; 10 adult pairs transferred to 2-l cylindrical cartons with plastic tops; kept at 15-30°C and 30-100% RH. Oviposition on pecan seedling in cage.

DEVELOPMENT Adult yield, 51%. Incubation period, 6 days; larval period, 14 days; pupal period, 8 days. Average pupal wt (mg): ♂, 23; ♀, 22. Sex ratio, 1:1. No effect on larval development, pupal weight or mating if pecan foliage incorporated in larval diet. Six generations reared.

Laspeyresia pomonella (Linnaeus) (previously *Carpocapsa* or *Cydia*), codling moth

Brinton, F.E.; Proverbs, M.D.; Carty, B.E. 1969. Artificial diet for mass production of the codling moth, *Carpocapsa pomonella* (Lepidoptera: Olethreutidae). CAN ENTOMOL 101: 577-584.

COMPOSITION (%): water, 71.70; wood pulp, air dry, 1.24; wood sawdust, air dry, 6.89; citric acid, 0.90; sucrose, 2.69; wheat bran, 1.80; whole wheat flour, 9.86; wheat germ, 0.90; Wesson's salt mixture, 0.62; casein, 2.69; choline chloride, 0.10; triturated ingredients, 0.61.

triturated ingredients (g): niacinamide, 5.00; calcium pantothenate, 5.00; riboflavin, 2.50; thiamine hydrochloride, 1.25; pyridoxine, 1.25; folic acid, 1.25; vitamin B_{12}, 0.1% in mannitol, 1.00; biotin, 0.10; Aureomycin, 5.5%, 810.00; ascorbic acid, 1804.00; and sorbic acid, 449.00.

PREPARATION Diet mixed in 130-l, steam-jacketted kettle fitted with two sets of counter-rotating paddles. About 80% of water boiled, then sheets of wood pulp mixed in until fibres separated and dispersed. Sawdust, citric acid and sucrose added, and mixture held at 90°C for 10 minutes. Bran, flour, wheat germ, salt mixture and casein pre-mixed and gradually stirred into hot mixture; again maintained with continuous stirring at 90°C for 10 minutes. Additional water then slowly added, until consistency that of porridge. Choline chloride added, and mixture cooled to 60°C, then triturate, containing vitamins, Aureomycin and sorbic acid, added and entire mixture stirred for 2 minutes.

Moisture balance improved and mould inhibited by replacing sawdust with wood chips ca. 2 x 2 x 20 mm in size.

REARING At 27 ± 1°C and 60-65% RH with 17 h photoperiod. Eggs surface sterilized at black-head stage by washing for 3 minutes in 3% NaClO and 0.1% wetting agent. Rinsed for 1 minute in water and air dried at 27°C. Egg paper placed on diet, and removed after 1-2 days when all eggs hatched. Light kept subdued initially, but once larvae established, intensity at diet surface increased to 3 ft-c. Trays removed after 15 days to adult emergence room at 27 ± 1°C and 50% RH. Adults collected in 4.5-l plastic jars and mated in revolving cylindrical drums lined with wax paper for oviposition at 26 ± 2°C and 70% RH in daylight.

DEVELOPMENT Egg hatch, 87%. Survival from egg to adult, 52%. Pupal wt (mg): ♂, 26-40; ♀, 30-50. Adult longevity: ♂, 15; ♀, 10 days. Seven generations reared.

In small-scale rearing, addition of peptone to diet @ 0.4 g/100 g diet resulted in increased average weight of newly emerged adult male and female moths to 20.7 and 31.9 mg respectively. Adults similar in size to those reared on apple.

See also: Batiste & Olson (1973).

→ Coutin, R. 1952. Alimentation des larves de *Laspeyresia pomonella* L. (Lépidoptère: Tortricidae) sur milieux artificiels. C R SEANCES SOC BIOL PARIS 146: 516-520. (In French)

COMPOSITION (g): dried, powdered pear pips, 20; glucose, 10; agar, 2.5; Nipagin, 0.2; and distilled water, 100 ml.

PREPARATION Pear pips dried with calcium chloride for 48 h at 50°C ground and sieved to 80 mesh. Glucose dissolved in 30 g water and added to pip powder. Agar and Nipagin added to remaining 70 g water and heated in water bath. Two portions then mixed and homogenised, cooled and dispensed to watch glasses of ca. 2 ml capacity. When cold, covered with another watch glass and sealed with tape. Final pH, 5.8.

REARING Larvae reared individually in watch glasses at 25°C and 60-70% RH in darkness. Diet renewed every 8-10 days or when necessary.

DEVELOPMENT From neonate to 3rd instar larvae only. 2nd or 3rd instar larvae reared on apples develop to 5th instar on diet.

→ Edelman, N.M. 1970. Principles of working out nutritive media for the pests of generative organs of fruit crops. (The codling moth taken as an example). ZOOL ZH 49: 1240-1246. (In Russian with English summary)

COMPOSITION (g): agar, 2; casein, 5; unsifted wheat germ, 4.0; dry apples, 4.0; crushed apple seeds, 1.5; ascorbic acid, 0.7; sucrose, 4.0; cellulose, 2.0; mineral salts, 0.84; cholesterol, 0.2; choline chloride, 0.040; folic acid, 0.001; nicotinic acid, 0.025; calcium pantothenate, 0.012; biotin, 0.00005; B_1, 0.0012; B_2, 0.0018; B_6, 0.0018; B_{12}, 0.0004; inositol, 0.008; cystine, 0.100; glycine, 0.100; formalin, 0.5; Metaben, 0.15; (ml): autolytic yeast, 8.0; water to 100 ml.

Hamilton, D.W.; Hathaway, D.O. 1966. Codling moths. pp. 339-354. *in* Smith, C.N. (Ed.). Insect Colonization and Mass Production. Academic Press, New York and London.

Two diets described.

COMPOSITION Diet 1: distilled water, 1100 ml; whole apple seeds, 3/cup and (g): ground apple seeds, 30; dried apple, 60; sucrose 60; soybean protein, 60; Wesson's salt mixture, 12.5; yeast, 1.25; glycine, 1.25; cysteine, 0.625; cholesterol, 0.625; α-cellulose, 30; ascorbic acid, 10; agar, 20; and (ml): linseed oil, 10; vitamin solution, 12.5; 10% KOH, 37.5; and mould inhibitor, 22.5.

Diet 2: as diet 1, except whole apple seeds, ground apple seeds and soybean protein replaced by 45 g wheat germ and 60 g of casein.

vitamins (g): niacinamide, 0.5; thiamine hydrochloride, 0.125; folic acid, 0.125; calcium pantothenate, 0.5; B_{12}, in mannitol, 1.0; riboflavin, 0.25; pyridoxine hydrochloride, 0.125; choline chloride, 25.0; and distilled water, 427.4 ml.

mould inhibitors (g): sorbic acid, 20; methyl *p*-hydroxybenzoate, 15; and 170 ml ethyl alcohol.

PREPARATION Dried apple and apple seeds blended with ca. 120 ml boiling water. Agar dissolved in remaining boiling water and added. Sucrose, soybean protein or casein, wheat germ, Wesson's salt mixture, yeast, glycine, cysteine and cholesterol premixed; then mixed with linseed oil, vitamin solution, KOH, and mould inhibitor; then blended into main mixture with α-cellulose as well. When cooled to 140° F, ascorbic acid, dissolved in cold water, added and diet finally blended. Dispensed to ⅝-oz cups @ 5-7 ml/cup, and if required, 3 whole apple seeds inserted into diet before diet solid. Larvae introduced after 12 h.

REARING At 82° F and 50% RH. Oviposition in 6 x 3" diam fibreboard tube covered at each end and internally coated with paraffin. 1" hole drilled near one end and 1 x 2½" water filled, cotton plugged, glass vial inserted. Three small branches of pear wood, ca. 5 cm thick and 70 cm long placed in each tube to increase egg laying. 100 moths placed in each tube under dim lighting, and removed after 5 days. Tubes refrigerated until needed; then placed in bright light, and water vial replaced by empty vial. Larvae collected in glass vials, transferred to diet @ 2 larvae/cup and kept at 82° F and 65% RH for adult emergence.

DEVELOPMENT Adult emergence, 29-54 days from seeding. Larval period, ca. 34 days. 35 generations reared.

See also: Hathaway *et al.* (1971, 1973); Hatmosoewarno & Butt (1974).

Howell, J.F. 1970. Rearing the codling moth on an artificial diet. J ECON ENTOMOL 63: 1148-1150.

COMPOSITION (g): agar, 90; casein, 126; sucrose, 96; wheat germ, 108; immature apple pulp, 108; Wesson's salt mixture, 36; Alphacel, 18; methyl *p*-hydroxybenzoate (dissolved in 95% ethanol), 5.4; Vanderzant's fortification mixture, 36; Aureomycin, 0.25; ascorbic acid, 15; and (ml): distilled water, 3400; 4M KOH, 18; formaldehyde (0.1 ml/ml water), 15.

PREPARATION Whole, fresh, immature apples surface sterilized in 10% formaldehyde and pulped. Agar dissolved in boiling water and blended with other ingredients when cooled to 60° C. Vitamin mixture dissolved in 200 ml, and ascorbic acid in 500 ml of water and blended into mixture. Dispensed to ½ or ¾ full 40 ml wide-mouth bottles, or 11 ml shell vials, or plastic cups. When gelled, refrigerated in uncovered containers for 12 h at 4° C. Warmed to ambient temperature before use.

REARING At 27 ± 3°C and 60 ± 7% RH. Oviposition cages 26 x 12.5 cm diam cylinder of 10 ml clear plastic fitted into top ¼ of two ½-gal paper cans; lined with waxed paper cylinder held in place by folding back 2 cm of each end under lid. Each lid punched with 3 cm hole and covered with screen; 25-50 paired moths in each tube. Moths transferred to new cages every 24 h. Egg sheets recovered, surface sterilized at black-head stage in formaldehyde (0.1 ml/ml water) and rinsed in sterile distilled water. Eggs or neonate larvae seeded on to diet, and containers inverted. On adult emergence, containers turned upright and adults collected in 61 x 61 x 92 cm box with front and back removed and filter and extract fan mounted on top; interior painted black, back covered with white cloth.

DEVELOPMENT Eight generations reared. 88.3% larvae matured to adults when confined in individual containers. Prepupal period, 18.9 days; fecundity, 131.1 eggs/♀, and pupal wt: ♂, 34.7; ♀, 44.5 mg on diet; data compared favourably with specimens reared on immature apples.

→ Howell, J.F. 1971. Problems involved in rearing the codling moth on diet in trays. J ECON ENTOMOL 64: 631-636.

COMPOSITION (g): agar, 45-60; casein, 126; sucrose, 96; wheat germ, 108; Wesson's salt mixture, 36; ascorbic acid, 15 (dissolved in 500 ml distilled water just before use); methyl *p*-hydroxybenzoate (dissolved in 95% ethanol), 5.4; sorbic acid (dissolved in 95% ethanol), 5.0; and (ml): water, 3600; formaldehyde (0.1 ml in 1 ml water), 15; propylene glycol, 30; and vitamin mixture, 45. pH, 5.7-6.0.

vitamins (part A, heat stable) (g): biotin, 0.24; nicotinic acid, 12.0; pyridoxine hydrochloride, 3.00; choline chloride (70%), 750.00 ml; α-tocopherol (unesterified), 96.00; Tween 80, 200.00; vitamin B_{12}, (0.1%), 24.00 ml; riboflavin, 6.00. (part B, heat labile) (g): thiamine hydrochloride, 3.00; *dl*-calcium pantothenate, 12.00; *i*-inositol, 240.00; folic acid, 3.00; adjusted to 12 l with distilled water.

PREPARATION Agar autoclaved with 2 l water for 15 minutes at 120°C, then mixed with 1-l cold distilled water. Solids, excluding antibiotics, added and blended for 30 seconds, then antibiotics, propylene glycol, vitamin mixture and ascorbic acid solution added. Water (0.8 l) added to fill 3.8-l Waring Blendor and diet blended for 30 seconds. Dispensed @ ca. 1.8 l of hot diet/tray; cooled overnight; then paraffin film applied as described by Howell (1967).

vitamin mixture (part A): nicotinic acid, and pyridoxine hydrochloride dissolved in 2500 ml boiling water. Biotin dissolved by first wetting with 900 ml 85% ethanol and then adding 1 l warm water. Added to hot solution of nicotinic acid and pyridoxine hydrochloride. Choline chloride added and stirred until dissolved. α-tocopherol mixed with Tween 80 and added; shaken thoroughly. Then vitamin B_{12} and riboflavin added - riboflavin pre-dissolved in 1 l water with about 300 ml 0.1 N NaOH; mixture adjusted to pH 6 with 0.1 N NaOH and total volume of 8 l autoclaved for 15 minutes at 120°C, then cooled to 50°C before being combined with part B. Part B: thiamine hydrochloride, calcium pantothenate, and inositol dissolved in 3 l distilled water and filtered through sterilized filter. Folic acid suspended in 600 ml of water and dissolved in 150 ml 0.1 N NaOH filtered through sterilizing filter. Using sterile techniques, part A and part B combined, mixed thoroughly, and stored in dark at room temperature (21-24°C) for maximum of 90 days.

REARING At 27 ± 3°C and 40 ± 7% RH with 16.5 h photoperiod. Eggs deposited on waxed paper sheets (see Howell*, 1970), surface sterilized, and 700-1000 eggs seeded evenly/tray. After 5 days, egg sheets removed from trays; on 14th day corrugated cardboard strips placed around edges of trays for pupation. When adults ready to emerge, trays transferred to collecting booth and adults removed by mechanical aspirator and transferred to oviposition cage.

DEVELOPMENT Egg hatch, 71.6%. Ca. 50.6% establishment on diet; survival to adult, ca. 20.9%. Average adult yield/tray 144.6; range 23-360. Adult emergence 28 days from seeding with peak at 31-32 days. Adults comparable to apple-reared specimens. Problems of microbial contamination and effects of antimicrobial agents discussed. 45 generations mass reared.

→ Howell, J.F. 1972b. Rearing the codling moth on a soya, wheat germ, starch medium. J ECON ENTOMOL 65: 636-637.

COMPOSITION (g): soya meal (ground, ether-extracted), 900; wheat germ (ground, natural), 300; apple pomace (dried, ground), 100; Wesson's salt mixture, 10; wheat starch, 150; invert sugar, 150; methyl *p*-hydroxybenzoate, 6; sorbic acid, 50; ascorbic acid, 16; vitamin stock (Howell*, 1971), 50; cellulose fibre, 15; and (ml): propylene glycol, 60; formalin (1 ml stock/10 ml water), 15; and distilled water, 6.5 l.

PREPARATION Apple pomace ground to powder and boiled in 3 l of water. Cellulose cut into 3-4 cm^2 pads and soaked in boiling water for 20 minutes. Wheat germ and soya meal ground. Then half of pomace solution, 900 ml water, mineral salts, and 710 ml flour blended at high speed until too thick for blender; repeated with remaining pomace, flour and softened cellulose. Vitamins, invert sugar, propylene glycol, formalin and starch (dispersed in 500 ml cold water) combined with soya-pomace paste in mixer. Methyl *p*-hydroxybenzoate and sorbic acid dissolved in 200 ml 95% ethanol, and ascorbic acid dissolved in 200 ml water before adding to diet. Diet mixed until homogenous, then heated to 92-95°C over water bath with intermittent stirring; cooled until just warm, dispensed to rearing trays, and left uncovered for further cooling. When surface no longer sticky, sprayed with paraffin as described by Howell (1967).

REARING At 31 ± 2°C and 50 ± 5% RH with 16.5 h photoperiod. 100-200 neonate larvae transferred to diet surface and trays covered with cloth. Corrugated strips provided for pupation on 12th day; pupae extracted after 20-24 days. Adults removed in aspirator to oviposition cages of inflated plastic bags in 27 x 13 cm diam plastic cylinders with paper lid. Screened 2.5 cm hole in lid for ventilation. Oviposition rooms maintained at 24 ± 2°C and 70 ± 5% RH with 16.5 h photoperiod.

DEVELOPMENT Above diet better than agar, wheat germ and wheat starch, and immature apple diets, giving 90.1% larval establishment; 57.4% pupation; pupal mortality, 6.2%; pupal wt: ♂, 39.9 mg; ♀, 52.2 mg; fecundity, 80 eggs/♀; egg hatch, 33.3%. Rate of development similar to that on apple diet. Cost about half that of apple diet.

→ Howell, J.F.; Clift, A.E. 1972. Rearing codling moths on an artificial diet in trays. J ECON ENTOMOL 65: 888-890.

COMPOSITION water, 160 l; agar, 2000 g; wheat germ, 8640 g; casein 5040 g; sucrose, 5040 g; Wesson's salt mixture, 1440 g; formalin, 60 ml; NaClO, 800 ml (0.5% available chloride); methyl *p*-hydroxybenzoate, 216 g (dissolved in 1 l 95% ethanol); sorbic acid, (200 g in 1.0 l 95% ethanol); carboxy methyl cellulose (CMC), 800 g (wetted with 2.5 l 70% ethanol); vitamin stock (Howell*, 1971), 1800 ml; and ascorbic acid, 600 g (in 2 l water).

PREPARATION Agar suspended in cold water, boiled, then cooled to 80 C; ingredients added in order listed and blended for 3 minutes. Diet dispensed and cooled overnight, then surface paraffin covered as in Howell (1967). All equipment sterilized with boiling water or autoclaved.

REARING Oviposition at 25 ± 3° C and 80 ± 5% RH; rearing at 28 ± 3° C and 40-80% RH; both with 17 h photoperiod; emergence at 24 ± 3° C with no humidity control but with continuous lighting. Reared in 73.7 x 40.6 x 6.3 cm galvanized steel trays covered with muslin lids, and lined with 2 ml polyethylene sheet. Adults collected in waxed paper bags and placed on open shelves for 5 days; rotated daily. Then bags cut open, adults discarded, and bag submerged in formalin (1 part formaldehyde to 10 parts water) for 5 minutes, dipped in 70% isopropyl alcohol and hung to dry. 1500-3000 eggs/bag; cut into 3 x 34 cm strips, folded to V shape and placed on diet. Paper removed after 5 days. Trays stacked 16 high, 32/pallet with 48.3 x 6.5 x 4 cm wood blocks separating trays. After 13 days corrugated paper strips placed around tray for pupation. Any mould treated with 1% sorbic acid in acetone. Pupal strips removed and placed in emergence box. Waxed paper bags fitted on to 3 emergence holes, and 75-125 adults collected/bag.

DEVELOPMENT Method produced 13000 moths/day. Average egg hatch, 46%. Eggs mature in 5 days.

See also: Howell (1972a); Hathaway *et al.* (1971).

→ Navon, A. 1968. Progress report on rearing of the codling moth. pp. 105-106. *in* Radiation, Radioisotopes and Rearing Methods in Control of Insect Pests. Proceedings of a Panel Organized by the Joint FAO/IAEA Division of Atomic Energy in Food and Agriculture.

COMPOSITION (g): full-fat soy meal (fine mesh), 24.0; sucrose, 48.0; brewers' yeast, 18.0; apple (var. Sarcocarp), 100.0; ascorbic acid, 0.3; *l*-malic acid, 4.0; methyl *p*-hydroxybenzoate, 1.5; agar (flakes), 24.0; cellulose powder, 60.0; and (ml): citrate buffer (pH 3.4), 120.0; distilled water, 480.0.

PREPARATION Soy meal, methyl *p*-hydroxybenzoate, sucrose, and brewers' yeast blended and autoclaved, thus destroying antitrypsin factor of soy meal. Agar dissolved in boiling water, and cellulose powder added. This fraction then shaken with sterilized homogenate. Just before jelling at ca. 45° C, apple blended to puree, and citrate buffer solution containing malic and ascorbic acids added. Refrigerated when cool till use.

REARING At 27 ± 1° C and 70-75% RH with constant lighting. Neonate larvae reared individually in test tubes with 5 g/diet; food added every 10 days; pupation in cotton plugs. Emerging adults transferred to cylindrical cage lined with wax paper. One end covered with cheese cloth, other with opaque paper. Centre hold cut in cheese cloth with water supply in plastic box above. Maintained at 23 ± 1° C and 70% RH with constant lighting to induce mating. Adults transferred to new cages every 2 days.

DEVELOPMENT Two successive generations reared; greatest yield in terms of adult/neonate larva, 20-30%; adults normal size.

→ Navon, A.; Moore, I. 1971. Artificial rearing of the codling moth (*Carpocapsa pomonella* L.) on calcium alginate gels. ENTOMOPHAGA 16: 381-387.

Best of 4 diets tested.

COMPOSITION Distilled water, 220.0 ml; (g): sodium alginate, 5.0; autoclaved soy germ, 50.0; vitamin-free casein, 20.0; sucrose, 20.0; torula yeast, 20.0; Nipagin, 0.4; sorbic acid, 0.2; $CaH_4\ (PO_4)_2$, 1.25; ascorbic acid, 2.0; and (ml): CH_3COOH (20% v/v), 20.0; autoclaved ground peanut shells, 70.0 g.

PREPARATION Sodium alginate dissolved with part of water. Nutrients (in order of listing), dry antimicrobial agents, and calcium phosphate added and blended. Resulting mixture, ascorbic acid, and acetic acid then homogenized. Resulting gel broken by further mixing, ground peanut shells mixed with gel particles, and dispensed to plastic bags for collective rearing or to glass vials for individual rearing.

Soy germs obtained by steeping soy beans under running water for 24 h, followed by actual germination on moist flannel overnight. Soy germs then drained, and ground (1000 mesh). Peanut shells mixed with 50 ml of water; autoclaved, and deep frozen until use.

Replacement of agar by sodium alginate reduced cost by factor of 5.

REARING At 27 ± 2° C and 70-80% RH with 16 h photoperiod; in thin polyethylene bags with diet, placed in plastic boxes and flattened to obtain 2 cm thick food layer; bags tied shut and surface punctured. Inoculated area surrounded by strip of corrugated paper for pupation. Or, individually in glass vials with 5 g diet plugs coated in Myvacet to prevent desiccation. Adults placed in mating/oviposition cages @ 5-10 pairs/cage. Food and oviposition lining renewed every 2 days.

DEVELOPMENT Survival to pupation, 50%; larval period, 14-18 days when reared collectively, and 15-21 days when reared singly; adult emergence, 100%; pre-oviposition period, 3 days; eggs viable.

When adults fed mixture of water, honey, 10.0%, ascorbic acid 0.5%, B-complex vitamins (thiamine 0.02%, riboflavin 0.001%, nicotinamide 0.002%, choline chloride 0.24%), fecundity increased from 48.5 eggs/♀ to 118.7 eggs/♀.

→ Redfern, R.E. 1964b. Concentrate medium for rearing the codling moth. J ECON ENTOMOL 57: 607-608.

COMPOSITION (g): apple seeds, 6.0; dried apple, 12.0; English walnut meats, 3.0; sugar, 12.0; soybean protein, 12.0; Wesson's salt mixture, 2.5; yeast powder, 2.5; glycine, 0.25; cysteine, 0.125; cholesterol, 0.125; Alphacel, 6.0; agar, 4.0; ascorbic acid, 2.0; and (ml): linseed oil, 2.0; vitamin solution, 2.5; mould inhibitor, 4.5; 10% KOH, 7.5; water, 220.0.

vitamins and mould inhibitor as in Redfern* (1963).

PREPARATION Apple and walnut blended with 120 ml boiling water. Agar dissolved in remaining boiling water. Remaining ingredients (except ascorbic acid and Alphacel), and apple/walnut solution added and thoroughly blended. Alphacel added, mixed and cooled to 140° F; then ascorbic acid added and diet mixed again. Dispensed to cups @ 0.25-oz/cup. 3 apple seeds imbedded in diet before solidification.

REARING At 80°F and 65% RH; @ 2 larvae/cup.

DEVELOPMENT Adult emergence, 60%.

Rock, G.C. 1967. Aseptic rearing of the codling moth on synthetic diets: ascorbic acid and fatty acid requirements. J ECON ENTOMOL 60: 1002-1005.

Two diets used.

COMPOSITION Diet 1 (g/100 g diet): vitamin-free casein, 4.0; cysteine, 0.2; glycine, 0.2; salt mixture, 0.471; sucrose, 5.0; safflower oil, 0.4; cholesterol, 0.2; vitamin mixture, 0.952; Tween 80, 0.3; agar, 3.0; and (ml): 2M KOH, 3.5; water to 100 g.

salts (mg/100 g diet): K_2HPO_4, 160; NaH_2PO_4, 80; NaCl, 20; $MgSO_4.7H_2O$, 100; $FeSO_4.7H_2O$, 40; $CuSO_4.5H_2O$, 2; $MnSO_4.H_2O$, 5; $ZnCl_2$, 5; KI, 1; and $CaCO_3$, 60.

vitamins (mg/100 g diet): folic acid, 2; riboflavin, 2; calcium pantothenate, 4; pyridoxine hydrochloride, 6; thiamine hydrochloride, 1.2; nicotinic acid, 12; biotin, 0.025; B_{12}, 0.004; choline chloride, 100; inositol, 10; ascorbic acid, 800; and α-tocopherol, 15.

Diet 2 (g/100 g of diet): casein, 3.50; cysteine, 0.20; wheat germ, 3.00; sucrose, 3.00; Wesson's salt mixture, 1.20; safflower oil, 0.30; vitamin mixture, 0.95; agar, 3.00; and (ml): 2M KOH, 3.00; mould inhibitor, 2.20; and water to 100 g.

vitamins: as for diet 1.

mould inhibitors: sorbic acid, 20.0 g; methyl parasept, 15.0 g; and ethyl alcohol, 170.0 ml.

PREPARATION Diet 1: all constituents, except ascorbic acid, mixed in Waring Blendor, and heat-sterilized in flasks at 15 lb pressure for 15 minutes. Cooled to ca. 40°C then aqueous solution of ascorbic acid, sterilized through Millipore filter (0.45 μm), added. Shaken then aseptically dispensed, in 5-6 g portions, to sterilized vials; stoppered with sterilized cotton. pH of diet, 6.0.

Diet 2: ingredients blended in order listed; agar dissolved in hot water before adding.

REARING Oviposition at 81 ± 1°F with constant lighting in 10" lantern globes lined with cellophane paper; 4-day-old eggs removed from paper, surface sterilized by immersion in 6.5% formalin for 30 minutes, washed 3 times in sterile distilled water and placed in sterile shell vials stoppered with sterile cotton. Neonate larvae transferred aseptically to sterile rearing vials, @ 1 larva/vial and reared at 81 ± 1°F with 18 h photoperiod.

DEVELOPMENT On diet 1, 3 successive generations reared. Growth and development similar to that on immature apples. On diet 2, 5 successive generations reared; larval survival to adult, 55-65%. Larval development, adult fecundity, and egg viability similar to those on immature apples.

→ Sender, C. 1969. Élevage permanent du carpocapse des pommes *Carpocapsa* (= *Laspeyresia*) *pomonella* L. sur milieu artificiel simplifié. ANN ZOOL ECOL ANIM 1: 321-326. (In French with English summary)

COMPOSITION (g): agar, 5.0; pulped apple powder, 9.0; corn semolina, 10.0; wheat germ, 12.0; brewers' yeast, 10.0; ascorbic acid, 1.0; methyl *p*-hydroxybenzoate, 0.3; benzoic acid, 0.3; and (ml): distilled water, 170.0; corn oil, 0.4; Aureomycin solution (1 g/50 ml), 1.0.

PREPARATION Agar dissolved in boiling water, cooled to 56° C, and blended with remaining ingredients for 2 minutes at 15000 r.p.m. Dispensed to boxes, cooled and stored at 8° C.

REARING Individually in tubes with 2 cm^3 plug of diet at 25° C and 70% RH with 18 h photoperiod. From ca. 15th day, tubes placed in darkness. As larvae surface from diet to pupate, transferred to 80 mm diam cylindrical boxes with rolls of transparent grooved paper; pupation in grooves. Adults paired singly or in groups in 80 x 100 mm diam plastic boxes, covered by mesh lid. Walls and base lined with glassine paper. Fed from dental rolls with 5% sugar solution. Oviposition on paper, renewed every 2 days. Adults held at 20° C and 75% RH with natural lighting.

DEVELOPMENT Incubation period, 7 days. Average survival to adults, 88%. Fecundity and fertility of adults normal.

Corollary

→ Sender, C. 1970. Élevage due carpocapse des pommes sur un nouveau milieu artificiel non spécifique. ANN ZOOL ECOL ANIM 2: 93-95. (In French with English summary)

Two diets compared.

COMPOSITION Diet 1: as in Sender* (1969).

Diet 2: as in Poitout & Bues* (1970) developed for noctuid larvae.

DEVELOPMENT Both diets satisfactory. Ten generations successfully reared with no detrimental effects. >80% survival to adult. Average larval and pupal period, 32 days; ♀ fertility and longevity of adults satisfactory.

→ Shumakov, E.M.; Edelman, N.M.; Borisova, A.E.; Yakimova, N.L. 1974. Mass rearing of the codling moth on artificial nutritive media. PROC ALL-UNION RES INST PLANT PROT 40: 7-17. (In Russian with English summary)

Best of five diets tested.

COMPOSITION (g): agar, 2.0; casein, 4.0; sifted wheat germ, 4.5; dry yeast, 1.5; ascorbic acid, 0.8; sucrose, 4.0; cellulose, 2.0; choline chloride, 0.050; folic acid, 0.001; nicotinic acid, 0.010; calcium pantothenate, 0.012; biotin, 0.00005; B_6, 0.0018; Metaben, 0.150; formalin, 0.5; water to 100 ml.

REARING In petri dish with 30-35 g diet to 1.5 cm depth. 125-160 larvae/dish. Tracing cloth with eggs incubated under inverted petri dish cover.

DEVELOPMENT Egg to adult, 29 days; average wt: ♀, 21.7 mg; ♂, 14.3 mg. Fecundity, 144 eggs/♀. Average number of adults/kg diet, 230. 32 generations reared.

See also: Bode (1971); Borisova (1972, 1973a b); Petrushova (1971); Pristavko & Yanishevskaya (1972a b); Shumakov (1959); Szalay-Marzsó (1972); White (1966); Yakimova (1973); Yanishevskaya (1969).

Rhyacionia buoliana (Schiffermüller), European pine shoot moth

→ Chawla, S.S.; Harwood, R.F. 1968. Artificial diets for the European pine shoot moth, *Rhyacionia buoliana* (Schiffermüller) (Lepidoptera: Olethreutidae). WASH AGRIC EXP STN TECH BULL 59: 13 pp.

Three diets tested.

COMPOSITION Diet 1. Modification of that of Berger* (1963) for *Heliothis* sp. (ml): water, 285.0; 22.5% KOH, 4.0; and (g): casein, 28.0; Wesson's salt mixture, 2.0; sucrose, 21.0; and (ml): 36.7% formaldehyde, 1.1; 15% methyl *p*-hydroxybenzoate in alcohol, 8.0; and (g): sorbic acid, 0.83; wheat germ, 24.0; Alphacel, 12.0; vitamin diet fortification mixture, 8.0; ascorbic acid, 2.8; agar, 16.0; dissolved in 300.0 ml of water.

Diet 2. Based on bean diet formulated by Shorey & Hale* (1965) for noctuid spp. (ml): water, 263.0; 22.5% KOH, 4.0; and (g): soaked beans, 177.8; brewers' yeast, 26.7; and (ml): 36.7% of formaldehyde, 1.7; and (g): methyl *p*-hydroxybenzoate, dry, 1.67; sorbic acid, 0.83; ascorbic acid, 2.67; agar, 10.7; dissolved in 270.0 ml of water.

Diet 3. Combination of diets 1 and 2 (ml): water, 285.0; 22.5% KOH, 4.0; and (g): casein, 14.0; soybean flour, 40.0; brewers' yeast, 13.3; Wesson's salt mixture, 1.0; sucrose, 10.5; and (ml): 36.7% formaldehyde, 1.4; 15% methyl *p*-hydroxybenzoate in alcohol, 8.0; and (g): sorbic acid, 0.83; wheat germ, 12.0; Alphacel, 6.0; vitamin diet fortification mixture, 4.0; ascorbic acid, 2.8; agar, 13.5; dissolved in 284.0 ml of water.

vitamins (ml/100 g diet): niacinamide, 1.00; calcium pantothenate, 1.0; thiamine hydrochloride, 0.25; riboflavin, 0.50; pyridoxine hydrochloride, 0.25; folic acid, 0.25; biotin, 0.02; vitamin B_{12}, 0.002; inositol, 20.0. Vitamin mixture of diet 2 without inositol.

PREPARATION Diet 1: see Berger* (1963).

Diet 2: see Shorey & Hale* (1965).

Diet 3: ingredients blended in order given, autoclaved at 10 lb pressure for 10 minutes then dispensed to rearing containers.

Surface treatment of diets to control mould as in Chawla *et al.* (1967):- two methods used; (i) routine light surface spray or dust applied weekly to diet surface using aqueous mixture of 1 ml each of 3 stock solutions in 100 ml of water. Stock solutions: aqueous preparations of Amphotericin B (Squibb fungizone: E.R. Squibb & Sons, Ltd, Montreal, P.Q.), 41 mg/5 ml; Penicillin G, 50 mg/5 ml; and Streptomycin, 50 mg/5 ml.

(ii) Antifungal sprays of NaClO, sorbic acid, and methyl *p*-hydroxybenzoate @ 0.01, 0.20, and 1.00% of NaClO in 95% ethyl alcohol; and 0.5, 1.0, and 5.0% of sorbic acid in acetone. For dust treatment, powdered sorbic acid used.

REARING At 25 C with 15 h photoperiod in clear plastic 7-dram vials. 2nd to 4th instar larvae dissected from infested pine shoots and introduced to vials @ 2-3/vial. Inside cap of vial lined with 100 mesh screen to prevent larval escape by boring. Mass rearing in 4 x 8" or 9.5 x 13.5" plastic containers; lids lined with aluminium foil. Diets offered in plugs or diet bored and sections of drinking straw containing larvae transferred to holes thus formed; or larvae directly put in holes and capped with small piece of diet obtained during boring.

DEVELOPMENT Development and growth on diet 1, normal; on diet 2, sub-normal. On diet 3, >70% survival from individual rearing vials.

Mass rearing successful. Moulds inhibited by surface treatments of diets with antifungal agents, particularly sorbic acid. Pupation, 64.5%. First instar larvae grew well until 60 days; <36.7% 1st instar larvae reached pupation; ca. 90% survival to pupae of field collected 2nd or 3rd instar larvae.

Corollary

Ross, R.H. Jr; Monroe, R.E.; Butcher, J.W. 1971. Studies on techniques for the xenic and aseptic rearing of the European pine shoot moth, *Rhyacionia buoliana* (Lepidoptera: Olethreutidae). CAN ENTOMOL 103: 1449-1454.

Best of 18 diets tested.

COMPOSITION See diet 1, Chawla & Harwood* (1968).

REARING Eggs sterilized in 0.1% NaClO solution for 5 minutes and individually placed in vials 24 h before hatching. (4/vial in non aseptic control vials).

DEVELOPMENT Under aseptic conditions, pupation, 7.4%; 9 adults emerged from 231 larvae.

Rhyacionia frustrana (Comstock), Nantucket pine tip moth

Creswell, M.J.; Sturgeon, E.E.; Eikenbary, R.D. 1971. Laboratory rearing of the Nantucket pine tip moth, *Rhyacionia frustrana*, on artificial diets. ANN ENTOMOL SOC AM 64: 1159-1163.

Best of 7 diets tested.

COMPOSITION (g): wheat germ, 24.0; sucrose, 10.5; casein, 28.0; Wesson's salt mixture, 2; Alphacel, 12.0; dextrin, 10.5; sorbic acid, 0.8; Vanderzant's fortification mixture, 8.0; ascorbic acid, 2.8; agar, 16; and (ml): 22.5% KOH, 4; 10% formaldehyde, 1.1; methyl *p*-hydroxybenzoate, (6 g in 400 ml of 95% ethanol), 8.0; water, 585.0.

PREPARATION See Chawla & Harwood* (1968). When cooled, diet cubed, and placed into 1-oz plastic medicine cups with plastic coated paper lids. pH 4.6. Stored at 1-3° C.

REARING At 23-26° C, individually or in groups. Pupation in cups. Adults transferred to 1-dram glass vials for oviposition, and vials set aside for development of eggs. Larvae removed within 2 h of hatching and placed on diet.

DEVELOPMENT Larval period, 51 days; pupation, 28.4%; adult emergence, 9.2%; pupal wt (mg): ♀, 6.1; ♂, 4.1, comparable to field-collected pupae.

Papilionidae

Papilio protenor demetrius Cramer

→Nagasawa, S.; Nakayama, I. 1969. Rearing of *Papilio protenor demetrius* Cramer (Lepidoptera: Papilionidae) on artificial diets with a special reference to the nutritional roles of yeast, soybean protein, cholesterol and choline chloride. KONTYÛ 37: 327-337.

COMPOSITION (g): agar, 6; cellulose powder, 10; mandarine orange leaf powder, 40; sucrose, 4; glucose, 2; sodium copper chlorophyllin, 0.1; sorbic acid (sodium salt), 0.3; sodium *l*-ascorbate, 0.6; Wesson's salt mixture, 0.6; brewers' yeast, 1, 2, 4, 8, or 16; soybean protein, 1, 2, 4, 8 or 16; cholesterol, 0.05, 0.1, 0.2, 0.4 or 0.8; choline chloride, 0.1, 0.2, 0.4, 0.8 or 1.6; and (ml): 40% formaldehyde, 4; water, 200.

PREPARATION Components except ascorbic acid and formaldehyde thoroughly blended and autoclaved for 15 minutes at 120° C and 1 kg/cm^2. Ascorbic acid in formaldehyde added when diet cooled to below 55° C. Diet stirred and refrigerated.

REARING At 25° C. Neonate larvae fed on young leaves of *Citrus unshiu*. 2nd instar larvae placed on diet in covered 9.0 x 1.5 cm petri dishes. Mature larvae transferred to individual glass vials lined with filter paper for pupation.

DEVELOPMENT Faster with increasing yeast or choline chloride. Best growth on diet with 0.8 g cholesterol, 0.4 g choline chloride, 1.0 g soybean protein, and 16.0 g yeast. Pupal wt: ♀, 2.09 g; ♂, 1.85 g. Developmental periods: ♀, 46.4 days; ♂, 44.2 days; pupation, 82.6%; emergence from pupae, 89%.

Papilio xuthus Linnaeus, swallow tail butterfly

→Kamano, S. 1965. Rearing of *Papilio xuthus* larvae on artificial diets. JPN J APPL ENTOMOL ZOOL 9: 133-135. (In Japanese)

Two diets tested.

COMPOSITION Diet 1: Kamano* (1964a) diet developed for *Barathra brassicae* (Lep.; Noctuidae), except that Na-Cu-chlorophene, 0.05 g, replaced by Cu-chlorophillin, 0.05 g; and dehydro-sodium-acetate, 0.05 g replaced by 2.0 ml 4% formaldehyde. Leaf content varied from nil to 20.0 g finely-chopped trifoliate orange (*Poncirus trifoliatum*) leaves, to 20.0 g finely chopped cabbage leaves.

Diet 2: as diet 1, except casein content doubled to 4.0 g, and both dehydro-sodium-acetate, 0.05 g, and 4% formaldehyde, 2.0 ml, used.

PREPARATION As in Kamano (1964a), leaving sodium ascorbate and 4% formaldehyde to be combined and added last.

REARING On diet 1: at 30° C in 9 cm petri dishes; 4-5 1st instar larvae, 2-3 later instar larvae. Diet renewed every 3-4 days. On diet 2: at 25° C. 1st to 4th instar larvae in 10 x 2 cm diam test tubes plugged with cotton. 5th instars in 6 x 3 cm vinyl containers. Diet renewed 3 times. Reared at 16 h and 12 h photoperiods; pupae from 12 h photoperiod kept at 4° C for 60 days then returned to 25° C.

DEVELOPMENT On diet 1: larvae reared from eclosion to adults with orange leaves; all adults from 3rd and 4th instar larvae introduced to diets with or without cabbage leaves, emerged with wing deformities. Neonate larvae reared from eclosion on diet with no leaf, all died in 5-10 days.

On diet 2: more rapid development; greater pupal weight and lower mortality with increased casein content. With 16 h photoperiod, emergence from 10-12 days after pupation; with 12 h photoperiod, 60 days at 4°C required, then emergence 15-27 days after return to 25°C.

Adults all slightly smaller than field sample average, though those from increased-casein diet with 12 h photoperiod nearly equal.

Pieridae

Colias eurytheme Boisduval, alfalfa caterpillar

→Tanada, Y.; Chang, G.Y. 1968. Resistance of the alfalfa caterpillar *Colias eurytheme*, at high temperatures to a cytoplasmic - polyhedrosis virus and thermal inactivation point of the virus. J INVERTEBR PATHOL 10: 79-83.

Diet modified from Shorey & Hale* (1965) developed for noctuid species.

COMPOSITION (g/100 ml diet): dried pinto beans, 12.0; dried brewers' yeast, 3.5; Wesson's salt mixture, 1.0; agar, 2.0; alfalfa leaf meal, 1.5; 10% formaldehyde, 0.05; methyl *p*-hydroxybenzoate (15% w/v in 95% ethyl alcohol), 0.15; sorbic acid (7.5% in 95% ethyl alcohol), 0.075; α-tocopherol, 0.025; soybean oil, 0.45; Tween 80, 0.3; ascorbic acid, 1.5; Aureomycin (pure), 1.5; distilled water, 77.4.

PREPARATION See Shorey & Hale* (1965).

REARING At 23 or 35°C. Individually in 2-oz plastic styrene jars with snap-on plastic covers containing 2-3 g diet.

Pieris brassicae (Linnaeus), cabbage worm or white butterfly

→Chun, M.W. 1972. Dynamics of feeding responses in *Pieris brassicae* Linn. as a function of chemosensory input: a behavioural, ultrastructural and electrophysiological study. MEDED LANDBOUWHOGESCH WAGENINGEN 72-11, 162 pp.

COMPOSITION Stock mixture (g): vitamin-free casein, 350; wheat germ, 300; Alphacel, 50; Wesson's salt mixture, 100; sucrose, 350; cholesterol + β-sitosterol, 50/50, 5; sorbic acid, 24; methyl *p*-hydroxybenzoate, 18; Aureomycin, 25; kanamycin sulphate, 1.4.

others (g): agar, 12.5; (ml): 10% formaldehyde, 2; 10% KOH, 5; linseed oil, 0.5; Vanderzant's vitamin mixture, 7.5; and water, 420.

PREPARATION Stock mixture ingredients blended and refrigerated. Agar dissolved in 300 ml water and blended with all remaining ingredients and 61 g stock mixture.

REARING Diet for routine rearing of 1st to 5th instar larvae, but not for maintenance of continuous cultures. Larvae reared at temperatures of 22.5 ± 2.5°C during day and 17.0 ± 2.5°C at night with 18 h photoperiod in 15 x 18 x 6 cm plastic boxes.

→ David, W.A.L.; Gardiner, B.O.C. 1965. Rearing *Pieris brassicae* L. larvae on a semi-synthetic diet. NATURE LONDON 207: 882-883.

COMPOSITION Diet compounded in 5 parts:- (a) casein (light white soluble), 12.6 g; and (ml): distilled water, 110; 4M KOH, 1.8.

(b) (g): sucrose, 12.6; wheat germ (Bemax), 10.8; cabbage (dried powder), 5.4; salt mixture, 3.6; Whatman, Chromedia cellulose powder, CF11 grade, 1.8.

(c) (ml): 10% choline chloride, 3.6; methyl *p*-hydroxybenzoate (15% in 95% ethyl alcohol), 3.6; formaldehyde solution (10% w/v), 1.5; vitamin stock, 0.8.

(d) Fine (Japanese powder) agar, 9 g; and 200 ml of distilled water.

(e) (g): *l*-ascorbic acid, 1.5; Aureomycin (veterinary grade), 0.8.

salts (g): $CaCO_3$, 120; K_2HPO_4, 129; $CaHPO_4.2H_2O$, 30; $MgSO_4.7H_2O$, 40.8; NaCl, 67; $FeC_6H_5O_7.6H_2O$, 11; KI, 0.32; $MnSO_4.4H_2O$, 2.0; $ZnCl_2$, 0.10; $CuSO_4.5H_2O$, 0.12.

vitamins (mg/100 ml water): nicotinic acid, 600; calcium pantothenate, 600; riboflavin, 300; aneurine hydrochloride, 150; pyridoxine hydrochloride, 150; folic acid, 150; *d*-biotin, 12; cyanocobalamine (B_{12}), 1.2.

veterinary grade Aureomycin soluble powder (Cyanamid of Great Britain) contained 25 g chlortetracycline hydrochloride/lb.

PREPARATION Cabbage leaves dried in ventilated oven at 105°C for 15-20 minutes, then ground to 60 mesh. Part a ingredients blended, and mixed with those of part b. Part c added. Agar dissolved in water, cooled to 70°C and added to abc. Ascorbic acid and Aureomycin added and diet blended. Warm diet dispensed to sterilized 1-lb jam jars to depth of ½" and jar rotated so as to coat sides. Diet cooled and jars inverted; stored at 12°C.

REARING Neonate larvae to jars covered with filter paper until 5th instar, then with 'Terylene' gauze. About 15, 5th instar larvae reared/jar. Smaller numbers of larvae reared in 2 x 1" specimen tubes with ½" diet. Diet renewed weekly.

DEVELOPMENT 1-2 days longer to pupation than on cabbage leaves. Pupal yield from neonate larvae, 80-100%; pupal wt, 0.37 g (0.38 g on cabbage); ca. 100% normal adult emergence.

See also: David & Gardiner (1966b c d); David *et al.*(1972a b); Geest & Sloog-Hoebe (1972).

→ Turunen, S. 1974b. Polyunsaturated fatty acids in the nutrition of *Pieris brassicae* (Lepidoptera). ANN ZOOL FENN 11: 300-303.

COMPOSITION (%): distilled water, 81.50; casein, 3.30; sucrose, 3.30; wheat-germ (lipid-free), 2.80; cabbage leaf powder (lipid-free), 1.40; trilinolein or trilinolenin, 0.50; cholesterol, 0.30; salt mixture, 0.95; vitamin stock, 0.25; *l*-ascorbic acid, 0.39; 10% choline chloride, 0.95; methyl *p*-hydroxybenzoate (15% in 94% ethanol), 0.95; 10% formaldehyde, 0.39; Aureomycin, 0.21; 4M KOH, 0.46; agar, 2.36.

salts and vitamins as in David & Gardiner* (1965).

PREPARATION As in David & Gardiner* (1965). Control diet with 0.25% linseed oil in addition to whole wheat germ and cabbage leaf powder without lipid extraction. In test diets, lipid extraction of wheat germ and cabbage carried out for 5 h with chloroform-methanol (2:1). Cholesterol added in 94% ethanol 2 minutes before diet dispensed into dishes.

REARING At 23°C and 70% RH with 18 h photoperiod. 1st-3rd instar larvae reared in inverted 100 ml glass dishes covered with filter paper. 3rd instars transferred to 250 ml dishes, and transferred again to fresh diet at 5th instar.

DEVELOPMENT From egg to adult, 27 days on control diet; 35 days on trilinolenin supplemented diet. Zero emergence from trilenolein supplemented diet, but 43% attempted to do so after 38 days. Of adults from trilinolenin diet, 69% ♂♂ normal; 21% with wing deformities; and 10% failed to emerge. Females normal. On control diet, 98% normal adult emergence.

See also: Turunen (1973a b, 1974a).

→ Wardojo, S. 1969b. Artificial diet without crude plant material for two oligophagous leaf feeders. ENTOMOL EXP APPL 12: 698-702.

COMPOSITION & PREPARATION Similar to diet for *Leptinotarsa decemlineata* (Say) (Col.; Chrysomelidae) in same paper, except: 0.22 g amino acid mixture replaced by 5.0 ml of 10% solution of enzymatic casein hydrolysate.

REARING At 25 ± 0.5°C and 50-70% RH with 18 h photoperiod. Ten neonate larvae in petri dish with 10 g diet; transferred to clean petri dish with fresh diet every 3 days. After 3rd moult, placed in 18.5 x 10.7 cm plastic box with gauze lid. Adults to rearing cages with cabbage plant and glass vial with 15% sucrose solution.

DEVELOPMENT Larval period, 14.8 ± 0.1 days; mean pupal wt, 421.1 ± 4.4 mg; adult emergence, >90%. Diet compared favourably with cabbage leaves. Two generations reared.

Pieris rapae crucivora Boisduval, cabbage white butterfly

→ Kono, Y. 1968. Rearing *Pieris rapae crucivora* Boisduval (Lepidoptera: Pieridae) on artificial diets. APPL ENTOMOL ZOOL 3: 96-98.

Best of six diets tested.

COMPOSITION (g): agar, 3.5; casein, 5.0; sucrose, 5.0; cellulose powder, 2.0; dried yeast powder (Ebios), 4.0; Beckman's salt mixture, 0.8; cholesterol, 0.1; choline chloride, 0.2; sodium ascorbate, 0.3; dried cabbage leaf powder, 4.0; sodium sorbate, 0.15; and (ml): distilled water, 120; rapeseed oil, 0.8; 4% formaldehyde solution, 2.0.

salts (%): $CaCO_3$, 120.0; K_2HPO_4, 129.0; $CaHPO_4.2H_2O$, 30.0; $MgSO_4.7H_2O$, 40.8; NaCl, 67.0; $FeC_6H_5O_7.6H_2O$, 11.0; KI, 0.32; $MnSO_4.6H_2O$, 2.0; $ZnCl_2$, 0.10; $CuSO_4.5H_2O$, 0.12.

PREPARATION Ingredients except sodium ascorbate and formaldehyde mixed thoroughly and autoclaved for 30 minutes. When cooled to 50°C sodium ascorbate dissolved in 4% formaldehyde and added; refrigerated until use. Cabbage leaves dried at 80-100°C for 2 h and ground in mortar. Rapeseed oil emulsified with Tween-20 (0.1 ml in 0.8 ml rapeseed oil).

REARING At 20°C with 16 h photoperiod. About 24 adults of both sexes placed in 25 x 25 x 45 cm cage with cabbage plant for oviposition, and sucrose solution on cotton pad. 24-h-old-1st instar larvae transferred to 9 x 2 cm petri dishes with 2 x 3 x 0.5 cm diet plug @ 10-15 larvae/dish. Numbers reduced to 5 or 6 from 3rd instar. Diet changed every 3-4 days until 5th instar when changed every other day.

DEVELOPMENT On diet, larval period, 24.5 ± 1.3 days; and mean pupal wt, 283 ± 31 mg; on cabbage leaf, 18.4 ± 1.0 days, and 234 ± 14 mg respectively. Normal adult emergence, >80%. Two generations reared.

Sato, Y. 1974. Rearing *Pieris rapae crucivora* larvae on artificial diets. KONTYÛ 42: 467-472. (In Japanese with English summary)

Two diets tested.

COMPOSITION Diet 1, as in Kono* (1968), except Beckman's salt mixture replaced by Wesson's salt mixture.

Diet 2 (g): dried cabbage leaf powder, 60; dried Chlorella powder, 60; sucrose, 5; agar, 13; inositol, 1.25; ascorbic acid, 2.5; vitamin mixture, 0.125; choline chloride, 0.4; sorbic acid, 1; Streptomycin, 0.125; gallic acid, 1.25; K_2HPO_4, 3.5; KH_2PO_4, 1.4; and distilled water, 700 ml.

vitamins (mg): biotin, 0.05; calcium pantothenate, 0.5; folic acid, 0.05; nicotinamide, 0.5; VB_6-HCl, 0.25; FMN-Na, 0.25; VB_1-CHl, 0.25; VB_T, 0.25; made up to 125 mg with starch.

PREPARATION Cabbage leaf powder dried for 1 h at 100°C and powdered in ball mill for 2 h. Agar, choline chloride, sorbic acid, K_2HPO_4 and KH_2PO_4 mixed, and dissolved in boiling water. Other mixtures added. Dispensed to either rearing jar @ 10 ml/jar, or to petri dishes to harden then cut into 2 x 4 x 0.3 cm plugs.

REARING At 25°C with 16 h photoperiod. Neonate larvae fed on cabbage for 28 h prior to transference on to diet. Diet 2: 40 larvae of 1st-2nd instar, or 20 of 3rd instar/jar. On cut-up diet, 10-20 larvae of 1st-2nd instar, 7-10 of 3rd-4th instar, 5-7 of 5th instar/segment. Diets stored at 5°C, renewed once in 2 days for 1st-2nd instars, daily for 3rd-5th.

DEVELOPMENT No differences between diets 1 and 2. Larval periods, 12.5 days and 12.2 days on diet and cabbage. Pupal wt, 191.3 mg and 188.9 mg on diet and cabbage.

Pyralidae

Cadra cautella (Walker) [previously *Ephestia*], almond moth

→ Waites, R.E.; Gothilf, S. 1969. Nutrition of the almond moth. I. Analysis and improvements of the experimental diet. J ECON ENTOMOL 62: 301-305.

Two diets used.

COMPOSITION Diet 1. For stock culture (parts): whole wheat flour, 4; cornmeal, 4; Purina dog food (ground fine), 2; brewers' yeast, 1; oatmeal, 1; wheat germ, 0.5; honey, 1; glycerine, 1.

Diet 2. Modification from Fraenkel & Blewett* (1946b) (g): casein, 48; sucrose, 48; McCollum-Davis salt mixture No. 185, 2; cholesterol, 1; wheat germ oil, 1; and 12 ml of vitamin solution.

vitamins (µg/g dry diet); thiamin hydrochloride, 50.0; riboflavin, 50.0; nicotinic acid, 50.0; pyridoxine hydrochloride, 50.0; *dl*-calcium pantothenate, 50.0; *p*-aminobenzoic acid, 50.0; *i*-inositol (meso), 500.0; choline chloride, 500.0; folic acid, 10.0; biotin, 0.1; and vitamin B_{12}, 0.4.

Diet later improved by addition of tribasic potassium phosphate @ 1 g/100 g dry diet; by replacing wheat germ oil with linolenic acid @ 160 mg/100 g dry diet; and by reducing cholesterol level to 400 mg/100 g dry diet.

PREPARATION Powdered sucrose mixed with casein, salt mixture, and pre-mixed cholesterol and oils. Vitamin solution added, transferred to closed jar and refrigerated overnight. Remixed in mortar before use. Vitamin solutions preserved at -90° C; stirred at ambient temperature before use.

REARING Adults at 27 ± 2° C and 50-60% RH; @ 100, 2-4-day-old/½-gal wide-mouthed mason jar. Shell vial with moistened cellucotton placed in each jar to feed adults and to provide high humidity. Jar covered with filter paper and sealed with metal jar ring. After 10 days water vials removed and corrugated cardboard added for pupation. After emergence, adult moths placed in quart mason jars containing water vials and sealed with screened jar ring. Jars then turned upsidedown in racks and eggs collected in petri dishes as dropped through screen cap. Eggs washed in distilled water, dried on filter paper. Neonate larvae transferred to shell vial with 1.5 g diet and reared at 30 ± 0.5° C and 70-80% RH in darkness.

DEVELOPMENT Larval periods on synthetic diet, improved diet and stock diet, 26.4, 22.8 and 16.1 days respectively. Adult emergence, 90%. Adults normal.

See also: Stevens (1971).

→ LeCato, G.L. 1976. Yield, development, and weight of *Cadra cautella* (Walker) and *Plodia interpunctella* (Hübner) on twenty-one diets derived from natural products. J STORED PROD RES 12: 43-47.

Best of 21 diets tested.

COMPOSITION (parts by volume): white cornmeal, 4.0; whole wheat flour, 4.0; Gaines dog food (granular or pellets), 2.0; yeast (brewer's or torula), 1.0; honey, 1.0; glycerine (96% white, chemically pure), 1.0; rolled oats, Quaker old fashioned, 1.0; wheat germ, 0.5.

PREPARATION Ingredients mixed. Diet frozen for 2 weeks at -20°C. Held at 60 ± 10% RH for 1-2 weeks before use.

REARING At 27 ± 2°C and 60 ± 10% RH with 12 h photoperiod. Three newly emerged pairs of either species placed into 0.47-1 jars containing 200 ml of diet.

DEVELOPMENT Quotient of log yield (number of adult progeny produced) divided by minimum number of days to adults (development) for *C. cautella* and *P. interpunctella* on diet, 0.0990 and 0.0912; mean live adult wt, 12.7 ± 0.1 and 12.0 ± 0.2 mg respectively; quotient and weights of both species greater on artificial diet than on most natural foods.

Chilo agamemnon Bleszynski

→ Isa, A.L. 1972. A standard technique for rearing *Chilo agamemnon* Bles. on an artificial diet. AGRIC RES REV 50: 61-64.

COMPOSITION (g): bacto-agar, 3; cellulose, 1.5; glucose, 1.5; sucrose, 1.5; casein, 1.5; yeast extract, autoclaved, 6.0; powdered rice bran, 9.0; ascorbic acid, 0.45; and (ml): 20% methyl paraben in alcohol, 1.5; 10% sorbic acid in alcohol, 1.5; distilled water, 150.

PREPARATION Agar and cellulose boiled in distilled water for 15-20 minutes. Other dry ingredients except rice bran added and stirred for 5 minutes. Antibiotic solutions stirred into mixture then rice bran added; cooled for 3-5 h. 4 g cooled diet dispensed to 2.5 x 8 cm glass vials.

REARING Individually in vials at 27°C in darkness. Vials plugged with cotton and aluminium caps. Pupae placed on moist cotton in clean vials for emergence. Pairs of adults placed in glass jars with muslin covers, lined with wax paper for oviposition, and with moist cotton to maintain humidity. Pieces of wax paper with eggs incubated in petri dishes.

DEVELOPMENT Larval stage, 27 days; on corn, 21.9 days; larval mortality, 9 out of 45 on diet; 15 out of 40 on corn. 80% pupation on diet; 62.5% on corn. ♀ pupal wt, 45.2 mg on diet; 40.0 mg on corn. Viable eggs laid.

Chilo auricilius (Dudgeon), stalk borer

→ Varma, A.; Avasthy, P.N. 1973. An artificial diet for the rearing of stalk borer, *Chilo auricilius* Dudg. EXPERIENTIA 29: 1161-1162.

COMPOSITION Larval diet (g): French bean (*Phaseolus vulgaris*), 100.0; yeast extract, 16.0; ascorbic acid, 1.0; casein, 50.0; Wesson's salt mixture, 5.0; sorbic acid, 0.5; methyl *p*-hydroxybenzoate, 1.0; vitamin diet fortification mixture (NBC), 8.0; agar-agar, 8.0; powder of dried sugar cane tops, 50.0; and (ml): formaldehyde, 1.0; water, 425.0.

PREPARATION Beans soaked in tap water for 24 h, boiled and seed coats removed. Peeled seeds mixed with other ingredients except agar, ascorbic acid, sugar cane powder and vitamins, and ground to paste in pestle and mortar. Agar dissolved in 155 ml water and sugar cane powder added whilst stirring. Bean paste, in 240 ml water, added, mixed and cooked for 5 minutes in water-bath (temp. not given). Vitamin mixture and ascorbic acid dissolved separately in 30 ml distilled water and added with formaldehyde to cooling diet. Dispensed to ¼ fill sterilized 9.5 x 2.5 cm glass tubes, slanted, and allowed to set. Tubes stored overnight for excess moisture to evaporate.

REARING At 22-27°C and 60-75% RH. Neonate larvae placed on diet @ 10/tube. After 2-3 weeks, transferred to fresh tubes @ 5/tube. Pupation from 33-58 days either near cotton plug or in pupal cell within diet. Pupae removed and incubated in 7.5 cm petri dishes; adult emergence from 6-10 days; mated in 15 x 10 cm diam glass jars. Oviposition on freshly cut 15 cm long sugar cane leaf sections.

DEVELOPMENT Mean larval period, 47.0 days; pupation, 82.0%; mean pupal period, 8.3 days; adult emergence, 65.8%; adult longevity (days): ♂, 5-6; ♀, 4-5; eggs laid/♀, 224.8; sex ratio of ♂:♀, 1:0.8. Longevity and fecundity not significantly different from wild population.

Chilo orichalcociliellus Strand

→ Delobel, A. 1975. *Chilo orichalcociliellus* Strand (Lepidoptera, Pyralidae), foreur des tiges du sorgho et du mais â Madagascar. I. Caractéristiques morphologiques. CAH ORSTOM SER BIOL 10: 3-9. (In French with English summary)

COMPOSITION (g): brewers' yeast, 5; wheat germ, 30; cornflour, 30; chickpea flour, 30; ascorbic acid, 1.4; methyl *p*-hydroxybenzoate, 1.0; levulose, 10.0; dried milled maize stem, 2.0; agar, 7.0; and (ml): 10% formaldehyde, 2.0; water, 340.0.

REARING At 26-31°C with 12 h photoperiod; or 18-21 C with 11 h photoperiod. Neonate larvae reared individually in 15-45 mm diam transparent plastic boxes. Diet renewed twice weekly. Pupae sexed and transferred to 208 x 50 mm diam plastic emergence boxes @ 10/box. Same size boxes lined with creased cellophane paper for oviposition; one pair/box; humidity maintained by damp cotton.

Chilo suppressalis Walker, rice stem borer

→ Guennelon, G.; Soria, F. 1973. Mise au point au laboratoire d'un élevage permanent de la pyrale du riz, *Chilo suppressalis* Walker (Lepidoptera, Pyralidae) sur milieu artificiel. ANN ZOOL ECOL ANIM 5: 547-558. (In French with English summary)

Modified from Poitout & Bues* (1970) for rearing noctuid larvae.

COMPOSITION (g): agar, 4.0; maize semolina, 28.0; wheat germ, 7.0; brewers' yeast, 7.5; ascorbic acid, 1.0; benzoic acid, 0.3; Nipagin, 0.25; powdered Aureomycin, 0.025; and (ml): water, 150.0.

PREPARATION Agar dissolved in water, cooled to 56°C and mixed with powdered ingredients in Waring mixer for 1-2 minutes. Diet dispensed to, and cooled in 6 x 8 cm diam plastic boxes to depth of 1.0-1.5 cm. Boxes covered with filter paper overnight to allow evaporation of surface moisture. Diet stored at 11°C for several weeks before use.

REARING At 25°C and 70% RH with 18 h photoperiod in sealed boxes. Diet scored for easier establishment of neonate larvae. Larval diapause avoided under these conditions. When larvae mature, boxes opened and placed within 8 x 10 cm diam boxes with corrugated lining to upper half for pupation of negatively geotrophic larvae. Emerging adults maintained in oviposition cages using technique similar to that perfected by Lazarevic (1961). Eggs incubated in 20 x 17 cm diam boxes at high humidity.

DEVELOPMENT Larval period, 26-28 days; adult yield, 85-90%. Generation completed every 56 days. Mean fecundity, ca. 200 eggs/♀. Ten generations reared.

Hormchong, T.; Srithunya, S.; Podhisritong, P. 1972. Mass rearing of rice stem borers. JPN PESTIC INF 10: 126-127.

COMPOSITION (g): sucrose, 12.0; Alphacel, 4.5; Wesson's salt mixture, 4.5; casein, 15.5; ascorbic acid, 2.0; cholesterol, 0.5; dried *Wolffia*, 13.5; agar, 8.0; and (ml): 22.5% KOH, 2.3; 10% formaldehyde, 2.0; 15% methyl *p*-hydroxybenzoate, 4.5; propionic acid, 0.5; Tetracycline, R-capsule, 0.5; Vanderzant's fortification mixture, 4.5; water, 400.0.

Wolffia arrhiza - small aquatic plant from Thailand.

REARING Flat whisky bottle (325 ml), clear plastic container, (7 x 4.5 x 2.5"), glass shell vials (4 drams) and medicine cups all used as rearing containers. Whisky bottle best.

Also suitable for other unspecified pyralids.

→ Ishii, S. 1971. Nutritional studies of the rice stem borer, *Chilo suppressalis* Walker, and its mass rearing. ENTOMOPHAGA 16: 165-173.

Best 4 of 8 diets tested.

COMPOSITION Diet 1, for aseptic rearing (g): agar, 0.7; cellulose, 1.0; glucose, 1.5; casein, 2.0; cholesterol, 0.02; minerals, 0.2; water extract of rice plant, 0.5-1.0; dried brewers' yeast, 1.0; and water, 42 ml.

Diet 2, basal diet to determine essential vitamins (g): agar, 0.3; cellulose, 1.0; glucose, 1.5; sucrose, 0.7; casein, 2.2; minerals, 0.2; cholesterol, 0.02; and water, 30 ml.

vitamins (μg): thiamine hydrochloride, 150; riboflavin, 75; nicotinic acid, 150; pyridoxine, 75; calcium pantothenate, 150; folic acid, 15; choline chloride, 3000; inositol, 1500; *p*-aminobenzoic acid, 50; biotin, 15.

For testing amino acids, carbohydrates, lipids, growth inhibitors of rice plants and attractants, composition of basal diet slightly varied.

Diet 3, semi-synthetic diet for mass rearing (g): agar, 1.0; cellulose, 1.0; glucose, 0.5; sucrose, 0.5; casein, 1.5; wheat bran 1.0; salt mixture, 0.4; sodium ascorbate, 0.3; cholesterol, 0.02; choline chloride, 0.05; *l*-cysteine hydrochloride, 0.03; Chlorella, 1.0; dried yeast, 1.0; and water, 50 ml.

Diet 4, simplified diet for mass rearing (g): wheat bran, 12.0; sucrose, 3.0; casein, 5.0; sodium ascorbate, 0.2; and (ml): 1% formalin, 0.3; water, 45.

PREPARATION All diets. Ingredients mixed and autoclaved at 1.2 kg/cm^2 for 15 minutes in flask.

REARING All diets. At 25 ± 1°C and 50-70% RH with 15 h photoperiod. Egg surface sterilized with 0.1% $HgCl_2$ solution, and transferred aseptically @ 50/flask. Matured larvae removed to petri dishes for pupation. Adults mated in glass pots; oviposition on paraffin paper.

DEVELOPMENT Diet 1: ca. 70% larvae pupated. Diet 2: used for nutritional requirements. Diet 3: ca. 97.5% larval survival and 89.5% adult emergence. Diet 4: ca. 91.5% larval survival and 87% adult emergence.

See also: Hirano (1963); Hirano & Ishii (1957, 1962); Ishii (1956); Ishii & Hirano (1955, 1957, 1964); Ishii & Urushibara (1954); Ishii *et al.* (1959); Koyama & Ishii (1950); Koyama *et al.* (1951); Yatomi *et al.* (1951).

→ Kamano, S. 1961. On the successive rearing of rice stem borer on the artificial diets under aseptic conditions. I. The effect of choline contained in the artificial diets on the oviposition and the hatching potency. JPN J APPL ENTOMOL ZOOL 5: 254-259. (In Japanese with English summary)

Two diets tested.

COMPOSITION Diet 1 (g): agar, 0.6; cellulose, 0.5; glucose, 0.7; sucrose, 0.3; casein, 1.0; brewers' yeast, 1.0; McCollum-Simmond's salt mixture, 0.2; cholesterol, 0.2; choline chloride, 0.2; barley juice (living stems and leaves pulped in blender and liquid squeezed through cloth bag), 20.0; and water, 40.0 ml.

Diet 2 (g): agar, 0.5; sucrose, 1.5; casein, 1.0; choline chloride, 0.3; barley (living stem and leaves cut o 0.5 cm lengths), 25.0; and water, 40.0 ml.

PREPARATION Cellulose, glucose, sucrose, yeast, casein, cholesterol, salt mixture and agar mixed; barley and choline chloride solution added, stirred and autoclaved at 112°C and 11-12 lb for 30 minutes.

REARING At 25°C or 30°C with long-day photoperiod. Mature larvae removed to tin with cellophane rolls, at 30°C and >90% RH.

DEVELOPMENT Larval period and weights, pupation and fecundity on diets inferior to wild strain. Shonai-ecotype strain reared aseptically on diet for 10 generations and Saigoku-ecotype strain for 6 generations under non-diapause conditions.

→ Kamano, S. 1964b. On the successive rearing of rice stem borer on the artificial diets under aseptic conditions. II. Ascorbic acid requirement of larvae. JPN J APPL ENTOMOL ZOOL 8: 101-105. (In Japanese with English summary)

COMPOSITION (g): agar, 0.6; cellulose (powdered filter paper), 1.0; sucrose, 0.5; glucose, 0.5; casein, 1.0; dried yeast, 1.0; McCollum-Simmond's salt mixture, 0.2; choline chloride, 0.1; cholesterol, 0.02; sorbic acid, 0.05; water, 50.0 ml. Ascorbic acid varied (mg): 0, 50, 100, 150, 200.

PREPARATION See Kamano* (1961). Sorbic acid dissolved in ether, absorbed by powdered filter paper; ether evaporated. Ascorbic acid (as sodium ascorbate) dissolved in 3 ml sterilized water, and added to diet aseptically immediately after autoclaving.

DEVELOPMENT 200-400 mg ascorbic acid/100 g diet increased larval weights, pupation, and emergence of normal female adults; also number of egg masses laid and hatching ratio increased. Larvae at late stages of development require more ascorbic acid than during early stages.

→ Kamano, S.; Fukaya, M. 1965. On the successive rearing of rice stem borer on the artificial diets under aseptic conditions. IV. The inbreeding and the rotational breeding of the rice stem borer. JPN J APPL ENTOMOL ZOOL 9: 89-93. (In Japanese with English summary)

COMPOSITION (g): powdered agar, 0.6; powdered filter paper, 1.0; glucose, 0.5; sucrose, 0.5; casein, 1.0; dried yeast, 1.0; McCollum-Simmond's salt mixture, 0.2; cholesterol, 0.02; choline chloride, 0.05; sodium ascorbate, 0.1; sodium sorbate, 0.1; rice-plant stems (1st-4th and 10th generation diet) or barley stems (5th-9th generation diet), finely chopped, 25.0; water, 50.0 ml.

REARING At 24-25°C with 16 h photoperiod. Inbreeding and rotational breeding methods compared.

DEVELOPMENT Larval growth, pupation, fecundity and hatching in 2nd or 3rd generation affected by inbreeding. Only 4 generations reared. Viability maintained for 10 generations during rotational breeding. Further improvement of diet necessary.

See also: Inoue & Kamano (1957).

→ Kamano, S.; Yushima, T. 1969. Rearing method of insects under aseptic and semi-aseptic conditions. PLANT PROT 23: 353-356. (In Japanese)

Also for *Adoxophyes privatana* (Walker) (Lep.; Tortricidae).

COMPOSITION For *C. suppressalis* (g): powdered gelatin, 0.6; powdered filter paper, 1.0; glucose, 0.5; cane sugar, 0.5; casein, 1.5; wheat bran, 2.0; Ebios, 1.0; mineral salts, 0.2; cholesterol, 0.02; choline chloride, 0.05; ascorbic acid, 0.3; dehydro-acetic acid, soda, 0.01; and (ml): formalin, 1%, 2.0; water, 50.0.

For *A. privatana* under semi-aseptic conditions (g): soybean flour, 10.0; powdered hydrocyanic acid, 6.0; Ebios, 2.0; gelatin, 1.4; propionic acid, 0.2; dehydro-acetic acid, soda, 0.025; 4N HCl, 1.0; and water, 47.0 ml.

PREPARATION For *C. suppressalis*. Solid ingredients mixed with water, autoclaved at 76 kPa (approx. 112°C) for 30 minutes and cooled to ca. 60°C; then ascorbic acid and formalin added and mixed.

For *A. privatana*. Gelatin dissolved in water and autoclaved at zero pressure. Soybean flour and Ebios added and mixed, and propionic acid and acetic acid solutions in water added when cooled. Finally HCl added, and diet blended.

REARING *C. suppressalis*. Glass flasks with cotton stoppers heated to 150-160°C and cooled; diet prepared in flasks and eggs inoculated aseptically. Eggs previously soaked in 70% alcohol for 10 seconds, then 0.1% corrosive sublimate solution for 4-5 minutes, washed in 70% alcohol and pasted to surface of flask. Adults mated by rotational breeding.

A. privatana in flasks or plastic containers washed in NaClO. Paraffin paper to support food.

DEVELOPMENT Life cycle of both species completed.

See also: Yushima & Kamano (1964).

Kamano, S. 1971. Studies on artificial diets of the rice stem borer, *Chilo suppressalis* Walker. BULL NATL INST AGRIC SCI JPN SER C 25: 1-45.

Reared on artificial, synthetic and chemically defined diets.

Artificial diets: based on fresh barley, rice bran, "Kinako" (parched soybean powder) and wheat bran.

COMPOSITION Fresh barley plant diet (g): agar, 0.6; sucrose, 1.5; casein, 1.0; choline chloride, 0.3; chopped barley plant, 25.0; water, 40.0 ml.

REARING For this, and all other diets described: larvae reared at 30°C with 16 h photoperiod in 200 ml flasks. Mature larvae transferred to wire-screen cages with cellophane tubes after 20-25 days and kept at 25°C and 90% RH. Pupae placed in petri dish with moist cotton. Adults kept in 15 x 12 cm diam glass containers with folded wax paper sheet for oviposition, and moist cotton. Eggs incubated in petri dishes with moist cotton at 25°C. Black head stage eggs sterilized in 0.1% $HgCl_2$ for 4 minutes, then washed with 70% ethyl alcohol and placed in wall of diet flask.

DEVELOPMENT Larval period, 17-19 days; mean larval wt, 99.0 mg; pupation, 98.4%; adult emergence, 87.8%; 5 generations reared.

Best of 3 diets based on rice bran.

COMPOSITION (g): agar, 1.0; cellulose, 1.0; sucrose, 1.0; casein, 0.5; dry yeast, 2.0; rice bran, 3.0; sodium ascorbate, 0.3; and (ml): 1% formaldehyde, 2.0; water, 50.0.

DEVELOPMENT Larval period, 20-22 days; mean larval wt, 83.9 mg; pupation, 98.6%; emergence, 80%; viable eggs laid. Two generations reared.

Best of 3 diets based on "Kinako".

COMPOSITION (g): agar, 1.0; sucrose, 1.0; casein, 1.0; Kinako, 3.0; dry yeast, 2.0; sodium ascorbate, 0.3; and (ml): 1% formaldehyde, 2.0; water, 50.

DEVELOPMENT Larval period, 20 days; mean larval wt, 89.1 mg; pupation, 98.8%; adult emergence, 87.1%; viable eggs laid. One generation reared.

Best of 4 wheat bran diets.

COMPOSITION (g): wheat bran, 12.0; sucrose, 3.0; casein, 5.0; sodium ascorbate, 0.2; and (ml): 1% formaldehyde, 3.0; water, 47.0.

DEVELOPMENT Larval period, 22 days; mean larval wt, 96.8 mg; pupation, 96%; adult emergence, 95%; viable eggs laid.

Synthetic diets:

Best of 5 diets tested.

COMPOSITION (g): agar, 1.5; glucose, 2.0; sucrose, 0.8; casein, 3.0; dry yeast, 2.0; McCollum-Simmond's salt mixture, 0.5; cholesterol, 0.05; choline chloride, 0.05; chopped rice plant leaf, 30.0; and 62.0 ml water.

DEVELOPMENT Larval period, 25 days; mean larval wt: ♀, 109.4; ♂, 77.6 mg. Pupation: ♂, 20.7%; ♀, 25.0%; adult emergence, 100.0%. Diet considerably improved by addition of further choline chloride and ascorbic acid.

Chemically defined diet:

Two diets tested.

COMPOSITION Diet 1 (g): agar, 0.6; cellulose, 1.0; glucose, 1.0; sucrose, 1.0; casein, 3.0; Wesson's salt mixture, 0.2; cholesterol, 0.02; choline chloride, 0.1; sodium ascorbate, 0.2; and (ml): 1% formaldehyde, 1.0; B-vitamin mixture, 2.0; water, 50.0.

Diet 2, as diet 1 except: agar increased from 0.6 to 1.0 g; protein reduced from 3.0 to 2.0 g; Wesson's salt mixture replaced by 0.4 g of salt mix of Beck *et al.** (1968); formaldehyde omitted; 0.03 g of *l*-cysteine hydrochloride added.

Beck's salts (%): K_2HPO_4, 49.75; KH_2PO_4, 18.00; $MgSO_4$, 16.00; $CaHPO_4$, 8.00; NaCl, 5.00; $Fe_2(SO_4)_3$, 2.00; $MnSO_4$, 0.50; $Zn(C_2H_3O_2)_2.2H_2O$, 0.50; $CuSO_4.5H_2O$, 0.025.

vitamins (mg/1000 ml water): thiamine hydrochloride, 200; riboflavin, 100; nicotinic acid, 200; pyridoxine hydrochloride, 100; calcium pantothenate, 200; folic acid, 20; biotin, 20; *p*-aminobenzoic acid, 200; inositol, 2000.

casein replaceable by *l*-amino acid mixture (mg): alanine, 60; arginine hydrochloride, 60; aspartic acid, 120; cystine, 30; glutamic acid, 380; glycine, 30; histidine, 60; hydroxyproline, 30; isoleucine, 120; leucine, 180; lysine hydrochloride, 180; methionine, 60; phenyl-alanine, 90; proline, 180; serine, 90; threonine, 90; tryptophan, 60; tyrosine, 60; valine, 120; total, 2000.

PREPARATION Sterilized in Koch's steam-sterilizer on 3 successive days. B-complex solution added after second sterilization and sodium ascorbate solution added aseptically after final sterilization.

DEVELOPMENT In protein comparison, best results obtained with 3.0 g casein. Larval feeding period, 22 days; larval survival, 74.7%; mean larval wt, 76.0 mg; pupation, 83.8%; adult emergence, 90.1%; viable eggs laid. On amino acid diet: larval feeding period, 18 days; larval survival, 80.3%; mean larval wt, 74.0 mg; pupation, 83.0%; adult emergence, 86.5%.

Corollary

→ Kamano, S. 1973. Studies on artificial diets and laboratory rearing methods suitable for successive generations of the rice stem borer, *Chilo suppressalis* Walker. BULL NATL INST AGRIC SCI JPN SER C 27: 1-51. (In Japanese with English summary)

Improved artificial diet.

COMPOSITION (g): agar, 1.0; cellulose, 1.0; sucrose, 1.5; corn starch, 2.0; casein, 1.5; salt mixture, 0.4; cholesterol, 0.02; choline chloride, 0.05; sodium ascorbate, 0.3; *l*-cysteine hydrochloride, 0.03; wheat germ, 1.0; and (ml): vitamin mixture, 1.0; water, 50.0 ml.

salt mixture of Beck *et al.** (1949).

DEVELOPMENT Improved egg production and hatching. Six generations successfully reared. Details given on further aspects, including: development of chemically defined diet, effects of antimicrobial agents on growth; growth and development on artificial diet under diapausing conditions; technique for breeding successive generations, and influence of environmental conditions on larval growth and development.

→ Salama, H.S.; Tolba, R.A. 1967. A simple artificial diet for rearing corn borers. BULL SOC ENTOMOL ÉGYPTE 51: 151-152.

COMPOSITION (g): dry beans, 150.0; ascorbic acid, 1.5; benzoic acid, 2.0; hydrochloric acid, 1.0; brewers' yeast, 16.0; agar, 6.5; formaldehyde, 1.0; and water, 325.

PREPARATION Beans soaked overnight and boiled in water for few minutes before use. Ingredients, except agar, blended with water; agar dissolved in water, cooled to 60°C then mixed with other blended ingredients. Diet dispensed to 2-oz glass tubes.

REARING At 30°C; paper with 20-25 eggs glued to surface of diet tube and plugged with cotton. 4th instar larvae removed in groups of 10 to 5 x 9 cm diam glass dishes with 6-oz diet. Pupae transferred to glass jars; emerging adults fed on 10% honey solution; oviposition on paper strips.

DEVELOPMENT Larval period, ca. 16 days; pupal period, 5 days; weights of ♂ and ♀ pupae: 21.8 and 42.3 mg respectively; fertility of emerging adults normal. Data compare well with larvae reared on sugar cane and maize stems.

See also: Chung *et al.* (1973); Kajita (1973); Yen (1972).

Chilo zonellus Swinhoe, sorghum stem borer

→ Chatterji, S.M.; Siddiqui, K.H.; Panwar, V.P.S.; Sharma, G.C.; Young, W.R. 1968. Rearing of the maize stem borer, *Chilo zonellus* Swinhoe on artificial diet. INDIAN J ENTOMOL 30: 8-12.

COMPOSITION & PREPARATION See Keaster & Harrendorf* (1965) diet for *Zeadiatraea grandiosella*.

Diet dispensed to sterile 3 x 1" glass specimen tubes, then plugged, slanted and kept at room temperature for 24 h for excess moisture to evaporate.

REARING At 27-30°C and 60-80% RH. Neonate larvae @ 3 larvae/tube; diet changed when required.

DEVELOPMENT Average of 33 days/generation; 55% larval survival to adult. Comparable to data from maize stem rearing.

→ Dang, K.; Anand, M.; Jotwani, M.G. 1970. A simple improved diet for mass rearing of sorghum stem borer, *Chilo zonellus* Swinhoe. INDIAN J ENTOMOL 32: 130-133.

Best of 4 diets tested.

COMPOSITION (g): agar, 51; yeast, 40; ascorbic acid, 13; methyl *p*-hydroxybenzoate, 8; sorbic acid, 4; *Kabuli* gram flour, (*Cicer arietinum*), 420; and (ml): 10% formaldehyde, 8; water, 3120.

PREPARATION Agar dissolved in ca. 1200 ml water. Other ingredients, except formaldehyde, blended with remaining water, then with agar solution; formaldehyde added and blended, then diet dispensed to 25 x 10 cm glass specimen tubes to about 4 cm depth and allowed to cool.

REARING See Chatterji *et al.** (1968).

DEVELOPMENT Larval period, 24.7 days; pupal period, 6.83 days; pupal wt, 85.30 mg; emergence, 90%; adult longevity, 2.9 days; average number of eggs laid/♀, 265.2; incubation period, 5.16 days; viability, 60.2%; yield of adults from larva, 70.0%. Two generations reared.

→ Pant, N.C.; Gupta, P.; Nayar, J.K. 1960a. Physiological studies of *Chilo zonellus* Swinh. a pest on maize crop. I. Growth on artificial diets. PROC NATL INST SCI INDIA 26B: 379-383.

Best of 4 diets tested.

COMPOSITION (g): casein, 5.50; glucose, 5.50; salt mixture, 0.44; yeast, 2.20; choline chloride, 0.09; cholesterol, 0.22; CMC cellulose, 2.50; leaf factor, 1.80; agar, 3.50; methyl *p*-hydroxybenzoate, 0.10; water, 125.

PREPARATION Dry ingredients ground; cholesterol dissolved in volatile solvent and added to dry mix with water; then CMC cellulose, dispersed in warm water, added. Autoclaved at 15 lb pressure for 20 minutes; diet transferred aseptically to 1 x 2" glass vials with cotton wool plug. Vials stood at room temperature for 1-2 days before use.

REARING Individually at 30°C and 40-80% RH. Diet scored for larval penetration; vials covered with black paper. Adults removed to 8 x 6" diam jars lined with waxed paper for oviposition; fed on sugar solution from cotton wick; jars covered with muslin. Eggs removed and sterilized when necessary.

DEVELOPMENT Egg to adult in 39 days.

→ Siddiqui, K.H.; Chatterji, S.M. 1972. Laboratory rearing of the maize stem borer, *Chilo zonellus* Swinhoe (Crambidae: Lepidoptera) on a semi-synthetic diet using indigenous ingredients. INDIAN J ENTOMOL 34: 183-185.

COMPOSITION (g): agar, 5.1; red rajmah grain powder (*Phaseolus vulgaris*); 74.8; wheat atta (*Triticum aestivum* var. *Sarbati sanora*), 20; yeast powder, 4; ascorbic acid, 1.3; sorbic acid, 0.4; methyl paraben, 0.8; vitamin E, 0.1; and (ml): 40% formaldehyde, 1; water, 380.

PREPARATION Agar dissolved in 130 ml water. Other ingredients blended with remaining water, and then with agar solution; dispensed to 3 x 1" glass tubes to depth of 2 cm and plugged with cotton wool; left for 24-48 h before use.

REARING In tubes @ 2 neonate larvae/tube.

DEVELOPMENT 80% larval survival to adult in 1st generation; 72.5% in 2nd. Fecundity, 272 eggs/♀, with 68.52% viability in 1st generation; 279 eggs/♀, with 62.9% viability in 2nd generation. Four generations reared.

Corcyra cephalonica Stainton, rice moth

→ Sarma, P.S. 1943. Riboflavin and pyridoxin (vitamin B_6) as growth promoting factors for rice moth larvae (*Corcyra cephalonica* St.). INDIAN J MED RES 31: 165-171.

COMPOSITION (g): extracted wheat, 100; McCollum-Davis salt mixture, 4; sugar, 4; and (µg): thiamin, 10; riboflavin, 5; nicotinic acid, 50; calcium pantothenate, 15; pyridoxin, 5.

REARING At 30°C and 70-75% RH. 8-day-old larvae placed on diet in petri dishes.

DEVELOPMENT Weight at 28 days, 220.5 mg. Pupation in 35 days. Data on diet and on whole wheat flour comparable.

→ Singh, K.R.P. 1955b. Studies on nutrition of *Corcyra cephalonica* Staint (Galleriidae, Lepidoptera) with special reference to carbohydrate requirements. SCI CULT 20: 339-340.

COMPOSITION (parts): vitamin-free casein, 50; glucose, 80; cholesterol, 1; McCollum's salt mixture, No. 185, 2; dried yeast, 5; water, 10.

yeast replaceable by vitamins (µg/g diet): thiamin, 50; riboflavin, 50; nicotinic acid, 50; pyridoxin, 50; pantothenic acid, 50; *p*-aminobenzoic acid, 50; folic acid, 4; choline chloride, 500; and biotin, 0.1 (or 2.5 parts of water insoluble fraction of yeast).

DEVELOPMENT Adult emergence in 34-56 days.

Sivaramakrishnan, R.; Sarma, P.S. 1957. Effect of soyabean inhibitor and vitamin B_{12} on growth of rice moth larva (*Corcyra cephalonica* St.). CURR SCI 26: 283-284.

COMPOSITION (parts): casein, 10; dextrose, 90; dried brewers' yeast, 5; cholesterol, 1; Philip & Hart salts mixture IV, 2; vitamin B_{12}, 0.05 µg/5g diet; raw inhibitor, 8000 x 10^{-6} T.I.U./5 g diet.

REARING 10-12-day-old larvae reared on diet for 3 weeks.

DEVELOPMENT Larval wt at 15 days, 258 mg.

→ Uberoi, N.K. 1956. Fatty acids in the lipid metabolism of rice moth *Corcyra cephalonica* Stainton. J ZOOL SOC INDIA 8: 85-90.

COMPOSITION (parts): casein, 50; glucose, 50; cholesterol, 1; McCollum's salt mixture, 2; water, 10; and (mg/g diet): aneurin, 25; riboflavin, 25; nicotinic acid, 25; pyridoxin, 25; pantothenic acid, 25; choline chloride, 500; *p*-aminobenzoic acid, 25; folic acid, 2.5; linoleic acid, 12; ether extract of yeast, 50.

PREPARATION Ingredients mixed and diet dispensed @ 2 g/2 x 1" glass tube.

REARING At 29°C and 70% RH. Eggs incubated in petri dishes. Larvae reared @ 5/tube.

DEVELOPMENT Larval wt after 12 days, 9.5 mg; after 24 days, 32 mg.

Crambus teterrellus (Zincken), bluegrass webworm

→ Ward, A.G.; Pass, B.C. 1969. Rearing sod webworms on artificial diets. J ECON ENTOMOL 62: 510-511.

Best of three diets. Modification of wheat germ diet for rearing *Pectinophora gossypiella* larvae, developed by Adkisson *et al.** (1960).

COMPOSITION (g): wheat germ, 3.2; vitamin-free casein, 3.8; sucrose, 3.0; Alphacel, 1.0; Wesson's salt mixture, 1.1; ascorbic acid, 0.4; agar, 3.0; and (ml): B-vitamin mixture, 1.1; methyl *p*-hydroxybenzoate (15% in 95% ethanol), 1.5; distilled water, 90.0; blue grass, powdered extract in water (10 g powdered bluegrass/500 ml of water), 10.0; corn oil, 1.0.

B-vitamin mixture as in Vanderzant *et al.** (1957).

PREPARATION Agar dissolved in water and homogenized with remaining ingredients when cooled to 60°C. Diet dispensed to 1-oz plastic cups @ 10 ml/cup.

Bluegrass extract prepared by steeping 100 g freshly cut bluegrass leaves in 500 ml acetone for 30 minutes at room temperature. Acetone then drawn off and bluegrass air dried, then powdered. 10 g powder added to 500 ml distilled water, boiled and filtered. Resulting liquid used in diet.

REARING Individual eggs incubated on moist filter paper and neonate larvae or eggs placed on diet. Fresh diet added 2 or 3 times during larval development.

DEVELOPMENT Survival to pupal stage, 40.8%; survival to adult, 35.8%. Duration of larval and pupal stages, 49 days compared with 47 days on bluegrass. Reared for 3 or 4 generations.

Crambus trisectus (Walker) and *Crambus mutabilis* Clemens also reared.

Crambus trisectus (Walker), sod webworm

→ Dupnik, T.D.; Kamm, J.A. 1970. Development of an artificial diet for *Crambus trisectus*. J ECON ENTOMOL 63: 1578-1581.

Modified from Riddiford* (1968) diet for *Hyalophora cecropia* and other saturniid silkworms.

COMPOSITION (g): wheat germ, 30.0; casein, 17.5; Wesson's salt mixture, 5.0; Alphacel, 7.5; β-sitosterol, 0.25; wheat-germ oil, 1.25; methyl *p*-hydroxybenzoate, 0.88; sorbic acid, 1.17; agar, 12.5; Aureomycin, 0.25; Kantrex, 0.07; and (ml): 37% formaldehyde, 0.5; 10% KOH, 5.0; vitamin suspension, 6.0; water, 420.

vitamins (mg/ml water): α-tocopherol, 8; ascorbic acid, 270; biotin, 0.02; calcium pantothenate, 1; choline chloride, 50; folic acid,

0.25; inositol, 20; niacinamide, 1; pyridoxine hydrochloride, 0.25; riboflavin, 0.5; thiamine hydrochloride, 0.25; vitamin B_{12}, 2.

PREPARATION Wheat germ, casein, salts, Alphacel, sitosterol, methyl *p*-hydroxybenzoate, sorbic acid, Aureomycin and Kantrex mixed. Agar and water boiled, cooled to 70° C. Then all ingredients blended for 2 minutes, and immediately dispensed to containers.

DEVELOPMENT Four generations successfully reared; mean survival to adult in each generation: 87, 82, 72, and 86% respectively.

Diatraea grandiosella (Dyar) [previously *Zeadiatraea*], southwestern corn borer

→ Chippendale, G.M. 1970b. Metamorphic changes in fat body proteins of the southwestern corn borer, *Diatraea grandiosella*. J INSECT PHYSIOL 16: 1057-1068.

Modified diet of Keaster & Harrendorf* (1965) for *Zeadiatraea grandiosella*.

COMPOSITION (g): vitamin-free casein, 3.29; dextrose, 3.29; whole wheat germ, 2.82; bacto-agar, 2.63; salt mixture, 0.94; cellulose powder, 0.47; sodium alginate, 0.47; ascorbic acid, 0.37; wheat-germ oil, 0.23; cholesterol, 0.18; choline chloride, 0.10; methyl *p*-hydroxybenzoate, 0.15; sorbic acid, 0.05; and (ml): 37% formalin, 0.05; 4M KOH, 0.46; vitamin solution, 9.40; distilled water, 75.10.

for salts and vitamins see Chippendale & Beck* (1968).

PREPARATION Agar and sodium alginate dissolved in boiling water and blended with remaining ingredients; solidified in shallow pan and dispensed to cups or vials.

REARING At 30° C and 60% RH with 16 h photoperiod. Larvae reared singly on ca. 6.6 g diet in 23 x 85 mm shell vials plugged with cotton, or @ 2/cup with 14 g diet.

DEVELOPMENT Larval maturity after ca. 15 days; 50% pupation by 16th day; pupal period, ca. 7 days; viable eggs laid.

See also: Chippendale & Reddy (1972a b, 1974); Jacob & Chippendale (1971); Reddy & Chippendale (1972).

→ Keaster, A.J.; Harrendorf, K. 1965. Laboratory rearing of the southwestern corn borer, *Zeadiatraea grandiosella*, on a wheat germ medium. J ECON ENTOMOL 58: 923-924.

COMPOSITION (g/500 g diet): wheat germ, 15.00; vitamin-free casein, 17.50; dextrose, 11.50; Wesson's No. 2 salt mixture, 5.00; cellulose, 2.50; vitamin fortification mixture (NBC), 7.00; ascorbic acid, 1.70; Aureomycin, 0.07; methyl paraben, 0.75; agar, 12.50; and (ml): 4M KOH, 2.50; formaldehyde, 1.00; and water, 450.00.

PREPARATION Agar dissolved in 350 ml water, and blended with other ingredients in 100 ml water. Diet dispensed from plastic squeeze bottle to ⅓ fill ½-oz cups.

REARING At 80° F. Neonate larvae transferred to diet, cap replaced and cup inverted. Pupae removed to oviposition cages on moist paper in petri dishes. Cages from 2½-gal cartons lined with waxed paper for oviposition with adult food supply of 5% sugar solution from cloth pad. Eggs collected, and incubated in cartons lined with moist paper towelling.

DEVELOPMENT Incubation period, 3-5 days; larval period, 16-20 days; larval mortality, <5%; pupal period, 7-10 days; fecundity/♀, 5 eggs/mass to 75 masses/♀ over 18 generations reared. Life cycle, 32 days.

Diatraea saccharalis (Fabricius), sugarcane borer

→ Dinther, J.B.M. van; Goossens, P.A. 1970. Rearing of *Diatraea saccharalis* on diets in Surinam. ENTOMOL EXP APPL 13: 320-326.

Best of 9 diets tested for mass-rearing.

COMPOSITION (g): kidney bean meal, 35; corn plant powder, 10; carrot powder, 5; brewers' yeast, 10; casein, 1; ascorbic acid, 0.5; agar, 3; water, 210; methyl *p*-hydroxybenzoate, 1; streptomycin sulphate, 0.1; Penicillin, 0.2; sorbic acid, 0.2.

PREPARATION Dry ingredients except agar blended in about 100 ml water. Agar dissolved in remaining water, cooled to 70° C, and blended with first mix. Dispensed to 10 x 2.5 cm vials or 25 ml plastic cups with 2-cm hole for aeration and stoppered with cotton wool.

REARING At 24-28° C and 65-90% RH with 12 h photoperiod. Pupae incubated on moist cotton in open petri dish in 30 x 23 cm diam cardboard cylinder lined with paper for oviposition. Adults fed on sugar solution from cotton wool. Eggs cut from paper and neonate larvae transferred to diet. Eggs or pupae refrigerated at 0° C for maximum of 10 days till use.

DEVELOPMENT Larval period, 32 ± 0.6 days; pupal period, 8 days; pupal wt: ♂, 59 ± 1.1; ♀, 112 ± 3.1 mg; eggs laid/♀, 394 ± 2.5; incubation period, 6 days; emergence, 94%. Growth rates comparable to those on rice or corn stalks.

See also: Brewer & Martin (1976).

→ Hensley, S.D.; Hammond, A.M. Jr 1968. Laboratory techniques for rearing the sugarcane borer on an artificial diet. J ECON ENTOMOL 61: 1742-1743.

Modified diet of Adkisson *et al.** (1960) for *Pectinophora gossypiella*.

COMPOSITION (ml): water, 3116.0; vitamin solution, 36.0; 37.2% formaldehyde, 1.8; and (g): Aureomycin, 1.0; Wesson's salts mixture, 36.0; casein, 108.0; sucrose, 180.0; wheat germ, 108; choline chloride, 3.6; ascorbic acid, 14.4; methyl *p*-hydroxybenzoate, 5.4; bacto-agar, 72.0.

vitamins (mg/1000 ml): niacinamide, 1000; calcium pantothenate, 1000; riboflavin, 500; thiamine hydrochloride, 250; pyridoxine hydrochloride, 250; folic acid, 100; biotin, 20; and 2 ml of vitamin B_{12} (1000 mg/ml).

PREPARATION Dry ingredients mixed with 900 ml water, then vitamin and formalin solutions added. Blended with agar dissolved in 2216 ml distilled water. Diet dispensed to ⅓rd fill 6-dr shell vials. Quantity sufficient for rearing ca. 500 larvae.

REARING At 25 ± 1° C with 14 h photoperiod. 20-30 pairs of pupae placed/1-gal paper carton covered by 16-mesh screen, lined with waxed paper and containing 2-3" layer of moist vermiculite for emergence. Oviposition on creased paper, removed 3 times/week. Eggs sterilized in 0.1% $HgCl_2$ for 3 minutes, and washed in 70% ethanol and distilled water. Dried on filter paper and incubated in 500 ml flasks. Neonate larvae removed individually to rearing vials stoppered with sterile cotton. After 10-12 days, or at 3rd instar, removed to insectary shelves to complete development.

DEVELOPMENT Average development from egg to pupa, 32 days; survival from 66-87%. Reared continuously for more than 3 years.

→ Miskimen, G.W. 1965. Non aseptic laboratory rearing of the sugarcane borer, *Diatraea saccharalis*. ANN ENTOMOL SOC AM 58: 820-823.

COMPOSITION (g): sugarcane tops, 450; bacto-agar, 3; methyl *p*-hydroxybenzoate, 1; sorbic acid, 2; ascorbic acid, 8; and (ml): distilled water, 550; 40% formaldehyde, 2.0.

PREPARATION Sugarcane tissue ether killed. Diet steam sterilized for 10 minutes at 95° C; pH to 5.8 with 0.1 M HCl buffered with 100 ml acetate.

REARING At 80° F and 75% RH in darkness for 1st and 2nd instar, and 90% RH with continuous lighting for 3rd instar to pupae. Reared in 80 x 25 mm glass vials. Mature larvae placed in quart jars with moist filter paper. Pupae placed on moist filter paper in separate glass jars for each sex. Adults collected twice daily and placed in ratio of 2♂:1♀ in 150 ml polyethylene sputum cups, with compression seal covers. Hole cut in centre of covers plugged with non absorbent cotton. Filter paper moistened with tap water placed in cup. Sections of cup bearing egg mass cut out, sterilized in alcohol for 10 seconds and placed in sterilized petri plates for hatching. Neonate larvae transferred with brush, and vials sealed with $HgCl_2$ treated cotton plugs.

DEVELOPMENT Larval period, 5-6 weeks; pupal period, 10 days.

→ Pan, Y.S.; Long, W.H. 1961. Diets for rearing the sugarcane borer. J ECON ENTOMOL 54: 257-261.

Best of 2 diets.

COMPOSITION (g): corn plant powder, 10.0; sucrose, 5.0; cholesterol, 0.2; Wesson's salt mixture, 1.0; yeast extract, 2.0; choline chloride, 0.1; bacto-agar, 3.5; and (ml): Dowicide-A (10% solution in 95% ethyl alcohol), 0.15; and distilled water, 125.0.

PREPARATION Agar dissolved in distilled water. Choline chloride added, then casein, sucrose, cholesterol, Wesson's salt mixture, and yeast extract stirred in. 0.15 ml of 10% solution of Dowicide-A, and finally corn plant powder added, and diet blended for 5 minutes. Kept at room temperature until set, then dispensed to 125 or 250 ml Erlenmeyer flasks, and sterilized at 17 psi for 20 minutes. Cooled at room temperature for 48 h.

Corn plant powder by grinding whole corn plants or leaves with ca. 3 volumes of acetone. Powdered extract recovered by filtration, resuspended in acetone, filtered again, and dried by evaporation. Sieved to 16 mesh.

REARING At 80° F in Erlenmeyer flasks stoppered with non-absorbent cotton. Eggs collected from adults in ½-gal cartons lined with wax paper and sterilized at black head stage by immersion in 0.1% $HgCl_2$ for 4 minutes; then rinsed in 70% ethyl alcohol and aseptically transferred to flaks @ 10 eggs or neonate larvae/flask. After 28-30 days larvae placed in vials on moist cotton for pupation.

DEVELOPMENT Mortality, 53%; pupation, 47%. Pupal wt: ♀, 116.7 mg; ♂ 59.5 mg. Fecundity, 390-496 eggs/♀. Sugar cane tops slightly superior than artificial diet.

→ Walker, D.W.; Alemañy, A.; Quintana, V.; Padovani, F.; Hagen, K.S. 1966. Improved xenic diets for rearing the sugarcane borer in Puerto Rico. J ECON ENTOMOL 59: 1-4.

Best 3 of 100 diets tested.

COMPOSITION Diet 1 (g): corn fibre, blended, 20.0; carrot powder, 10.0; agar, 2.0; brewers' yeast, 15.0; enzymatic hydrolyzed casein, 1.0; sodium benzoate, 0.3; methyl *p*-hydroxybenzoate, 0.05; ascorbic acid, 0.32; and (ml): corn filtrate, 125.0; and 1N HCl, 10.0. pH 5.8.

Diet 2: as diet 1, except corn filtrate replaced by 125.0 ml distilled water.

Diet 3: as diet 1, except brewers' yeast decreased to 10.0 g; enzymatically hydrolyzed casein increased to 2.0 g; ascorbic acid increased to 0.50 g; and corn filtrate replaced by 125.0 ml distilled water.

PREPARATION Agar dissolved in water or corn filtrate. Carrot powder and brewers' yeast added and mixed, temperature reduced to 70° C, and remaining dry ingredients added. Corn fibre or corn filtrate then added and stirred vigorously for 2 minutes. HCl added and mixture again stirred thoroughly for at least 2 minutes. Dispensed to 8-dr shell vials @ 20 g/vial, or to baby food jars.

Corn fibre and corn filtrate prepared according to Pan & Long* (1961).

REARING At 85 ± 2° F and 70-90% RH in darkness. Neonate larvae transferred to diet surface @ 5/vial or 25/jar. Adults not fed.

DEVELOPMENT Little difference on any diet. Average larval period and adult longevity, 22-25 and 8-10 days respectively. Five larval instars. Fecundity ca. 500 eggs/♀. On diet 3, 90% survival from egg to adult. Pupal weights less than wild population, but adult more vigorous.

See also: Walker (1968).

→ Wongsiri, T.; Randolph, N.M. 1962. A comparison of the biology of the sugarcane borer on artificial and natural diets. J ECON ENTOMOL 55: 472-473.

COMPOSITION As wheat germ diet No. 1 developed for *Pectinophora gossypiella* by Adkisson *et al.** (1960) plus addition of 2 g ascorbic acid dissolved in 20 ml water.

PREPARATION Agar dissolved in water and blended with dry ingredients. Liquids added, blended again, and diet dispensed to 8-dram shell vials to depth of 1" from plastic squeeze bottle. Vials wrapped in paper and heat sterilized in oven before use. One drop of ascorbic acid added to each vial after cooling below 90° F.

REARING At 78° F. Neonate larvae transferred to vials plugged with non-absorbent cotton. Adults kept in 1-gal glass jars with 1" layer moist sand covered with paper towel. Sugar solution from jar via cotton wick; oviposition on folded waxed paper.

DEVELOPMENT Average duration in days: egg, 6.1; larva, 29.7; pupa, 7.2; adult, 5.9. Average life cycle, 42.8 days; fecundity, 333.5 eggs/♀. Life cycle on grain sorghum, 40.0 days; and fecundity, 227.6 eggs/♀.

Dioryctria abietella (Denis & Schiffermüller)

→ Fatzinger, C.W. 1970a. Rearing successive generations of *Dioryctria abietella* (Lepidoptera: Pyralidae (Phycitinae)) on artificial media. ANN ENTOMOL SOC AM 63: 809-814.

Best diet tested.

COMPOSITION (g): vitamin-free casein, 17.50; wheat germ, 20.00; Wesson's salt mixture, 5.00; levulose, 5.30; glucose, 5.30; sucrose, 6.40; raffinose, 0.50; stachyose, 0.05; choline chloride, 0.50; ascorbic acid, 2.00; methyl *p*-hydroxybenzoate (in 95% ethanol), 0.75; agar, 8.00; and (ml): vitamin solution, 5.00; 37% formalin, 0.25; 4M KOH, 2.50; distilled water, 420.00.

vitamins (mg/100 ml water): nicotinamide, 100; calcium pantothenate, 100; riboflavin, 50; thiamine hydrochloride, 25; pyridoxine hydrochloride, 25; folic acid, 25; biotin, 2; and vitamin B_{12}, 0.2.

PREPARATION Sugars, vitamin solution, choline chloride, ascorbic acid, formalin, and methyl *p*-hydroxybenzoate solution blended in 110 ml water. Casein and wheat germ then added. Agar, KOH, and salt mixture dissolved in remaining 310 ml of water, cooled to 60°C, and blended with other ingredients. Diet dispensed while hot as described in Burton *et al.* (1966).

REARING At 30°C with 11½ h photoperiod in test tubes with non-absorbent cotton plugs, or 22 x 94 mm glass petri dishes with porcelain covers. Neonate larvae transferred to top of slant in tubes; pupae placed in 25 x 95 mm shell vials with 1 x 6 cm strip of plastic screen for emergence. Adults kept in 30 cm^3 wire screen cages and fed 5% sucrose solution via cotton wick. Mated ♀♀ placed in individual vials with moistened plug of absorbent cotton wrapped in cheesecloth. Plastic screen for oviposition coated 4 times in 1.3% methyl cellulose. Screen soaked in water for 45 minutes with air bubble action to dislodge eggs; washed for 1 minute in 0.1% sodium alkylaryl sulphonate in 250 ml funnel, washed for 5 minutes in 0.1% peracetic acid, then transferred to sterile filter paper in Buchner funnel and washed 3 times in sterile distilled water. Incubated on moist filter paper.

DEVELOPMENT Life cycle over 4 generations from 32.7-38.9 days; survival, 66-74%. Number of eggs laid/♀, 104.7-133.8. Higher survival, fecundity, and growth on artificial diet than on natural food of pine cones.

See also: Fatzinger (1970b).

→ Fatzinger, C.W. 1973. A wheat germ-free, meridic medium for rearing *Dioryctria abietella* (Lepidoptera: Pyralidae: Phycitinae). FLA ENTOMOL 56: 327-332.

COMPOSITION Group I (g): α-tocopherol, 0.10; cholesterol, 1.50; ergosterol, 1.50; methyl *p*-hydroxybenzoate, 0.75; and (mg): linolenic acid, 250.0.

group II (g): vitamin-free casein, 22.40; glycine, 1.0; cystine, 1.0; and (mg): folic acid, 1.250.

group III (g): agar, 8.0; Wesson's salt mixture, 5.0; and (ml): distilled water, 325.0; 4M KOH, 2.50.

group IV (g): levulose, 7.10; glucose, 7.10; sucrose, 6.40; raffinose, 0.50; stachyose, 0.050; choline chloride, 0.60; ascorbic acid, 2.0; and (ml): 37% formalin, 0.250; vitamin solution, 100.0.

vitamins (mg/175 ml water): inositol, 320; niacin, 250; calcium pantothenate, 250; riboflavin-5'-phosphate, sodium, 159; pyridoxine hydrochloride, 63; thiamine hydrochloride, 63; biotin, 5; vitamin B_{12}, 0.5.

PREPARATION Ingredients of group I dissolved in 100 ml 95% ethanol at 60° C. Groups I and II mixed, ethanol evaporated, and mix reground. Group III autoclaved at 15 psi for 30 minutes and cooled to 60° C. Group IV ingredients dissolved in water; then powder from groups I and II, and groups III and IV all blended and dispensed to rearing containers.

REARING At 27° C with 12 h photoperiod. Neonate larvae reared individually to pupal stage in 18 x 150 mm test tubes each containing ca. 7 ml diet. Pupae removed and held in vials until emergence.

DEVELOPMENT Over 3 generations: survival to adult, 71, 67, and 69%; average pupal wt: 56.3, 56.2, and 57.3 mg; larval period: 27.7, 32.6, and 28.7 days respectively.

Ectomyelois ceratoniae (Zeller), carob moth

→ Gothilf, S. 1968. The biology of carob moth (*Ectomyelois ceratoniae* (Zell)) in Israel. I. Mass culture on artificial diet. ISR J ENTOMOL 3: 109-118.

COMPOSITION (%): soybean flour, 43.5; sucrose, 43.5; water, 13.0.

PREPARATION Sucrose dissolved in water, mixed with soybean meal and stored till use.

REARING At 25 ± 1° C and 75 ± 5% RH in 31 x 23.5 x 7.5 cm plastic trays, each with 650 g diet. 500 eggs placed on diet and covered with paper. Mature larvae bore through paper before pupation. Rearing trays removed to 70 x 80 x 180 cm screened emergence cages after 28 days. Mated ♀♀ removed to 3.5 x 3 cm diam plastic cups inverted on to rough paper for oviposition.

DEVELOPMENT Larval period on diet, 29.3 days; on carob and acacia pods, 29.5 and 27.5 days respectively. Larval survival, size of adults, and fecundity on artificial diet significantly greater than on natural foods; 16 generations reared.

→ Levinson, H.Z.; Gothilf, S. 1965. A semisynthetic diet for axenic growth of the carob moth, *Ectomyelois ceratoniae* (Zell.). REV PARASITOL 26: 19-26.

Best of 14 diets tested.

COMPOSITION (%): α-soy protein, 49.0; sucrose, 23.5; vitamin solution, 25.0; linoleic acid, 1.4; β-sitosterol, 0.1; salt mixture (Glaxo BS2407), 1.0.

vitamins (mg/ml aqueous solution): thiamine, 0.25; riboflavin, 0.12; nicotinic acid, 0.49; pyridoxine, 0.12; calcium pantothenate, 0.25; *i*-inositol, 2.47; biotin, 0.002; ammonium folate, 0.005; choline chloride, 4.95.

PREPARATION Dry ingredients powdered and mixed with sterol dissolved in soybean oil or linoleic acid. B-vitamins added shortly before starting experiments. 6 g diet placed in 15 x 1.5 cm test tubes and autoclaved at 15 psi for 25 minutes.

REARING Individually at 25 ± 1° C and 75 ± 5% RH. Neonate larvae from surface sterilized eggs (10 minutes in 0.1% $HgCl_2$) added to tubes.

DEVELOPMENT Pupation, 97%; total development period, 34.4 days. Adult wt: ♂, 12.9 mg; ♀, 20.8 mg. Two generations reared.

Elasmopalpus lignosellus Zeller, lesser cornstalk borer

➞ Chalfant, R.B. 1975. A simplified technique for rearing the lesser cornstalk borer (Lepidoptera: Phycitidae). J GA ENTOMOL SOC 10: 33-37.

COMPOSITION (% by weight): beans (pinto or southern peas), 12; torula yeast, 3.8; wheat germ, 6; 55% linolenic acid, 0.3; ascorbic acid, 0.38; Vanderzant's vitamin mixture, 0.044; methyl paraben, 0.23; sorbic acid, 0.12; Aureomycin or Kanamycin, 0.044; water, 76; agar, 1.

PREPARATION See Burton* (1969). Diet dispensed to 1-oz plastic cups.

REARING At 13 h/30° C light and 11 h/24° C dark, and 40-80% RH. Ten pairs of adults in 6 x 6" diam cages, with screen wall and 6 x 9" plastic dish ends, with centre replaced by screen wire. Fed on 8% sugar solution from paper tissue wick; whole cage wrapped with paper towel for oviposition through screen wall. Renewed every 2-3 days. Eggs incubated in plastic bags. Neonate larvae mixed with fine vermiculite @ 1 larva/ 2 ml vermiculite. Diet covered by thin layer of vermiculite then larvae/ vermiculite added @ 5 larva/cup.

DEVELOPMENT Egg to adult, 25 days; adult longevity: ♀, 5 days; ♂, 8 days; pre-oviposition period, 3 days; average fecundity, 47 eggs/♀. Adult emergence, 46-50%. Reared for 25 generations.

Galleria mellonella Linnaeus, greater wax moth

➞ Balázs, A. 1958. Nutritional and nervous factors in the adaptation of *Galleria mellonella* to artificial diet. ACTA BIOL ACAD SCI HUNG 4: 47-69.

COMPOSITION (%): maize flour, 22.0; wheat flour, 11.0; dried skimmed milk, 11.0; dried powdered yeast, 5.5; wheat grain, 11.0; beeswax, 17.5; honey, 11.0; glycerine, 11.0.

REARING At 29° C and 60-70% RH. 50-100 larvae in 500 ml flasks plugged with cotton. After 4 weeks flask capped with drum-like structure for pupation. Drums changed after 3 days, pupae placed in petri dishes for eclosion. Pairs of moths placed in petri dish with cellophane and dark paper strip; eggs laid under paper. Larvae also reared individually in sealed watchglasses until 4-5th instar, then transferred to cotton plugged test tubes for pupation.

DEVELOPMENT Incubation period, 8-10 days; larval period, 42 days; mature larval wt, 214 mg. 5-6 generations/year.

→ Beck, S.D. 1960b. Growth and development of the greater wax moth, *Galleria mellonella* (L.) (Lepidoptera: Galleriidae). TRANS WIS ACAD ARTS LETT 49: 137-148.

COMPOSITION (mg/g): honey (strained), 236; glycerol, 207; water, 94; Pablum (mixed cereal), 322; brewers' yeast powder, 94; beeswax (yellow), 47.

PREPARATION Pablum, yeast powder, and beeswax mixed with sufficient diethyl ether to dissolve wax, then ether evaporated off. Honey, glycerol, and water mixed and added to dry ingredients and mixed again. Resulting diet friable and sticky: texture changes if refrigerated for ca. 24 h.

REARING At 35°C in darkness in 190 x 100 mm crystallizing dishes covered with 210 mm diam glass plate with 15 mm cotton plugged hole in centre. 100 eggs seeded on to 30 mm layer of diet in dish. Mature larvae or cocoons transferred to 2.5-l battery jars covered with cheesecloth for emergence. Oviposition on 100 mm folded waxed paper squares; eggs collected daily.

DEVELOPMENT Increase in larval weight from 0.02 mg at eclosion to 200 mg at 15 days. Development and growth comparable with control on diet of (honeybee) brood combs and pollen.

Corollary

→ Campadelli, G. 1973. Allevamento di *Galleria mellonella* L. (Lepidoptera: Galleriidae) con dieta semiartificiale. BOLL IST DEI ENTOMOL UNIV STUDI BOLOGNA 32: 11-25. (In Italian with English summary)

COMPOSITION Modification of Beck's (1960) diet: brewers' yeast replaced by bacto-yeast extract, and following added (g/100 g Pablum): cholesterol, 0.6; ascorbic acid, 0.6; saccharose, 0.6; and methyl *p*-hydroxybenzoate, 0.1.

REARING At 27 ± 1°C and 75 ± 5% RH in darkness.

DEVELOPMENT Incubation period, 7-9 days; larval period (6 instars), 32 days; pupal period, 9-13 days. Generation completed in 55 days.

→ Bronskill, J.F. 1962. A cage to simplify the rearing of the greater wax moth, *Galleria mellonella* (Pyralidae). J LEPID SOC 15: 102-104.

COMPOSITION (g): Pablum, 420; ground brood comb, 20; and (ml): white honey, 150; glycerine, 150; tap water, 30.

PREPARATION Liquids blended with Pablum and brood comb.

REARING At 82 ± 1°F and 37 ± 3% RH. One ♀ and 2 ♂ adults transferred to lantern globe cages, ends covered with copper screen. Eggs laid on walls, and on folded card. Larval diet on screen rack placed within cage when eggs mature. Cocoons spun from rack. 200-250 larvae/cage. Illustrated.

DEVELOPMENT Incubation period, 8-10 days; larval period, 42 days; pupal wt, 207.8 ± 13.5 mg.

Dadd, R.H. 1964. A study of carbohydrate and lipid nutrition in the wax moth, *Galleria mellonella* (L.), using partially synthetic diets. J INSECT PHYSIOL 10: 161-178.

Best of several diets described.

COMPOSITION (g): cellulose powder, 23.0; linoleic acid, 0.1; Wesson's salt mixture, 2.0; casein (light white soluble), 15.0; glycerol, 20.0; and vitamin solution, 30 ml; and glucose or dextrin, 30-40%.

vitamins (μg/g): thiamin, 25; riboflavin, 25; nicotinic acid, 100; pyridoxine, 25; folic acid, 5; calcium pantothenate, 50; *meso*-inositol, 500; choline chloride, 1000; biotin, 1; B_{12}, 1.

PREPARATION Lipids dissolved in ether in mortar, and cellulose powder ground in as ether evaporated. Casein added gradually and mixed. Dispensed to 5 x 4 cm diam, wide-necked jars. Diet stored at 5°C till used.

REARING At 35 ± 0.5°C and 50-60% RH in darkness. Oviposition through gauze lid of diet jar on to filter paper. Incubated for 4 days, then filter paper cut up and placed on diet @ 30 eggs/jar. Fresh diet every 3 days. Numbers reduced to 5/jar when larval weight >50 mg; diet sprinkled over fragmented filter paper for pupation.

DEVELOPMENT Complete. Growth improved by addition of 40% glucose to diet. Effects of other nutritional factors discussed.

Corollary

Dadd, R.H. 1966. Beeswax in the nutrition of the wax moth, *Galleria mellonella* (L.). J INSECT PHYSIOL 12: 1479-1492.

Two diets tested.

COMPOSITION Diet 1: see Dadd* (1964).

Diet 2 (g): cellulose powder, 27; linoleic acid, crude, 0.1; cholesterol, 0.05; Wesson's salt mixture, 2; glucose, 30; casein (light white soluble), 20; glycerol, 20; and vitamin solution, 20 ml.

DEVELOPMENT Optimum growth with 10-15% beeswax, compared with 30-40% glucose.

Dutky, S.R.; Thompson, J.V.; Cantwell, G.E. 1962. A technique for mass rearing the greater wax moth (Lepidoptera: Galleriidae). PROC ENTOMOL SOC WASH 64: 56-58.

COMPOSITION (g): dry Pablum (Mead-Johnson), 225; sterilized sucrose-glycerol-water mixture, 319; and vitamin mixture (Meads Deca-Vi-Sol), 0.6 ml.

sucrose-glycerol-water mixture (in parts): sucrose, 1.00; glycerol, 1.19; distilled water, 0.94.

PREPARATION Pablum added to sucrose, glycerol, water and vitamin mixture in separate container and mixed rapidly.

REARING In darkness; 30 mg or 1000 mature eggs placed in wide-mouthed gallon jar, diet poured over eggs and jars covered with 20 mesh wire screen, filter paper, cardboard gasket and lid. 4" diam holes for ventilation provided in cardboard and lid. Jars kept at 30° C for 6 weeks or 34° for 4 weeks until larvae mature. Removed to pint emergence jars when just at cocooning stage @ 50/jar. Oviposition on accordion-folded wax paper; 300 mg eggs/jar collected by 16th day. Eggs incubated at 25° C.

DEVELOPMENT Incubation period, 8 days at 25° C; 4 days at 30° C and 30 days at 18° C; larval period, 33 days at 34° C.

Good, M.E.; Morrison, F.O.; Mankiewicz, N.E. 1953. Lipidolytic enzymes extracted from *Galleria mellonella* L. (Lepidoptera: Pyralidae) reared on natural and artificial media. CAN ENTOMOL 85: 252-253.

COMPOSITION (g): fine corn meal, 20; whole wheat flour, 40; skim milk powder, 20; powdered dry yeast, 10; honey and glycerine, equal amounts until mixture wet sand consistency.

PREPARATION Ingredients mixed well.

REARING Diet placed in screen covered jar to ca. 2". Seeded with 700 eggs laid on folded wax paper.

DEVELOPMENT Larval period, ca. 4 weeks.

Haydak, M.H. 1941. Nutrition of the wax moth larvae. Vitamin requirement. I. Requirement for vitamin B_1. PROC MINN ACAD SCI 9: 27-29.

COMPOSITION Dry portion (g): vitamin-free casein, 40; agar, 2; dried yeast, 20; salt (Hawk-Oser), 4; cottonseed oil, 4; dextrine, 30; liquid portion: equal volumes of invert sugar (Nulomoline brand) and glycerine.

PREPARATION One g dry portion mixed with 1 ml wet. Small vials half filled and plugged with cotton. Kept at constant temperature for 2 days.

REARING Individually in vials at 32° C and 75% RH. Cotton plugs exchanged for corks protected with wire cloth and small holes plugged with cotton wool.

DEVELOPMENT From neonate larva to adult, 67 days.

See also: Haydak (1936a b, 1940).

Marston, N.; Campbell, B.; Boldt, P.E. 1975. Mass producing eggs of the greater wax moth, *Galleria mellonella* (L.). U S DEP AGRIC RES SERV TECH BULL 1510: 15 pp.

COMPOSITION (g): bran, 260; Wheast, 65; wheat flour, 162; cornmeal, 162; glycerine, 193; water, 158.

PREPARATION Liquids mixed with bran. Remaining mixed dry ingredients added and diet mixed in 56.6-l cement mixer. Aureomycin added to diet for disease control with no adverse affect on egg production.

REARING At 35 C and 65 ± 5% RH in 4.4-l glass jars for first 2 weeks of larval period with 2.2-l diet. Then larvae transferred to 11.5-l galvanized iron pans with fresh diet; optimum density, 1100 eggs/kg, producing 773 cocoons/kg. Mature larvae and pupae transferred to 122 x 101 x 71 cm emergence boxes. Adults attracted through transparent plastic tube by light into 76 x 61 x 61 cm oviposition cage of 32 mesh Saran screen.

Oviposition on plastic sheeting sprayed with 10% sugar and coated with granulated sugar. Sheeting renewed every 7 days, eggs removed by washing and passed through hose to 60 mesh sieve. 20 ml detergent added to rinsing water to remove scales. Eggs suspended in water and poured on to velvet cloth, separated by rubbing cloth, and dried by fan. Rearing methods illustrated.

DEVELOPMENT Larval period, 4 weeks; pupal period, 3 weeks; adult longevity, 7-10 days. Fecundity in 3 days at 35°C, 83 eggs/♀. Cost of diet, $0.1866/kg.

See also: Marston & Brown (1974); Marston & Campbell (1975).

→Waterhouse, D.F. 1959. Axenic culture of wax moths for digestion studies. ANN N Y ACAD SCI 77: 283-289.

Also for lesser wax moth, *Achroia grisella* (Fabricius).

COMPOSITION (g): yeast, 0.5; Pablum, 1.5; cholesterol, 10 mg and water, 1 ml.

PREPARATION Yeast powder and Pablum dried in vacuo over phosphorous pentoxide, then 24 h Soxhlet extraction with light petroleum. Solids weighed into 50 ml Erlenmeyer flask, water added and stirred in. Flasks plugged and sterilized by tyndallization (exposure to steam at 100°C for 20 minutes on each of 3 consecutive days), with incubation at 30°C between. Small pieces of comb extracted with light petroleum, autoclaved, and placed in flask.

REARING Eggs surface sterilized in 0.1% Zephiran and incubated at 30°C and 80% RH. 25 eggs transferred aseptically to each flask.

DEVELOPMENT Egg to adult in 5-9 weeks.

See also: Beard (1958); Benson & Oberlander (1974); Ishikawa & Newburgh (1971); Mitsui *et al.* (1973); Morrison & Perron (1963); Pipa (1963); Smith (1937); Stairs (1965); Yendol (1970); Young (1961).

Hypsipyla grandella Zeller, meliaceous shoot borer

→Berrios, F.; Hidalgo-Salvatierra, O. 1971. Estudios sobre el barrenador *Hypsipyla grandella* Zeller. VI. Susceptibilidad de la larva al hongo *Metarrhizium anisopliae* (Metch). TURRIALBA 21: 214-219. (In Spanish with English summary)

COMPOSITION (g): soy bean, 80.7; agar, 8.0; wheat germ, 36; Wesson's salt mixture, 10; vitamin diet fortification mixture (NBC), 20; methyl *p*-hydroxybenzoate, 1.8; sorbic acid, 1.1; ground corn cob, sieved to 0.21 mm, 80; and (ml): 10% formalin, 5; 25% acetic acid, 13.3; Aureomycin (actual 166.7 mg), 20; 4M KOH, 6; and water, 850.

PREPARATION Soya beans blended with 450 ml water. Agar dissolved in 400 ml boiling water and cooled to 60°C. Mixed and remaining ingredients stirred in.

See also: Grijpma (1971).

Hypsipyla robusta Moore, cedar shoot borer

→ Achan, P.D. 1968. Preliminary observations on the development of an artificial diet for *Hypsipyla robusta* Moore. COMMONW INST BIOL CONT TECH BULL 10: 23-26.

COMPOSITION (g): part 1, vitamin-free casein, 3.5; Alphacel, 0.5; Wesson's salt mixture, 1; sucrose, 3.5; wheat flour, 3.0; choline chloride, 0.1; soy bean powder, 1; ascorbic acid, 0.4; methyl *p*-hydroxybenzoate, 0.15; Aureomycin, 0.03; agar, 2.5; and (ml): distilled water, 22; KOH, 0.5; vitamin solution, 1; formalin, 0.05; part 2, agar, 2.5 g; and water, 62.0 ml. Best diet also with egg-yolk @ ca. 8.0 ml/250 g of diet.

salts (g): $MgSO_4.7H_2O$, 14.6; $C_6H_5O_7Fe.5H_2O$, 3.16; Na_2HPO_4, 20.19; $CaHPO_4$, 28.18; NaCl, 4.15; $Ca[CH_3CH(OH)COO]_2$, 18.02; K_2HPO_4, 11.7.

DEVELOPMENT Average larval period, 31.3 days; pupal period: ♂, 15.5 days; ♀, 17 days. First instar to adult in 46-51 days on diet, and 23-33 days on natural host.

Ostrinia nubilalis (Hübner) [previously *Pyrausta*], European corn borer

→ Beck, S.D.; Lilly, J.H.; Stauffer, J.F. 1949. Nutrition of the European corn borer, *Pyrausta nubialis* (Hbn.). I. Development of a satisfactory purified diet for larval growth. ANN ENTOMOL SOC AM 42: 483-496.

COMPOSITION (g): bacto agar, 3.00; cellu-cotton, 1.625; Butaben, 0.20; glucose, 4.90; casein, 4.38; cholesterol, 0.175; corn oil, 0.175; Wesson's salt mixture, 0.35; brewers' yeast, 1.75; choline chloride, 0.07; dried corn leaves, 0.875; distilled water, 100 ml. pH to 6.7.

PREPARATION Corn leaves from 3' high plant homogenized 30 minutes after collection and freeze dried; leaf suspension kept cold by adding crushed ice; transferred to flask at -70° C in mixture of dry ice and 95% ethyl alcohol, flask revolved within freezing mixture, and attached to vacuum pump until dry leaf left. Dry leaf placed in brown bottle with $CaCl_2$ and sealed. Agar dissolved in 50 ml water in Erlenmeyer flask. Other ingredients added, remaining water last. Stirred for 15 minutes in water bath, then plugged with cotton and autoclaved at 15 lb for 20 minutes. Cooled and shaken. Flasks refrigerated until use.

REARING Individually in sterile 25 ml shell vials at 85 ± 5° F; diet smeared on to side of vial; vial plugged with cotton. Larvae transferred to clean vials and new diet every 3 days.

DEVELOPMENT Survival to pupa, 90%; larval wt after 30 days with 5% leaf content, 90 mg; with 10% leaf content, 100 mg.

→ Beck, S.D.; Stauffer, J.F. 1950. An aseptic method for rearing European corn borer larvae. J ECON ENTOMOL 43: 4-6.

COMPOSITION (g): bacto-agar, 3.35; cellulose, 2.7; glucose, 5.475; casein, 5.475; cholesterol, 0.219; linoleic acid (with 1% α-tocopherol), 0.219; Wesson's salt mixture, 0.437; brewers' yeast, 2.188; choline chloride, 0.087; leaf factor, 1.75; distilled water, 125 ml.

PREPARATION Agar dissolved in 60 ml water in water bath. Other ingredients and remaining water added, stirred for 15 minutes removed from water bath and dispensed to 125 ml Erlenmeyer flasks plugged with cotton, or to 50 x 90 mm diam crystallizing dish covered with filter paper lined petri dish lid. Diet autoclaved for 25 minutes at 17 lb pressure, and shaken while cooling.

REARING At 30° C, individually in vials or @ 10-15/flask. Eggs laid on waxed paper in wire screen cage. For egg sterilization, black head stage eggs removed from paper and placed in 2% solution of commercial 1:300 trypsin solution at pH 8 with 0.05 M phosphate buffer. Trypsin solution and eggs kept at 35° C with vigorous agitation every 10-15 minutes for 60 minutes, then rinsed in distilled water. Last rinse in 40% ethyl alcohol for 1 minute then distilled water. Sterilized for 10 minutes in 2% NaOH, and 2% formaldehyde solution, then to 70% ethyl alcohol solution. Separated eggs transferred to diet.

DEVELOPMENT Larval mortality, <30%; larval period, 17-25 days. Normal pupation and emergence; viable eggs.

See also: Beck (1950, 1953, 1956a b c d, 1960a); Beck & Hanec (1958); Beck & Kapadia (1957); Wressell (1955).

Corollary

→ Beck, S.D.; Chippendale, G.M.; Swinton, D.E. 1968. Nutrition of the European corn borer, *Ostrinia nubilalis*. VI. A larval rearing medium without crude plant fractions. ANN ENTOMOL SOC AM 61: 459-462.

COMPOSITION Improved diet (g/100 g diet): vitamin-free casein, 4.13; dextrose, 3.23; powdered cellulose, 1.94; fibrous cellulose, 1.29; bacto-agar, 2.32; salt mixture, 0.50; sodium alginate, 0.39; *l*-ascorbic acid, 0.39; *l*-cysteine hydrochloride, 0.08; cholesterol, 0.15; corn oil + 1% α-tocopherol, 0.21; sorbic acid, 0.09; choline chloride, 0.07; B-vitamin mixture, 0.04; distilled water, 85.17.

salts (%): K_2HPO_4, 49.75; KH_2PO_4, 18.00; $MgSO_4$, 16.00; $Ca(H_2PO_4)_2 . H_2O$, 8.00; NaCl, 5.00; $Fe_2(SO_4)_3 . 6H_2O$, 2.00; $MnSO_4 . H_2O$, 0.50; $Zn(OOCCH_3)_2 . 2H_2O$, 0.50; $CuSO_4 . 5H_2O$, 0.25.

B-vitamins (mg): *myo*-inositol, 29.50; calcium pantothenate, 5.45; niacin, 2.12; *p*-aminobenzoic acid, 1.00; riboflavin, 0.31; pyridoxine hydrochloride, 0.24; thiamine hydrochloride, 0.18; folic acid, 0.08; biotin, 0.05; and cobalamine, 0.01.

DEVELOPMENT Eight consecutive generations reared without reduction in average body weight, rate of growth, adult longevity, or fecundity.

→ Becton, A.J.; George, B.W.; Brindley, T.A. 1962. Continuous rearing of European corn borer larvae on artificial medium. IOWA STATE COLL J SCI 37: 163-172.

COMPOSITION (g): distilled water, 255; agar, 6.6; glucose, 10.5; vitamin-free casein, 10.5; cholesterol, 0.85; corn oil (containing 1% α-tocopherol), 0.5; salt mixture, U.S.P. XIII No. 2, 1.3; choline chloride, 0.12; brewers' yeast, 6.9; leaf factor, 13.8; n-butyl *p*-hydroxybenzoate (in 95% ethanol), 0.6; sorbic acid (in 95% ethanol), 0.6.

PREPARATION See Beck *et al.** (1949).

REARING Eggs hatched in moist container and neonate larvae transferred to vials @ 2/vial. Pupae removed to emergence cages; eggs laid on waxed paper placed on top of wire screen cages.

DEVELOPMENT Larval period, 20 days. No. of egg masses/♀ in 11th generation, 7.1. 11 generations reared.

→ Bottger, G.T. 1942. Development of synthetic food media for use in nutrition studies of the European corn borer. J AGRIC RES 65: 493-500.

Best of 20 diets tested.

COMPOSITION (g): cellulose, 3.5; agar, 3; casein, 2; glucose, 2; sucrose, 1; total mineral salts, 0.30; fat, 0.3; and water, 85 ml. pH, 4.5.

salts (g): KH_2PO_4, 1; $CaSO_4.2H_2O$, 1; $FeCl_3.6H_2O$, 1; $MgSO_4.7H_2O$, 1; and $Cu(NH_3)_4(OH)_2$, 0.13 ml. Copper ammonium hydroxide saturated solution prepared by mixing ammonium hydroxide with excess cuprous oxide.

Fats supplied in form of corn oil, wheat-germ oil, and halibut-liver oil.

PREPARATION Agar dissolved in hot water and mixed with cellulose previously soaked in water; boiled for 10 minutes, then dry ingredients and fats added and homogenized. Vitamins added when cool.

REARING At 80 ± 1° F and 70% RH. Diet spread to ¼" thickness around sides of 1.5 x 2 cm diam glass crystallizing dishes, with copper mesh lids. 20 larvae reared/dish. Transferred to fresh diet every 2-3 days.

DEVELOPMENT Survival to adult, 36%.

→ Chippendale, G.M.; Beck, S.D. 1964. Nutrition of the European corn borer, *Ostrinia nubilalis* (Hübn.). V. Ascorbic acid as the corn leaf factor. ENTOMOL EXP APPL 7: 241-248.

Two basic diets tested.

COMPOSITION Diet 1 (g): whole wheat germ, 5.0; vitamin-free casein, 3.5; dextrose, 3.5; powdered cellulose, 3.0; bacto-agar, 2.5; Wesson's salt mixture, 1.0; cholesterol, 0.30; sorbic acid, 0.12; choline chloride, 0.10; ascorbic acid, 0.50; sodium alginate, 0.50; and (ml): vitamin solution, 10.0; distilled water, 75.0.

vitamins (mg/l): *myo*-inositol, 2000; calcium pantothenate, 100; niacin, 100; riboflavin, 50; pyridoxine hydrochloride, 25; thiamine hydrochloride, 25; folic acid, 25; biotin, 2; and B_{12}, 0.2.

Diet 2 (g): spray-dried alfalfa juice, 4.17; vitamin-free casein, 4.17; dextrose, 4.17; fibrous cellulose, 1.7; bacto-agar, 2.8; Wesson's salt mixture, 0.38; corn oil (containing 1% α-tocopherol acetate), 0.215; cholesterol, 0.19; sorbic acid, 0.15; choline chloride, 0.069; and (ml): vitamin solution, 8.3; distilled water, 96.

vitamins (mg/l): *myo*-inositol, 4600; calcium pantothenate, 850; niacin, 330; *p*-aminobenzoic acid, 155; riboflavin, 48; pyridoxine hydrochloride, 37; thiamine hydrochloride, 28; folic acid, 11; biotin, 8; and B_{12}, 2.

PREPARATION Components, except ascorbic acid, thoroughly blended and autoclaved for 20 minutes at 121° C and 17 psi. Ascorbic acid added as solution in sterile distilled water after diet cooled to 55° C. Diet stirred and dispensed to 5 x 9 cm diam crystallization dishes to depth of 2 cm.

REARING Individually at 30° C in 23 x 85 mm sterile steel vials, with 5 g diet, and plugged with cotton.

DEVELOPMENT Diet 1 superior; mean days to 50% pupation, 12.5 ± 0.29; average pupal wt, 79.4 mg; viable eggs laid.

See also: Beck & Smissman (1960, 1961).

Corollary

→ Chippendale, G.M.; Beck, S.D. 1968. Biochemical requirements for mass rearing of plant feeding lepidopterans. pp. 31-39. *in* Radiation, Radiosotopes and Rearing Methods in the Control of Insect Pests. Proceedings of a Panel Organized by the joint FAO/IAEA Division of Atomic Energy in Food and Agriculture.

Two synthetic diets given. Diet 1, modification of that of Chippendale & Beck* (1964) for *O. nubilalis*.

COMPOSITION Diet 1 (g): whole-wheat germ, 20.0; vitamin-free casein, 25.0; dextrose, 25.0; powdered cellulose, 8.0; fibrous cellulose, 10.0; bacto-agar, 18.0; Wesson's salt mixture, 2.5; sodium alginate, 3.0; *l*-ascorbic acid, 3.0; cysteine, 0.6; cholesterol, 1.2; corn oil + 1% α-tocopherol, 1.6; sorbic acid, 0.72; choline chloride, 0.56; and (ml): vitamin solution, 50.0; distilled water, 610.0.

Diet 2a (g): wheat germ residue, 4.0; vitamin-free casein, 5.0; dextrose, 5.0; agar, 3.6; fibrous cellulose, 2.0; powdered cellulose, 1.6; sodium alginate, 0.6; *l*-ascorbic acid, 0.6; Wesson's salt mixture, 0.5; corn oil + 1% α-tocopherol, 0.32; cholesterol, 0.24; sorbic acid, 0.14; cysteine, 0.12; choline chloride, 0.11; and (ml): vitamin solution, 10.0; distilled water, 122.0.

2b: as 2a, except: wheat-germ residue omitted; casein increased to 6.4 g; cellulose increased to 3.0 g.

2c: as 2b, except: 0.5 g Wesson's salt mixture replaced by 0.5 g salt mixture of Beck *et al.** (1968).

DEVELOPMENT Percent pupation on 17th day at 30° C on diets 1, 2 and 3, 91, 21 and 90 respectively.

Also diet for cabbage looper, *Trichoplusia ni* (Hübner) as given in Chippendale & Beck* (1965).

→ Guthrie, W.D.; Raun, E.S.; Dicke, F.F.; Pesho, G.R.; Carter, S.W. 1965. Laboratory production of European corn borer egg masses. IOWA STATE J SCI 40: 65-83.

COMPOSITION (g): agar, 68.0; dextrose, 98.0; casein, 97.0; cholesterol, 7.5; USP salt mixture No. 2, 13.0; brewers' yeast, 50.0; leaf and whorl powder, 135.0; ascorbic acid, 30.0; vitamin supplement, 15.0; and (ml): distilled water, 3000.0; wheat germ oil, 7.5; mould inhibitor, 40.0.

vitamins (%): vitamin A (200000 units/g), 0.45; vitamin D (400000 units/g), 0.025; α-tocopherol, 0.5; ascorbic acid, 4.5; inositol, 0.5; choline chloride, 7.5; riboflavin, 0.1; menadione, 0.225; *p*-aminobenzoic acid, 0.5; niacin, 0.45; pyridoxine hydrochloride, 0.1; thiamine hydrochloride, 0.1; calcium pantothenate, 0.3; biotin, 0.002; folic acid, 0.009; B_{12}, 0.00135; and dextrose, 84.738865.

mould inhibitor: 90.0 g of sorbic acid dissolved in 400.0 ml of 95% ethanol.

PREPARATION Leaf and whorl powder prepared from corn borer susceptible inbred WF 9, or single cross WF 9 x M 14. Field grown plants cut at first collar of leaves when tassel just entering boot, sun or heat dried in forced-air grain-drying bin at 130° F. Dried leaf material then ground and deep frozen.

Agar dissolved in half of water, then cooked with corn powder for 15 minutes. Added to premix of remaining water, vitamins and other ingredients and blended. Diet dispensed to depth of ¾" in 3-dram vials, or to ½" depth in 12" diam plastic containers, to solidify.

REARING At 80° F and 70% RH individually in vials. Pupae removed to small cups for emergence, and adults paired in small oviposition cages. Or, mass reared in plastic containers with ventilated snap-on lids, with 200-300 eggs/box. Pupae sterilized in 1.5% NaClO, rinsed in water, dried, and transferred to 6' cube emergence and oviposition cages @ 100 pairs/cage.

DEVELOPMENT From neonate larva to adult, 20 days. Fecundity, 1.3-16.9 egg masses/♀; minimum cost/moth, 1.6 cents.

See also: Conway & Harding (1967); Harding *et al.* (1967); Raun (1961, 1966, 1967).

→ Gahukar, R.T. 1975. Nouvelles techniques adoptées pour l'élevage d'*Ostrinia nubilalis* Hübner sur milieu artificiel. ANN ZOOL ECOL ANIM 7: 491-498. (In French with English summary).

COMPOSITION As in Poitout & Bues* (1970), except agar reduced to 4 g; benzoic acid and Nipagin increased to 0.3 and 0.25 g respectively.

PREPARATION As in Poitout & Bues* (1970).

REARING Adults at 24 ± 1° C and 60-70% RH with 16 h photoperiod. Eggs, larvae and pupae at 28° C and 70-80% RH with 16 h photoperiod. 30 adult pairs confined in 20 x 16 cm diam gauzed cages. Oviposition on paper sheets. For large numbers, oviposition cages as described by Guthrie *et al.* (1965). Adults fed on 5% sugar solution. Eggs sterilized in 2% sorbic acid in 95% ethanol for 2-3 minutes, then rinsed in running water. Egg sheets incubated between damp filter papers in 12 x 9 cm box. Neonate larvae transferred to 20 x 8 x 2.5 cm box with long strips of diet, and covered with black paper for 2 days to reduce mortality. After 7 days transferred to larger box with fresh diet. Pupation in corrugated paper strips treated with sorbic acid as in Reed *et al.** (1972).

DEVELOPMENT Eclosion in 3-4 days; larval period, ca. 17 days; larval survival, 72-92%; pupal survival, 65-80%; mean pupal wt: ♀, 63.6 mg; ♂, 56.5 mg; fecundity, 12 egg masses/♀.

Isa, A.L.; Khadr, G.D. 1973. Rearing of *Ostrinia nubilalis* (Hbn) on artificial diets. AGRIC RES REV 51: 15-19.

Three diets tested.

COMPOSITION Diet 1 (g): agar, 3; cellulose, 1.5; glucose, 1.5; sucrose, 1.5; casein, 1.5; yeast extract, 6.0; rice bran, 9; ascorbic acid, 0.45; and (ml): methyl paraben (20% solution in alcohol), 1.5; sorbic acid (10% solution in alcohol), 1.5; distilled water, 150.

Diet 2: as diet 1, except rice bran replaced by 9 g corn plant powder.

Diet 3: as diet 1, plus (g): rice bran, 6.75; and corn plant powder, 2.25.

PREPARATION Agar and cellulose boiled in distilled water. Blended for 5 minutes with other dry ingredients then antibiotics added. Diet cooled for 3-5 h, then dispensed to 3 x 1" glass vials @ 4 g/vial.

REARING At 27°C in darkness, individually in vials. Pupae removed to clean vials on moist cotton for emergence. Adults paired in glass jars lined with moist cotton, and wax paper for oviposition. Eggs incubated in petri dishes.

DEVELOPMENT On diet 1: larval period, 24 days; pupation, 83-96%; pupal period, 10 days; pupal wt: ♀, 86-89 mg; ♂, 65-67 mg. On diet 2: larval period, 23-27 days; pupation, 80-100%; pupal period, 10 days; pupal wt: ♀, 82-93 mg; ♂, 60-65 mg. On diet 3: larval period, 21-25 days; pupation, 84-92%; pupal period, 10 days; pupal wt: ♀, 87-94 mg; ♂, 65 mg.

Klun, J.A.; Tipton, C.L.; Brindley, T.A. 1967. 2,4-Dihydroxy-7-methoxy-1,4-benzoxazin-3-one (DIMBOA), an active agent in the resistance of maize to the European corn borer. J ECON ENTOMOL 60: 1529-1533.

COMPOSITION (ml): water, 200; wheat germ oil, 0.50; and (g): agar, 4.50; sucrose, 6.50; casein, 6.50; cholesterol, 0.50; U.S.P. salt mixture No. 2, 0.85; corn leaf, 9.00; sorbic acid, 0.40.

vitamins (mg): choline chloride, 52.5; riboflavin, 0.70; pyridoxine hydrochloride, 0.70; folic acid, 0.06; calcium pantothenate, 2.10; thiamine hydrochloride, 0.70; vitamin B_{12}, 9.4×10^{-3}; and (g): ascorbic acid, 2.0.

PREPARATION See Guthrie *et al.** (1965).

REARING At 80°F and 75% RH with 24 h photoperiod. Eggs incubated in screw cap jars; larvae reared individually in shell vials.

DEVELOPMENT Pupal wt: ♀, 111-117 mg; ♂, 83-86 mg.

Lewis, L.C.; Lynch, R.E. 1969b. Rearing the European corn borer, *Ostrinia nubilalis* (Hübner), on diets containing corn leaf and wheat germ. IOWA STATE J SCI 44: 9-14.

Best 2 of 5 diets tested.

COMPOSITION Diet 1, wheat diet (ml): water, 3000; and (g): agar, 70.0; dextrose, 100.0; casein, 110.0; cholesterol, 8.0; salt mixture No. 2, 36.0; vitamin supplement (NBC), 23.0; ascorbic acid, 30.0; sorbic acid (dissolved in 7.5 ml of 95% ethyl alcohol), 1.5; and wheat germ, 130.0.

Diet 2, corn leaf diet described by Guthrie *et al.** (1965).

PREPARATION See Guthrie *et al.** (1965).

REARING Individually in shell vials at 80° F and 75% RH with 24 h photoperiod. Pupae removed to jelly cups for emergence. Adults mated singly in wire cages at 80° F and 75% RH with 16 h photoperiod then 8 h darkness at 65° F. Eggs incubated as larvae.

DEVELOPMENT Pupal weight of ♂♂ on diet 2 significantly greater than on diet 1: 75.47 mg and 72.68 mg respectively. No other difference after 32 and 62 generations on diets 1 and 2 respectively.

See also: Lewis & Lynch (1969a, 1970); Lewis & Raun (1966); Lewis *et al.* (1971); Lynch & Lewis (1971); Reed *et al.* (1972).

Corollary

→ Lynch, R.E.; Lewis, L.C.; Robinson, J.C. 1973. Lyophilized rearing medium for bioassays. J ECON ENTOMOL 66: 1224-1225.

PREPARATION Diet plugs frozen in vials, then freeze-dried at -55° C with 5 μ Hg vacuum for 24 h. Vials plugged with cotton and stored at -5° C. Prepared for bioassay by rehydrating with water or test-material suspension; performance equal to that of standard diet.

→ Nagy, B. 1970. Rearing of the European corn borer (*Ostrinia nubilalis* Hbn.) on a simplified artificial diet. ACTA PHYTOPATHOL ACAD SCI HUNG 5: 73-79.

COMPOSITION (g): agar, 20; saccharose (crystallized), 40; wheat germ, 150; lucerne meal, 100; yeast (household, fresh), 40; ascorbic acid, 4; Nipagin, 4; and (ml): acetic acid, 5; and water, 1000.

PREPARATION Agar dissolved in ca. 500 ml water. Wheat germ, lucerne meal, yeast and sugar dissolved in remaining water, added to agar solution, and cooked for 5-10 minutes. Nipagin (in 20 ml of 96% alcohol) added, mixture cooled to 60° C, and glacial acetic acid and ascorbic acid in 50 ml of water stirred in. Diet dispensed to jars or vials. Final pH, 4.0-4.5.

REARING At 28 ± 0.5° C with 17 h photoperiod; humidity not regulated. 300-350 g diet smeared on bottom and sides of 1.36-l glass jars; 75-300 eggs seeded on to diet. Pupation in folded filter paper (illustrated). Adults to 8.2-l gauze oviposition cages. Oviposition on dark blue paper strips, renewed every 2 days. Bunch of cotton attached to cover and moistened every day to stimulate oviposition. Egg clusters incubated at 18-28° C in petri dishes on moist cotton.

DEVELOPMENT Generation completed in 34-37 days; 39 generations reared.

→ Salama, H.S. 1970. Rearing the corn borer, *Ostrinia nubilalis* (Hübn.), on a semi-artificial diet. Z ANGEW ENTOMOL 65: 216-218.

COMPOSITION (g): dry kidney beans, 150.0; brewers' yeast, 16.0; ascorbic acid, 1.5; agar, 6.5; water, 320; benzoic acid or sodium benzoate, 2; and (ml): formaldehyde, 1; 0.1 M HCl, 1.

PREPARATION Agar dissolved in water, cooled to 60 C, then mixed with other blended ingredients. Beans soaked overnight, boiled for few minutes in water before use. Diet dispensed to 15 x 10 cm glass jars @ 2 oz/jar.

REARING At 30° C. Oviposition on paper strips. 20-30 eggs seeded to jars with diet. More diet added when 4th instar. Pupae removed individually to small jars for emergence. Adults fed 10% honey solution.

DEVELOPMENT Larval period, 12.3 days; pupal period, 3.3 days; adult longevity, 7 days. Viable eggs laid. Development retarded and fecundity reduced with excess or more concentrated mould inhibitor.

Spodoptera littoralis (Boisduval) (Noctuidae) also reared.

Plodia interpunctella (Hübner), Indian meal moth

→ Morère, J.-L. 1970a. Nutrition d'une chenille de Lépidoptère *Plodia interpunctella* (Pyralidae). I. Essai d'alimentation semi-synthétique permettant l'obtention d'adultes féconds. ARCH SCI PHYSIOL 24: 97-107. (In French)

Best of 4 diets tested.

COMPOSITION Dry part (%): sucrose, 7.0; dextrose, 18.0; vitamin-free casein (Hammersten), 55.0; corn germ oil, 17.0; cholesterol, 0.7; α-tocopherol, 0.035; ascorbic acid, 0.35; Tanton mineral salt mix, 2.0.

aqueous part, vitamins (%): thiamine hydrochloride, 0.0025; riboflavin, 0.0012; pyridoxine hydrochloride, 0.0012; nicotinic acid, 0.0025; calcium pantothenate, 0.0025; folic acid, 0.0001; biotin, 0.00001; inositol, 0.025; choline, 0.025.

PREPARATION Folic acid and biotin dissolved and mixed with 0.005 N NaOH solution, dried for 24 h in desiccator and added to remainder of vitamins in water. Casein, wheat germ oil, salts and cholesterol mixed. Dextrose ground and mixed with vitamin solution. Whole then pulverised in ball mill for >6 h at >90% RH.

REARING At 28 ± 1° C and 60% RH with 12 h photoperiod.

DEVELOPMENT On artificial diet and control on pistachio nuts (in brackets):- mean larval period, 34.3 (37.2) days; mean adult emergence (%), 69.0 (45.2); fecundity (eggs/♀), 158 (124); fertility (%), 96 (96). Seven generations reared.

Corollary

→ Morère, J.-L. 1974. Nutrition d'une chenille de Lépidoptère *Plodia interpunctella* Hübner (Lépidoptère, Pyralidae) sur un milieu chimiquemement défini. ARCH INT PHYSIOL BIOCHEM 82: 625-630. (In French with English summary)

COMPOSITION (g): amino acids, 40.0; RNA, 1.0; DNA, 1.0; glucose, 17.0; sucrose, 5.0; Tanton mix mineral salts, 2.0; cholesterol, 0.6; oleic acid, 7.0; linoleic acid, 6.0; palmitic acid, 1.0; stearic acid, 0.5; α-tocopherol, 0.2; β-carotene, 0.05; ascorbic acid, 0.3; Nipagin, 0.5; cellulose, 14.0; and (mg): vitamins, 60.01.

l-amino acids (g): glutamic acid, 8.0; aspartic acid, 2.0; histidine, 1.0; lysine, 2.0; arginine, 1.0; glycine, 1.0; alanine, 1.0; valine, 2.5; leucine, 3.5; isoleucine, 2.0; proline, 4.0; phenylalanine, 2.0; cysteine, 2.0; methionine, 2.0; tryptophane, 0.5; tyrosine, 2.0; serine, 2.0; threonine, 1.5.

vitamins (mg): thiamine hydrochloride, 2.5; pyridoxine hydrochloride, 1.2; nicotinic acid, 2.5; calcium pantothenate, 2.5; folic acid, 0.1; biotin, 0.01; inositol, 25.0; choline, 25.0.

DEVELOPMENT On host (pistachio nuts), semi-synthetic diet, and chemically defined diet respectively: larval and pupal period (days), 37.2, 33.0, and 55.3; emergence, 45.2, 83.8, 15.0%; fecundity, (eggs/♀), 158, 205, 157; egg fertility, 96.0, 91.0, 94.7.

See also: Morère (1970b, 1971a b, 1973 a b)

→ Silhacek, D.L.; Miller, G.L. 1972. Growth and development of the Indian meal moth, *Plodia interpunctella* (Lepidoptera: Phycitidae), under laboratory mass-rearing conditions. ANN ENTOMOL SOC AM 65: 1084-1087.

COMPOSITION (g): Gaines dog meal (ground), 620; rolled oats, 255; white cornmeal, 1665; whole wheat flour, 1480; wheat germ, 160; brewers' yeast, 325; glycerol, 1000; and honey, 900.

PREPARATION Dry and liquid components mixed separately then briefly blended, held at room temperature for 1-2 days in closed container, then ground in hammer mill through 0.48 cm screen. Passed through No. 8 sieve before use; dispensed to 13 x 19 x 10.5 cm or 12 x 22.5 x 7.5 cm boxes @ 400 g/box.

REARING At 30 ± 0.5°C and 70 ± 5% RH with 16 h or 24 h photoperiod. About 1800 (14 mg) 0-1 h old eggs mixed with diet. Adults removed to cylindrical oviposition cages with 20 mesh sieve base through which eggs fall on to black paper. Surface sterilized as in Boles & Marzke* (1966).

DEVELOPMENT Larval period, 12-13 days; pupal period, 6 days; adult emergence from 18-19 days, with peak from 20-22 days. Mature larval wt, 16 mg; pupal wt, 10 mg; wt of newly emerged ♀, 11-12 mg; of ♂, 8-9 mg. Maximum oviposition at 23-24 days. Fecundity, 1 g eggs/ 10000 moths.

Corollary

→ Bruce, W.A.; McCray, T.C. 1975. A simple method of dispensing eggs of *Plodia interpunctella*. J GA ENTOMOL SOC 10: 74-75.

REARING Egg dispenser made of shell vial; 10 evenly spaced holes drilled in base with No. 62 bit. Pressure-sensitive tape placed outside vial over holes. Vial filled with eggs, cover replaced, vial agitated, inverted and tape containing eggs removed. Tape secured to 25 x 75 mm glass microscope slides by double-sided pressure-sensitive tape. Eggs incubated at 27°C and 60% RH. Black electrical tape best.

Method also used with eggs of *Tribolium castaneum* (Herbst) and *Aedes aegypti* (Linnaeus); hole size varied for use with other insect species.

Corollary

→ Rutter, R.R.; Ferkovich, S.M. 1973. A rapid separation of larvae of the Indian meal moth from rearing medium. ANN ENTOMOL SOC AM 66: 919-920.

REARING Larvae separated from 400 g loosely packed diet in 20.5 cm sieve mounted 6" above collecting pan. Diet temperature gradually increased by 250-W reflector IR heat lamp mounted 17 cm above sieve; larvae move in to stainless steel pan; collected in jar by 150-W projector flood lamp mounted 43 cm behind sieve. Larvae collection completed within 18 minutes.

DEVELOPMENT From thermal collection: adult emergence, 99% in 20.2 days; fecundity, 2.2 mg/♀. From hand collection: adult emergence, 97% in 20.1 days; fecundity, 2.5 mg/♀.

Scirpophaga nivella (Fabricius), sugarcane top borer

→ Wahid, A.; Akhtar, M. 1971. Nutritional requirements of sugarcane top borer, *Scirpophaga nivella* Fab. PAK J ZOOL 3: 15-21.

Best of 35 casein and amino acid diets.

COMPOSITION Casein diet (g): bacto-agar, 1.116; cellulose, 1.0; sucrose, 3.0; casein, 1.825; cholesterol, 0.073; Wesson's salt mixture, 0.145; fresh sugarcane leaves, 3.0; choline chloride, 0.035; ascorbic acid, 0.25; and distilled water, 42.0 ml.

Amino acid diet, as above, except casein replaced by (mg/g): alanine, 0.0693; arginine, 0.0784; aspartic acid, 0.1533; cystine, 0.0073; glutamic acid, 0.4106; glycine, 0.0419; histidine, 0.0529; isoleucine, 0.1168; leucine, 0.1441; lysine, 0.1624; methionine, 0.0456; phenylalanine, 0.0839; proline, 0.1368; serine, 0.1149; threonine, 0.894; tryptophan, 0.0292; tyrosine, 0.1478; valine, 0.1149.

PREPARATION Agar dissolved in 14 ml water. Dry ingredients, except choline chloride and ascorbic acid, mixed with agar solution for 15-20 minutes, then when cool, choline chloride and ascorbic acid added and thoroughly mixed. Diet dispensed to 50 ml beakers, solidified and autoclaved.

REARING 5 larvae/beaker.

DEVELOPMENT Larval survival: on casein diet, 60.0 ± 4.6%; on amino acid diet, 70.6 ± 1.0%. No difference in larval weight on casein or amino acid diets.

Vitula edmandsae serratilineella Ragonot, dried-fruit moth

→ Miller, G.J.; Blankenship, J.W. 1973. Influence of dietary lipids upon lipids in larvae and adults of the dried-fruit moth. J INSECT PHYSIOL 19: 65-74.

Three diets tested.

COMPOSITION Diet 1 (g): yeast, 80; corn starch, 80; and (ml): glycerine, 50; honey, 50; α-tocopherol acetate, 0.06% by weight.

Diet 2: as diet 1, with addition of 5% beef tallow.

Diet 3: as diet 1, with addition of 5% corn oil.

PREPARATION Honey and glycerine blended with yeast and corn starch. Beef tallow or corn oil blended with yeast and corn starch first.

REARING In 1-gal glass jars ventilated by holes in metal lid. In closets at room temperature along with open jars of water containing filter paper wicks to provide humidity.

DEVELOPMENT Three generations reared.

See also: Blankenship & Miller (1971).

Miscellaneous - Pyralids

→ Boles, H.P.; Marzke, F.O. 1966. Lepidoptera infesting stored products. pp. 259-270. *in* Smith C.N. (Ed.). Insect Colonization and Mass Production. Academic Press, New York and London.

Rearing of Indian-meal moth, *Plodia interpunctella* (Hübner), Mediterranean flour moth, *Anagasta kühniella* (Zeller), and almond moth, *Cadra cautella* (Walker) on artificial diet described.

COMPOSITION (pints): cornmeal, 4; whole wheat flour, 4; Gaines Meal [General Foods Ltd, Toronto], 2; dried yeast, 1; honey, 1; glycerine, 1; wheat germ, 0.5; and oat meal, 1.

Gaines Meal: Hominy feed, soybean meal, wheat middlings, meat and bone meal, wheat red dog, animal fat preserved in BHA, fish meal, iodized salt, riboflavin supplement, U.S. certified food colour, vitamin B_{12} supplement, vitamin A supplement, and vitamin D_2 supplement.

PREPARATION Whole wheat flour and corn meal autoclaved separately for 90 minutes at 12-16 lb pressure at 240°F. Cooled to room temperature and mixed. Stored at 40°F in sealed containers. Mixed as required with remaining ingredients.

REARING At 80°F and 60% RH in 1-gal wide-mouthed jars with two-piece screw tops, centre lid replaced by filter paper to provide light cover and permit air exchange. 600 g diet dispensed to jars, and 50, 1-3 day-old adults added. Adults removed and eggs incubated.

DEVELOPMENT Of *P. interpunctella:* egg to adult from 23-30 days, maximum from 23-27 days. Larval period, 17 days; pupal period, 7 days; fecundity, 90 eggs/♀; incubation period, 4 days. Adult longevity, 13 days. Data for other 2 spp. similar.

→ Gonen, M.; Donahay, E. 1973. An improved technique for rearing the tropical warehouse moth *Ephestia cautella* (Walker) (Lepidoptera: Phycitidae). ISR J ENTOMOL 8: 179-180.

Improved rearing method for *Ephestia cautella.*

REARING In 31 x 23 x 6 cm plastic containers, with 800 g diet spread to 3 cm depth. Funnels attached to holes in lids to collect mature larvae in plastic cups.

DEVELOPMENT 2.1 larvae/g diet if 800 g in container and 6000 eggs seeded. Plastic containers better than glass jars.

See also: Navarro & Gonen (1970).

Fraenkel, G.; Blewett, M. 1946b. The dietetics of the caterpillars of three *Ephestia* species, *E. kuehniella, E. elutella,* and *E. cautella,* and of a closely related species, *Plodia interpunctella.* J EXP BIOL 22: 162-171.

All 4 spp. reared on diet.

COMPOSITION Basal diet (parts): casein, vitamin- and fat-free (E or A/E, Glaxo Lab.), 50; glucose (or starch), 50; dried debitterised brewers' yeast (Glaxo Lab.), 5; cholesterol, 1; McCollum's salt mixture, 2; water, 10; wheat-germ oil, 1.

PREPARATION See Fraenkel & Blewett* (1943a b). Modified by mixing salt mix in casein with water; mixture dried and ground in mortar after 24 h at 25°C and 70% RH.

REARING At 25°C and 70% RH. 5 larvae/2 x 1" tube with 3 g diet. 2 x 1" corrugated paper strip placed in tube for pupation as in Fraenkel & Blewett* (1943a b).

DEVELOPMENT Optimal growth of *E. kuehniella* with 80% glucose in diet; for other species optimal survival with 50% glucose. Replacement of glucose by starch highly satisfactory for *E. kuehniella,* but unsuitable for other 3 spp.

See also: Fraenkel & Blewett (1946d).

Corollary

Fraenkel, G.; Blewett, M. 1946c. Linoleic acid, vitamin E and other fat-soluble substances in the nutrition of certain insects, *Ephestia kuehniella, E. elutella, E. cautella* and *Plodia interpunctella* (Lep.). J EXP BIOL 22: 172-190.

COMPOSITION Basal diet for *E. kuehniella* (parts): casein (A/E, E or 'extracted', Glaxo Lab.), 20; glucose (anhydrous), 80; yeast (dried brewers', debitterised, Glaxo Lab.), 5; cholesterol, 1; McCollum's salt mixture, No. 185, 2; wheat germ oil, <1; water, 10.

Basal diet for *E. elutella, E. cautella* and *Plodia interpunctella* (Hübner) as above, except that casein increased to 50 parts, and glucose reduced to 50 parts.

In absence of wheat-germ oil: growth slow, mortality high, and adults failed to emerge from pupae. With suboptimal quantities of wheat-germ oil, adults emerged with wings lacking in scales.

See also: Perron *et al.* (1966).

Strong, R.G.; Partida, G.J.; Warner, D.N. 1968. Rearing stored-product insects for laboratory studies: six species of moths. J ECON ENTOMOL 61: 1237-1249.

Systems developed for continuous maintenance of almond moth, *Cadra cautella* (Walker); dried-fruit moth, *Vitula edmandsae serratilineella* Ragonot; Indian meal moth, *Plodia interpunctella* (Hübner); Mediterranean flour moth *Anagasta kuehniella* (Zeller); raisin moth, *Ephestia figulilella* (Gregson); and tobacco moth, *Ephestia elutella* (Hübner).

COMPOSITION For all 6 spp. (quarts): poultry laying mash, 15; chick starter mash, 15; rolled barley, 10; raisins (lightly fluffed, not packed), 5; and glycerine, 4.

PREPARATION Dry ingredients held at 5° F for 2 weeks before use. Diet prepared by blending ingredients in concrete mixer as follows: (a) rolled barley into mixer; (b) 1 qt glycerine poured over barley; (c) ca. 5 qt of laying or starter mash added; (d) 1 pt glycerine added slowly; (e) steps c and d repeated, alternating laying and starter mash until all poultry feed in mixer; (f) mixed for further 10 minutes, lumps broken apart by hand; (g) finally raisins added, and entire mixture thoroughly blended. Diet stored for 2-3 days before use to allow thorough adsorption of glycerine.

REARING At 80 ± 2° F and 55 ± 5% RH in 1-gal ventilated jars. Corrugated cardboard placed on top of food when larvae mature for pupation; pupae removed to gallon jars for emergence. Eggs collected every 2-3 days from moths in 1-pint egg laying jars covered with screen and inverted over filter paper in bottom of 10 cm glass petri dish. Eggs cleaned through sieve and sprinkled on ½-pint food placed on bottom of jars, then covered with 3½-pints food. Eggs incubation varied from 110 mg to 620 mg depending on species.

DEVELOPMENT Egg production (mg/adult): *C. cautella*, 1.03; *V. edmandsae*, 1.57; *P. interpunctella*, 0.97; *A. kuehniella*, 2.37; *E. figulilella*, 0.74; *E. elutella*, 0.91. Development period from egg to adult (days) for above spp.: 23, 35, 21, 32, 35 and 35 respectively. Numbers of adults/culture: 2188, 901, 2547, 1078, 2104, 1365 respectively.

See also: Bell (1975).

Saturniidae

Hyalophora cecropia (Linnaeus), cecropia moth

→ Levengood, W.C. 1968. Rearing a giant silkworm moth, *Hyalophora cecropia*, on an artificial culture medium. ANN ENTOMOL SOC AM 61: 1594-1596.

COMPOSITION Group A (ml): water, 293.0; 22.5% KOH, 6.0; and (g): Wesson's salt mixture, 12.0; sucrose, 42.0; and (ml): 37% formaldehyde, 1.2; inhibitor solution (40 g sorbic acid, and 30 g methyl *p*-hydroxybenzoate dissolved in 340 ml ethyl alcohol), 24.0; and (g): wheat germ, 36.0; casein, 42.0; Alphacel, 6.0.

group B (ml): water, 733.0; and (g): agar, 30.0; ascorbic acid, 4.8; and (ml): choline chloride (10 g/100 ml of water), 12.0; vitamin suspension, 2.0; and (g): Aureomycin, 3.0; kanamycin sulphate, 0.166.

vitamins (mg/ml water): nicotinic acid, 6.0; calcium pantothenate, 6.0; riboflavin, 3.0; thiamine hydrochloride, 1.5; folic acid, 1.5; pyridoxine hydrochloride, 1.5; vitamin B_{12}, 0.012; biotin (added to 1 ml of distilled water and thus used as liquid base for vitamin suspension), 0.12.

PREPARATION Agar dissolved in water, cooled to 60° C and blended first with group A components, then remainder of group B. Diet dispensed to 6-dram vials and refrigerated until use.

REARING At 22 ± 2° C with natural lighting. Air exchange via cotton plug in lid of vial. Larvae transferred to pint jars with fresh diet after 3-4 weeks for pupation; cocoons spun on cross-support of applicator sticks.

DEVELOPMENT Cocoon spinning from 2 months; pupae returned to ambient temperature from winter environment after 5 months. Adult emergence 35 days later, but with deformed wings. Eggs non viable.

Philosamia cynthia ricini (Boisduval), eri-silkworm or oak-silkworm

→ Fukuda, T.; Higuchi, Y. 1963. Artificial food for oak-silkworm raising. AGRIC BIOL CHEM 27: 99-102.

COMPOSITION For 1st to 3rd instar larvae (g): powdered dry oak-tree leaves, 6.0; parched soybean powder, 1.0; sucrose, 1.0; agar-agar, 1.0; and antiseptic solution, 27 ml.

For 4th and 5th instars; as above, except antiseptic solution reduced to 21 ml.

antiseptic solution; 2 g sodium dehydroacetate dissolved in 1-1 of water.

PREPARATION Hot solution of agar, sucrose and antiseptic added to powdered leaves of oak trees (*Quercus dentata*) and soybean powder. Mixture stirred and refrigerated when cool. Dry powder prepared at 60°C by air-blasting, pulverizing and screening through 80 mesh.

REARING At 26°C and 70-80% RH in petri dishes in cabinet. Diet on paraffin paper until 3rd instar. Fourth instar in wooden box with wire netting; with diet on bottom. Diet provided in 0.1-1 mm thick slices twice daily.

DEVELOPMENT Larval period, 29-45 days; pupal period, 22 days; 23% to cocoons yielding 0.34 g silk fibre; fecundity, 203 eggs/ .

See also: Fukuda (1963); Fukuda *et al.* (1961).

→ Fukuda, T. 1964. Chemically refined diets for eri-silkworm raising. BULL SERIC EXP STN JPN 19: 201-221. (In Japanese with English summary)

Best of 4 diets tested.

COMPOSITION (g): agar-agar, 1.5; casein, 2.0; sucrose, 1.5; Wesson's salt mixture, 0.3; β-sitosterol, 0.2; ether extracts from dry leaves of castor-oil plant, etc., 0.4; cellulose powder, 2.9; and (mg): vitamin mixture, 93; choline chloride, 16.5; sodium hydroacetate, 53; and distilled water, 27 ml.

l-amino acids (g): alanine, 1.0; arginine, 1.0; cystine, 1.0; glycine, 1.2; histidine hydrochloride, 0.7; isoleucine, 1.2; leucine, 1.8; lysine hydrochloride, 1.5; methionine, 0.3; phenylalanine, 1.1; proline, 1.1; serine, 1.2; sodium aspartate, 2.1; sodium glutamate, 2.3; threonine, 1.1; tryptophan, 0.3; tyrosine, 0.7; valine, 1.4; total, 21 g.

vitamins (mg): *p*-aminobenzoic acid, 10; ascorbic acid, 2500; biotin, 1; calcium pantothenate, 30; β-carotene, 20; ergosterol, 400; folic acid, 1; inositol, 100; nicotinic acid, 10; pyridoxine hydrochloride, 30; riboflavin, 10; thiamin hydrochloride, 10; total, 3.122 mg.

PREPARATION In 2 parts. 1. Casein, vitamin mixture, salt mixture, cellulose powder portion with β-sitosterol, cellulose, powder portion with ether extracts from dried leaves of castor-oil plant, and remaining cellulose powder weighed and mixed. 2. Sucrose, agar and choline chloride dissolved in 26.5 ml hot water and sodium dehydroacetate. Part 1 mixture added at ca. 77° C, stirred, cooled and refrigerated for maximum of 4 days.

REARING At 24-26° C and 95% RH. 20 1st-4th instar larvae/9 cm diam petri dish and 20/21 cm diam petri dish for 5th instars. Thin slices of diet fed twice daily.

DEVELOPMENT Larval period, 25-31 days; cocoon rating, 95%; weight of cocoon fibre produced/larva, 116 mg; and eggs laid/♀, 142; adult, 80%.

Fukuda, T.; Higuchi, Y. 1967. A semi-synthetic diet for eri-silkworm raising (continued). BULL SERIC EXP STN JPN 21: 401-423. (In Japanese with English summary).

Best of many diets tested.

COMPOSITION (g): agar, 1.5; soybean powder, 2.5; casein, 1.3; sugar, 1.0; β-sitosterol, 0.1; Wesson's salt mixture, 0.3; dry brewers' yeast, 1.0; cellulose powder, 1.1; sorbic acid, 0.1; and (mg): ascorbic acid, 25; and water, 27 ml.

PREPARATION Casein, ascorbic acid, Wesson's salt mixture, and cellulose powder containing β-sitosterol, brewers' yeast and soybean powder mixed in order. Agar and sucrose in 0.25 ml water added to water and sorbic acid. Boiled, and casein fraction stirred in. Cooled and refrigerated for maximum of 4 days.

REARING See Fukuda* (1964).

DEVELOPMENT Growth, total development, and fecundity comparable with that on fresh leaves of *Ailanthus glandulosa*.

Riddiford, L.M. 1968. Artificial diet for cecropia and other saturniid silkworms. SCIENCE 160: 1461-1462.

Species reared:- *Hyalophora cecropia* (Linnaeus), *Hyalophora gloveri* Streek, *Samia cynthia* (Drury), *Antheraea pernyi* Guerin-Meneville, *A. polyphemus* (Cramer).

COMPOSITION Group 1 (g): Wesson's salt mixture, 10.0; sucrose, 35.0; wheat germ, 30.0; vitamin-free casein, 35.0; Alphacel, 5.0; cholesterol or β-sitosterol, 0.5; and (ml): 37% formaldehyde, 1.0; inhibitor solution, 20; and 10% KOH, 10.

group 2 (g): Aureomycin, 2.5; kanamycin sulphate, 0.14; ascorbic acid, 4.0; choline chloride, 1.0; inositol (if required), 0.15; and (ml): linseed oil, 1.0; and vitamin suspension, 2.0.

group 3 (g): agar, 25; and water, 840 ml.

inhibitor solution: 40 g sorbic acid and 30 g methyl *p*-hydroxybenzoate dissolved in 340 ml of 95% ethanol.

vitamins (mg/ml): biotin, 0.12; calcium pantothenate, 6.00; folic acid, 1.50; niacinamide, 6.00; pyridoxine hydrochloride, 1.50; riboflavin, 3.00; thiamine hydrochloride, 1.50; and vitamin B_{12}, 0.012.

PREPARATION Agar dissolved in water and blended with ingredients of group 1. Cooled to 50° C, then vitamins and inhibitors added and blended. Final mixture poured into beakers for solidifcation, then transferred to polyethylene bags and stored at 2° C.

REARING At 25°C and 65-80% RH in perforated aluminium foil covered glass jars; wooden stick applicators provided for pupation.

DEVELOPMENT Of *H. cecropia:* cocooning from 41-44 days; pupal wt: ♀, 6.0-7.5 g; ♂, 4.0-5.5 g; normal adults after 15 weeks at 5°C or after injection of α-ecdysone. Fertile eggs laid.

H. gloveri: cocoons spun after 26-30 days at 25°C.

S. cynthia: cocoons spun after 26-30 days at 25°C if inositol added to diet.

A. polyphemus: 25% pupation only on addition of 0.015-0.03% inositol and 0.001-0.1% trans-2-hexonal. Wing defects in emerging adults eliminated by addition of 0.1% linseed oil.

A. pernyi: 100% mortality before cocooning.

See also: Garrison *et al.* (1963); Riddiford (1967).

Sesiidae

Sanninoidea exitiosa (Say), peach tree borer

→Smith, H.R.; Richardson, E.G.; Yonce, C.E.; Jacklin, S.W. 1969. A method of rearing the peach tree borer on artificial diet. J ECON ENTOMOL 62: 961-962.

COMPOSITION (g): vitamin-free casein, 25.0; wheat germ, 15.0; dextrose, 15.0; salt mixture U.S.P. No. 2, 5.0; choline chloride, 1.0; agar, 40.0; vitamin diet fortification mixture (NBC), 5.0; cholesterol, 1.0; glycine, 0.3125; cysteine, 0.156; Alphacel, 7.5; ascorbic acid, 2.0; methyl *p*-hydroxybenzoate, 4.0; sorbic acid (dissolved in 10 ml 95% ethanol), 2.0; and distilled water, 650 ml.

PREPARATION Agar dissolved in 325 ml boiling water. Ascorbic acid and vitamin mixture dissolved in 325 ml water and remaining dry ingredients added and blended; blended with agar solution, and then warm diet dispensed to 1-oz paper medicine cups. Caps on after 24 h.

REARING At ca. 24°C and 60% RH with 10 h photoperiod @ 4 larvae/cup; pupation in soil.

DEVELOPMENT Survival to adult, 42.5%; 513 viable eggs from ♀ of one mating.

Synanthedon pictipes (Grote & Robinson), lesser peach tree borer

→Antonio, A.Q.; McLaughlin, J.R.; Leppla, N.C.; Green, C.W. 1975. Culturing the lesser peach tree borer. J ECON ENTOMOL 68: 309-310.

COMPOSITION (g): pinto beans (ground), 412; wheat germ, 200; brewers' yeast, 125; casein, 100; gelcarin, 46; methyl *p*-hydroxybenzoate, 8; sorbic acid, 4; ascorbic acid, 12; and (ml): vitamin mixture (400 g Hoffmann-La Roche in 1000 ml water), 30; demineralized water, 2700; 40% formalin, 15; and Tetracycline, 250 mg.

PREPARATION In 11.6 l batches. Dry ingredients except ascorbic acid added to warm distilled water in 18.9-l steam-jacketed kettle. Heated at 78°C for 10 minutes, formalin added after 5 minutes. Vitamins, tetracycline, and ascorbic acid added to hot diet to high speed blender, mixed for 30 seconds and dispensed @ 0.3 l diet/16-oz plastic cup with vented lid.

REARING At 28 ± 1°C and 70 ± 5% RH. Mated ♀♀ caged in 1-oz plastic cups with moist cotton gauze for oviposition. After 6 days eggs washed on to fine nylon screen with water and sterilized with NaClO solution. Dried eggs sprinkled on 2 cm^2 discs of sterilized paper towelling covered lightly with dilute casein glue @ 60-80 eggs/disc. Discs pinned to sponge wrapped in moist towelling and held for 2 days suspended on pin 12 mm above substrate in each cup. When larvae mature, diet transferred to 30 x 17 x 10 cm plastic boxes covered with 25-50 mm autoclaved sawdust for pupation. Pupae removed and stored in emergence cages at 28-35°C.

DEVELOPMENT Larval development, 51 days; giving 17 ± 1 pupae/cup; mean pupal wt, 48 ± 0.5 mg. Fecundity, 147 ± 18 eggs/♀ with 90% viability. Five generations reared.

→ Meyer, H.J.; Norris, D.M.; Koval, C.F. 1974. Laboratory rearing of the lesser peach tree borer, *Synanthedon pictipes*, on artificial diets. ANN ENTOMOL SOC AM 67: 721-722.

Best of two diets tested; modified from that of Chippendale & Beck* (1964) for *Ostrinia nubilalis*.

COMPOSITION (g): dextrose, 25; casein, 25; Wesson's salt, 2.5; cholesterol, 1.2; sorbic acid, 0.72; Alphacel, 8; agar, 18; cellulose (fibre), 10; wheat germ, 20; sodium alginate, 3; *l*-ascorbic acid, 3; *l*-cysteine hydrochloride, 0.6; and (ml): distilled water, 585; vitamin solution, 50; and corn oil with α-tocopherol, 50 drops. (1 g α-tocopherol/100 ml).

vitamins: see Norris & Baker* (1969).

PREPARATION As in Chippendale & Beck* (1964) except diet dispensed into 150 mm crystallizing dishes to 1" depth, autoclaved, cooled and refrigerated. Plugs of diet ca. 1" thick used.

REARING At 25°C and 55% RH in darkness; 1st instar larvae placed on diet plugs in sterile vials; emerging adults transferred to 14 x 16 x 36" screen cage. Mated ♀♀ isolated in small jelly cups at 25°C and 65% RH with 12 h photoperiod; oviposition on moist filter paper in cup. Eggs incubated at 100% RH in cups on tray of wet silica sand.

DEVELOPMENT Eclosion from 8th day; pupation in 38 days; pupal period, 11 days. Adult emergence, 50%; ♀♀ fecund and eggs viable.

→ Russell, W.G. 1968. A medium for laboratory rearing of the peach tree and lesser peach tree borers. J ECON ENTOMOL 61: 1450-1451.

COMPOSITION (ml): apple juice, 132.0; 22.5% KOH, 2.7; and (g): casein hydrolysate, 19.0; Wesson's salt mixture, 5.5; sucrose, 14.0; and (ml): 10% formaldehyde, 2.0; 15% methyl *p*-hydroxybenzoate in 95% ethyl alcohol, 5.5; and (g): apple (var. Winesap), 25.0; Alphacel, 2.5; vitamin mixture, 5.5; agar (hot, dissolved in 330 ml apple juice), 13.5; ascorbic acid, 2.0; and (mg): Aureomycin, 75.0; and (ml): cholesterol in ether (5 mg/ml ether), 5.0. Total ca. 565 g.

vitamins (mg): niacin, 236.3; calcium pantothenate, 236.3; thiamine hydrochloride, 59.0; riboflavin, 118.2; pyridoxine, 59.0; folic acid, 59.0; biotin, 4.7; vitamin B_{12}, 0.5; inositol, 4726.7.

PREPARATION First 10 ingredients, up to and including vitamin mixture, thoroughly mixed in order given. Hot agar cooled slightly, then with remaining 3 ingredients, blended with first mix.

REARING At 72 ± 5° F under diffuse light in cabinet. Fourth instar larvae reared in 72 x 22 mm vials half full of diet.

DEVELOPMENT Survival rates to adult from 4th instar larvae of *S. pictipes* and *S. exitiosa* 55 and 45% respectively. Nil survival of neonate larvae.

Sphingidae

Celerio euphorbiae (Linnaeus)

→ House, H.L. 1965d. Effects of low levels of the nutrient content of a food and of nutrient imbalance on the feeding and the nutrition of a phytophagous larva, *Celerio euphorbiae* (Linnaeus) (Lepidoptera: Sphingidae). CAN ENTOMOL 97: 62-68.

Modified Berger* (1963) diet for *Heliothis* spp.

COMPOSITION (g): vitamin-free casein, 3.5; Wesson's salt mixture, 1.0; sucrose, 3.5; wheat embryo, 3.0; choline chloride, 0.1; ascorbic acid, 0.4; agar, 2.5; cellulose flour, 0.5; and (ml): vitamin solution, 1.0; 4M KOH, 0.5; water (with concentration of ether extract of 4.5 g of fresh spurge, *Euphorbia* sp.), 84.0; formalin, 0.05.

vitamins (g/100 ml water): niacin, 0.1; calcium pantothenate, 0.1; riboflavin, 0.05; thiamine hydrochloride, 0.025; pyridoxine, 0.025; folic acid, 0.025; biotin, 0.002; cyanocobalamin (B_{12}), 0.0002.

PREPARATION See Berger* (1963).

REARING At room temperature.

DEVELOPMENT Weight gain in 4th instar larvae.

Celerio lineata (Fabricius), white-lined sphinx

→ Keeley, L.L. 1970. Diapause metabolism and rearing methods for the white-lined sphinx, *Celerio lineata* (Lepidoptera: Sphingidae). ANN ENTOMOL SOC AM 63: 905-907.

COMPOSITION (g): pre-mixed diet (NBC), 230.00; agar, 38.00; Aureomycin, 2.00; methyl paraben, 3.00; Acromycin, 2.00; Kanamycin, 0.28; cholesterol, 2.00; and (ml): raw linseed oil, 20; 36% formaldehyde, 2; 20% acetic acid, 20; water, 1680; and Vanderzant's vitamins, 60.

pre-mix diet (%): wheat germ, 24; sucrose, 28; Wesson's salt mixture, 8; Alphacel, 12; casein, 28.

vitamin mixture: 177 g Vanderzant's/1000 ml water; then (mg/60 ml solution): α-tocopherol, 85; biotin, 0.21; calcium pantothenate, 10; choline chloride, 531; folic acid, 2.66; inositol, 210; niacinamide, 10; pyridoxine hydrochloride, 2.66; riboflavin, 5.31; thiamine hydrochloride, 2.66; vitamin B_{12} (triturate), 20; and ascorbic acid, 2870.

REARING At 26-29°C with 10 h photoperiod in ¾ x 2½" plastic vials @ 2 larvae/vial. Mature larvae reared individually in 2-oz plastic cups with diet. Prepupa removed to filter paper lined petri dishes for pupation.

DEVELOPMENT Larval survival, 50-66%; pupal wt, ca. 2 g. Three generations reared.

Manduca sexta Linnaeus [previously *Protoparce*], tobacco hornworm

→ Hoffman, J.D.; Lawson, F.R.; Yamamoto, R.T. 1966. Tobacco hornworms. pp. 479-486. *in* Smith C.N. (Ed.). Insect Colonization and Mass Production. Academic Press, New York and London.

COMPOSITION (g): wheat germ, 50.00; sucrose, 30.00; casein, high nitrogen, 35.00; torula or brewers' yeast, 15.00; Wesson's salt mixture, 10.00; ascorbic acid, 4.00; cholesterol, 1.00; sorbic acid, 1.50; methyl *p*-hydroxybenzoate, 1.00; streptomycin sulphate, 0.20; Aureomycin, 0.10; agar, 2.00; and (ml): vitamin solution, 5.0; water, for agar, 600.0; and water for dietary mixture, 250.0.

vitamins (mg/200 ml water): nicotinic acid, 300; calcium pantothenate, 300; riboflavin, 150; thiamine, 75; pyridoxine, 75; folic acid, 75; biotin, 6; and choline chloride, 20 g.

PREPARATION Dry ingredients except agar ball-milled for 12 h then refrigerated or deep frozen until use, when suspended in 250 ml water; vitamins added. Agar dissolved in water and both mixtures blended, dispensed to trays @ 3.5 l/tray, and refrigerated.

REARING At 80°F and 80% RH; 300 larvae/tray. Prepupa placed in screened 1 x 1 x 4" wooden cells and kept at 75°F and 90-100% RH. After 7 days pupae removed to 2 x 2 x 1' screen wire cages @ 200/cage. Adults kept at 75 F and 80% RH; fed 10% sucrose solution; oviposition on tobacco plant or on artificial blossoms made from 2 x 2" diam plastic 'stem' with light coloured paper around rim painted with synthetic attractant (isoamyl salicylate) to stimulate oviposition.

DEVELOPMENT Larval period, 2 days longer on tobacco leaves than on artificial diet. 5th instar larval weights equal on host and diet. Pupal wt: 2170-4700 mg and 2650-4710 mg on host and diet respectively. Eight generations reared.

Manduca quinquemaculata (Haworth), tomato hornworm, also reared on diet, if 2% of tobacco leaf powder or water extract of tobacco leaves added. 80% pupation in 14-17 days.

See also: Stewart & Baker (1970).

→ Waldbauer, G.P.; Yamamoto, R.T.; Bowers, W.S. 1964. Laboratory rearing of the tobacco hornworm, *Protoparce sexta* (Johnson) (Lepidoptera: Sphingidae). J ECON ENTOMOL 57: 93-95.

COMPOSITION (g): bacto-agar, 35.0; cellulose powder (standard grade, Whatman), 30.0; sucrose, 50.0; vitamin-free casein, 25.0; bacto-peptonel (Difco), 25.0; corn oil (Mazola + 1% α-tocopherol), 2.00; cholesterol, 2.0; Wesson's salt mixture, 4.0; brewers' yeast, 20.0; choline chloride, 1.0; methyl *p*-hydroxybenzoate, 4.0; *l*-sorbose, 1.0; and (ml): leaf homogenate, 500; distilled water, 500.0.

leaf homogenate: 150 g fresh tobacco or tomato leaves homogenized in 300 ml distilled water.

PREPARATION Agar dissolved in water. Remaining ingredients added and mixed, poured into Erlenmeyer flasks plugged with cotton and autoclaved for 20 minutes at 15 lb pressure. Diet dispensed to petri dishes to depth of 1 cm and refrigerated until use.

REARING At 27-30°C in 10¼ x 12½ x 4" ventilated plastic trays, with paper towelling on base. Diet placed on platform of 4 mesh screen 1" above towelling. 5th instars transferred to large metal tubs; prepupae placed in battery jars with 6" layer moist sand @ 10/jar. Pupae removed after 7-8 days to 3 x 3 x 6' cheesecloth covered cages for emergence. Dusk simulated by 25 W lamp held 15' from cage. Adults hand fed sucrose solution from bulb pipette. Oviposition on potted tomato or tobacco plants, or cut stems in jars of water; eggs incubated in petri dishes.

DEVELOPMENT Larval period, 15-18 days; prepupal to adult period, 19-21 days; egg incubation period, 3 days. Generation completed in ca. 40 days.

→Yamamoto, R.T. 1969. Mass rearing of the tobacco hornworm. II. Larval rearing and pupation. J ECON ENTOMOL 62: 1427-1431.

Diet modified from that of Hoffman *et al.** (1966) for *M. sexta*.

COMPOSITION (g): wheat germ (pre-ground), 100.0; casein, 45.0; sucrose, 40.0; torula yeast, 30.0; Wesson's salt mixture, 15.0; ascorbic acid, 1.5; methyl *p*-hydroxybenzoate, 1.0; cholesterol, 0.5; agar, 20.0; and (ml): 10% formaldehyde, 20.0; water, 980.0.

vitamins (mg): nicotinic acid, 10.0; calcium pantothenate, 10.0; riboflavin, 5.0; thiamine hydrochloride, 2.3; pyridoxine hydrochloride, 2.3; folic acid, 0.2; and biotin, 0.2.

PREPARATION Dry ingredients ball-milled for 24 h and refrigerated till use; then blended with formaldehyde and about ⅓rd of water. Agar dissolved in remaining water, and both mixtures blended. Diet dispensed to aluminium pans, cooled and refrigerated.

REARING At 26°C and 55% RH. Young larvae reared in 61 x 28 x 10 cm metal trays with climbing support ring, feeding platform, lid and light shield (illustrated) @ 300/tray. 5th instar larvae and prepupae reared in 72 x 61 x 10 cm tray; diet renewed every 3 days. Prepupae transferred to wooden cells. Pupae stored for >10 months between 8-15°C.

DEVELOPMENT Neonate larva to 5th instar from 12-14 days; 5th instar to prepupa, 4 days; to pupa on 5th day; pupal period, 21 days; 5th instar wt, ca. 8.0 g; pupal yield from eggs, 45-65%.

See also: Yamamoto (1968).

→Bell, R.A.; Joachim, F.G. 1975. Techniques for rearing laboratory colonies of tobacco hornworms and pink bollworms. ANN ENTOMOL SOC AM 69: 365-373.

COMPOSITION For *M. sexta:* see Yamamoto* (1969) except 0.4% raw linseed oil added and water reduced by 25%.

For *P. gossypiella:* water content reduced to 500 ml/l.

PREPARATION Dry ingredients poured into plastic bag and deep frozen until use. Agar weighed into separate plastic bag and stored dry at room temperature. Vitamin solution (Hoffman *et al.* 1960) prepared and stored at 2° C. Boiling water poured into pre-warmed, 1-gal Waring Blendor and blended with agar, then remaining ingredients. Diet dispensed to greased baking pans, allowed to harden, and cut into cubes.

REARING *M. sexta:* at 26 ± 1° C and 40-50% RH with 15 h photoperiod. Eggs placed individually in 120 ml polystyrene cups containing few cubes of larval diet suspended above eggs on perforated plastic rack (illustrated). Diet cup placed within another, painted black so that light enters only through perforated lid. Tissue towelling sandwiched between lid and cup to prevent neonate larval escape through vent holes. After 24 h, larvae transferred to 30 ml polystyrene cups with ca. 8 g cubes of diet and plastic coated lid. Final instar larvae transferred to 120 ml polystyrene cups containing ca. 25 g diet. Cups covered with tight-fitting polyethylene lids with 3-4 reinforced vent holes. Upon termination of feeding transferred to pupation sites in 40 x 12 x 4 cm wooden blocks containing individual cells formed by drilling 2.5 cm diam holes to depth of 12 cm. Blocks saturated with water and larvae placed individually in each hole. Pupation blocks placed on shelves. Pupae removed from blocks after 6-7 days, sexed and placed on wire racks in oviposition cage 2-3 days prior to adult emergence.

P. gossypiella: Diet dispensed to 2.5 ml polystyrene vials @ ca. 1 ml/vial. Best results obtained with diet stored at room temperature for 2 days or less. Adult feeding, mating and oviposition cage from 200 ml dairy cups with base replaced by window screen. 10% sucrose solution fed via cotton wick. Oviposition on moist paper disc placed over double screen cover to cup. Eggs collected daily and placed in 30 ml jelly cups capped with plastic coated lids and stored at 25° C and 50% RH. Neonate larvae transferred individually to diet vials for development to adult.

DEVELOPMENT *M. sexta:* over 50 generations reared. Eclosion, 91-99%; mature larval wt, 8-9 g; pupal wt, 4-5 g; survival, >90%; fecundity, >100 eggs/♀/day. Cost, $30/1000 larvae.

P. gossypiella: several generations reared. Eclosion, 82.0%; survival to adult, 85.0%; fecundity, 413 eggs/♀.

Also reared successfully: *Anthonomus grandis* Boheman, (Col., Curculionidae), *Antheraea pernyi* Buerin-Meneville (Lep., Saturniidae); *Tetanops myaeformis (= myopaeformis)* Röder (Dipt., Oitidae).

Tineidae

Tineola bisselliella Hummel, webbing clothes moth

➞ Crowell, M.F.; McCay, C.M. 1937. Nutritional studies of the webbing clothes moth, *Tineola bisselliella* Hum. PHYSIOL ZOOL 10: 368-372.

Best 2 diets of many tested.

COMPOSITION Diet 1 (%): vitamins A- and B-free casein, 94; fish meal ash, 5.0; Harris B concentrate, 0.5; cystine, 0.5.

Diet 2 (%): vitamins A- and B-free casein, 89.5; vitamins A- and B-free lactalbumin, 10; Haris B concentrate (details not given), 0.5.

PREPARATION Diets ground, ball-milled and mixed.

REARING At 25° C and 75% RH individually from neonate larvae in 20 x 40 or 80 mm shell vials.

DEVELOPMENT Diet 1: adult emergence, 30%; average life cycle, 57.9 ± 0.53 days.

Diet 2: adult emergence, 20%; average life cycle, 48.2 ± 0.2 days.

On fish meal: adult emergence, 29%; average life cycle, 56.5 ± 1.06 days.

Fraenkel, G.; Blewett, M. 1946a. The dietetics of the clothes moth, *Tineola bisselliella* Hum. J EXP BIOL 22: 156-161.

Two diets tested.

COMPOSITION Diet 1 (parts): casein (A/E, Glaxo Lab.), 50; glucose, 50; yeast (brewers', debittered, Glaxo Lab.), 5; cholesterol, 1; McCollum's salt mixture, 2; water, 10.

Diet 2 (parts): casein, 50; glucose, 50; 'insoluble' yeast, 5; cholesterol, 1; salt mixture, 2; water, 10; and (µg/g dry diet): thiamin, 50; riboflavin, 50; nicotinic acid, 50; pyridoxin, 50; pantothenic acid, 50; inositol, 500; choline chloride, 500; *p*-aminobenzoic acid, 500.

PREPARATION See Fraenkel & Blewett* (1943a).

REARING At 25° C and 70% RH; 10 larvae in 2 x 1" tubes with 2 g food.

DEVELOPMENT Larval period, 25 days; pupal period, 10 days. Adult emergence, 33 days from inoculation. Growth similar on both diets.

See also: Griswald (1944).

Tortricidae

Adoxophyes orana Fischer von Röslerstamm, smaller tea tortrix

Tamaki, Y. 1959. Studies on nutrition and metabolism of the smaller tea tortrix, *Adoxophyes orana* Fischer von Röslerstamm. I. Aseptic rearing of the larva on synthetic diets. JPN J APPL ENTOMOL ZOOL 3: 286-289.

Best of many diets tested.

COMPOSITION (g): agar, 1.0; cellulose, 1.3; glucose, 3.0; milk casein, 3.1; cholesterol, 0.1; choline chloride, 0.025; Wesson's salt mixture, 0.3; dried yeast, 2.9; tea leaf powder, 0.1; and distilled water, 40 ml.

PREPARATION Ingredients placed in 300 ml Erlenmeyer flask and sterilized for 30 minutes on 3 successive days. To prepare powder, tea leaves steamed at 0.8 kg wt/cm^2 for 20-30 seconds, dried at 80° C for 2 h, then ground and sifted through 100 mesh sieve.

REARING At 27 ± 3°C in flasks. Adults in 3 x 12 cm oviposition cylinders lined with paraffin paper. Fed from sugar solution on cotton; oviposition on paraffin. Egg masses washed in 70% ethanol, sterilized in 0.1% $HgCl_2$ for 4 minutes and washed in alcohol. Egg mass (20-50 eggs) inoculated aseptically to diet. Pupae placed in small glass tubes for emergence.

DEVELOPMENT 15th day larval wt, 17.7 mg. Emergence poor until level of tea leaf powder increased to >10%.

Tamaki, Y. 1964. Studies on nutrition and metabolism of the smaller tea tortrix, *Adoxophyes orana* Fischer von Röslerstamm. IV. Some chemical properties of an unknown dietary factor responsible for larval growth. JPN J APPL ENTOMOL ZOOL 8: 55-61.

Two basal diets tested.

COMPOSITION Diet 1 (g): agar, 0.7; cellulose, 0.9; glucose, 2.0; casein, 2.5; cholesterol, 0.1; Wesson's salt mixture, 0.2; linseed oil, 0.1; and (ml): distilled water, 20.0; B-vitamin solution, 1.0.

vitamins (µg/ml water): choline chloride, 4000; inositol, 2000; thiamine hydrochloride, 200; nicotinic acid, 200; calcium pantothenate, 200; *p*-aminobenzoic acid, 200; riboflavin, 100; pyridoxine, 100; folic acid, 20; biotin, 20.

Diet 2: as diet 1, with addition of 0.2 g sodium propionate and 0.015 g HCl.

PREPARATION Ingredients mixed, transferred to 300 ml Erlenmeyer flasks, then autoclaved for 30 minutes daily for 3 successive days.

REARING At 27 ± 3°C; aseptically and semi-aseptically.

DEVELOPMENT On diet 2, larval period, 18 days; survival, 52%. Normal adults produced. Similar results when reared aseptically on diet 1.

See also: Tamaki (1961a b, 1962).

Tamaki, Y. 1966a. Rearing of *Adoxophyes orana* Fischer von Röslerstamm and *Homona magnanima* Diakonoff on simplified artificial food. JPN J APPL ENTOMOL ZOOL 10: 46-48. (In Japanese)

Four diets tested.

COMPOSITION Diet 1 (g): glucose, 2; casein, 2.5; cholesterol, 0.1; Wesson's salt mixture, 0.2; linseed oil, 0.1; powdered cellulose, 0.9; powdered tea (sieved through 1 mm mesh), 2.0; dried yeast, 1.0; powdered agar, 0.7; sodium propionate, 0.2; and (ml): 4N HCl, 0.3; water, 23.

Diet 2 (g): soybean meal, 0.8; dried yeast, 1.0; powdered agar, 0.7; sodium propionate, 0.2; and (ml): 4N HCl, 0.3; water, 24.

Diet 3 (g): as diet 2, with addition of (g): powdered tea, 3.0; soybean meal, 5.0.

Diet 4 (g): as diet 3, with addition of (g): sodium dehydroacetate, 0.0125; sodium propionate, 0.1; and 0.5 ml 4N HCl.

PREPARATION Ingredients mixed, autoclaved for 10 minutes to dissolve agar, cooled, and dispensed to 100 x 15 mm test tubes. Divided into 20 or 30-15 x 100 mm test tubes, plugged with cotton.

REARING At 25° or 28°C with 16 h photoperiod; diet in tube sufficient for complete larval development.

DEVELOPMENT Of *A. orana:* compared with controls on natural hosts - leaves of tea and soybean larval development on diet 1 slower, but pupal weight greater; on diet 2, growth rate and pupal weight both greater; on diets 3 and 4, growth rate about that on diet 2, but pupal weight less. Can be used for mass rearing, but many abnormal pupae produced.

Of *H. magnanima:* only diets 2 and 3 compared with natural host of tea and persimmon leaves. Both diets give greater growth rate and far greater pupal weight, but abnormal pupae result from mass rearing on diet 2.

→ Tamaki, Y. 1966b. Mass rearing of the smaller tea tortrix, *Adoxophyes orana* Fischer von Röslerstamm, on a simplified artificial diet for successive generations (Lepidoptera: Tortricidae). JPN J APPL ENTOMOL ZOOL 10: 120-124.

COMPOSITION (g): soybean powder (Kinako), 10.0; tea leaf powder, 6.0; yeast (Ebios), 2.0; agar, 1.4; sodium propionate, 0.2; sodium dehydroacetate, 0.025; and (ml): 4N HCl, 1.0; deionized water, 47.0.

PREPARATION Ingredients thoroughly mixed and steamed for 10 minutes, mixed again and cooled at room temperature. Tea leaf ground to pass through 1 mm sieve.

REARING 0.8 g diet placed on 3 x 6 cm sterilized paraffin paper. 250 sheets of paper placed in 10 x 20 cm diam glass rearing container with glass lid, egg masses introduced, and when emergence complete, egg shells removed; pupae transferred to 30 x 25 x 3 cm plastic oviposition chamber with paraffin sheet on top and moist filter on base with illumination for 16 h. Resulting egg masses cut off paraffin sheets and incubated in petri dishes at 87% RH.

DEVELOPMENT Larval period, ca. 20 days; pupal wt: ♀, 34.7 mg; ♂, 23.8 mg. Average number of eggs/egg mass, 98.1; egg masses/♀, 2.2. Twelve generations mass-reared using rotational breeding system.

→ Yamaya, K.; Tamaki, Y. 1972. Mass rearing method for Tortricidae. SHOKUBUTSU BOEKI 26: 165-168. (In Japanese)

COMPOSITION (g): soybean meal, 400; powdered tea leaves, 240; dried yeast, 80; powdered agar, 56; and (ml): mould inhibitor solution, 80; 4N HCl, 40; water, 1800.

mould inhibitor solution (g): sodium propionate, 50; sodium dehydroacetate, 6.3; water, 500 ml.

PREPARATION Agar dissolved in water, and blended with soybean, tea and yeast. Mould inhibitor and HCl added slowly; blended, and dispensed to 28 x 19.5 x 6 cm plastic dishes @ 90 g/dish to cool.

REARING At 25°C with 16 h photoperiod. Eggs surface sterilized in 0.1% $HgCl_2$ at black-head stage and placed on diet @ 500/dish. Egg cases removed after 2-3 days. Adults removed to dish ⅓ full of wet cotton wool with wax paper strips for oviposition. Eggs incubated at 87% RH in petri dishes lined with wet filter paper.

DEVELOPMENT Pupal period, 18-20 days; both *A. orana* and *H. maganima* reared.

Adoxophyes reticulana (Hübner)

→Ankersmit, G.W. 1968. The photoperiod as a control agent against *Adoxophyes reticulana* (Lepidoptera: Tortricidae). ENTOMOL EXP APPL 11: 231-240.

Diet modified from Adkisson *et al.** (1960) for *P. gossypiella*.

COMPOSITION (g): casein, 35.0; wheat germ, 30.0; sugar, 35.0; Wesson's salt mixture, 10.0; brewers' yeast, 30.0; choline chloride, 4.0; agar, 25.0; carboxymethylcellulose Hz 858 (Algemene Kunstzijde Unie, Arnhem, Netherlands), 5.0; and (ml): linseed oil, 5.0; inhibitor, 20.0; water, 850.0.

mould inhibitor solution (g): sorbic acid, 8.0; and methyl *p*-hydroxybenzoate, 8.0; both dissolved in 95% ethanol, 100 ml.

PREPARATION See Adkisson *et al.** (1960).

REARING At 15, 20 and 25°C with 18 h photoperiod. Larvae reared individually in 15 x 50 mm vials partially filled with diet. Pupae transferred to large cage for emergence. Oviposition on plastic sheet covering jars; adults fed 30% sugar solution. Eggs rinsed in 1% NaClO solution and incubated in glass vials.

DEVELOPMENT Larval period at 15, 20 and 25°C, 50, 25 and 20 days respectively.

Argyrotaenia citrana (Fernald), orange tortrix

→Coudriet, D.L. 1970. Rearing the orange tortrix on a synthetic diet. J ECON ENTOMOL 63: 1004-1005.

Best diet tested.

COMPOSITION (g): agar, 5.0; wheat germ, 62.5; ascorbic acid, 2.0; Tetracycline, 0.125; and (ml): vitamin solution, 3.0; water, 500.0; choline chloride (0.1 g/ml water), 9.0; methyl *p*-hydroxybenzoate (15% w/v in 95% ethanol), 13.0.

vitamins (g/500 ml water): niacinamide, 1.0; thiamine hydrochloride, 0.025; pyridoxine hydrochloride, 0.25; biotin, 0.02; calcium pantothenate, 1.0; riboflavin, 0.5; folic acid, 0.25; vitamin B_{12}, 0.002.

PREPARATION Agar dissolved in water and when cool blended with remaining ingredients. Dispensed from 'squeeze' bottle to 18 ml cups @ 0.25 oz/cup. Dried overnight. Diet scored, cups capped and stored until use.

REARING At 21 ± 2°C and 70 ± 5% RH with 14 h photoperiod. 6 larvae/cup. Pupae removed to individual cups; 25 adult pairs placed in 30 x 27 cm cylinder of 16-mesh screen, one end pressed into 2.5 cm of sand; other covered with plywood, roughened on inside, with 2.5 cm hole for introduction of adults or pupae. Oviposition on waxed paper suspended from plywood. Eggs removed after 72 h, and incubated in 18 ml cups.

DEVELOPMENT Larval period, 32 days; pupal wt: ♀, 30 mg; ♂, 14 mg. Emergence to adult, 93%. Fecundity, 2273 eggs/♀, with 82% hatch. Twenty-three generations reared.

Other species reared: bulb mite, *Rhizoglyphus echinopus* (Fumouze & Robin) (Acarina, Acaridae), for more than 3 years; varigated cutworm, *Peridroma saucia* (Hübner) (Lep., Noctuidae); lesser bulb fly, *Eumerus tuberculatus* Rondani (Dipt., Syrphidae); narcissus bulb fly, *Merodon equestris* (Fabricius) (Dipt., Syrphidae) reared by replacing alfalfa meal with 27 g of narcissus bulb powder.

Argyrotaenia velutinana (Walker), red banded leaf-roller

→Karpel, M.A.; Hagmann, L.E. 1968. Medium and techniques for mass rearing the red banded leaf-roller. J ECON ENTOMOL 61: 1452-1454.

COMPOSITION (g): alfalfa leaf and stem meal, 160; water, 200; sorbic acid, 3; methyl *p*-hydroxybenzoate, 3; and sucrose, 10.

PREPARATION Ingredients mixed, and autoclaved at 20 psi for >30 minutes. Diet dispensed to 12 x 15 x 3½" plastic boxes to depth of 1".

REARING >1000 larvae reared/box. Lid lined with autoclaved paper towelling for pupation after 18-20 days. Pupae placed in 8 x 12" polystyrene foam cages @ 200/cage for emergence; trough cut in bottom for adult food wick; oviposition on wax paper introduced through slit near top. Cage covered with cheesecloth. Adults fed 0.01% tocopherol solution; and subjected to 17 h photoperiod; eggs removed daily and washed in 1% sorbic acid in 70% ethyl alcohol, rinsed in distilled water and incubated at 100% RH.

DEVELOPMENT Larval period, 18-20 days. Fecundity, 125 eggs/♀. Egg hatch, 91.75%. Six generations reared.

→Redfern, R.E. 1963. Concentrate media for rearing red-banded leaf roller. J ECON ENTOMOL 56: 240-241.

COMPOSITION (g): agar, 5.0; apple seeds, 25.0; English walnut meats, 12.0; yeast, 5.0; glycine, 0.5; cysteine, 0.25; cholesterol, 0.25; sugar, 25.0; Wesson's salt mixture, 5.0; soybean protein, 25.0; dried apple, 25.0; and (ml): vitamin solution, 5.0; water, 350.0; 10% KOH, 14.0; mould inhibitor, 10.0; apple juice, 50.0.

vitamin solution (g): niacinamide, 0.5; calcium pantothenate, 0.5; thiamin hydrochloride, 0.125; riboflavin, 0.25; pyridoxine hydrochloride, 0.125; folic acid, 0.125; biotin, 0.01; vitamin B_{12} in mannitol, 1.0; choline chloride, 50.0; and 447.4 ml of distilled water.

mould inhibitor: 20 g sorbic acid, 15 g methyl *p*-hydroxybenzoate, and 170 ml ethyl alcohol.

PREPARATION Agar dissolved in water and blended with vitamins then remaining ingredients; diet dispensed to cups @ 0.25 oz/cup from plastic squeeze bottles. When set, surface scored and stored overnight.

REARING At 76°F and 35-40% RH; 4 neonate larvae/cup. Pupae removed to rearing cages; 20-30 adults placed in oviposition cans.

DEVELOPMENT Egg incubation, 8.7 days; larval period, 25.3 days; pupal period, 9.8 days. Larval survival to pupation, 75%; with 92% emergence.

Corollary

Redfern, R.E. 1964a. An improved medium for rearing red-banded leafroller. J ECON ENTOMOL 57: 296-297.

COMPOSITION (g): alfalfa leaf meal, 62.5; agar, 50.0; and (ml): mould inhibitor, 13.0; vitamin solution, 6.0; water, 500. Details of vitamin solution and mould inhibitor in Redfern* (1963).

PREPARATION Agar dissolved in water and blended with other ingredients; mould inhibitor added last and mixture homogenized. Diet dispensed to cups.

DEVELOPMENT Ca. 75% larval survival to adult; 15 generations reared.

Rock, G.C.; Glass, E.H.; Patton, R.L. 1964. Axenic rearing of the red-banded leaf roller, *Argyrotaenia velutinana*, on meridic diets. ANN ENTOMOL SOC AM 57: 617-621.

Two diets described.

COMPOSITION Diet 1 (g): Fava bean leaf (*Vicia faba*), 7.0; yeast extract (Difco), 1.0; dextrose, 2.0; Wesson's salt mixture, 1.0; bacto-peptone, 0.5; bacto-agar granules, 0.5; water to 100 g.

Diet 2 (g): vitamin-free casein, 3.00; glycine, 0.15; *l*-cystine, 0.10; *l*-arginine, 0.13; *l*-lysine, 0.08; dextrose, 5.00; cholesterol, 0.20; linseed oil, 0.30; RNA, 0.10; Wesson's salt mixture, 1.00; Tween 80, 0.35; bacto-agar granules, 3.00; and (mg): $FeSO_4.7H_2O$, 20.00; $ZnCl_2$, 15; niacin, 6.25; calcium pantothenate, 2.00; riboflavin, 1.00; thiamine hydrochloride, 6.00; folic acid, 1.20; choline chloride, 100.00; α-tocopherol, 25.00; biotin (aqueous solution), 0.025; B_{12} (aqueous solution), 0.002; and water to 100 g.

PREPARATION Diet 1: bean leaves lyophilized and thoroughly blended; heated with remaining ingredients, dispensed to petri dishes and cooled. 4-5 g pieces transferred to 20 x 90 mm vials, stoppered, and autoclaved for 15 minutes at 15 lb pressure. Slanted while cooling; left for 3 days before use.

Diet 2: cystine dissolved in 150 ml distilled water with 4 ml HCl. 2N NaOH added to pH 6. RNA dissolved in 30 ml warm distilled water with 8 ml 2N NaOH and added to cystine solution. Casein, dextrose, mineral salts, amino acids, and vitamins dissolved in cystine-RNA solution. 1 g cholesterol dissolved in 30 ml hot 95% ethyl alcohol and pipetted into 170 ml distilled water with Tween 80; α-tocopherol and linseed oil added to 3 ml hot ethyl alcohol and mixture pipetted into cholesterol suspension; reduced to 150 ml volume by evaporation over steam bath, blended, and then mixed with other ingredients. pH to 6.3 with 2N KOH. Diet autoclaved for 15 minutes at 15 lb pressure then dispensed to sterilized 20 x 90 mm vials @ 4-5 g/vial. Vials slanted while cooling and kept for 4-5 days before use.

REARING At 80 ± 1°F individually in vials. Pupae removed to other vials capped with aluminium foil punctured to hold cotton wad moistened with 20% sucrose solution for adult feeding. Also 50 pupae placed in oviposition box; eggs laid on waxed paper. Five days old eggs sterilized in 6.5% formalin solution.

DEVELOPMENT On diet 1: larval period (days): ♀, 16.8; ♂, 15.9; pupal period, 10 days; pupal wt: ♀, 24.7 mg; ♂, 16.3 mg. Adult longevity, 14.1 days.

On diet 2: larval period (days): ♀, 21.7; ♂, 19.2; pupal period, 12 days; pupal wt: ♀, 27 mg; ♂, 17.1 mg. Adult longevity, 14.9 days.

See also: Rock *et al.* (1965).

→ Rock, G.C.; King, K.W. 1966. Amino acid composition in hydrolysates of the red-banded leaf roller, *Argyrotaenia velutinana* (Lepidoptera: Tortricidae) during development. ANN ENTOMOL SOC AM 59: 273-277.

COMPOSITION (g): alfalfa leaf meal, 7.0; powdered cerophyl (dehydrated grass leaves), 1.0; bacteriological peptone, 0.5; dextrose, 2.0; Wesson's salt mixture, 1.0; agar, 3.0; and (ml): mould inhibitor, 2.2; water, 85.0.

mould inhibitor (g): sorbic acid, 20.0; methyl parasept, 15.0; and 170 ml ethyl alcohol.

PREPARATION Agar dissolved in water and blended with dry ingredients and mould inhibitor. Hot mixture dispensed to cups @ 0.25 oz/cup.

REARING See Redfern* (1964b). At 80 ± 1°F with 16 h photoperiod.

DEVELOPMENT Larval period, ca. 23 days; pupal period, 10 days.

→ Rock, G.C.; King, K.W. 1967a. Qualitative amino acid requirements of the red-banded leaf roller, *Argyrotaenia velutinana* (Lepidoptera: Tortricidae). J INSECT PHYSIOL 13: 59-68.

COMPOSITION (g/100 g diet): amino acids, 2.025; dextrose, 5.000; linseed oil, 0.300; salts, 0.471; vitamins, 0.132; cholesterol, 0.200; Tween 80, 0.300; agar, 3.00; 2M KOH to pH, 6.2; water to 100 g.

l-amino acids (mg/100 g diet): arginine, 311; histidine, 75; isoleucine, 113; leucine, 175; lysine hydrochloride, 113; methionine, 75; phenylalanine, 150; threonine, 100; tryptophan, 75; valine, 138; alanine, 150; aspartic acid, 100; glutamic acid, 100; glycine, 150; proline, 50; serine, 50; cystine, 50; tyrosine, 50.

vitamins (mg/100 g diet): folic acid, 1.0; riboflavin, 1.0; calcium pantothenate, 2.0; pyridoxine hydrochloride, 6.0; thiamine hydrochloride, 0.6; nicotinic acid, 6.0; biotin, 0.025; B_{12}, 0.002; choline chloride, 100; and α-tocopherol, 15.

PREPARATION In 400-600 g batches; heat sterilized at 15 lb pressure for 15 minutes. Dispensed to 20 x 90 mm vials @ 4-5 g/vial, slanted to cool; stored for 3-4 days before use.

REARING Axenically at 81 ± 1°F with 18 h photoperiod. 5-day-old egg masses surface sterilized in 10% formalin solution for 30 minutes, then washed 3 times in sterile distilled water. Eggs incubated in sterile shell vials; neonate larva transferred aseptically and individually to vials

DEVELOPMENT After 10 days, larvae in 2nd and 3rd instar; larval period (days): ♀, 23.3; ♂, 20.8; pupal wt (mg): ♀, 27.5; ♂, 17.1; adult longevity (days): ♀, 9.0; ♂, 11.2. Survival from larva to adult, 89%. Three generations reared.

Corollary

Rock, G.C.; King, K.W. 1967b. Quantitative amino acid requirements of the red-banded leaf roller, *Argyrotaenia velutinana* (Lepidoptera: Tortricidae). J INSECT PHYSIOL 13: 175-186.

COMPOSITION As Rock & King* (1967a), except that vitamins (except choline chloride) doubled; carbohydrate reduced from 5% to 4%; sodium bicarbonate buffer added to amino-acids; and amino-acids altered as follows (mg/100 g diet): arginine, 125; histidine, 100; isoleucine, 92; leucine, 156; lysine hydrochloride, 190; methionine, 72; phenylalanine, 107; threonine, 100; tryptophan, 90; valine, 95; alanine, 125; aspartic acid, 18; *l*-glutamic acid, 220; glycine, 97; proline, 99; serine, 57; cystine, 55; cysteine, 10; tyrosine, 102.

PREPARATION See Rock *et al.** (1964), except that amino acids, carbohydrate, water-soluble vitamins, and salts mostly sterilized by autoclaving and thoroughly mixed with sterile-filtered solution.

DEVELOPMENT Larval growth on diets containing amino acid mixture at 3% dietary level similar to growth on intact protein diets and on synthetic diet containing dehydrated plants used for mass rearing.

Rock, G.C. 1971. Utilization of *d*-isomers of the dietary indispensable amino acids by *Argyrotaenia velutinana* larvae. J INSECT PHYSIOL 17: 2157-2168.

COMPOSITION (g/100 g diet): *l*-amino acids, 3.0; dextrose, 5.0; cholesterol, 0.20; safflower oil, 0.30; Wesson's salt mixture, 1.0; $ZnCl_2$ + $CuSO_4$, 0.005; $FeSO_4.7H_2O$, 0.020; B-vitamin mixture, 0.142; Tween 80, 0.30; agar, 3.0; and 2M KOH, pH to 6.2; and distilled water to 100 g.

amino acids (mg/100 g diet): arginine, 180; histidine, 145; isoleucine, 130; leucine, 225; lysine hydrochloride, 280; methionine, 95; phenylalanine, 155; threonine, 145; tryptophan, 130; valine, 137; alanine, 180; aspartic acid, 265; glutamic acid, 325; glycine, 135; proline, 144; serine, 80; cystine, 98; tyrosine, 145.

vitamins (mg/100 g diet): folic acid, 2; riboflavin, 2; calcium pantothenate, 4; pyridoxine hydrochloride, 6; thiamine hydrochloride, 1.2; nicotinic acid, 12; biotin, 0.025; vitamin B_{12}, 0.004; choline chloride, 100; α-tocopherol, 15.

PREPARATION As in Rock & King* (1967a).

REARING At 27 ± 0.5 C with 16 h photoperiod. As in Rock & King* (1967a).

DEVELOPMENT On complete diet, larval period (days): ♀, 27.0; ♂, 24.5; adult survival, ca. 73%. Three generations reared.

See also: Chou *et al.* (1973); Lii *et al.* (1975); Rock (1969a b); Rock & Sharma (1974); Rock *et al.* (1973); Sharma *et al.* (1972a b).

Roelofs, W.L. 1967. Agarless medium for mass rearing the red-banded leaf roller. J ECON ENTOMOL 60: 1477-1478.

COMPOSITION (kg): apple cider, 26; alfalfa leaf meal, 17.5; and (g): yeast extract, 875; sorbic acid, 315; Tegosept M, 280.

PREPARATION Apple cider poured into bowl and ingredients stirred in as follows: yeast extract, sorbic acid, Tegosept M, and alfalfa leaf meal. Dispensed to ½" depth in 10 x 12½ x 4" plastic boxes. Storable at 5°C or -12°C with moisture content of 55% for several months.

REARING Eggs surface sterilized in 10% formalin, washed in distilled water and transferred to diet @ 500/10 x 12 x 4" box. Lid lined with cloth towel to provide light seal. After 2 weeks lid replaced by plastic bag to collect moths.

DEVELOPMENT Adult emergence, 27 days from eclosion.

Choristoneura fumiferana (Clemson), spruce budworm

→ McMorran, A. 1965. A synthetic diet for the spruce budworm, *Choristoneura fumiferana* (Clem.) (Lepidoptera: Tortricidae). CAN ENTOMOL 97: 58-62.

COMPOSITION (/100 g diet): water (distilled), 22.0 ml; vitamin-free casein, 3.5 g; 4M KOH, 0.5 ml; Alphacel, 0.5 g; Wesson's salt mixture, 1.0 g; sucrose, 3.5 g; wheat embryo, 3.0 g; choline chloride, 0.1 g; ascorbic acid, 0.4 g; nutrient agar 4% w/v, 62.0 ml; antimicrobial agents: formalin (37% formaldehyde), 0.05 ml; methyl *p*-hydroxybenzoate, 0.15 g; Aureomycin, 0.03 g; and vitamin solution, 1.0 ml.

vitamins (mg/100 ml water): niacin, 100.0; calcium pantothenate, 100.0; riboflavin, 50.0; thiamine hydrochloride, 25.0; pyridoxine hydrochloride, 25.0; folic acid, 25.0; biotin, 2.0; and vitamin B_{12}, 0.2.

PREPARATION Agar dissolved in water and blended with premix of remaining ingredients. Diet dispensed to 23 x 57 mm vials @ 6 ml/vial; excess moisture evaporated and vials sealed.

REARING Adults at 74 ± 1°F and 78-80% RH with natural lighting. Larvae at 72 ± 1°F and 70-74% RH with 18 h photoperiod. Pupae held at 72°F (♂) and 74°F (♀) to synchronise emergence. Adults mated singly in 5-oz paper containers with punctured tab lids for ventilation. Oviposition on balsam twig in container. Eggs collected and incubated in petri dishes. Larvae transferred individually to cotton plugged diet vials.

DEVELOPMENT Survival to adult, 75%; pupal wt: ♀, 121.7 mg; ♂, 77.4 mg. 2nd generation pupal wt: ♀, 90.7 mg; ♂, 63.0 mg. Of two successive laboratory generations reared on synthetic diet and on frozen balsam fir buds, those on diet exhibited greater survival, faster development to greater weights, more successful mating and higher fecundity.

See also: Grisdale (1970, 1972); Harvey (1974, 1975); Shaw (1973); Wilson (1974).

Corollary

→ Grisdale, D.G. 1973. Large volume preparation and processing of a synthetic diet for insect rearing. CAN ENTOMOL 105: 1553-1557.

COMPOSITION As McMorran* (1965) except: nutrient agar replaced by granulated agar; Aureomycin reduced to 18 g/68 l; formalin omitted.

PREPARATION Ascorbic acid and choline chloride stirred into vitamin solution. Aureomycin capsules placed in small amount water and crushed when soft. Agar dissolved in half of water in electrically-heated steam-jacket kettle and mixed at 36 rpm. Premixed dry ingredients divided into 2 or 3 parts and blended in 1-gal blender with minimum water;

wheat embryo prepared similarly. These, plus vitamins, mixed with agar solution at 60 rpm. Cooled to 58°C and dispensed to creamer cups using suction hose from kettle. When solidified, antifungal solution (Grisdale, 1970) sprayed on to diet surface. Placed in plastic bags, sealed and refrigerated till use.

Choristoneura occidentalis Freeman, western spruce budworm

Lyon, R.L.; Richmond, C.E.; Robertson, J.L.; Lucas, B.A. 1972. Rearing diapause and diapause-free western spruce budworm (*Choristoneura occidentalis*) (Lepidoptera: Tortricidae) on an artificial diet. CAN ENTOMOL 104: 417-426.

Diet modified from that of McMorran* (1965).

COMPOSITION Group I: distilled water, 19 ml; 4M KOH, 0.6 ml; wheat embryo, 3.3 g.

group II: vitamin mixture, 1.2 g; ascorbic acid, 0.4 g.

group III: methyl *p*-hydroxybenzoate, 0.18 g; potassium sorbate, 0.14 g; Aureomycin (5.51% chlorotetracycline hydrochloride), 0.08 g.

group IV: distilled water, 9.0 ml; vitamin-free casein, 3.9 g; Alphacel, 0.6 g; sucrose, 2.9 g; Wesson's salt mixture, 1.0 g.

group V: distilled water, 55 ml; agar, 2.7 g.

antimicrobial agent quantities doubled when rearing field collected larvae, or if high contamination.

PREPARATION Ingredients in groups I to III mixed then blended with group IV. Group V autoclaved at 103 kPa for 30 minutes, cooled to 60°C, and mixed with other ingredients. Diet poured into trays, gelled, then cut into slabs, wrapped in aluminium foil, and stored at 4°C. 6-oz plastic cups half filled with 1 cm squares of diet; or diet cubes to 1 pint containers.

REARING Larvae at 23-26°C and 33-52% RH with 24 h photoperiod; 16 h at 1080-1400 lux and 8 h at 108 lux; @ 50-100 larvae/container, reduced to 10-15/container at 5th instar. Pupae removed to paper bags for mating, with wax paper strips for oviposition, @ 40-100 pairs/bag, and kept at 20-24°C and 39-68% RH for 10 days.

DEVELOPMENT Larval period (6 instars), 21-23 days; pupal period, ca. 7 days; average of 136 eggs/♀. Elimination of diapause gave 7½ generations/year instead of normal 2¼.

Choristoneura pinus Freeman, jack pine budworm

Allen, D.C.; Knight, F.B.; Foltz, J.L. 1968. A technique for rearing *Choristoneura pinus* on artificial diet. ANN ENTOMOL SOC AM 61: 362-364.

COMPOSITION & PREPARATION See McMorran* (1965).

REARING Newly laid eggs incubated in small vials lined with cheesecloth at 72 ± 2°F. After 3 weeks, vials placed at 40 ± 2°F for 4 months then removed and left 3 days at room temperature. Larvae reared individually on diet in plastic petri dishes. Pupae removed to empty containers. Adults paired in plastic vials or in 16-oz wide bottom jars lined with wax paper, and plugged with moist cotton.

DEVELOPMENT Larval size equal to that of field collected specimens. Survival on diet, ca. 70%. Average life cycle, 40-50 days; adult longevity: ♂, 14 days; ♀, 18 days; fecundity, 264 ± 2.19 days/♀.

Epichoristodes acerbella (Walker)

→Cavalloro, R.; Piana, A. 1972. Prove di radiosensibilita' alle radiazioni ionizzanti dei tortricidi del garofano con particolare riguardo ad *Epichoristodes acerbella* (Walker). REDIA 53: 281-302. (In Italian with English summary)

Best of 5 diets tested.

COMPOSITION (g/kg): 'medica' grass flour, 26.0; wheat germ, 26.0; cellulose, 24.0; agar, 22.0; casein, 35.0; dextrose, 35.0; "olio di arachidi", 3.0; cholesterol, 1.5; choline chloride, 1.0; vitamin mix (NBC), 16.0; salt mix No. 2 (NBC), 3.2; Nipagin, 2.5; sorbic acid, 0.8; ascorbic acid, 4.0.

Epiphyas postvittana (Walker), light brown apple moth

→Dunwoody, J.E.; Hooper, G.H.S. 1967. An artificial medium for rearing *Epiphyas postvittana*. J ECON ENTOMOL 60: 1753-1754.

Diet modified from that of Redfern* (1963) for *A. velutinana*.

COMPOSITION (g): soybean protein, 25.0; ground apple seeds, 25.0; dehydrated apple, 25.0; sucrose, 25.0; ground walnuts, 12.0; agar, 5.0; yeast, 5.0; Wesson's salt mixture, 5.0; glycine, 0.5; cysteine, 0.25; cholesterol, 0.25; Nipagin, 2.2; and (ml): vitamin solution, 7.0; apple juice, 55.0; water, 400.0.

vitamins: as in Vanderzant and Davich* (1958).

PREPARATION See Redfern* (1963).

REARING At 24 ± 1.5°C and 60 ± 5% RH. Diet surface scored and 4 larvae inoculated/cup. Adults mated singly in 50 x 25 mm diam, or in groups in 150 x 50 mm diam glass tubes; oviposition on cellophane lining. Eggs incubated at 24°C and 40-60% RH.

DEVELOPMENT Larval period, 21 days; pupal weights equal to those of stock culture. Early larval mortality, 20-30%. Five generations reared.

See also: Bartell & Shorey (1969).

→Singh, P. 1974a. Aseptic rearing of *Epiphyas postvittana* (Lepidoptera: Tortricidae) on a meridic diet. N Z J ZOOL 1: 111-117.

COMPOSITION (g): agar, 3.0; cellulose powder, fibrous, 1.0; vitamin- and fat-free casein, 3.5; sodium alginate, 0.5; salt mixture USP XIV (NBC), 1.0; wheat germ, 2.5; cholesterol, 0.2; wheat germ oil, 0.2; Tween 80, 0.05; dextrose, 3.0; and (mg): niacinamide, 4.0; calcium pantothenate, 4.0; thiamine hydrochloride, 1.0; riboflavin, 2.0; pyridoxine hydrochloride, 1.0; folic acid, 1.0; biotin, 0.075; B_{12}, 0.75; choline chloride, 400; ascorbic acid, 400; and (ml): ethanol, 2.8; 4M KOH, 0.25; distilled water to 100 ml.

PREPARATION Agar, cellulose powder, casein, sodium alginate, salt mixture and wheat germ homogenized. Cholesterol/lipid emulsion prepared by dissolving cholesterol in warm 95% ethanol, then mixing wheat germ oil and 9 ml distilled water with Tween 80 heated to 70°C, and emulsifying both solutions. 10 ml emulsion added to dry mix, followed by 58.5 ml water. Diet blended, pH adjusted to 6.5 by KOH, and autoclaved at 100 kPa for 20 minutes; cooled to 60°C and 20 ml of microbiologically filtered dextrose/vitamin solution added: (dextrose and vitamins dissolved in 2.8 ml 95% ethanol and distilled water to 20 ml). Diet finally mixed and aseptically dispensed to sterile 75 x 12 mm test tubes @ 2 g/tube. Tubes slanted and set aside for 24 h before use.

Non aseptic rearing by adding mould inhibitor to diet and adjusting water accordingly. Mould inhibitor: methyl *p*-hydroxybenzoate, 15; sorbic acid, 20; ethyl alcohol 95%, 175 ml. Used @ 1.25 ml/100 g diet.

REARING Axenically and individually in 75 x 12 mm test tubes at 25° ± 1°C and 50-60% RH with 12 h photoperiod. Egg masses surface sterilized in 5% formalin for 30 minutes and hatched in petri dishes. Neonate larvae individually transferred to tubes, until pupation. Adults mated in open ended 15 x 3 cm diam glass tubing lined with scored polyethylene for oviposition. Ends stoppered with cotton wool soaked in 10% honey solution; renewed daily.

DEVELOPMENT Average larval period, 28 days; prepupal and pupal period, 11-12 days; preoviposition period, 2 days. Egg incubation, 7 days. Egg to adult, 49 days. Survival from larva to adult, 89%. Sex ratio 1:1; 3 successive generations reared.

→Singh, P. 1974b. A chemically defined medium for rearing *Epiphyas postvittana* (Lepidoptera: Tortricidae). N Z J ZOOL 1: 241-243.

COMPOSITION (mg): alanine, 180; arginine, 180; aspartic acid, 265; cysteine, 98; glutamic acid, 325; glycine, 135; histidine, 145; isoleucine, 130; leucine, 225; lysine, 280; methionine, 95; phenylalanine, 155; proline, 144; serine, 80; threonine, 145; tryptophane, 130; tyrosine, 145; valine, 137; and cholesterol, 200; safflower oil, 300; Tween 80, 300; α-tocopherol, 15; sucrose, 3000; glucose, 500; ascorbic acid, 600; choline chloride, 300; inositol, 40; niacinamide, 4; calcium pantothenate, 4; riboflavin, 2; folic acid, 1; thiamine hydrochloride, 1; pyridoxine hydrochloride, 1; biotin, 0.04; B_{12}, 0.004; Wesson's salt mixture, 1000; agar, 3000; cellulose powder (fibrous), 2000; ribonucleic acid, 100; and (ml): 4M KOH, 0.3; distilled water to 100 g.

PREPARATION Amino acids ground in mortar in bulk and stored in air-tight bottles. Cholesterol dissolved in warm ethanol. α-tocopherol and safflower oil dissolved in 95% ethanol. 100 ml distilled water with Tween 80 heated to 60-70°C. Cholesterol, and safflower oil/α-tocopherol solutions added and blended, heated to 80-90°C, cooled, volume adjusted to 125 ml with distilled water, and refrigerated till use.

To prepare diet, amino acids, RNA, salt mixture, agar, and cellulose powder dissolved in 62.9 ml distilled water, 12.5 ml warm cholesterol emulsion added, pH adjusted to 6.5 with KOH, mixture thoroughly stirred, autoclaved at 103.35 kPa for 20 minutes, and cooled to 60°C. Sugars and vitamins dissolved in distilled water and sterilized by passing through 45 μm Millipore filter. Then both parts combined aseptically, thoroughly mixed and aseptically dispensed to sterile test tubes. Tubes slanted and set aside 24 h before use.

REARING See Singh (1974a).

DEVELOPMENT Incubation period, 7 days; larval period, 31.5 days; pre-pupal/pupal period, 10 days; preoviposition period, 2 days. Average life-cycle, 49 days. Average fecundity, 347 eggs/♀. Larval survival to adult, 78%. Sex ratio 1.3♂:1♀. Three successive generations reared.

→Thomas, W.P. 1968. An artificial diet for rearing the light brown apple moth *Epiphyas postvittana* (Walk.) (Lepidoptera: Tortricidae). N Z ENTOMOL 4: 31-32.

COMPOSITION (g): "sheep nuts" (D.H. Brown & Sons Ltd, Christchurch, New Zealand), 50; lucerne meal, 5; and (ml): mould inhibitor, 2.5; distilled water, 85.

mould inhibitor: 15 g of methyl *p*-hydroxybenzoate and 20 g of sorbic acid in 170 ml of 95% alcohol.

basic components of commercially manufactured "sheep nuts":- barley meal, linseed meal, pea meal, lucerne meal, bran, pollard, lime, salt, urea, copper sulphate, potassium iodide and cobalt sulphate.

PREPARATION Sheep nuts partially crushed and mixed thoroughly with other components. Stored at 5°C for 48 h before use; then thoroughly remixed.

REARING At 22°C in plastic containers.

DEVELOPMENT Final instar larvae in 22 days. Viable eggs produced.

Lobesia botrana Dennis et Schiffermüller

→Guennelon, G.; Sender, C.; D'Arcier, F.; Audemard, H. 1970. Mise au point d'un milieu artificiel pour l'élevage au laboratoire des larves de l'eudémis de la vigne, *Lobesia botrana* Den. et Schiff. (Lepidoptera, Tortricidae). ANN ZOOL ECOL ANIM 2: 51-77. (In French with English summary)

COMPOSITION (g): agar, 5.0; cellulose, 6.0; vine leaves, 20.0; ripe seedless grapes, 70.0; proteolysed casein, 7.0; glucose, 3.5; cholesterol, 0.6; Wesson's salt mixture, 2.0; brewers' yeast, 6.0; vitamin mix (NBC), 4.0; ascorbic acid, 1.0; Nipagin, 0.1; benzoic acid, 0.25; and (ml): grape seed oil, 0.4; distilled water, 100.

PREPARATION Agar dissolved in water and cooled to 56°C. Grapes blended for two minutes and heated to 56°C in water bath. Other powdered ingredients blended for 1 minute. All mixtures then blended together; final pH ca. 4.5. Diet dispensed to rearing tubes or boxes. Boxes of 2 sizes; 15 x 45 mm diam with 14 g diet to depth of 7 mm for 5 larvae; and 30 x 80 mm diam with 56 g diet to depth of 9 mm for 50 larvae.

REARING At 25 ± 1°C and 70% RH with 18 h photoperiod. Pupation in or on diet or on lids. Adults mated singly, or in groups of 5 pairs, in 80 x 100 mm diam polystyrene boxes at 20°C and 80% RH with natural lighting. Fed on 5% sugar solution; oviposition on box surfaces. Adults transferred to new boxes twice weekly.

DEVELOPMENT Egg incubation period at 20°C, 6-7 days. Mean adult yield, 70.8% from larvae reared in groups of 5, and 52.6% from those in groups of 50. Sex ratio, 1:1. Mean development from larva to adult, 27 days. 95% adults hatch 21-33 days from eclosion. Fertility and fecundity sufficient for 16:5 proliferation rate. Diet uneconomical for mass production.

Moreau, J.P. 1965b. Comportement des vers de la grappe vis-á-vis de divers cépages et essais d'alimentation artificielle. REV ZOOL AGRIC APPL 64: 13-16. (In French)

Also for *Clysiella ambiguella* Hübner [previously *Eupoecilia*] (Cochylidae).

COMPOSITION (g): dried powdered vine leaf, 1.0; cellulose powder, 1.0; agar, 0.2; yeast extract, 0.12; benzoic acid, trace; and (ml): distilled water, 8.5-9.0.

PREPARATION Dried leaf, cellulose and yeast blended. Agar and benzoic acid dissolved in water at 85-90°C. Cooled to 40°C and blended rapidly with dry ingredients.

REARING At 25°C and high humidity. Food changed every 4-5 days.

DEVELOPMENT Complete; worse than growth on inflorescences or young grapes, but better than on tendrils or leaves.

Tzanakakis, M.E.; Savopoulou, M.C. 1973. Artificial diets for larvae of *Lobesia botrana* (Lepidoptera: Tortricidae). ANN ENTOMOL SOC AM 66: 470-471.

Two diets tested.

COMPOSITION Diet 1 (g): tap water, 50; fine agar powder, 2; potassium sorbate, 0.1; methyl *p*-hydroxybenzoate, 0.15; tomato juice, 25; carrot baby food (Gerber), 5; brewers' yeast, 10; vitamin mixture (Vanderzant's modification), 2; alfalfa meal, 10; Wesson's salt mixture, 0.5.

Diet 2: as diet 1, plus (g): ascorbic acid, 1; wheat germ, 3; Alphacel, 5.

PREPARATION Water, agar, potassium sorbate and methyl *p*-hydroxybenzoate, mixed and heated for 15 minutes; blended with tomato juice, carrot baby food, and salt mixture and then vitamins and yeast and finally alfalfa meal. Diet dispensed to 9 cm glass petri dishes to depth of 8 mm, cooled for 30 minutes, then dishes covered and refrigerated. After 24 h diet surface sprayed with wax. Holes punched for larval entry.

REARING At 26°C and 70% RH with 16 h photoperiod; 30 neonate larvae/dish, covered with glass lid; corrugated paper strips for pupation. Adults mated in ca. 7 x 9 cm diam transparent cups covered with white cloth @ 4 pairs/cup. Two pieces of vine cane for oviposition. Fed on dilute sucrose solution and 7:10:40:50 wt/wt mixture of egg yolk:yeast hydrolysate: sucrose:water solution on cotton wads. Liquids renewed every 2 days. Kept at 25°C and 70% RH with natural lighting for oviposition.

DEVELOPMENT Diet 1: pre-oviposition period, 4-6 days. On last of 7 generations reared: larval survival to adult, 47%; pupal wt: ♀, 10.0 mg; ♂, 8.3 mg; fecundity, 32.6 eggs/♀; eclosion, 79%.

Diet 2: similar data.

Merophyas divulsana (Walker), lucerne leafroller

Galloway, I.D. 1970. Food medium and techniques for the laboratory rearing of *Merophyas divulsana* (Walker). QUEENSL J AGRIC ANIM SCI 27: 325-327

COMPOSITION See Karpel & Hagmann* (1968) diet developed for *Argyrotaenia velutinana*.

PREPARATION Diet not autoclaved if individually reared. Lucerne leaves and stalks oven dried, finely ground and mixed with sorbic acid, sucrose and methyl *p*-hydroxybenzoate. Distilled water added until diet of moist, fragmented consistency; dispensed to tubes or thinly spread on layer of plaster of Paris in 105 x 70 mm diam plastic containers.

REARING Adults mated singly in 85 x 40 mm diam tubes or in groups in polystyrene cages. Oviposition on wax paper; egg masses collected daily and incubated separately. Dark head stage eggs transferred to diet. Larvae reared in plastic containers with small hole bored through bottom and then layer of plaster of Paris poured to ca. 10 mm deep. Tubes and containers placed in shallow tray of water for 5-10 minutes every day to maintain high humidity. Diet renewed as required. Pupae removed to emergence tubes. Adults fed 2% sugar solution on dental pad. Oviposition cage and rearing tube illustrated.

DEVELOPMENT 116 adults reared from 150 1st instar larvae.

Platynota stultana Walsingham, omnivorous leaf roller

Aliniazee, M.T.; Stafford, E.M.; Fukushima, C. 1971. Rearing of the omnivorous leaf roller in the laboratory on artificial diet. ANN ENTOMOL SOC AM 64: 1172-1173.

Diet developed for *Trichoplusia ni* by Henneberry & Kishaba* (1966) except wheat germ replaced by wheat bran to produce coarser texture.

REARING At 24 ± 1°C. 1-3 day-old adults mated in 1-gal icecream carton @ 20/carton. Fed on 10% sugar solution; oviposition on wax paper lining. Eggs collected every 4 days. Neonate larvae reared individually in 4-5-oz Lily cups with 2-oz diet. Pupae removed to open mouthed cardboard cartons for emergence.

DEVELOPMENT Average egg to adult period, 32.7 days; compared with 46 days on carnations, 43 days on oranges, 48.6 days on roses, and 35.9 days on cotton. Ten generations reared.

Sparganothis pilleriana Schiffermüller

Russ, K. 1971. Investigations with a view to mass rearing of *Sparganothis pilleriana* Schiffermüller for genetic control. pp. 295-302. *in* Sterility Principle for Insect Control or Eradication. Proceedings of a Panel Organized by the Joint FAO/IAEA Division of Atomic Energy in Food and Agriculture.

COMPOSITION (g): agar, 3-5; wheat germ, 11; maize semolina, 11; powdered *Vicia faba*, 11; yeast, 11; ascorbic acid, 1; Nipagin, 0.35; benzoic acid, 0.35; and Aureomycin, 5 mg; and water, 170 ml.

REARING Individually in small polystyrene containers; food changed daily; or mass reared in 150 x 180 x 30 mm dishes and 1000 ml buckets. Pupation in 'Perlite'.

DEVELOPMENT Larval wt (mg) on diet, *Vicia faba* and grape vine: ♀, 78.74, 36.07, 48.73; ♂, 50.68, 33.01, 46.22 respectively. Fecundity: on diet, 298 eggs/♀; on *V. faba*, 242 eggs/♀. 64% hatch on diet; 88% on *V. faba*.

Zeiraphera diniana (Guenée), larch bud moth

Altwegg, P. 1971. Ein semisynthetisches Nährmedium und Ersatzsubstrate für die Oviposition zur von der Jahreszeit unabhängigen Zucht des grauen Lärchenwicklers, *Zeiraphera diniana* (Gn.) (Lep.; Tortricidae). Z ANGEW ENTOMOL 69: 135-170. (In German with English summary)

COMPOSITION Larval diet (g): powdered larch needle (*Larix decidua*), 7; dry yeast, 2; wheat germ, 1; sucrose, 3; milk powder, low fat (perhaps casein), 3; cellulose powder, 3; cholesterol, 0.1; ascorbic acid, 0.5; choline chloride, 0.05; agar, 2.5; and (ml): Nipagin, 20% in 96% ethanol, 2; distilled water, 7775; and Streptomycin, 1 mg.

Oviposition medium, part a (g): powdered larch needle, 7; dry yeast, 2; wheat germ, 2; sucrose, 3; milk powder, low fat, 4; cellulose, 3; cholesterol- water, 25 ml.

part b: agar, 2.5 g; and water, 50 ml.

part c: ascorbic acid, 0.5 g; choline chloride, 0.05 g; linseed oil, 10 ml; and antimicrobial mixture, 2 ml.

antimicrobial mixture: 20 g sorbic acid and 15 g Nipagin M in 170 ml ethanol.

PREPARATION Larval diet: powdered larch needle, dry yeast, wheat germ, sugar, milk powder, and cellulose powder mixed in 2775 ml cold water. Agar dissolved in 500 ml water and cooled to 40-50° C and both solutions blended; then cholesterol dissolved in ethanol, ascorbic acid, choline chloride, streptomycin dissolved in water, and Nipagin M dissolved in 96% ethanol added. Whole diet finally blended and dispensed to 20 ml transparent polystyrene boxes.

Oviposition medium: group 'a' mixed with agar dissolved in water, cooled to 40-50° C; group 'c' added and mixed.

REARING At 20° C and 80-85% RH; 1st and 2nd instar larvae in total darkness; 3rd-5th instar with 15 h photoperiod. 10 larvae/box with 8 mm ventilation hole until 3rd instar, then individually in 2 x 6.5 cm diam boxes with 10 mm ventilation hole. Pupation and emergence in containers.

DEVELOPMENT Pupal wt (mg) on diet and larch: ♀, 40.2, 47.4; ♂, 32.3, 35.4 respectively. Adult emergence on diet and larch, 94.4% and 91.4% respectively.

Yponomeutidae

Acrolepia assectella Zeller, leek moth

Pralavorio, R.; Arambourg, Y.; Guennelon, G. 1973. Essai de mise au point d'un élevage permanent d'*Acrolepia assectella* Zeller (Lepidoptera, Hyponomeutidae) sur milieu artificiel. ANN ZOOL ECOL ANIM 5: 569-580. (In French with English summary)

COMPOSITION (g): agar, 10; corn semolina, 56; wheat germ, 14; powdered leek, 10; brewers' yeast, 15; ascorbic acid, 2; Nipagin, 0.4; benzoic acid, 0.5; and (ml): distilled water, 340.

PREPARATION Agar dissolved in water and cooled to 56° C. Dry contents added and blended for 2 minutes at 15000 rpm. Diet dispensed to depth of 1 cm in 2.7 x 8.2 cm diam plastic boxes, dried for 7-8 h under paper filters, and boxes covered with tight lids. Stored at 8° C for maximum of 2-3 weeks before use.

REARING At 25-26° C and 60-80% RH with 16 h photoperiod; 150 eggs seeded/box, diet scored with scalpel to help larval establishment. 80 μm mesh cover for limited ventilation. Pupation on cover or sides of box. Adults mated in groups of 200-500 in 20 ml transparent plexiglass sided cages. Fed on honey and water solution on cotton plugs. Natural lighting in summer and 14 h photoperiod in winter. Oviposition on paper gauze sachet containing leek leaf; removed as required. After 24 h, eggs sterilized in 2% Merseptyl (0.1% mercurothiolic acid and 1% mono ethanol amine) and rinsed with distilled water.

DEVELOPMENT Eclosion, 72.8-91.0%. Pupation from neonate larvae, 31-67%; survival from egg to adult, 23.1-55.3%; incubation period, ca. 4 days; larval period, ca. 14-18 days; pupal period, 9 days. >7 generations reared.

Plutella xylostella (Linnaeus), diamond back moth

Biever, K.D.; Boldt, P.E. 1971. Continuous laboratory rearing of the diamond back moth and related biological data. ANN ENTOMOL SOC AM 64: 651-655.

Diet modified from that of Berger* (1963) for *Heliothis* spp.

COMPOSITION As in Berger* (1963) except dried ground rape leaves added @ 50 g/gal of diet, and vitamin mixture prepared by Hoffman-La Roce Inc., Nutley, N.J. and contained vitamin A and E in addition to those of Berger's diet.

PREPARATION 2.5 g nutrient agar dissolved in 62 ml water, and blended with dietary ingredients while hot.

REARING At 23 ± 1° C and 60 ± 5% RH. Sections of paper towelling with ca. 100 eggs attached to lids of 6-oz waxed cups with diet. Pupation on lids. Adults removed to oviposition cage @ 100-150/cage. Oviposition on paper strips renewed every 2-4 days. Cage illustrated.

DEVELOPMENT Incubation period, 3 days; larval period, 11 days; pupal period, 5 days; survival from egg to adult, 70%; average adult longevity, 8 days (6-21). Average fecundity, 139 eggs/♀ (55-266) 50% oviposition within 3 days of emergence.

Miscellaneous

Bailey, D.L.; Chada, H.L. 1968. Effects of natural (sorghum) and artificial (wheat germ) diets on development of the corn earworm, fall armyworm and southwestern corn borer. J ECON ENTOMOL 61: 257-260.

Species reared: *Zeadiatraea grandiosella* (Dyar), southwestern corn borer (Pyralidae); *Heliothis zea* (Boddie), corn earworm (Noctuidae); *Spodoptera frugiperda* (J.E. Smith), fall armyworm (Noctuidae).

COMPOSITION (g): agar, 90; casein, 126; sucrose, 126; Wesson's salt mixture, 36; Alphacel, 18; wheat germ, 108; ascorbic acid, 15; and (ml): methyl *p*-hydroxybenzoate (15% in ethyl alcohol), 36; formaldehyde (0.1 g/ml water), 18; choline chloride, (0.1 g/ml water), 36; 4M KOH, 18; vitamin stock solution, 6; vitamin E and corn oil, (1 ml α-tocopherol in 99 ml corn oil), 10; water, 3100.

REARING At 80° F. 25 neonate larvae of *H. zea* and *S. frugiperda* reared individually on diet in 1¾ x 1¼ x 1" cells of polyethylene ice cube trays covered with plywood. *Z. grandiosella* reared in 1 x 3¾" glass shell vials, plugged with cotton. 10 pairs adults/oviposition cage.

DEVELOPMENT *H. zea*: egg incubation period, 3 days; larval period, 15.3 days; pupal period, 10.3 days; adult longevity, 8 days; generation completed in 36.9 days; control on sorghum in 36.6 days.

S. frugiperda: larval period, 15 days; pupal period, 10.2 days; adult longevity, 7.5 days; generation completed in 35.6 days; control in 35.0 days.

Z. grandiosella: egg incubation period, 5 days; larval period, 19.3 days; pupal period, 10.1 days; adult longevity, 6.0; generation completed in 40.4 days; control in 39.7 days.

Weight, fecundity and development of all 3 species comparable to data from control on sorghum.

See also: Bailey (1964).

Edelman, N.M. 1974. The rearing of phytophagous insects with different types of feeding on nutrient media containing leguminous sprouts. TR VSES INST ZASHCH RAST 40: 128-135. (In Russian with English summary)

Species reared: *Trichoplusia ni* (Hübner), cabbage looper (Noctuidae); *Porthetria dispar* (Linnaeus), gypsy moth (Lymantriidae); *Agrotis ipsilon* (Hufnagel), black cutworm (Noctuidae).

Best of 3 diets tested for noctuids.

COMPOSITION (g): sprouting seeds of mung bean, 87.5; vitamin C, 2.0; agar-agar, 10.0; dry medicinal (bakers or brewers' ?) yeast, 2.0; Nipagin, 0.150; cellulose, 5.0; and (ml): water, 50; 16% formalin, 0.5.

Best of 8 diets tested for lymantriids.

(g): sprouts of chick pea seeds, 87.5; saccharose, 5; cellulose, 20; dry medicinal yeasts, 7.5; vitamin C, 2.0; Wesson's salt mixture, 2.5; choline chloride, 0.04; linseed oil, 2.5; agar-agar , 12; Nipagin, 0.75; and (ml): water, 70; 16% formalin, 0.5.

REARING *P. dispar*: first to 3rd instar at 18-20°C; mature larvae at 25-28°C with 16 h photoperiod. 1st-3rd instar @ 15-20/petri dish; subsequently 10/500 ml jars.

T. ni: to 4th instar in 30 ml jars @ 5-8/jar; larvae separated in jar by compressed tracing paper to prevent cannibalism; from 4th instar, individually in test tubes. Adults @ 10/500 ml jar.

A. ipsilon: at 24-25°C.

DEVELOPMENT *T. ni* and *A. ipsilon*: normal development; adults fecund. Diet containing cellulose more acceptable, giving increased larval weights. Addition of saccharose (to change carbohydrate-protein ratio) did not affect the physiological condition. Four generations reared.

P. dispar: life cycle, 55 days. Fecundity, 350 eggs/♀.

➞ Shumakov, E.M.; Edelman, N.M.; Borisova, A.E. 1960. An attempt to rear phytophagous insects on synthetic media. DOK AKAD NAUK SSSR SER BIOL 130: 237-240. (In Russian)

Species reared:- *Chloridea obsoleta* Fabricius, cotton moth (Pyralidae); *Loxostege sticticalis* Linnaeus, meadow moth (Noctuidae); *Gelechia malvella* (Hübner), mallow moth (Gelechiidae).

Best 4 of several diets tested.

COMPOSITION Diet 1 (g): casein, 8; agar, 4; cellulose, 5; saccharose, 6; cottonseed oil, 1; mineral salts, 2; yeast, 8; and (mg): cholesterol, 100; acetylcholine chloride, 60; α-tocopherol, 36; vitamin B_1, 2.4; vitamin B_2, 3.6; vitamin B_{12}, 0.06; vitamin B_6, 3.2; *p*-aminobenzoic acid, 4; pantothenic acid, 8; nicotinic acid, 21; folic acid, 1; ascorbic acid, 30; inositol, 4; valine, 10; arginine, 10; leucine, 10; lysine, 10; tyrosine, 10; cysteine, 4; cystine, 100; glycine, 150; methionine, 4; phenylalanine, 6; alanine, 4; serine, 4.

Diet 2 (g): casein, 8; agar, 4; cellulose, 5; saccharose, 6; cottonseed oil, 1; mineral salts, 2; yeast, 8; and (mg): peptone, 100; cholesterol, 100; acetylcholine chloride, 60; α-tocopherol, 36; vitamin B_1, 1.2; vitamin B_2, 1.8; vitamin B_{12}, 0.06; vitamin B_6, 1.6; *p*-aminobenzoic acid, 2; pantothenic acid, 4; nicotinic acid, 24; folic acid, 1; ascorbic acid, 30; inositol, 2; valine, 15; arginine, 15; leucine, 15; lysine, 15; tyrosine, 15; cysteine, 10; cystine, 150; glycine, 150; methionine, 10; phenylalanine, 10; alanine, 10; serine, 10.

Diet 3: as diet 1, except (g): mineral salt, 1.5; yeast, 4; (mg): α-tocopherol, 60; vitamin B_1, 1.2; B_2, 1.8; B_6, 1.6; *p*-aminobenzoic acid, 2; pantothenic acid, 4; nicotinic acid, 12; folic acid, 0.8; inositol, 2.

Diet 4: as diet 3, except (g): saccharose, 5; mineral salt, 2; (mg): peptone, 100; acetylcholine chloride, 40; pantothenic acid, 2; nicotinic acid, 18.

PREPARATION Yeast autolyzed for 48 h at 50°C. pH 6.2. Casein dissolved in 2M NaOH and remaining ingredients added. Agar dissolved in water and blended with dietary mixture. Diet dispensed to rearing vessels and sterilized.

REARING Larvae of *C. obsoleta* and *L. sticticalis* reared in wide test tubes with cotton plugs. *G. malvella* larvae reared in 30-40 ml glass jars with plastic screw-on caps. Eggs sterilized for 10 minutes in 2% NaOH solution and rinsed in distilled water. *C. obsoleta* and *G. malvella* larvae reared individually; those of *L. sticticalis* @ 3-5/tube.

DEVELOPMENT *C. obsoleta*: on diet 1: larval period, 14.2 days; mature larval wt, 385 mg; pupal wt, 250 mg; pupation, 93%; adult emergence, 98%. On diet 2: larval period, 14.0 days; mature larval wt, 380 mg; pupal wt, 243 mg; pupation, 95%; emergence, 98%. Viable eggs laid on both diets.

G. malvella: 2 generations reared. Growth normal on both diets.

L. sticticalis: larval period, 21 days; larval wt, 42-47 mg; growth also normal on diets 3 and 4.

Vanderzant, E.S.; Pool, M.C.; Richardson, C.D. 1962. The role of ascorbic acid in the nutrition of three cotton insects. J INSECT PHYSIOL 8: 287-297.

Species reared: *Anthonomus grandis* Boheman, boll weevil (Curculionidae); *Heliothis zea* (Boddie), boll worm (Noctuidae); *Estigmene acrea* (Drury), salt-marsh caterpillar (Arctiidae).

COMPOSITION Larval diet for *H. zea* and *A. grandis* (g): vitamin-free casein, 5.0; cysteine hydrochloride, 0.1; sucrose, 5.0; Wesson's salt mixture, 1.0; corn oil (Mazola), 0.5; cholesterol, 0.05; choline chloride, 0.05; inositol, 0.02; cellulose (Whatman ashless powder), 5.0; agar (Difco Bacto granules), 3.0; and (ml): cobalt, molybdenum and zinc salts, 0.5; B-vitamins, 1.0; and water, 80.0.

Larval diet for *E. acrea* (g): vitamin-free casein, 4.0; cysteine hydrochloride, 0.1; sucrose, 4.0; Wesson's salt mixture, 1.0; corn oil (Mazola), 0.5; cholesterol, 0.1; choline chloride, 0.05; inositol, 0.02; β-carotene-protein complex (Hoffman-La Roche), 0.025; cellulose (Whatman ashless powder), 10.0; agar (Difco Bacto granules), 2.0; and (ml): cobalt, molybdenum, and zinc salts, 0.5; fat-soluble vitamins, 0.5; B-vitamins, 1.0; and water, 80.0.

Adult diet for *A. grandis* (g): vitamin-free casein, 2.0; (enzymatic hydrolysate) casein, 2.0; fructose, 3.0; galactose, 1.0; Wesson's salt mixture, 0.75; corn oil (Mazole), 0.5; cholesterol, 0.05; choline chloride, 0.05; inositol, 0.02; agar (Difco Bacto granules), 3.0; and (ml): cobalt, molybdenum, and zinc salts, 1.0; B-vitamins, 2.0; and water, 100.0.

salts (mg): $Zn(C_2H_3O_2)_2.2H_2O$, 0.5; $CoCl_2.6H_2O$, 0.25; and $Na_2MoO_4.2H_2O$, 0.25.

B-vitamins (mg): niacinamide, 1.0; calcium pantothenate, 1.0; thiamine hydrochloride, 0.25; riboflavin, 0.5; pyridoxine hydrochloride, 0.25; folic acid, 0.25; biotin, 0.02; vitamin B_{12}, 0.0002.

fat soluble vitamins: α-tocopherol, 10 mg; vitamin D_3, 0.5 mg; menadione, 0.5 mg in ethanol.

PREPARATION Casein dissolved in water and alkali. Cholesterol triturated with sucrose, choline chloride, inositol and vitamins, added to remaining water and mixed with dry ingredients. Both portions combined in Waring Blendor. Diet heated in water bath and dispensed from polyethylene squeeze bottle. Two g diet in 17 x 60 mm glass vials sterilized for 15 minutes at 121°C. Ascorbic acid solution sterilized by filtration and added. Diet cooled for 5 minutes.

REARING At 29 ± 1°C with 13.5 h photoperiod. *A. grandis* eggs sterilized with $HgCl_2$ solution or solution of alkyldimethylbenzylammonium chloride. Rinsed in distilled water; eggs transferred individually to hole cut in diet. Adults placed in 6" cube screened cages and fed twice daily. Food given in aluminium cups or small cylinders coated with paraffin.

H. zea eggs similarly sterilized. Transferred aseptically 1 week after hatching to 20 x 70 mm vials or 17 x 150 mm tubes with 5 g diet.

E. acrea eggs surface sterilized for 5-30 minutes with 0.25% NaClO. Eggs placed in sterile vial for hatching. Larvae transferred to vials of sterile diet; 3 or more aseptic transfers to diet in larger containers. Mature larvae placed in glass jars with diet and shredded paper or sand for pupation. Pupae removed to emergence cages.

DEVELOPMENT *A. grandis* on larval diet containing 0.2 g ascorbic acid, 74% adult yield; ♀ wt, 13.0 mg; ♂ wt, 11.6 mg. 47 eggs laid in 20 days.

H. zea on diet with 0.4 g ascorbic acid, 79% adult yield.

E. acrea on diet with 0.4 g ascorbic acid, 32% adult yield.

Wellington, E.F. 1949. Artificial media for rearing some phytophagous Lepidoptera. NATURE LONDON 163: 574.

Species reared: *Archips fervidana* (Clemens) (Tortricidae); *Archips rosaceana* (Harris) (Tortricidae); *Choristoneura fumiferana* (Clemens), spruce budworm (Tortricidae); *Dioryctria reniculella* (Grote) (Pyralidae).

COMPOSITION (parts): balsam fir foliage suspension, (*Abies balsamea*), 100; 8% agar, 50; 10% Mycoban, 1.5; and autolyzed yeast, 1.5 g.

PREPARATION Foliage placed in blender and mixed with minimal distilled water. Agar added and diet hardened in petri dishes.

DEVELOPMENT *C. fumiferana*, complete. Survival and fecundity good.

D. reniculella, A. rosaceana complete.

A. fervidana reared on diet based on red oak (*Quercus rubra*).

Wilkinson, J.D.; Morrison, R.K.; Peters, P.K. 1972. Effect of Calco Oil Red N-1700 dye incorporated into a semiartificial diet of the imported cabbage worm, corn earworm, and cabbage looper. J ECON ENTOMOL 65: 264-268.

Species reared: *Pieris rapae* (Linnaeus), imported cabbage worm (Pieridae); *Heliothis zea* (Boddie), corn earworm (Noctuidae); *Trichoplusia ni* (Hübner), cabbage looper (Noctuidae).

Modified diet of Berger* (1963) for *Heliothis* spp.

COMPOSITION Group 1: water, 2000 ml; agar, 90 g.

group 2: water, 1000 ml; 45% KOH, 9 ml; casein, 126 g; attractant (dried, ground rape leaves), 50 g; Tenox, 0.8 g; sucrose, 135 g; Wesson's salt mixture, 36 g; wheat germ, 175 g; Alphacel, 25 g; potassium sorbate (in sufficient water to dissolve), 4 g; methyl *p*-hydroxybenzoate (in 95% ethanol, 15%), 36 ml; 40% formaldehyde, 30 ml; Vanderzant's vitamin mixture, 36 g; Aureomycin (chorotetracycline hydrochloride 25 g/lb), 10 g; and raw linseed oil, 26 ml. Sufficient for ca. 1-gal diet.

PREPARATION See Berger* (1963). Calco Oil Red N-1700 incorporated by dissolving 1.6 g in 5 ml of linseed oil and adding to diet at time of blending, just before pouring into rearing cups.

REARING Larvae and pupae of 3 species reared at 27 ± 3° C and 50 ± 10% RH with constant lighting. *P. rapae* and *T. ni* reared in 6-oz waxed paper cups @ 25/cup. *H. zea* reared singly in 1-oz clear plastic jelly cups. Containers filled with 1.75 cm diet and capped with cardboard lids. *P. rapae* pupae placed in oviposition cage at 32 ± 9° C. 10 adults placed in 31 x 30 cm cylindrical sleeve cages covered with hardware cloth. Cotton pad saturated with honey-yeast solution and fresh mustard leaf provided daily. Eggs kept in individual compartment of plastic block at 24 ± 4° C. For *H. zea* and *T. ni*, 20 pairs of adults held at 27 ± 3° C and 55 ± 10% RH with 12 h photoperiod in 28 x 15 cm cylindrical hardware cloth cages with solid metal lids. Fed 10% sucrose solution on cotton pad; eggs laid on Teflon cloth covering of cage. Eggs incubated as for *P. rapae*.

DEVELOPMENT *P. rapae:* larval period, 13.7 days; pupal period, 7.7 days; adult longevity, 5-6 days; pupal survival, 97%; fecundity, 171 eggs/♀.

H. zea: larval period, 20 days; pupal period, 17 days; larval survival, 78%; adult longevity, 15 days; mean fecundity, 473 eggs/♀.

T. ni: larval period, 13 days; pupal period, 8.5 days; pupal survival, 96%; adult longevity, 17-19 days; fecundity, 855 eggs/♀.

Calco Oil Red affected larval period, pupal weights, and longevity of adults, but not to an extent to prevent effective use of dye for marking released specimens for field studies.

→ Zlotin, A.Z.; Lymareva, M.A. 1966. Rearing insects in winter for biological tests of insecticides. ZOOL ZH 45: 1100-1102. (In Russian with English summary)

Porthetria dispar (Linnaeus), gypsy moth (Lymantriidae).

COMPOSITION (g): casein, 6.0; tannin, 2.5; Wesson's salt mixture, 1.5; Hephephytin (?), 2.0; agar, 4.0; gelatin, 1.0; acorn flour, 7.0; glucose, 6.425; sodium alginate, 1.0; distilled water, 68; and (mg): arginine hydrochloride, 15; β-phenyl α-alanine, 15; valine, 15; histidine hydrochloride, 15 leucine, 15; lysine hydrochloride, 15; methionine, 15; threonine, 15; tryptophan, 15; cysteine, 15; cystine, 200; glycine, 100; vitamin B_1, 5; vitamin B_2, 5; vitamin B_6, 5.0; vitamin B_{12}, 0.5; PP, 50; C, 50; calcium pantothenate, 10.

PREPARATION Acorn flour prepared by drying cotyledons of acorns at 45° C and grinding. Agar and gelatin dissolved in water. Vitamin B_{12} and dry finely ground components added. Flask shaken continuously for 10 minutes. Diet poured into petri dishes and refrigerated.

REARING Diet given daily in acorn-size pieces. Eggs immersed in 0.1% $KMnO_4$.

DEVELOPMENT Normal from 1st to 4th instar.

Barathra brassicae Linnaeus, cabbage moth (Noctuidae)

COMPOSITION (g): agar, 1; casein, 1; glucose, 1.425; gelatin, 0.3; flour from sunflower achenes, 1.0; cabbage leaf flour, 1.0; Hephephytin, 1.5; Wesson's salt mixture, 0.5; benzoic acid, 0.07; sodium alginate, 0.18; distilled water, 16.9; and (mg): vitamins B_1, 5.0; B_2, 5.0; B_6, 5.0; PP, 50; C, 50; calcium pantothenate, 10.

PREPARATION Agar and gelatin dissolved in water; ground dry elements added. Flask shaken and heated for 10 minutes, then diet cooled.

REARING At 23 ± 2°C and 80 ± 5% RH with 24 h photoperiod. Egg sac (15-20 eggs) immersed in 0.1% $KMnO_4$ placed in flask stoppered with cotton wool. Adults paired in petri dishes and fed 5% sugar solution.

DEVELOPMENT Fourth instar larvae in 15-17 days. Pupae storeable for 45 days at 5°C. Normal development from 1st to 4th instar.

Pieris brassicae Linnaeus and *Pieris rapae* Linnaeus reared on *B. brassicae* diet to 2nd or 3rd instar.

Neuroptera

Chrysopidae

Chrysopa carnea Stephens, common green lacewing

→Hagen, K.S.; Tassan, R.L. 1965. A method of providing artificial diets to *Chrysopa* larvae. J ECON ENTOMOL 58: 999-1000.

COMPOSITION Adult diet (g): enzymatic protein hydrolysate of yeast, 4.0; fructose, 7.0; choline chloride, 0.01; and distilled water, 10.0 ml.

Larval diet (g): enzymatic protein hydrolysate of yeast, 5; choline chloride, 0.0125; ascorbic acid, 0.5; fructose, 8.75; and water, 12.5 ml. Mixture diluted 4-fold for complete larval development.

PREPARATION Food droplets prepared by adding ca. 0.1 g paraffin (mp 43-47° C) to 20 ml diluted diet in 30 ml beaker. 1-2 mm layer formed on diet surface. Heated to 43-57° C; glass applicator rod dipped into warm mixture, and then touched to 16 cm^2 parafilm square to form droplets of ca. 1 mm diam. Applicator rod (illustrated) made from 5 mm diam glass tube drawn to taper of 1-2 mm over 30 mm length. About 25 droplets placed on parafilm food strips. Smaller droplets made for 1st instar larvae. Deeper dip of rod into mixture gives larger droplets and cooler mixture gives thicker paraffin layer over droplets.

REARING At 25 ± 7° C and 28-58% RH. One egg or neonate larva into half pint cage with masking tape around lip to prevent escape. Food strip placed on equal-sized square of cheesecloth on cotton in bottom of cage. Food and cotton changed every 2 days and exuviae removed.

DEVELOPMENT *C. carnea:* 50 days from newly hatched larva to adult emergence. Developmental period longer on diet than on aphids; and adults smaller but fertile. *C. plorabunda* Fitch and *C. californica* Coquillett: development completed.

See also: Hagen & Tassan (1966a b).

→Hagen, K.S.; Tassan, R.L. 1966c. The influence of protein hydrolysates of yeasts and chemically defined diets upon the fecundity of *Chrysopa carnea* Stephens (Neuroptera). VESTN CESK SPOL ZOOL 30: 219-227.

Many adult diets tested.

COMPOSITION Diet 1 (g): yeast autolysate, 4; choline chloride, 0.01; and (ml): distilled water, 10.

Diets 2 and 3: as diet 1, except different types of yeast autolysate used.

Diet 4: as diet 1, but 7 g fructose.

Diet 5: as diet 1, but 7 g sucrose.

Diet 6 (g): fructose, 3.5; amino acids, 1.5; Wesson's salt mixture, 0.16; cholesterol, 0.003; and vitamin mix plus water, 5 ml.

Diet 7: as diet 6, except half amino acid quantity.

l-amino acids (g): alanine, 0.25; arginine hydrochloride, 0.20; aspartic acid, 0.30; cystine, 0.70; glutamic acid, 1.10; glycine, 0.15; histidine hydrochloride, 0.10; hydroxyproline, 0.10; isoleucine, 0.27; leucine, 0.33; lysine hydrochloride, 0.33; methionine, 0.12; phenylalanine, 0.20; proline, 0.42; serine, 0.37; threonine, 0.20; tryptophane, 0.07; tyrosine, 0.22; valine, 0.22.

vitamins (mg): nicotinic acid amide, 10; calcium pantothenate, 5; thiamine hydrochloride, 5; riboflavin, 5; pyridoxine, 5; folic acid, 5; inositol, 20; choline chloride, 100; biotin, 0.2; B_{12}, 0.001.

PREPARATION Diets 1-5 mixed and kept at 7°C. All ingredients of diets 6 and 7 except vitamins ground; 1 mg B_{12} added to 5 l distilled water; and remaining vitamins added to 100 ml of B_{12} solution, and blended with first mix.

REARING At 27 ± 1° and 65 ± 2% RH with 24 h photoperiod from fluorescent light. Cocoons weighed and held in individual vials till emergence. Adults sexed and fed 70% fructose solution until diet test. Oviposition units of 1-pint or ½-gal cylindrical, waxed cardboard cartons. Cheesecloth cover lined with heavy wax paper streaked with test diet. Water saturated cotton wad on bottom. Adults CO_2 anaesthetized every 2 days and transferred. For detailed description of oviposition cage see Hagen (1950).

DEVELOPMENT Fecundity over 21 days whilst feeding on protein hydrolysate of yeast/carbohydrate/choline chloride/distilled water mix, 786 eggs/♀; on chemically defined diet, 391 eggs/♀.

See also: Hagen *et al*. (1970).

→Hagen, K.S.; Tassan, R.L. 1970. The influence of food Wheast and related *Saccharomyces fragilis* yeast products on the fecundity of *Chrysopa carnea* (Neuroptera: Chrysopidae). CAN ENTOMOL 102: 806-811.

Best of 12 diets for adults.

COMPOSITION (g): Food Wheast, 4.8; sucrose, 5.8; and water, 10 ml.

PREPARATION Food Wheast autolyzed for 17 h; ingredients mixed and stored at 7°C.

REARING At 27 ± 1°C and 65 ± 5% RH with 24 h photoperiod. See Hagen (1950).

DEVELOPMENT Mean eggs/♀/day, 32.0 ± 6.6; total eggs/♀, 8071 over 28 days.

See also: Butler & Ritchie (1971); Hagen & Tassan (1972); Tulisalo & Korpela (1973).

→Ridgway, R.L.; Morrison, R.K.; Badgley, M. 1970. Mass rearing a green lacewing. J ECON ENTOMOL 63: 834-836.

COMPOSITION Adult food (parts): food Wheast, 5; sucrose, 6; water, 10.

PREPARATION Mixed and painted on to sides of adult cage.

REARING At 27 ± 2° C. Adults in 35 cm diam fibreglass oviposition cage, 10 cm deep, with nylon net insert in centre of screened bottom. Saturated cotton placed under nylon insert. Top of cage covered with brown paper for oviposition. 400 adults/cage produced 2000-2500 eggs/day, collected daily. Food renewed every 2 days. Eggs removed by gently rubbing paper with ball of nylon netting. 1 egg placed in cardboard hexacel, (27.9 x 30.5 cm) divided into 600 cells and glued to cotton organdie. Piece of glass placed over cells. Young larvae fed eggs of *Sitotroga cerealella* (Olivier) @ 3 g/unit. After 3 days eggs stuck to glass top criss-crossed with lines of honey. Glass changed daily until 13th day. Before food changed, rack placed over CO_2 to anaesthetize larvae. Larvae not pupated after 14 days removed from rack to prevent cannibalism. Rearing unit (without top) and pupae placed in emergence cage. Hexacel unit and oviposition cage illustrated.

DEVELOPMENT Larval stage, 14 days. Adult emergence, 95%, 15-20 days from egg hatch. Eggs laid after 3-4 days.

Corollary

→Morrison, R.K.; House, V.S.; Ridgway, R.L. 1975. Improved rearing unit for larvae of a common green lacewing. J ECON ENTOMOL 68: 821-822.

REARING Improved rearing method described.

DEVELOPMENT 82% of pupae emerged as adults; 62% (range 41-82%) of cells produced adults. Units produced ca. 15% more adults than methods of Ridgway *et al.* (1970).

See also: Hassan (1975).

→Sundby, R.A. 1967. Influence of food on the fecundity of *Chrysopa carnea* Stephens (Neuroptera, Chrysopidae). ENTOMOPHAGA 12: 475-497.

COMPOSITION Adult diet (g): liver, 2.0; hydrolysate of yeast, 5.0; hydrolysate of casein, 3.0; fructose, 5.0; glucose, 2.0; sucrose, 2.0; sorbitol, 0.5; mannitol, 0.5; leucine assay, 1.0; choline chloride, 2.0; ascorbic acid, 0.5; galactose, 0.5; riboflavin, 0.5; cholesterol, 0.5; honey, 12.0; Bacto-Soyton, 12.0; and distilled water, 15.0.

REARING See Sundby (1966). At 21° C with natural lighting. Paired adults kept in glass jars covered with cloth gauze smeared with synthetic diet.

DEVELOPMENT Artificial diet alone inferior to artificial diet plus pollen of *Matricaria indora*, or honey and flower. Fecundity, 135 eggs/♀ on artificial diet; 475/♀ on artificial diet and flowers. Duration of oviposition period about equal. % hatch on honey and flower, artificial diet and flower and artificial diet, 7.3, 67.1, and 41.0 respectively.

See also: Sundby (1966).

→ Vanderzant, E.S. 1969b. An artificial diet for larvae and adults of *Chrysopa carnea*, an insect predator of crop pests. J ECON ENTOMOL 62: 256-257.

COMPOSITION Larval and adult diet (g): casein hydrolysate, enzymatic, 5.00; soy hydrolysate, enzymatic, 5.00; fructose, 15.00; K_2HPO_4, 0.16; $NaH_2PO_4.H_2O$, 0.08; $MgSO_4.7H_2O$, 0.05; $FeSO_4.7H_2O$, 0.005; soybean lecithin and oil, 0.5; cholesterol, 0.05; choline, 0.05; inositol, 0.02; ascorbic acid, 0.10; and (ml): B-vitamins, 2.0; water to 120.0.

B-vitamins (mg/ml): nicotinamide, 1.0; calcium pantothenate, 1.0; thiamine hydrochloride, 0.25; riboflavin, 0.5; pyridoxine hydrochloride, 0.25; folic acid, 0.25; biotin, 0.02; vitamin B_{12}, 0.002.

PREPARATION Cholesterol first mixed with lecithin and oil, then blended with water and other ingredients, except ascorbic acid. Divided into 10 ml portions and stored in freezer. Ascorbic acid added to each portion just before use. Portion used for daily feeding refrigerated for maximum of 4 days before replacement with fresh diet.

REARING Neonate larvae placed individually in 14 ml plastic jelly cups, with plastic covers, and 5 mm^3 cellulose sponge saturated with test diet. Fresh diet daily until larvae mature. Adults sexed and transferred to cups covered with cheesecloth. Maximum of 12 adults/cup.

DEVELOPMENT Larvae spinning by 17-20 days (average 14) compared with 8 days for larvae on host (*Sitotroga cerealella* (Olivier)) eggs. Adult yield, 56-65% on diet; 85% on eggs; emergence, 61.76% on diet; 87% on eggs. Oviposition, 220-579 eggs/♀ in 30 days from diet-fed adults. Seven generations reared.

Corollary

→ Vanderzant, E.S. 1973. Improvements in the rearing diet for *Chrysopa carnea* and the amino acid requirements for growth. J ECON ENTOMOL 66: 336-338.

COMPOSITION Basic diet. See Vanderzant* (1969b). Enzymatic casein and soy hydrolysate replaced by amino acid mix: essential *l*-amino acids (g/100 ml diet): arginine hydrochloride, 0.30; histidine hydrochloride, 0.15; isoleucine, 0.25; leucine, 0.40; lysine hydrochloride, 0.40; methionine, 0.20; phenylalanine, 0.25; threonine, 0.25; tryptophan, 0.15; valine, 0.30; non-essential: alanine, 0.40; aspartic acid, 0.30; cysteine hydrochloride, 0.10; glutamic acid, 0.90; glycine, 0.35; proline, 0.40; serine, 0.25; tyrosine, 0.15.

PREPARATION As in Vanderzant* (1969b), except adult diet made with ca. 5 ml water, to form homogenous paste. Additional water provided in sponge.

DEVELOPMENT Egg production, 586-717 eggs/♀ in 30 days on concentrated diet; 220-579 eggs/♀ on liquid diet. Pupation, 70% comparable to standard diet but only 5% emergence.

Diets (sucrose 15 g + yeast hydrolysate 2 g + casein 1 g) with or without autoclaving gave similar results e.g. emergence, 54 and 51%, and yields 34 and 39% respectively. Eighteen generations reared on diet containing soy and casein hydrolysate with 75% average emergence.

Chrysopa septempunctata Wesmael

→ Okada, I.; Niijima, K. 1971. Artificial rearing of lacewing, *Chrysopa septempunctata* Wesmael, with special reference to a new diet using drone honey bee brood. HEREDITY 25: 41-44. (In Japanese with English summary)

COMPOSITION Drone honey bee brood; see Okada (1970).

PREPARATION Various: frozen (-9° C); pulped in blender; freeze-dried and powdered; set with agar, etc.

REARING At 25° C with 12 h photoperiod (60 W white fluorescent light) in 7 cm petri dishes @ 20 larvae/dish; or individually.

DEVELOPMENT Best growth on fresh or frozen brood. Maximum emergence rate 22% in F_1 declining to 6.1% in F_4. 5th generation not attained. Adults smaller than normal with pale coloration. Mean number of eggs/♀ 986 in F_1 and 104 in F_4. Larval period, 9.0-12.3 days; pupal period, 12.2-14.7 days. Larval development faster when reared individually - comparable to natural diet.

Corollary

→ Okada, I.; Matsuka, M.; Tani, M. 1974. Rearing a green lacewing, *Chrysopa septempunctata* Wesmael on pulverized drone honeybee brood. BULL FAC AGRIC TAMAGAWA UNIV 14: 26-32.

REARING At 23-25° C. Female adult fed pulverized drone honey bee in 9 cm petri dish. Young honey bee pupae blended and placed on small glass plate. Water supplied on sponge. Eggs laid on dish. 10 neonate larvae reared in inverted petri dish with drone powder scattered over soft folded tissue paper. 2 glass pieces with powder provided for 3rd instars, replaced daily. Larvae reared with 9 h photoperiod until 5th generation then 12 h photoperiod.

DEVELOPMENT Life span: ♂, 71 days; ♀, 49 days; after 5 generations, fecundity, 825 eggs/♀; 6 generations reared.

Chrysopa oculata Say, golden eye lacewing

Chrysopa rufilabris Burmeister

Chrysopa plorabunda Fitch

→ Burke, H.R.; Martin, D.F. 1956. The biology of three chrysopid predators of the cotton aphid. J ECON ENTOMOL 49: 698-700.

COMPOSITION Adult diet (g): "Delmors nutrient powder", 1; honey and water (equal parts), 10 ml.

REARING In 25 x 95 mm glass vials. Larvae fed on *Aphis gossypii* Glover.

DEVELOPMENT *C. oculata:* larval period, 15.7 days; pupal period, 11 days; fecundity, 921 eggs/♀; life span: ♀, 22.9 days; ♂, 26.4 days.

C. plorabunda: larval period, 13.7 days; pupal period, 8.1 days; fecundity, 85 eggs/♀; life span: ♀, 62.5 days; ♂, 34.6 days.

C. rufilabris: larval period, 10.4 days; pupal period, 6.5 days; fecundity, 87 eggs/♀; life span: ♀, 23.6 days; ♂, 23.5 days.

See also: Ru *et al.* (1976).

Orthoptera

Acrididae

Locusta migratoria cinerascens Fabricius

→ Petavy, G. 1966. Rapports entre alimentation et physiologie chez un insecte: *Locusta migratoria cinerascens* Fab. (Orthoptera: Acrididae). I. Utilisation de milieux nutrifis artificiels. ANN NUTR ALIMENT 20: 279-299. (In French)

COMPOSITION Adult diet (g/100 g): bran, 50.0; "Gerblé", 25.0; "Superlevure Gayelord", 25.0; Mineral solution "de Tatum" (see Albritton, 1954).

Gerblé contains stabilized wheat germ and various (mainly unsaturated) fatty acids, lecithin, proteins, carotene and ascorbic acid. Superlevure is commercial yeast brand.

PREPARATION & REARING at 33 ± 2° C and 60-70% RH. Egg incubation at 25° C.

DEVELOPMENT Fecundity and fertility lower than on control.

Locusta migratoria manilensis (Meyen), Oriental locust

Locusta migratoria migratoria Linnaeus, Asian migratory locust

→ Borisova, A.E. 1966. Rearing *Locusta migratoria manilensis* (Mey.) and *Locusta migratoria migratoria* L. on semi-synthetic media. ZOOL ZH 45: 858-864. (In Russian with English summary)

COMPOSITION (g): agar, 3; casein, 4; bran, 4; sucrose, 8; Wesson's salt mixture, 2.5; dry reed flour, 2; and (mg): autolyzed brewers' yeast, 16; ascorbic acid, 150; cholesterol, 300.

minerals (g): $CaCO_3$, 200; K_2HPO_4, 322.5; $CaHPO_4.2H_2O$, 75; $MgSO_4.7H_2O$, 102; NaCl, 167.5; $FeC_6H_5O_7.6H_2O$, 27.5; KI, 0.8; $MnSO_4$, 5; $ZnCl_2$, 0.28; $CuSO_4.5H_2O$, 0.30.

PREPARATION Casein hydrolyzed in 2 N NaOH for 8-10 h in water bath until pink. Ground flour, reeds mixed with previously scalded bran, cholesterol, ascorbic acid, Wesson's salt mixture, sucrose and autolyzed brewers' yeast all crushed in mortar and added to casein. Yeast kept for 2 days at 50° C. Water and dissolved agar added last to give 75-88% moisture content and pH of 6.2.

REARING At 35-37° C during day, 30-32° C at night under continuous light. Petri dish of water placed in each cage. Bran given as well as diet.

DEVELOPMENT Best development of *L. m. migratoria* with ascorbic acid/cholesterol ratio of 1:2; of *L. m. manilensis*, ratio of 1:1.5. Fertility of *L. m. manilensis* increased with addition of 1 mg vitamin D_2. Three successive generations of *L. m. manilensis* reared.

→ Borisova, A.E. 1969. The role of carbohydrates in the fertility of the Asian migratory locust, *Locusta migratoria migratoria* L. (Orthoptera, Acrididae), when reared on semi synthetic media. ENTOMOL REV USSR 48: 141-144.

Best of nine diets tested.

COMPOSITION (g): dry reeds, 2.0; dried cow's milk, 2.0; dry brewers' yeast, 1.0; bran, 2.0; soluble starch, 0.30; and (mg): ascorbic acid, 0.30; cholesterol, 0.15.

dried milk components (%): fat, 26.5; protein, 28.5; lactose, 36.5; mineral salts, 5.5; water, 3.0.

DEVELOPMENT Over two generations, colouration, weight and fertility comparable with control. Nymphal period, 18 days; sexual maturation period, 14-15 days; mean number of eggs laid/♀, 351-391; average weight of newly hatched nymphs, 19.0 mg.

See also: Le Berre & Lenoir (1965, 1969); Le Berre & Tira (1976); Sinoir (1968).

Locusta migratoria Linnaeus

Schistocerca gregaria (Forskal), desert locust

→ Dadd, R.H. 1960a. The nutritional requirements of locusts - I. Development of synthetic diets and lipid requirements. J INSECT PHYSIOL 4: 319-347.

Best of 5 diets tested.

COMPOSITION (g): cellulose powder, 15.0; cholesterol, 0.2; salt mixture, 1.5; sucrose, 5.0; white dextrin, 5.0; casein (fat and vitamin free), 6.0; bacteriological peptone, 2.0; egg albumen powder, 2.0; and (mg): β-carotene, 25; ascorbic acid, 100; choline chloride, 50; and (ml): wheat germ oil, 1.0; vitamin solution, 10.0.

vitamin solution in 20% ethanol (μg/g diet): thiamine, 25; riboflavin, 25; nicotinic acid, 100; pyridoxine, 25; folic acid, 25; *meso*-inositol, 250; calcium pantothenate, 50; *p*-aminobenzoic acid, 25; biotin, 1.

salt mixture (Glaxo's salt mixture DL 6) (parts): NaCl, 22.0; calcium phosphate, 130.0; $K_3C_6H_5O_7.H_2O$, 125.0; $MgSO_4.7H_2O$, 30.0; iron citrate, 5.0; trace mixture, 0.7. Trace mixture (parts): KI, 12; NaF, 10; $MnSO_4$, 2; CuI, 1; anhydrous potassium alure, 1; $ZnSO_4.7H_2O$, 1.

PREPARATION Lipids weighed and dissolved in ether or chloroform. Casein, cellulose and yeast added, and ground in mortar and pestle as ether evaporated off in oven at 60°C. Salts, sugars, albumen, peptone, nucleic acids and other dry components ground together to fine powder then mixed with first fraction. Water soluble factors added in solution and all ground to friable mixture. Calcium pantothenate and *p*-aminobenzoic acid dissolved in 20% alcohol, pH to 7.0; other B-vitamins made up in 20% alcohol, pH to 3.5.

REARING At 32 ± 0.5°C and 55 ± 5% RH with 24 h photoperiod. Hoppers in 1.5-l glass jars with plastic gauze and screw-on lids. First instar @ 15-30 hoppers/jar; 2nd and 3rd @ 10-15, and 5-10 thereafter. 12-l celluloid cages (Watkin & Doncaster type) also used with 20-50 hoppers/cage. Strips of perforated zinc for perches. Diet in small petri dishes with cotton wool water pads to increase humidity to 60-70%. Diet supplied every 2 days.

DEVELOPMENT Digestibility coefficient of 45-50% obtained. *S. gregaria* achieved 30-50% survival of adult compared with 30-65% on control (grass). Mean weight at end of nymphal development on synthetic diet 1300-1500 mg; on control 1500-1700 mg.

Growth rate of *L. migratoria* on diet less than on grass.

See also: Cavanagh (1963); Dadd (1957, 1960b c d e, 1961a b c).

→ Howden, G.F.; Hunter-Jones, P. 1958. An artificial diet for the laboratory rearing of locusts. NATURE LONDON 182: 1527-1528.

COMPOSITION (parts by volume): bran, 2; dried whole milk, 2; grass-meal, 2; dried brewers' yeast, 1.

PREPARATION Ingredients ground and thoroughly mixed.

REARING At 28°C night, 34-35°C day, and 55-60% RH mornings and 45-50% RH later in day. 20 insects in 12-l cages of plastic and metal with twigs in cage for perches and dish containing wet cotton wool.

DEVELOPMENT Growth and weights of final instars similar for both experimental and control groups (given only fresh green grass daily) of *S. gregaria*, but coloration differed. Mortalities also differed, being 30% in control and 80% in experimental group. Poor results with *L. migratoria*.

→ Srivastava, P.D.; Kumar, S.S.; Bhamburkar, M.W.; Pradhan, S. 1964. An artificial diet for rearing the desert locust. INDIAN J ENTOMOL 26: 352-356.

Best of many diets tested.

COMPOSITION (g): dried cabbage leaf powder, 2; wheat bran, 2; whole milk powder, 2; dried brewers' yeast, 1.

PREPARATION Ingredients mixed and water added until of crumbly consistency.

REARING In sunlight and at 32-35°C at night. Initially reared in glass jars with wood perches, then transferred to 15 x 15 x 22" wire gauze cages; cleaned daily; diet given every other day.

DEVELOPMENT Adult survival 30% and 53% on diet and control respectively. Development on dried cabbage leaf powder 9 days longer than on diet. Eggs not viable if yeast not included in diet. Colony maintained for 1 year.

See also: Ashby (1972).

Melanoplus bivattatus (Say), two striped grasshopper

→ Nayar, J.K. 1964a. The nutritional requirements of grasshoppers. I. Rearing of the grasshopper, *Melanoplus bivattatus* (Say), on a completely defined synthetic diet and some effects of different concentrations of B-vitamin mixture, linoleic acid, and β-carotene. CAN J ZOOL 42: 11-22.

COMPOSITION (g): Alphacel, 15; sucrose, 5; soluble starch, 5; anhydrous glucose, 5; Wesson's salt mixture, 1.5; casein, 10.0; cholesterol, 0.2; *l*-ascorbic acid, 0.2; and vitamin B mix; (mg): thiamine, 1.25; riboflavin, 1.25; nicotinic acid, 2.5; pyridoxine, 1.25; folic acid, 1.25; *i*-inositol, 15.00; calcium pantothenate, 2.50; *p*-aminobenzoic acid, 1.25; choline chloride, 50.00; and biotin, 50 μg.

PREPARATION Casein, sucrose, glucose, soluble starch, cellulose and salt mixture ground and mixed with cholesterol and lipids dissolved in ethyl ether. Ether evaporated at room temperature and mixture put in bottles with screw-on lids. Shaken or rolled for 2-4 h. B-vitamin mix and ascorbic acid made in 10 ml of 20% ammonical ethanol and mixed with remaining ingredients. Diet ground in mortar and pestle and stored.

REARING At 30 ± 2° C and 75% RH with 24 h photoperiod. Eggs stored in Mason jars under moist sand at 4-5° C for 90 days; hatch 8 days after transfer to 30° C. 20 larvae reared in 3 x 2 x 2" screen cages; or 5 larvae reared in 5 cm diam plastic dishes with 2.5 cm diam ventilation hole covered with 40 mesh plastic screen. Inside of dishes roughened for purchase by softening with acetone then scratching surface. Diet provided in 1.5 cm diam dish with small wet cotton pad in each dish, initially on alternate days, then daily after 3rd moult.

DEVELOPMENT On natural diet of lettuce and bran mean weight after 21 days in plastic dish, 349.0 mg; in cage, 206.6 mg. Larval development in plastic dish, 22-27 days; in cage, 36-41 days. On synthetic diet mean weight after 21 days in plastic dish, 164.7 mg; in cage, 79.4. Larval development in plastic dish, 24-34 days; in cage, 35-45 days. Some adults with wing deformities. Better growth in smaller dishes, attributed to higher RH.

See also: Harley & Thorsteinson (1967); Nayar (1964b).

Melanoplus femurrubrum (De Geer), red legged grasshopper

→ Mulkern, G.B.; Toczek, D.R. 1970. Bioassays of plant extracts for growth-promoting substances for *Melanoplus femurrubrum* (Orthoptera: Acrididae). ANN ENTOMOL SOC AM 63: 272-284.

COMPOSITION (g): Alphacel, 8.0; sucrose, 8.0; vitamin-free casein, 5.0; brewers' yeast, 1.5; corn, 1.0; Wesson's salt mixture, 0.7; ascorbic acid, 0.5; chlorophyllin powder, 0.5; cholesterol, 0.3; glycine, 0.15; choline chloride, 0.1; *l*-cystine, 0.1.

PREPARATION Chlorophyllin added to serve as indicator for calculating digestibility and to mask slight variations in colour among different extracts. Plant extract components added as required; for most preparations liquid extract added to cellulose and dried. Dry cellulose-extract mixture then blended with other ingredients. Freeze-dried extracts added as dry powder. Ingredients blended to viscous mixture by addition of 30% ethanol, then pressed into titration mould and air dried. Tablets (ca. 0.05 g) removed from mould, dried to constant moisture level in warm oven, weighed individually and stored at -18° C.

REARING Neonate hoppers placed in groups of 5 in small rearing cages; 85 x 30 mm polystyrene vial fitted into lid of 78 x 75 x 30 mm container and separated by screen. Strip of screen fastened on side of vial for hoppers to cling especially during ecdysis. Water provided in glass tube inserted in "O" neoprene stopper from top of vial. Food supplied as needed. Plastic cage illustrated.

DEVELOPMENT On basic diet, nil survival to 4th instar. Preparations containing ethanol and petroleum ether extracts of *Ambrosia psilostachya*, *Amorpha canescens*, *Bouteloua gracilis*, or *Poa pratensis* no more effective than basic diet.

Melanoplus bivittatus (Say), two striped grasshopper

Melanoplus sanguinipes (Fabricius), migratory grasshopper

→ Kreasky, J.B. 1962. A growth factor in romaine lettuce for the grasshoppers *Melanoplus sanguinipes* (F.) and *M. bivittatus* (Say). J INSECT PHYSIOL 8: 493-504.

COMPOSITION (g): casein, 5.0; *l*-cystine, 0.1; glycine, 0.15; sucrose, 8.0; corn oil, 1.0; cholesterol, 0.3; Wesson's salt mixture, 0.7; choline chloride, 0.1; powdered cellulose, 4.0; brewers' yeast, 1.5; ascorbic acid, 0.5; agar, 4.0; and (ml): water, 125.0; water extract of dried lettuce, 12, 24 or 36 g.

PREPARATION Sucrose dissolved in portion of water and added to rest of ingredients except agar. Blended thoroughly then agar and experimental additives added with remainder of water. Autoclaved at 20 lb/in^2 for 20 minutes. Stirred whilst cooling, then refrigerated until needed. Water extracts prepared by adding 12-36 g pulverized, dried romaine lettuce (*Lactuca sativa* cv. 'Romana') to 1 l distilled water and boiled for 1 h. Mixture filtered through several layers of cheesecloth and then muslin. Concentration effected by gentle boiling. Precipitates removed by frequent filtration through filter paper. Evaporated to 25 ml before adding to basic diet.

REARING At 30° C with 24 h photoperiod. Neonate hoppers into 6 x 1" clean glass tubes stoppered with gauze-covered cotton plugs. Portions of diet aseptically transferred from flask to petri dish, and transferred to rearing tube every 24 h.

DEVELOPMENT Complete development, but adults smaller with some wing deformities. With water extract preparations of *Cleome serrulata*, *Asclepias syriaca*, *Amorpha canescens*, *Cyperus schweinitzii*, and *Calamovilfa longifolia*, 90-100% survival to adult.

See also: Henry & Oma (1975); McGinnis & Kasting (1966, 1967); Roberts and Smith (1971a b).

Gryllidae

Acheta domesticus (Linnaeus) [previously *Gryllus*], house cricket

→ Ashby, G.J. 1972a. The house cricket. pp. 588-589. *in* Universities Federation for Animal Welfare (Ed.), The UFAW Handbook on the Care and Management of Laboratory Animals. Fourth Edition. Churchill Livingstone. Edinburgh and London.

COMPOSITION (g): biscuit meal, 50; dried milk, 5; Bemax, 40; dried yeast, 5.

PREPARATION Ingredients mixed thoroughly.

REARING At 25-35° C in large, uncovered glass jars at least 23 cm deep. Bottom covered with 1 cm clean sawdust or sand and folded paper towels or cardboard for shelter. Water provided from cotton wool pad in petri dish. Eggs laid on damp cotton wool.

DEVELOPMENT Eclosion in 11 days at 32° C. 5-6 weeks to adult stage at 35° C.

→ Busvine, J.R. 1955. Simple methods for rearing the cricket (*Gryllus domesticus* L.) with some observations on speed of development at different temperatures. PROC R ENTOMOL SOC LONDON SER A 30: 15-18.

COMPOSITION (%): yellow maize meal, 35; wheat feed, 20; "Vitamix", 15; dried milk products, 10; animal protein, 5; grass meal, 5; arachis oil, 5; dried yeast, 3; linseed cake meal, 2; salt, 0.5. Fresh green stuff added to containers weekly.

REARING In large, glass, open containers, bottom covered with dry sand, and pieces of folded cardboard for shelter. Food in shallow trays, water from soaked cotton wool pad in petri dish. Eggs laid on damp cotton pads.

DEVELOPMENT Best between 25-35° C, 11 larval stages in 52 days at 30° C.

→ Ghouri, A.S.K.; McFarlane, J.E. 1958. Observations on the development of crickets. CAN ENTOMOL 90: 158-165.

COMPOSITION (parts by wt): ground wheat, 2; ground oat groats, 3; skim milk, 37; fish meal, 20; brewers' yeast, 5; cane molasses, 10; corn oil, 5; vitamin, mineral, and antibiotic supplement, 3.

vitamin, mineral, antibiotic supplement (mg): iodized salt, 124.0; bone flour, 645.0; $FeSO_4$, 36.0; vitamin B_{12} plus penicillin, 70.0; CrO_3, 72.5; vitamin B_2, 0.13; vitamin B_6, 0.13; calcium pantothenate, 0.26; vitamin A (250000 I.U.), 1.16; and vitamin D_2, 38.5.

REARING From 23-38° C and 50% RH. 10 larvae in muslin covered 2-lb jar, with food in plastic dishes or on strong, non-absorbent paper, water tube, and folded paper. When adult, ♀♀ removed, paired within 24 h, and kept in 2-lb jar with food and water. 2" diam plastic lids with ½" layer of fine, sterilized, moistened sand, placed in cage for 12 h every other day for oviposition. Eggs emptied on to paper towel in air-tight jar with moist cotton to maintain 100% RH.

DEVELOPMENT Synthetic diet inferior to control diet of rabbit pellets. At 28° C, longevity and mean eggs laid for *A. domesticus*, 57 days and 1060 eggs; for *A. configuratus* (Walker), 51 days and 667 eggs; for *Gryllodes sigillatus* (Walker), 61 days and 864 eggs.

McFarlane, J.E.; Neilson, B.; Ghouri, A.S.K. 1959. Artificial diets for the house cricket, *Acheta domesticus* (L.). CAN J ZOOL 37: 913-916.

Three diets tested.

COMPOSITION Diet 1 (g): dextrose, 20; casein, 40; yeast, 10; cholesterol, 1; salt mixture, 2; cellulose powder, 30.

Diet 2 (g): dextrose, 45; casein, 45; yeast, 10; cholesterol, 1; salt mixture, 2.

Diet 3: as diet 1, but no yeast.

Three diets tested in conjunction with 3 vitamin mixtures, A (mg): thiamine, 2.5; riboflavin, 1.25; nicotinic acid, 5.0; pyridoxine, 1.25; calcium pantothenate, 2.5; choline chloride, 50; inositol, 25; folic acid, 0.25; biotin, 0.025; *p*-aminobenzoic acid, 2.5.

B: as A, except *p*-aminobenzoic acid omitted.

C: quantities in B doubled.

PREPARATION Casein, glucose, salts and yeast mixed together; vitamins added in aqueous solution. Diet dried and ground, cellulose powder added, followed by cholesterol in chloroform solution. Vitamin mixture added to 103 g of diet 1, 104 g of diet 2, and 93 g of diet 3.

REARING At 33 ± 1° C and 55 ± 5% RH. 10 nymphs/16-oz jar. 32 ml vial filled with water, plugged with aseptic cotton, and inverted in jar. Dry diet placed on filter paper away from water. Jars changed weekly.

DEVELOPMENT Synthetic diets inferior to control diet of rabbit pellets. On diets 1 and 2, nymphal weights after 28 days, 32.1 and 21.9 mg; on rabbit pellets, 202 mg.

Diet 2 also satisfactory for *Gryllodes sigillatus* (Walker).

Corollary 1

Ritchot, C.; McFarlane, J.E. 1961. The B-vitamin requirements of the house cricket. CAN J ZOOL 39: 11-15.

Best vitamin mixture: mg/93 g diet: thiamine, 1.2; riboflavin, 1.8; nicotinic acid, 10.0; pyridoxine 1.6; calcium pantothenate, 4.0; choline chloride, 100; inositol, 200; *p*-aminobenzoic acid, 5.0; folic acid, 0.5; biotin, 0.06.

Corollary 2

McFarlane, J.E. 1974. The functions of copper in the house cricket and the relation of copper to vitamin E. CAN ENTOMOL 106: 441-446.

COMPOSITION (g): casein, 40; glucose, 20; cellulose, 30; cholesterol, 1; Wesson's salt mixture, 2; vitamin mixture as in corollary 1. Also added:- vitamin E, 86 μg/g diet; Cu (as $CuSO_4$), 10 μg/g diet.

DEVELOPMENT Improved by copper and vitamin E at 10 μg/g for ♂♂ and 2 μg/g for ♀♀. Mean weights without and with copper and vitamin E in diet (mg): 216.6 ± 9.5, 295.3 ± 8.4, 246.9 ± 9.2 and 377.4 ± 8.4 for ♂♂ and 303.6 ± 9.5, 326.8 ± 10.6, 383.0 ± 7.7, and 408.4 ± 11.7 for ♀♀. Similarly survival improved from 55 to 63% and from 55 to 77%.

See also: McFarlane (1964, 1972a b c d, 1976a b); Meikle & McFarlane (1965); Ritchot & McFarlane (1962).

Neville, P.F.; Stone, P.C.; Luckey, T.D. 1961. Cricket nutrition. II. An unidentified factor in the nutrition of *Acheta domesticus*. J NUTR 74: 265-273.

Best of several diets described.

COMPOSITION (%): casein, 30.0; corn oil, 8.0; corn starch, 30.0; cellulose, 12.0; sucrose, 12.0; total salts, 8.0; ascorbic acid, 1.0; α-tocopherol, 0.01; vitamin K_3 (menadione), 0.001; B-vitamin mix, 0.423; cholesterol, 0.4; vitamin A (10000 I.U./kg); vitamin D_3 (2000 I.U./kg).

salts (g/kg): CH_3COOK, 20.0; $CaCO_3$, 18.0; Na_2HPO_4, 12.0; $CaHPO_4$, 13.5; NaCl, 3.0; KI, 0.045; $MgSO_4.7H_2O$, 4.5; MgO, 4.0; $MnSO_4.4H_2O$, 0.75; $Fe(C_6H_5O_7)_2$, 4.5; $CuSO_4.5H_2O$, 0.23; $CoCl_2.6H_2O$, 0.03; $NaMoO_4.2H_2O$, 0.30; $ZnSO_4.7H_2O$, 0.06; $Na_2B_4O_7.10H_2O$, 0.03; $KAl(SO_4)_2.10H_2O$, 0.045.

B-vitamins (g/kg): thiamine hydrochloride, 0.02; nicotinamide, 0.1; calcium *d*-pantothenate, 0.05; inositol, 2.0; choline hydrochloride, 2.0; pyridoxine hydrochloride, 0.02; biotin, 0.01; folic acid, 0.02; vitamin B_{12}, 0.00005.

PREPARATION Dry ingredients less vitamins and cholesterol ground separately in ball mill to 200 mesh, and mixed together in ball mill for 6 h. Corn oil, fat soluble vitamins and cholesterol added. Ether added then mixed with electric mixer until ether evaporated. Materials to be assayed added and mixed for further 15 minutes.

REARING At 37 ± 1°C and 58 ± 2% RH. At 1 week, 6 crickets placed in 10 cm^3 jars with 2 g diet and cotton stoppered flask with distilled water. Jars covered with cheesecloth.

DEVELOPMENT Average weight, 29.7 mg; 21 out of 24 crickets survived. Growth improved with addition of ca. 1.2% arginine.

See also: Luckey & Stone (1968); Neville & Luckey (1962, 1971).

➡ Patton, R.L. 1967. Oligidic diets for *Acheta domesticus* (Orthoptera: Gryllidae). ANN ENTOMOL SOC AM 60: 1238-1242.

Best four of 16 diets.

COMPOSITION Diet 1 (g/100 g): dry brewers' yeast, 5; corn meal, 20; liver powder, 5; dried skim milk, 15; soybean meal, 20; and wheat middlings, 35. Used for routine rearing for more than three years.

Diet 2 (g/100 g): dry brewers' yeast, 5; corn meal, 35; liver powder, 5; dried skim milk, 15; soybean meal, 10; and wheat middlings, 30.

Diet 3 (g/100 g): dry brewers' yeast, 5; corn meal, 20; fish meal, 10; dried skim milk, 15; soybean meal, 20; and wheat middlings, 30.

Diet 4 (g/100 g): dry brewers' yeast, 10; corn meal, 10; liver powder, 5; dried skim milk, 15; soybean meal, 30; and wheat middlings, 25.

PREPARATION Mixed and finely ground.

REARING & DEVELOPMENT See Patton (1963b). 10 x 12 x 4.5" plastic refrigerator hydrators covered with screened air vents. Food in 4-oz icecream cups and water from small jars covered with folded facial tissue and capped with disc of aluminium foil. Foil crimped over top of jar; inverted. 100 newly hatched nymphs reared in each cage at 26.5°C. Most colonies exposed to incandescent and influorescent light for 16 h photoperiod approximating to natural cycle. Regardless of photoperiod (16 h or 12 h), mixture of long- and short-winged adults produced on diet 2, all long-winged adults produced on diets 1, 3 and 4.

See also: Patton (1963b).

Acheta domesticus (Linnaeus), house cricket

Gryllus bimaculatus (De Geer)

➡ Vago, C.; Fosset, J. 1965. Élevage aseptique des gryllides. 90 CONGR SOC SAVANTES NICE II: 501-505. (In French)

COMPOSITION (g): bran, 25; crushed lettuce leaves, 25; crushed carrots, 25; guinea-pig food, 25; and (ml): water, 70.

PREPARATION Ingredients mixed, placed in tubes or flasks and autoclaved at 110°C for 20 minutes.

REARING 2-4-days-old eggs rinsed in sterile water, transferred to 4% solution of NaClO at 18°C and stirred gently for 3 minutes. Rinsed in sterile water and laid on nutritive agar surface to verify asepsis. Eclosion on agar, then neonate nymphs transferred to rearing containers. Flasks sealed with foil covered cotton plugs and maintained at 28°C.

DEVELOPMENT After 14 trials, average of 82% survival to adults in 50 days. Remainder exhibit moulting problems. Survival slightly increased to 88% by addition of sterile yeast.

See also: Charpentier *et al.* (1972).

Plebeiogryllus guttiventris (Walker)

Dakshayani, K.; Mathad, S.B. 1973. Growth, development, and survival of the cricket, *Plebeiogryllus guttiventris* on various diets. ANN ENTOMOL SOC AM 66: 228-229.

Best of 9 diets tested.

COMPOSITION Concentrated poultry feed (Royal Products, 11 South Parade, Bangalore-1, India).

REARING See Mathad & Dakshayani (1972). Nymphs reared in groups of 10 in 20-oz jars at 33 ± 1°C and 50 ± 5% RH; most suitable rearing temperature 35°C.

DEVELOPMENT After about 1 month nymphal survival, 95%; wt, 101.9 ± 4.3 mg; nymphal period, 39 ± 0.7 days; survival to adult, 83%; adult wt: ♂, 197 ± 4.4 mg; ♀, 209 ± 5.3 mg.

See also: Bentur & Mathad (1974); Mathad & Dakshayani (1972).

Siphonaptera

Pulicidae

Ctenocephalides felis Bouché, cat flea

→ Bruce, W.N. 1948. Studies on the biological requirements of the cat flea. ANN ENTOMOL SOC AM 41: 346-352.

Many diets tested.

COMPOSITION Dried beef blood, 1 g; brewers' yeast, 50 mg. Other successful diets:- dried mammalian blood; desiccated red blood cells, dried blood albumin, haemoglobin and fibrin.

Synthetic diet (%): amino acids, 85; Mazola corn oil, 1.5; glucose, 0.5; agar, 2; salt mixture, 4; and vitamins.

amino acids (mg): glycine, 20; alanine, 60; valine, 80; leucine, 180; isoleucine, 80; proline, 80; hydroxyproline, 10; phenylalanine, 78; glutamic acid, 220; aspartic acid, 41; serine, 20; tyrosine, 65; cystine, 23; histidine hydrochloride, 50; arginine hydrochloride, 64; lysine hydrochloride, 193; tryptophane, 23; methionine, 35; *dl*-threonine, 120; hydroxyglutamic acid, 40; cysteine, 20; 3,5-1-diiodo tyrosine, 20; *dl*-amino-n-valeric acid, 20; *dl*-amino-n-butyric acid, 20.

vitamins (mg/g): thiamin, 0.033; riboflavin, 0.047; nicotinic acid, 0.33; pantothenic acid, 0.14; pyridoxine, 0.021; ascorbic acid, 1.07; ergosterol, 0.0029; β-carotene, 0.043; α-tocopherol, 0.02. Folic acid, biotin and choline added if available.

PREPARATION Ingredients powdered and mixed.

REARING At 80 -90°F and 65-90% RH in 5-10 cm petri dish with diet and 4 g sterile sand. 100 viable eggs in each petri dish.

DEVELOPMENT 92-100% maturity in minimum of 4 days; 61-98% maturity in 7-8 days on synthetic diet when 0.25 g brewers' yeast added.

Xenopsylla cheopis (Rothschild), oriental rat flea

→ Pausch, R.D.; Fraenkel, G. 1966. The nutrition of the larva of the oriental rat flea, *Xenopsylla cheopis* (Rothschild). PHYSIOL ZOOL 39: 202-222.

COMPOSITION Stock diet of dried beef blood and brewers' yeast. Artificial diet (parts): casein, 100; cholesterol, 1; Wesson's salt mixture, 2; and (μg/g dry diet): thiamin, 25; riboflavin, 12.5; nicotinic acid, 50; pyridoxin, 12.5; pantothenic acid, 25; choline, 500; biotin, 1.0; folic acid, 2.5; inositol, 250; (Peanut protein, dry beef blood and fibrin could replace protein).

PREPARATION Dry ingredients finely ground. Transferred to mixing bottle on mechanical roller and rotated for 3 h. 2 g diet dispensed/ 2 x 1" vials. Vitamins added in solution @ 0.2 ml water/vial. Vials kept at 30°C and 75% RH for 3 days. Diet reground in mortar and halved into new vials.

REARING At 30°C and 75% RH. 20 neonate larvae/vial covered with tissue paper. Left for 1 week. Cocoons placed in separate vials; vials immersed in ice-water bath for 2-3 minutes to immobilize nymphs or adults for counting. Adults fed on rat and eggs laid on cellulose layer on bottom of rat cage.

DEVELOPMENT Larval weight after 7 days, 0.335 mg; larval period, 10-14 days; ca. 100% pupation. Similar results when casein plus 5% yeast used: when 5% brewers' yeast added to basic diet instead of vitamin mix, then larval weight, 0.555 mg; larval period, 9-11 days.

See also: Fraenkel (1959b).

Acarina

Acaridae

Acarus siro Linnaeus, grain mite

Griffiths, D.A. 1966. Nutrition as a factor influencing hypopus formation in the *Acarus siro* complex (Acarina, Acaridae). J STORED PROD RES 1: 325-340.

Best of 3 diets.

COMPOSITION Dried yeast powder supplying 35.5% carbohydrates and 49.0% protein (amounts not specified, ample food), supplemented with vitamin-B complex (mg/100 g): thiamine, 10.0; riboflavin, 5.0; niacin, 35.0.

PREPARATION Ingredients thoroughly mixed.

REARING At 20 ± 1° C and 90% RH, in microcells similar to Solomon and Cunnington (1964). Microcells made by drilling tapered hole through strip of plastic sheeting, preferably black ebonite or perspex, so that sides of hole visible from above. Sealed beneath by black filter paper affixed with shellac. Glass lid held with magazine clips. Could rear thousands of mites in cells 3 mm deep and 25 mm maximum diameter. Smaller version of cell for single pair rearing, 1.5 mm deep and 12 mm maximum diam, with microscope cover-slip lid held by ticket clip. Cells stacked on small racks in desiccator. Virgin females placed in cells with males and diet. Both types cells illustrated.

DEVELOPMENT Reproduction and hypopi formation satisfactory; also for *Acarus farris* (Oudemans) and *Acarus immobilis* Griffiths.

See also: Curry (1971); Griffiths (1969).

Caloglyphus anomalus Nesbitt

Pillai, P.R.P.; Winston, P.W. 1968. Culture and nutrition of *Caloglyphus anomalus* (Acarina: Acaridae). ANN ENTOMOL SOC AM 61: 56-60.

Mites axenically reared on microbes (*Bacillus cereus*, *Eschericia coli*, and *Penicillium* sp.) grown on artificial media, such as:- (i) ground beef inoculated with original substrate, (ii) decaying cheddar cheese, (iii) commercial chicken feed (Purina Lay Mash, Raulston Purina Company), (iv) *Neurospora* medium-Difco No. 322, and (v) Bonner's medium. Ground beef and Bonner's medium best two.

COMPOSITION Bonner's medium (g): dextrose, 10; proteose peptone No. 3 (Difco), 10; $Na_2HPO_4.12H_2O$, 0.96; KH_2PO_4, 1.45; agar, 20; and water, 1000 ml.

PREPARATION See Bonner (1947). Ingredients dissolved in water and autoclaved.

REARING Individual mites reared within 2 x 1.5 cm diam glass rings, glued to base microscope slide. Two ventilation holes, 2.5 mm diam, bored in sides of glass ring and sealed with fine mesh glass cloth. Agar poured into cells and bacteria streaked. After 24 h, 1 or 2 sterilized eggs introduced; cells sealed with greased cover slip and autoclaved shell vial inverted over each to reduce contamination. Another cell-type for individual rearing made of 1 x 1 cm ring, cut from unglazed porcelain filter, placed on agar surface in petri dish. Cell sealed as before. Air exchange through porous walls of cell. Both cells illustrated.

Cultures kept in total darkness; desiccators immersed in water bath at 21 ± 2°C and saturated humidity. Eggs sterilized in 2% solution of NaClO for 20 minutes. Mass rearing in petri dishes possible.

DEVELOPMENT First generation completed in 8-10 days on 3 different microbes. Second generation completed sooner on *Penicillium* than on *E. coli*. Those fed on *B. cereus* died during the second generation due to rapid growth of bacteria. After 2 months, mites on *Penicillium* died, but those on *E. coli* did well, completing 4 generations on single substrate.

See also: Woodring & Cutcher (1968).

Lardoglyphus konoi Sasa & Asanuma, grain mite

Matsumoto, K. 1970. Studies on the environmental factors for the breeding of grain mites. Part X. Effect of temperature and relative humidity on the breeding and hypopus formation in *Lardoglyphus konoi*. JPN J SANIT ZOOL 21: 213-219. (In Japanese with English summary)

COMPOSITION Brewers' yeast (Ebios) and fish meal powder (Niboshi) in ratio 4:6.

PREPARATION Ingredients thoroughly mixed.

REARING At series of relative humidities and temperatures.

DEVELOPMENT Optimum breeding at 30°C and 87% RH. In Matsumoto (1968) at 25°C and 82% RH, 256 mites/0.5 g diet multiplied to 912 mites after 7 days.

See also: Matsumoto (1968, 1973).

Rhizoglyphus echinopus (Fumouze & Robin), bulb mite

→ Bot, J.; Meyer, M.K.P. 1967. An artificial rearing medium for acarid mites. J ENTOMOL SOC SOUTH AFR 29: 199.

COMPOSITION (g): wheat germ, 24.0; vitamin-free casein, 5.0; yeast (finely-ground, dry, live *Saccharomyces utywale africana*), 22.0; agar, 6.0; ascorbic acid, 2.5; choline chloride, 0.2; inositol, 0.1; cholesterol, 0.1; Nipagin M, 1.2; water, 250.0.

PREPARATION Wheat germ, yeast, and casein thoroughly mixed. Nipagin and cholesterol dissolved in 25 ml of 25-75% acetone-ethyl ether mixture, and solution stirred into first mixture. Then thinly spread to allow evaporation of solvents. Agar boiled in 200 ml water; ascorbic acid, choline, and inositol dissolved in remaining 50 ml water, and added, together with wheat germ-yeast-casein mixture, to agar solution; boiled again. Resulting mixture then poured into 4" diam petri dishes, cooled, and covered ready for use.

REARING At 25°C and 70-75% RH in 4" diam petri dishes.

DEVELOPMENT *R. echinopus* and *Tyrophagus* sp.: complete life cycles from 8-10 days. Survival of cultures started with 5 females and 5 males, 5-6 weeks. One colony maintained for >1 year.

Tyrophagus putrescentiae (Schrank), mould mite

→ Rodriguez, J.G.; Lasheen, A.M. 1971. Axenic culture of *Tyrophagus putrescentiae* in a chemically defined diet and determination of essential amino acids. J INSECT PHYSIOL 17: 979-985.

COMPOSITION Chemically defined diet (g/100 ml diet): *l*-amino acids: alanine, 0.26; arginine, 0.34; aspartic acid, 0.60; cysteine, 0.09; glutamic acid, 0.94; glycine, 0.28; histidine, 0.18; isoleucine, 0.23; leucine, 0.40; lysine, 0.31; methionine, 0.07; phenylalanine, 0.25; proline, 0.33; serine, 0.26; threonine, 0.20; tryptophan, 0.08; tyrosine, 0.15; valine, 0.25; total, 5.22 g.

other ingredients: RNA, 0.10; agar, 3.0; sucrose, 8.0; Alphacel, 1.0; Wesson's salt mixture, 2.5; myristic acid, 0.056; palmitic acid, 0.494; stearic acid, 0.268; oleic acid, 0.494; linoleic acid, 0.070; linolenic acid, 0.028; cholesterol, 0.500; Tween 80, 0.01; Vanderzant's fortification mixture, 0.40; Streptomycin, 0.05; and (ml): methyl *p*-hydroxybenzoate (15% in 95% ethanol), 1.0; formalin (10%), 0.1; and KOH (22.5% in water), 0.5.

PREPARATION See Rodriguez & Hampton (1966). Mineral and non water-soluble fractions autoclaved; water-soluble fraction sterilized by filtration. Diet poured into sterile 3-dram glass vials @ 2.5 ml/vial; plugged with sterilized cotton. Preparation of radioactive diet similar to above, except sucrose content reduced to 1%. Diet labelled by addition of 250 μC glucose-UL^{14}C to 5 ml aliquot of completed diet.

REARING On soft moist dog food in environmental chamber at 25°C and 83 ± 3% RH. Eggs sterilized in 2% formalin for 5 minutes and seeded into 3-dram vials containing diet. After hatching, immediately transferred by aseptic technique to radioactive or ordinary diet.

DEVELOPMENT After 28 days on basic diet inoculated with 100 neonate larvae, 68 adults and 364 larvae obtained. Essentiality of amino acids determined by using glucose-UL^{14}C, and also be deletion of some individual amino acids from diet.

Corollary

→ Rodriguez, J.G. 1972. Inhibition of acarid mite development by fatty acids. pp. 637-650. *in* Rodriguez, J.G. (Ed.), Insect and Mite Nutrition. North-Holland Publishing Company, Amsterdam.

Two diets.

COMPOSITION Chemically defined diet: see Rodriguez & Lasheen* (1971). Casein diet: as above, except amino acids and lipid fraction replaced by 8 g vitamin-free casein and 7 g wheat germ respectively.

REARING At 27 ± 1°C and 80.0 ± 2.5% RH in microcups.

DEVELOPMENT Casein diet: after 28 days, 200 neonate larvae reproduced to 284 adults and 370 larvae and nymphs. Both diets also supported growth of wet grain mite, *Caloglyphus berlesei* (Michael) (Acarina: Acaridae).

See also: Rodriguez & Potts (1974).

→ Fleurat-Lessard, F. 1974. Contribution a l'étude des relations trophiques entre *Tyrophagus putrescentiae* Schrank (Acarina Acaridae) et la mycoflore des graines de colza stockées. Mise au point d'une technique d'élevage de l'acarien sur moisissures. ANN ZOOL ECOL ANIM 6: 97-109. (In French with English summary)

REARING Cages illustrated. Constructed from transparent plastic "altuglass". Funnel shaped hollow (rearing chamber) drilled in centre of parallelepiped (4.5 x 4.5 x 1.0 cm). 4.5 x 4.5 x 0.15 cm plastic cover held in place during experiments by grease layer. Holes (0.7 cm diam) drilled laterally on each of two sides to within 2-3 mm of rearing chamber. Final connections with rearing chamber then drilled; on one side by 2 mm diam hole, on other by 1 mm diam hole, both blocked by cotton plugs during experiments. Food of fungal spores and mycelium added as follows: rearing chamber half filled with agar (20 g agar/1000 ml distilled water). When gelled, second layer of agar poured on to first to depth of 2 mm, and when set, 6 mm diam disc removed from this second layer and replaced by similar disc cut from pure colony of fungus cultivated in petri dish. Mites then deposited on agar, near to fungal colony and lid fitted. Cells joined together by communication tunnel fitting into one of lateral holes.

DEVELOPMENT Eclosion, 72.8-91.0%; pupae from neonate larvae, 31-67%; adults from eggs laid, 23.1-55.3%. Eclosion after ca. 4 days; and majority of larvae pupated, 14-18 days later. Pupal development, 9 days. Over 7 generations reared.

See also: Hana & Hibbs (1970).

→ Radinovsky, S.; Krantz, G.W. 1961. The biology and ecology of granary mites of the Pacific Northwest. II. Techniques for laboratory observation and rearing. ANN ENTOMOL SOC AM 54: 512-518.

Best 2 of many rearing cells tested.

REARING Plexiglass microcells (illustrated); series of 1.5-2.0 x 0.5 cm diam glass tubes embedded in charcoal-yeast-plaster food mixture poured into syracuse watch glasses. Electric field to repel mites created by 2 brass plates of 1 mm thickness cut to 115 x 45 mm. Thin sheet of celluloid placed between two plates and glued together. Series of 18 holes, 6 mm diam drilled through sandwich, embedded in charcoal-yeast-plaster mixture. Resulting microcells deepened by drilling into food mixture. Electric potential created by wire connected to each plate and to transformer. Sides of microcells filed. Mites introduced into cells by means of fine brush. Potential difference of 7 volts sufficient to repel mites. However no avoidance response achieved so only useful for temporary confinement.

Fluon coating of test tubes only effective means. Dipped into Fluon 3 times, dried slowly at 21°C. Stock cultures maintained in transparent lid slave tins $2\frac{3}{4}$ x $\frac{7}{8}$ inches with holes in lids for ventilation and kept in trays of paraffin.

Species reared: *Acarus siro* Linnaeus, *Tyrophagus putrescentiae* Schrank, *Caloglyphus berlesi* (Michael), *Leiodinychus krameri* Canestrini.

Ceratozetidae

Ceratozetes cisalpinus Berlese

Scheloribates laevigatus Koch

Oppia neerlandica Oudemans (Fam., Oppiidae)

Woodring, J.P.; Cook, E.F. 1962. The biology of *Ceratozetes cisalpinus* Berlese, *Scheloribates laevigatus* Koch, and *Oppia neerlandica* Oudemans (Oribatei) with a description of all stages. ACAROLOGIA 4: 109-137.

COMPOSITION (g): glucose (ground and sifted), 360; vitamin-free casein, 300; salt mixture, 40; vitamin mixture, 40.22; choline mixture, 25; corn oil, 30; cholesterol, 10.

salts (g): $CaCO_3$, 2; $MgSO_3$, 4.8; K_2SO_4, 4.4; KH_2PO_4, 13.6; $NaHCO_3$, 1.7; $CaSO_4.5H_2O$, 0.1; $Fe(NH_3)(SO_4)_2.12H_2O$, 0.960; or $Fe(NH_3)(SO_4)_2.24H_2O$, 1.29; $MnSO_4$ (anhydrous), 0.223; $ZnSO_4.7H_2O$, 0.290.

vitamins (mg): thiamin, 68; nicotinic acid, 98; pyridoxine hydrochloride, 82; calcium pantothenate, (quantity missing from original paper); cyanocobalamin (B_{12}), 2; biotin, 12; folic acid, 88; riboflavin, 7.6; inositol, 3.6 g.

Also fed on mushrooms (*Coprinus* sp.) and lichen (*Physica* sp.).

PREPARATION Corn oil and cholesterol dissolved in acetone in hot water bath. Poured on to dry mixture of all other ingredients in shallow pan. Mixed while fan blowing to remove acetone. 174.78 g celluflour added to 1 kg total. 2.8 g choline chloride dissolved in 25 ml hot absolute ethanol, and 22.2 g celluflour added to make choline mixture. Dried at 120°C, while stirring for 4-6 h. Smaller quantities of salts dissolved in 15 ml double distilled water, 12 g celluflour added, mixed and oven dried at 120°C. Larger amounts of salts ground in mortar and pestle, mixed well and added to other dried salts.

Inositol dissolved in 15 ml hot water, 16.4 g celluflour added, mixed and dried at 120°C for 4-6 h. Then riboflavin added. Thiamin, nicotinic acid, pyridoxine hydrochloride, calcium pantothenate dissolved in 3 or 4 ml hot water and 9.8 g celluflour added. Biotin dissolved in few ml hot 95% ethanol, 5 g celluflour added. Folic acid added to 5 ml hot water with 1 drop concentrated (28%) ammonia, 4.9 g celluflour added. All dried at 50°C in vacuum oven for 2 h and mixed.

Mushrooms oven dried at 100°C and ground slightly. Lichen scraped from bark and oven dried at 100°C. Artificial diet passed through 100 mesh screen and stored in deep freeze until use.

REARING At 25°C and 90-95% RH in individual plastic culture tubes (illustrated) kept in 6.5 x 12.5 cm diam glass dishes covered with another dish. Moisture supplied by 4-5 layers of filter paper changed every 2 weeks. Each dish with 10 tubes arranged around circumference with open centre for cleaning and watering daily. Top half of No. 5 rubber stopper used as culture tube lid. *C. cisalpinus* females oviposited in dried leaves or cast nymphal skins. *S. laevigatus* oviposited in decayed grass tissue. *O. neerlandica* oviposited in fungal hyphae. Artificial diet removed every other day; unused mushroom and lichen weekly. Given less food than could be consumed to prevent fungal growth. Neonate *C. cisalpinus* need fungal hyphae for food.

Aseptic rearing possible in sterile box. Eggs sterilized by immersion in 1.0% NaClO for 10-15 seconds, washed with sterile water.

DEVELOPMENT Life cycle (egg to adult emergence): *C. cisalpinus*, 32 days; *S. laevigatus*, 64 days; *O. neerlandica*, 23 days.

Phytoseiidae

Amblyseius hibisci (Chant)

→ McMurtry, J.A.; Scriven, G.T. 1962. The use of agar media in transporting and rearing phytoseiid mites. J ECON ENTOMOL 55: 412-414.

COMPOSITION (%): agar, 2; water, 98. Addition of 6% honey and 10% yeast hydrolysate increased reproduction.

PREPARATION Agar dissolved in hot water and poured to depth of ca. ⅜" in rearing or shipping vials.

REARING Rearing container: constructed from plastic dome and 2 x 2" glass plate. Both parts sealed together with asphalt emulsion (Tree Seal). Dome cut from egg storage tray (Lustro Ware No. L-31-S) and 3 holes cut with cork borer: 5 mm hole on one side for introducing host mites or transferring predators; 12 mm hole on opposite side covered by piece of felt for ventilation and for oviposition site; and 18 mm hole in top into which vial with agar diet inserted; 10 x 19 mm diam vial cut from standard 3-dram shell vial, filled to top with diet heated to liquid state, and 15 mm round cover glass immediately placed on surface, leaving exposed, narrow area of periphery. RH within cell 80-90%. Cell illustrated.

Shipping container: One-dram shell vial equipped with stopper of rubber tubing and piece of fine mesh nylon cloth. Cloth tightly stretched over one end of tubing, then pushed into vial. Care needed to eliminate excess folds in cloth to prevent mite escape. Container illustrated.

DEVELOPMENT Satisfactory in both shipping and rearing container. Also adopted for citrus red mite, *Panonychus citri* (McGregor), and Pacific spider mite, *Tetranychus pacificus* McGregor (Tetranychidae).

McMurtry, J.A.; Scriven, G.T. 1966. Effects of artificial foods on reproduction and development of four species of phytoseiid mites. ANN ENTOMOL SOC AM 59: 267-269.

Several foods in water solution tested.

COMPOSITION (%): sucrose, 20; sucrose + yeast hydrolysate, 20 + 20; molasses, 20; molasses + yeast hydrolysate, 20 + 20.

REARING Mites confined to washed, excised avocado leaves. Small droplets of food placed directly on leaf surface and fresh diet supplied every 3 days.

DEVELOPMENT Both sucrose and molasses increased survival of adult females of *Amblyseius limonicus*; also *A. hibisci* (Chant), *Typhlodromus occidentalis* Nesbitt, and *T. rickeri* Chant. Yeast hydrolysate combined with either molasses or sucrose induced oviposition by *A. limonicus* and *A. hibisci*. Latter food combination provided sufficient nutrients for both *Amblyseius* species to develop from egg to maturity, but developmental rate slow and mortality high. No food induced significant oviposition or development of either 2 *Typhlodromus* species.

Phytoseiulus persimilis Athias-Henriot

Shehata, K.K.; Weismann, L. 1972. Rearing the predaceous mite *Phytoseiulus persimilis* Athias-Henriot on artificial diet (Acarina Phytoseiidae). BIOLOGIA BRATISLAVA 27: 609-615.

Three diets evaluated.

COMPOSITION Diet 1: honey, 15 g; and (mg): ascorbic acid, 100; vitamin complex, 300; and (ml): antiseptic, 1; distilled water, 100; pangamin extract, 50.

Diet 2: as diet 1, except only 50 ml of vitamin extract used.

Diet 3: 5 ml distilled water added to 10 ml of diet 2 and mixed well To this solution, 40 mg of following amino acids added: *dl*-serine, *dl*-*meso*-homocystine, *l*-cystine, *dl*-methionine, and *l*-threonine.

PREPARATION Pangamin extract prepared by dissolving 5 g pangamin in 50 ml distilled water, then heating for 2 h at 50°C. Solution cooled and filtered. Honey added, and stirred until dissolved. B-vitamin complex dissolved in remaining 50 ml distilled water and filtered. Ascorbic acid added to vitamin complex and stirred well. Antiseptic prepared by dissolving 15 g sorbic acid and 20 g Nipagin in 170 ml water. From this solution, 1 ml taken and added to vitamin complex with ascorbic acid which then added to pangamin extract with honey. Whole solution mixed well. pH adjusted to 6.8-7.0 with KOH solution. Diet stored in tightly closed flasks, and refrigerated until use.

REARING At 24 ± 1°C, using sachet of stretched parafilm. Illustrated description of rearing sachet and its preparation given; 15 x 22 mm diam glass cell open at both ends. Plastic cylinder holder 20 mm diam placed on dish, and divided by ridge of filter paper. Glass cell then fitted over holder forming tight seal. 1.5 cm parafilm square stretched around top end of glass cell, then covered completely with diet dispensed by micropipette. Second square of parafilm of same dimensions stretched and pressed over first thus sandwiching food droplets between thin layers of film and completely enclosing feeding pocket. Disc of moistened filter paper put on base of internal part of holder for maintaining RH.

DEVELOPMENT Best results obtained on diet 3. Egg hatch, 2-2.5 days; larval period, 0.5-1.5 days; protonymph, 1.0-1.5 days; deutonymph, 2.0-2.5 days; life cycle, 3.0-4.5 days (average 3.75 days). 62% of newly hatched larvae completed development to adult stage; no viable eggs obtained. However, when preyed on *Tetranychus urticae*, viable eggs laid. Average survival 16 days; oviposition period, ca. 7 days.

Further improvements to diet needed.

Pyroglyphidae

Dermatophagoides farinae Hughes, American house dust mite

➞ Larson, D.G.; Mitchell, W.F.; Wharton, G.W. 1969. Preliminary studies on *Dermatophagoides farinae* Hughes, 1961 (Acari) and house dust allergy J MED ENTOMOL 6: 295-299.

COMPOSITION House dust and crumbled dog food (Gaines burger).

REARING At 25° C and 75% RH in desiccators with saturated NaCl solution. Mites for testing in 9 x 30 mm shell vials stoppered by nylon bolting silk over rubber tubing. Food added 3 times/week.

DEVELOPMENT 85% adults after 30 days. Mean range of 0.8-1.4 eggs laid/♀/day.

➞ Waki, S.; Matsumoto, K. 1973a. Studies on the environmental requirements for the breeding of the dust mite, *Dermatophagoides farinae* Hughes, 1961. Part 1. Observations on the mode of breeding under various temperature and humidity conditions. EISEI DOBUTSU 23: 159-163. (In Japanese with English summary)

COMPOSITION Mixture of equal amount of dried yeast and fish meal.

PREPARATION Commercial dried sardines (Niboshi) heated at 105° C for 5 h then powdered. Mixed with yeast.

REARING At various temperatures and humidities in 50 ml beaker.

DEVELOPMENT Maximum mite count at 25° C and 57% RH, 4800 after 23 weeks; at 25° C and 60% RH, 4460 after 13 weeks and 5665 after 23 weeks; at 30° C and 60% RH, 2445 after 11 weeks. Initial inoculation 900 mites/25 g diet.

See also: Waki & Matsumoto (1973b).

Dermatophagoides pteronyssinus (Troussart), European house dust mite

Miyamoto, J.; Ishii, A.; Sasa, M. 1975. A successful method for mass culture of the house dust mite, *Dermatophagoides pteronyssinus* (Troussart, 1897). JPN J EXP MED 45: 133-138.

COMPOSITION (parts): powdered animal food for laboratory mouse and rat (CE-2, Clea Co. Ltd, Tokyo), 2; dried yeast (Ebios, Ebios Pharmaceutical Co. Ltd, Tokyo), 2; dried fish powder (Niboshi), 1.

PREPARATION Dried fish powder thrashed through graining machine, ("AGA" type, Ishikawa Manufacturing Co. Ltd, Tokyo). All ingredients heated dry at 90° C for 30 minutes. Measured volumes of water added to obtain desired humidities.

REARING At 25° C and various humidities (best 75%), either in 20 x 10 mm petri dishes or in 5 x 8 cm diam glass vials. Diet spread in 1 cm layer to provide large surface area and some degree of ventilation. Vial opened weekly and diet mixed by shaking. Constant RH maintained by keeping containers in desiccators containing saturated solutions of various salts.

DEVELOPMENT Average yield after 12 weeks, 2-3 g wet wt/100 g diet after inoculation with 15000 mites.

Tetranychidae

Tetranychus urticae (Koch), two-spotted spider mite

Ekka, I.; Rodriguez, J.G.; Davis, D.L. 1971. Influence of dietary improvement on oviposition and egg viability of the mite, *Tetranychus urticae*. J INSECT PHYSIOL 17: 1393-1399.

COMPOSITION Improved chemically defined diet: *l*-amino acids (mg): alanine, 150; arginine, 200; aspartic acid, 200; cysteine, 50; glutamic acid, 180; glycine, 200; histidine, 100; isoleucine, 100; leucine, 160; lysine, 120; methionine, 60; phenylalanine, 60; proline, 60; serine, 80; threonine, 120; tryptophan, 40; tyrosine, 50; valine, 120; RNA, 100.

sugars (g): sucrose, 3.0; levulose, 0.5; dextrose, 0.5.

fat-soluble vitamins (mg): vitamin A, 3.0; vitamin E, 5.0.

water-soluble vitamins (mg): folic acid, 0.34; inositol, 6.00; *p*-aminobenzoic acid, 0.50; pyridoxine hydrochloride, 0.10; riboflavin, 0.20; thiamine hydrochloride, 0.20; nicotinic acid, 0.30; biotin, 0.0001; calcium pantothenate, 0.50; choline chloride, 10.00; B_{12}, 0.002; ascorbic acid, 100.00.

lipids (mg): β-sitosterol, 0.50; stigmasterol, 0.44; stearic acid, 1.00; palmitic acid, 0.50; oleic acid, 0.50; linoleic acid, 0.50; linolenic acid, 3.00.

salts (mg): K_2HPO_4, 37.50; $Na_2HPO_4.12H_2O$, 10.60; $MgSO_4.7H_2O$, 15.60; $CaCl_2.2H_2O$, 3.62; $CoCl_2.6H_2O$, 0.08; (sequestrenes); Fe, 1.00; Mn, 0.50; Cu, 0.08; Zn, 0.12.

Water to 100 ml.

PREPARATION Heat stable nutrients autoclaved, others Millipore filtered. Parafilm membrane sterilized by exposure to ethylene oxide for 4-6 h. Quiescent deutonymphs surface-sterilized with 2% formalin, and sachets prepared under sterile conditions. Mites transferred to fresh diets every second day. Details of preparation described by Walling *et al.* (1968); amino acids and RNA mixed with 40 ml warm sterile distilled water. Solution cooled, titrated to pH 5.0 with 4 N KOH, brought to 40 ml volume, and again mixed. Lipids made up in stock solutions in ethanol and pipetted into 50 ml beaker. Water (5 ml) added until mixture cloudy, then solution evaporated to dryness in nitrogen stream and made up to 40 ml volume with sterile distilled water. Tween 80 solution (10 mg in 1 ml) added and lipid fraction then sonicated. Some water-soluble vitamins in stock solutions; others dissolved in sterile distilled water and brought to final volume of 10 ml. Salts in stock solutions made up to 10 ml volume. Diet mixed in sequence given above.

REARING Liquid diet in parafilm sachet at end of 25 x 25 mm diam glass tube. Lower membrane unstretched but slightly depressed, served to hold diet; top membrane fully stretched, served as membrane through which mites fed. Parafilm covering wrinkled to give anchorage for mites. Mites confined on feeding surface by thin line tanglefoot placed around rim of tube. Rearing cages housed in 3-l Nalgene jars with 300 ml KOH solution to provide 80% RH. Jars covered to protect against contamination. Transparent tops permit 24 h photoperiod provided by a GE 40 W cool-white fluorescent tube, giving 1000 lux. Room temperature maintained at 27° C.

DEVELOPMENT Pre-oviposition period shortened from 3.5 to 1.5 days; oviposition rate increased from 0.343 to 0.649 eggs/♀/day; oviposition period extended from 25 to 36 days; egg viability increased from 3.6 to 19.2%.

See also: Ekka (1969); Rodriguez (1969); Rodriguez & Hampton (1966); Rodriguez & Singh (1967); Rodriguez *et al.* (1967); Seay (1967); Sun (1963); Sun *et al.* (1963); Walling *et al.* (1968).

→ Fritzsche, R. 1960. Morphologische, biologische und physiologische Variabilität und ihre Bedeutung für Epidemilogie und Bekämpfung von *Tetranychus urticae* Koch. BIOL ZENTRALB 79: 521-576. (In German)

COMPOSITION (g): sucrose, 0.7; asparagine, 0.001; aspartic acid, 0.001; KCl, 0.15; $CaHPO_4$, 0.2; vitamins B_1, B_2, C, traces; double distilled water, 50 ml.

PREPARATION Solutions autoclaved for 30 minutes and vitamins added. pH 5.1-6.0.

REARING Collodion membrane prepared using 7.5% amyl acetate solution as solvent. Membrane stretched over end of 10 x 17 mm diam glass ring with 1 mm thick wall. Smaller, 5 x 8 mm diam, glass ring placed on top in centre of membrane and attached to upper rim to serve as cage. Upper edge greased to confined mites. Diet hangs as drop on underside of membrane. Whole assembly placed on saturated filter paper inside petri dish cover. 1 ♀/cage.

DEVELOPMENT Adult longevity on diet, 22-26 days; on control of bean leaves, 21-28 days. Reproduction higher on host plants than on artificial diet.

→ Storms, J.J.H.; Noordink, J.Ph.W. 1970. Nutritional requirements of the two-spotted spider mite, *Tetranychus urticae* (Acarina: Tetranychidae). 7 EUR MITE SYMP POL AKAD 29-39.

COMPOSITION Chemically defined diet: amino acids (mg): asparagine, 19; aspartic acid, '; glutamine, 9; arginine, 9; histidine, 9; isoleucine, 9; methionine, 9 leucine, 6; lysine, 6; phenylalanine, 6; tyrosine, 6; valine, 6; threonine, 5; proline, 5; alanine, 3; cysteine, 3; cystine, 2; amino-butyric acid, 2; glutamic acid, 3; glycine, 2; homoserine, 2; serine, 2; tryptophan, 3; ornithine, 6; citrulline, 2.

fatty acids and sterols (mg): palmitic acid, 25; stearic acid, 10; ergosterol, 15; sitosterol, 10; cholesterol benzoate, 12; sucrose, 1500.

vitamins (mg x 10^{-3}): vitamin E, 60; aneurine, 625; choline chloride, 625; folic acid, 60; inositol, 1875; nicotinic acid, 625; *p*-aminobenzoic acid, 625; calcium pantothenate, 310; pyridoxine hydrochloride, 200; riboflavin, 250; thiamine hydrochloride, 310; lipoic acid, 125; cholic acid, 625; biotin, 60; vitamin B_{12}, 60; niacinamide, 60; ascorbic acid, 6000.

salts (mg): KH_2PO_4, 12; $MgCl_2.6H_2O$, 3; $NaH_2PO_4.12H_2O$, 3; $CaCl_2.6H_2O$, 3; (sequestrenes); Fe, 0.1; Zn, 0.05; Mn, 0.05; Cu, 0.05.

water to 100 ml.

PREPARATION Ingredients in solution mixed together and pH adjusted to 10 with 0.1 N KOH. Diets boiled for 3 minutes and parafilm membranes dipped in 96% ethanol before use.

REARING At 20-24°C and 80% RH under diffuse light. Preparation of sachet using Parafilm 'M' described. Parafilm stretched in 2 directions and mounted on plastic cup of 3 cm^2 area; 0.3 ml diet injected between two layers of membrane. Mites confined on membrane by applying sticky barrier. Sachets prepared axenically.

DEVELOPMENT Highest diet uptake and egg production of adults at pH 10 on standard diet. Adult females and larvae kept alive for 3 weeks. Not confirmed if whole life cycle completed.

Trombiculidae

→ Lipovsky, L.J. 1953. Improved technique for rearing chigger mites (Acarina: Trombiculidae). ENTOMOL NEWS 64: 4-7.

REARING In 3- or 5-dram vials lined with powdered charcoal and plaster of Paris mixed in dry ratio of 12:1. 7 ml dry mix placed in 3-dram vial and mixed with 3 ml water. Plastic stopper placed tightly on vial and end tapped to get uniform 15 mm layer of wet plaster. 4-5 longitudinal slits made in plaster on sides of tube for light. Pin holes made in tube if shipping. Few drops water added if tube opened frequently.

20 x 65 mm diam stender dishes also used; lined with plaster mixture, upper 6 mm plaster free and kept smooth as barrier. Coating outside of dish and lid with paraffin, and covering lid with wax paper recommended. Two mm deep grooves cut on surface. 50 adults reared/dish. Food available continuously to prevent cannibalism. Culture dish also used for mites of Oribatidae, Tyroglyphidae, Trombidiidae, Phytoseiidae, Macrochelidae, Anoetidae, Uropodidae, and Cheyletidae; and insects of families Cryptostemmatidae (Hemiptera), Staphylinidae and other Coleoptera; and Stratiomyiidae, Sciaridae, Mycetophylidae and other Diptera.

Tydeidae

Lorryia ferrulus Baker

Tydeus bakeri Brickhill

Brickhill, C.D. 1958. Biological studies of two species of tydeid mites from California. HILGARDIA 27: 601-620.

COMPOSITION (g): enzymatic protein hydrolysate, 3; fructose, 6; water added to make paste.

REARING At 18-20° C in winter, 25-28° C in summer. Diet placed on ½" square of black waxed paper floated on water in 2" diam stender dish. 3-4 drops of diet given/week. Oviposition increased if finely cut splinter of box wood placed on fine kapok dusted on substrate.

DEVELOPMENT *L. ferrulus:* egg hatch in 6-7 days; larval period, 3-4 days; protonymph, 6-8 days; deutonymph and tritonymph, 10-12 days; oviposition begins 3-4 days after adult emergence; life cycle, 30-34 days. Vigorous colony in ca. 23 days from 10 mites.

T. bakeri: egg hatch in 7-8 days; larval period, 4 days; protonymphs, 7-10 days; deutonymph and tritonymph, 11-14 days. Oviposition 3-4 days after adult emergence; life cycle, 32-38 days. Vigorous colony in 30 days from 10 mites.

Araneida

Clubionidae

Chiracanthium inclusum (Hentz)

Lycosidae

Lycosa gulosa Walckenaer

→ Peck, W.B.; Whitcomb, W.H. 1968. Feeding spiders on artificial diet. ENTOMOL NEWS 79: 233-236.

COMPOSITION For *C. inclusum*, (parts): egg yolk, 4; dye, 1.

For *L. gulosa*. (parts): homogenized milk, 2; egg yolk, 10; dye, 1.

REARING At 55% RH in glass tubing, one end plugged with cotton, other end corked, with hollow-stemmed cotton swab inserted through cork. Cotton swab saturated with diet with long needled syringe through plastic stem of swab.

DEVELOPMENT Period between ecdyses for *L. gulosa* 28 days compared with 10 days in control. *C. inclusum* maintained for 6 weeks; and *L. gulosa*, 90 days. Most specimens moulted once or twice.

See also: Peck & Whitcomb (1967).

Addendum

COLEOPTERA

SCARABAEIDAE

→ Donaldson, J.M.I. 1976. An artificial diet for flower- and fruit-eating cetonid beetles (Coleoptera: Scarabaeidae). PHYTOPHYLACTICA 8: 17-18.

COMPOSITION (g): dehydrated peaches, 40; maize meal, 40; brown sugar, 10; vitamin- and fat-free casein, 4; brewers' yeast (ground dry live *Saccharomyces utywala-africana*), 15; methyl *p*-hydroxybenzoate, 1.8; cholesterol, 0.1; ascorbic acid, 2.5; choline chloride, 0.2; and (ml): acetone and ethyl ether 3:1, 20; water (including peach juice), 120.

PREPARATION Peaches soaked in sufficient water for 4-5 h, then machine-minced to soft pulp (ca. 40 g dried peaches give ca. 100 g of pulp). Maize meal, sugar, casein and yeast mixed. Methyl *p*-hydroxybenzoate and cholesterol dissolved in acetone and ethyl ether mixture, and added to dry ingredients, stirred and solvents evaporated. Ascorbic acid and choline chloride dissolved in ca. 60 ml of water and/or peach juice (left over water after soaking peaches) then added to dried ingredients to form paste. Paste stirred into additional 60 ml of boiling water or peach juice until dry enough not to stick to sides of container; removed from heat, minced peaches added and stirred. Dispensed to pairs of compartments of plastic ice-cube trays, wrapped in tin foil and stored at ca. 4°C until required.

REARING At 27 ± 1°C and 60-90% RH with 13 h photoperiod. Adults in 75 x 48 x 48 cm wooden framed, wire gauze, cages. Bowls of damp soil supplied for oviposition. Food renewed twice weekly.

DEVELOPMENT Viable eggs laid by *Dischista cincta* De Geer, *Dyspilophora trivittata* ssp. *nigricans* Kraatz and *Genyodonta flavomaculata* ssp. *egregia* Boheman when adults fed on diet. *Pachnoda sinuata* ssp. *flaviventris* Gory and Percheron reared for 2 generations. Mean eggs/♀/day varied from 0.17 to 0.75 on diet and 0.46-50 on rose blossom.

LEPIDOPTERA

NOCTUIDAE

Pseudaletia separata Walker

Neelgund, Y.F.; Mathad, S.B. 1974. Comparative studies of developments of the armyworm, *Pseudaletia separata* Walker, reared on Napier grass and artificial diet. KARNATAK UNIV J SCI 19: 7-13.

COMPOSITION (g): beans (*Dolichos lablab*), 133.2; casein, 13.6; sucrose, 13.6; yeast, 16.0; wheat germ oil, 16.0; ascorbic acid, 4.0; agar, 8.0; methyl *p*-hydroxybenzoate (dissolved in 9.0 ml of 95% ethyl alcohol), 0.9; and (ml): formalin, 2.0; water (sterile), 400.

PREPARATION Beans washed and soaked in 250 ml sterile water for 18-24 h, then blended. Casein, sucrose, yeast, wheat germ oil and ascorbic acid added. Agar dissolved in 150 ml boiling sterile water, cooled to 80° C then stirred in. Inhibitor solution added after diet cooled for 5-10 minutes. Dispensed to clean sterilized beakers and refrigerated until use.

REARING At 26 ± 1° C and 70-80% RH in 9.5 ml plastic cup. Diet renewed weekly.

DEVELOPMENT On Napier grass and artificial diet respectively: average larval duration (days), 27.5; 25.4; larval weight 20 days after hatching, 249.7, 395.5 mg; % pupation, 72.0, 84.0; % adult emergence, 80.6, 94.2; average eggs laid/♀, 524.7, 604.0; % egg viability, 91.7, 97.7; % reaching maturity, 58.0, 79.0; nutritional index, 116.9, 189.5.

LEPIDOPTERA

NOCTUIDAE

Spodoptera frugiperda (J.E. Smith), fall armyworm

Burton, R.L. 1967. Mass rearing the fall armyworm in the laboratory. U S DEP AGRIC PRESENTATION PAP 33-117, 12 pp.

COMPOSITION Group A: agar, 720 g; distilled water, 17.6 l.

Group B (g): casein, 1008; powdered cellulose, 144; sucrose, 1008; wheat germ, 900; Wesson's salt mixture, 288; and distilled water, 6.4 l.

Group C: microbial inhibitor stock solution (g): sorbic acid, 117.6; methyl *p*-hydroxybenzoate, 88.2 g; dissolved in 1 l 95% ethyl alcohol; added @ 576 ml.

Group D (ml): 10% formaldehyde, 120; 22.5% KOH solution, 240.

Group E (g): ascorbic acid, 120; kanamycin sulphate, 4; Streptomycin, 4; choline chloride, 29; and (ml): linseed oil plus α-tocopherol (99:1), 48; vitamin stock suspension, 48.

vitamins (g): niacin, 3.00; calcium pantothenate, 3.00; riboflavin, 1.50; thiamine hydrochloride, 0.75; pyridoxine hydrochloride, 0.75; folic acid, 2.50; vitamin B_{12} (0.1% triturated with mannitol), 7.50; choline chloride, 15.00 (dissolved in 500 ml biotin stock solution) [0.1 g biotin in 833 ml distilled water].

PREPARATION Agar mixed with water, divided equally into 2 large covered enamel pans and autoclaved for 4 minutes at 121°C and 17 lb/inch^2. Removed, stirred and autoclaved for further 4 minutes, then cooled to 70°C. Group B blended for 2 minutes, then group C added. BC mixture blended for ½ minute then group D added in order listed with ½ minute stirring between them. Whole mixture blended for 2 minutes then agar solution added. Mixture blended again for 1 minute before addition of group E, then whole diet mixed and blended for ca. 4 minutes until homogenous. Diet dispensed by machine to 1-oz plastic cups as described by Burton & Cox (1966).

REARING At 24°C and 78% RH with 14 h photoperiod in 1 cubic foot cages (described). Ca. 40 pupae placed in cage; emerging moths fed daily from folded square of cheesecloth or absorbent cotton saturated with beer. Oviposition on paper towelling. Neonate larvae transferred to rearing cups; kept at 26.5°C at constant low intensity lighting.

DEVELOPMENT Adult emergence, 87%; cost US$16.50/1000 adults.

Appendix

Composition of salt and vitamin mixtures

Salt Mixtures

Wesson's Salt Mixture	(%)
Calcium carbonate	21.000
Copper sulphate ($5H_2O$)	0.039
Ferric phosphate	1.470
Manganous sulphate (anhydrous)	0.020
Magnesium sulphate (anhydrous)	9.000
Potassium aluminium sulphate	0.009
Potassium chloride	12.000
Potassium dihydrogen phosphate	31.000
Potassium iodide	0.005
Sodium chloride	10.500
Sodium fluoride	0.057
Tricalcium phosphate	14.900

Salt Mixture No. 2 USP XIII	(%)
Calcium biphosphate	13.58
Calcium lactate ($5H_2O$)	32.70
Ferric citrate ($5H_2O$)	2.97
Magnesium sulphate	13.70
Potassium phosphate (dibasic)	23.98
Sodium biphosphate ($2H_2O$)	8.72
Sodium chloride	4.35

McCollum-Davis Salt Mixture No. 185	(%)
Calcium lactate ($5H_2O$)	35.19
Calcium biphosphate (H_2O)	14.60
Potassium phosphate (mono)	25.78
Sodium phosphate (mono)	9.38
Sodium chloride	4.67
Magnesium sulphate (anhydrous)	7.19
Ferric citrate	3.19

Salt Mixture USP XIV

			(g)
Cupric sulphate	0.48 g	)	
Ferric ammonium		)	
citrate	94.33 g	)	
Manganese sulphate	1.24 g	)	16.2
Ammonium alum	0.57 g	)	
Potassium iodide	0.25 g	)	
Sodium fluoride	3.13 g	)	
To make	100 g		
Calcium carbonate			68.6
Calcium citrate			308.3
Calcium biphosphate			112.8
Magnesium carbonate			35.2
Magnesium sulphate			38.3
Potassium chloride			124.7
Dibasic potassium phosphate			218.8
Sodium chloride			77.1
			1000 g

Vitamin Mixtures

Vanderzant's Fortification Mixture (Composition/1000 g).

α-tocopherol	8 g
Ascorbic acid	270 g
Biotin	20 mg
Calcium pantothenate	1 g
Choline chloride	50 g
Folic acid, crystalline	250 mg
Inositol	20 g
Niacinamide	1 g
Pyridoxine hydrochloride	250 mg
Riboflavin	500 mg
Thiamine hydrochloride	250 mg
Vitamin B_{12} trituration in Mannitol	2 g

Vitamin Diet Fortification Mixture

A mixture of the following vitamins triturated in dextrose

	g
Vitamin A Concentrate (200000 units per gram)	4.5
Vitamin D Concentrate (400000 units per gram)	0.25
α-tocopherol	5.0
Ascorbic acid	45.0
Inositol	5.0
Choline chloride	75.0
Menadione	2.25
p-aminobenzoic acid	5.0
Niacin	4.5
Riboflavin	1.0
Pyridoxine hydrochloride	1.0
Thiamine hydrochloride	1.0
Calcium pantothenate	3.0
	mg
Biotin	20
Folic acid	90
Vitamin B_{12}	1.35

Bibliography

To consult bibliography see Format, briefly:-

(i) * indicate main references abstracted.

(ii) ** indicate reviews and papers on related work.

(iii) References with no asterisk are supplementary references given under "See also".

Abdel-Malek, A.A.; Adham, F.K. 1974 Larval and adult diets as factors influencing autogeny in *Aedes caspius*, (Diptera: Culicidae). ACTA ENTOMOL BOHEMOSLOV 71: 294-297.

*Aboul-Nasr, A.E.; El-Miniawi, S.F. 1974 Successful mass-rearing of *Megaselia scalaris* Loew, on a low cost pablum in Egypt. *in* FAO/IAEA Information Circular on Radiation Techniques and their Application to Insect Pests No. 18.

*Achan, P.D. 1968 Preliminary observations on the development of an artificial diet for *Hypsipyla robusta* Moore. COMMONW INST BIOL CONT TECH BULL 10: 23-26.

*Addy, N.D. 1969 Rearing the forest tent caterpillar on an artificial diet. J ECON ENTOMOL 62: 270-271.

Adkisson, P.L. 1961a Effect of larval diet on the seasonal occurrence of diapause in the pink bollworm. J ECON ENTOMOL 54: 1107-1112.

*Adkisson, P.L. 1961b Fecundity and longevity of the adult female pink bollworm reared on natural and synthetic diets. J ECON ENTOMOL 54: 1224-1227.

*Adkisson, P.L.; Bull, D.L.; Allison, W.E. 1960 A comparison of certain artificial diets for laboratory cultures of the pink bollworm. J ECON ENTOMOL 53: 791-793.

*Adkisson, P.L.; Vanderzant, E.S.; Bull, D.L.; Allison, W.E. 1960 A wheat germ medium for rearing the pink bollworm. J ECON ENTOMOL 53: 759-762.

Afrikian, E.G. 1960 Causal agents of bacterial diseases of the silkworm and the use of antibiotics in their control. J INSECT PATHOL 2: 299-304.

Agarwal, H.C. 1970 Sterol requirements of the beetle *Trogoderma*. J INSECT PHYSIOL 16: 2023-2026.

*Agui, N.; Ogura, N.; Okawara, M. 1975 Rearing of the cabbage armyworm, *Mamestra brassicae* L. (Lepidoptera: Noctuidae) and some lepidopterous larvae on artificial diets. JPN J APPL ENTOMOL ZOOL 19: 91-96. (In Japanese with English summary)

*Akey, D.H.; Beck, S.D. 1971 Continuous rearing of the pea aphid, *Acyrthosiphon pisum*, on a holidic diet. ANN ENTOMOL SOC AM 64: 353-356.

*Akey, D.H.; Beck, S.D. 1972 Nutrition of the pea aphid, *Acyrthosiphon pisum:* requirements for trace metals, sulphur and cholesterol. J INSECT PHYSIOL 18: 1901-1914.

*Akey, D.H.; Beck, S.D. 1975 Programmable automated system for continuous rearing of the pea aphid, *Acyrthosiphon pisum,* on holidic diets. ENTOMOL EXP APPL 18: 1-16.

Akhtar, M.; Ali, A. 1972 Nutritional requirements of the silkworm *Bombyx mori* Linn. PAK J ZOOL 4: 101-107.

*Akov, S. 1962a A qualitative and quantitative study of the nutritional requirements of *Aedes aegypti* L. larvae. J INSECT PHYSIOL 8: 319-335.

Akov, S. 1962b Antimetabolites in the nutrition of *Aedes aegypti* L. larvae. The substitution of choline by related substances and the effect of choline inhibitors. J INSECT PHYSIOL 8: 337-348.

**Akov, S. 1964 The aseptic rearing of *Aedes aegypti.* BULL W H O 31: 463-464.

Akov, S. 1967 Effect of folic acid antagonists on larval development and egg production in *Aedes aegypti.* J INSECT PHYSIOL 13: 913-923.

**Albritton, E.C. (Ed.) 1954 Standard Values in Nutrition and Metabolism. W.B. Saunders Co, Philadelphia, PA.

*Aliniazee, M.T.; Stafford, E.M.; Fukushima, C. 1971 Rearing of the omnivorous leaf roller in the laboratory on artificial diet. ANN ENTOMOL SOC AM 64: 1172-1173.

*Allen, D.C.; Knight, F.B.; Foltz, J.L. 1968 A technique for rearing *Choristoneura pinus* on artificial diet. ANN ENTOMOL SOC AM 61: 362-364.

*Allen, W.R.; Askew, W.L. 1970 A simple technique for mass-rearing the onion maggot (Diptera: Anthomyiidae) on an artificial diet. CAN ENTOMOL 102: 1554-1558.

**Altman, P.L.; Dittmer, D.S. (Ed.) 1968 Nutrient requirements and utilization; insects, pp. 148-163; Synthetic diets: insects, pp. 164-167. *in* Metabolism. Federation of American Societies of Experimental Biology. Bethesda, Maryland.

*Altwegg, P. 1971 Ein semisynthetisches Nährmedium und Ersatzsubstrate für die Oviposition zur von der Jahreszeit unabhängigen Zucht des grauen Lärchenwicklers, *Zeiraphera diniana* (Gn.) (Lep; Tortricidae). Z ANGEW ENTOMOL 69: 135-170. (In German with English summary)

*Ankersmit, G.W. 1968 The photoperiod as a control agent against *Adoxophyes reticulana* (Lepidoptera; Tortricidae). ENTOMOL EXP APPL 11: 231-240.

*Anonymous 1971 Diet for *Piophila casei:* rearing method. INSECT TOXICOL INF SERV 14: 47-48.

*Anthon, E.; Smith, L.O.; Garrett, S.D. 1971 Artificial diet and pheromone studies with peach twig borer. J ECON ENTOMOL 64: 259-262.

*Antonio, A.Q.; McLaughlin, J.R.; Leppla, N.C.; Green, C.W. 1975 Culturing the lesser peach tree borer. J ECON ENTOMOL 68: 309-310.

Anwar, M.; Bhatti, M.A.; Qureshi, Z.A.; Ahmad, S. 1966 Rearing of *Dacus zonatus* (Saunders) for their control by sterile male technique. 1. Formulation of an artificial diet. PROC AGRIC SYMP DACCA 152-154.

Applebaum, S.W.; Konijn, A.M. 1966 The utilization of starch by larvae of the flour beetle, *Tribolium castaneum.* J NUTR 85: 275-282.

*Applebaum, S.W.; Konijn, A.M.; Menco, B. 1971 Growth and biochemical adaptation of larvae of the beetle *Dermestes maculatus* to carbohydrate-free diets. INSECT BIOCHEM 1: 1-13.

*Applebaum, S.W.; Lubin, Y. 1967 The comparative effects of vitamin deficiency on development and on adult fecundity of *Tribolium castaneum.* ENTOMOL EXP APPL 10: 23-30.

Arai, N.; Ito, T. 1964 Amino acid requirements of the silkworm, *Bombyx mori* L. J SERIC SCI TOKYO 33: 107-110.

Arai, N.; Ito, T. 1967 Nutrition of the silkworm, *Bombyx mori.* XVI. Quantitative requirements of essential amino acids. BULL SERIC EXP STN JPN 21: 373-384. (In Japanese)

Arevad, K. 1963 Various investigations on houseflies. Rearing houseflies on agar medium. ANNU REP DAN PEST INFESTATION LAB (1959-1960) pp. 53-54.

Arevad, K. 1974 Juvenile hormone mimics against housefly larvae. ANNU REP DAN PEST INFESTATION LAB (1973-1974) pp. 42-43.

Argauer, R.J.; Gilliam, M. 1974 A fluorometric method for determining oxytetracycline in treated colonies of the honey bee, *Apis melifera.* J INVERT PATHOL 23: 51-54.

*Arnold, M.-L. 1975 Zur Sexuales - Produktion von *Dysaphis plantaginea* bei Haltung auf Künstlichem Diät (Homoptera: Aphididae). ENTOMOL GERM 1: 253-257. (In German with English summary)

Arthur, A.P.; Coppel, H.C. 1953 Studies on dipterous parasites of the spruce budworm, *Choristoneura fumiferana* (Clem.) (Lepidoptera: Tortricidae). I. *Sarcophaga aldrichi* Park (Diptera: Sarcophagidae). CAN J ZOOL 31: 374-391.

Arthur, A.P.; Hegdekar, B.M.; Batsch, W.W. 1972 A chemically defined, synthetic medium that induces oviposition in the parasite *Itoplectus conquisitor* (Say) (Hymenoptera: Ichneumonidae). CAN ENTOMOL 104: 1251-1258.

Arthur, D.R. 1965 Feeding in ectoparasitic acari with special reference to ticks. ADV PARASITOL 3: 249-295.

Asahina, S. 1964 Food material and feeding procedures for mosquito larvae. BULL W H O 31: 465-466.

Ascher, K.R.S.; Levinson, Z.H. 1956 The influence of protein addition to the larval diet on oviposition of the housefly. RIV PARASSITOL 17: 217-222.

*Ashby, G.J. 1972a The House Cricket. pp. 588-589. *in* Universities Federation for Animal Welfare (Ed.), The UFAW Handbook on the Care and Management of Laboratory Animals. Fourth Edition, Churchill Livingstone, Edinburgh and London.

Ashby, G.J. 1972b Locusts. pp. 582-587. *in* Universities Federation for Animal Welfare (Ed.), The UFAW Handbook on the Care and Management of Laboratory Animals. Fourth Edition, Churchill Livingstone, Edinburgh and London.

*Ashby, G.J. 1972c Blowflies. pp. 554-555. *in* Universities Federation for Animal Welfare (Ed.), The UFAW Handbook on the Care and Management of Laboratory Animals. Fourth Edition, Churchill Livingstone, Edinburgh and London.

*Ashby, G.J. 1972d The Housefly. pp. 549-553. *in* Universities Federation for Animal Welfare (Ed.), The UFAW Handbook on the Care and Management of Laboratory Animals. Fourth Edition, Churchill Livingstone, Edinburgh and London.

*Ashrafi, S.H. 1964 The cultivation and nutritional requirements of *Stomoxys calcitrans*. BULL W H O 31: 519-520.

Atallah, Y.H.; Killebrew, R. 1967 Ecological and nutritional studies on *Coleomegilla maculata* (Coleoptera: Coccinellidae). IV. Amino acid requirements of the adult determined by the use of C^{14}-labelled acetate. ANN ENTOMOL SOC AM 60: 186-188.

*Atallah, Y.H.; Newsom, L.D. 1966 Ecological and nutritional studies on *Coleomegilla maculata* De Geer (Coleoptera: Coccinellidae). I. The development of an artificial diet and a laboratory rearing technique. J ECON ENTOMOL 59: 1173-1179.

Atkin, E.E.; Bacot, A. 1917 The relation between the hatching of the eggs and the development of the larvae of *Stegomyia fasciata* (*Aedes calopus*), and the presence of bacteria and yeast. PARASITOLOGY 9: 482-536.

**Auclair, J.L. 1963 Aphid feeding and nutrition. ANNU REV ENTOMOL 8: 439-490.

*Auclair, J.L. 1965 Feeding and nutrition of the pea aphid, *Acyrthosiphon pisum* (Homoptera: Aphidae), on chemically defined diets of various pH and nutrient levels. ANN ENTOMOL SOC AM 58: 855-875.

*Auclair, J.L. 1967a Effects of pH and sucrose on rearing the cotton aphid, *Aphis gossypii*, on a germ-free and holidic diet. J INSECT PHYSIOL 13: 431-446.

Auclair, J.L. 1967b Effects of light and sugars on rearing the cotton aphid, *Aphis gossypii*, on a germ-free and holidic diet. J INSECT PHYSIOL 13: 1247-1268.

**Auclair, J.L. 1969 Nutrition of plant sucking insects on chemically defined diets. ENTOMOL EXP APPL 12: 623-641.

Auclair, J.L.; Cartier, J.J. 1963 Pea aphid: rearing on a chemically defined diet. SCIENCE 142: 1068-1069.

Auclair, J.L.; Cartier, J.J. 1964 L'élevage du puceron du pois, *Acyrthosiphon pisum* (Harr.), sur un régime nutritif de composition chimique connue. ANN SOC ENTOMOL QUE 10: 68-72. (In French)

Auclair, J.L.; Raulston, J.R. 1966 Feeding of *Lygus hesperus* (Hemiptera: Miridae) on a chemically defined diet. ANN ENTOMOL SOC AM 59: 1016-1017.

Auclair, J.L.; Srivastava, P.N. 1972 Some mineral requirements of the pea aphid, *Acyrthosiphon pisum* (Homoptera: Aphididae). CAN ENTOMOL 104: 927-936.

Auclair, J.L.; Srivastava, S.T.; Srivastava, P.N. 1973 Utilization of sugars by the large milkweed bug, *Oncopeltus fasciatus* (Hemiptera: Lygaeidae) maintained on chemically defined diets. ENTOMOL EXP APPL 16: 525-540.

**Audy, J.R.; Lavoipierre, M.M.J. 1964 The laboratory rearing of parasitic Acarina. BULL W H O 31: 583-586.

*Awadallah, A.; Fares, F. 1973 A low-cost artificial medium for the mass rearing of the Mediterranean fruit fly, *Ceratitis capitata* (Wied.). AGRIC RES REV 51: 63-67.

Bacot, A.W.; Harden, A. 1922 Vitamin requirements of *Drosophila*. I. Vitamins B and C. BIOCHEM J 16: 148-152.

Bailey, D.L. 1964 Development of techniques for mass production of lepidopterous larvae for use in connection with sorghum insect resistance studies. DISS ABSTR 26: 1253-1254.

*Bailey, D.L.; Chada, H.L. 1968 Effects of natural (sorghum) and artificial (wheat germ) diets on development of the corn earworm, fall armyworm, and southwestern corn borer. J ECON ENTOMOL 61: 257-260.

*Bailey, D.L.; Whitfield, T.L.; LaBrecque, G.C. 1975 Laboratory biology and techniques for mass production of the stable fly, *Stomoxys calcitrans* (L.) (Diptera: Muscidae). J MED ENTOMOL 12: 189-193.

*Bailey, P. 1973 Respiration and nutrition of radiation-sterilized female *Dacus cucumis* (Diptera: Tephritidae). ENTOMOL EXP APPL 16: 433-444.

*Bailey, P.; Shipp, E. 1970 Corpus allatum size and ovarian environment in irradiated cucumber fly, *Dacus cucumis*. J INSECT PHYSIOL 16: 1293-1299.

Baker, H. 1964 Mating problems as related to the establishment and maintenance of laboratory colonies of mosquitoes. BULL W H O 31: 467-468.

*Baker, J.E. 1974a Influence of nutrients on the utilization of woollen fabric as a food for larvae of *Attagenus megatoma* (F.) (Coleoptera: Dermestidae). J STORED PROD RES 10: 155-160.

Baker, J.E. 1974b Differential net food utilization by larvae of *Sitophilus oryzae* and *Sitophilus granarius*. J INSECT PHYSIOL 20: 1937-1942.

*Baker, J.E. 1974c Differential sterol utilization by larvae of *Sitophilus oryzae* and *Sitophilus granarius*. ANN ENTOMOL SOC AM 67: 591-594.

*Baker, J.E. 1975a Protein utilization by larvae of the black carpet beetle, *Attagenus megatoma*. J INSECT PHYSIOL 21: 613-621.

*Baker, J.E. 1975b Vitamin requirements of larvae of *Sitophilus oryzae*. J INSECT PHYSIOL 21: 1337-1342.

Baker, J.E.; Mabie, J.M. 1973a Growth responses of larvae of *Sitophilus granarius* (Coleoptera: Curculionidae) on a meridic diet. ANN ENTOMOL SOC AM 66: 681-683.

*Baker, J.E.; Mabie, J.M. 1973b Growth and development of larvae of the granary weevil, *Sitophilus granarius* (Coleoptera: Curculionidae), on natural and meridic diets. CAN ENTOMOL 105: 249-256.

*Baker, J.E.; Mabie, J.M. 1973c Growth responses of larvae of the rice weevil, maize weevil, and granary weevil on a meridic diet. J ECON ENTOMOL 66: 723-726.

*Baker, J.E.; Schwalbe, C.P. 1975 Food utilization by larvae of the furniture carpet beetle, *Anthrenus flavipes*. ENTOMOL EXP APPL 18: 213-219.

*Balázs, A. 1958 Nutritional and nervous factors in the adaptation of *Galleria mellonella* to artificial diet. ACTA BIOL ACAD SCI HUNG 9: 47-69.

*Balboni, E.R. 1964 Influence of chloramphenicol on some aspects of the metabolism of the parasitoid *Encarsia formosa* Gahan. J INSECT PHYSIOL 10: 887-896.

*Balogun, R.A. 1969 Rearing of *Ips cembrae* larvae on artificial diets: some essential amino acids and sugars. J INSECT PHYSIOL 15: 141-148.

*Banegas, A.D. 1968 Progress Report on the rearing of torsalo larvae *Dermatobia hominis* (Diptera: Cuterebridae). pp. 7-18. *in* Radiation, Radioisotopes and Rearing Methods in the Control of Insect Pests. Proceedings of a Panel Organized by the Joint FAO/IAEA Division of Atomic Energy in Food and Agriculture.

**Banks, C.J. 1965 Aphid nutrition and reproduction. ROTHAMSTED EXP STN REP (1964): 299-309.

*Bardner, R.; Kenten, J. 1957 Notes on laboratory rearing and biology of the wheat bulb fly, *Leptohylemyia coarctata* (Fall.). BULL ENTOMOL RES 48: 821-831.

Barlow, J.S. 1965 Effects of diet on the composition of body fat in *Agria affinis* (Fallén). CAN J ZOOL 43: 337-340.

Barlow, J.S.; House, H.L. 1956 Ethylene oxide for sterilizing diets. SCIENCE 123: 229.

Barlow, J.S.; House, H.L. 1960 Effects of dietary glucose on haemolymph carbohydrates of *Agria affinis* (Fall.). J INSECT PHYSIOL 5: 181-189.

*Barlow, J.S.; Kollberg, S. 1971 An improved chemically defined diet for *Lucilia sericata* (Diptera: Calliphoridae). CAN ENTOMOL 103: 1341-1345.

*Barnes, E.E.; Bass, M.H. 1973 Whitefringed beetles: an artificial diet for adults. J ECON ENTOMOL 66: 581.

*Barnes, B.N. 1976 Mass rearing the Natal fruit-fly *Pterandrus rosa* (Ksh.) (Diptera: Trypetidae). J ENTOMOL SOC S AFR 39: 121-124.

*Barnes, F.J.; Goodfellow, R.D. 1971 Mevalonate kinase: localization and variation in activity during the development of *Sarcophaga bullata*. J INSECT PHYSIOL 17: 1415-1427.

Barnett, J.W.; Berger, R.S. 1970a Dietary fats and fatty-acid composition of bollworms as factors influencing insecticide susceptibility. J ECON ENTOMOL 63: 1430-1433.

Barnett, J.W.; Berger, R.S. 1970b Growth and fatty acid composition of bollworms, *Heliothis zea* (Lepidoptera: Noctuidae), as affected by dietary fats. ANN ENTOMOL SOC AM 63: 917-924.

*Barras, S.J.; Norris, D.M. Jr 1965 *In vitro* establishment and development of *Eucosma* sp. larvae from cones of *Pinus resinosa* on an artificial nutrient medium. J ECON ENTOMOL 58: 1033-1034.

Bartell, R.J.; Shorey, H.H. 1969 A quantitative bioassay for the sex pheromone of *Epiphyas postvittana* (Lepidoptera) and factors limiting male responsiveness. J INSECT PHYSIOL 15: 33-40.

*Barton-Browne, L. 1956 The effect of light on the fecundity of the Queensland fruit-fly, *Strumeta tryoni* (Frogg.). AUST J ZOOL 4: 125-145.

*Barton-Browne, L. 1975 Regulatory mechanisms in insect feeding. ADV INSECT PHYSIOL 11: 1-116.

*Basden, E.B. 1947 Breeding the housefly (*Musca domestica* L.) in the laboratory. I. Introduction and technique. BULL ENTOMOL RES 37: 381-387.

Bass, M.H.; Barnes, E.E. 1969a Toxicities of antimicrobial agents to white-fringed beetle larvae and the effectiveness of certain of these agents against microbial growth. J ECON ENTOMOL 62: 718-719.

Bass, M.H.; Barnes, E.E. 1969b A laboratory rearing technique for the white-fringed beetle. J ECON ENTOMOL 62: 1512-1513.

Batiste, W.C.; Olson, W.H. 1973 Codling moth: mass production in controlled - environment rearing units. J ECON ENTOMOL 66: 383-388.

Bauer, J.W.; Monroe, R.E. 1969 Metabolism of cholesterol-4-^{14}C in insecticide-resistant and susceptible strains of aseptically reared house flies, *Musca domestica*. ANN ENTOMOL SOC AM 62: 1021-1025.

Baumberger, J.P. 1917 The food of *Drosophila melanogaster* Meigen. PROC NATL ACAD SCI U S A 3: 122-126.

**Baumberger, J.P. 1919 A nutritional study of insects with special reference to microorganisms and their substrate. J EXP ZOOL 28: 1-81.

Baumberger, J.P.; Glaser, R.W. 1917 The rearing of *Drosophila ampelophila* Loew on solid media. SCIENCE 45: 21-22.

*Baumhover, A.H.; Husman, C.N.; Graham, A.J. 1966 Screw-worms. pp. 533-554. *in* Smith, C.N. (Ed.), Insect Colonization and Mass Production. Academic Press, New York and London.

Beard, R.L. 1958 Secondary physiological effects of DDT in *Galleria* larvae. ENTOMOL EXP APPL 1: 260-267.

Beck, S.D. 1950 Nutrition of the European corn borer, *Pyrausta nubilalis* (Hbn). II. Some effects of diet on larval growth characteristics. PHYSIOL ZOOL 23: 353-361.

Beck, S.D. 1953 Nutrition of the European corn borer, *Pyrausta nubilalis* (Hbn.). III. An unidentified dietary factor required for larval growth. J GEN PHYSIOL 36: 317-325.

Beck, S.D. 1956a A biomodal response to dietary sugars by an insect. BIOL BULL 110: 219-228.

Beck, S.D. 1956b The European corn borer, *Pyrausta nubilalis* (Hubn.) and its prepupal host plant. I. Orientation and feeding behaviour of the larva of the corn plant. ANN ENTOMOL SOC AM 49: 552-558.

Beck, S.D. 1956c The European corn borer, *Pyrausta nubilalis* (Hubn.), and its principal host plant. II. The influence of nutritional factors on larval establishment and development on the corn plant. ANN ENTOMOL SOC AM 49: 582-588.

Beck, S.D. 1956d Nutrition of the European corn borer, *Pyrausta nubilalis* (Hbn.). IV. Feeding reactions of first instar larvae. ANN ENTOMOL SOC AM 49: 399-405.

Beck, S.D. 1960a The European corn borer, *Pyrausta nubilalis* (Hubn.), and its principal host plant. VII. Larval feeding behaviour and host plant resistance. ANN ENTOMOL SOC AM 53: 206-212.

*Beck, S.D. 1960b Growth and development of the greater wax moth, *Galleria mellonella* (L.) (Lepidoptera: Galleriidae). TRANS WIS ACAD SCI ARTS LETT 49: 137-148.

*Beck, S.D. 1971 Growth and retrogression in larvae of *Trogoderma glabrum* (Coleoptera: Dermestidae). I. Characteristics under feeding and starvation conditions. ANN ENTOMOL SOC AM 64: 149-155.

**Beck, S.D.; Chippendale, G.M. 1968 Environmental and behavioural aspects of the mass rearing of plant-feeding lepidopterans. pp. 19-30. *in* Radiation, Radioisotopes and Rearing Methods in the Control of Insect Pests. Proceedings of a Panel Organized by the Joint FAO/IAEA Division of Atomic Energy in Food and Agriculture.

*Beck, S.D.; Chippendale, G.M.; Swinton, D.E. 1968 Nutrition of the European corn borer, *Ostrinia nubilalis*. VI. A larval rearing medium without crude plant fractions. ANN ENTOMOL SOC AM 61: 459-462.

*Beck, S.D.; Edwards, C.A.; Medler, J.T. 1958 Feeding and nutrition of the milkweed bug, *Oncopeltus fasciatus*. ANN ENTOMOL SOC AM 51: 283-288.

Beck, S.D.; Hanec, W. 1958 Effect of amino acids on feeding behaviour of the European corn borer, *Pyrausta nubilalis* (Hubn.). J INSECT PHYSIOL 2: 85-96.

Beck, S.D.; Kapadia, G.G. 1957 Insect nutrition and metabolism of sterols. SCIENCE 126: 258-259.

*Beck, S.D.; Lilly, J.H.; Stauffer, J.F. 1949 Nutrition of the European corn borer, *Pyrausta nubilalis* (Hbn.). I. Development of a satisfactory purified diet for larval growth. ANN ENTOMOL SOC AM 42: 483-496.

Beck, S.D.; Smissman, E.E. 1960 The European corn borer, *Pyrausta nubilalis*, and its principal host plant. VIII. Laboratory evaluation of host resistance to larval growth and survival. ANN ENTOMOL SOC AM 53: 755-762.

Beck, S.D.; Smissman, E.E. 1961 The European corn borer, *Pyrausta nubilalis*, and its principal host plant. IX. Biological activity of chemical analogues of corn resistance factor A (6-methoxybenzoxazolinone). ANN ENTOMOL SOC AM 54: 53-61.

*Beck, S.D.; Stauffer, J.F. 1950 An aseptic method for rearing European corn borer larvae. J ECON ENTOMOL 43: 4-6.

Becker, G. 1965 Über haltung von termiten im laboratorium. Z ANGEW ZOOL 52: 385-398. (In German with English summary)

Becker, P. 1958 The behaviour of larvae of *Culicoides circumscriptus* Kieff. (Dipt., Ceratopogonidae) towards light stimuli as influenced by feeding, with observations on the feeding habits. BULL ENTOMOL RES 49: 785-802.

*Beckman, H.F.; Bruckart, S.M.; Reiser, R. 1953 Laboratory culture of the pink bollworm on chemically defined media. J ECON ENTOMOL 46: 627-630.

*Becton, A.J.; George, B.W.; Brindley, T.A. 1962 Continuous rearing of European corn borer larvae on artificial medium. IOWA STATE COLL J SCI 37: 163-172.

*Bedard, W.D. 1966 A ground phloem medium for rearing immature bark beetles (Scolytidae). ANN ENTOMOL SOC AM 59: 931-938.

Begg, M. 1949 Nutritional requirements of *Drosophila*. NATURE LONDON 163: 881.

*Begg, M. 1956 Some observations on water-soluble factors which accelerate the growth in aseptic culture of larvae of *Drosophila melanogaster*. J EXP BIOL 33: 142-154.

Begg, M.; Robertson, F.W. 1948 Nutritional requirements of *Drosophila melanogaster*. NATURE LONDON 161: 769-770.

Begg, M.; Robertson, F.W. 1949 The nutritional requirements of *Drosophila melanogaster*. J EXP BIOL 26: 380-387.

Begg, M.; Sang, J.H. 1950 A method for collecting and sterilizing large numbers of *Drosophila* eggs. SCIENCE 112: 11-12.

Bell, C.H. 1975 Effects of temperature and humidity on development of four pyralid moth pests of stored products. J STORED PROD RES 11: 167-175.

Bell, R.A.; Joachim, F.G. 1976 Techniques for rearing laboratory colonies of tobacco hornworms and pink bollworms. ANN ENTOMOL SOC AM 69: 365-373.

Benschoter, C.A. 1967 Effect of dietary biotin on reproduction of the house fly. J ECON ENTOMOL 60: 1326-1328.

Benschoter, C.A.; Paniagua, G.R. 1966 Reproduction and longevity of Mexican fruit flies fed biotin in the diet. ANN ENTOMOL SOC AM 59: 298-300.

Bentur, J.S.; Mathad, S.B. 1974 Effects of surplus dietary biotin on the egg production of the cricket, *Plebeiogryllus guttiventris* Walker (Orth. Gryllidae). KARNATAK UNIV J SCI 19: 3-6.

Benson, J.; Oberlander, H. 1974 Protein synthesis in wing discs of *Galleria mellonella*: effects of ecdysone and fat body *in vitro*. INSECT BIOCHEM 4: 423-428.

Beratlief, C.; Boguleau, G. 1974 Experiments with rearing the fall webworm (*Hyphantria cunea* Drury) on artificial media. AN INST CERCET PROT PLANT 10: 302-307. (In Rumanian with English summary)

*Berger, R.S. 1963 Laboratory techniques for rearing *Heliothis* species on artificial medium. U S DEP AGRIC PRESENTATION PAP ARS 33-84, 4 pp.

Bergman, E.D.; Levinson, Z.H,; Mechoulam, R. 1958 The toxicity of *Veratrum* and *Solanum* alkaloids to housefly larvae. J INSECT PHYSIOL 2: 162-177.

*Berrios, F.; Hidalgo-Salvatierra, O. 1971 Estudios sobre el barrenador *Hypsipyla grandella* Zeller. VI. Susceptibilidad de la larva al hongo *Metarrhizium anisopliae* (Metch). TURRIALBA 21: 214-219. (In Spanish)

*Bhatkar, A.; Whitcomb, W.H. 1970 Artificial diet for rearing various species of ants. FLA ENTOMOL 53: 229-232.

Bhattacharya, A.K.; Pant, N.C. 1968 Dietary efficiency of natural, semi-synthetic and synthetic diets with special reference to qualitative amino acid requirements of khapra beetle, *Trogoderma granarium* Everts (Coleoptera: Dermestidae). J STORED PROD RES 4: 249-257.

Bieber, L.L.; Monroe, R.E. 1969 The relation of carnitine to the formation of phosphatidyl-β-methylcholine by *Tenebrio molitor* L. larvae. LIPIDS 4: 293-298.

*Biever, K.D. 1971 Effect of diet and competition in laboratory rearing of chironomid midges. ANN ENTOMOL SOC AM 64: 1165-1169.

*Biever, K.D.; Boldt, P.E. 1971 Continuous laboratory rearing of the diamond back moth and related biological data. ANN ENTOMOL SOC AM 64: 651-655.

Blankenship, J.W.; Miller, G.J. 1971 *In vivo* synthesis of phosphatidyl choline and ethanolamine in larvae of the dried fruit moth. J INSECT PHYSIOL 17: 2061-2067.

Blewett, M.; Fraenkel, G. 1944 Intracellular symbiosis and vitamin requirements of two insects, *Lasioderma serricorne* and *Sitoderpa panicea*. PROC R SOC LONDON SER B 132: 212.

Bloch, K.; Langdon, R.G.; Clark, A.J.; Fraenkel, G. 1956. Impaired steroid biogenesis in insect larvae. BIOCHIM BIOPHYS ACTA 21: 176.

Bode, W.M. 1971 The codling moth *Laspeyresia pomonella* (Lepidoptera: Olethreutidae): effects of an introduced genalasis virus on a field population and laboratory rearing on artificial diets. DISS ABSTR INT B 32: 348.

**Bodenstein, O.F.; Fales, J.H. 1969 A list of references on the rearing of cockroaches for experimental purposes. PEST CONTROL 37: 30, 32, 34.

Bodnaryk, R.P. 1968 Dietary amino acid and nucleic acid imbalance leading to altered tumour frequency in the adult housefly, *Musca domestica*. J INSECT PHYSIOL 14: 223-242.

*Bogdanow, E.A. 1908 Über das Abhängigkeit des Wachstums der Fliegenlarven von Bakterien und Fermenten und über Variabilität und Vererbung bei den Fleischfliegen. ARCH ANAT PHYSIOL ABT SUPPL: 173-200.

*Boles, H.P.; Marzke, F.O. 1966 Lepidoptera infesting stored products. pp. 259-270. *in* Smith, C.N. (Ed.), Insect Colonization and Mass Production. Academic Press, New York and London.

Boller, E.F. 1968 An artificial oviposition device for the European cherry fruit fly, *Rhagoletis cerasi*. J ECON ENTOMOL 61: 850-852.

**Boller, E.F. 1972 Behavioural aspects of mass-rearing of insects. ENTOMOPHAGA 17: 9-25.

*Boller, E.F.; Remund, U. 1975 Application of sit on the European cherry fruit fly, *Rhagoletis cerasi* L. in northwest Switzerland. pp. 77-82. *in* Controlling Fruit Flies by the Sterile-insect Technique. Proceedings of a Panel Organized by the Joint FAO/IAEA Division of Atomic Energy in Food and Agriculture.

*Bond, E.J.; Monro, H.A. 1954 Rearing the cadelle *Tenebroides mauritanicus* (L.) as a test insect for insecticidal research. CAN ENTOMOL 86: 402-408.

**Boness, M. 1968 Insektenzuchten auf Künstlichen Futter. UMSCH WISS TECH 18/68: 562. (In German)

**Boness, M. 1969 Insektenzuchten auf Künstlichen Medien. ANZ SCHAEOLINGSKD PFLANZENSCHUTZ 42: 26-30. (In German)

**Boness, M. 1970 Nahrung und Ernährung der Insekten, ihre Erforschung und praktische Bedeutung. Z ANGEW ENTOMOL 65: 223-230. (In German with English summary)

Bonner, J.T. 1947 Evidence for the formation of cell aggregates by chemotaxis in the development of the slime mould, *Dictyostelium discoideum*. J EXP ZOOL 106: 1-26.

*Boorman, J. 1967 Aseptic rearing of *Aedes aegypti* Linn. NATURE (LONDON) 213: 197-198.

*Boorman, J. 1974 The maintenance of laboratory colonies of *Culicoides variipennis* (Coq.), *C. nubeculosus* (Mg.) and *C. riethi* Kieff. (Diptera, Ceratopogonidae). BULL ENTOMOL RES 64: 371-377.

*Borisova, A.E. 1966 Rearing *Locusta migratoria manilensis* Mey and *Locusta migratoria migratoria* L. on semi-synthetic media. ZOOL ZH 45: 858-864. (In Russian with English summary)

*Borisova, A.E. 1969 The role of carbohydrates in the fertility of the Asian migratory locust, *Locusta migratoria migratoria* L. (Orthoptera, Acrididae), when reared on semisynthetic media. ENTOMOL REV USSR 48: 141-144.

Borisova, A.E. 1972 A semisynthetic medium for rearing the codling moth. BYULL VSES NAUCHNO ISSLED INST ZASHCH RAST 23: 9-12. (In Russian)

Borisova, A.E. 1973a Search for media for mass rearing of the codling moth. pp. 136-137. *in* Papers of a Conference on Progressive Methods of Pest Control in Agricultural Crops, Moscow. (In Russian)

Borisova, A.E. 1973b Studies on diets for the mass rearing of codling moth. *in* Papers of a Conference on Progressive Methods of Pest Control in Agricultural Crops, Moscow. (In Russian)

*Bot, J. 1965 An aseptic artificial rearing medium for the false codling moth, *Argyroploce leucotreta* Meyr. S AFR J AGRIC SCI 8: 1075-1078.

*Bot, J. 1966a An artificial diet for the plusia looper, *Plusia acuta* Walker. S AFR J AGRIC SCI 9: 67-70.

*Bot, J. 1966b Rearing *Heliothis armigera* Hubn. and *Prodenia litura* F. on an artificial diet. S AFR J AGRIC SCI 9: 535-538.

*Bot, J. 1967a An artificial rearing medium for three noctuids of economic importance belonging to the genus *Spodoptera* (Lepidoptera). J ENTOMOL SOC SOUTH AFR 29: 157-160.

*Bot, J. 1967b Rearing three species of fruit-sucking moths on artificial diets. S AFR J AGRIC SCI 10: 1009-1114.

*Bot, J.; Meyer, M.K.P. 1967 An artificial rearing medium for acarid mites. J ENTOMOL SOC SOUTH AFR 29: 199.

*Bottger, G.T. 1942 Development of synthetic food media for use in nutrition studies of the European corn borer. J AGRIC RES 65: 493-500.

Bouletreau, M. 1968 Premiers résultats de l'élevage des larves d'un hyménoptère chalcidien (*Pteromalus puparum* L.) sur hémolymphe de lépidoptère. ENTOMOPHAGA 13: 217-222. (In French with English summary)

*Boush, G.M.; Baerwald, R.J.; Miyazaki, S. 1969 Development of a chemically defined diet for adults of the apple maggot based on amino acid analysis of honeydew. ANN ENTOMOL SOC AM 62: 19-21.

Boush, G.M.; Dunkel, F.V.; Burkholder, W.E. 1968 Progeny suppression of *Attagenus megatoma* and *Trogoderma parabile* by dietary factors. J ECON ENTOMOL 61: 644-646.

*Bowling, C.C. 1967 Rearing of two lepidopterous pests of rice on a common artificial diet. ANN ENTOMOL SOC AM 60: 1215-1216.

*Bracken, G.K. 1965 Effects of dietary components on fecundity of the parasitoid *Exeristes comstockii* (Cress.) (Hymenoptera: Ichneumonidae). CAN ENTOMOL 97: 1037-1041.

Bracken, G.K. 1966 Role of ten dietary vitamins on fecundity of the parasitoid *Exeristes comstockii* (Cresson) (Hymenoptera: Ichneumonidae). CAN ENTOMOL 98: 918-922.

*Bracken, G.K. 1969 Effects of dietary amino acids, salts, and protein starvation on fecundity of the parasitoid *Exeristes comstockii* (Hymenoptera: Ichneumonidae). CAN ENTOMOL 101: 91-96.

*Branson, T.F.; Guss, P.L.; Krysan, J.L.; Sutter, G.R. 1975 Corn rootworms: laboratory rearing and manipulation. U S DEP AGRIC AGRIC RES SERV ARS-NC-28, 17 pp.

*Brazzel, J.R. 1970 Second conference on test methods for resistance in insects of agricultural importance. Standard method for detection of insecticide resistance in *Heliothis zea* (Boddie) and *H. virescens* (F.). Tentative methods for detection in *Diabrotica* and *Hypera*. BULL ENTOMOL SOC AM 16: 147-153.

Brazzel, J.R.; Chambers, H.; Hamman, P.J. 1961 A laboratory rearing method and dosage-mortality data on the boll worm *Heliothis zea*. J ECON ENTOMOL 54: 949-952.

Brazzel, J.R.; Davich, T.B.; Raven, K. 1959 Rearing boll weevils on an artificial diet. TEX AGRIC EXP STN MISC PUBL 353.

*Bready, J.K.; Friedman, S. 1963a The nutritional requirement of termites in axenic cultures. 1. Sterilization of eggs of *Reticulitermes flavipes* and the requirements of first instar nymphs. ANN ENTOMOL SOC AM 56: 703-706.

*Bready, J.K.; Friedman, S. 1963b The nutritional requirements of termites in axenic cultures. 2. Studies on the effectiveness of antibiotics in the sterilization of workers of *Reticulitermes flavipes*. ANN ENTOMOL SOC AM 56: 706-708.

Brewer, F.D.; Martin, D.F. 1976 Substitutes for agar in a wheat germ diet used to rear the corn earworm and the sugarcane borer. ANN ENTOMOL SOC AM 69: 255-256.

Brewer, F.D.; Tidwell, A.L. Jr 1975 Supplemental B-vitamins necessary to rear *Heliothis zea* on a wheat-germ based diet. ANN ENTOMOL SOC AM 68: 365-366.

*Brickhill, D.D. 1958 Biological studies of the species of tydeid mites from California. HILGARDIA 27: 601-620.

Bridges, C.B. 1932 Apparatus and methods for *Drosophila* culture. AM NAT 66: 250-273.

Bridges, C.B.; Darby, H.H. 1933 Culture media for *Drosophila* and the pH of media. Part I. Improvements in culture methods: the role of yeast. AM NAT 67: 437-471.

Bridges, R.G. 1972 Choline metabolism in insects. ADV INSECT PHYSIOL 9: 51-110.

Bridges, R.G.; Ricketts, J.; Cox, J.T. 1965 The replacement of lipid-bound choline by other bases in the phospholipids of the housefly. J INSECT PHYSIOL 11: 225-236.

Brinton, F.E.; Proverbs, M.D.; Carty, B.E. 1969 Artificial diet for mass production of the codling moth, *Carpocapsa pomonella* (Lepidoptera: Olethreutidae). CAN ENTOMOL 101: 577-584.

Bronnimann, H. 1964 Rearing anthocorids on an artificial medium. COMMONW INST BIOL CONTROL TECH BULL 4: 147-150.

Bronskill, J.F. 1962 A cage to simplify the rearing of the greater wax moth, *Galleria mellonella* (Pyralidae). J LEPID SOC 15: 102-104.

Bronskill, J.F.; House, H.L. 1957 Notes on rearing a pupal endoparasite, *Pimpla turionellae* (L.) (Hymenoptera: Ichneumonidae), on unnatural food. CAN ENTOMOL 89: 483.

*Brookes, V.J.; Fraenkel, G. 1958 The nutrition of larvae of the housefly, *Musca domestica* L. PHYSIOL ZOOL 31: 208-223.

Brooks, M.A. 1960 Some dietary factors that affect ovarial transmission of symbiotes. PROC HELMINTHOL SOC WASH 27: 212-220.

*Brooks, M.A. 1964 Symbiotes and the nutrition of medically important insects. BULL W H O 31: 555-559.

Brooks, M.A.; Kringen, W.B. 1972 Polypeptides and proteins as growth factors for aposymbiotic *Blattella germanica* (L.). pp. 353-374. *in* Rodriguez, J.G. (Ed.), Insect and Mite Nutrition. North-Holland Publishing Company, Amsterdam and London.

Brooks, M.A.; Richards, A.G. 1955 Intracellular symbiosis in cockroaches. I. Production of aposymbiotic cockroaches. BIOL BULL 108: 22-39.

Brooks, M.A.; Richards, A.G. 1956 Intracellular symbiosis in cockroaches. III. Re-infestation of aposymbiotic cockroaches with symbiotes. J EXP ZOOL 132: 447-465.

*Brown, J.H.; Chippendale, G.M. 1975 Survival of the adult maize weevil, *Sitophilus zeamais:* role of nutrients larval reserves and symbionts. COMP BIOCHEM PHYSIOL 50A: 83-90.

*Bruce, W.A.; McCray, T.C. 1975 A simple method of dispensing eggs of *Plodia interpunctella.* J GA ENTOMOL SOC 10: 74-75.

*Bruce, W.N. 1948 Studies on the biological requirements of the cat flea. ANN ENTOMOL SOC AM 41: 346-352.

*Bruen, J.P.; Goldberg, L.J.; Tanabe, A.M.; Watkins, H.M.S.; Ford, I.J. 1976 Quantitative rearing of small numbers of mosquito larvae. MOSQ NEWS 36: 95-97.

**Brues, C.T. 1946 Insect Dietary; an account of the food habits of insects. Harvard University Press, Cambridge Mass 466 pp.

*Brust, M.; Fraenkel, G. 1955 The nutritional requirements of the larvae of a blowfly, *Phormia regina* (Meig.). PHYSIOL ZOOL 28: 186-204.

Brust, R.A. 1971 Laboratory mating in *Aedes dianteus* and *Aedes communis* (Diptera: Culicidae). ANN ENTOMOL SOC AM 64: 243-247.

Bryan, D.E.; Jackson, C.G.; Stoner, A. 1969 Rearing cotton insect parasites in the laboratory. U S DEP AGRIC PROD RES REP 109, 13 pp.

Bryant, P.J.; Sang, J.H. 1969 Physiological genetics of melanotic tumors in *Drosophila melanogaster.* VI. The tumorigenic effects of juvenile hormone-like substances. GENETICS 62: 321-336.

**Büchner, P. 1953 Endosymbiose der Tiere unit pflanzlichen Mikroorganismen. Birkhäuser, Basel. (In German)

*Buddington, A.R. 1941 The nutrition of mosquito larvae. J ECON ENTOMOL 34: 275-281.

Budowski, P.; Ishaaya, I.; Katz, M. 1967 Growth inhibition of *Dermestes maculatus* by phytosterols. J NUTR 91: 201-207.

Bull, D.L.; Adkisson, P.L. 1962 Fat content of the larval diet as a factor influencing diapause and growth-rate of the pink bollworm. ANN ENTOMOL SOC AM 55: 499-502.

Bullock, H.R.; Mangum, C.L.; Guerra, A.A. 1969 Treatment of eggs of the pink bollworm *Pectinophora gossypiella,* with formaldehyde to prevent infection with a cytoplasmic polyhedrosis virus. J INVERTEBR PATHOL 14: 271-273.

Burchfield, H.P.; Hilchey, J.D.; Storrs, E.E. 1952 An objective method for insecticide bioassay based on photo-migration of mosquito larvae. CONTRIB BOYCE THOMSON INST 17: 57-86.

Burchfield, H.P.; Redder, A.M.; Storrs, E.E.; Hilchey, J.D. 1953 Improved methods for rearing larvae of *Aedes aegypti* (L.) for use in insecticide bioassay. CONTRIB BOYCE THOMSON INST 17: 317-331.

*Burditt, A.K. Jr; Lopez D, F.; Steiner, L.F.; Windeguth, D.L. von; Baranowski, R.; Anwar, M. 1974 Application of sterilization techniques to *Anastrepha suspensa* Loew in Florida, United States of America. PP. 93-101. *in* Sterility Principle for Insect Control 1974. Proceedings of a Panel Organized by the Joint FAO/IAEA Division of Atomic Energy in Food and Agriculture.

*Burke, H.R.; Martin, D.F. 1956 The biology of three chrysopid predators of the cotton aphid. J ECON ENTOMOL 49: 698-700.

*Burkholder, W.E.; Schwalbe, C.P.; Boush, G.M. 1973 Antimicrobial food additives and their effects on *Trogoderma variabile* and *Attagenus megatoma* (Coleoptera, Dermestidae). J STORED PROD RES 9: 205-211.

Burnet, B.; Sang, J.H. 1963 Dietary utilization of DNA and its derivatives by *Drosophila melanogaster* (Meig.). J INSECT PHYSIOL 9: 553-562.

Burnet, B.; Sang, J.H. 1968 Physiological genetics of melanotic tumors in *Drosophila melanogaster*. V. Amino acid metabolism and tumor formation in the tu bw; st su-tu strain. GENETICS 59: 211-235.

*Burton, R.L. 1967 Mass rearing the fall armyworm in the laboratory. U S DEP AGRIC PRESENTATION PAP 33-117, 12 pp.

*Burton, R.L. 1969 Mass rearing the corn earworm in the laboratory. U S DEP AGRIC PRESENTATION PAP 33-134, 8 pp.

*Burton, R.L. 1970 A low-cost artificial diet for the corn earworm. J ECON ENTOMOL 63: 1969-1970.

Burton, R.L.; Cox, H.C. 1966 An automated packaging machine for lepidopterous larvae. J ECON ENTOMOL 59: 907-909.

Burton, R.L.; Harrell, E.A. 1966 Modification of a lepidopterous larvae dispenser for a packaging machine. J ECON ENTOMOL 59: 1544-1545.

Burton, R.L.; Harrell, E.A.; Cox, H.C.; Hare, W.W. 1966 Devices to facilitate rearing of lepidopterous larvae. J ECON ENTOMOL 59: 594-596.

*Burton, R.L.; Perkins, W.D. 1972 WSB, A new laboratory diet for the corn earworm and the fall armyworm. J ECON ENTOMOL 65: 385-386.

Burton, R.L.; Snow, J.W. 1970 A marker dye for the corn earworm. J ECON ENTOMOL 63: 1976-1977.

**Bush, G.L. 1975 Genetic variation in natural insect populations and its bearing on mass rearing programmes. pp. 9-17. *in* Controlling Fruit Flies by the Sterile-insect Technique. Proceedings of a Panel Organized by the Joint FAO/IAEA Division of Atomic Energy in Food and Agriculture.

*Busvine, J.R. 1955 Simple methods for rearing the cricket (*Gryllus domesticus* L.) with some observations on speed of development at different temperatures. PROC R SOC LONDON SER A 30: 15-18.

Butler, G.D. Jr; Ritchie, P.L. Jr 1971 Feed Wheast and the abundance and fecundity of *Chrysopa carnea*. J ECON ENTOMOL 64: 933-934.

**Butt, B.A. 1974 Survey of synthetic diets for codling moths. pp. 565-578. *in* Sterility Principle for Insect Control. Proceedings of a Panel Organized by the Joint FAO/IAEA Division of Atomic Energy in Food and Agriculture.

*Byers, R.A. 1974 Artificial diets for maintaining the adult clover root borer. J ECON ENTOMOL 67: 806.

Calabrese, E.J.; Stoffolano, J.G. Jr 1974 The influence of age and diet on respiration in adult male and female blowflies, *Phormia regina*. J INSECT PHYSIOL 20: 383-393.

*Campadelli, G. 1973 Allevamento di *Galleria mellonella* L. (Lepidoptera: Galleriidae) con dieta semiartificiale. BOLL IST DEI ENTOMOL UNIV STUDI BOLOGNA 32: 11-25. (In Italian with English summary)

**Campbell, F.L.; Moulten, F.R. 1943 Laboratory procedures in studies of the chemical control of insects. AM ASSOC ADV SCI PUBL 20: 206 pp.

Cantarino, M.L.; Rey, J.M. 1970 Improvements in the artificial rearing of *Dacus oleae*. BOL R SOC ESP HIST NAT SECC BIOL 68: 181-188.

**Cantwell, G.E. (Ed.) 1974 Insect Diseases. Vol I 1-330, Vol II 301-595. Marcel Dekker, Inc, New York.

Cartier, J.J. 1965 Action des radiations orangées sur le puceron du pois, *Acyrthosiphon pisum* (Harr.) et sur le puceron de la pomme de terre, *Macrosiphum euphorbiae* (Thos.) en élevage sur diète synthétique. PHYTOPROTECTION 46: 65-73.

**Cartier, J.J. 1966 Aphid responses to colours in artificial rearings. BULL ENTOMOL SOC AM 12: 378-380.

**Cartier, J.J. 1968 Factors of host plant specificity and artificial diets. BULL ENTOMOL SOC AM 14: 18-21.

Cartier, J.J.; Auclair, J.L. 1964 Pea aphid behaviour: colour preference on a chemical diet. CAN ENTOMOL 96: 1240-1243.

Cartier, J.J.; Auclair, J.L. 1965 Effets des couleurs sur le comportement de diverses races du puceron du pois, *Acyrthosiphon pisum* (Harris), en élevage sur um régime nutritif de composition chimique connue. PROC 12 INT CONGR ENTOMOL LONDON (1964): 414. (In French)

Cartier, J.J.; Morin, St. G. 1965 L'élevage du puceron de la pomme de terre, *Macrosiphum euphorbiae* (Thos.) (Homoptera: Aphididae), sur diète synthétique. ANN SOC ENTOMOL QUE 10: 69-79. (In French)

Castillón, M.P.; Catalán, R.E.; Jimenez, C.; Madariaga, M.A.; Muncio, A.M.; Suarez, A. 1974. Effect of dietary lipids on phospholipid composition of *Ceratitis capitata*. J INSECT PHYSIOL 20: 507-512.

*Cavalloro, R. 1967 Orientamenti sull'allevamento permanente di *Dacus oleae* Gmelin (Diptera, Trypetidae) in laboratorio. REDIA 50: 337-344. (In Italian with English summary)

Cavalloro, R.; Delrio, G. 1973 Incremento della fertilità delle uova di *Dacus oleae* Gmelin negli allevamenti permanenti. (Increase of the fertility of the eggs of *Dacus oleae* Gmelin in permanent cultures). NOTE APPUNTI SPER ENTOMOL AGRAR 14: 1-12. (In Italian with English summary)

Cavalloro, R.; Girolami, V. 1969a Nuove tecniche di allevamento in laboratorio del *Dacus oleae* Gmel. I. Adulti. REDIA 51: 127-152. (In Italian with English summary)

*Cavalloro, R.; Girolami, V. 1969b Miglioramenti nell'allevamento in massa di *Ceratitis capitata* Wiedemann (Diptera, Trypetidae). REDIA 51: 315-327. (In Italian with English summary)

*Cavalloro, R.; Piana, A. 1972 Prove di radiosensibilita' alle radiazioni ionizzanti dei tortricidi del garofano con particolare riguardo ad *Epichoristodes acerbella* (Walker). REDIA 53: 281-302. (In Italian with English summary)

*Cavalloro, R.; Prota, R. 1972 Method di allevamento di *Sesamia nonagrioides* (Lef.) (Lepidoptera, Noctuidae) con diete artificiali. REDIA 53: 395-411. (In Italian with English summary)

Cavanagh, G.G. 1963 The use of the Dadd synthetic diet as a food for adult *Schistocerca gregaria* (Forsk.) and the effects of some additions and modifications to it. J INSECT PHYSIOL 9: 759-775.

Cavicchi, S. 1971 Lo studio della *Ceratitis capitata* in laboratorio. I. Metodi di allevamento. BOLL SOC ENTOMOL ITAL 103: 117-124. (In Italian with English summary).

*Chalfant, R.B. 1975 A simplified technique for rearing the lesser cornstalk borer (Lepidoptera: Phycitidae). J GA ENTOMOL SOC 10: 33-37.

*Chamberlain, W.F. 1970 *In vitro* culture of the common cattle grub for six weeks after hatching. ANN ENTOMOL SOC AM 63: 1465-1466.

**Chambers, D.L. 1975 Quality in mass-produced insects. Definition and Evaluation. pp. 19-32. *in* Controlling Fruit Flies by the Sterile-insect Technique. Proceedings of a Panel Organized by the Joint FAO/IAEA Division of Atomic Energy in Food and Agriculture.

Chambliss, J.E.; Jones, C.M. 1966 Cucurbitacins: specific insect attractants in Cucurbitaceae. SCIENCE 153: 1392-1393.

*Chang, J.T.; Wang, M.Y. 1958 Nutritional requirements of the common house fly, *Musca domestica vicina* Macq. NATURE LONDON 181: 566.

*Chang, V.C.S.; Jensen, L. 1972 A diet for studying clonal resistance of sugarcane to the New Guinea sugarcane weevil. J ECON ENTOMOL 65: 1197-1199.

*Chaniotis, B.N. 1975 A new method for rearing *Lutzomyia trapidoi* (Diptera: Psychodidae), with observations on its development and behaviour in the laboratory. J MED ENTOMOL 12: 183-188.

**Chapman, R.F. 1969a Feeding. pp. 21-37. *in* The Insects: Structure and Function. The English University Press Ltd, London.

**Chapman, R.F. 1969b Nutrition. pp. 70-82. *in* The Insects: Structure and Function. The English University Press Ltd, London.

**Chapman, R.F. 1974 The chemical inhibition of feeding by phytophagous insects: a review. BULL ENTOMOL RES 64: 339-363.

Chapman, R.N. 1924 Nutritional studies on the confused flour beetle, *Tribolium confusum* Duval. J GEN PHYSIOL 6: 565-585.

Charas, C. 1960 L'attractivite exercée par *Fraxinus excelsior* L. a l'egard de *Leperesinus fraxini* Panz. (Coléoptèra: Scolytidae) et les modifications physiologique de la plante-hote. C R ACAD SCI SER D 250: 3872-3874. (In French)

Charbonneau, R.; Lemonde, A. 1960a Unidentified growth factors in brewers' yeast. I. Necessity of these factors for *Tribolium confusum* Duval. CAN J ZOOL 38: 87-90.

Charbonneau, R.; Lemonde, A. 1960b Unidentified growth factors in brewers' yeast. II. Some chemical and physical properties of these factors. CAN J ZOOL 38: 443-448.

*Charpentier, R. 1968 Élevage aseptique d'un Coléoptère coprophage: *Aphodius constans* Duft (Col., Scarabaeidae). ANN EPIPHYT 19: 533-538. (In French with English summary)

Charpentier, R.; Larsson, R.; Olofsson, E. 1972 A method for axenic rearing of the house cricket *Acheta domesticus* L. (Orth.). ENTOMOL SCAND 3: 239-243.

*Chatterji, S.M.; Sharma, G.C.; Siddiqui, K.H.; Panwar, V.P.S.; Young, W.R. 1969 Laboratory rearing of the pink stem borer, *Sesamia inferens* Walker, on artificial diet. INDIAN J ENTOMOL 31: 75-77.

*Chatterji, S.M.; Siddiqui, K.H.; Panwar, V.P.S.; Sharma, G.C.; Young, W.R. 1968 Rearing of the maize stem borer, *Chilo zonellus* Swinhoe on artificial diet. INDIAN J ENTOMOL 30: 8-12.

*Chaudhary, K.D.; Lemonde, A. 1962 Phosphorus in the nutrition of *Tribolium confusum* Duval. CAN J ZOOL 40: 375-380.

Chauthani, A.R.; Adkisson, P.L. 1964 Comparison of two artificial diets for rearing *Heliothis zea* (Boddie) larvae for insecticide tests. J ECON ENTOMOL 58: 1163-1164.

*Chauthani, A.R.; Claussen, D. 1968 Rearing Douglas-fir tussock moth larvae on synthetic media for the production of nuclear-polyhedrosis virus. J ECON ENTOMOL 61: 101-103.

*Chauthani, A.R.; Snideman, M.; Rehnborg, C.S. 1971 Comparison of commercially produced *Bacillus thuringiensis* var. *thuringiensis* with two bioassay techniques based on toxicity units. J ECON ENTOMOL 64: 1291-1293.

**Chauvin, R. 1956 Physiologie de l'Insecte; le comportement, les grandes fonctions, écophysiologie. Masson et cie, Paris, 917 pp. (In French)

**Chauvin, R. 1968a Digestion et nutrition des adultes. pp. 347-377. *in* Chauvin, R. (Ed.), Traité de Biologie de l'Abeille. Masson et cie, Paris. (In French)

**Chauvin, R. (Ed.). 1968b Traité de Biologie de l'Abeille. Masson et cie, Paris, 547 pp. (In French)

Chawla, S.S. 1966a Studies on the nutritional requirements of the fruit fly, *Dacus cucurbitae* Coquillett (Diptera: Trypetidae). I. Responses of *D. cucurbitae* larvae to various foodstuffs. RES BULL PANJAB UNIV SCI 17: 111-120.

Chawla, S.S. 1966b Studies on the nutritional requirements of the fruit fly, *Dacus cucurbitae* Coquillett (Diptera: Trypetidae). II. Responses of *D. cucurbitae* larvae to various semi-natural diets under aseptic conditions. RES BULL PANJAB UNIV SCI 17: 141-147.

*Chawla, S.S. 1966c Studies on the nutritional requirements of the fruit fly, *Dacus cucurbitae* Coquillett (Diptera: Trypetidae). III. Development of a synthetic medium for *D. cucurbitae* larvae. RES BULL PANJAB UNIV SCI 17: 243-250.

Chawla, S.S. 1966d Studies on the nutritional requirements of the fruit fly, *Dacus cucurbitae* Coquillett (Diptera: Trypetidae). IV. Responses of *D. cucurbitae* larvae to vitamin deficient diets. RES BULL PANJAB UNIV SCI 17: 327-334.

*Chawla, S.S.; Harwood, R.F. 1968 Artificial diets for the European pine shoot moth, *Rhyacionia buoliana* (Schiffermüller) (Lepidoptera: Olethreutidae). WASH AGRIC EXP STN TECH BULL 59: 13 pp.

Chawla, S.S.; Howell, J.F.; Harwood, R.F. 1967 Surface treatment to control fungi on wheat germ diets. J ECON ENTOMOL 60: 307-308.

*Chawla, S.S.; Perron, J.M.; Cloutier, M. 1974 Effect of different growth factors on the potato aphid, *Macrosiphum euphorbiae* (Aphididae: Homoptera) fed on an artificial diet. CAN ENTOMOL 106: 273-280.

*Chawla, S.S.; Perron, J.M.; Cloutier, M. 1975 Preliminary observations on an artificial diet for *Hydroecia micacea* (Esp.) (Lepidoptera: Noctuidae). ANN SOC ENTOMOL QUE 20: 8-10.

*Cheldelin, V.H.; Newburgh, R.W. 1959 Nutritional studies on the blowfly. ANN N Y ACAD SCI 77: 373-383.

**Chen, P.S. 1966 Amino acid and protein metabolism in insect development. pp. 53-132. *in* Beament, J.W.L.; Treherne, J.E.; Wigglesworth, V.B. (Eds), Insect Physiology Vol 3. Academic Press, New York and London.

*Childress, D.; Williams, P.P. 1973 Control of a bacterial contaminant of boll weevil diet. J ECON ENTOMOL 66: 554-555.

*Chippendale, G.M. 1970a Development of artificial diets for rearing the Angoumois grain moth. J ECON ENTOMOL 63: 844-848.

*Chippendale, G.M. 1970b Metamorphic changes in fat body proteins of the southwestern corn borer, *Diatraea grandiosella*. J INSECT PHYSIOL 16: 1057-1068.

*Chippendale, G.M. 1971a Growth and development of the Angoumois grain moth, *Sitotroga cerealella* in an artificial diet. J INSECT PHYSIOL 17: 109-115.

*Chippendale, G.M. 1971b Observations on the physical and chemical composition of diets for the Angoumois grain moth. J INSECT PHYSIOL 17: 1257-1266.

Chippendale, G.M. 1971c Lipid requirements of the Angoumois grain moth, *Sitotroga cerealella.* J INSECT PHYSIOL 17: 2169-2177.

*Chippendale, G.M. 1972 Dietary carbohydrates: role in survival of the adult rice weevil, *Sitophilus oryzae.* J INSECT PHYSIOL 18: 949-957.

*Chippendale, G.M. 1973 Metabolic reserves of larvae and pupae of the Angoumois grain moth, *Sitotroga cerealella.* INSECT BIOCHEM 3: 1-10.

*Chippendale, G.M.; Beck, S.D. 1964 Nutrition of the European corn borer, *Ostrinia nubilalis* (Hübn.). V. Ascorbic acid as the corn leaf factor. ENTOMOL EXP APPL 7: 241-248.

*Chippendale, G.M.; Beck, S.D. 1965 A method for rearing the cabbage looper, *Trichoplusia ni,* on a meridic diet. J ECON ENTOMOL 58: 377-378.

**Chippendale, G.M.; Beck, S.D. 1968 Biochemical requirements for mass rearing of plant-feeding lepidopterans. pp. 31-39. *in* Radiation, Radioisotopes and Rearing Methods in the Control of Insect Pests. Proceedings of a Panel Organized by the Joint FAO/IAEA Division of Atomic Energy in Food and Agriculture.

Chippendale, G.M.; Beck, S.D.; Strong, F.M. 1964 Methyl linolenate as an essential nutrient for the cabbage looper, *Trichoplusia ni* (Hübner). NATURE LONDON 204: 710-711.

*Chippendale, G.M.; Beck, S.D.; Strong, F.M. 1965 Nutrition of the cabbage looper, *Trichoplusia ni* (Hübn.). I. Some requirements for larval growth and wing development. J INSECT PHYSIOL 11: 211-223.

Chippendale, G.M.; Mann, R.A. 1972 Feeding behaviour of Angoumois grain moth larvae. J INSECT PHYSIOL 18: 87-94.

Chippendale, G.M.; Reddy, G.P.V. 1972a Dietary sterols: role in larval feeding behaviour of the southwestern corn borer, *Diatraea grandiosella.* EXPERIENTIA 28: 485-486.

Chippendale, G.M.; Reddy, G.P.V. 1972b Polyunsaturated fatty acid and sterol requirements of the southwestern corn borer, *Diatraea grandiosella.* J INSECT PHYSIOL 18: 305-316.

Chippendale, G.M.; Reddy, G.P.V. 1974 Dietary carbohydrates: role in feeding behaviour and growth of the southwestern corn borer, *Diatraea grandiosella.* J INSECT PHYSIOL 20: 751-759.

*Chirigos, M.A.; Meiss, A.N.; Pisano, J.J.; Taylor, M.W. 1960 Growth response of the confused flour beetle, *Tribolium confusum* (Duval) to six selected protein sources. J NUTR 72: 121-130.

*Chiu, S.F.; McCay, C.M. 1939 Nutritional studies of the confused flour beetle (*Tribolium confusum* Duval) and the bean weevil (*Acanthoscelides obtectus* Say) ANN ENTOMOL SOC AM 32: 164-170.

Chou, Y.M.; Rock, G.C.; Hodgson, E. 1973 Consumption and utilization of chemically defined diets by *Argyrotaenia velutinana* and *Heliothis virescens.* ANN ENTOMOL SOC AM 66: 627-632.

**Choumakov, E.M. 1970 Problèmes essentiels posés par les élevages de masse, en laboratoire, d'insectes utilisés pour la lutte biologique. pp. 9-16. *in* Colloque Franco-sovietique sur l'utilisation des Entomophages. Institute National de la Recherche Agronomique. (In French with English and Russian summary)

Christenson, L.D.; Maeda, S.; Halloway, J.R. 1956 Substitution of dehydrated for fresh carrots in medium for rearing fruit flies. J ECON ENTOMOL 49: 135-136.

*Christmas, P.E. 1970 Laboratory rearing of the biting fly *Stomoxys calcitrans* (Diptera: Muscidae). N Z ENTOMOL 4: 45-49.

*Chu, H.M.; Norris, D.M.; Kok, L.T. 1970 Pupation requirement of the beetle *Xyleborus ferrugineus:* sterols other than cholesterol. J INSECT PHYSIOL 16: 1379-1387.

*Chu, Y.I. 1969 On the bionomics of *Lyctocoris beneficus* (Hiura) and *Xylocoris galactinus* (Fieber) (Anthocoridae, Heteroptera). J FAC AGRIC KYUSHU UNIV 15: 1-136.

*Chun, M.W. 1972 Dynamics of feeding responses in *Pieris brassicae* Linn. as a function of chemosensory input: a behavioural ultrastructural and electrophysiological study. MEDED LANDBOUWHOGESCH WAGENINGEN 72-11, 162 pp.

Chung, K.H.; Ryu, J.; Kaim, Y.P.; Kwan, S.H. 1973 Studies on the artificial rearing of the rice stem borer (*Chilo suppressalis* Walker). KOREAN J PLANT PROT 12: 29-32 (In Korean with summary in English)

Clark, A.J.; Bloch, K. 1959a The absence of sterol synthesis in insects. J BIOL CHEM 234: 2578-2582.

Clark, A.J.; Bloch, K. 1959b Function of sterols in *Dermestes vulpinus.* J BIOL CHEM 234: 2583-2588.

*Clark, E.W. 1965 An artificial diet for the southern pine beetle and other bark beetles. U S FOR SER RES NOTE SE-45: 3 pp.

Clark, E.W. 1971 Chemically defined diet for pales weevil larvae. J ELISHA MITCHELL SCI SOC 87: 161.

*Clark, E.W. 1973 *Hylobius pales:* a meridic larval diet using sitosterol. J ECON ENTOMOL 66: 841-843.

*Clark, E.W.; Richmond, C.A.; McGough, J.M. 1961 Artificial media and rearing techniques for the pink bollworm. J ECON ENTOMOL 54: 4-9.

Clayton, R.B. 1959 A simplified method for the culture of *Blattella germanica* under aseptic conditions. NATURE LONDON 184: 1166-1167.

**Clayton, R.B. 1964 The utilization of sterols by insects. J LIPID RES 5: 3-19.

*Cline, L.D. 1973 Rearing the lesser grain borer on ground dog biscuits to obtain mature larvae and pupae. ANN ENTOMOL SOC AM 66: 551-552.

*Cloutier, C.; MacKauer, M. 1975 A modified technique for rearing of the pea aphid on artificial diets. ENTOMOL EXP APPL 18: 395-396.

*Cobb, P.P.; Bass, M.H. 1975 Beet armyworm (Lepidoptera: Noctuidae): a laboratory rearing technique. J GA ENTOMOL SOC 10: 190-191.

*Cohen, B. 1968 The influence of larval crowding on the development time of populations of *Drosophila melanogaster* on chemically defined medium. CAN J ZOOL 46: 493-497.

Cohen, R.A.; Stürckow, B. 1971 Gustatory behavioural responses of the adult blow fly, *Phormia regina* reared on sugar - containing media. ANN ENTOMOL SOC AM 64: 1269-1272.

**Coker, W.Z.; Frizzi, G.; Kitzmiller, J.B.; Laven, H.; Mason, G.F.; Milani, R.; Pal, R.; Sharma, G.P.; Wright, J.W. 1967 Standardized strains of insects of public health importance. pp. 735-766. *in* Wright, J.W.; Pal, R. (Eds), Genetics of Insect Vectors of Disease. Elsevier Publishing Company, Amsterdam.

Coluzzi, M. 1964 Maintenance of laboratory colonies of *Anopheles* mosquitoes. BULL W H O 31: 441-443.

**Coluzzi, M. 1971 Problèmes théoriques et pratiques liés à l'élevage et à la production de masse des Culicides. ANN PARASITOL HUM COMP 46: 91-101. (In French)

Conway, S.C.; Harding, J.A. 1967 Daily weight gain of the European corn borer reared on artificial media (Lepidoptera: Pyraustidae). J KANS ENTOMOL SOC 40: 1-3.

*Cook, S.F.; Scott, K.G. 1933 The nutritional requirements of *Zootermopsis (Termopsis) angusticollis.* J CELL COMP PHYSIOL 4: 95-110.

Cooke, J.; Sang, J.H. 1970 Utilization of sterols by larvae of *Drosophila melanogaster.* J INSECT PHYSIOL 16: 801-812.

*Cooper, M.I.; Fraenkel, G. 1952 Nutritive requirements of the small-eyed flour beetle, *Palorus ratzeburgi* Wissman (Tenebrionidae, Coleoptera). PHYSIOL ZOOL 25: 20-28.

*Coppel, H.C.; House, H.L.; Maw, M.G. 1959 Studies on dipterous parasites of the spruce budworm, *Choristoneura fumiferana* (Clem.) (Lepidoptera: Tortricidae). VII. *Agria affinis* (Fall.) (Diptera: Sarcophagidae). CAN J ZOOL 37: 817-830.

*Coudriet, D.L. 1970 Rearing the orange tortrix on a synthetic diet. J ECON ENTOMOL 63: 1004-1005.

*Coutin, R. 1952 Alimentation des larves de *Laspeyresia pomonella* L. (Lépidoptèra: Tortricidae) sur milieux artificiels. C R SEANCES SOC BIOL PARIS 146: 516-520. (In French)

**Craig, R.; Hoskins, W.M. 1940 Insect Biochemistry. ANNU REV BIOCHEM 9: 617-640.

*Creighton, C.S.; Cuthbert, E.R. Jr 1968 A semisynthetic diet for adult banded cucumber beetles. J ECON ENTOMOL 61: 337-338.

*Cress, D.C.; Chada, H.L. 1971a Development of synthetic diet for the greenbug, *Schizaphis graminum.* 1. Greenbug development on two synthetic diets. ANN ENTOMOL SOC AM 64: 1237-1240.

*Cress, D.C.; Chada, H.L. 1971b Development of a synthetic diet for the greenbug, *Schizaphis graminum.* 2. Greenbug development as affected by zinc, iron, manganese and copper. ANN ENTOMOL SOC AM 64: 1240-1244.

Cress, D.C.; Chada, H.L. 1971c Development of a synthetic diet for the greenbug, *Schizaphis graminum.* 3. Responses of greenbug biotypes A and B to the same diet medium. ANN ENTOMOL SOC AM 64: 1245-1247.

*Creswell, M.J.; Sturgeon, E.E.; Eikenbary, R.D. 1971 Laboratory rearing of the Nantucket pine tip moth, *Rhyacionia frustrana,* on artificial diets. ANN ENTOMOL SOC AM 64: 1159-1163.

*Crowell, M.F.; McCay, C.M. 1937 Nutritional studies of the webbing clothes moth, *Tineola bisselliella* Hum. PHYSIOL ZOOL 10: 368-372.

Curry, J.P. 1971 Development of populations of *Acarus siro* (Acarina, Acaridae) on various foodstuffs. ANN ENTOMOL SOC AM 64: 531-532.

Cuthbert, F.P. Jr; Creighton, C.S.; Cuthbert, R.B. II. 1968 Mass rearing banded cucumber beetles, with notes on rearing spotted and striped cucumber beetles. J ECON ENTOMOL 61: 288-292.

*Cymorek, S. 1965 Verfahren zur Beschleunigten Züchtung des *Anobium punctatum* im laboratorium. PROC 12 INT CONGR ENTOMOL LONDON (1964) 687-689. (In German)

Dadd, R.G. 1957 Ascorbic acid and carotene in the nutrition of the desert locust, *Schistocerca gregaria* Forsk. NATURE LONDON 179: 427-428.

*Dadd, R.H. 1960a The nutritional requirements of locusts - I. Development of synthetic diets and lipid requirements. J INSECT PHYSIOL 4: 319-347.

Dadd, R.H. 1960b The nutritional requirements of locusts - II. Utilization of sterols. J INSECT PHYSIOL 5: 161-168.

Dadd, R.H. 1960c The nutritional requirements of locusts - III. Carbohydrate requirements and utilization. J INSECT PHYSIOL 5: 301-316.

Dadd, R.H. 1960d Observations on the palatability and utilization of food by locusts, with particular reference to the interpretation of performances in growth trials using synthetic diets. ENTOMOL EXP APPL 3: 283-304.

Dadd, R.H. 1960e Some effects of dietary ascorbic acid on locusts. PROC R SOC LONDON SER B 153: 128-143.

Dadd, R.H. 1961a The nutritional requirements of locusts - IV. Requirements for vitamins of the B-complex. J INSECT PHYSIOL 6: 1-12.

Dadd, R.H. 1961b The nutritional requirements of locusts - V. Observations on essential fatty acids, chlorophyll, nutritional salt mixtures, and the protein or amino acid components of synthetic diets. J INSECT PHYSIOL 6: 126-145.

Dadd, R.H. 1961c Observations on the effects of carotene on the growth and pigmentation of locusts. BULL ENTOMOL RES 52: 63-81.

**Dadd, R.H. 1963 Feeding behaviour and nutrition in grasshoppers and locusts. pp. 47-109. *in* Beament, J.W.L.; Treherne, J.E.; Wigglesworth, V.B. (Eds), Insect Physiology Vol I. Academic Press, New York and London.

*Dadd, R.H. 1964 A study of carbohydrate and lipid nutrition in the wax moth, *Galleria mellonella* (L.), using partially synthetic diets. J INSECT PHYSIOL 10: 161-178.

*Dadd, R.H. 1966 Beeswax in the nutrition of the wax moth, *Galleria mellonella* (L.). J INSECT PHYSIOL 12: 1479-1492.

Dadd, R.H. 1967 Improvement of synthetic diet for the aphid *Myzus persicae* using plant juices, nucleic acids, or trace metals. J INSECT PHYSIOL 13: 763-778.

Dadd, R.H. 1968a Dietary amino acids and wing determination in the aphid *Myzus persicae*. ANN ENTOMOL SOC AM 61: 1201-1210.

**Dadd, R.H. 1968b Problems connected with inorganic components of aqueous diets. BULL ENTOMOL SOC AM 14: 22-26.

**Dadd, R.H. 1970a Arthropod nutrition. pp. 35-95. *in* Florkin, M.; Scheer, B.T. (Eds), Chemical Zoology Vol V, part A, Arthropoda. Academic Press, New York and London.

**Dadd, R.H. 1970b Digestion in Insects. pp. 117-145. *in* Florkin, M.; Scheer, B.T. (Eds), Chemical Zoology Vol V, part A, Arthropoda. Academic Press, New York and London.

Dadd, R.H. 1972 Ambiguities in the interpretation of growth experiments with mosquito larvae in semi-synthetic dietary media. pp. 199-209. *in* Rodriguez, J.G. (Ed.), Insect and Mite Nutrition. North-Holland Publishing Company, Amsterdam.

**Dadd, R.H. 1973 Insect nutrition: current developments and metabolic implications. ANNU REV ENTOMOL 18: 381-420.

*Dadd, R.H.; Gomez, I.; Nambe, M. 1973 Requirement for ribonucleic acid in a semi-synthetic larval diet for the mosquito *Culex pipiens*. J MED ENTOMOL 10: 47-52.

*Dadd, R.H.; Krieger, D.L. 1967 Continuous rearing of aphids of the *Aphis fabae* complex on sterile synthetic diet. J ECON ENTOMOL 60: 1512-1514.

Dadd, R.H.; Krieger, D.L. 1968 Dietary amino acid requirements of the aphid, *Myzus persicae*. J INSECT PHYSIOL 14: 741-764.

Dadd, R.H.; Krieger, D.L.; Mittler, T.E. 1967 Studies on the artificial feeding of the aphid *Myzus persicae* (Sulzer) - IV. Requirements for water-soluble vitamins and ascorbic acid. J INSECT PHYSIOL 13: 249-272.

Dadd, R.H.; Mittler, T.E. 1965 Studies on the artificial feeding of the aphid *Myzus persicae* (Sulzer). III. Some major nutritional requirements. J INSECT PHYSIOL 11: 717-743.

*Dadd, R.H.; Mittler, T.E. 1966 Permanent culture of an aphid on a totally synthetic diet. EXPERIENTIA 22: 832-833.

*Dahlman, D.L. 1963 Survival of *Empoasca fabae* (Harris) (Cicadellidae, Homoptera) on synthetic media. PROC IOWA ACAD SCI 70: 498-504.

*Dakshayani, K.; Mathad, S.B. 1973 Growth development, and survival of the cricket *Plebeiogryllus guttiventris* on various diets. ANN ENTOMOL SOC AM 66: 228-229.

*Dambre-Raes, H. 1974 Nutrition of the cabbage root fly *Hylemya brassicae* (Bouché). I. An artificial diet and axenic culture method. MEDED FAC LANDBOUWWET RIJKSUNIV GENT 39: 1603-1610.

*Dambre-Raes, H. 1975 Axenic rearing method of cabbage root maggots *Hylemya brassicae* (Bouché) on an artificial diet. MEDED FAC LANDBOUWWET RIJKSUNIV GENT 40: 751-758.

*Dang, K.; Anand, M.; Jotwani, M.G. 1970 A simple improved diet for mass rearing of sorghum stem borer *Chilo zonellus* Swinhoe. INDIAN J ENTOMOL 32: 130-133.

*Dang, K.; Doharey, K.L.; Srivastava, B.G.; Jotwani, M.G. 1971 Artificial diet for mass rearing of *Atherigona varia soccata* Rondani. ENTOMOL NEWSL 1: 65.

Danneel, I. Von 1969 Untersuchungen zur Nahrungsaufnahme und Phosphatausscheidung durch *Aphis fabae* Scop. (Insecta, Homoptera, Aphididae) mit Hilfe der Tracer-Methode. I. Untersuchungen über die Faktoren, die Nahrungsaufnahme von *Aphis fabae* Scop. bei künstlicher Fütterung durch Membranen beeinflussen. Z ANGEW ZOOL 56: 229-251. (In German)

Daum, R.J.; McKibben, G.H.; Davich, T.B.; McLaughlin, R. 1969 Development of the bait principle for boll weevil control: Calco Oil Red N-1700 dye for measuring ingestion. J ECON ENTOMOL 62: 370-375.

David, J. 1959 Étude quantitative de developpement de la Drosophile élevée en milieu axénique. BULL BIOL 93: 472-505. (In French)

David, J. 1962 A new medium for rearing *Drosophila* in axenic conditions. DROSOPH INF SERV 36: 128.

**David, J. 1967 Méthodes d'évaluation des besoins nutritionnels des insectes élevés sur milieux artificiels. ANN NUTR ALIMENT 21: 25-54. (In French)

*David, J.; Clavel, M.F. 1965 Interaction entre le génotype et le milieu d'élevage conséquences sur les caractéristiques du développement de la Drosophile. BULL BIOL FR BELG 99: 369-378. (In French with English summary)

**David, J.; Croissant, J. 1956 Recherches sur la fécondité et la fertilité chez *Drosophila melanogaster* Meig. ANN UNIV LYON SCI SECT C 9: 73-128. (In French)

David, J.; Fouillet, P.; Van Herrewege, J. 1970 Sous-alimentation quantitative chez la Drosophile. I. Action sur le développement larvaire et sur la taille des adultes. ANN SOC ENTOMOL FR 6: 367-378. (In French with English summary)

David, W.A.L.; Ellaby, S. 1975 The viability of the eggs of the African armyworm, *Spodoptera exempta* in laboratory cultures. ENTOMOL EXP APPL 18: 269-280.

*David, W.A.L.; Ellaby, S.; Taylor, G. 1972a The fumigant action of formaldehyde incorporated in a semisynthetic diet on the granulosis virus of *Pieris brassicae* and its evaporation from the diet. J INVERTEBR PATHOL 19: 76-82.

David, W.A.L.; Ellaby, S.; Taylor, G. 1972b The effect of reducing the content of certain ingredients in a semisynthetic diet on the incidence of granulosis virus disease in *Pieris brassicae*. J INVERTEBR PATHOL 20: 332-340.

*David, W.A.L.; Ellaby, S.; Taylor, G. 1975 Rearing *Spodoptera exempta* on semi-synthetic diets and on growing maize. ENTOMOL EXP APPL 18: 226-237.

*David, W.A.L.; Gardiner, B.O.C. 1965 Rearing *Pieris brassicae* L. larvae on a semi-synthetic diet. NATURE LONDON 207: 882-883.

*David, W.A.L.; Gardiner, B.O.C. 1966a Rearing *Mamestra brassicae* (L.) on semi-synthetic diets. BULL ENTOMOL RES 57: 137-142.

David, W.A.L.; Gardiner, B.O.C. 1966b Breeding *Pieris brassicae* apparently free from granulosis virus. J INVERTEBR PATHOL 8: 325-333.

David, W.A.L.; Gardiner, B.O.C. 1966c The effect of sinigrin on the feeding of *Pieris brassicae* L. larvae transferred from various diets. ENTOMOL EXP APPL 9: 95-98.

David, W.A.L.; Gardiner, B.O.C. 1966d Rearing *Pieris brassicae* (L.) on semi-synthetic diets with and without cabbage. BULL ENTOMOL RES 56: 581-593.

**Davidson, G. 1974 The principles and dynamics involved in the sterile insect technique. pp. 5-30. *in* Davidson, G. (Ed.), Genetic Control of Insect Pests. Academic Press, London and New York.

Davies, D.M.; Anderson, E.L. 1967 Egg production of house-fly adults fed human blood containing various supplements. PROC ENTOMOL SOC ONT 97: 111-114.

Davies, D.M.; Morrison, P.E.; Taylor, G.C.; Goodman, T. 1965 Larval nutrient stores, adult diets and reproduction in Diptera. PROC 12 INT CONGR ENTOMOL LONDON (1964): 174.

*Davis, G.R.F. 1956 Amino acid requirements of *Oryzaephilus surinamensis* (L.) (Coleoptera: Silvanidae) for pupation. CAN J ZOOL 34: 82-85.

*Davis, G.R.F. 1959a Alanine and proline in the diet of larvae of *Oryzaephilus surinamensis* (L.) (Coleoptera: Silvanidae). ANN ENTOMOL SOC AM 52: 164-167.

*Davis, G.R.F. 1959b A method for rearing larvae of *Ctenicera aeripennis destructor* (Brown) (Coleoptera: Elateridae) aseptically in test tubes. ANN ENTOMOL SOC AM 52: 173-175.

Davis, G.R.F. 1961a Effects of variations in casein and dextrin content of a synthetic diet on larvae of *Ctenicera aeripennis destructor* (Brown) (Coleoptera: Elateridae). ENTOMOL EXP APPL 4: 273-276.

Davis, G.R.F. 1961b The biting response of larvae of the prairie grain wireworm, *Ctenicera aeripennis destructor* (Brown) (Coleoptera: Elateridae) to various extracts of germinating rye seed. CAN J ZOOL 39: 299-303.

Davis, G.R.F. 1961c Lysine requirements of larvae of *Oryzaephilus surinamensis* (L.) (Coleoptera: Silvanidae). J INSECT PHYSIOL 6: 122-125.

Davis, G.R.F. 1961d Sulfur-containing amino acids in the nutrition of the saw-toothed grain beetle, *Oryzaephilus surinamensis* (L.) (Coleoptera: Silvanidae). J NUTR 75: 275-278.

Davis, G.R.F. 1962 Quantitative *l*-arginine requirements of larvae of the saw-toothed grain beetle, *Oryzaephilus surinamensis* (Coleoptera: Silvanidae). J INSECT PHYSIOL 8: 377-382.

Davis, G.R.F. 1964 The importance of carnitine in the diet of larvae of *Oryzaephilus surinamensis* (L.) (Coleoptera: Silvanidae). ARCH INT PHYSIOL BIOCHIM 72: 70-75.

Davis, G.R.F. 1965 The effect of structural analogs of methionine and glutamic acid on larvae of the prairie grain wireworm, *Ctenicera destructor* (Brown) (Coleoptera: Elateridae). ARCH INT PHYSIOL BIOCHIM 73: 177-187.

Davis, G.R.F. 1966a Survival and development of *Oryzaephilus surinamensis* (L.) (Coleoptera: Silvanidae) on wheat-flour diets and on casein diets supplemented with biotin and ascorbic acid. CAN ENTOMOL 98: 263-267.

Davis, G.R.F. 1966b Effects of including yeast ribonucleic acid in a synthetic diet for larvae of *Oryzaephilus surinamensis* (L.) (Coleoptera: Silvanidae). CAN J ZOOL 44: 345-346.

Davis, G.R.F. 1966c Replacement of RNA in the diet of *Oryzaephilus surinamensis* (L.) (Coleoptera: Silvanidae) by purines, pyrimidines, and ribose. CAN J ZOOL 44: 781-785.

*Davis, G.R.F. 1966d Growth and development of prairie grain wireworm *Ctenicera destructor* Brown (Coleoptera: Elateridae), on synthetic diets. CAN J ZOOL 44: 995-1001.

Davis, G.R.F. 1966e Putrescine and spermidine as growth-promoting substances for the saw-toothed grain beetle, *Oryzaephilus surinamensis* (L.) (Coleoptera: Silvanidae). COMP BIOCHEM PHYSIOL 19: 619-627.

Davis, G.R.F. 1967a Effects of dietary lipids on survival and development of the saw-toothed grain beetle, *Oryzaephilus surinamensis* (L.) (Coleoptera: Silvanidae). REV CAN BIOL 26: 119-124.

Davis, G.R.F. 1967b Reassessment of some amino acid requirements of larvae of *Oryzaephilus surinamensis* (L.) (Coleoptera: Silvanidae). J NUTR 91: 255-260.

Davis, G.R.F. 1967c Nutritional relationships in larvae of *Oryzaephilus surinamensis*. J INSECT PHYSIOL 13: 1737-1743.

Davis, G.R.F. 1968a Quantitative dietary requirements of *Oryzaephilus surinamensis* (L.) (Coleoptera: Silvanidae) for phenylalanine and tyrosine. CAN J ZOOL 46: 469-473.

*Davis, G.R.F. 1968b Effects of 2,4-D, maleic acid hydrazide, and 2,4,5-T on survival and growth of the prairie grain wireworm, *Ctenicera destructor* Brown (Coleoptera: Elateridae). CAN J ZOOL 46: 747-750.

**Davis, G.R.F. 1968c Phagostimulation and consideration of its role in artificial diets. BULL ENTOMOL SOC AM 14: 27-30.

Davis, G.R.F. 1968d Glutamic acid requirements of the saw-toothed grain beetle, *Oryzaephilus surinamensis* (L.) (Coleoptera: Silvanidae). COMP BIOCHEM PHYSIOL 24: 395-401.

Davis, G.R.F. 1968e Quantitative requirements of the saw-toothed grain beetle, *Oryzaephilus surinamensis*, for dietary *l*-tryptophan. J INSECT PHYSIOL 14: 1693-1696.

Davis, G.R.F. 1969a Effect of dietary serine on survival and rate of development of *Oryzaephilus surinamensis* (Coleoptera: Silvanidae). CAN ENTOMOL 101: 27-31.

*Davis, G.R.F. 1969b Protein nutrition of *Tenebrio molitor* L. X. Improvement of the nutritional value of lactalbumin by supplementation with amino acids. ARCH INT PHYSIOL BIOCHIM 77: 741-748.

*Davis, G.R.F. 1970a Protein nutrition of *Tenebrio molitor* L. XI. Effect on growth of partial replacement of dietary casein by lactalbumin or lactalbumin hydrolysates and of partial replacement of these two proteins by casein. ARCH INT PHYSIOL BIOCHIM 78: 29-36.

*Davis, G.R.F. 1970b Protein nutrition of *Tenebrio molitor* L. XII. Effects of dietary casein concentration and of dietary cellulose on larvae of race F. ARCH INT PHYSIOL BIOCHIM 78: 37-41.

Davis, G.R.F. 1970c Protein nutrition of *Tenebrio molitor* L. XIII. Consideration of some dietary factors of casein, lactalbumin, and lactalbumin hydrolysate. ARCH INT PHYSIOL BIOCHIM 78: 467-473.

Davis, G.R.F. 1971a Protein nutrition of *Tenebrio molitor* L. XIV - Further investigations of effects of components of vitamin-free casein, lactalbumin and lactalbumin hydrolysate on growth of larvae of race F. ARCH INT PHYSIOL BIOCHIM 79: 1-9.

*Davis, G.R.F. 1971b Protein nutrition of *Tenebrio molitor* L. XV - Amino acid mixtures as replacements for protein of the artificial diet. ARCH INT PHYSIOL BIOCHIM 79: 11-17.

**Davis, G.R.F. 1972a Application of insect nutrition in solving general nutrition problems. pp. 33-39. *in* Rodriguez, J.G. (Ed.), Insect and Mite Nutrition. North-Holland Publishing Company, Amsterdam and London

**Davis, G.R.F. 1972b Refining diets for optimal performance. pp. 171-181. *in* Rodriguez, J.G. (Ed.), Insect and Mite Nutrition. North-Holland Publishing Company, Amsterdam and London.

*Davis, G.R.F. 1972c Quantitative requirements of the saw-toothed grain beetle, *Oryzaephilus surinamensis*, for dietary glycine and *l*-threonine. J INSECT PHYSIOL 18: 1287-1294.

Davis, G.R.F. 1973a Quantitative dietary requirements of the saw-toothed grain beetle, *Oryzaephilus surinamensis*, for *l*-leucine. J INSECT PHYSIOL 19: 273-276.

Davis, G.R.F. 1973b Quantitative requirements of the saw-toothed grain beetle, *Oryzaephilus surinamensis*, for dietary *l*-isoleucine and *l*-valine. J INSECT PHYSIOL 19: 1657-1661.

Davis, G.R.F. 1974 Protein nutrition of *Tenebrio molitor* L. XVII - Improved amino acid mixture and interaction with dietary carbohydrate. ARCH INT PHYSIOL BIOCHIM 82: 631-637.

*Davis, G.R.F.; Leclercq, J. 1969 Protein nutrition of *Tenebrio molitor* L. IX. Replacement caseins for the reference diet and a comparison of the nutritional values of various lactalbumins and lactalbumin hydrolysates. ARCH INT PHYSIOL BIOCHIM 77: 687-693.

Davis, G.R.F.; Sosulski, F.W. 1973a Improvement of basic diet for use in determining the nutritional value of proteins with larvae of *Tenebrio molitor* L. ARCH INT PHYSIOL BIOCHIM 81: 495-500.

Davis, G.R.F.; Sosulski, F.W. 1973b Protein nutrition of *Tenebrio molitor* L. XVI. Effects of dietary protein concentration on larvae of race F. ARCH INT PHYSIOL BIOCHIM 81: 661-665.

*Davis, G.R.F.; Sosulski, F.W. 1974a Protein nutrition of *Tenebrio molitor* L. XVIII. Growth of larvae of race F fed protein blends of cereals and legumes. ARCH INT PHYSIOL BIOCHIM 82: 861-871.

Davis, G.R.F.; Sosulski, F.W. 1974b Nutritional quality of oilseed protein isolates as determined with larvae of the yellow mealworm, *Tenebrio molitor* L. J NUTR 104: 1172-1177.

Delcourt, A.; Guyénot, E. 1910 De la possibilité d'étudier certains Diptères en milieu défini. C R ACAD SCI SER D 151: 255-257. (In French)

Delcourt, A.; Guyénot, E. 1912 Génétique et milieu. Nécessité de la determination des conditions. Sa possibilité chez les Drosophiles technique. BULL SCI FR BELG 45: 249-332. (In French)

*Delobel, A. 1975 *Chilo orichalcociliellus* Strand (Lepidoptera, Pyralidae), foreur des tiges du sorgho et du mais à Madagascar. I. Caractéristiques morphologiques. CAH ORSTOM SER BIOL 10: 3-9. (In French with English summary)

Demal, J. 1961 Problèmes concernant la morphogenèse in vitro chez les insectes. BULL SOC ZOOL FR 86: 522-533. (In French)

*De Meillon, B.; Golberg, L.; Lavoipierre, M. 1945 The nutrition of the larva of *Aedes aegypti* L. I. J EXP BIOL 21: 84-89.

**Demerec, M.; Kaufmann, B.P. 1962 Drosophila Guide. 7th ed. CARNEGIE INST WASHINGTON

*Deoras, P.J. 1954 Breeding the Indian house-fly (*Musca domestica nebulo* Fabr.) for experimental studies. PARASITOLOGY 44: 305-309.

Depner, K.R. 1962 Continuous propagation of the horn fly, *Haematobia irritans* (L.) (Diptera: Muscidae). CAN ENTOMOL 94: 893-895.

Dethier, V.G. 1953 Host plant perception in phytophagous insects. TRANS 9 INT CONGR ENTOMOL AMSTERDAM (1951) 2: 81-89.

Dethier, V.G.; Barton-Browne, L.; Smith, C.N. 1960 The designation of chemicals in terms of responses they elicit from insects. J ECON ENTOMOL 53: 134-136.

Dethier, V.G.; Goldrich, N. 1971 Blowflies: alteration of adult taste responses by chemicals present during development. SCIENCE 173: 242-244.

*Dietz, A. 1973 Longevity and survival of honey bee larvae on artificial diets. J GA ENTOMOL SOC 8: 59-63.

Dietz, A.; Haydak, M.H. 1971 Caste determination in honey bees. I. The significance of moisture in larval food. J EXP ZOOL 177: 353-358.

*Dimetry, N.Z. 1970 A successful technique for mass rearing of the cotton leaf worm, *Spodoptera littoralis* Boisd. on an artificial diet. Z ANGEW ENTOMOL 66: 408-411.

*Dimond, J.B.; Lea, A.O.; Hahnert W.F. Jr; DeLong, D.M. 1956 The amino acids required for egg production in *Aedes aegypti*. CAN ENTOMOL 88: 57-62.

*Dinther, J.B.M. Van; Goossens, P.A. 1970 Rearing of *Diatraea saccharalis* on diets in Surinam. ENTOMOL EXP APPL 13: 320-326.

*Doane, W.W. 1970 A quick and easy method for rearing large quantities of 'clean' larvae. DROSOPH INF SERV 45: 189.

*Dodgshun, T.J. 1970 A semi-synthetic maintenance diet suitable for hepialid and melolonthid larvae. N Z ENTOMOL 4: 78-79.

Doll, J.P.; Trexler, P.C.; Reynolds, L.I.; Bernard, G.R. 1963 The use of peracetic acid to obtain germfree invertebrate eggs for gnotobiotic studies. AM MIDL NAT 69: 231-239.

*Donaldson, J.M.I. 1976 An artificial diet for flower- and fruit-eating cetoniid beetles (Coleoptera: Scarabeidae). PHYTOPHYLACTIA 8: 17-18.

**Dougherty, E.C. 1959 Introduction to axenic culture of invertebrate metazoa: a goal. ANN N Y ACAD SCI 77: 27-54.

*Drake, E.F. 1971 Life cycle and laboratory diet for *Atrichopogon jacobsoni* (de Meijere) (Diptera: Ceratopogonidae). PROC HAWAII ENTOMOL SOC 21: 63-66.

Dunkel, F.V.; Boush, G.M. 1968 Biology of the gregarine *Pyxinia frenzeli* in the black carpet beetle, *Attagenus megatoma*. J INVERTEBR PATHOL 11: 281-288.

*Dunwoody, J.E.; Hooper, G.H.S. 1967 An artificial medium for rearing *Epiphyas postvittana*. J ECON ENTOMOL 60: 1753-1754.

*Dupnik, T.D.; Kamm, J.A. 1970 Development of an artificial diet for *Crambus trisectus*. J ECON ENTOMOL 63: 1578-1581

Dutky, R.C.; Robbins, W.E.; Shortino, T.J.; Kaplanis, J.N.; Vroman, H.E. 1967 The conversion of cholestanone to cholestanol by the housefly, *Musca domestica* L. J INSECT PHYSIOL 13: 1501-1510.

*Dutky, S.R.; Thompson, J.V.; Cantwell, G.F. 1962 A technique for mass rearing the greater wax moth (Lepidoptera: Galleriidae). PROC ENTOMOL SOC WASH 64: 56-58.

Earle, N.W.; Gaines, R.C.; Roussel, J.S. 1959 A larval diet for the boll weevil containing an acetone powder of cotton squares. J ECON ENTOMOL 52: 710-712.

*Earle, N.W.; MacFarlane, J. 1968 A unisexual strain of the salt-marsh caterpillar, *Estigmene acrea*. ANN ENTOMOL SOC AM 61: 949-953.

Earle, N.W.; Padovani, I.; Thompson, M.J.; Robbins, W.E. 1970 Inhibition of larval development and egg production in the boll weevil following ingestion of ecdysone analogues. J ECON ENTOMOL 63: 1064-1069.

Earle, N.W.; Walker, A.B.; Burks, M.L. Jr 1965 Storage and excretion of steroids in the adult boll weevil. COMP BIOCHEM PHYSIOL 16: 277-288.

*Earle, N.W.; Walker, A.B.; Burks, M.L. 1966 An artificial diet for the boll weevil, *Anthonomus grandis* (Coleoptera: Curculionidae), based on the analysis of amino acids in cotton squares. ANN ENTOMOL SOC AM 59: 664-669.

Earle, N.W.; Walker, A.B.; Burks, M.L.; Slatten, B.H. 1967 Sparing of cholesterol by cholestanol in the diet of the boll weevil, *Anthonomus grandis* (Coleoptera: Curculionidae). ANN ENTOMOL SOC AM 60: 599-603.

Earle, W.R. 1943 Production of malignancy *in vitro*. IV. The mouse fibroblast cultures and changes in the living cells. J NAT CANCER INST 4: 165-212.

Economopoulos, A.P.; Tzanakakis, M.E. 1967 Egg yolk and olive juice as supplement to the yeast hydrolysate-sucrose diet for adults of *Dacus oleae*. LIFE SCI 6: 2409-2416.

**Edelman, N.M. 1961 The feeding of insects on artificial media. USP SOVREM BIOL 51: 204-219.

*Edelman, N.M. 1970 Principles of working out nutritive media for the pests of generative organs of fruit crops. (The codling moth as an example). ZOOL ZH 49: 1240-1246. (In Russian with English summary)

*Edelman, N.M. 1974 The rearing of phytophagous insects with different types of feeding on nutrient media containing leguminous sprouts. TR VSES INST ZASHCH RAST 40: 128-135. (In Russian with English summary)

*Ehrhardt, P. 1965 Speicherung anorganischer Substanzen in den Mitteldarmzellen von *Aphis fabae* Scop. und ihre Bedeutung für die Ernährung. Z VGL PHYSIOL 50: 293-312. (In German with English summary)

Ehrhardt, P. 1968a Nachweis einer durch symbiontische Mikroorganismen bewirkten Sterinsynthese in Künstlich ernahrten Aphiden (Homoptera, Rhynchota, Insecta). EXPERIENTIA 24: 82-83. (In German with English summary)

Ehrhardt, P. 1968b Der Vitaminbedarf einer siebröhrensaugenden Aphidae, *Neomyzus circumflexus* Buckt. (Homoptera: Insecta). Z VGL PHYSIOL 60: 416-426. (In German with English summary)

*Ehrhardt, P. 1968c Einfluss von Ernährungsfaktoren auf die Entwicklung von Säfte saugenden Insekten unter besonderer Berücksichtigung von Symbionten. Z PARASITENKD 31: 38-66. (In German with English summary)

*Ehrhardt, P. 1968d Die Wirkung verschiedener Spurenelemente auf Wachstum, Reproduktion und Symbionten von *Neomyzus circumflexus* Buckt. (Aphidae, Homoptera, Insecta) bei künstlicher Ernährung. Z VGL PHYSIOL 58: 47-75. (In German with English summary)

Ehrhardt, P. 1969 Die Rolle von Methionin, Cystin, und Sulfat bei der künstlichen Ernährung von *Neomyzus (Aulacorthum) circumflexus* Buckt. (Aphidae, Homoptera, Insecta). BIOL ZENTRALBL 88: 335-348. (In German with English summary)

Ehrhardt, P.; Schmutterer, H. 1966 Die Wirkung Verschiedener antibiotica auf Entwicklung und Symbionten künstlich ernäharter Bohnenblattläuse (*Aphis fabae* Scop.). Z MORPHOL OEKOL TIERE 56: 1-20.

Ekka, I. 1969 The rearing of the two-spotted spider mite (*Tetranychus urticae* Koch) on artificial diets. Doctoral Thesis, MacDonald College, McGill University, Canada, 116 pp.

*Ekka, I.; Rodriguez, J.G.; Davis, D.L. 1971 Influence of dietary improvement on oviposition and egg viability of the mite *Tetranychus urticae*. J INSECT PHYSIOL 17: 1393-1399.

Ellinger, P.; Fraenkel, G.; Abdel Kader, M.M. 1947 The utilization of nicotinamide derivatives and related compounds by mammals, insects and bacteria. BIOCHEM J 41: 59-68.

Elliot, K.R. 1955 Studies on the nutrition of the southern armyworm *Prodenia eridania* (Cramer). REP ENTOMOL SOC ONT 86: 17-19.

*Ellis, J.F. 1959 Reversal of an adenine and a cytidine requirement in axenic *Drosophila* culture. PHYSIOL ZOOL 32: 29-39.

*El-Miniawi, S.F.; Moustafa, M.A. 1965 On the biology of *Megaselia scalaris* Loew (Dipt. Phoridae). BULL SOC ENTOMOL EGYPTE 49: 81-91.

*El-Minshawy, A.M.; Zeid, M. 1972 Rearing the larvae of the cotton leafworm *Spodoptera littoralis* (Boisd.) on semi-artificial diet. Z ANGEW ENTOMOL 70: 101-104.

*Erickson, E.H.Jr.; Werner, F.G. 1974. Bionomics of nearctic bee-associated Meloidae (Coleoptera); life histories and nutrition of certain Meloinae. ANN ENTOMOL SOC AM 67: 394-400.

*Erk, F.C.; Sang, J.H. 1966 The comparative nutritional requirements of two sibling species, *Drosophila simulans* and *D. melanogaster*. J INSECT PHYSIOL 12: 43-51.

*Eschle, J.L.; Schmidt, C.D.; Matter, J.J. 1973 Horn fly: preliminary evaluation of media for rearing larvae. ANN ENTOMOL SOC AM 66: 352-354.

*Étienne, J. 1973 Conditions artificielles nécessaires à l'élevage massif de *Ceratitis rosa* (Diptera: Trypetidae). ENTOMOL EXP APPL 16: 380-388. (In French with English summary)

*Fales, J.H.; Bodenstein, O.F.; Keller, J.C. 1961 Face fly laboratory rearing. First successful rearing in the laboratory of *Musca autumnalis*, a serious pest of livestock, is major step forward in developing controls. SOAP CHEM SPEC 37: 81-83.

Falk, D.R.; Nash, D. 1972 The search for auxotrophic mutants in *Drosophila melanogaster*. pp. 19-33. *in* Rodriguez, J.G. (Ed.), Insect and Mite Nutrition. North-Holland Publishing Company, Amsterdam and London.

**Fast, P.G. 1964 Insect lipids: a review. MEM ENTOMOL SOC CAN 37: 1-50.

*Fatzinger, C.W. 1970a Rearing successive generations of *Dioryctria abietella* (Lepidoptera: Pyralidae (Phycitinae)) on artificial media. ANN ENTOMOL SOC AM 63: 809-814.

Fatzinger, C.W. 1970b Aseptic techniques for rearing *Dioryctria abietella* (Lepidoptera: Pyralidae, Phycitinae) on artificial medium. ANN ENTOMOL SOC AM 63: 1716-1718.

*Fatzinger, C.W. 1973 A wheat germ-free, meridic medium for rearing *Dioryctria abietella* (Lepidoptera: Pyralidae: Phycitinae). FLA ENTOMOL 56: 327-332.

*Feeny, P.P. 1968 Effect of oak leaf tannins on larval growth of the winter moth *Operophtera brumata*. J INSECT PHYSIOL 14: 805-817.

Feir, D.; Beck, S.D. 1963 Feeding behaviour of the large milkweed bug, *Oncopeltus fasciatus*. ANN ENTOMOL SOC AM 56: 224-229.

Féron, M.; Delanoue, P.; Soria, F. 1968 L'élevage massif artificiel de *Ceratatis capitata* Wied. ENTOMOPHAGA 3: 45-53. (In French with English summary)

*Ferran, A.; Laforge, J.P. 1975 L'alimentation artificielle des larves de la coccinelle aphidiphage *Adonia notata* Schn. (Col. Coccinellidae). ANN ZOOL ECOL ANIM 7: 1-12. (In French with English summary)

Finney, G.L. 1956 A fortified carrot medium for mass-culture of the oriental fruit fly and certain other tephritids. J ECON ENTOMOL 49: 134.

**Finney, G.L.; Fisher, T.W. 1964 Culture of entomorphagous insects and their hosts. pp. 328-355. *in* DeBach, P.; Schlinger, E.I. (Eds), Biological Control of Insect Pests and Weeds. Chapman and Hall, London.

Fisher, R.W.; Jursic, F. 1958 Rearing houseflies and roaches for physiological research. CAN ENTOMOL 90: 1-7.

**Fisher, T.W. 1964 Shipping techniques for entomophagous insects. BULL W H O 31: 503-504.

**Flanders, S.E.; Bay, E.C. 1964 Standardization of mass rearing procedures for Entomophaga. BULL W H O 31: 505-507.

*Fleming, W.E.; Parker, L.B.; Maines, W.W.; Plasket, E.L.; McCabe, P.J. 1962 Bioassay of soil containing residues of chlorinated hydrocarbon insecticides with special reference to control of Japanese beetle grub. U S DEP AGRIC TECH BULL 1266: 44 pp.

*Fleurat-Lessard, F. 1974 Contribution a l'étude des relations trophiques entre *Tyrophagus putrescentiae* Schrank (Acarina Acaridae) et la mycoflore des graines de colza stockées. Mise au point d'une technique d'élevage de l'acarien sur moisissures. ANN ZOOL ECOL ANIM 6: 97-109.

*Forgash, A.J. 1958 The effect of inositol on growth, survival, and maturation in *Periplaneta americana* (L.). ANN ENTOMOL SOC AM 51: 406-409.

*Forgash, A.J.; Moore, R.F. 1960 Dietary inositol requirements of *Periplaneta americana* (L.). ANN ENTOMOL SOC AM 53: 91-94.

**Fraenkel, G. 1943 Insect nutrition. J R COLL SCI 13: 59-69.

Fraenkel, G. 1951a Effect and distribution of vitamin B_T. ARCH BIOCHEM BIOPHYS 34: 457-467.

Fraenkel, G. 1951b Vitamin B_T. I. Deficiency symptoms, method of testing, distribution, isolation procedures and some properties. FED PROC FED AM SOC EXP BIOL 10: 183-184.

**Fraenkel, G. 1952 The role of symbionts as sources of vitamins and growth factors for their insects hosts. TIJDSCHR ENTOMOL 95: 183-195.

**Fraenkel, G. 1953a The nutritional requirements of insects for known and unknown vitamins. TRANS 9 INT CONGR ENTOMOL AMSTERDAM (1951) 1: 277-280.

**Fraenkel, G. 1953b The nutritional value of green plants for insects. TRANS 9 INT CONGR ENTOMOL AMSTERDAM (1951) 2: 90-100.

*Fraenkel, G. 1953c Studies on the distribution of vitamin B_T (carnitine). BIOL BULL 104: 359-371.

Fraenkel, G. 1954 The distribution of vitamin B_T (carnitine) throughout the animal kingdom. ARCH BIOCHEM BIOPHYS 50: 486-595.

Fraenkel, G. 1955 Inhibitory effects of sugars on the growth of the mealworm, *Tenebrio molitor* L. J CELL COMP PHYSIOL 45: 393-408.

Fraenkel, G. 1957 The *Tenebrio* assay for carnitine. METHODS ENZYMOL 3: 662-667.

**Fraenkel, G. 1958 The effect of zinc and potassium in the nutrition of *Tenebrio molitor*, with observations of the expression of a carnitine deficiency. J NUTR 65: 361-395.

**Fraenkel, G. 1959a The physiology of insect nutrition. SYMP GENET BIOL ITAL 9: 407-415.

**Fraenkel, G. 1959b A historical and comparative survey of the dietary requirements of insects. ANN N Y ACAD SCI 77: 267-274.

**Fraenkel, G. 1969 Evaluation of our thoughts on secondary plant substances. ENTOMOL EXP APPL 12: 473-486.

Fraenkel, G.; Blewett, M. 1942a Biotin as a possible growth factor for insects. NATURE LONDON 149: 301.

Fraenkel, G.; Blewett, M. 1942b Biotin, B_1, riboflavin, nicotinic acid, B_6 and pantothenic acid as growth factors for insects. NATURE LONDON 150: 177-178.

*Fraenkel, G.; Blewett, M. 1943a The basic food requirements of several insects. J EXP BIOL 20: 28-34.

Fraenkel, G.; Blewett, M. 1943b The vitamin B-complex requirements of several insects. BIOCHEM J 37: 686-692.

Fraenkel, G.; Blewett, M. 1943c The sterol requirements of several insects. BIOCHEM J 37: 692-695.

Fraenkel, G.; Blewett, M. 1943d Vitamins of the B-group required by insects. NATURE LONDON 151: 703-704.

Fraenkel, G.; Blewett, M. 1943e The natural foods and the food requirements of several species of stored products insects. TRANS R ENTOMOL SOC LONDON 93: 457-490.

Fraenkel, G.; Blewett, M. 1944 Stages in the recognition of biotin as a growth factor for insects. PROC R ENTOMOL SOC LONDON SER A 19: 30-35.

Fraenkel, G.; Blewett, M. 1945 Linoleic acid, α-tocopherol and other soluble substances as nutritional factors for insects. NATURE LONDON 155: 392-393.

*Fraenkel, G.; Blewett, M. 1946a The dietetics of the clothes moth, *Tineola bisselliella* Hum. J EXP BIOL 22: 156-161.

*Fraenkel, G.; Blewett, M. 1946b The dietetics of the caterpillars of three *Ephestia* species, *E. kuehniella, E. elutella,* and *E. cautella,* and of a closely related species, *Plodia interpunctella.* J EXP BIOL 22: 162-171.

*Fraenkel, G.; Blewett, M. 1946c Linoleic acid, vitamin E and other fat-soluble substances in the nutrition of certain insects, *Ephestia kuehniella, E. elutella, E. cautella* and *Plodia interpunctella* (Lep.). J EXP BIOL 22: 172-190.

Fraenkel, G.; Blewett, M. 1946d Folic acid in the nutrition of certain insects. NATURE LONDON 157: 697.

Fraenkel, G.; Blewett, M. 1947a The importance of folic acid and unidentified members of the vitamin B-complex in the nutrition of certain insects. BIOCHEM J 41: 469-475.

Fraenkel, G.; Blewett, M. 1947b Linoleic acid and arachidonic acid in the metabolism of two insects, *Ephestia kuehniella* (Lep.) and *Tenebrio molitor* (Col.). BIOCHEM J 41: 475-478.

Fraenkel, G.; Blewett, M.; Coles, M. 1948 B_T, a new vitamin of the B-group and its relation to the folic acid group, and other anti-anaemia factors. NATURE LONDON 161: 981-983.

*Fraenkel, G.; Blewett, M.; Coles, M. 1950 The nutrition of the mealworm, *Tenebrio molitor* L. (Tenebrionidae, Coleoptera). PHYSIOL ZOOL 23: 92-108.

Fraenkel, G.; Chang, P-I. 1954 Manifestations of vitamin B_T (carnitine) deficiency in the larvae of the mealworm, *Tenebrio molitor* L. PHYSIOL ZOOL 27: 40-56.

**Fraenkel, G.S.; Friedman, S.; Hinton, T.; Laszlo, S.; Noland, J.L. 1954 The effect of substituting carnitine for choline in the nutrition of several organisms. ARCH BIOCHEM BIOPHYS 54: 432-439.

Fraenkel, G.; Leclercq, J. 1956 Nouvelles recherches sur les besoins nutritifs de la larve du *Tenebrio molitor* L. (Insecte Coléoptère). ARCH INT PHYSIOL BIOCHIM 64: 601-622.

*Fraenkel, G.; Printy, G.E. 1954 The amino acid requirements of the confused flour beetle, *Tribolium confusum* Duval. BIOL BULL 106: 149-157.

*Fraenkel, G.; Reid, J.A.; Blewett, M. 1941 The sterol requirements of the larva of the beetle, *Dermestes vulpinus* Fabr. BIOCHEM J 35: 712-720.

*Fredeen, F.J.H.; Glen, G.S. 1970 The survival and development of *Leptocera caenosa* (Diptera: Sphaeroceridae) in laboratory cultures. CAN ENTOMOL 102: 164-171.

French, E.W.; Fraenkel, G. 1954 Carnitine (vitamin B_T) as a nutritional requirement for the confused flour beetle. NATURE LONDON 173: 173.

**Friedman, S. 1970 Metabolism of carbohydrates in insects. pp. 167-197. *in* Florkin, M.; Scheer, B.T. (Eds), Chemical Zoology Vol V, part A, Arthropoda. Academic Press, New York and London.

**Friedman, S.; Fraenkel, G. 1972 III. Carnitine. pp. 329-355. *in* Sebrell, W.H.; Harris, R.S. (Eds), The Vitamins Vol V. Academic Press, New York and London.

Friedman, S.J.; Skehan, P.J.; Polan, M.L.; Fausto-Sterling, A.; Brown, P.R. 1974 DNA synthesis during early development of *Drosophila melanogaster*. INSECT BIOCHEM 4: 381-394.

**Friend, W.G. 1955 Problems in nutritional studies on phytophagous insects. REP ENTOMOL SOC ONT 86: 13-17.

**Friend, W.G. 1958 Nutritional requirements of phytophagous insects. ANNU REV ENTOMOL 3: 57-74.

Friend, W.G. 1965 The gorging response in *Rhodnius prolixus* Stahl. CAN J ZOOL 43: 125-132.

**Friend, W.G. 1968 The nutritional requirements of Diptera. pp. 41-57 *in* Radiation, Radio-isotopes and Rearing Methods in the Control of Insect Pests. Proceedings of a Panel Organized by the Joint FAO/IAEA Division of Atomic Energy in Food and Agriculture.

*Friend, W.G.; Backs, R.H.; Cass, L.M. 1957 Studies on amino acid requirements of larvae of the onion maggot, *Hylemya antiqua* (Mg.), under aseptic conditions. CAN J ZOOL 35: 535-543.

Friend, W.G.; Cartwright, E. 1963 A practical apparatus for feeding artificial diet to all stages of *Rhodnius prolixus* Stål. CAN ENTOMOL 95: 362-364.

*Friend, W.G.; Patton, R.L. 1956 Studies on vitamin requirements of larvae of the onion maggot, *Hylemya antiqua* (Mg.), under aseptic conditions. CAN J ZOOL 34: 152-162.

*Friend, W.G.; Salkeld, E.H.; McClanahan, R.J. 1958 A chemically defined diet and axenic rearing method for larvae of the seedcorn maggot, *Hylemya cilicrura* (Rond.). CAN J ZOOL 36: 931-936.

*Friend, W.G.; Salkeld, E.G.; Stevenson, I.L. 1959 Nutrition of onion maggots, larvae of *Hylemya antiqua* (Meig.), with reference to other members of the genus *Hylemya*. ANN N Y ACAD SCI 77: 384-393.

Friend, W.G.; Smith, J.J.B. 1971 Feeding in *Rhodnius prolixus:* potenencies of nucleotide phosphate in initiating gorging. J INSECT PHYSIOL 17: 1315-1320.

Friend, W.G.; Smith, J.J.B. 1975 Feeding in *Rhodnius prolixus:* increasing sensitivity to ATP during prolonged food deprivation. J INSECT PHYSIOL 21: 1081-1084.

Frings, H. 1948 Rearing houseflies and blow flies on dog biscuit. SCIENCE 107: 629-630.

Frings, H.; Frings, M. 1953 Dog biscuit as larval medium for *Sarcophaga bullata*. J ECON ENTOMOL 46: 183.

*Fritzsche, R. 1960 Morphologische, biologische und physiologische Variabilität und ihre Bedeutung für die Epidemiologie und Bekämpfung von *Tetranychus urticae* Koch. BIOL ZENTRALB 79: 521-576. (In German)

*Frobrich, G.; Offhaus, K. 1953 Der qualitative Vitamintest mit dem Reismehilkäfer *Tribolium confusum* Duv. (Tenebrionidae, Coleoptera) als Testorganismus. Z VITAM HORM FERMENTFORSCH 5: 358-369. (In German with English summary)

*Frost, F.M.; Herms, W.B.; Hoskins, W.M. 1936 The nutritional requirements of the larvae of the mosquito, *Theobaldia incidens* (Thom.). J EXP ZOOL 73: 461-479.

Fukuda, T. 1963 A semi-synthetic diet for eri-silk raising. AGRIC BIOL CHEM 27: 601-609.

*Fukuda, T. 1964 Chemically defined diets for eri-silkworm raising. BULL SERIC EXPT STN JPN 19: 201-221. (In Japanese with English summary)

*Fukuda, T.; Higuchi, Y. 1963 Artificial food for oak-silkworm raising. AGRIC BIOL CHEM 27: 99-102.

*Fukuda, T.; Higuchi, Y. 1967 A semi-synthetic diet for eri-silkworm raising (continued). BULL SERIC EXP STN JPN 21: 401-423. (In Japanese with English summary)

Fukuda, T.; Higuchi, Y.; Matsuda, M. 1961 Artificial food for eri-silkworm raising. AGRIC BIOL CHEM 25: 417-420.

*Fukuda, T.; Suto, M.; Higuchi, Y. 1960a Silkworm raising on the artificial food. J SERIC SCI JPN 29: 1-3. (In Japanese)

Fukuda, T.; Suto, M.; Higuchi, Y. 1960b Silkworm raising on artificial food. NATURE LONDON 187: 669-670.

*Fukuda, T.; Suto, M.; Kameyama, T.; Kawasugi, S. 1962 Synthetic diet for silkworm raising. NATURE LONDON 196: 53-54.

Gahan, J.B.; Smith, C.N. 1964 Problems connected with raising mosquitoes in the laboratory. BULL W H O 31: 445-448.

*Gahukar, R.T. 1975 Nouvelles techniques adoptées pour l'élevage d'*Ostrinia nubilalis* Hubner sur milieu artificiel. ANN ZOOL ECOL ANIM 7: 491-498. (In French with English summary)

*Galford, J.R. 1967a A technique for rearing larvae of the smaller European elm bark beetle on an artificial medium. J ECON ENTOMOL 60: 1192.

Galford, J.R. 1967b Emergence of *Entedon leucogramma* from smaller European elm bark beetle larvae reared on artificial media. J ECON ENTOMOL 60: 1482-1483.

*Galford, J.R. 1969a Artificial rearing of 10 species of wood-boring insects. U S FOR SERV RES NOTE NE-102, 6 pp.

*Galford, J.R. 1969b A larval and ovipositional medium for the smaller European elm bark beetle. J ECON ENTOMOL 62: 1166-1168.

*Galford, J.R. 1971 Improved techniques for rearing the smaller European elm bark beetle on artificial media. J ECON ENTOMOL 64: 1327-1328.

*Galford, J.R. 1972 Some basic nutritional requirements of smaller European elm bark beetle larvae. J ECON ENTOMOL 65: 681-684.

*Galford, J.R. 1974 Some physiological effects of temperature on artificially reared red oak borers. J ECON ENTOMOL 67: 709-710.

*Galloway, I.D. 1970 Food medium and techniques for the laboratory rearing of *Merophyas divulsana* (Walker). QUEENSL J AGRIC ANIM SCI 27: 325-327.

**Galun, R. 1967 Feeding stimuli and artificial feeding. BULL W H O 36: 590-593.

**Galun, R. 1971 Recent developments in the biochemistry and feeding behaviour of haematophagous arthropods as applied to their mass rearing. pp. 273-282. *in* Sterility Principle for Insect Control or Eradication. Proceedings of a Panel Organized by the Joint FAO/IAEA Division of Atomic Energy in Food and Agriculture.

*Garcia, C. Jr; Frick, K.E. 1975 *Bactra verutana,* a possible biological control agent of purple and yellow nutsedge: large scale rearing on artificial diet. ANN ENTOMOL SOC AM 68: 15-18.

*Gardiner, L.M. 1970 Rearing wood-boring beetles (Cerambycidae) on artificial diet. CAN ENTOMOL 102: 113-117.

Garrison, M.; Wittwer, F.; Ziegler, R.; Bucklin, D. 1963 Artificial diet for *Hyalophora cecropia.* AM ZOOL 3: 531.

Gast, R.T. 1961 Some shortcuts in laboratory rearing of boll weevils. J ECON ENTOMOL 64: 395-396.

Gast, R.T. 1965 Mass producing artificial diet pellets for adult boll weevils. J ECON ENTOMOL 58: 1024-1025.

Gast, R.T. 1966a Oviposition and fecundity of boll weevils in mass-rearing laboratory cultures. J ECON ENTOMOL 59: 173-176.

Gast, R.T. 1966b A spray technique for implanting boll weevil eggs on artificial diets. J ECON ENTOMOL 59: 239-240.

**Gast, R.T. 1968 Mass rearing of insects: its concept, methods and problems. pp. 59-67. *in* Radiation, Radioisotopes and Rearing Methods in the Control of Insect Pests. Proceedings of a Panel Organized by the Joint FAO/IAEA Division of Atomic Energy in Food and Agriculture.

*Gast, R.T.; Davich, T.B. 1966 Boll weevils. pp. 405-418. *in* Smith, C.N. (Ed.), Insect Colonization and Mass Production. Academic Press, New York and London.

Gast, R.T.; Landin, M. 1966 Adult boll weevils and eggs marked with dye fed in larval diet. J ECON ENTOMOL 59: 474-75.

*Gay, F.J. 1938 A nutritional study of the larva of *Dermestes vulpinus* F. J EXP ZOOL 79: 93-107.

*Geer, B.W. 1963 A ribonucleic acid-protein relationship in *Drosophila* nutrition. J EXP ZOOL 154: 353-364.

Geer, B.W. 1966 Utilization of *d*-amino acids for growth by *Drosophila melanogaster* larvae. J NUTR 90: 31-39.

Geer, B.W.; Olander, R.M.; Sharp, P.L. 1970 Quantification of dietary choline utilization in adult *Drosophila melanogaster* by radioisotope methods. J INSECT PHYSIOL 16: 33-43.

*Geer, B.W.; Vovis, G.F. 1965 The effect of choline and related compounds on the growth and development of *Drosophila melanogaster.* J EXP ZOOL 158: 223-236.

Geest, L.P.S. van der; Sloog-Hoebe, A.A.M. 1972 Effect of various vitamins on the development of *Pieris brassicae* larvae. MEDED FAC LANDBOUWWET RIJKSUNIV GENT 37: 713-716.

**Gelperin, A. 1971 Regulation of feeding. ANNU REV ENTOMOL 16: 365-378.

*Gemetchu, T. 1971 Liver and yeast as larval diets in colonization of a sandfly (*Phlebotomus longipes*). TRANS R SOC TROP MED HYG 65: 682-683.

Gemrich, E.G. 1972 Effect of polyene antibiotics on the growth and development of the house fly. J ECON ENTOMOL 65: 1552-1554.

*George, B.W.; Raun, E.S.; Peters, D.C.; Mendoza, C. 1960 Artificial medium for rearing some lepidopterous corn insects. J ECON ENTOMOL 53: 318-319.

George, J.A.; Howard, M.G. 1965 A waxed-paper laboratory cage for sterilization studies with the oriental fruit moth, *Grapholitha molesta* (Busck). PROC ENTOMOL SOC ONT 95: 146-147.

*Gerberg, E.J. 1970 Manual for mosquito rearing and experimental techniques. AM MOSQ CONTROL ASSOC BULL 5: 109 pp.

Gerberich, J.B. 1948 Rearing house-flies on common bacteriological media. J ECON ENTOMOL 41: 125-126.

*Gershenson, S. 1928 A new sex-ratio abnormality in *Drosophila obscura*. GENETICS 13: 488-507.

Getzin, L.W. 1962 Mass rearing of virus-free cabbage loopers on an artificial diet. J INSECT PATHOL 4: 486-488.

*Ghouri, A.S.K.; McFarlane, J.E. 1958 Observations on the development of crickets. CAN ENTOMOL 90: 158-165.

Gier, H.T. 1947 Growth rate in the cockroach *Periplaneta americana* (Linn). ANN ENTOMOL SOC AM 40: 303-317.

Gifawesen, C.; Funke, B.R.; Proshold, F.I. 1975 Control of antifungal-resistant strains of *Aspergillus niger* mould contaminants in insect rearing media. J ECON ENTOMOL 68: 441-444.

**Gilbert, I.H. 1964 Laboratory rearing of cockroaches, bed-bugs, human lice and fleas. BULL W H O 31: 561-563.

*Gilliam, M.; Argauer, R.J. 1975a How long is Teramycin stable in diets fed to honey bee colonies for disease control? AM BEE J 115: 230-234.

Gilliam, M.; Argauer, R.J. 1975b Stability of oxytetracycline in diets fed to honeybee colonies for disease control. J INVERTEBR PATHOL 26: 383-386.

**Gilmour, D. 1961 The Biochemistry of Insects. Academic Press, New York and London. 343 pp.

**Gilmour, D. 1965 The Metabolism of Insects. W.H. Freeman and Company, San Francisco. 195 pp.

**Ginet, R. 1966 Influence de la nutrition sur la durée de la vie des poecilothermes. ANN BIOL 5: 75-93. (In French)

*Gingrich, R.E. 1960 Development of a synthetic medium for aseptic rearing of larvae of *Stomoxys calcitrans* (L.). J ECON ENTOMOL 53: 408-411.

*Gingrich, R.E. 1964 Nutritional studies on screw worm larvae with chemically defined media. ANN ENTOMOL SOC AM 57: 351-360.

Gingrich, R.E. 1972 Nutrient studies: their bearing on the development of practical oligidic diets for mass rearing larvae of the screw worm, *Cochliomyia hominivorax*. pp. 257-270. *in* Rodriguez, J.G. (Ed.), Insect and Mite Nutrition. North Holland Publishing Company, Amsterdam.

*Gingrich, R.E.; Graham, A.J.; Hightower, B.G. 1971 Media containing liquified nutrients for mass-rearing larvae of the screw worm. J ECON ENTOMOL 64: 678-683.

Glaser, R.W. 1924 Rearing flies for experimental purposes with biological notes. J ECON ENTOMOL 17: 486-496.

Glaser, R.W. 1938 Method for the sterile culture of houseflies. J PARASITOL 24: 177-179.

**Glaser, R.W. 1943 The germ-free culture of certain invertebrates. pp. 164-187. *in* Reyniers, J.A. (Ed.), Micrurgical and Germfree Methods; their Application to Experimental Biology and Medicine. Thomas, Springfield, Illinois.

Gochnauer, T.A.; Furgala, B. 1962 The *Nosema* inhibitory activity of alcohol-1, a component of Fumagillin. J INSECT PATHOL 4: 489-491.

Golanski, K. 1961 The effectiveness of formalin in controlling jaundice (nuclear polyhedrosis) of the silkworm in Poland. J INSECT PATHOL 3: 11-14.

Golberg, L.; De Meillon, B. 1947 Further observations on the nutritional requirements of the larva of *Aedes aegypti* L. NATURE LONDON 160: 582-583.

*Golberg, L.; De Meillon, B. 1948 The nutrition of the larvae of *Aedes aegypti* Linnaeus. 4. Protein and amino-acid requirements. BIOCHEM J 43: 379-387.

Golberg, L.; De Meillon, B.; Murray, J.F. 1947 Pteroylglutamic acid in liver extracts and urine. NATURE LONDON 160: 22-23.

*Gonen, M.; Donahay, E. 1973 An improved technique for rearing the tropical warehouse moth *Ephestia cautella* (Walker) (Lepidoptera; Phycitidae). ISR J ENTOMOL 8: 179-180.

*Good, M.E.; Morrison, F.O.; Mankiewicz, N.E. 1953 Lipidolytic enzymes extracted from *Galleria mellonella* L. (Lepidoptera: Pyralidae) reared on natural and artificial media. CAN ENTOMOL 85: 252-253.

*Goodfellow, R.D.; Barnes, F.J.; Graham, W.J. Jr 1971 Dietary sterol effects on mevalonate kinase activity in axenic and non-axenic *Sarcophaga bullata* (Diptera). J INSECT PHYSIOL 17: 1625-1635.

**Gooding, R.H. 1972 Digestive processes of haematophages insects. 1. A literature review. QUAEST ENTOMOL 8: 5-60.

*Goodnight, K.C.; Kircher, H.W. 1971 Metabolism of lathosterol by *Drosophila pachea*. LIPIDS 6: 166-169.

*Goodwin, R.H. 1966 Axenic culture of the armyworm *Pseudaletia unipuncta* (Haworth) on a meridic diet. NATURE LONDON 212: 799-800.

*Goonewardene, H.F.; McKay, J.E. 1969 An artificial diet for the adult Japanese beetle. J ECON ENTOMOL 62: 964.

*Goonewardene, H.F.; Zepp, D.B.; McGuire, J.U. Jr 1974 Evaluation of artificial diets for 3rd stage larvae of Japanese beetles. ANN ENTOMOL SOC AM 67: 413-415.

**Gordon, H.T. 1959 Minimal nutritional requirements of the German roach, *Blattella germanica* L. ANN N Y ACAD SCI 77: 290-351.

**Gordon, H.T. 1961 Nutritional factors in insect resistance to chemicals. ANNU REV ENTOMOL 6: 27-54.

Gordon, H.T. 1968a Intake rates of various solid carbohydrates by male German cockroaches. J INSECT PHYSIOL 14: 41-52.

**Gordon, H.T. 1968b Quantitative aspects of insect nutrition. AM ZOOL 8: 131-138.

Gordon, H.T. 1974 Cohort rearing and prolonged cold-storage of the sunflower strain of *Oncopeltus fasciatus*. ANN ENTOMOL SOC AM 67: 976-980.

*Gothilf, S. 1968 The biology of the carob moth (*Ectomyelois ceratoniae* (Zell)) in Israel. I. Mass culture on artificial diet. ISR J ENTOMOL 3: 109-118.

*Gothilf, S.; Beck, S.D. 1966 Rearing the citrus mealy bug, *Planococcus citri* (Risso), on a defined diet. J ECON ENTOMOL 59: 489-490.

*Gothilf, S.; Beck, S.D. 1967 Larval feeding behaviour of the cabbage looper, *Trichoplusia ni*. J INSECT PHYSIOL 13: 1039-1053.

*Gouger, R.J.; Bray, D.F. 1968 Two semi-artificial oviposition media for the smaller European elm bark beetle. J ECON ENTOMOL 61: 564-565.

Graf, E.; Benz, G. 1970 Toxicity of Streptomycin and Terramycin, and influence on growth and development time of *Drosophila melanogaster*. EXPERIENTIA 26: 1339-1341.

*Graham, A.J.; Dudley, F.H. 1959 Culture methods for mass rearing of screw worm larvae. J ECON ENTOMOL 52: 1006-1008.

Graham, H.M.; Mangum, C.L. 1971 Larval diets containing dyes for tagging pink bollworm moths internally. J ECON ENTOMOL 64: 376-379.

Greenberg, B. 1954 A method for the sterile culture of house fly larvae, *Musca domestica* L. CAN ENTOMOL 86: 527-528.

Greenberg, B. 1959 House fly nutrition. I. Quantitative study of the protein and sugar requirements of males and females. J CELL COMP PHYSIOL 53: 169-177.

Greenberg, B. 1960a House fly nutrition. II. Comparative survival values of sucrose and water. ANN ENTOMOL SOC AM 53: 125-128.

*Greenberg, B. 1960b Host contaminant biology of muscoid flies. I. Bacterial survival in the pre-adult stages and adults of four species of blow flies. J INSECT PATHOL 2: 44-54.

**Greenberg, B. 1967 Gnotobiotic insects in biomedical research. pp. 410-416. *in* Miyakawa, M.; Luckey, T.D. (Eds), Advances in Germfree Research and Gnotobiology. Chemical Rubber Company, Cleveland, Ohio.

*Greenberg, B. 1969 Sterile culture of *Musca sorbens*. ANN ENTOMOL SOC AM 62: 450.

**Greenberg, B. 1970 Sterilizing procedures and agents, antibiotics and inhibitors in mass rearing of insects. BULL ENTOMOL SOC AM 16: 31-36.

*Greenberg, B.; Burkman, A.M. 1963 Effect of B-vitamins and a mixed flora on the longevity of germ-free adult houseflies, *Musca domestica* L. J CELL COMP PHYSIOL 62: 17-22.

Greene, G.L. 1970 Hatching of bean leaf roller eggs as influenced by insect diet components. J ECON ENTOMOL 63: 321.

Grenier, S.; Bonnot, G.; Delobel, B. 1974 Définition et mise au point de milieux artificiels pour l'élevage *in vitro* de *Phryxe caudata* Rond. (Diptera: Tachinidae). I. Survie du parasitoide sur milieux dont la composition est basée sur celle de l'hémolymphe de l'hôte. ANN ZOOL ECOL ANIM 6: 511-520.

*Grenier, S.; Bonnot, G.: Delobel, B. 1975 Définition et mise au point de milieux artificiels pour l'élevage *in vitro* de *Phryxe caudata* Rond. (Diptera, Tachinidae). II. Croissance et mues larvaires du parasitoide en milieux définis. ANN ZOOL ECOL ANIM 7: 13-25.

*Grieve, E.G. 1937 The culture of *Muscina stabulans*. pp. 432-434. *in* Needham, J.G. *et al.* (Eds), Culture Methods for Invertebrate Animals. Comstock Publishing Company, Ithaca N.Y. (Dover edition, 1959).

Griffin, J.G.; Lindig, O.H. 1973 Ultraviolet lamp pass-through cabinet for use in boll weevil mass-rearing facility. J ECON ENTOMOL 66: 1063-1066.

Griffin, J.G.; Lindig, O.H. 1974 Mechanized production of boll weevil diet pellets. TRANS ASAE 17: 15-19.

Griffin, J.G.; Lindig, O.H. 1975 Granular materials used in implanting boll weevil eggs on artificial diet. J ECON ENTOMOL 68: 433-434.

Griffin, J.G.; Lindig, O.H.; McLaughlin, R.E. 1974 Flash sterilizers: sterilizing artificial diets for insects. J ECON ENTOMOL 67: 689.

*Griffiths, D.A. 1966 Nutrition as a factor influencing hypopus formation in the *Acarus siro* species complex (Acarina, Acaridae). J STORED PROD RES 1: 325-340.

Griffiths, D.A. 1969 The influence of dietary factors on hypopus formation in *Acarus immobilis* Griffiths (Acari, Acaridae). PROC 2 INT CONGR ACAROL (1967): 419-432.

Griffiths, D.C.; Greenway, A.R.; Lloyd, S.L. 1975 An artificial diet with improved acceptability to three strains of *Myzus persicae* (Sulz.) (Hemiptera: Aphididae). BULL ENTOMOL RES 65: 349-358.

Grijpma, P. 1971 Studies on the shoot borer, *Hypsipyla grandella* Zeller. V. Observations on a rearing technique and on host selection of adults in captivity. TURRIALBA 21: 202-213.

*Grisdale, D.G. 1963 Rearing insects on artificial diet. Interim Research Report of the Insect Pathology Research Institute, Sault Ste Marie, Ontario, Canada.

Grisdale, D.G. 1970 An improved laboratory method for rearing large numbers of spruce budworm, *Choristoneura fumiferana* (Lepidoptera: Tortricidae). CAN ENTOMOL 102: 1111-1117.

Grisdale, D.G. 1972 An improved method for producing large numbers of second-instar spruce budworm larvae, *Choristoneura fumiferana* (Lepidoptera: Tortricidae). CAN ENTOMOL 104: 1955-1957.

*Grisdale, D.G. 1973 Large volume preparation and processing of a synthetic diet for insect rearing. CAN ENTOMOL 105: 1553-1557.

*Grisdale, D.G. 1976 Laboratory methods for rearing the forest tent caterpillar. BIMON RES NOTES CAN FOR SERV 32: 1.

**Grison, P. 1948 Alimentation artificielle des insectes phytophages. ANN SCI NAT ZOOL 10: 59-65. (In French)

Griswald, G.H. 1944 Studies on the biology of the webbing clothes moth (*Tineola bisselliella* Hum.). MEM CORNELL UNIV AGRIC EXP STN 262, 59 pp.

**Groot, A.P. De 1953 Protein and amino acid requirements of the honey bee (*Apis mellifica* L.). PHYSIOL COMP OECOL 3: 197-285.

Gueldner, R.C.; Hedin, P.A.; Woodard, D.N. 1975 Mineral content of boll weevils, cotton buds and synthetic diets. J ECON ENTOMOL 68: 428-430.

**Guennelon, G. 1967 L'alimentation artificielle des insectes. REV ZOOL AGRIC APPL 66: 20-28. (In French)

**Guennelon, G. 1968 L'alimentation artificielle des larves de lépidoptères phytophages. ANN EPIPHYT 19: 539-570. (In French)

*Guennelon, G.; Sender, C.; D'Arcier, F.; Audemard, H. 1970 Mise au point d'un milieu artificiel pour l'élevage au laboratoire des larves de l'eudémis de la vigne, *Lobesia botrana* Den.et Schiff. (Lepidoptera: Tortricidae). ANN ZOOL ECOL ANIM 2: 51-77. (In French with English summary)

*Guennelon, G.; Soria, F. 1973 Mise au point au laboratoire d'un élevage permanent de la pyrale du riz, *Chilo suppressalis* Walker (Lepidoptera: Pyralidae) sur milieu artificiel. ANN ZOOL ECOL ANIM 5: 547-558. (In French with English summary)

*Guss, P.L.; Krysan, J.L. 1973 Maintenance of the southern corn rootworm on a dry diet. J ECON ENTOMOL 66: 352-353.

*Guthrie, W.D.; Raun, E.S.; Dicke, F.F.; Pesho, G.R.; Carter, S.W. 1965 Laboratory production of European corn borer egg masses. IOWA STATE COLL J SCI 40: 65-83.

Guyénot, E. 1913 Études biologique sur une mouche *Drosophila ampelophila* Low. V. Nutrition des adultes et fecondite. C R SOC BIOL 74: 332-334. (In French)

*Guyénot, E. 1917 Recherches expérimentiles sur la vie aseptique et la développement d'un organisme (*Drosophila ampelophila*) en fonction du milieu. BULL BIOL FR BELG 51: 1-330. (In French)

Hadlington, P.W.; Johnston, J.A. 1973 Mass rearing of *Plagiohammus spinipennis* (Thomas.) (Col: Cerambycidae) for the biological control of *Lantana camara* (Verbenaceae). J ENTOMOL SOC AUST 8: 32-36.

*Haeger, S.; Provost, W. 1964 The laboratory colonization of *Opifex*. BULL W H O 31: 451-452.

Hafez, M. 1948 A simple method for breeding the house-fly, *Musca domestica* L., in the laboratory. BULL ENTOMOL RES 39: 385-386.

*Hafez, M.; Attia, M.A. 1958 Rearing and culturing *Musca sorbens* Wied. in the laboratory. BULL ENTOMOL RES 49: 633-635.

*Hafez, M.; Salama, H.S.; Saleh, M.R. 1971 Survival and development of *Lecanium acuminatum* Sign. (Coccoidea) on a host plant and artificial diet. Z ANGEW ENTOMOL 69: 182-186.

*Hafez, M.; Shoukry, A. 1972 Effect of irradiation on adult fecundity and longevity of the Mediterranean fruit fly, *Ceratitis capitata* Wied., in Egypt (Diptera: Tephritidae). Z ANGEW ENTOMOL 72: 59-66.

*Hafez, M.; Zein-el-Dine, K. 1964 Culturing *Phlebotomus papatasii* (Scopali) in the laboratory. BULL ENTOMOL RES 54: 657-659.

Hagen, K.S. 1950 Fecundity of *Chrysopa californica* Coquillet as affected by synthetic foods. J ECON ENTOMOL 43: 101-104.

Hagen, K.S. 1958 Honeydew as an adult fruit fly diet affecting reproduction. PROC 10 INT CONGR ENTOMOL MONTREAL (1956) 3: 25-30.

*Hagen, K.S. 1964 Nutrition of entomophagous insects and their hosts. pp. 356-380. *in* de Bach, P. (Ed.), Biological Control of Insect Pests and Weeds. Chapman & Hall, London.

Hagen, K.S. 1966 Dependence of the olive fly *Dacus oleae*, larvae on symbiosis with *Pseudomonas savastanoi* for the utilization of olive. NATURE LONDON 209: 423-424.

Hagen, K.S.; Finney, G.L. 1950 A food supplement for effectively increasing the fecundity of certain tephritid species. J ECON ENTOMOL 43: 735.

*Hagen, K.S.; Santas, L.; Tsecouras, A. 1963 A technique for culturing the olive fly, *Dacus oleae* Gmel., on synthetic media under xenic conditions. pp. 333-356. *in* Radiation and Radioisotopes Applied to Insects of Agricultural Importance. Proceedings of a Panel Organized by the Joint FAO/IAEA Division of Atomic Energy in Food and Agriculture.

*Hagen, K.S.; Tassan, R.L. 1965 A method of providing artificial diets to *Chrysopa* larvae. J ECON ENTOMOL 58: 999-1000.

Hagen, K.S.; Tassan, R.L. 1966a Artificial diet for *Chrysopa carnea* Stephens. pp. 83-87. *in* Ecology of Aphidophagous Insects. Proceedings of Symposium held in Liblice Prague, Sept 27-Oct 1, 1965.

Hagen, K.S.; Tassan, R.L. 1966b A method of coating droplets of artificial diets with paraffin for feeding *Chrysopa* larvae. pp. 89-90. Proceedings of Symposium held in Liblice Prague, Sept 27-Oct 1, 1965.

*Hagen, K.S.; Tassan, R.L. 1966c The influence of protein hydrolysates of yeasts and chemically defined diets upon the fecundity of *Chrysopa carnea* Stephens (Neuroptera). VESTN CESK SPOL ZOOL 30: 219-227.

*Hagen, K.S.; Tassan, R.L. 1970 The influence of food Wheast and related *Saccharomyces fragilis* yeast products on the fecundity of *Chrysopa carnea* (Neuroptera: Chrysopidae). CAN ENTOMOL 102: 806-811.

**Hagen, K.S.; Tassan, R.L. 1972 Exploring nutritional roles of extracellular symbiotes on the reproduction of honeydew feeding adult chrysopids and tephritids. pp. 323-351. *in* Rodriguez, J.G. (Ed.), Insect and Mite Nutrition. North-Holland Publishing Company, Amsterdam.

Hagen, K.S.; Tassan, R.L.; Sawall, E.F. Jr 1970 Some ecophysiological relationships between certain *Chrysopa*, honeydews and yeasts. BOLL LAB ENTOMOL AGRAR "FILIPPO SILVESTRI" PORTICI 28: 113-134.

*Hagley, E.A.C. 1967 Artificial diet for the adult froghopper. NATURE LONDON 213: 414-415.

*Haisch, A. 1968 Preliminary results in rearing the cherry fruit-fly (*Rhagoletis cerasi* L.) on a semi-synthetic medium. pp. 69-78. *in* Radiation, Radioisotopes and Rearing Methods in the Control of Insect Pests. Proceedings of a Panel Organized by the Joint FAO/IAEA Division of Atomic Energy in Food and Agriculture.

*Haisch, A. 1970 Rearing experiment with the European cherry fruit fly. pp. 107-109. *in* Sterile-male Technique for Control of Fruit Flies. Proceedings of a Panel Organized by the Joint FAO/IAEA Division of Atomic Energy in Food and Agriculture.

Haisch, A. 1974 Genetic control of *Rhagoletis cerasi* L.: preconditions, present situation and prospects. pp. 27-35. *in* The Sterile-insect Technique and its Field Application. Proceedings of a Panel Organized by the Joint FAO/IAEA Division of Atomic Energy in Food and Agriculture.

*Haisch, A. 1975 Raising and multiplying of the European cherry fruit fly, *Rhagoletis cerasi* L, pp. 120-121. *in* Controlling Fruit Flies by the Sterile-insect Technique. Proceedings of a Panel Organized by the Joint FAO/IAEA Division of Atomic Energy in Food and Agriculture.

*Haisch, A.; Boller, E.F. 1971 Genetic control of the European fruit fly, *Rhagoletis cerasi* L. Progress report on rearing and sterilization. pp. 67-76. *in* Sterility Principle for Insect Control or Eradication. Proceedings of a Panel Organized by the Joint FAO/IAEA Division of Atomic Energy in Food and Agriculture.

Hamamura, Y. 1970 The substances that control the feeding behaviour and the growth of the silkworm, *Bombyx mori* L. pp. 55-80 *in* Wood, D.C.; Silverstein, R.M.; Nakajima, M. (Eds), Control of Insect Behaviour by Natural Products. Academic Press, New York and London.

Hamamura, Y.; Kuwata, K.; Masuda, H. 1966 Effect of gallic acid on the growth of the silkworm larvae *Bombyx mori* L. NATURE LONDON 212: 1386-1387.

*Hamilton, D.W.; Hathaway, D.O. 1966 Codling moths. pp. 339-354. *in* Smith, C.N. (Ed.), Insect Colonization and Mass Production. Academic Press, New York and London.

Hamilton, E.W. 1972 Pollen substitute and honey in diet for *Diabrotica*. J ECON ENTOMOL 65: 887.

Hammen, C.S. 1956 Nutrition of *Musca domestica* in single-pair culture. ANN ENTOMOL SOC AM 49: 365-368.

Hammen, C.S. 1957 A growth-promoting effect of cholesterol in the diet of larvae of the house fly, *Musca domestica* L. ANN ENTOMOL SOC AM 50: 125-127.

Hanna, M.A.; Hibbs, E.T. 1970 Feeding phytophagous mites on liquid formulations. J ECON ENTOMOL 63: 1672-1673.

Harding, J.A.; Brindley, T.A.; Dyar, R.C. 1967 Survival and development of European corn borers fed gossypol in artificial diets. J ECON ENTOMOL 60: 1764-1765.

*Hardwick, D.F. 1965 The corn earworm complex. MEM ENTOMOL SOC CAN 40: 247 pp.

Hare, W.W.; Harrell, E.A.; Perkins, W.D. 1973 Moisture removed from rearing cells during incubation of corn earworm on an artificial diet. J ECON ENTOMOL 66: 284-285.

Harley, K.L.S.; Thorsteinson, A.J. 1967 The influence of plant chemicals on the feeding behaviour, development, and survival of the two-striped grasshopper, *Melanopus bivittatus* (Say), Acrididae: Orthoptera. CAN J ZOOL 45: 305-319.

*Harley, K.L.S.; Willson, B.W. 1968 Propagation of a cerambycid borer on a meridic diet. CAN J ZOOL 46: 1265-1266.

*Harlow, P.M. 1956 A study of ovarial development and its relation to adult nutrition in the blowfly, *Protophormia terrae-novae* (R.D.). J EXP BIOL 3: 777-797.

*Harper, J.D. 1970 Laboratory production of *Peridroma saucia* and its nuclear polyhedrosis virus. J ECON ENTOMOL 63: 1633-1634.

Harrell, E.A.; Sparks, A.N.: Perkins, W.D.; Hare, W.W. 1973 An insect diet filler for an inline form - fill - seal machine. J ECON ENTOMOL 66: 1340-1341.

Harrewijn, P. 1973 Functional significance of indole alkylamines linked to nutritional factors in wing development of the aphid *Myzus persicae*. ENTOMOL EXP APPL 16: 499-513.

Harrewijn, P.; Noordink, J.Ph.W. 1971 Taste reception of *Myzus persicae* in relation to food uptake and developmental processes. ENTOMOL EXP APPL 14: 413-419.

*Harris, C.R.; Svec, H.J.; Begg, J.A. 1966 Mass rearing of root maggots under controlled environmental conditions: seed corn maggot, *Hylemya cilicrura;* bean seed fly *H. liturata, Euxesta notata;* and *Chaetopsis* sp. J ECON ENTOMOL 59: 407-410.

*Harris, R.L. 1962 Laboratory colonization of the horn fly, *Haematobia irritans* (L.). NATURE LONDON 196: 191-192.

*Harris, R.L.; Frazer, E.D.; Grossman, P.D. 1967 Artificial media for rearing larvae of horn flies. J ECON ENTOMOL 60: 891-892.

*Hartzell, A.; Burchfield, H.P.; Coty, V.F. 1958 Improvements in culturing larvae of *Aedes aegypti* (L.) for use in insecticide bioassay. CONTRIB BOYCE THOMSON INST 19: 321-326.

Harvey, G.T. 1974 Nutritional studies of eastern spruce budworm (Lepidoptera: Tortricidae). I. Soluble sugars. CAN ENTOMOL 106: 353-365.

Harvey, G.T. 1975 Nutritional studies of Eastern spruce budworm (Lepidoptera: Tortricidae). II. Starches. CAN ENTOMOL 107: 717-728.

Harvey, T.L.; Howell, D.E. 1965 Resistance of house fly strains resistant to *Bacillus thuringiensis* Berliner. J INVERTEBR PATHOL 7: 92-100.

Haskins, K.P.F. 1962 A new system for rearing the American cockroach. ENTOMOLOGIST LONDON 95: 27-29.

Hassan, S.A. 1975 Über die Massenzucht von *Chrysopa carnea* Steph. (Neuroptera, Chrysopidae). Z ANGEW ENTOMOL 79: 310-315. (In German with English summary)

*Hatchett, J.H.; Jackson, R.D.; Barry, R.M. 1973 Rearing a weed cerambycid, *Dectes texanus*, on an artificial medium, with notes on biology. ANN ENTOMOL SOC AM 66: 519-522.

Hathaway, D.O.; Clift, A.E.; Butt, B.A. 1971 Development and fecundity of codling moths reared on artificial diets or immature apples. J ECON ENTOMOL 64: 1088-1090.

Hathaway, D.O.; Lydin, L.V.; Butt, B.A.; Morton, L.J. 1973 Monitoring mass rearing of the codling moth. J ECON ENTOMOL 66: 390-393.

**Hatmosoewarno, S.; Butt, B.A. 1974 Larval diets for the codling moth (*Laspeyresia pomonella* L. Lepidoptera: Olethreutidae). pp. 579-584. *in* Sterility Principle for Insect Control. Proceedings of a Panel Organized by the Joint FAO/IAEA Division of Atomic Energy in Food and Agriculture.

Hayashiya, K.; Kato, M.; Hamamura, Y. 1965 Acetylcholine as a growth factor in early larval development of the silkworm. NATURE LONDON 205: 620-621.

*Haydak, M.H. 1936a A food for rearing laboratory insects. J ECON ENTOMOL 29: 1026.

Haydak, M.H. 1936b Is wax a necessary constituent of the diet of the wax moth larvae? ANN ENTOMOL SOC AM 29: 581-588.

Haydak, M.H. 1940 The length of development of the greater wax moth. SCIENCE 91: 525.

*Haydak, M.H. 1941 Nutrition of the wax moth larvae. Vitamin requirement. I. Requirement for vitamin B_1. PROC MINN ACAD SCI 9: 27-29.

**Haydak, M.H. 1968 Nutrition des larves d'abeilles. pp. 302-333. *in* Chauvin, R. (Ed.), Traité de Biologie de l'abeille. Masson et die Paris. (In French)

**Haydak, M.H. 1970 Honeybee nutrition. ANNU REV ENTOMOL 15: 143-156.

Hays, S.B.; Amerson, G.M. 1966 New devices for rearing and handling house flies in the laboratory. J ECON ENTOMOL 59: 1523-1524.

Hedin, P.A.; Thompson, A.C.; Gueldner, R.C.; Henson, R.D. 1974 Analysis of the antimicrobial agents potassium sorbate and methyl *p*-hydroxybenzoate, in boll weevil diets. J ECON ENTOMOL 67: 147-149.

*Hedlin, A.F. 1965 Rearing cone insects in the laboratory on an artificial diet. BI-MON NOTES CAN FOR RES SERV 21: 4.

Heed, W.B.; Kricher, H.W. 1965 Unique sterol in the ecology and nutrition of *Drosophila pachea*. SCIENCE 149: 758-761.

*Hegdekar, B.M. 1970 Amino acid analogues as inhibitors of insect reproduction. J ECON ENTOMOL 63: 1950-1956.

**Helms, J.T.; Raun, E.S. 1971 Perennial laboratory culture of disease-free insects. pp 639-654. *in* Burges, H.D.; Hussey, N.W. (Eds), Microbial Control of Insects and Mites. Academic Press, New York and London.

*Henneberry, T.J.; Kishaba, A.N. 1966a Cabbage loopers. pp. 461-478. *in* Smith, C.N. (Ed.), Insect Colonization and Mass Production. Academic Press, New York and London.

Henneberry, T.J.; Kishaba, A.N. 1966b Pupal size and mortality, longevity and reproduction of cabbage loopers reared at several densities. J ECON ENTOMOL 59: 1490-1493.

Henneberry, T.J.; Kishaba, A.N. 1967 Mating and oviposition of the cabbage looper in the laboratory. J ECON ENTOMOL 60: 692-696.

Henry, J.E. 1968 *Malameba locustae* and its antibiotic control in grasshopper cultures. J INVERTEBR PATHOL 11: 224-233.

Henry, J.E.; Oma, E.A. 1975 Sulphonamide antibiotic control of *Malameba locustae* (King & Taylor) and its effect on grasshoppers. ACRIDA 4: 217-226.

**Henry, S.M. 1962 The significance of microorganisms in the nutrition of insects. TRANS N Y ACAD SCI 24: 676-683.

Henry, S.M.; Cotty, V.F. 1957 The rearing of aseptic adult house flies for physiological studies. CONTR BOYCE THOMSON INST PL RES 19: 227-229.

*Hensley, S.D.; Hammond, A.M. Jr 1968 Laboratory techniques for rearing the sugarcane borer on an artificial diet. J ECON ENTOMOL 61: 1742-1743.

Herger, P. 1973 Künstliche Ernährung von Blattläusen: Eine neue Technik zur Herstellung der Parafilm-Sachet. EXPERIENTIA 29: 1312-1314. (In German with English summary)

*Herger, P. 1975 Einfluss von Farbe und Nahrungszusammensetzung auf das Saugerhalten der künstlich ernährten Ampferblattlaus, *Aphis rumicics* (Homoptera: Aphididae). ENTOMOL GERM 2: 149-166. (In German with English summary)

**Herger, P.; Lampel, G. 1975 Zuchtversuche mit mono-bis polyphagen Röhrenblattlaus-Arten auf Künstlichem Nährmedium; (Homoptera: Aphidina: Aphididae). ENTOMOL GERM 1: 258-278. (In German with English summary)

*Hertel, R. 1974 Einfluss von ATP in einer holidischen Diät auf *Myzus persicae* (Sulz.) (Aphidina). EXPERIENTIA 30: 775-776. (In German with English summary)

Hertig, M. 1964 Laboratory colonization of central American *Phlebotomus* sandflies. BULL W H O 31: 569-570.

*Herzog, G.A.; Phillips, J.R. 1974 Selection for a nondiapause strain of the bollworm, *Heliothis zea* (Lepidoptera: Noctuidae). ENVIRON ENTOMOL 3: 525-527.

Hilchey, J.D. 1953 Studies on the qualitative requirements of *Blattella germanica* (L.) for amino acids under aseptic conditions. CONTR BOYCE THOMSON INST PL RES 17: 203-219.

*Hill, D.L.; Bell, V.A.; Chadwick, L.E. 1947 Rearing of the blowfly, *Phormia regina* Meigen, on a sterile synthetic diet. ANN ENTOMOL SOC AM 40: 213-216.

*Hill, S.T. 1964 Axenic culture of the foreign grain beetle, *Ahasverus advena* (Waltl) (Coleoptera: Silvanidae) and the role of fungi in its nutrition. BULL ENTOMOL RES 55: 681-690.

Hinton, H.E. 1956 Dietary requirements of insects. Amino acids and vitamins. SCI PROG LONDON 44: 292-309.

Hinton, T. 1952 A quantitative study of folic acid requirements and reversal of aminoprotein inhibition in *Drosophila*. SCIENCE 116: 708-710.

Hinton, T. 1955 The genetic basis of a nutritional requirement in *Drosophila*. GENETICS 40: 224-234.

Hinton, T. 1956a Nucleic acid utilization by *Drosophila*. PHYSIOL ZOOL 29: 20-26.

Hinton, T. 1956b The effects of arginine, ornithine and citrulline on the growth of *Drosophila*. ARCHS BIOCHEM BIOPHYS 62: 78-85.

*Hinton, T.; Dunlap, A. 1958 The genetical basis for nutritional requirements in insects. PROC 10 INT CONGR ENTOMOL MONTREAL (1956) 2: 123-126.

Hinton, T.; Noyes, D.T.; Ellis, J. 1951 Amino acids and growth factors in a chemically defined medium for *Drosophila*. PHYSIOL ZOOL 64: 335-353.

Hirano, C. 1963 Effect of dietary unsaturated fatty acids on the growth of larvae of *Chilo suppressallis* Walker (Lepidoptera: Pyralidae). JPN J APPL ENTOMOL ZOOL 7: 59-62.

Hirano, C.; Ishii, S. 1957 Nutritive values of carbohydrates for the growth of larvae of the rice stem borer, *Chilo suppressallis* Walker. BULL NATL INST AGRIC SCI TOKYO 7: 89-99.

Hirano, C.; Ishii, S. 1962 Utilization of dietary carbohydrates and nitrogen by rice stem borer larvae under axenic conditions. ENTOMOL EXP APPL 5: 53-60.

Hobson, R.P. 1933 Growth of blowfly larvae on blood and serum. I. Responses of aseptic larvae to vitamin B. BIOCHEM J 27: 1899-1909.

*Hobson, R.P. 1935a On a fat-soluble growth factor required by blowfly larvae. I. Distribution and properties. BIOCHEM J 29: 1292-1296.

Hobson, R.P. 1935b On a fat-soluble growth factor required by blowfly larvae. II. Identity of the growth factor with cholesterol. BIOCHEM J 29: 2023-2026.

Hodgson, E.; Cheldelin, V.H.; Newburgh, R.W. 1956 Substitution of choline by related compounds and further studies on amino acid requirements in nutrition of *Phormia regina* (Meig.). CAN J ZOOL 34: 527-532.

Hodgson, E.; Dauterman, W.C. 1960 Nutrition and metabolism of methyl donars and related compounds in the blow fly, *Phormia regina* (Meigen). ARCH BIOCHEM BIOPHYS 87: 48-54.

Hodgson, E.; Dauterman, W.C. 1964 The nutrition of choline, carnitine and related compounds in the blow fly, *Phormia regina* Meigen. J INSECT PHYSIOL 10: 1005-1008.

*Hodgson, E.; Rock, G.C. 1971 Insect Nutrition. pp. 105-145. *in* Kerkut, G.A. (Ed.), Experiments in Physiology and Biochemistry Vol 4. Academic Press, New York and London.

*Hoffman, J.D.; Ignoffo, C.M. 1974 Growth of *Pteromalus puparum* in a semisynthetic medium. ANN ENTOMOL SOC AM 67: 524-525.

*Hoffman, J.D.; Ignoffo, C.M.; Dickerson, W.A. 1975 *In vitro* rearing of the endoparasitic wasp, *Trichogramma pretiosum*. ANN ENTOMOL SOC AM 68: 335-336.

*Hoffman, J.D.; Ignoffo, C.M.; Long, S.H. 1973 *In vitro* cultivation of an endoparasitic wasp, *Pteromalus puparum*. ANN ENTOMOL SOC AM 66: 633-634.

*Hoffman, J.D.; Lawson, F.R.; Yamamoto, R. 1966 Tobacco hornworms. pp. 479-486. *in* Smith, C.N. (Ed.), Insect Colonization and Mass Production. Academic Press, New York and London.

*Hogan, G.R. 1972 Development of *Tribolium castaneum* (Coleoptera: Tenebrionidae) in media supplemented with pyrimidines and purines. ANN ENTOMOL SOC AM 65: 631-636.

Hollingsworth, M.J.; Burcombe, J.V. 1970 The nutritional requirements for longevity in *Drosophila*. J INSECT PHYSIOL 16: 1017-1025.

Hoog, E.G. van t' 1935a Aseptische Drosophila-kultur ein Reagenz auf Sterole. ACTA BREVIA NEERL PHYSIOL PHARMACOL MICROBIOL 5: 160-162. (In German)

*Hoog, E.G. van t' 1935b Aseptic culture of insects in vitamin research. I. Z VITAMINFORSCH 4: 300-324.

Hoog, E.G. van t' 1936 Aseptic culture of insects in vitamin research. II, Z VITAMINFORSCH 5: 118-126.

*Hori, K. 1972 Utilization of sucrose and starch by the bug, *Lygus disponsi* Linnavuori (Hemiptera: Miridae). APPL ENTOMOL ZOOL 7: 79-82.

Horie, Y. 1962 Effects of various fractions of mulberry leaves on feeding of the silkworm, *Bombyx mori* L. J SERIC SCI JPN 31: 258-264. (In Japanese with English summary)

*Horie, Y.; Inokuchi, T.; Watanabe, K. 1970 Effects of dietary amino acid balances on growth and cocoon quality in the silkworm, *Bombyx mori* L. BULL SERIC EXP STN JPN 24: 345-365. (In Japanese with English summary)

Horie, Y.; Ito, T. 1963 Vitamin requirements of the silkworm. NATURE LONDON 197: 98-99.

Horie, Y.; Ito, T. 1965 Nutrition of the silkworm, *Bombyx mori*. X. Vitamin B requirements and the effects of several analogues. J INSECT PHYSIOL 11: 1585-1593.

Horie, Y.; Nakasone, S. 1968 Effect of dietary biotin on fatty acid composition of the silkworm, *Bombyx mori* L. J INSECT PHYSIOL 14: 1381-1387.

Horie, Y.; Nakasone, S. 1971 Effects of the levels of fatty acids and carbohydrates in a diet on the biosynthesis of fatty acids in larvae of the silkworm, *Bombyx mori*. J INSECT PHYSIOL 17: 1441-1450.

Horie, Y.; Watanabe, K.; Ito, T. 1966 Nutrition of the silkworm, *Bombyx mori*. XIV. Further studies on the requirements for B-vitamins. BULL SERIC EXP STN JPN 20: 393-409.

*Hormchong, T.; Srithunya, S.; Podhisritong, P. 1972 Mass rearing of rice stem borers. JPN PESTIC INF 10: 126-127.

Hosoi, T. 1964 Problems of establishing and maintaining colonies of subspecies of *Culex pipiens*. BULL W H O 31: 453.

Hostounský, Z.; Abrahám, J.; Dvorácek, J. 1972 A simple dosimeter of synthetic food for caterpillars. ACTA ENTOMOL BOHEMOSLOV 69: 69-70.

*Hou, R.F.; Brooks, M.A. 1975 Continuous rearing of the aster leafhopper, *Macrosteles fascifrons*, on a chemically defined diet. J INSECT PHYSIOL 21: 1481-1483.

*House, H.L. 1949a Nutritional studies with *Blattella germanica* (L.) reared under aseptic conditions. II. A chemically defined diet. CAN ENTOMOL 81: 105-112.

House, H.L. 1949b Nutritional studies with *Blattella germanica* (L.) reared under aseptic conditions. III. Five essential amino acids. CAN ENTOMOL 81: 133-139.

*House, H.L. 1954a Nutritional studies with *Pseudosarcophaga affinis* (Fall.), a dipterous parasite of the spruce budworm, *Choristoneura fumiferana* (Clem.). I. A chemically defined medium and aseptic-culture technique. CAN J ZOOL 32: 331-341.

House, H.L. 1954b Nutritional studies with *Pseudosarcophaga affinis* (Fall.), a dipterous parasite of the spruce budworm, *Choristoneura fumiferana* (Clem.). II. Effects of eleven vitamins on growth. CAN J ZOOL 32: 343-350.

House, H.L. 1954c Nutritional studies with *Pseudosarcophaga affinis* (Fall.), a dipterous parasite of the spruce budworm, *Choristoneura fumiferana* (Clem.). III. Effects of nineteen amino acids on growth. CAN J ZOOL 32: 351-357.

House, H.L. 1954d Nutritional studies with *Pseudosarcophaga affinis* (Fall.), a dipterous parasite of the spruce budworm, *Choristoneura fumiferana* (Clem.). IV. Effects of ribonucleic acid, glutathione, dextrose, a salt mixture, cholesterol and fats. CAN J ZOOL 32: 358-365.

**House, H.L. 1955 Nutritional requirements and artificial diets for insects. REP ENTOMOL SOC ONT 86: 5-23.

**House, H.L. 1958a Nutritional requirements of insects associated with animal parasitism. EXP PARASITOL 7: 555-609.

**House, H.L. 1958b The nutrition of insects with particular reference to the entomophagous parasites. PROC 10 INT CONGR ENTOMOL MONTREAL (1956) 2: 139-143.

**House, H.L. 1959 Nutrition of the parasitoid *Pseudosarcophaga affinis* (Fall.) and of other insects. ANN N Y ACAD SCI 77: 394-405.

**House, H.L. 1961a Insect nutrition. ANNU REV ENTOMOL 6: 13-26.

**House, H.L. 1961b Insect diseases resulting from malnutrition. PROC ENTOMOL SOC ONT 91: 13-22.

**House, H.L. 1962 Insect nutrition. ANNU REV BIOCHEM 31: 653-672.

**House, H.L. 1963 Nutritional diseases. pp. 133-160. *in* Steinhaus, E.A. (Ed.), Insect Pathology Vol I. Academic Press, New York and London.

House, H.L. 1964a Effects of dietetic nucleic acids and components on growth of *Agria affinis* (Fallén) (Diptera: Sarcophagidae). CAN J ZOOL 42: 801-806.

**House, H.L. 1964b Contributions of nutritional research to entomology. PROC NORTH CENT BRANCH ENTOMOL SOC AM 19: 13-20.

**House, H.L. 1965a Insect Nutrition. pp. 769-813. *in* Rockstein, M. (Ed.), The Physiology of Insecta Vol II. Academic Press, New York and London.

**House, H.L. 1965b Digestion. pp. 814-858. *in* Rockstein, M. (Ed.), The Physiology of Insecta Vol II. Academic Press, New York and London.

House, H.L. 1965c Effects of vitamin A acetate and structurally related substances on growth and reproduction of *Agria affinis* (Fallén) (Diptera: Sarcophagidae). J INSECT PHYSIOL 11: 1039-1045.

*House, H.L. 1965d Effects of low levels of the nutrient content of a food and of nutrient imbalance of the feeding and the nutrition of a phytophagous larva, *Celerio euphorbiae* (Linnaeus) (Lepidoptera: Sphingidae). CAN ENTOMOL 97: 62-68.

House, H.L. 1966a Effects of varying the ratio between the amino acids and the other nutrients in conjunction with a salt mixture on the fly, *Agria affinis* (Fall.). J INSECT PHYSIOL 12: 299-310.

**House, H.L. 1966b The role of nutritional principles in biological control. CAN ENTOMOL 98: 1121-1134.

*House, H.L. 1967a The nutritional status and larvicidal activities of C_{6}- to C_{14}- saturated fatty acids in *Pseudosarcophaga affinis* (Diptera: Sarcophagidae). CAN ENTOMOL 99: 384-392.

*House, H.L. 1967b The role of nutritional factors in food selection and preference as related to larval nutrition of an insect, *Pseudosarcophaga affinis* (Diptera: Sarcophagidae), on synthetic diets. CAN ENTOMOL 99: 1310-1321.

**House, H.L. 1967c Artificial diets for insects: a compilation of references with abstracts. INF BULL RES INST CAN DEP AGRIC BELLEVILLE ONTARIO 5, 163 pp.

House, H.L. 1969a The diet pattern of feeding activity of larval *Pseudosarcophaga affinis* (Diptera: Sarcophagidae). CAN ENTOMOL 101: 409-412.

House, H.L. 1969b Effects of different proportions of nutrients on insects. ENTOMOL EXP APPL 12: 651-669.

House, H.L. 1970 Choice of food by larvae of the fly, *Agria affinis* related to dietary proportions of nutrients. J INSECT PHYSIOL 16: 2041-2050.

House, H.L. 1971a Changes from initial food choice in a fly larva, *Agria affinis* as related to dietary proportions of nutrients. J INSECT PHYSIOL 17: 1051-1059.

House, H.L. 1971b Relations between dietary proportions of nutrients, growth rate, and choice of food in the fly larvae, *Agria affinis*. J INSECT PHYSIOL 17: 1225-1238.

**House, H.L. 1972a Insect Nutrition. pp. 513-573. *in* Fiennes, R.N. (Ed.), Biology of Nutrition. I.E.F.N. Vol 18. Pergamon Press, Oxford and New York.

*House, H.L. 1972b Inversion in the order of food superiority between temperatures effected by nutrient balance in the fly larva *Agria housei* (Diptera: Sarcophagidae). CAN ENTOMOL 104: 1559-1564.

House, H.L.; Barlow, J.S. 1956 Nutritional studies with *Pseudosarcophaga affinis* (Fall.), dipterous parasite of the spruce budworm, *Choristoneura fumiferana* (Clem.). V. Effects of various concentrations of the amino acid mixture, dextrose, potassium ion, the salt mixture, and lard on growth and development, and a substitute for lard. CAN J ZOOL 34: 182-189.

House, H.L.; Barlow, J.S. 1957a New equipment for rearing small numbers of *Pseudosarcophaga affinis* (Fall.) (Diptera: Sarcophagidae) for experimental purposes. CAN ENTOMOL 89: 145-150.

House, H.L.; Barlow, J.S. 1957b Effects of nucleic acids on larval growth, pupation and adult emergence of *Pseudosarcophaga affinis* (Fall.). NATURE LONDON 180: 44.

*House, H.L.; Barlow, J.S. 1958 Vitamin requirements of the house fly, *Musca domestica* L. (Diptera: Muscidae). ANN ENTOMOL SOC AM 51: 299-302.

*House, H.L.; Barlow, J.S. 1960 Effects of oleic and other fatty acids on the growth rate of *Agria affinis* (Fall.) (Diptera: Sarcophagidae) J NUTR 72: 409-414.

House, H.L.; Barlow, J.S. 1964 Effects on the parasitoid *Agria affinis* (Fall.) (Diptera: Sarcophagidae) of small molecules in diets. J INSECT PHYSIOL 10: 225-260.

House, H.L.; Barlow, J.S. 1965 The effects of a new salt mixture developed for *Agria affinis* (Fallén) (Diptera: Sarcophagidae) on the growth rate, body weight and protein content of the larvae. J INSECT PHYSIOL 11: 915-918.

House, H.L.; Patton, R.L. 1949 Nutritional studies with *Blattella germanica* (L.) reared under aseptic conditions. Equipment and techniques. CAN ENTOMOL 81: 94-100.

**House, H.L.; Singh, P.; Batsch, W.W. 1971 Artificial diets for insects: a compilation of references with abstracts. INFORMATION BULL RES INST CAN DEP AGRIC BELLVILLE ONTARIO 7, 156 pp.

House, H.L.; Traer, M.G. 1948 An artificial food for rearing *Pseudosarcophaga affinis* (Fall.), a parasite of the spruce budworm *Choristoneura fumiferana* (Clem.). REP ENTOMOL SOC ONT 79: 1-4.

*Howden, G.F.; Hunter-Jones, P. 1958 An artificial diet for the laboratory rearing of locusts. NATURE LONDON 182: 1527-1528.

Howell, J.F. 1967 Paraffin films to control dehydration of an artificial rearing medium for codling moth. J ECON ENTOMOL 60: 289-290.

*Howell, J.F. 1970 Rearing the codling moth on an artificial diet. J ECON ENTOMOL 63: 1148-1150.

*Howell, J.F. 1971 Problems involved in rearing the codling moth on diet in trays. J ECON ENTOMOL 64: 631-636.

Howell, J.F. 1972a Modifications of the artificial diet for codling moths to improve larval acceptance and production of moths. J ECON ENTOMOL 65: 57-59.

*Howell, J.F. 1972b Rearing the codling moth on a soya, wheat germ, starch medium. J ECON ENTOMOL 65: 636-637.

*Howell, J.F.; Clift, A.E. 1972 Rearing codling moths on an artificial diet in trays. J ECON ENTOMOL 65: 888-890.

Hsiao, T.H. 1967 The host plant specificity of the Colorado potato beetle, *Leptinotarsa decemlineata* (Say). DISS ABSTR 27: 2917B.

*Hsiao, T.H. 1972 Chemical feeding requirements of oligophagous insects. pp. 225-240. *in* Rodriguez, J.G. (Ed.), Insect and Mite Nutrition. North-Holland Publishing Company, Amsterdam.

*Hsiao, T.H.; Fraenkel, G. 1968 The role of secondary plant substances in the food specificity of the Colorado potato beetle. ANN ENTOMOL SOC AM 61: 485-493.

Hsiao, T.H.; Hsiao, C. 1974a Benomyl: a novel drug for controlling a microsporidan disease of the alfalfa weevil. J INVERTEBR PATHOL 22: 303-304.

*Hsiao, T.H.; Hsiao, C. 1974b A practical artificial diet for laboratory rearing of the alfalfa weevil, *Hypera postica* (Gyllenhal). ANN ENTOMOL SOC AM 67: 149-150.

*Hsiao, T.H.; Hsiao, C. 1974c Feeding requirements and artificial diets for the alfalfa weevil. ENTOMOL EXP APPL 17: 83-91.

**Hubert-Dahl, M.L. 1973 Nahrungswahl und Bedentung der Nahrungsqualität bei der künsllichen Ernährung von Aphiden. WISS Z UNIV ROSTOCK MATH NATURWISS REIHE 22: 785-797. (In German with English summary)

Hudspeth, W.N.; Jenkins, J.N.; Maxwell, F.G. 1969 Ascorbic acid impractical as a character for resistance in cotton to the boll weevil. J ECON ENTOMOL 62: 583-584.

Huot, L.; Bernard, R.; Lemonde, A. 1957 Aspects quantitatifs des besoins en minéraux des larves de *Tribolium confusum* Duval. I. Pourcentage optimum d'un mélauge salin. CAN J ZOOL 35: 513-518. (In French with English summary)

Huot, L.; Bernard, R.; Lemonde, A. 1958 Aspects quantitatifs des besoins en minéraux de *Tribolium confusum* Duval. II. Pourcentage optimum des cations Mg, Ca, Na, et K. CAN J ZOOL 36: 7-13. (In French with English summary)

**Hurpin, B. 1962 Alimentation, développement et fécondité chez les insectes. ANN NUTR ALIMENT 16: 153-200. (In French)

*Ignoffo, C.M. 1963 A successful technique for mass-rearing cabbage loopers on a semisynthetic diet. ANN ENTOMOL SOC AM 56: 178-182.

Ignoffo, C.M. 1965 The nuclear-polyhedrosis virus of *Heliothis zea* (Boddie) and *Heliothis virescens* (Fabricius). II. Biology and propagation of diet-reared *Heliothis*. J INVERTEBR PATHOL 7: 217-226.

*Ignoffo, C.M. 1966 Insect Viruses. pp. 501-530. *in* Smith, C.N. (Ed.), Insect Colonization and Mass Production. Academic Press, New York and London.

Ignoffo, C.M.; Boening, O.P. 1970 Compartmented disposable plastic trays for rearing insects. J ECON ENTOMOL 63: 1696-1697.

Ignoffo, C.M.; Garcia, C. Jr 1968 Formalin interaction of nuclear polyhedrosis virus. J INVERTEBR PATHOL 10: 430-432.

Ignoffo, C.M.; Gard, I. 1970 Use of an agar-base diet and house fly larvae to assay β-exotoxin activity of *Bacillus thuringiensis*. J ECON ENTOMOL 63: 1987-1989.

*Injac, M.; Veyrunes, J-C.; Kuhl, G. 1973 Gajenje dudovca (*Hyphantria cunea* Drury) veštačkom hranom u aseptičnim uslovima. ZAST BILJA 24: 111-117. (In Yugoslav with French summary)

*Inokuchi, T. 1969a Nutritional studies of amino acids in the silkworm, *Bombyx mori*. I. Effect of dietary cystine on methionine requirement and its metabolism. BULL SERIC EXP STN JPN 23: 371-387.

Inokuchi, T. 1969b Nutritional studies of amino acids in silkworm, *Bombyx mori*. II. Effect of several amino acids on proline and arginine requirements. BULL SERIC EXP STN JPN 23: 389-410.

Inokuchi, T. 1970 Nutritional studies of amino acids in the silkworm, *Bombyx mori*. III. Effect of dietary amino acid on the concentration of protein and free amino acids in the haemolymph of larvae. BULL SERIC EXP STN JPN 24: 389-408. (In Japanese)

Inokuchi, T. 1971 Effects of dietary conditions on free amino acid patterns in haemolymph of the silkworm, *Bombyx mori*. J SERIC SCI JPN 40: 281-287. (In Japanese with English summary)

Inokuchi, T.; Horie, Y.; Ito, T. 1967 Nutrition of the silkworm, *Bombyx mori* - XIX. Effects of omission of essential amino acids in each of the larval instars. BULL SERIC EXP STN JPN 22: 195-205. (In Japanese)

Inoue, H.; Kamano, S. 1957 The effect of photoperiod and temperature on the induction of diapause in the rice stem borer, *Chilo suppressalis* Walker. JPN J APPL ENTOMOL ZOOL 1: 100-105.

*Isa, A.L. 1972 A standard technique for rearing *Chilo agamemnon* Bles. on an artificial diet. AGRIC RES REV 50: 61-64.

*Isa, A.L.; Khadr, G.D. 1973 Rearing of *Ostrinia nubilalis* (Hbn) on artificial diets. AGRIC RES REV 51: 15-19.

Ishaaya, I.; Moore, I.; Joseph, D. 1971 Protease and amylase activity in larvae of the Egyptian cotton worm, *Spodoptera littoralis*. J INSECT PHYSIOL 17: 945-953.

Ishii, S. 1952 Some problems on the rearing method of rice stem borer by synthetic media under aseptic conditions. OYO-KONTYU 8: 93-98. (In Japanese with English summary)

**Ishii, S. 1956 Culture of phytophagous insects on artificial diets. SHOKUBUTSU BOEKI 10: 7-10. (In Japanese)

**Ishii, S. 1959 Methods of experiment with insects. pp. 126-143. KONCHU JIKKAN KAKU. (In Japanese)

*Ishii, S. 1971 Nutritional studies of the rice stem borer, *Chilo suppressalis* Walker, and its mass rearing. ENTOMOPHAGA 16: 165-173.

Ishii, S.; Azim, A.; Hirano, C. 1959 A further experiment on the effect of dietary levels of protein and carbohydrate on the growth of the rice stem borer, *Chilo suppressalis*, larvae. JPN J APPL ENTOMOL ZOOL 3: 143-145.

Ishii, S.; Hirano, C. 1955 Qualitative studies on the essential amino acids for the growth of larvae of the rice stem borer, *Chilo simplex* Butler, under aseptic conditions. BULL NATL INST AGRIC SCI TOKYO (C) 5: 35-48.

Ishii, S.; Hirano, C. 1957 Effect of various concentrations of protein and carbohydrate in a diet on the growth of the rice stem borer larva. JPN J APPL ENTOMOL ZOOL 1: 75-79.

Ishii, S.; Hirano, C. 1964 Inability of *Chilo suppressalis* larvae (Lepidoptera: Pyralidae) to utilize farnesol for synthesizing cholesterol. JPN J APPL ENTOMOL ZOOL 8: 84-86.

Ishii, S.; Urushibara, H. 1954 On fat soluble and water soluble growth factors required by the rice stem borer, *Chilo simplex* Butler. BULL NATL INST AGRIC SCI TOKYO (C) 4: 109-133.

Ishikawa, H.; Newburgh, R.W. 1971 Changes in RNA during metamorphosis of the central nervous tissue of *Galleria*. J INSECT PHYSIOL 17: 1113-1124.

Ito, T. 1960 Effect of sugars on feeding of larvae of the silkworm *Bombyx mori*. J INSECT PHYSIOL 5: 95-107.

*Ito, T. 1961a Nutrition of silkworm, *Bombyx mori*. I. Nutritive effects of soybean oil for the larva. BULL SERIC EXP STN JPN 16: 311-348. (In Japanese with English summary)

Ito, T. 1961b Nutrition of the silkworm, *Bombyx mori*. III. Requirements for sterols and their effects on feeding. BULL SERIC EXP STN JPN 17: 91-117.

Ito, T. 1961c Nutrition of silkworm, *Bombyx mori*. IV. Effects of ascorbic acid. BULL SERIC EXP STN JPN 17: 119-136. (In Japanese with English summary)

Ito, T. 1961d Sterol requirements of the silkworm, *Bombyx mori*. NATURE LONDON 191: 882-883.

Ito, T. 1962 An artificial diet for the silkworm, *Bombyx mori* and the effect of soybean oil on its growth. PROC 11 INT CONGR ENTOMOL WIEN (1960) 3: 157-162.

**Ito, T. 1967 Nutritional requirements of the silkworm, *Bombyx mori* L. PROC JPN ACAD 43: 57-61.

*Ito, T. 1969 Rearing of the silkworm under aseptic condition. JPN AGRIC RES Q 4: 33-35.

**Ito, T. 1970 Lipid nutrition of the silkworm, *Bombyx mori* L. PROC JPN ACAD 46: 1036-1040.

**Ito, T. 1972a An approach to nutritional control mechanisms in the silkworm, *Bombyx mori* (Lepidoptera: Bombycidae). ISR J ENTOMOL 7: 1-6.

**Ito, T. 1972b Amino acid nutrition of the silkworm, *Bombyx mori*. PROC JPN ACAD 48: 613-618.

*Ito, T.; Arai, N. 1965a Nutrition of the silkworm, *Bombyx mori*. VIII. Amino acid requirements and nutritive effects of various proteins. BULL SERIC EXP STN JPN 19: 345-373. (In Japanese)

Ito, T.; Arai, N. 1965b Nutrition of the silkworm, *Bombyx mori*. IX. Further studies on the nutritive effects of ascorbic acid. BULL SERIC EXP STN JPN 20: 1-19. (In Janapese with English summary)

Ito, T.; Arai, N. 1966a Nutrition of the silkworm, *Bombyx mori*. XI. Requirements for aspartic and glutamic acids. J INSECT PHYSIOL 12: 861-869.

*Ito, T.; Arai, N. 1966b An amino-acid diet for the silkworm, *Bombyx mori* L. J AGRIC CHEM SOC JPN 40: 110-112. (In Japanese with English summary)

Ito, T.; Arai, N. 1967 Nutritive effects of alanine, cystine, glycine, serine and tyrosine on the silkworm, *Bombyx mori*. J INSECT PHYSIOL 13: 1813-1824.

Ito, T.; Arai, N. 1973 Artificial diets for the silkworm, *Bombyx mori*, with special references to a semi-synthetic diet for fifth-instar larvae, which includes a large amount of defatted soybean meal. J AGRIC CHEM SOC JPN 47: 397-401. (In Japanese with English summary)

Ito, T.; Arai, N.; Inokuchi, T. 1967 Nutrition of the silkworm, *Bombyx mori*. XVII. Effect of dietary levels of amino acids on growth of the fifth-instar larvae and on cocoon quality. BULL SERIC EXP STN JPN 21: 385-400. (In Japanese with English summary)

Ito, T.; Fraenkel, G. 1966 The effect of nitrogen starvation on *Tenebrio molitor* L. J INSECT PHYSIOL 12: 803-817.

Ito, T.; Horie, Y. 1962 Nutrition of the silkworm, *Bombyx mori*. VII. An aseptic culture of larvae on semi-synthetic diets. J INSECT PHYSIOL 8: 569-578.

Ito, T.; Horie, Y. 1966 Utilization of sterols and related compounds by the silkworm, *Bombyx mori* L. ANNOT ZOOL JPN 39: 1-6.

Ito, T.; Horie, Y.; Kurta, K. 1967 Rearing of the silkworm, *Bombyx mori* L., by the germ-free method and its application to silkworm pathology. pp. 424-429. *in* Miyakawa, M.; Luckey, T.D. (Eds), Advances in Germfree Research and Gnotobiology. Chemical Rubber Co., Cleveland, Ohio.

Ito, T.; Horie, Y.; Nakasone, S. 1975 Deterrent effect of soybean meal on feeding of the silkworm, *Bombyx mori*. J INSECT PHYSIOL 21: 995-1006.

Ito, T.; Horie, Y.; Tanaka, M. 1961 Nutrition of silkworm, *Bombyx mori*. II. Some effects of artificial diets on feeding and nutrition. BULL SERIC EXP STN JPN 16: 349-373. (In Japanese with English summary)

Ito, T.; Horie, Y.; Watanabe, K. 1967 Rearing germ-free silkworm by means of vinyl-isolater. J SERIC SCI JPN 36: 409-412.

Ito, T.; Horie, Y.; Watanabe, K. 1972 Rearing of the silkworm, *Bombyx mori*, on semi-synthetic diets, with special reference to effects of dietary levels of some ingredients on larval growth. BULL SERIC EXP STN JPN 25: 219-234. (In Japanese with English summary)

*Ito, T.; Horie, Y.; Watanabe, K.; Takamiya, K.; Furuyama, M.; Miyabayashi, M.; Yamamoto, K.; Nagashima, M. 1974 Rearing of larvae of the silkworm, *Bombyx mori*, entirely on semi-synthetic diets. J AGRIC CHEM SOC JPN 48: 403-407. (In Japanese with English summary)

**Ito, T.; Inokuchi, T. 1972 Nutrition and metabolism of amino acids in the silkworm, *Bombyx mori*. pp. 517-529. *in* Rodriguez, J.G. (Ed.), Insect and Mite Nutrition. North-Holland Publishing Company, Amsterdam.

Ito, T.; Kawashima, K.; Nakahara, M.; Nakanishi, K.; Terahara, A. 1964 Effects of sterols on feeding and nutrition of the silkworm, *Bombyx mori* L. J INSECT PHYSIOL 10: 225-238.

*Ito, T.; Mizuta, Y.; Takamiya, K.; Ueda, S.; Kimura, R.; Higuchi, T.; Takahashi, S. 1975 Growth, development, and cocoon production in artificial-diet-rearing of the silkworm, *Bombyx mori*, with or without application of a synthetic juvenile hormone analog. J AGRIC CHEM SOC JPN 49: 39-48. (In Japanese with English summary)

Ito, T.; Mukaiyama, F. 1964 Relationship between protein content of diets and xanthine oxidase activity in the silkworm, *Bombyx mori* L. J INSECT PHYSIOL 10: 789-796.

Ito, T.; Nakasone, S. 1966 Nutrition of the silkworm, *Bombyx mori*. XIII. Nutritive effects of fatty acids. BULL SERIC EXP STN JPN 20: 375-391. (In Japanese with English summary)

Ito, T.; Nakasone, S. 1967 Nutrition of the silkworm, *Bombyx mori*. XV. Utilization of sterols in the presence of dietary fatty acids. J INSECT PHYSIOL 13: 281-287.

Ito, T.; Nimura, M. 1966 Nutrition of the silkworm, *Bombyx mori*. XII. Nutritive effects of minerals. BULL SERIC EXP STN JPN 20: 361-374. (In Japanese with English summary)

Ito, T.; Tanaka, M. 1960 Rearing of the silkworm on an artificial diet and the segregation of pentamolters. J SERIC SCI JPN 29: 191-196. (In Japanese with English summary)

*Ito, T.; Tanaka, M. 1962 Rearing of the silkworm by means of aseptic technique. J SERIC SCI JPN 31: 7-10. (In Japanese with English summary)

*Ivaldi-Sender, C. 1974 Techniques simples pour un élevage permanent de la tordeus orientale, *Grapholita molesta* (Lépidoptèra: Tortricidae) sur milieu artificiel. ANN ZOOL ECOL ANIM 6: 337-343. (In French with English summary)

Jacob, D.; Chippendale, G.M. 1971 Growth and development of the southwestern corn borer *Diatraea grandiosella* on a meridic diet. ANN ENTOMOL SOC AM 64: 485-488.

*Jensen, R.L.; Newsom, L.D.; Gibbens, J. 1974 The soybean looper: effects of adult nutrition on oviposition, mating frequency, and adult longevity. J ECON ENTOMOL 67: 467-470.

**Johansson, A.S. 1964 Feeding and nutrition in reproductive processes in insects. pp. 43-55. Symp R Entomol Soc London 2, Insect Reproduction.

*Jones, R.G.; Brindley, W.A. 1970 Test of eight rearing media for the mountain pine beetle, *Dendroctonus ponderosae* (Coleoptera: Scolytidae), from lodgepole pine. ANN ENTOMOL SOC AM 63: 313-316.

Jones, R.H. 1964 Mass production methods in rearing *Culicoides variipennis* (Coquillett). BULL W H O 31: 571-572.

*Jones, R.H.; Potter, H.W. Jr; Baker, S.K. 1969 An improved larval medium for colonized *Culicoides variipennis*. J ECON ENTOMOL 62: 1483-1486.

*Jones, W.R.; Cochran, D.G.; Graham, P.P.; Kelly, R.F. 1973 Growth and development of the cheese skipper under sterile, unsterile and sterile-inoculated conditions. J ECON ENTOMOL 66: 1252-1254.

Jurzitza, G. 1966 Harnsäure als zusätzliche Stickstoffquelle für Larven von *Lasioderma serricorne*. Z MORPHOL OEKOL TIERE 57: 244-248. (In German with English summary)

*Jurzitza, G. 1969a Die Rolle der hefeartigen Symbionten von *Lasioderma serricorne* F. (Coleoptera, Anobiidae) im Proteinmetabolismus ihrer Wirte. I. Das Wachstum normaler und aposymbiontischer Larven in Diäten mit proteinen, Proteinderivaten und Aminosäuregemischen als N-Quellen. Z VGL PHYSIOL 63: 165-181. (In German with English summary)

Jurzitza, G. 1969b Aufzuchtversuche an *Lasioderma serricorne* F. in Drogen-und Holzpulvern im Hinblick auf die Rolle der hefeartigen Symbionten. Z NATURFORSCH 24: 760-763. (In German with English summary)

Jurzitza, G. 1969c Der Vitaminbedarf normaler und aposymbiontischer *Lasioderma serricorne* F. (Coleoptera, Anobiidae) und die Bedeutung der symbiontischen Pilze als Vitaminquelle für ihre Wirte. OECOLOGIA 3: 70-83. (In German with English summary)

Jurzitza, G. 1970 Untersuchungen über die Wirkung sekundärer Pflanzeninhaltsstoffe auf die Pilzsymbiose des Tabakkäfers *Lasioderma serricorne* F. 3. Mitteilung. Die wirkung von Sulfanilamid in Vitaminmangeldiäten. Z ANGEW ENTOMOL 66: 96-100. (In German with English summary)

Jurzitza, G. 1971 Untersuchungen über die Wirkung sekundärer Pflanzeninhaltsstoffe auf die Pilzsymbiose des Tabakkäfers *Lasioderma serricorne* F. 4. Mitteilung. Wirkung einiger Alkaloide und Saponine auf den Vitaminhaushalt des symbiontischen Systems. Z ANGEW ENTOMOL 67: 432-440. (In German with English summary)

*Jurzitza, G. 1972 Die Funktion der endosymbiontischen Pilze des Tabakkäfers *Lasioderma serricorne* F. (Col.; Anobiidae) bei einer Aufzucht unter Proteinmangelbedingungen. Z ANGEW ENTOMOL 72: 198-205. (In German with English summary).

Jurzitza, G. 1974 Über die Lieferung von Sterinen durch die hefeartigen Endosymbionten von *Lasioderma serricorne* F. (Coleoptera, Anobiidae) und die ökologische Bedeutung dieser Leistung für den Wirt. OECOLOGIA 16: 163-172. (In German with English summary)

Jurzitza, G.; Wolters, B. 1967 Untersuchungen über die Wirkung sekundärer Pflanzeninhaltss-toffe auf die Pilzsymbiose des Tabakkäfers *Lasioderma serricorne* F. Z ANGEW ENTOMOL 59: 397-402. (In German with English summary)

*Kadner, C.H.; LaFleur, F.M. 1951 The vitamin requirements of *Phaenicia sericata* (Meigen) larvae (Diptera: Calliphoridae). WASMANN J BIOL 9: 129-136.

Kajita, H. 1973 Rearing of *Apanteles chilonis* Munakata on the rice stem borer, *Chilo suppressalis* Walker, bred on a semi-artificial diet. JPN J APPL ENTOMOL ZOOL 17: 5-9. (In Japanese with English summary)

*Kamano, S. 1961 On the successive rearing of rice stem borer on the artificial diets under aseptic conditions. I. The effect of choline contained in the artificial diets on the oviposition and the hatching potency. JPN J APPL ENTOMOL ZOOL 5: 254-259. (In Japanese with English summary)

*Kamano, S. 1964a Rearing the cabbage army-worm, *Barathra brassicae* Linné, on artificial diets. JPN J APPL ENTOMOL ZOOL 8: 340-342. (In Japanese with English summary)

*Kamano, S. 1964b On the successive rearing of rice stem borer on the artificial diets under aseptic conditions. II. Ascorbic acid requirement of larvae. JPN J APPL ENTOMOL ZOOL 8: 101-105. (In Japanese with English summary)

*Kamano, S. 1965 Rearing of *Papilio xuthus* larvae on artificial diets. JPN J APPL ENTOMOL ZOOL 9: 133-135. (In Japanese with English summary)

*Kamano, S. 1971 Studies on artificial diet of the rice stem borer, *Chilo suppressalis* Walker. BULL NATL INST AGRIC SCI JPN SER C 25: 1-45.

*Kamano, S. 1973 Studies on artificial diets and laboratory rearing methods suitable for successive generations of the rice stem borer, *Chilo suppressalis* Walker. BULL NATL INST AGRIC SCI JPN SER C 27: 1-51. (In Japanese with English summary)

*Kamano, S.; Fukaya, M. 1965 On the successive rearing of rice stem borer on the artificial diets under aseptic conditions. IV. The inbreeding and the rotational breeding of the rice stem borer. JPN J APPL ENTOMOL ZOOL 9: 89-93. (In Japanese with English summary)

*Kamano, S.; Yushima, T. 1969 Rearing method of insects under aseptic and semi-aseptic conditions. PLANT PROT 23: 353-356. (In Japanese)

*Kamasaki, H.; Sutton, R.; Lopez, D.F.; Selhime, A. 1970 Laboratory culture of the Caribbean fruit fly, *Anastrepha suspensa*, in Florida. ANN ENTOMOL SOC AM 63: 639-642.

Kamioka, S.; Mukaiyama, F.; Takei, T.; Ito, T.;1971 Digestion and utilization of artificial diet by the silkworm *Bombyx mori*, with special reference to the efficiency of the diet at varying levels of dietary soybean meal. J SERIC SCI JPN 40: 473-483.

*Karpel, M.A.; Hagmann, L.E. 1968 Medium and techniques for mass rearing the red-banded leaf roller. J ECON ENTOMOL 61: 1452-1454.

Kasting, R.; Davis, G.R.F.; McGinnis, A.J. 1962 Nutritionally essential and non-essential amino acids for the prairie grain wireworm, *Ctenicera destructor* Brown, determined with glucose-U-C^{14}. J INSECT PHYSIOL 8: 589-596.

*Kasting, R.; McGinnis, A.J. 1958 Note on a method of artificially sustaining larvae of the wheat stem sawfly, *Cephus cinctus* Nort. (Hymen-optera: Cephidae). CAN ENTOMOL 90: 63-64.

**Kasting, R.; McGinnis, A.J. 1966 Radio-isotopes and the determination of nutrient requirements. ANN N Y ACAD SCI 139: 98-108.

*Kasting, R.; McGinnis, A.J. 1967 An artificial diet and some growth factor require-ments for the pale western cutworm. CAN J ZOOL 45: 787-796.

Kasting, R.; McGinnis, A.J. 1973 Construction and operation of a 'clean room' to avoid infectious disease when rearing the pale western cutworm. LAB PRACT 368-369.

Kato, M.; Sumimoto, K. 1968 Nutritional studies on excessive larval moulting in the silkworm *Bombyx mori*. ANNOT ZOOL JPN 41: 140-147.

Kato, M.; Yamada, H. 1966a Chlorogenic acid promotes the utilization of oil in the growth of silkworm. PROC JPN ACAD 42: 1185-1188.

Kato, M.; Yamada, H. 1966b Silkworm requires 3, 4-dihydroxy-benzene structure of chlorogenic acid as a growth factor. LIFE SCI 5: 717-722.

*Katz, M.; Budowski, P.; Bondi, A. 1971 The effect of phytosterols on the growth and sterol composition of *Dermestes maculatus*. J INSECT PHYSIOL 17: 1295-1303.

Kaul, S.; Saxena, S.C. 1974a Vitamin B_2 (riboflavin) sparing activity of ascorbic acid in the larval nutrition of *Oryzaephilus mercator* Fauvel (Coleoptera: Silvanidae). ARCH INT PHYSIOL BIOCHIM 82: 639-644.

Kaul, S.; Saxena, S.C. 1974b Quantitative requirements for some vitamins of the B-complex for the growth and development of larvae of *Oryzaephils mercator* (Coleoptera: Cucujidae). ACTA ENTOMOL BOHEMOSLOV 71: 80-85.

*Kaul, S.; Saxena, S.C. 1975a Pantothenic acid sparing activity of ascorbic acid in the larval nutrition of *Oryzaephilus mercator* (Coleoptera). ACTA ENTOMOL BOHEMOSLOV 72: 236-238.

Kaul, S.; Saxena, S.C. 1975b Beneficial effects of *l*-methionine in choline deficient diets of *Oryzaephilus mercator* (Fauvel) (Coleoptera: Silvanidae). ARCH INT PHYSIOL BIOCHIM 83: 299-302.

*Keaster, A.J.; Chippendale, G.M.; Pill, B.A. 1975 Feeding behaviour and growth of the wireworms *Melanotus depressus* and *Limonius dubitans*: effect of host plants, temperature photoperiod and artificial diets. ENVIRON ENTOMOL 4: 591-595.

*Keaster, A.J.; Harrendorf, K. 1965 Laboratory rearing of the southwestern corn borer, *Zeadiatraea grandiosella*, on a wheat germ medium. J ECON ENTOMOL 58: 923-924.

*Keeley, L.L. 1970 Diapause metabolism and rearing methods for the white-lined sphinx, *Celerio lineata* (Lepidoptera: Sphingidae). ANN ENTOMOL SOC AM 63: 905-907.

Keiding, J.; Arevad, K. 1964 Procedure and equipment for rearing a large number of housefly strains. BULL W H O 31: 527-528.

*Kennedy, M.K. 1973 A culture method for *Bradysia impatiens* (Diptera: Sciaridae). ANN ENTOMOL SOC AM 66: 1163-1164.

*Kennedy, M.K. 1974 Survival and development of *Bradysia impatiens* (Diptera: Sciaridae) on fungal and non-fungal food sources. ANN ENTOMOL SOC AM 67: 745-749.

Kerr, R.W. 1954 Rearing *Drosophila melanogaster* Mg. for insecticide investigations. BULL ENTOMOL RES 45: 313-316.

*Kettle, D.S.; Wild, C.H.; Elson, M.M. 1975 A new technique for rearing individual *Culicoides* larvae (Diptera: Ceratopogonidae). J MED ENTOMOL 12: 263-264.

*Khalaf, K.T. 1974 Non aseptic wheat germ diet for *Megalopyge opercularis* (Lepidoptera: Megalopygidae). FLA ENTOMOL 57: 377-381.

*Khalifa, A.; Salama, H.S.; El-sharaby, A.F. 1973 Rearing the cotton leafworm *Spodoptera littoralis* Boisd. on a semiartificial diet. Z ANGEW ENTOMOL 73: 129-132.

Khlistovskiy, E.D. 1965 Artificial media for the rearing of the bollworm *Heliothis armigera*. ZASHCH RAST VREDIT BOLEZ 12: 44. (In Russian)

Khlistovskiy, E.D. 1970 The rearing of the bollworm, *Chloridea obsoleta* F. (Lepidoptera, Noctuidae), on an artificial nutrient medium. ENTOMOL REV USSR 49: 55-57.

*Khlistovskiy, E.D.; Uspenskaya, N.V. 1969 A formulation for a semisynthetic food medium and the procedures for rearing larvae of *Agrotis segetum* on it. ZOOL ZH 48: 599-604. (In Russian with English summary)

**Kidder, G.W. 1953 The nutrition of invertebrate animals. pp. 162-195. *in* Bourne, G.H.; Kidder, G.W. (Eds), Biochemistry and Physiology of Nutrition. Academic Press, New York and London.

*Kieckhefer, R.W.; Derr, R.F. 1967 Rearing three species of cereal aphids on artificial diets. J ECON ENTOMOL 60: 663-665.

King, R.C. 1965 Genetics, 2nd Edition. Oxford University Press, London. 429 pp.

*Kirollos, J.Y.; Hibbs, E.T. 1971 Viability of *Empoasca fabae* (Homoptera: Cicadellidae) eggs in sterile media with added monosaccharides, amino acids, or phorate. ANN ENTOMOL SOC AM 64: 32-36.

Kishaba, A.N.; Henneberry, T.J. 1966 Effects of cold storage on egg viability of the cabbage looper and some aspects of the biology of the progeny survivors. J ECON ENTOMOL 59: 1169-1171.

Kishaba, A.N.; Henneberry, T.J.; Pangaldan, R.; Tsao, P.H. 1968 Effects of mold inhibitors in larval diet on the biology of cabbage looper. J ECON ENTOMOL 61: 1189-1194.

*Kiyoku, M.; Tsukuda, P. 1966 Successive laboratory cultures of the tobacco-cutworms *Prodenia litura* Fabricius on some simple semisynthetic media. REP FAC AGRIC OKAYAMA UNIV 28: 1-11.

*Klun, J.A.; Tipton, C.L.; Brindley, T.A. 1967 2, 4-Dihydroxy-7-methoxy-1, 4-benzoxazin-3-one (DIMBOA), an active agent in the resistance of maize to the European corn borer. J ECON ENTOMOL 60: 1529-1533.

*Knierim, J.A.; Lea, A.O.; Dimond, J.B.; DeLong, D.M. 1955 Feeding adult mosquitoes on preserved blood to maintain egg production. MOSQ NEWS 15: 176-179.

**Knipling, E.F. 1966 Introduction. pp. 1-12. *in* Smith, C.N. (Ed.), Insect Colonization and Mass Production. Academic Press, New York and London.

**Koch, A. 1954 Symbionten als Vitaminquellen der Insekten. FORSCH FORTSCHR 28: 33-37. (In German)

*Kogan, M. 1971 Feeding and nutrition of insects associated with soybeans. I. Growth and development of the Mexican bean beetle, *Epilachna varivestis*, on artificial media. ANN ENTOMOL SOC AM 64: 1044-1050.

Kogan, M.; Cope, D. 1974 Feeding and nutrition of insects associated with soybeans. 3. Food intake, utilization, and growth in the soybean looper, *Pseudoplusia includens*. ANNALS ENTOMOL SOC AM 67: 66-72.

*Kogan, M.; Goeden, R.D. 1971 Feeding and host-selection behaviour of *Lema trilineata daturaphila* larvae (Coleoptera: Chrysomelidae). ANN ENTOMOL SOC AM 64: 1435-1448.

Kok, L.T.; Norris, D.M.; Chu, H.M. 1970 Sterol metabolism as a basis for a mutualistic symbiosis. NATURE LONDON 225: 661-662.

*Kolya, A.K.; Pant, N.C. 1961 Effect of qualitative and quantitative vitamin deficiencies in larval development of *Oryzaephilus surinamensis* L. INDIAN J ENTOMOL 23: 285-292.

*Komai, T. 1927 The culture medium for *Drosophila*. SCIENCE 65: 42-43.

*Kono, Y. 1968 Rearing *Pieris rapae crucivora* Boisduval (Lepidoptera: Pieridae) on artificial diets. APPL ENTOMOL ZOOL 3: 96-98.

*Kouni, M.H.; Nash, D. 1974 Biology of 5-flouro-2-deoxyuridine sensitivity of *Drosophila melanogaster* larvae. J INSECT PHYSIOL 20: 1481-1490.

*Kovitvadhi, K.; Kerr, S.H. 1968 Artificial diet for the zoysiagrass billbug, *Sphenophorus venatus vestitus* (Coleoptera: Curculionidae), and notes on its biology. FLA ENTOMOL 51: 247-250.

Koyama, K. 1973a Preference on colour of the diets in *Inazuma dorsalis* (Hemiptera: Deltocephalidae). JPN J APPL ENTOMOL ZOOL 17: 49-53. (In Japanese with English summary)

*Koyama, K. 1973b Rearing of *Inazuma dorsalis* and *Nephotettix cincticeps* on a synthetic diet. JPN J APPL ENTOMOL ZOOL 17: 163-166. (In Japanese with English summary)

*Koyama, K.; Mitsuhashi, J. 1969 Artificial feeding of smaller brown plant-hopper, *Laodelphax striatellus* Fallén (Hemiptera: Delphacidae). JPN J APPL ENTOMOL ZOOL 13: 89-90.

Koyama, K.; Mitsuhashi, J. 1973 A dietary factor affecting the wing form of *Laodelphax striatellus* Fallén (Hemiptera: Delphacidae). JPN J APPL ENTOMOL ZOOL 17: 111-112. (In Japanese)

Koyama, K.; Mitsuhashi, J. 1975 Essential amino acids for the growth of the smaller brown plant-hopper, *Laodelphax striatellus* Fallén (Hemiptera: Delphacidae). APPL ENTOMOL ZOOL 10: 208-215.

Koyama, T.; Ishii, S. 1950 Rearing method of rice stem borer by artificial media. KONTYU 18: 160-161. (In Japanese)

Koyama, T.; Yasuda, S.; Ishii, S. 1951 On the rearing method of rice stem borer by artificial media. OYO-KONTYU 6: 198-201. (In Japanese with English summary)

*Kreasky, J.B. 1962 A growth factor in romaine lettuce for the grasshoppers *Melanoplus sanguinipes* (F.) and *M. bivittatus* (Say). J INSECT PHYSIOL 8: 493-504.

*Krieger, D.L. 1971 Rearing several aphid species on a synthetic diet. ANN ENTOMOL SOC AM 64: 1176-1177.

*Kunimi, Y.; Aruga, H. 1974 Susceptibility to infection with nuclear- and cytoplasmic-polyhedrosis viruses of the fall webworm, *Hyphantria cunea* Drury reared on several artificial diets. JPN J APPL ENTOMOL ZOOL 18: 1-4. (In Japanese with English summary)

Kunkel, H.M.; Mittler, T.E. 1971 Einfluss der Ernährung bei Junglarven von *Myzus persicae* (Sulz.) (Aphididae) auf ihre Entwicklung zu Geflügelten oder Ungeflügelten. OECOLOGIA 8: 110-134. (In German with English summary)

Kunz, S.E.; Eschle, J.L. 1971 Possible use of the sterile-male technique for eradication of the horn fly. pp. 145-156. *in* Sterility Principle for Insect Control or Eradication. Proceedings of a Panel Organized by the Joint FAO/IAEA Division of Atomic Energy in Food and Agriculture.

Kuwata, K.; Hirayama, K.; Hamamura, Y. 1970 The effect of putrescine on the ability of the fibroin synthesis in the posterior silk gland of *Bombyx mori*. REVUE VER SOIE ÉDITION SPÉCIALE (August, 1970): 1-6.

Kuwata, K.; Yamamoto, E.; Hamamura, Y. 1968 The effect of β-carotene and α-tocopherol on the growth and egg formation of silkworm *Bombyx mori*. J SERIC SCI JPN 37: 63-65. (In Japanese)

LaBrecque, G.C.; Keller, J.C. (Ed.) 1965 Advances in Insect Population Control by the Sterile-male Technique. Technical Report Series 44. Report of a Panel held in Vienna 20-24 July 1964. International Atomic Energy Agency, Vienna.

*Lafon, M. 1939 Recherches sur quelques aspects du besoin qualitatif d'azote. 2e Memoire Essai sur le besoin qualitatif d'azote chez un insecte: *Drosophila melanogaster* Meig. ANN PHYSIOL PHYSICOCHIM BIOL 15: 215-260. (In French)

*Laing, D.R.; Hagen, K.S. 1970 Axenic, partially synthetic diet for the oriental fruit moth, *Grapholitha molesta* (Lepidoptera: Olethreutidae). CAN ENTOMOL 102: 250-252.

*Lake, P.; Friend, W.G. 1968 The use of artificial diets to determine some of the effects of *Nocardia rhodnii* on the development of *Rhodnius prolixus*. J INSECT PHYSIOL 14: 543-562.

Lall, B.S. 1958 Studies on the dietary requirements of the hide beetle (*Dermestes maculatus* De Geer; Dermestidae: Coleoptera) with particular reference to amino acids and vitamins. CURR SCI 27: 399-400.

Lambremont, E.N.; Bennett, A.F.; Stanley, R.F. 1967 Modified laboratory techniques for preparing sterile larval diets for the boll weevil. J ECON ENTOMOL 60: 1473.

*Landes, D.A.; Strong, F.E. 1965 Feeding and nutrition of *Lygus hesperus* (Hemiptera: Miridae). I. Survival of bugs fed on artificial diets. ANN ENTOMOL SOC AM 58: 306-309.

*Lang, C.A. 1963 The effect of temperature on the growth and chemical composition of the mosquito. J INSECT PHYSIOL 9: 279-286.

Lang, C.A.; Lau, H.Y.; Jefferson, D.J. 1965 Protein and nucleic acid changes during growth and aging in the mosquito. BIOCHEM J 95: 372-377.

Lang, C.A.; Stephan, J.K. 1967 Nicotinamide-ademine dinucleotide phosphate enzymes in the mosquito during growth and aging. BIOCHEM J 102: 331-337.

*Larson, D.G.; Mitchell, W.F.; Wharton, G.W. 1969 Preliminary studies on *Dermatophagoides farinae* Hughes, 1961 (Acari) and house dust allergy. J MED ENTOMOL 6: 295-299.

Laurence, B.R. 1964 The problem of stabilizing conditions of larval culture in mosquito colonies. BULL W H O 31: 457-458.

Lazarevic, B. 1961 Inquiries into the resistance of some hybrid corns towards the corn borer. ANN EPIPHYT 12: 405-411.

Lea, A.O. 1957 Studies on the nutritional requirements of *Aedes aegypti* L. for growth, development and reproduction. Thesis No. 58-2083. Ohio State University Microfilms Inc, Ann Arbor, Mich.

*Lea, A.O.; Dimond, J.B.; DeLong, D.M. 1956a A chemically defined medium for rearing *Aedes aegypti* larvae. J ECON ENTOMOL 49: 313-315.

Lea, A.O.; Dimond, J.B.; DeLong, D.M. 1956b Role of diet in egg development by mosquitoes (*Aedes aegypti*). SCIENCE 123: 890-891.

Le Berre, J.R.; Lenoir, A. 1969 Incidence de régimes alimentaires de type méridique et de la duree d'éclairement sur le développement larvaire du criquet migrateur *Locusta migratoria*. ENTOMOL EXP APPL 12: 670-697. (In French with English summary)

Le Berre, J.R.; Petavy, G. 1965 Action comparée de quelques milieux nutritifs artificiels sur la longévité et la reproduction du criquet migrateur *Locusta migratoria* L. C R ACAD SCI SER D 260: 5877-5880.

Le Berre, J.R.; Tira, R. 1976 Influence d'une alimentation artificielle de type méridique sur divers caractères biologiques du criquet pélerin *Schistocerca gregaria* Forsk. C R ACAD SCI PARIS D 282: 473-475. (In French with English summary)

*LeCato, G.L. 1976 Yield, development, and weight of *Cadra cautella* (Walker) and *Plodia interpunctella* (Hübner) on twenty-one diets derived from natural products. J STORED PROD RES 12: 43-47.

Leckstein, P.M. 1974 Sulphur-amino acid metabolism in *Aphis fabae*. COMP BIOCHEM PHYSIOL 49 B: 743-747.

Leckstein, P.M.; Llewellyn, M. 1973 Effect of dietary amino acids on the size and alary polymorphism of *Aphis fabae*. J INSECT PHYSIOL 19: 973-980.

Leckstein, P.M.; Llewellyn, M. 1974 The role of amino acids in diet intake and selection and the utilization of dipeptides by *Aphis fabae*. J INSECT PHYSIOL 20: 877-885.

Leclercq, J. 1954 Action vitaminique de la "*dl*-di carnitine" chez "*Tenebrio molitor* L." (Insecte, Coléoptère). ARCH INT PHYSIOL 62: 101-108. (In French)

*Leclercq, J.; DeBast, D. 1965 Projet d'utilisation des larves de "*Tenebrio molitor*" pour comparer la valeur nutritive des protéines. ANN NUTR ALIMENT 19: 19-25. (In French)

Leclercq, J.; Lopez-Francos, L. 1966 Nutrition protidique chez *Tenebrio molitor* L. VII. Nouveaux essais de remplacement de la caséine par des préparations d'acides aminés. ARCH INT PHYSIOL BIOCHIM 74: 397-415. (In French)

Leclercq, J.; Lopez-Francos, L. 1967 Nutrition protidique chez *Tenebrio molitor* L. VIII. Sur la valeur nutritive des fractions de la caséine. ARCH INT PHYSIOL BIOCHIM 75: 89-95. (In French with English summary)

*Leclercq, J.; Magis, N.; Rey, C. 1954 Sur les besoins nutritifs du *Gnathocerus cornutus* F. (Coleoptère, Ténébrionidae). Recherche de l' optimum glucidique et de l'optimum protidique dans un régime alimentaire artificiel. ARCH INT PHYSIOL 62: 264-271. (In French)

*Lee, B.L.; Bass, M.H. 1969 Rearing technique for the granulate cutworm and some effects of temperature on its life cycle. ANN ENTOMOL SOC AM 62: 1216-1217.

**Legay, J.M. 1958 Recent advances in silkworm nutrition. ANNU REV ENTOMOL 3: 75-86.

*Legay, J.M.; Pascal, M. 1963-1964 Essais préliminaires à la mise au point d'une alimentation preparée à partir de feuilles de mûrier cuites pour le ver à soie *Bombyx mori* L. REV VER SOIE 15-16: 101-104. (In French)

Legge, J.B.B.; Palmer, J. 1968 Behaviour of *Myzus persicae* (Sulz.) on tobacco. III. Choice by alatae of sites for feeding and larviposition on leaves and artificial substrates of differing nutrient status. BULL ENTOMOL RES 57: 479-493.

*Legner, E.F.; Bay, E.C. 1965 Culture of *Hippelates pusio* (Diptera: Chloropidae) in the West Indies for natural enemy exploration and some notes on behaviour and distribution. ANN ENTOMOL SOC AM 58: 436-440.

*Lemonde, A.; Bernard, R. 1951a Nutrition des larves de *Tribolium confusum* Duval. I. Recherche d'un régime synthétique basal satisfaisant leurs besoins nutritifs. CAN J ZOOL 29: 71-79. (In French with English summary)

Lemonde, A.; Bernard, R. 1951b Nutrition des larves de *Tribolium confusum* Duval. II. Importance des acides aminés. CAN J ZOOL 29: 80-83. (In French with English summary)

*Lemonde, A.; Bernard, R. 1953 Aspects nutritifs des larves de *Stegobium paniceum* L. (Anobiidae) et d'*Oryzaephilus surinamensis* L. (Cucujidae). NAT CAN 80: 125-142. (In French)

Lemonde, A.; Bernard, R. 1955 Importance de la choline et de certaines substances analogues dans la nutrition des larves de *Tribolium confusum* Duval. REV CAN BIOL 1: 8-13. (In French with English summary)

*Lennox, F.G. 1939 Studies of the physiology and toxicology of blowflies. 1. The development of a synthetic medium for aseptic cultivation of larvae of *Lucilia cuprina*. AUST COMMONW COUNC SCI IND RES PAM 90: 7-24.

*Leonard, D.E.; Doane, C.C. 1966 An artificial diet for the gypsy moth, *Porthetria dispar* (Lepidoptera: Lymantriidae). ANN ENTOMOL SOC AM 59: 462-464.

Leppla, N.C.; Carlyle, S.L.; Carlysle, T.C. 1974 Effects of surface sterilization and automatic collection on cabbage looper eggs. J ECON ENTOMOL 67: 33-36.

Leppla, N.C.; Carlyle, S.L.; Pons, W.J.; Mitchell, E.R. 1975 Use of a rodent rearing container for culturing insects. J GA ENTOMOL SOC 10: 326-327.

*Leppla, N.C.; Solomon, J.D.; Doolittle, R.E. 1975 Culturing the carpenterworm. ANN ENTOMOL SOC AM 68: 683-685.

*Leppla, N.C.; Standifer, L.N.; Erickson, E.H. Jr 1974 Culturing larvae of blister beetles on diets containing different pollens collected by honeybees. J APIC RES 13: 243-247.

*Levengood, W.C. 1968 Rearing a giant silkworm moth, *Hyalophora cecropia*, on an artificial culture medium. ANN ENTOMOL SOC AM 61: 1594-1596.

**Levinson, Z.H. 1955 Nutritional requirements of insects. RIV PARASSITOL 16: 113-183.

*Levinson, Z.H. 1960 Food of housefly larvae. NATURE LONDON 188: 427-428.

**Levinson, Z.H. 1962 The function of dietary sterols in phytophagous insects. J INSECT PHYSIOL 8: 191-198.

**Levinson, Z.H. 1963 Production of metabolic deficiencies as a possible approach to insect control. BULL RES COUN ISR 10: 116-124.

*Levinson, Z.H.; Barelkovsky, J.; Bar Ilan, A.R. 1967 Nutritional effects of vitamin omission and antivitamin administration on development and longevity of the hide beetle *Dermestes maculatus* Deg. (Coleoptera, Dermestidae). J STORED PROD RES 3: 345-352.

Levinson, Z.H.; Bergmann, E.D. 1957 Steroid utilization and fatty acid synthesis by the larvae of the housefly, *Musca vicina* (Macq.). BIOCHEM J 65: 254-260.

Levinson, Z.H.; Bergmann, E.D. 1959 Vitamin deficiencies in the housefly produced by antivitamins. J INSECT PHYSIOL 3: 293-305.

*Levinson, Z.H.; Cohen, E. 1973 The action of overdosed biotin on reproduction of the hide beetle, *Dermestes maculatus*. J INSECT PHYSIOL 19: 551-558.

*Levinson, Z.H.; Gothilf, S. 1965 A semi-synthetic diet for axenic growth of the carob moth, *Ectomyelois ceratoniae* (Zell.). RIV PARASSITOL 26: 19-26.

*Levinson, Z.H.; Navon, A. 1969 Ascorbic acid and unsaturated fatty acids in the nutrition of the Egyptian cotton leafworm, *Prodenia litura*. J INSECT PHYSIOL 15: 591-595.

Levinson, Z.H.; Silverman, P.H. 1954 Studies on the lipids of *Musca vicina* (Macq.) during growth and metamorphosis. BIOCHEM J 58: 294-297.

*Lewallen, L.L. 1954 Biological and toxicological studies of the little house fly. J ECON ENTOMOL 47: 1137-1141.

Lewis, E.B. 1960 A new standard food medium. DROSOPH INF SERV 34: 117-118.

Lewis, L.C.; Lynch, R.E. 1969a Use of drugs to reduce *Perezia pyraustae* infections in the European corn borer. PROC NORTH CENT BRANCH ENTOMOL SOC AM 24: 84-87.

*Lewis, L.C.; Lynch, R.E. 1969b Rearing the European corn borer, *Ostrinia nubilalis* (Hübner), on diets containing corn leaf and wheat germ. IOWA STATE COLL J SCI 44: 9-14.

Lewis, L.C.; Lynch, R.E. 1970 Treatment of *Ostrinia nubilalis* larvae with Fumidil B to control infections caused by *Perezia pyraustae*. J INVERTEBR PATHOL 15: 43-48.

Lewis, L.C.; Lynch, R.E.; Guthrie, W.D. 1971 Biology of European corn borers reared continuously on a diet containing Fumidil B. ANN ENTOMOL SOC AM 64: 1264-1269.

Lewis, L.C.; Raun, E.S. 1966 Consumption and utilization of laboratory diets by European corn borers. IOWA STATE COLL J SCI 41: 173-180.

**Ley, F.J.; Bleby, J.; Coates, M.E.; Paterson, J.S. 1969 Sterilization of laboratory animal diets using *gamma* radiation. LAB ANIM 3: 221-254.

*Lichtenstein, E.P. 1948 Growth of *Culex molestus* under sterile conditions. NATURE LONDON 162: 227.

Lii, G.Y.; Garlich, J.D.; Rock, G.C. 1975 Protein and energy utilization by the insect, *Argyrotaenia velutinana* (Walker) (Lep., Tortricidae), fed diets containing graded levels of an amino acid mixture. COMP BIOCHEM PHYSIOL 52A: 615-618.

Lindig, O.H.; Malone, O.L. 1973 Sodium benzoate and sodium propionate as anticontamination diets fed to adult boll weevil. J ECON ENTOMOL 66: 577.

**Lipke, H. 1957 Nutritional considerations for rearing insects. PROC NORTH CENT BRANCH ENTOMOL SOC AM 12: 90-91.

**Lipke, H.; Fraenkel, G. 1956 Insect nutrition. ANNU REV ENTOMOL 1: 17-44.

*Lipovsky, L.J. 1953 Improved technique for rearing chigger mites (Acarina: Trombiculidae). ENTOMOL NEWS 64: 4-7.

Lloyd, E.P.; Daum, R.J.; McLaughlin, R.E.; Tingle, F.C.; McKibben, G.H.; Burt, E.C.; McCoy, J.R.; Bell, M.R.; Cleveland, T.C. 1968 A red dye to evaluate bait formulations and to mass mark field populations of boll weevils. J ECON ENTOMOL 61: 1440-1444.

*Loaharanu, S.; Chiravatanapong, S.; Sutantawong, M.; Kaochnong, P. 1974 Mating competitiveness of the raiosterilized male armyworm, *Spodoptera exigua* Hb., and the male mosquito, *Aedes aegypti* L., in field cages. pp. 317-323. *in* Sterility Principle for Insect Control 1974. Proceedings of a Panel Organized by the Joint FAO/IAEA Division of Atomic Energy in Food and Agriculture.

Loeb, J. 1915a The simplest constituents required for growth and the completion of the life cycle in an insect (*Drosophila*). SCIENCE 41: 169-170.

Loeb, J. 1915b The salts required for the development of insects. J BIOL CHEM 23: 431-434.

Loeb, J.; Northrop, J.H. 1916 Nutrition and evolution. J BIOL CHEM 27: 309-312.

*Lopez, D.F. 1970 Sterile-male technique for eradication of the Mexican and Caribbean fruit flies. Review of current status. pp. 111-117. *in* Sterile Male Technique for Control of Fruit Flies. Proceedings of a Panel Organized by the Joint FAO/IAEA Division of Atomic Energy in Food and Agriculture.

**Louw, B.K. 1964 Physical aspects of laboratory maintenance of muscoid fly colonies. BULL W H O 31: 529-533.

**Luckey, T.D. 1954 A single diet for all living organisms. SCIENCE 120: 396-398.

*Luckey, T.D. 1961 A study in comparative nutrition. COMP BIOCHEM PHYSIOL 2: 100-124.

**Luckey, T.D. 1963 Germfree Life and Gnotobiology. Academic Press, New York and London. 512 pp.

Luckey, T.D.; Stone, P.C. 1968 Cricket nutrition: supplementation of grass by known nutrients. J INSECT PHYSIOL 14: 1533-1538.

Ludvik, G.F. 1972 A device for producing artificial squares for boll weevil oviposition and feeding. J ECON ENTOMOL 65: 1200-1201.

Lukefahr, M.J.; Martin, D.F. 1964 The effects of various larval and adult diets on the fecundity and longevity of the bollworm, tobacco budworm, and cotton leafworm. J ECON ENTOMOL 57: 233-235.

Lynch, R.E.; Lewis, L.C. 1971 Reoccurrence of the microsporidan *Perezia pyraustae* in the European corn borer, *Ostrinia nubilalis*, reared on diet containing Fumidil B. J INVERTEBR PATHOL 17: 243-246.

*Lynch, R.E.; Lewis, L.C.; Robinson, J.C. 1973 Lyophilized rearing medium for bioassays. J ECON ENTOMOL 66: 1224-1225.

*Lyon, R.L.; Brown, S.J. 1970 Contact toxicity of insecticides applied to fall cankerworm reared on artificial diet. J ECON ENTOMOL 63: 1970-1971.

*Lyon, R.L.; Flake, H.W. Jr 1966 Rearing Douglas-fir tussock moth larvae on synthetic media. J ECON ENTOMOL 59: 696-698.

*Lyon, R.L.; Richmond, C.E.; Robertson, J.L.; Lucas, B.A. 1972 Rearing diapause and diapause-free western spruce budworm (*Choristoneura occidentalis*) (Lepidoptera: Tortricidae) on an artificial diet. CAN ENTOMOL 104: 417-426.

McCay, C.M. 1938 The nutritional requirements of *Blattella germanica*. PHYSIOL ZOOL 11: 89-103.

McCay, C.M.; Melampy, R.M. 1937 Care and rearing of *Blattella germanica*. pp. 283-284. *in* Needham, J.G. *et al.*, (Eds), Culture Methods for Invertebrate Animals. Comstock Publishing Company, New York (Dover Edition 1959 Unaltered and Unabridged).

McClanahan, R.J. 1966 A synthetic diet for the onion maggot, *Hylemya antiqua* (Meigen) (Diptera: Anthomyiidae). CAN J ZOOL 44: 1089-1090.

*McClanahan, R.J.; Miller, L.A. 1958 Laboratory rearing of the seed-corn maggot, *Hylemya cilicura* (Rond.) (Diptera: Anthomyiidae). CAN ENTOMOL 90: 372-374.

*McClanahan, R.J.; Niemczyk, H.D. 1963 Continuous rearing of the carrot rust fly, *Psila rosae* (Fab.). CAN ENTOMOL 95: 827-830.

McFarlane, J.E. 1964 The protein requirements of the house cricket, *Acheta domesticus* L. CAN J ZOOL 42: 645-647.

McFarlane, J.E. 1972a Vitamin E, tocopherol quinone and selenium in the diet of the house cricket, *Acheta domesticus* (L.). ISR J ENTOMOL 7: 7-14.

McFarlane, J.E. 1972b Studies on vitamin E in the house cricket, *Acheta domesticus* (Orthoptera: Gryllidae). I. Nutritional albinism. CAN ENTOMOL 104: 511-514.

McFarlane, J.E. 1972c Studies on vitamin E in the house cricket, *Acheta domesticus* (Orthoptera: Gryllidae). II. *In vivo* inhibition by vitamin E of phenolose system in the egg. CAN ENTOMOL 104: 515-518.

McFarlane, J.E. 1972d Studies on vitamin E in the house cricket, *Acheta domesticus* (Orthoptera: Gryllidae). III. Tarsal absorption of vitamin E and fatty acid methyl esters by larvae. CAN ENTOMOL 104: 519-522.

*McFarlane, J.E. 1974 The functions of copper in the house cricket and the relation of copper to vitamin E. CAN ENTOMOL 106: 441-446.

McFarlane, J.E. 1976a Influence of dietary copper and zinc on growth and reproduction of the house cricket (Orthoptera: Gryllidae). CAN ENTOMOL 108: 387-390.

McFarlane, J.E. 1976b Vitamin K: A growth factor for the house cricket (Orthoptera: Gryllidae). CAN ENTOMOL 108: 391-394.

*McFarlane, J.E.; Neilson, B.; Ghouri, A.S.K. 1959 Artificial diets for the house cricket, *Acheta domesticus* (L.). CAN J ZOOL 37: 913-916.

McGhehey, J.H.; Nagel, W.P. 1966 A technique for rearing larvae of *Medetera aldrichii* (Diptera: Dolichopodidae). ANN ENTOMOL SOC AM 59: 290-292.

McGinnis, A.J.; Kasting, R. 1959 Nutrition of the pale western cutworm, *Agrotis orthogonia* Morr. (Lepidoptera: Noctuidae). I. Effects of underfeeding and artificial diets on growth and development, and a comparison of wheat sprouts of Thatcher, *Triticum aestivum* L., and Golden Ball, *T. durum* Desf., as food. CAN J ZOOL 37: 259-266.

McGinnis, A.J.; Kasting, R. 1960 Nutrition of the pale western cutworm, *Agrotis orthogonia* Morr. (Lepidoptera: Noctuidae). III. Lyophilized sprouts and leaves of wheat as a basal diet for larvae, and effects of supplementation with *l*-leucine. CAN J ZOOL 38: 585-592.

McGinnis, A.J.; Kasting, R. 1962 A method of rearing larvae of the wheat stem fly, *Cephus cinctus* Nort. (Hymenoptera: Cephidae) under artificial conditions. CAN ENTOMOL 94: 573-574.

**McGinnis, A.J.; Kasting, R. 1964a Nutritional methods for phytophagous insects. CAN ENTOMOL 96: 130.

McGinnis, A.J.; Kasting, R. 1964b Comparison of gravimetric and chromic oxide methods for measuring percentage utilization and consumption of food by phytophagous insects. J INSECT PHYSIOL 10: 989-995.

McGinnis, A.J.; Kasting, R. 1966 Comparison of tissues from solid- and hollow-stemmed spring wheats during growth. IV. Apparent dry matter utilization and nitrogen balance in the two-striped grasshoppers, *Melanoplus bivittatus* (Say). J INSECT PHYSIOL 12: 671-678.

McGinnis, A.J.; Kasting, R. 1967 Dietary cellulose: effect on food consumption and growth of a grasshopper. CAN J ZOOL 45: 365-367.

**McGinnis, A.J.; Kasting, R. 1972 Quantitative nutrition and evaluation of protein in food of phytophagous insects. pp. 57-61. *in* Rodriguez, J.G. (Ed.), Insect and Mite Nutrition. North-Holland Publishing Company, Amsterdam.

*McGinnis, A.J.; Newburgh, R.W.; Cheldelin, V.H. 1956 Nutritional studies on the blow-fly, *Phormia regina* (Meig.). J NUTR 58: 309-324.

*McKibben, G.H.; Hedin, P.A.; Davich, T.B.; Daum, R.J.; Laseter, M.W. 1971 Addition of food acidulants to increase attractiveness to boll weevils of bait containing cottonseed oil. J ECON ENTOMOL 64: 583-585.

McKibben, G.H.; Mitlin, N.; Haynes, J.W.; Davich, T.B. 1972 Feeding stimulants added to diet of adult boll weevil. J ECON ENTOMOL 65: 1190-1191.

*McKinley, D.J. 1970 The laboratory culture and biology of *Spodoptera littoralis* Boisduval. REP ANTI-LOCUST RES CENT 18, 12 pp.

**McKinley, D.J. 1971 An introduction to the use and preparation of artificial diets with special emphasis on diets for phytophagous Lepidoptera. PANS 17: 421-424.

*McKinley, D.J. 1974 An attempt to develop a completely artificial diet for rearing the red bollworm, *Diparopsis castanea*. CENT OVERSEAS PEST RES MISC REP 16: 1-4.

McLaughin, R.E.; Keller, J.C. 1964 Antibiotic control of an epizootic caused by *Serratia marcescens* Bizio in the boll weevil, *Anthonomus grandis* Boheman. J INSECT PATHOL 6: 481-485.

*McLean, D.L. 1971 Probing behaviour of the pea aphid, *Acyrthosiphon pisum*. V. Comparison of *Vicia faba*, *Pisum sativum*, and a chemically defined diet as food sources. ANN ENTOMOL SOC AM 64: 499-503.

McLeod, D.G.R. 1964 Nutrition and feeding behaviour of the adult onion maggot, *Hylemya antiqua*. J ECON ENTOMOL 57: 845-847.

*McMorran, A. 1965 A synthetic diet for the spruce budworm, *Choristoneura fumiferana* (Clem) (Lepidoptera: Tortricidae). CAN ENTOMOL 97: 58-62.

*McMurtry, J.A.; Scriven, G.T. 1962 The use of agar media in transporting and rearing phytoseiid mites. J ECON ENTOMOL 55: 412-414.

*McMurtry, J.A.; Scriven, G.T. 1966 Effects of artificial foods on reproduction and development of four species of phytoseiid mites. ANN ENTOMOL SOC AM 59: 267-269.

**Mackauer, M. 1972 Genetic aspects of insect production. ENTOMOPHAGA 17: 27-48.

**Mackauer, M. 1976 Genetic problems in the production of biological control agents. ANNU REV ENTOMOL 21: 369-385.

*Mackerras, M.J.; Freney, M.R. 1933 Observations on the nutrition of maggots of Australian blow-flies. J EXP BIOL 10: 237-246.

Madariaga, M.A.; Mata, F.; Municio, A.M.; Ribera, A. 1972a Effect of cholesterol feeding on the relative fatty acid distribution during development of *Ceratitis capitata*. J INSECT PHYSIOL 18: 1-8.

Madariaga, M.A.; Mata, F.; Municio, A.M.; Ribera, A. 1972b Effect of the lipid composition of the larval diet on the fatty acid composition during development of *Ceratitis capitata*. INSECT BIOCHEM 2: 249-256.

Madariaga, M.A.; Mata, F.; Municio, A.M.; Ribera, A. 1974 Changes in the fatty acid patterns of glycerolipids of *Dacus oleae* during metamorphosis and development. INSECT BIOCHEM 4: 151-160.

*Maeda, S.; Hagen, K.S.; Finney, G.L. 1953 Artificial media and the control of microorganisms in the culture of tephritid larvae (Diptera: Tephritidae). PROC HAWAII ENTOMOL SOC 15: 177-185.

**Magalhaes, S.C. de 1970 Sterile-male technique for control of the olive fly. pp. 119-130. *in* Sterile-male Tĕchnique for Control of Fruit Flies. Proceedings of a Panel Organized by the Joint FAO/IAEA Division of Atomic Energy in Food and Agriculture.

Magis, N. 1954a Nutrition comparee des *Tribolium*. II. - Importance du cholestérol et des facteurs de la levure dans la nutrition des larves de *Tribolium confusum* Duv., *T. castaneum* Herbst et *T. destructor* Uytt. (Insectes, Coléoptères). BULL SOC CHIM BIOL 36: 681-690. (In French)

*Magis, N. 1954b Nutrition comparée des *Tribolium*. III. - Action vitaminique de la *dl*-carnitine et de *dl*-dicarnitine chez *Tribolium castaneum* Herbst (Insecte, Coléoptêra). ARCH INT PHYSIOL 62: 505-511. (In French)

*Magnoler, A. 1970 A wheat germ medium for rearing of the gypsy moth, *Lymantria dispar* L. ENTOMOPHAGA 15: 401-406.

*Maleki-Milani, H. 1971 Élevage de la noctuelle du chou, *Mamestra (Barathra) brassicae* L. sur milieu artificiel. ANN SOC ENTOMOL FR 7: 39-56. (In French with English summary)

*Malphettes, C.B. 1968 Premiers essais d' élevage des larves d'*Hylobius abietis* sur milieu synthétique. ANN SCI FOR 25: 189-192. (In French)

Maltais, J.B. 1952 A simple apparatus for feeding aphids aseptically on chemically defined diets. CAN ENTOMOL 84: 291-294.

*Mangat, B.S. 1970 Rearing the black cutworm in the laboratory. J ECON ENTOMOL 63: 1325-1326.

Mangat, B.S. 1971 Development of the black cutworm. J ECON ENTOMOL 64: 766.

Mangum, C.L.; Ridgway, W.O.; Brazzel, J.R. 1969 Large-scale laboratory production of the pink bollworm for sterilization programs. U S DEP AGRIC PRESENTATION PAP ARS 81-35, 7 pp.

Manoukas, A.G. 1974 Protein hydrolysate-free larval diets for rearing the olive fruit fly, *Dacus oleae*, and the nutritional role of Brewer's yeast. pp. 219-228. *in* Proceedings Symposium on the Sterility Principle for Insect Control. Proceedings of a Panel Organized by the Joint FAO/IAEA Division of Atomic Energy in Food and Agriculture.

*Manoukas, A.G. 1975 Low-cost larval diets for mass production of the olive fruit fly. J ECON ENTOMOL 68: 22-24.

Mansingh, A.; Pratt, J.J. Jr 1975 Effects of RNA, purines and pyrimidines on the development of *Tribolium confusum* (Duval). J STORED PROD RES 11: 155-160.

Markkula, M.; Laurema, S. 1967 The effect of amino acids, vitamins and trace elements on the development of *Acyrthosiphon pisum* Harris (Hom., Aphididae). ANN AGRIC FENN 6: 77-80.

Marlowe, R.H. 1934 An artificial food medium for the Mediterranean fruit fly (*Ceratitis capitata* Wied.). J ECON ENTOMOL 27: 1100.

Marston, N.; Brown, B. 1974 Constituents in diets for *Galleria mellonella*. J ECON ENTOMOL 67: 497-500.

Marston, N.; Campbell, B. 1973 Comparison of nine diets for rearing *Galleria mellonella*. ANN ENTOMOL SOC AM 66: 132-136.

*Marston, N.; Campbell, B.; Boldt, P.E. 1975 Mass producing eggs of the greater wax moth, *Galleria mellonella* (L.). U S DEP AGRIC TECH BULL 1510: 15 pp.

*Martin, D.F. 1966 Pink bollworms. pp. 355-366. *in* Smith, C.N. (Ed.), Insect Colonization and Mass Production. Academic Press, New York and London.

Martínez-Palacios, A. 1964 Maintenance of *Aedes aegypti* in the insectary in Mexico city. BULL W H O 31: 461-462.

Marucci, P.E.; Clancy, D.W. 1950 The artificial culture of fruit flies and their parasites. PROC HAWAII ENTOMOL SOC 14: 163-166.

**Massonié, G. 1971a L'élevage des aphides sur milieu synthétique. ANN ZOOL ECOL ANIM 3: 103-123. (In French with English summary)

Massonié, G. 1971b Élevage d'un biotype de *Myzus persicae* Sulzer sur milieu synthétique. I. Influence quantitative de l'Azote sur la fecondite. REV ZOOL AGRIC PATHOL VEG 4: 83-88. (In French with English summary)

Massonié, G. 1973 Élevage d'un biotype de *Myzus persicae* Sulzer sur milieu synthétique. II. Influence de la qualité de l'alimentation azotée sur la biologie au puceron. ANN ZOOL ECOL ANIM 5: 559-568. (In French with English summary)

Massonié, G. 1974 Élevage d'un biotype de *Myzus persicae* Sulzer sur milieu synthétique. III. Valeur alimentaire respective de la *l*. méthionine, de la *l*. cystéine et des sulfates. ANN NUTR ALIMENT 29: 45-50. (In French with English summary)

Mathad, S.B.; Dakshayani, K. 1972 Laboratory rearing of the cricket *Plebeiogryllus guttivantris*. ANN ENTOMOL SOC AM 65: 282-283.

Matsubara, F. 1965 Studies on the disease in the aseptically reared silkworms, *Bombyx mori* L. BULL FAC TEXT FIBRES KYOTO UNIV IND ARTS 4: 291-297.

*Matsubara, F.; Kato, M.; Hayashiya, K.; Kodama, R.; Hamamura, Y. 1967 Aseptic rearing of silkworm with prepared food. J SERIC SCI JPN 36: 39-45. (In Japanese with English summary)

*Matsuda, M.; Matsuura, Y. 1967 The germ free rearing of the silkworm (*Bombyx mori* L.) in the flexible isolator. J SERIC SCI JPN 36: 403-408. (In Japanese with English summary)

*Matsuka, M.; Okada, I. 1975 Nutritional studies of an aphidophagous coccinellid, *Harmonia axyridis*. (I). Examination of artificial diets for the larval growth with special reference to drone honeybee powder. BULL FAC AGRIC TAMAGAWA UNIV 15: 1-9.

Matsuka, M.; Shimotori, D.; Senzaki, T.; Okada, I. 1972 Rearing some coccinellids on pulverized drone honeybee brood. BULL FAC AGRIC TAMAGAWA UNIV 12: 28-38.

Matsumoto, K. 1968 Studies on the environmental factors for the breeding of grain mites. Part IX. The effect of relative humidity on the age composition of the population of *Lardoglyphus konoi*. JPN J SANIT ZOOL 19: 196-203. (In Japanese with English summary)

*Matsumoto, K. 1970 Studies on the environmental factors for the breeding of grain mites. Part X. Effect of the temperature and relative humidity on the breeding and hypopus formation in *Lardoglyphus konoi*. JPN J SANIT ZOOL 21: 213-219. (In Japanese with English summary)

Matsumoto, K. 1973 Studies on environmental factors for the breeding of grain mites. XI. The effect of nutrient on hypopus formation in *Lardoglyphus konoi*. JPN J SANIT ZOOL 24: 1-7. (In Japanese)

*Matsumoto, K. 1975 Studies on the environmental requirement for breeding the dust mites, *Dermatophagoides farinae* Hughes, 1961. Part 3. Effect of the lipids in the diet on the population growth of the mites. JPN J SANIT ZOOL 26: 121-127. (In Japanese with English summary)

Matsumoto, Y. 1954 An aseptic rearing of the oriental fruit moth on synthetic food media. BER OHARA INST LANDWIRTSCH BIOL 10: 66-72.

*Matsumoto, Y.; Yasue, Y. 1952 Artificial sterile breeding of the 'Princess' pear core borer, *Laspeyresia (= Grapholitha) molesta* Busck, in an artificial medium. J AGRIC RES 40: 205-208. (In Japanese)

*Mattingly, E.; Ogle, S. 1969 A new culture method for *Rhynchosciara*. ANN ENTOMOL SOC AM 62: 94-96.

*Mauldin, J.K.; Lambremont, E.N.; Graves, J.B. 1971 Principal lipid classes and fatty acids synthesized during growth and development of the beetle *Lyctus planicollis*. INSECT BIOCHEM 1: 316-326.

*Mauldin, J.K.; Rich, N.M. 1975 Rearing two subterranean termites, *Reticulitermes flavipes* and *Coptotermes formosanus*, on artificial diet. ANN ENTOMOL SOC AM 68: 454-456.

Medici, J.C. 1964 Nutritional studies, including quantitative amino acid and mineral requirements, with the flour beetle, *Tribolium confusum* Duval. DISS ABSTR 25: 2207.

*Medici, J.C.; Taylor, M.W. 1966 Mineral requirements of the confused flour beetle, *Tribolium confusum* (Duval). J NUTR 88: 181-186.

Medici, J.C.; Taylor, M.W. 1967 Interrelationships among copper, zinc, and cadmium in the diet of the confused flour beetle. J NUTR 93: 307-309.

Meikle, J.E.S.; McFarlane, J.E. 1965 The role of lipid in the nutrition of the house cricket, *Acheta domesticus* L. (Orthoptera: Gryllidae). CAN J ZOOL 43: 87-98.

Melampy, R.M.; Maynard, L.A. 1937 Nutrition studies with the cockroach (*Blattella germanica*). PHYSIOL ZOOL 10: 36-44.

Melvin, R.; Bushland, R.C. 1936 A method of rearing *Cochliomyia americana* C & P on artificial media. U S DEP AGRIC BUR ENTOMOL PLANT QUANT E7-88, 2 pp.

*Melvin, R.; Bushland, R.C. 1941 The nutritional requirements of screw-worm larvae. J ECON ENTOMOL 33: 850-852.

*Meyer, H.J.; Norris, D.M.; Koval, C.F. 1974 Laboratory rearing of the lesser peach tree borer, *Synanthedon pictipes*, on artificial diets. ANN ENTOMOL SOC AM 67: 721-722.

**Meynadier, G. 1971 Aseptic rearing of invertebrates for tissue culture. pp. 141-167. *in* Vago, C. (Ed.), Invertebrate Tissue Culture Vol I, Academic Press, New York and London.

Michael, C.A.; Meiss, A.N.; Pisano, J.J.; Taylor, M.W. 1960 Growth response of the confused flour beetle, *Tribolium confusum* (Duval) to six selected protein sources. J NUTR 72: 121-130.

*Michelbacher, A.E.; Hoskins, W.M.; Herms, W.B. 1932 The nutrition of flesh fly larvae, *Lucilia sericata* (Meig.). 1. The adequacy of sterile synthetic diets. J EXP ZOOL 64: 109-131.

*Miller, G.J.; Blankenship, J.W. 1973 Influence of dietary lipids upon lipids in larvae and adults of the dried-fruit moth. J INSECT PHYSIOL 19: 65-74.

*Miskimen, G.W. 1965 Non aseptic laboratory rearing of the sugarcane borer, *Diatraea saccharalis*. ANN ENTOMOL SOC AM 58: 820-823.

**Misra, N.S. 1973 Nutritional requirements of stored grain insect pests. pp. 144-166. *in* Pant, N.C.; Ghai, S. (Eds), Insect Physiology and Anatomy. Indian Council Agricultural Research, New Delhi, India.

Mitchell, S.; Tanaka, N.; Steiner, L.F. 1965 Methods of mass culturing melon flies and oriental and Mediterranean fruit flies. U S DEP AGRIC PRESENTATION PAP ARS 33-104, 22 pp.

Mitsuhashi, J. 1970 A device for collecting planthopper and leafhopper eggs (Hemiptera: Delphacidae) and Deltocephalidae). APPL ENTOMOL ZOOL 5: 47-49.

*Mitsuhashi, J. 1974 Methods for rearing leafhoppers and planthoppers on artificial diets. REV PLANT PROT RES 7: 57-67.

Mitsuhashi, J.; Koyama, K. 1971 Rearing of planthoppers on a holidic diet. ENTOMOL EXP APPL 14: 93-98.

*Mitsuhashi, J.; Koyama, K. 1972 Artificial rearing of the smaller brown planthopper, *Laodelphax striatellus* Fallén, with special reference to rearing conditions for the first instar nymphs. JPN J APPL ENTOMOL ZOOL 16: 8-17. (In Japanese with English summary)

Mitsuhashi, J.; Koyama, K. 1974 Folic acid as a dietary factor affecting the wing morph of the planthopper, *Laodelphax striatellus* (Hemiptera: Delphacidae). ENTOMOL EXP APPL 17: 77-82.

Mitsuhashi, J.; Koyama, K. 1975 Oviposition of smaller brown planthopper, *Laodelphax striatellus*, into various carbohydrate solutions through a parafilm membrane (Hemiptera: Delphacidae). APPL ENTOMOL ZOOL 10: 123-129.

*Mitsuhashi, J.; Maramorosch, K. 1963 Aseptic cultivation of four virus transmitting species of leafhoppers (Cicadellidae). CONTRIB BOYCE THOMPSON INST 22: 165-173.

Mitsui, T.C.; Nobusawa, C.; Fukami, J.; Mori, K.; Fukunaga K. 1973 Compounds with juvenile hormone activity. I. Inhibitory effects of the synthetic juvenile hormone and its analogues of the metamorphosis of several insects. APPL ENTOMOL ZOOL 8: 27-35.

Mittler, T.E. 1967a Effect of amino acid and sugar concentrations on the food uptake of the aphid *Myzus persicae*. ENTOMOL EXP APPL 10: 39-51.

Mittler, T.E. 1967b Gustation of dietary amino acids by the aphid *Myzus persicae*. ENTOMOL EXP APPL 10: 87-96.

Mittler, T.E. 1970a Effects of dietary amino acids on the feeding rate of the aphid *Myzus persicae*. ENTOMOL EXP APPL 13: 432-437.

Mittler, T.E. 1970b Uptake rates of plant sap and synthetic diet by the aphid *Myzus persicae*. ANN ENTOMOL SOC AM 63: 1701-1705.

Mittler, T.E. 1971a Some effects on the aphid *Myzus persicae* of ingesting antibiotics incorporated into artificial diets. J INSECT PHYSIOL 17: 1333-1347.

Mittler, T.E. 1971b Dietary amino acid requirements of the aphid *Myzus persicae* affected by antibiotic uptake. J NUTR 101: 1023-1028.

Mittler, T.E.; Dadd, R.H. 1962 Artificial feeding and rearing of the aphid, *Myzus persicae* (Sulzer), on a completely chemically defined diet. NATURE 195: 404.

Mittler, T.E.; Dadd, R.H. 1963a Studies on the artificial feeding of the aphid *Myzus persicae* (Sulzer) - I. Relative uptake of water and sucrose solutions. J INSECT PHYSIOL 9: 623-645.

Mittler, T.E.; Dadd, R.H. 1963b Studies on the artificial feeding of the aphid, *Myzus persicae* (Sulzer) - II. Relative survival, development, and larviposition on different diets. J INSECT PHYSIOL 9: 741-757.

Mittler, T.E.; Dadd, R.H. 1964 An improved method for feeding aphids on artificial diets. ANN ENTOMOL SOC AM 57: 139-140.

Mittler, T.E.; Dadd, R.H. 1965 Assimilation by apterous adult *Myzus persicae* maintained on a synthetic diet. ANN ENTOMOL SOC AM 58: 396-401.

Mittler, T.E.; Dadd, R.H.; Daniels, S.C. Jr 1970 Utilization of different sugars by the aphid *Myzus persicae*. J INSECT PHYSIOL 16: 1873-1890.

Mittler, T.E.; Kleinjan, J.E. 1970 Effect of artificial diet composition on wing-production by the aphid *Myzus persicae*. J INSECT PHYSIOL 16: 833-850.

Mittler, T.E.; Kleinjan, J.E.; Kunkel, H. 1970 Apteriform development induced by radish seedlings in larvae of the aphid *Myzus persicae* reared on artificial diets. J INSECT PHYSIOL 16: 2119-2125.

*Mittler, T.E.; Koski, P. 1974 Meridic artificial diets for rearing aphids. ENTOMOL EXP APPL 17: 524-525.

Mittler, T.E.; Kunkel, H. 1971 Wing production by grouped and isolated apterae of the aphid, *Myzus persicae* on artificial diet. ENTOMOL EXP APPL 14: 83-92.

*Mittler, T.E.; Tsitsipis, J.A. 1973 Economical rearing of larvae of the olive fruit fly, *Dacus oleae*, on a liquid diet offered on cotton towelling. ENTOMOL EXP APPL 16: 292-293.

Mittler, T.E.; Tsitsipis, J.A.; Kleinjan, J.E. 1970 Utilization of dehydroascorbic acid and some related compounds by the aphid *Myzus persicae* feeding on an improved diet. ENTOMOL EXP APPL 16: 2315-2326.

*Miura, K.; Takaya, T.; Koshiba, K. 1967 The effect of biotin deficiency on the biosynthesis of the fatty acids in a blowfly, *Aldrichina grahami* during metamorphosis under aseptic conditions. ARCH INT PHYSIOL BIOCHIM 75: 65-76.

**Miyakawa, M.; Luckey, T.D. (Eds) 1967 Advances in Germfree Research and Gnotobiology. Chemical Rubber Company, Cleveland, Ohio. 439 pp.

*Miyamoto, J.; Ishii, A.; Sasa, M. 1975 A successful method for mass culture of the house dust mite, *Dermatophagoides pteronyssinus* (Troussant). JPN J EXP MED 45: 133-138.

*Miyashita, K. 1971 Effects of constant and alternating temperatures on the development of *Spodoptera litura* F. (Lepidoptera: Noctuidae). APPL ENTOMOL ZOOL 6: 105-111.

*Mochida, O.; Miyahara, Y. 1974 Effect of *gamma* radiation on male pupae and adults of *Spodoptera litura* (F.) (Lepidoptera: Noctuidae) following pupal exposure. APPL ENTOMOL ZOOL 9: 41-47.

Monro, J. 1965 IAEA laboratory activities. TECH REP SER 55: 63-65.

*Monro, J. 1968 Improvements in mass rearing the Mediterranean fruit-fly *Ceratitis capitata* Wied. pp. 91-104. *in* Radiation, Radioisotopes and Rearing Methods in Control of Insect Pests. Proceedings of a Panel Organized by the Joint FAO/IAEA Division of Atomic Energy in Food and Agriculture.

*Monro, J.; Osborn, A.W. 1967 The use of sterile males to control populations of Queensland fruit-fly, *Dacus tryoni* (Frogg.) (Diptera: Tephritidae). I. Methods of mass-rearing, transporting, irradiating and releasing sterile flies. AUST J ZOOL 15: 461-473.

*Monroe, R.E. 1960 Effect of dietary cholesterol on housefly reproduction. ANN ENTOMOL SOC AM 53: 821-824.

*Monroe, R.E. 1962 A method for rearing housefly larvae aseptically on a synthetic medium. ANN ENTOMOL SOC AM 55: 140.

Monroe, R.E.: Hopkins, T.L.; Valder, A.S. 1967 Metabolism and utilization of cholesterol-4-C^{14} for growth and reproduction of aseptically reared houseflies, *Musca domestica* L. J INSECT PHYSIOL 13: 219-233.

Monroe, R.E.; Lamb, N.J. 1968 Effect of commercial proteins on house fly reproduction. ANN ENTOMOL SOC AM 61: 456-459.

Monroe, R.E.; Leyrer, R.L. 1973 A device for the collection of $^{14}CO_2$ from housefly larvae reared aseptically on a synthetic diet. ANN ENTOMOL SOC AM 66: 476-478.

Monty, J. 1973 Rearing the Natal fruit fly *Ceratitis (Pterandrus) rosa* Karsch (Diptera, Trypetidae) in the laboratory. REVUE AGRIC SUCR ILE MAURICE 52: 133-135.

*Moore, C.G.; Fisher, B.R. 1969 Competition in mosquitoes. Densities and species ratio effects on growth, mortality, fecundity and production of growth retardant. ANN ENTOMOL SOC AM 62: 1325-1331.

Moore, C.G.; Whitacre, D.M. 1972 Competition in mosquitoes. 2. Production of *Aedes aegypti* larval growth retardant at various densities and nutrition levels. ANN ENTOMOL SOC AM 65: 915-918.

Moore I. 1959 A method for artificially culturing the olive fly (*Dacus oleae* Gmel.) under aseptic conditions. KTAVIM ENG ED 9: 295-296.

*Moore, I. 1962 Further investigations on the artificial breeding of the olive fly *Dacus oleae* Gmel. under aseptic conditions. ENTOMOPHAGA 7: 53-57.

*Moore, I.; Navon, A. 1966 The rearing and some bionomics of the leopard moth, *Zeuzera pyrina* L., on an artificial medium. ENTOMOPHAGA 11: 285-296.

*Moore, I.; Navon, A. 1969 Calcium alginate: a new approach in the artificial culturing of insects, applied to *Spodoptera littoralis* (Boisduval). EXPERIENTIA 25: 221-222.

*Moore, I.; Navon, A. 1971 Rearing of the cotton worm, *Spodoptera littoralis* (Boisduval), on a calcium alginate medium. pp. 81-86. *in* Application of Induced Sterility for Control of Lepidopterous Populations. Proceedings of a Panel Organized by the Joint FAO/IAEA Division of Atomic Energy in Food and Agriculture.

*Moore, I.; Navon, A. 1973 A new gelling technique and improved calcium-alginate medium for rearing *Spodoptera littoralis*. J ECON ENTOMOL 66: 565-566.

*Moore, R.F. Jr; Whisnant, F.F.; Taft, H.M. 1967 A laboratory diet containing egg albumin for larval and adult boll weevils. J ECON ENTOMOL 60: 237-241.

*Moore, W. 1943 Some vitamin requirements of the black carpet beetles, *Attagenus* (?) spp. ANN ENTOMOL SOC AM 36: 483-485.

*Moore, W. 1946 Nutrition of *Attagenus* (?) sp. II. (Coleoptera: Dermestidae). ANN ENTOMOL SOC AM 39: 513-521.

Moreau, J.P. 1965a A propos de la biologie d'*Arctia caja* L. (Lépidoptères-Arctiidae). PROC 12 INT CONGR ENTOMOL LONDON (1964): 539. (In French)

*Moreau, J.P. 1965b Comportement des vers de la grappe vis-à-vis de divers cépages et essais d'alimentation artificielle. REV ZOOL AGRIC APPL 64: 13-16. (In French)

Moreland, C.R.; McLeod, W.S. 1957 Studies on rearing the house fly on a bran-alfalfa medium. J ECON ENTOMOL 50: 146-150.

*Morère, J.L. 1970a Nutrition d'une chenille de Lépidoptère *Plodia interpunctella* (Pyralidae) I. Essai d'alimentation semi-synthétique permettant l'obtention d'adultes féconds. ARCH SCI PHYSIOL 24: 97-107. (In French)

Morère, J.L.1970b Développement et reproduction de *Plodia interpunctella* (Hbn) (Lépidoptère: Pyralidae) sur un régime méridique (= semi-synthétique) a base d'acides aminés. C R ACAD SCI SER D 271: 1019-1022. (In French)

Morère, J.L. 1971a Remplacement de l'huile de germe de mais pars des acides gras dans l'alimentation de type méridique (= semi-synthétique) de *Plodia interpunctella* (Hbn) (Lépidoptère: Pyralidae). C R ACAD SCI SER D 272: 133-136. (In French)

Morère, J.L. 1971b Le carotène: substance indispensable pour la nutrition de *Plodia interpunctella* (Lep: Pyralidae). C R ACAD SCI SER D 272: 2229-2231. (In French)

Morère, J.L. 1973a Caractères physiologiques de *Plodia interpunctella* (Pyralidae) élevé sur un milieu semi-synthétique durant vingt générations. ENTOMOL EXP APPL 16: 358-364. (In French with English summary)

Morère, J.L. 1973b Importance des glucides dans la nutrition de la chenille de *Plodia interpunctella* (Lep: Pyralidae). ARCH INT PHYSIOL BIOCHIM 81: 709-716. (In French)

*Morère, J.L. 1974 Nutrition d'une chenille de Lépidoptère *Plodia interpunctella* Hübner (Lépidoptère, Pyralidae) sur un milieu chimiquemement défini. ARCH INT PHYSIOL BIOCHIM 82: 625-630. (In French with English summary)

Morgan, J.F.; Morton, H.J.; Parker, R.D. 1950 Nutrition of animal cells in tissue culture. I. Initial studies on a synthetic medium. PROC SOC EXP BIOL MED 73: 1-8.

Morisaki, M.; Ohotaka, H.; Awata, N.; Ikekawa, N.; Horie, Y.; Nakasone, S. 1974 Nutritional effect of possible intermediates of phytosterol dealkylation in the silkworm, *Bombyx mori*. STEROIDS 24: 165-176.

Morrison, F.O.; Perron, J.M. 1963 Sensibilité des différents stades larvaires de *Galleria mellonella* L. à *Bacillus thuringiensis* Berliner. PHYTOPROTECTION 44: 106-115. (In French)

*Morrison, P.E.; Davies, D.M. 1964a Feeding of dry chemically defined diets, and egg production in the adult house-fly. NATURE LONDON 201: 104-105.

Morrison, P.E.; Davies, D.M. 1964b Repeated ovarian cycles with ribonucleic acid in the diet of adult house-flies. NATURE LONDON 201: 948-949.

*Morrison, R.K.; House, V.S.; Ridgway, R.L. 1975 Improved rearing unit for larvae of a common green lacewing. J ECON ENTOMOL 68: 881-882.

*Mourad, A. 1974 Preliminary experiments in reducing the cost of medfly larval diets. pp. 237-240. *in* Sterility Principle for Insect Control 1974. Proceedings of a Panel Organized by the Joint FAO/IAEA Division of Atomic Energy in Food and Agriculture.

Mourad, A. 1975 Liquid larval diet for the olive fly, *Dacus oleae* Gmelin. pp. 145. *in* Controlling Fruit Flies by the Sterile-insect Technique. Proceedings of a Panel Organized by the Joint FAO/IAEA Division of Atomic Energy in Food and Agriculture.

*Mourikis, P.A.; Vassilaina-Alexopoulou, P. 1969 Notes on the rearing on semi-synthetic diet of the cotton bollworm *Heliothis armigera* (Hbn). ANN INST PHYTOPATHOL BENAKI 9: 46-53.

Mukaiyama, F.; Ito, T. 1962 Digestion experiments in the silkworm *Bombyx mori*, by means of artificial diets. II. Effect of cellulose powder added in the diets. J SERIC SCI JPN 31: 398-406. (In Japanese)

*Mukherjee, A.B.; SomChoudhury, A.K. 1971 Artificial diets for rearing of *Xylocoris flavipes* (Reuter), a predator on some pests of stored cereals. INDIAN J ENTOMOL 33: 356-358.

*Mulkern, G.B.; Toczek, D.R. 1970 Bioassay of plant extracts for growth-promoting substances for *Melanoplus femurrubrum* (Orthoptera: Acrididae). ANN ENTOMOL SOC AM 63: 272-284.

*Mulla, M.S. 1962 Mass rearing of three species of *Hippelates* eye gnats (Diptera: Chloropidae). ANN ENTOMOL SOC AM 55: 253-258.

*Mulla, M.S.; Barnes, M.M. 1957 On laboratory colonization of the eye gnat, *Hippelates collusor* Townsend. J ECON ENTOMOL 50: 813-816.

Municio, A.M.; Odrizolla , J.M.; Ramos, J.A. 1973 Effect of diet on lipogenesis of larvae of *Ceratitis capitata*. INSECT BIOCHEMISTRY 3: 359-366.

*Murdoch, F.F.; Smart, T.L. 1931 A method of producing sterile blowfly larvae for surgical use. U S NAV MED BULL 29: 406-417.

Mushi, A.M.; Chiang, H.C. 1974 Laboratory observations on the effect of common salt on *Dermestes maculatus* Deg. (Coleoptera: Dermestidae) infesting dried fresh water fish, *Roccus chrysops*. J STORED PROD RES 10: 57-60.

*Nachiappan, R.M.; Subramaniam, T.R. 1973 Studies on mass-culturing of *Heliothis armigera* Hbn. (Noctuidae: Lepidoptera) on semi-synthetic diets. I. Testing with various types of semi-synthetic diets. MADRAS AGRIC J 60: 561-564.

*Nadel, D.J. 1970 Current mass-rearing techniques for the Mediterranean fruit fly. pp. 13-18. *in* Sterile-male Technique for Control of Fruit Flies. Proceedings of a Panel Organized by the Joint FAO/IAEA Division of Atomic Energy in Food and Agriculture.

Nadel, D.J.; Peleg, B.A. 1968 Mass-rearing techniques for the Mediterranean fruit-fly in Israel. pp. 87-90. *in* Radiation, Radioisotopes and Rearing Methods in Control of Insect Pests. Proceedings of a Panel Organized by the Joint FAO/IAEA Division of Atomic Energy in Food and Agriculture.

Nagasawa, S. 1972 Comparison of rates of emergence of offsprings in *Drosophila melanogaster* Meigen given Hempa with saccharose and that with dried milk. Studies on the chemosterilants of insects. XVI. APPL ENTOMOL ZOOL 7: 1-8.

*Nagasawa, S.; Nakayama, I. 1969 Rearing of *Papilio protenor demetrius* Cramer on artificial diets with a special reference to the nutritional roles of yeast, soybean protein, cholesterol and choline chloride. KONTYÛ 37: 327-337.

*Nagy, B. 1970 Rearing of the European corn borer (*Ostrinia nubilalis* Hbn.) on a simplified artificial diet. ACTA PHYTOPATHOL ACAD SCI HUNG 5: 73-79.

*Napompeth, B.; Nishida, T.; Mitchell, W.C. 1972 Biology and rearing methods of the New Guinea sugarcane weevil, *Rhabdoscelus obscurus*. HAWAII AGRIC EXP STN TECH BULL 85: 51 pp.

*Nash, R.R.; Tombes, A.S. 1966 Evaluation of five artificial diets for the laboratory rearing of alfalfa weevil larvae. J ECON ENTOMOL 59: 220-221.

Nation, J.L.; Robinson, F.A. 1966 Gibberellic acid: effects of feeding in an artificial diet for honey bees. SCIENCE 152: 1765-1766.

Nation, J.L.; Robinson, F.A. 1968 Brood rearing by caged honey bees in response to inositol and certain pollen fractions in their diet. ANN ENTOMOL SOC AM 61: 514-517.

*Naton, V.E. 1960 Über die Entwicklung des schwarzbraunen Mehlkäfers *Tribolium destructor* Uyttenb. Teil II. Beitrag zur metamorphose des Darmtraktus bei natürlicher und "synthetischer" Ernährung. Z ANGEW ENTOMOL 47: 256-274. (In German with English summary)

Naton, V.E. 1961 Über die Entwicklung des schwarzbraunen Mehlkäfers, *Tribolium destructor* Uyttenb. Tiel III. Die Aufzucht in "synthetischen", carnitin-unterdosierten Diäten und die darauf beruhenden Schädigungen. Z ANGEW ENTOMOL 48: 58-74. (In German with English summary)

Navarro, S.; Gonen, M. 1970 Some techniques for laboratory rearing and experimentation with *Ephestia cautella* (Wlk) (Lepidoptera, Phycitidae). J STORED PROD RES 6: 187-189.

*Navon, A. 1968 Progress report on rearing of the codling moth. pp. 105-106. *in* Radiation, Radioisotopes and Rearing Methods in Control of Insect Pests. Proceedings of a Panel Organized by the Joint FAO/IAEA Division of Atomic Energy in Food and Agriculture.

*Navon, A.; Moore, I. 1971 Artificial rearing of the codling moth (*Carpocapsa pomonella* L.) on calcium alginate gels. ENTOMOPHAGA 16: 381-387.

*Nayar, J.K. 1964a The nutritional requirements of grasshoppers. I. Rearing of the grasshoppers, *Melanoplus bivittatus* (Say), on a completely defined synthetic diet and some effects of different concentrations of B-vitamin mixture, linoleic acid and β-carotene. CAN J ZOOL 42: 11-22.

Nayar, J.K. 1964b The nutritional requirements of grasshoppers. II. Effects of plant phospholipids and extracts of bran on growth, development, and survival of the grasshoppers *Melanoplus bivittatus* (Say) and *Camnula pellucida* (Scudder). CAN J ZOOL 42: 23-38.

*Nayar, J.K. 1966 A method of rearing saltmarsh mosquito larvae in defined sterile medium. ANN ENTOMOL SOC AM 59: 1283-1285.

Nayar, J.K.; Fraenkel, G. 1963a The chemical basis of the host selection in the catalpa sphinx, *Ceratomia catalpae* (Lepidoptera: Sphingidae). ANN ENTOMOL SOC AM 56: 119-122.

Nayar, J.K.; Fraenkel, G. 1963b The chemical basis of the host selection in the Mexican bean beetle, *Epilachna varivestes* (Coleoptera: Coccinellidae). ANN ENTOMOL SOC AM 56: 174-178.

Nayar, J.K.; Sauerman, D.M. Jr 1968 Larval aggregation formation and population density interrelations in *Aedes taeniorhynchus*, their effects on pupal ecdysis and adult characteristics at emergence. ENTOMOL EXP APPL 11: 423-442.

Nayar, J.K.; Sauerman, D.M. Jr 1969 Flight behaviour and phase polymorphism in the mosquito *Aedes taeniorhynchus*. ENTOMOL EXP APPL 12: 365-375.

*Nayar, J.K.; Sauerman, D.M. Jr 1971 The effects of diet on life span fecundity and flight potential of *Aedes taeniorhynchus* adults. J MED ENTOMOL 8: 506-513.

*Naylor, A.F. 1964 Possible value of casein, gluten, egg albumin, or fibrin as whole proteins in the diet of two strains of the flour beetle, *Tribolium confusum* (Tenebrionidae). CAN J ZOOL 42: 1-9.

**Needham, J.G. *et al.* (Eds) 1959 Culture methods for invertebrate animals. Dover Publications Inc, New York. 590 pp.

*Neelgund, Y.F.; Mathad, S.B. 1974 Comparative studies of developments of the armyworm, *Pseudaletia separata* Walker, reared on Napier grass and artificial diet. KARNATAK UNIV J SCI 19: 7-13.

Neilson, W.T.A. 1965 Culturing of the apple maggot, *Rhagoletis pomonella*. J ECON ENTOMOL 58: 1056-1057.

Neilson, W.T.A. 1969 Rearing larvae of the apple maggot on an artificial medium. J ECON ENTOMOL 62: 1028-1031.

*Neilson, W.T.A. 1973 Improved method for rearing apple maggot larvae on artificial media. J ECON ENTOMOL 66: 555-556.

Neilson, W.T.A.; McAllan, J.W. 1964a Artificial diets for the apple maggot, *Rhagoletis pomonella*. I. Mass rearing on certain diets. J ECON ENTOMOL 57: 333-335.

Neilson, W.T.A.; McAllan, J.W. 1964b Artificial diets for the apple maggot, *Rhagoletis pomonella*. II. Reproductive potential. J ECON ENTOMOL 57: 904-905.

Neilson, W.T.A.; McAllan, J.W. 1965 Artificial diets for the apple maggot. III. Improved defined diets. J ECON ENTOMOL 58: 542-543.

*Nelson, E.V.; Roberts, R.B.; Stephen, W.P. 1972 Rearing larvae of the leaf-cutter bee *Megachile rotundata* on artificial diets. J APIC RES 11: 153-156.

Nettles, W.C. Jr; Betz, N.L. 1966 Surface sterilization of eggs of the boll weevil with cupric sulfate. J ECON ENTOMOL 59: 239.

Neville, P.F.; Luckey, T.D. 1962 Carbohydrate and roughage requirement of the cricket, *Acheta domesticus*. J NUTR 78: 139-146.

Neville, P.F.; Luckey, T.D. 1971 Bioflavonoids as a new growth factor for the cricket *Acheta domesticus*. J NUTR 101: 1217-1224.

*Neville, P.F.; Stone, P.C.; Luckey, T.D. 1961 Cricket nutrition. II. An unidentified factor in the nutrition of *Acheta domesticus*. J NUTR 74: 265-273.

*Nihmura, M. 1972 Rearing of the silkworm, *Bombyx mori*, on the artificial feed by adapting the changes in feeding palatability according to larval growth. J SERIC SCI JPN 41: 375-382. (In Japanese with English summary)

Nihmura, M. 1973a Higher content of defatted soybean meal and exclusion of agar in an artificial feed of the silkworm. (Preparation of a practical artificial feed of the silkworm, *Bombyx mori* Part I.) J AGRIC CHEM SOC JPN 47: 241-249. (In Japanese with English summary)

Nihmura, M. 1973b Application of the feed materials and the feeds of the livestock to an artificial feed of the silkworm. (Preparation of a practical artificial feed of the silkworm, *Bombyx mori* Part II.) J AGRIC CHEM SOC JPN 47: 251-257. (In Japanese with English summary)

Nihmura, M. 1973c On the hardness of the artificial feed for the silkworm, *Bombyx mori*. J SERIC SCI JPN 42: 173-177. (In Japanese with English summary)

Nihmura, M.; Aomori, S.; Mori, K.; Matsui, M. 1972 Utilization of synthetic compounds with juvenile hormone activity for the silkworm rearing. AGRIC BIOL CHEM 36: 889-892.

Nihmura, M.; Kirimura, J. 1974 Studies on the antiseptic in the artificial feed for the silkworm. J SERIC SCI JPN 43: 163-170. (In Japanese with English summary)

Noland, J.L. 1954a Sterol metabolism in insects. I. Utilization of cholesterol derivatives by the cockroach *Blattella germanica* (L.). ARCH BIOCHEM BIOPHYS 48: 370-379.

Noland, J.L. 1954b Sterol metabolism in insects. II. Inhibition of cholesterol utilizations by structural analogs. ARCH BIOCHEM BIOPHYS 52: 323-330.

*Noland, J.L. 1956 An improved method for rearing cockroaches. J ECON ENTOMOL 49: 411-412.

Noland, J.L.; Baumann, C.A. 1949 Requirement of the German cockroach for choline and related compounds. PROC SOC EXP BIOL MED 70: 198-201.

Noland, J.L.; Baumann, C.A. 1951 Protein requirements of the cockroach *Blattella germanica* (L.). ANN ENTOMOL SOC AM 44: 184-188.

*Noland, J.L.; Lilly, J.H.; Baumann, C.A. 1949a A laboratory method for rearing cockroaches and its application to dietary studies on the German roach. ANN ENTOMOL SOC AM 42: 63-70.

Noland, J.L.; Lilly, J.H.; Baumann, C.A. 1949b Vitamin requirements of the cockroach, *Blattella germanica* (L.). ANN ENTOMOL SOC AM 42: 154-164.

*Norris, D.M. 1972 Dependence of fertility and progeny development of *Xyleborus ferrugineus* upon chemicals from its symbiotes. pp. 299-310. *in* Rodriguez, J.G. (Ed.), Insect and Mite Nutrition. North-Holland Press, Amsterdam.

*Norris, D.M.; Baker, J.M. 1968 A minimal nutritional substrate required by *Fusarium solani* to fulfill its mutualistic relationship with *Xyleborus ferrugineus*. ANN ENTOMOL SOC AM 61: 1473-1475.

*Norris, D.M.; Baker, J.M. 1969 Nutrition of *Xyleborus ferrugineus*. I. Ethanol in diets as a tunnelling (feeding) stimulant. ANN ENTOMOL SOC AM 62: 592-594.

Norris, D.M.; Baker, J.M.; Chu, H.M. 1969 Symbiotic interrelationships between microbes and ambrosia beetles. III. Ergosterol as the source of sterol to the insect. ANN ENTOMOL SOC AM 62: 413-414.

Norris, D.M.; Chu, H.M. 1970 Nutrition of *Xyleborus ferrugineus*. II. A holidic diet for the aposymbiotic insect. ANN ENTOMOL SOC AM 63: 1142-1145.

Northrop, J.H. 1926 Duration of life of an aseptic *Drosophila* culture inbred in the dark for 230 generations. J GEN PHYSIOL 9: 763-765.

*Odell, T.M.; Rollinson, W.D. 1966 A technique for rearing the gypsy moth, *Porthetria dispar* (L.), on an artificial diet. J ECON ENTOMOL 59: 741-742.

Okada, I. 1970 A new method of artificial rearing of coccinellids, *Harmonia axyridis* Pallas. HEREDITY 24: 32-35. (In Japanese with English summary)

Okada, I.; Hoshiba, H.; Maehara, T. 1972 An artificial rearing of a coccinellid beetle, *Harmonia axyridis* Pallas, on pulverized drone honeybee brood. BULL FAC AGRIC TAMAGAWA UNIV 12: 39-47. (In Japanese with English summary)

Okada, I.; Hoshiba, H.; Maruoka, T. 1971 An artificial rearing of a coccinellid beetle, *Harmonia axyridis* Pallas, on drone honeybee brood. BULL FAC AGRIC TAMAGAWA UNIV 11: 91-97. (In Japanese with English summary)

*Okada, I.; Matsuka, M. 1973 Artificial rearing of *Harmonia axyridis* on pulverized drone honeybee brood. ENVIRON ENTOMOL 2: 301-302.

Okada, I.; Matsuka, M.; Hoshiba, H. 1971 Utilization of drone larvae and pupae as semi-artificial diet for aphidophagous insects. APIACTA 6: 119-120.

*Okada, I.; Matsuka, M.; Tani, M. 1974 Rearing a green lacewing, *Chrysopa septempunctata* Wesmael on pulverized drone honeybee brood. BULL FAC AGRIC TAMAGAWA UNIV 14: 26-32.

*Okada, I.; Niijima, K. 1971 Artificial rearing of lace wing, *Chrysopa septempunctata* Wesmael, with special reference to a new diet using drone honeybee brood. HEREDITY 25: 41-44. (In Japanese with English summary)

Okada, I.; Sakai, T.; Tanigishi, K.; Yoshida, T. 1973 Some observations on oviposition habits of a coccinellid beetle (*Harmonia axyridis* Pallas). BULL FAC AGRIC TAMAGAWA UNIV 13: 63-71. (In Japanese with English summary)

*Oliver, B.F.; Maxwell, F.G.; Jenkins, J.N. 1970 Utilization of glanded and glandless cotton diets by the bollworm. J ECON ENTOMOL 63: 1965-1966.

**Orphanidis, P.S.; Pétsikou, N.A.; Patsakos, P.G. 1970 Élevage du *Dacus oleae* (Gmel.) sur substrat artificiel. ANN INST PHYTOPATHOL BENAKI 9: 147-169. (In French)

*Orr, C.W.M. 1964 The influence of nutritional and hormonal factors on egg development in the blow fly, *Phormia regina* (Meig.). J INSECT PHYSIOL 10: 53-64.

*Ouye, M.T. 1962 Effect of antimicrobial agents on micro-organisms and pink bollworm development. J ECON ENTOMOL 55: 854-857.

*Ouye, M.T.; Vanderzant, E.S. 1964 B-vitamin requirements of the pink bollworm. J ECON ENTOMOL 57: 427-430.

Paech, K.; Tracey, M.V. 1955 Modern Methods of Plant Analysis. Springer-Verlag, Berlin, 761 p.

Paik, W.H. 1956 Longevity of aphids (*Macrosiphum pisi*) on different nutrient solutions. Commemoration Theses Fiftieth Aniversary, pp. 105.

**Pain, J. 1968 Nutrition et developpement des organes sexuels adultes. pp. 410-435. *in* Chauvin, R. (Ed.), Traité de Biologie de l'abeille. Masson et cie Paris. (In French)

*Pan, Y.S.; Long, W.H. 1961 Diets for rearing the sugarcane borer. J ECON ENTOMOL 54: 257-261.

Panda, N. 1966 Effect of larval nutrition on the tolerance of adult *Musca nebula* F. to insecticides. INDIAN J ENTOMOL 28: 94-97.

*Pant, J.C.; Anand, M. 1972 An artificial diet for the spotted bollworm, *Earias fabia* Stoll. ENTOMOL NEWSL 2: 45-46.

*Pant, J.C.; Anand, M. 1975 An artificial diet for rearing castor semi looper, *Achaea janata* Linnaeus. ENTOMOL NEWSL 5: 49-50.

Pant, N.C. 1953 Nutritional requirements of *Trogoderma granaria* Everts (Coleoptera: Dermestidae). CURR SCI 22: 379-380.

*Pant, N.C. 1957 Nutritional studies on *Trogoderma granarium* Everts. Basic food and vitamin requirements. INDIAN J ENTOMOL 18: 259-266.

**Pant, N.C. 1973a Insect nutrition in economic entomolgy. pp. 139-143. *in* Pant, N.C.; Ghai, S. (Eds), Insect Physiology and Anatomy. Indian Council Agricultural Research, New Delhi, India.

**Pant, N.C. 1973b Nutrition of phytophagous insects. pp. 167-175. *in* Pant, N.C.; Ghai, S. (Eds), Insect Physiology and Anatomy. Indian Council Agricultural Research, New Delhi, India.

**Pant, N.C. 1973c Physiology of symbiotes in insects. pp. 176-188. *in* Pant, N.C.; Ghai, S. (Eds), Insect Physiology and Anatomy. Indian Council Agricultural Research, New Delhi, India.

Pant, N.C.; Dang, K. 1965 Effect of different carbohydrates on the growth and development of *Tribolium castaneum* Herbst. INDIAN J ENTOMOL 27: 432-441.

Pant, N.C.; Dang, K. 1967 Importance of minerals in the growth and development of the larvae of *Tribolium castaneum* Hbst. INDIAN J ENTOMOL 29: 275-280.

*Pant, N.C.; Fraenkel, G. 1950 The function of the symbiotic yeasts of two insect species, *Lasioderma serricorne* F. and *Stegobium (Sitodrepa) paniceum* L. SCIENCE 112: 498-500.

*Pant, N.C.; Fraenkel, G. 1954a On the function of the intracellular symbionts of *Oryzaephilus surinamensis* L. (Cucujidae, Coleoptera). J ZOOL SOC INDIA 6: 173-177.

Pant, N.C.; Fraenkel, G. 1954b Studies on the symbiotic yeasts of two insect species, *Lasioderma serricorne* F. and *Stegobium paniceum* L. BIOL BULL WOODS HOLE MASS 107: 420-432.

*Pant, N.C.; Gabrani, K. 1963 Qualitative vitamin B requirements of *Tribolium castaneum* Herbst. INDIAN J ENTOMOL 25: 110-115.

*Pant, N.C.; Ghai, S.; Chawla, S.S. 1959 A simple unfortified rice medium for mass-rearing of the melon fruit-fly, *Dacus cucurbitae* Coquillett under laboratory conditions. CURR SCI 28: 288-289.

*Pant, N.C.; Gupta, P.; Nayar, J.K. 1960a Physiological studies on *Chilo zonellus* Swinh. a pest on maize crop. I. Growth on artificial diets. PROC NATL INST SCI INDIA 26B: 379-383.

*Pant, N.C.; Gupta, P.; Nayar, J.K. 1960b Physiology of intracellular symbiotes of *Stegobium paniceum* L. with special reference to amino acid requirements of the host. EXPERIENTIA 16: 311-312.

Pant, N.C.; Kapoor, S. 1963 Physiology of intracellular symbiotes of *Lasioderma serricorne* F. with special reference to the cholesterol requirements of the host. INDIAN J ENTOMOL 25: 311-315.

*Pant, N.C.; Nayar, J.K.; Gupta, P. 1958 On the significance of amino acids in the larval development of khapra-beetle, *Trogoderma granarium* Everts (Coleoptera: Dermestidae). EXPERIENTIA 14: 176-177.

Pant, N.C.; Pant, J.C. 1960 Nutritional studies on *Trogoderma granarium* Everts. IV. Further studies on vitamin requirements of larvae grown on different casein samples. INDIAN J ENTOMOL 22: 115-120.

Paschke, J.D. 1964 Disposable containers for rearing cabbage loopers. J INSECT PATHOL 6: 248-251.

Patana, R. 1967 A pressure paint tank modified for use as a dispenser for insect diets. J ECON ENTOMOL 60: 1755-1756.

*Patana, R. 1969 Rearing cotton insects in the laboratory. U S DEP AGRIC PROD RES REP 108, 6 pp.

*Patana, R.; McAda, W.C. 1973 Tobacco budworms : use of dry diet flakes in rearing. J ECON ENTOMOL 66: 817-818.

*Patel, R.C.; Patel, J.C.; Patel, J.K. 1973 Biology and mass breeding of the tobacco caterpillar, *Spodoptera litura* (F.). ISR J ENTOMOL 8: 131-142.

*Patel, R.C.; Patel, J.K.; Patel, P.B.; Singh, R. 1968 Mass breeding of *Heliothis armigera* (Hbn.). INDIAN J ENTOMOL 30: 272-280.

*Patton, R.L. 1963a Insect nutrition. pp. 7-29. *in* Introductory Insect Physiology. W.B. Saunders Co, Philadelphia.

Patton, R.L. 1963b Rearing the house cricket, *Acheta domesticus*, on commercial feed. ANN ENTOMOL SOC AM 56: 250-251.

*Patton, R.L. 1967 Oligidic diets for *Acheta domesticus* (Orthoptera: Gryllidae). ANN ENTOMOL SOC AM 60: 1238-1242.

*Pausch, R.; Fraenkel, G. 1966 The nutrition of the larvae of the oriental rat flea, *Xenopsylla cheopis* (Rothschild). PHYSIOL ZOOL 39: 202-222.

*Payne, J.A.; Lowman, H.; Pate, R.R. 1975 Artificial diets for rearing the tilehorned *Prionus*. ANN ENTOMOL SOC AM 68: 680-682.

Pearl, R. 1926 A synthetic food medium for the cultivation of *Drosophila*. J GEN PHYSIOL 9: 513-519.

Pearl, R.; Allen, A.; Penniman, B.D. 1926 Culture media for *Drosophila*. II. A new synthetic medium and its influence on fertility at different densities of population. AM NAT 60: 357-366.

Peck, W.B.; Whitcomb, W.H. 1967 An adaptable method for rearing spider and cannibalistic insects. TURTOX NEWS 45: 242-244.

*Peck, W.B.; Whitcomb, W.H. 1968. Feeding spiders an artificial diet. ENTOMOL NEWS 79: 233-236.

*Peleg, B.A.; Rhode, R.H. 1970 New larval medium and improved pupal recovery method for the Mediterranean fruit fly in Costa Rica. J ECON ENTOMOL 63: 1319-1321.

Peleg, B.A.; Rhode, R.H.; Calderson, W. 1968 Mass rearing of the Mediterranean fruit fly in Costa Rica. pp. 107-110. *in* Radiation, Radioisotopes and Rearing Methods in Control of Insect Pests. Proceedings of a Panel Organized by the Joint FAO/IAEA Division of Atomic Energy in Food and Agriculture.

Perron, J.M.: Huot, L.; Smirnoff, W.A. 1966 Toxicité comparée des eléments Al, As, B, Co, Cu F, Fe, I, Li, Mg, Mo, Zn, pour *Plodia interpunctella* (Hbn.) (Lépidoptères). COMP BIOCHEM PHYSIOL 18: 869-879. (In French with English summary)

*Perry, A.S.; Miller, G.; Buckner, A.J. 1965 The essential role of folic acid and the effect of antimetabolites on growth and metamorphosis of housefly larvae *Musca domestica* L. J INSECT PHYSIOL 11: 1277-1287.

*Petavy, G. 1966 Rapports entre alimentation et physiologie chez un insecte: *Locusta migratoria cinerascens* Fab. (Orthoptera: Acridiae). I. Utilization de milieux nutritifs artificiels. ANN NUTR ALIMENT 20: 279-299. (In French)

Peters, L.L.; Fairchild, M.L.; Zuber, M.S. 1972 Effect of corn endosperm containing different levels of amylose on Angoumois grain moth biology. 3. Interrelationship of amylose levels and moisture content of diets. J ECON ENTOMOL 65: 1168-1169.

**Peterson, A. 1964 Entomological Techniques. How to Work with Insects. Edwards Brothers, Inc., Ann Arbor, Michigan, 435 pp.

Petrushova, N.I. *et al.* 1971 A laboratory method of rearing the codling moth (*Carpocapsa pomonella* L.) on fruit and artificial nutritive media. GNBS YALTA 24. (In Russian)

*Petterson, M.A.; Debolt, J.W. 1975 Rearing the beet armyworm in paraffined paper bags. J GA ENTOMOL SOC 10: 165-168.

*Petterson, M.A.; Wielgus, R.S. 1973 Acceptance of artificial diet by *Megathymus streckeri* (Skinner) (Megathymidae). J RES LEPID 12: 197-198.

Phelps, A. 1963 Growth of a thermophilic mite in axenic culture. AM ZOOL 3: 533.

Pillai, M.K.K.; Madhukar, B.V.R. 1969 Effect of biotin on the fertility of the yellow fever mosquito, *Aedes aegypti* (L.). NATURWISSENSCHAFTEN 56: 218-219.

*Pillai, P.R.P.; Winston, P.W. 1968 Culture and nutrition of *Caloglyphus anomalus* (Acarina: Acaridae). ANN ENTOMOL SOC AM 61: 56-60.

Pipa, R.L. 1963 Studies on the hexapod nervous system. VI. Ventral nerve cord shortening; a metamorphic process in *Galleria mellonella* (L.) (Lepidoptera, Pyrallidae). BIOL BULL WOODS HOLE MASS 124: 293-302.

*Podmanická, D.; Weismann, L. 1974 Effects of gradually increased doses of B-group vitamins on the development and vitality of *Scotia segetum* Den. & Schiff. BIOLOGIA BRATISLAVA 29: 641-648. (In Slovak with English summary)

*Podoler, H.; Applebaum, S.W. 1971 Basic nutritional requirements of larvae of the bruchid beetle, *Callosobruchus chinensis* L. J STORED PROD RES 7: 187-193.

*Poitout, S.; Bues, R. 1970 Élevage de plusieurs espèces de Lépidoptères Noctuidae sur milieu artificiel riche et sur milieu artificiel simplifié. ANN ZOOL ECOL ANIM 2: 79-91. (In French with English summary)

*Poitout, S.; Bues, R. 1972 Nutrition des insectes. Mise en évidence de besoins différents en acide linolénique entre Lépidoptères Noctuidae-Trifinae, appartenant à différentes sous-familles, et Lépidoptères Noctuidae - Quadrifinae de la sous-famille des Plusiinae. C R ACAD SCI PARIS 274: 3113-3115. (In French)

*Poitout, S.; Bues, R.; Le Rumer, C. 1972 Élevage zur milieu artificiel simple de deux noctuelles parasites du coton *Earias insulana* et *Spodoptera littoralis*. ENTOMOL EXP APPL 15: 341-350. (In French with English summary)

*Power, R.J.B.; Singh, P. 1974 Laboratory rearing method for the stem weevil, *Hyperodes bonariensis* (Coleoptera: Curculionidae). N Z J ZOOL 1: 531-536.

*Pralavorio, R.; Arambourg, Y.; Guennelon, G. 1973 Essai de mise au point d'un élevage permanent d'*Acrolepia assectella* Zeller (Lepidoptera: Hyponomeutidae) sur milieu artificiel. ANN ZOOL ECOL ANIM 5: 569-580. (In French with English summary)

*Pranter, W. 1960 Untersuchungen über die Ernährungsphysiologie von *Bostrychus capucinus* L. (Coleopt., Fam. Bostrychidae). Z ANGEW ZOOL 47: 385-430. (In German with English summary)

*Preiss, F.J.; Davidson, J.A. 1971 Adult longevity, pre-oviposition period and fecundity of *Alphitobius diaperinus* in the laboratory (Coleoptera: Tenebrionidae). J GA ENTOMOL SOC 6: 105-109.

**Pristavko, V.P. 1974 Insect rearing as a system. pp. 131-133. *in* The Pathology of Arthropods and the Biological Methods of Controlling Harmful Organisms. Kiev City Conference Abstract. (In Russian)

**Pristavko, V.P.; Yanishevskaya, L.V. 1971 Rearing of insects on artificial nutrient media. SELSK KHOZ RUBENZHON RAST 8: 60-62. (In Russian)

Pristavko, V.P.; Yanishevskaya, L.V. 1972a Microflora of the nutrient media for breeding the codling moth and effectiveness of suppressing growth of microbes by some antiseptics. BIOL NANK. 15: 114-118. (In Russian)

Pristavko, V.P.: Yanishevskaya, L.V. 1972b The method of all year culture of the codling moth on artificial nutritional media. ZAK ROSLYN (KIEV) 15: 22-26. (In Ukranian)

**Proceedings of a Panel Organized by the Joint FAO/IAEA Division of Atomic Energy in Food and Agriculture. 1970 Sterile-male Technique for Control of Fruit Flies, 175 pp.

**Proceedings of a Panel Organized by the Joint FAO/IAEA Division of Atomic Energy in Food and Agriculture. 1974 The Sterile-insect Technique and its Field Application, 134 pp.

**Proceedings of Planning and Training Conference for Insect Nutrition and Rearing held at Excelsior Springs, Missouri, March 4-8, 1963, United States Department of Agriculture.

*Prokopy, R.J. 1967 Artificial diet for apple maggot larvae. J ECON ENTOMOL 60: 1161-1162.

*Prokopy, R.J.; Boller, E.F. 1970 Artificial egging system for the European cherry fruit fly. J ECON ENTOMOL 63: 1413-1417.

**Prosser, L.C. 1952 Nutrition. pp. 112-143 *in* Comparative Animal Physiology. W.B. Saunders Co., U.S.A.

**Ptitsyna, N.V.; Gress, P.Y. 1973 Diets for the mass rearing of codling moths. pp. *in* Conference Proceedings on Progressive Methods of Agriculture Pest Control. (In Russian)

Punj, G.K. 1971 The effect of certain artificial diets on the susceptibility to ethylene oxide of larvae of *Trogoderma granarium* Everts. (Coleoptera: Dermestidae). J STORED PROD RES 6: 347-349.

*Punj, G.K.; Girish, G.K. 1968 Specific nutritional requirements of stored grain insect pests. BULL GRAIN TECHNOL 6: 143-152.

*Qureshi, Z.A.; Anwar, M.; Ashraf, M.; Chata, N.U.; Arif, M.D. 1972 Rearing, biology and sterilization of the pink rice borer, *Sesamia inferens* Walker. JPN PESTIC INF 10: 129.

*Qureshi, Z.A.; Ashraf, M.; Bughio, A.R.; Hussain, S. 1974 Rearing, reproductive behaviour and *gamma* sterilization of fruit fly, *Dacus zonatus* (Diptera: Tephritidae). ENTOMOL EXP APPL 17: 504-510.

*Rabbani, M.G.; Seawright, J.A.; Leatherwood, L.B. 1976 A method for culturing single families of *Anopheles albimanus*. MOSQ NEWS 36: 100-102.

**Radiation, Radioisotopes and Rearing Methods in the Control of Insect Pests. 1968 Proceedings of a Panel Organized by the Joint FAO/IAEA Division of Atomic Energy in Food and Agriculture.

*Radinovsky, S.; Krantz, G.W. 1961 The biology and ecology of granary mites of the pacific northwest. II. Techniques for laboratory observation and rearing. ANN ENTOMOL SOC AM 54: 512-518.

*Ramanankasina-Razafialisoa, E. 1970 Alimentation artificielle de la larvae d'*Earias insulana* Boisd. Noctuidae - Westermaniinae. C R ACAD SCI PARIS SER D 271: 408-410. (In French)

Ramazzotto, L.J.; Ludwig, D. 1967 Differences in composition of the fat body of normal and aposymbiotic cockroaches (*Periplaneta americana*). ANN ENTOMOL SOC AM 60: 227-230.

*Rasso, S.C.; Fraenkel, G. 1954 The food requirements of the adult female blow-fly, *Phormia regina* (Meigen), in relation to ovarian development. ANN ENTOMOL SOC AM 47: 636-645.

*Raulston, J.R. 1971 A practical diet containing cottonseed for rearing the pink bollworm. J ECON ENTOMOL 64: 1021-1023.

Raulston, J.R. 1975 B-vitamin supplements required for soy flour wheat germ diet used in rearing tobacco budworms. ANN ENTOMOL SOC AM 68: 387-388.

*Raulston, J.R.; Auclair, J.L. 1968 Responses of *Lygus hesperus* to chemically defined diets. ANN ENTOMOL SOC AM 61: 1495-1500.

*Raulston, J.R.; Lingren, P.D. 1969 A technique for rearing larvae of the bollworm and tobacco budworm in large numbers. J ECON ENTOMOL 62: 959-961.

*Raulston, J.R.; Shaver, T.N. 1970 A low agar casein-wheat germ diet for rearing tobacco budworms. J ECON ENTOMOL 63: 1743-1744.

Raun, E.S. 1961 Elimination of microsporidiosis in laboratory-reared European corn borers by the use of heat. J INSECT PATHOL 3: 446-448.

Raun, E.S. 1966 European corn borer. pp. 323-338. *in* Smith, C.N. (Ed.), Insect Colonization and Mass Production. Academic Press, New York and London.

Raun, E.S. 1967 Coloured European corn borers and eggs from dye-containing diet. PROC NORTH CENT BRANCH ENTOMOL SOC AM 22: 162-163.

*Rawlins, W.A. 1967 Oviposition by onion maggot adults fed on a chemically defined diet. J ECON ENTOMOL 60: 1747-1748.

Reddy, G.P.V.; Chippendale, G.M. 1972 Observations on the nutritional requirements of the south western corn borer, *Diatraea grandiosella*. ENTOMOL EXP APPL 15: 51-60.

*Redfern, R.E. 1963 Concentrate media for rearing red-banded leaf roller. J ECON ENTOMOL 56: 240-241.

*Redfern, R.E. 1964a An improved medium for rearing red-banded leaf roller. J ECON ENTOMOL 57: 296-297.

*Redfern, R.E. 1964b Concentrate medium for rearing the codling moth. J ECON ENTOMOL 57: 607-608.

Redfern, R.E. 1967 Laboratory techniques for rearing the southern armyworm. J ECON ENTOMOL 60: 308.

*Redfern, R.E.; Raulston, J.R. 1970 Improved rearing techniques for the southern armyworm. J ECON ENTOMOL 63: 296-297.

Reed, G.L.; Showers, W.B.; Huggans, J.L.; Carter, S.W. 1972 Improved procedures for mass rearing the European corn borer. J ECON ENTOMOL 65: 1472-1476.

*Reese, J.C.; English, L.M.; Yonke, T.R.; Fairchild, M.L. 1972 A method for rearing black cutworms. J ECON ENTOMOL 65: 1047-1050.

*Rehr, S.S.; Feeny, P.P.; Janzen, D.H. 1973 Chemical defense in central American non-ant-acacias. J ANIM ECOL 42: 405-416.

*Rejesus, R.S. 1975 Control of fruit flies by the sterile-insect technique. Studies on sterilization, population dynamics, fruit infestation sequence and dispersal of *Dacus dorsalis* Hendel. pp. 124-125. *in* Controlling Fruit Flies by the Sterile-insect Technique. Proceedings of a Panel Organized by the Joint FAO/IAEA Division of Atomic Energy in Food and Agriculture.

*Retnakaran, A.; Beck, S.D. 1967 Aspects of mineral nutrition of the pea aphid, *Acyrthosiphon pisum* (Harris). J NUTR 92: 43-52.

Retnakaran, A.; Beck, S.D. 1968 Amino acid requirements and sulfur amino acid metabolism in the pea aphid, *Acyrthosiphon pisum* (Harris). COMP BIOCHEM PHYSIOL 24: 611-619.

Retnakarn, A.; French, J. 1971 A method for separating and surface sterilizing the eggs of the spruce budworm, *Choristoneura fumiferana* (Lepidoptera: Tortricidae). CAN ENTOMOL 103: 712-716.

*Rexrode, C.O.; Krause, C.R. 1968 Rearing larvae of *Pseudopityophthorus pruinosus* and *P. minutissimus* on ground phloem media. ANN ENTOMOL SOC AM 61: 814-816.

Rey, J.M. 1970 Development of a diet for rearing olive fly larvae. ENTOMOPHAGA 15: 157-167.

Rhode, R.H. 1957 A diet for Mexican fruit flies. J ECON ENTOMOL 50: 215.

*Rhode, R.H.; Spishakoff, L.M. 1965 Tecnicas usadas en la cultivo de *Anastrepha ludens* (Loew). II. Memorias del dia Parasitologo, Departo. de Parasitologia. ESC NAC AGRI (CHAPINGO) (1964): 23-28. (In Spanish)

**Richards, G.A.; Brooks, M.A. 1958 Internal symbiosis in insects. ANNU REV ENTOMOL 3: 37-56.

Richardson, H.H. 1932 An efficient medium for rearing houseflies throughout the year. SCIENCE 76: 350-351.

*Richeson, J.S.; Nation, J.L.; Wilkinson, R.C. 1971 Fatty acid composition in *Ips calligraphus* (Coleoptera: Scolytidae) during postembryonic development. ANN ENTOMOL SOC AM 64: 251-254.

*Richeson, J.S.; Wilkinson, R.C.; Nation, J.L. 1970 Development of *Ips calligraphus* on foliage-based diets. J ECON ENTOMOL 63: 1797-1799.

Richmond, C.A.; Martin, D.F. 1966 Technique for mass rearing of the pink bollworm by infesting diet medium with eggs. J ECON ENTOMOL 59: 762-763.

Richmond, J.A.; Thomas, H.A. 1975 *Hylobius pales:* effect of dietary sterols on development and on sterol content of somatic tissue. ANN ENTOMOL SOC AM 68: 329-332.

Riddiford, L.M. 1967 Antibiotics in the laboratory-rearing of cecropia silkworms. SCIENCE 157: 1451-1452.

*Riddiford, L.M. 1968 Artificial diet for cecropia and other saturniid silkworms. SCIENCE 160: 1461-1462.

*Ridet, J.M. 1972 Étude des conditions optimales d'élevage et d'alimentation de *Lymantria dispar* L. ANN SOC ENTOMOL FR 8: 653-668. (In French with English summary)

*Ridgway, R.L.; Morrison, R.K.; Badgley, M. 1970 Mass rearing a green lacewing. J ECON ENTOMOL 63: 834-836.

*Ripley, L.B.; Hepburn, G.A.; Dick, J. 1939 Mass breeding of false codling-moth, *Argyroploce leucotreta* Meyr., in artificial media. SCI BULL DEP AGRIC FOR UNION S AFR 207, 18 pp.

*Ritchot, C.; McFarlane, J.E. 1961 The B vitamin requirements of the house cricket. CAN J ZOOL 39: 11-15.

Ritchot, C.; McFarlane, J.E. 1962 The effects of wheat germ oil and linoleic acid on growth and reproduction of the house cricket. CAN J ZOOL 40: 371-374.

*Rivas, A.M.; Buchanan, W.D. 1958 A new technique for rearing carpenterworms. J ECON ENTOMOL 51: 406-407.

**Robbins, W.E. 1963 Studies on the utilization, metabolism, and function of sterols in the house fly, *Musca domestica*. pp. 269-280. *in* Radiation and Radioisotopes Applied to Insects of Agricultural Importance. Proceedings of a Panel Organized by the Joint FAO/IAEA Division of Atomic Energy in Food and Agriculture.

Robbins, W.E.; Thompson, M.J.; Yamamoto, R.T.; Shortino, T.J. 1965 Feeding stimulants for the housefly, *Musca domestica* Linnaeus. SCIENCE 147: 628-630.

Roberts, J.A.; Kitching, R.L. 1974 Ingestion of sugar, protein and water by adult *Lucilia cuprina* (Wied.) (Diptera, Calliphoridae). BULL ENTOMOL RES 64: 81-88.

Roberts, R.B.; Smith, H.W. 1971a Free amino acids in the eggs and haemolymph of the grasshopper *Melanoplus sanguinipes* (Orthoptera: Acarididae). 1. Identification and quantitative changes throughout development. ANN ENTOMOL SOC AM 64: 693-702.

Roberts, R.B.; Smith, H.W. 1971b Free amino acids in the eggs and hemolymph of the grasshopper *Melanoplus sanguinipes* (Orthoptera: Acrididae). 2. Changes in composition as a result of diet and population and artificial stress. ANN ENTOMOL SOC AM 64: 703-708.

Robertson, J.L.; Noble, L.W. 1968 Rearing the tobacco budworm in honeycomb-like cells. J ECON ENTOMOL 61: 331-332.

*Robinson, F.A.; Nation, J.L. 1966 Artificial diets for honey bees, *Apis mellifera*. FLA ENTOMOL 49: 175-184.

*Robinson, F.A.; Nation, J.L. 1968 Substances that attract caged honeybees colonies to consume pollen supplements and substitutes. J APIC RES 7: 83-88.

*Rochford, P.J. 1972 The American cockroach. pp. 575-581. *in* Universities Federation for Animal Welfare (Ed.), The UFAW Handbook on the Care and Management of Laboratory Animals. Fourth Edition. Churchill Livingstone, Edinburgh and London.

*Rock, G.C. 1967 Aseptic rearing of the codling moth on synthetic diets: ascorbic acid and fatty-acid requirements. J ECON ENTOMOL 60: 1002-1005.

Rock, G.C. 1969a Sterol and water-soluble vitamin requirements for larvae of the red-banded leaf roller, *Argyrotaenia velutinana*. ANN ENTOMOL SOC AM 62: 611-613.

Rock, G.C. 1969b Nutritional evidence for the absence of the complete ornithine-urea cycle in the insect, *Argyrotaenia velutinana* (Lepidoptera: Tortricidae). J NUTR 98: 153-158.

*Rock, G.C. 1971 Utilization of *d*-isomers of the dietary indispensable amino acids by *Argyrotaenia velutinana* larvae. J INSECT PHYSIOL 17: 2157-2168.

*Rock, G.C.; Glass, E.H.; Patton, R.L. 1964 Axenic rearing of the red-banded leaf roller, *Argyrotaenia velutinana* on meridic diets. ANN ENTOMOL SOC AM 57: 617-621.

*Rock, G.C.; Hodgson, E. 1971 Dietary amino requirements for *Heliothis zea* determined by dietary deletion and radiometric techniques. J INSECT PHYSIOL 17: 1087-1097.

Rock, G.C.; Khan, A.; Hodgson, E. 1975 The nutritional value of seven *d*-amino acids and α-ketoacids for *Argyrotaenia velutinana*, *Heliothis zea*, and *Phormia regina*. J INSECT PHYSIOL 21: 693-703.

*Rock, G.C.; King, K.W. 1966 Amino acid composition in hydrolysates of the red-banded leaf roller, *Argyrotaenia velutinana* (Lepidoptera: Tortricidae) during development. ANN ENTOMOL SOC AM 59: 273-277.

*Rock, G.C.; King, K.W. 1967a Qualitative amino acid requirements of the red-banded leaf roller, *Argyrotaenia velutinana* (Lepidoptera: Tortricidae). J INSECT PHYSIOL 13: 59-68.

*Rock, G.C.; King, K.W. 1967b Quantitative amino acid requirements of the red-banded leaf roller, *Argyrotaenia velutinana* (Lepidoptera: Tortricidae). J INSECT PHYSIOL 13: 175-186.

Rock, G.C.; Ligon, B.G.; Hodgson, E. 1973 Utilization of methionine analogues by *Argyrotaenia velutinana* larvae. ANN ENTOMOL SOC AM 66: 177-179.

Rock, G.C.; Patton, R.L.; Glass, E.H. 1965 Studies of the fatty acid requirements of *Argyrotaenia velutinana* (Walker). J INSECT PHYSIOL 11: 91-101.

Rock, G.C.; Sharma, G.K. 1974 Comparison of the nutritive values of carbohydrates to *Argyrotaenia velutinana*. ANN ENTOMOL SOC AM 67: 391-393.

**Rockstein, M. 1957a Some aspects of intermediary metabolism of carbohydrates in insects. ANNU REV ENTOMOL 2: 19-36.

Rockstein, M. 1957b Longevity of male and female house flies. J GERONTOL 12: 253-256.

**Rodriguez, J.G. 1964 Nutritional studies in the Acarina. ACAROLOGIA 6: 324-337.

**Rodriguez, J.G. 1966 Axenic arthropoda: current status of research and further possibilities. ANN N Y ACAD SCI 139: 53-64.

Rodriguez, J.G. 1969 Dietetics and nutrition of *Tetranychus urticae* Koch. PROC 2 INT CONGR ACAROL NOTTINGHAM ENGLAND (1967): 469-475.

*Rodriguez, J.G. 1972a Inhibition of acarid mite development by fatty acids. pp. 637-650. *in* Rodriguez, J.G. (Ed.), Insect and Mite Nutrition. North-Holland Publishing Company, Amsterdam, London.

**Rodriguez, J.G. (Ed.) 1972b Insect and mite nutrition. Significance and implications in ecology and pest management. North-Holland Publishing Co., Amsterdam, London. 702 pp.

Rodriguez, J.G.; Hampton, R.E. 1966 Essential amino acids determined in the two-spotted spider mite, *Tetranychus urticae* Koch. (Acarina: Tetranychidae) with glucose U-C^{14}. J INSECT PHYSIOL 12: 1209-1216.

*Rodriguez, J.G.; Lasheen, A.M. 1971 Axenic culture of *Tyrophagus putrescentiae* in a chemically defined diet and determination of essential amino acids. J INSECT PHYSIOL 17: 979-985.

Rodriguez, J.G.; Potts, M.F. 1974 Nutritional requirements and metabolic inhibition of some acarid mites. pp. 160-166. PROC 1 INT WORK CONF STORED PROD ENTOMOL SAVANNAH.

Rodriguez, J.G.; Singh, P. 1967 A chemically defined diet for *Tetranychus urticae* Koch. PROC NORTH CENT BRANCH ENTOMOL SOC AM 22: 143.

Rodriguez, J.G.; Singh, P.; Seay, T.N.; Walling, M.V. 1967 Ingestion in the two-spotted spider mite, *Tetranychus urticae* Koch. as influenced by wavelength of light. J INSECT PHYSIOL 13: 925-932.

*Roelofs, W.L. 1967 Agarless medium for mass rearing the red-banded leaf roller. J ECON ENTOMOL 60: 1477-1478.

Rollinson, W.D. 1964 An artificial diet for the gypsy moth. U S FOR SERV NORTHEAST FOR EXP STN TECH PROG REP (1963-1964): 7-8.

*Romoser, W.S.; Tucker, D.D. 1975 The effect of various constant temperatures on the growth of the ventral diverticulum in *Aedes triseriatus* (Diptera: Culicidae). ANN ENTOMOL SOC AM 68: 771-776.

*Rosales-Ronquillo, M.C.; Simons, R.W.; Silverman, P.H. 1973 Aseptic rearing of *Anopheles stephensi* (Diptera: Culicidae). ANN ENTOMOL SOC AM 66: 949-954.

Rose, R.I. 1971 Laboratory rearing of the southern corn rootworm on a meridic diet and preliminary evaluation of the fungus, *Beauveria bassiana*, as a biological control agent. DISST ABSTR INT B 32: 115 pp.

*Rose, R.I.; McCabe, J.M. 1973 Laboratory rearing techniques for the southern corn rootworms. J ECON ENTOMOL 66: 398-400.

*Ross, R.H. Jr; Monroe, R.E.; Butcher, J.W. 1971 Studies on techniques for the xenic and aseptic rearing of the European pine shoot moth, *Rhyacionia buoliana* (Lepidoptera: Olethreutidae). CAN ENTOMOL 103: 1449-1454.

Roy, P.; Pant, N.C. 1972 Lipid requirements for *Tribolium castaneum* (Herbst). INDIAN J ENTOMOL 34: 35-41.

*Royes, W.V.; Robertson, F.W. 1964 The nutritional requirements and growth relations of different species of *Drosophila*. J EXP BIOL 156: 105-135.

Ru, N.; Whitcomb, W.H.; Murphey, M. 1976 Culturing of *Chrysopa rufilabris* (Neuroptera: Chrysopidae). FLA ENTOMOL 59: 21-26.

*Russ, K. 1971 Investigations with a view to mass rearing of *Sparganothis pilleriana* Schiffermüller for genetic control. pp. 295-302. *in* Sterility Principle for Insect Control or Eradication. Proceedings of a Panel Organized by the Joint FAO/IAEA Division of Atomic Energy in Food and Agriculture.

*Russell, W.G. 1968 A medium for laboratory rearing of the peach tree and lesser peach tree borers. J ECON ENTOMOL 61: 1450-1451.

*Ruth, D.S.; Hedlin, A.F. 1969 Rearing the Douglas-fir cone moth, *Barbara colfaxiana* (Kearfott) on an artificial diet in the laboratory. J ENTOMOL SOC B C 66: 22-25.

*Rutter, R.R.; Ferkovich, S.M. 1973 A rapid separation of larvae of the Indian meal moth from rearing medium. ANN ENTOMOL SOC AM 66: 919-920.

*Růžička, Z. 1974 A note on substitutes of plant pollen in the adult diet of *Metasyrphus corollae* (Fabr.) (Diptera: Syrphidae). VESTN CESK SPOL ZOOL 38: 148-151.

Safyanova, V.M. 1964 Laboratory cultivation of sandflies (Diptera: Phlebotominae). BULL W H O 31: 573-576.

*Saksena, C.; Perti, S.L. 1970 Development of a chemically defined diet for studies on the nutritional requirements of the cockroach, *Periplaneta americana* Linn. LABDEV PART B 8: 155-160.

Saksena, C.; Perti, S.L. 1971 Studies on vitamin requirements of the cockroach, *Periplaneta americana* Linn. LABDEV PART B 9: 126-131.

*Salama, H.S. 1970 Rearing the corn borer, *Ostrinia nubilalis* (Hübn.) on a semi-artificial diet. Z ANGEW ENTOMOL 65: 216-218.

Salama, H.S.; Saleh, M.R. 1971 A technique for artificial feeding of coccids. PROC 13 INT CONGR ENTOMOL MOSCOW (1968)1: 434-435.

*Salama, H.S.; Tolba, R.A. 1967 A simple artificial diet for rearing corn borers. BULL SOC ENTOMOL ÉGYPTE 51: 151-152.

*Salama, H.S.; Tolba, R.A. 1971 Development of the sugarcane borer, *Sesamia cretica* Led. on a semi-artificial diet. Z ANGEW ENTOMOL 68: 74-75.

*Sang, J.H. 1956 The quantitative nutritional requirements of *Drosophila melanogaster*. J EXP BIOL 33: 45-72.

Sang, J.H. 1957 XVII. Utilization of dietary purines and pyrimidines by *Drosophila melanogaster*. PROC R SOC EDINBURGH 66: 339-359.

Sang, J.H. 1959 Circumstances affecting the nutritional requirements of *Drosophila melanogaster*. ANN N Y ACAD SCI 77: 352-365.

Sang, J.H. 1962a Relationships between protein supplies and B-vitamin requirements, in axenically cultured *Drosophila*. J NUTR 77: 355-368.

Sang, J.H. 1962b Selection for rate of larval development using *Drosophila melanogaster* cultured axenically on deficient diets. GENET RES 3: 90-109.

Sang, J.H. 1964 Nutritional requirements of inbred lines and crosses of *Drosophila melanogaster*. GENET RES 5: 50-67.

**Sang, J.H. 1972a The use of mutants in nutritional research. pp. 9-16. *in* Rodriguez, J.G. (Ed.), Insect and Mite Nutrition. North-Holland Publishing Company, New York and London.

Sang, J.H. 1972b Utilization of lipids by *Drosophila melanogaster*. pp. 493-503. *in* Rodriguez, J.G. (Ed.), Insect and Mite Nutrition. North-Holland Publishing Company, New York and London.

**Sang, J.H. 1976 The nutritional requirements of *Drosophila*. *in* Wright, T.R.F.; Ashburner, M. (Eds), Genetics and Biology of *Drosophila*. Vol 2a, Academic Press, New York (In press).

Sang, J.H.; Burnet, B. 1963 Physiological genetics of melanotic tumors in *Drosophila melanogaster*. I. The effects of nutrient balance on tumor penetrance in the tu^k strain. GENETICS 48: 235-253.

Sang, J.H.; King, R.C. 1961 Nutritional requirements of axenically cultured *Drosophila melanogaster* adults. J EXP BIOL 38: 793-809.

Santas, L.A. 1965 On a new synthetic nutritive substrate for rearing larvae of *Dacus oleae* Gmel. (Diptera: Trypetidae). BULL AGRIC BANK GREECE 144: 8-15.

*Sarma, P.S. 1943 Riboflavin and pyridoxin (vitamin B_6) as growth promoting factors for rice moth larvae (*Corcyra cephalonica* St.). INDIAN J MED RES 31: 165-171.

*Sato, Y. 1965 The artificial diet for mass rearing of the tobacco cutworm, *Prodenia litura* Fabricius, and the common armyworm, *Leucania separata* Walker. JPN J APPL ENTOMOL ZOOL 9: 99-105.

*Sato, Y. 1974 Rearing *Pieris rapae crucivora* larvae on artificial diets. KONTYÛ 42: 467-472. (In Japanese with English summary)

*Saunders, J.L.; Knoke, J.K. 1967 Diets for rearing the ambrosia beetle, *Xyleborus ferrugineus* (Fabricius) *in vitro*. SCIENCE 157: 460-463.

**Sawicki, R.M. 1964 Some general considerations on housefly rearing techniques. BULL W H O 31: 535-537.

**Sawicki, R.M.; Holbrook, D.V. 1961 The rearing, handling and biology of house flies (*Musca domestica* L.) for assay of insecticides by the application of measured drops. PYRETHRUM POST 6: 3-18.

Saxena, S.C.; Kaul, S. 1974 Qualitative vitamin requirements of *Oryzaephilus mercator* Fauvel and their deficiency effects on the survival and growth of F_1-progeny. ARCH INT PHYSIOL BIOCHIM 82: 49-54.

Saxena, S.C.; Kaul, S. 1975 Effect of aminoprotein on the growth and metamorphosis of *Oryzaephilus mercator* Fauvel (Coleoptera: Silvanidae) larvae. APPL ENTOMOL ZOOL 10: 315-317.

Schader, L. 1941 The effect of vitamin B_1 (thiamin) on the body length and duration of the larval period in *Drosophila melanogaster*. GROWTH 5: 19-26.

Schaefer, C.H. 1968 The relationship of the fatty acid composition of *Heliothis zea* larvae to that of its diet. J INSECT PHYSIOL 14: 171-178.

Scheel, C.A.; Beck, S.D.; Medler, J.T. 1957 Nutrition of plant-sucking Hemiptera. SCIENCE 125: 444-445.

*Scheel, C.A.; Beck, S.D.; Medler, J.T. 1958 Feeding and nutrition of certain Hemiptera. PROC 10 INT CONGR ENTOMOL MONTREAL (1956) 2: 303-308.

Scherney, F.; Haisch, A. 1967 Über Massenzucht und Sterilisation der Mittelmeerfruchtfliege *Ceratitis capitata* Wied. (Ein Beitrag Zur Autozitmethode). BAYER LANDWIRTSCH JAHRB 44: 748-756. (In German)

*Schmidt, F.H. 1966 Two artificial (oligidic) media for the Douglas-fir beetle, *Dendroctonus pseudotsugae* Hopkins (Coleoptera: Scolytidae). CAN ENTOMOL 98: 1050-1055.

Schmidt, M.L. 1964 Laboratory culture of two *Phlebotomus* species, *P. papatasi* and *P. orientalis*. BULL W H O 31: 577-578.

**Schoof, H.F. 1964 Laboratory culture of *Musca, Fannia,* and *Stomoxys*. BULL W H O 31: 539-544.

**Schoonhoven, L.M. 1968 Chemosensory bases of host plant selection. ANNU REV ENTOMOL 13: 115-136.

*Schroeder, W.J. 1970 Rearing of the pecan bud moth on artificial diet. J ECON ENTOMOL 63: 650-651.

*Schroeder, W.J.; Miyabara, R.; Chambers, D.L. 1970 A reusable diet support for rearing the melon fly on fluid larval medium. J ECON ENTOMOL 63: 2002.

*Schroeder, W.J.; Miyabara, R.Y.; Chambers, D.L. 1972 Protein products for rearing three species of larval Tephritidae. J ECON ENTOMOL 65: 969-972.

*Schroeder, W.J.; Miyabara, R.Y.; Tanaka, N.; Chambers, D.L. 1971 A fluid larval medium for rearing the melon fly. J ECON ENTOMOL 64: 1221-1223.

Schultz, J.; St. Lawrence, P.; Newmeyer, D. 1946 A chemically defined medium for the growth of *Drosophila melanogaster*. ANAT REC 96: 540. (Summary only; details not published; hence not abstracted)

*Schwartz, P.H.; Turner, R.B. 1966 A dietary study of the adult eye gnat, *Hippelates pusio* (Diptera: Chloropidae). ANN ENTOMOL SOC AM 59: 277-280.

*Schwarz, I.; Koch, A. 1962 Der *Tribolium*-Test als quantitativer Test für 8 B-Vitamine I. Z VITAM HORM FERMENTFORSCH 12: 291-306. (In German)

Schwarz, I.; Koch, A. 1963 Der *Tribolium*-Test als quantitativer Test für 8 B-vitamine II. Z VITAM HORM FERMENTFORSCH 13: 16-21. (In German)

Schwarz, I.; Koch, A. 1966 Der *Tribolium*-Test als quantitativer Test für 8 B-Vitamine III. Über die Errechnung der "Kennzahl" für die quantitative Auswertung der Versuchsprotokolle. Z ERNAEHRUNGSWISS 7: 1-8. (In German)

**Scoggin, J.K.; Tauber, O.E. 1950 Survey of literature on insect lipids. IOWA STATE COLL J SCI 25: 99-124.

*Scott, B.A. Jr; Berryman, A.A. 1971 Laboratory rearing techniques for *Scolytus ventralis* (Coleoptera: Scolytidae). BULL WASH AGRIC EXP STN 741: 9 pp.

Scudo, F.M.; Peschiera, G. 1971 A technique for regulating artificial populations of *Drosophila*. ECOLOGY 52: 165-168.

Seay, T.N. 1967 Studies on the rearing of the two-spotted spider mite, *Tetranychus urticae* (Koch) on chemically defined diets. Doctoral Thesis, Department of Entomology, University of Kentucky, Lexington, USA.

*Sedee, J.W. 1952 Qualitative vitamin requirements for growth of larvae of *Calliphora erythrocephala* (Meig.). EXPERIENTIA 9: 142-143.

*Sedee, J.W. 1954 Qualitative amino acid requirements of larvae of *Calliphora erythrocephala* (Meigen). ACTA PHYSIOL PHARMACOL NEERL 3: 262-269.

*Sedee, J.W. 1956 Dietetic requirements and intermediary protein metabolism of the larvae of *Calliphora erythrocephala* (Meig.). Van Gorcum & Co. Ltd, Assen, Netherlands, 130 pp.

Sedee, J.W. 1958 Qualitative B vitamin requirements of the larvae of a blowfly, *Calliphora erythrocephala* (Meig.). PHYSIOL ZOOL 31: 310-316.

Sedee, J.W. 1961 Intermediary metabolism in aseptically reared blowfly larvae. I. Biosynthesis of squalane and cholesterol. ARCH INT PHYSIOL BIOCHIM 69: 284-294.

Sehgal, S.S.; Agarwal, H.C. 1973 Quantitative dietary requirements of the beetle *Trogoderma* for cholesterol and choline. J INSECT PHYSIOL 19: 419-425.

Sell, D.K.; Bodznick, D.A. 1971 Effects of dietary $ZnSO_4$ on the growth and feeding of the tobacco budworm, *Heliothis virescens*. ANN ENTOMOL SOC AM 64: 850-855.

*Sender, C. 1969 Élevage permanent du carpocapse des pommes *Carpocapsa (= Laspeyresia) pomonella* L. sur milieu artificiel simplifié. ANN ZOOL ECOL ANIM 1: 321-326. (In French with English summary)

*Sender, C. 1970 Élevage du carpocapse des pommes sur un nouveau milieu artificiel non spécifique. ANN ZOOL ECOL ANIM 2: 93-95. (In French with English summary)

Shankland, D.L.; Stark, J.H.; Whistler, R.L. 1968 The effect of 5-thio-*d*-glucose on insect development and its adsorption by insects. J INSECT PHYSIOL 14: 63-72.

*Shanks, C.H. Jr; Finnigan, B.F. 1969 Behaviour, survival, and reproduction of the strawberry aphid on holidic diets and host and nonhost plants. ANN ENTOMOL SOC AM 62: 1490-1492.

*Shanks, C.H. Jr; Finnigan, B.F. 1973 An artificial diet for *Otiorhynchus sulcatus* larvae. ANN ENTOMOL SOC AM 66: 1164-1166.

*Shannon, R.C. 1923 Rearing dipterous larvae on nutrient agar. PROC ENTOMOL SOC WASH 25: 103-104.

Sharma, G.K.; Hodgson, E.; Rock, G.C. 1972a Nutrition and metabolism of sulphur amino acid in *Argyrotaenia velutinana* larvae. J INSECT PHYSIOL 18: 9-18.

Sharma, G.K.; Rock, G.C.; Hodgson, E. 1972b Metabolism of sulphur amino acids via the cystathionine pathway in *Argyrotaenia velutinana* larvae. J INSECT PHYSIOL 18: 1333-1341.

Shaver, T.N.; Lukefahr, M.J. 1969 Effect of flavonoid pigments and gossypol on growth and development of the bollworm, tobacco budworm and pink bollworm. J ECON ENTOMOL 62: 643-646.

*Shaver, T.N.; Raulston, J.R. 1971 A soybean-wheat germ diet for rearing the tobacco budworm. ANN ENTOMOL SOC AM 64: 1077-1079.

Shaw, G.G. 1973 Importance of starches to spruce budworm (Lepidoptera: Tortricidae). CAN ENTOMOL 105: 129-132.

*Shehata, K.K.; Weismann, L. 1972 Rearing the predaceous mite *Phytoseiulus persimilis* Athias-Henriot on artificial diet (Acarina Phytoseiidae). BIOLOGIA BRATISLAVA 27: 609-615.

*Shorey, H.H. 1963 A simple artificial rearing medium for the cabbage looper. J ECON ENTOMOL 56: 536-537.

Shorey, H.H. 1966 The biology of *Trichoplusia ni* (Lepidoptera: Noctuidae). IV. Environmental control of mating. ANN ENTOMOL SOC AM 59: 502-506.

*Shorey, H.H.; Hale, R.L. 1965 Mass-rearing of the larvae of nine noctuid species on a simple artificial medium. J ECON ENTOMOL 58: 522-524.

Shumakov, E.M. *et al.* 1959 Rearing of phytophaga on artificial media. 4 CONGR ALL-UNION ENTOMOL SOC 1: 197-199. (In Russian)

*Shumakov, E.M. 1968 Difficulties and successes in the mass rearing of insects in the laboratory, and the possibility of autocidal control of some harmful species. pp. 219-232. (In Russian with English summary) *in* Isotopes and Radiation in Entomology. International Atomic Energy Agency, Vienna.

*Shumakov, E.M.; Edelman, N.M.; Borisova, A.E. 1960 An attempt to rear phytophagocytic insects on synthetic media. DOKL AKAD NAUK SSSR SER BIOL 130: 237-240. (In Russian)

*Shumakov, E.M.; Edelman, N.M.; Borisova, A.E. 1967 Propagation of phytophagous insects in artificial media. ZASHCH RAST MOSCOW 6: 21-22. (In Russian)

**Shumakov, E.M.; Edelman, N.M.; Borisova, A.E.; Yakimova, N.L. 1974 Mass rearing of the codling moth on artificial nutritive media. PROC ALL UNION RES INST PLANT PROT 40: 7-17. (In Russian with English summary)

**Shyamala, M.B.; Bhatt, J.V. 1962 Recent developments in silkworm nutrition. J SCI IND RES INDIA 21: 26-32.

*Siddiqui, K.H.; Chatterji, S.M. 1972 Laboratory rearing of the maize stem borer, *Chilo zonellus* Swinhoe (Crambidae: Lepidoptera) on a semi-synthetic diet using indigenous ingredients. INDIAN J ENTOMOL 34: 183-185.

Sieburth, J.F.; Bonsall, M.G.; McLaren, B.A. 1951 A simplified biological assay method using the cockroach, *Periplaneta americana* (Linn.), for protein utilization. ANN ENTOMOL SOC AM 44: 463-468.

*Sieburth, J.F.; McLaren, B.A. 1953 Growth studies with the cockroach, *Periplaneta americana* (Linn.), fed vitamin-deficient diets substituted with corresponding antivitamins. ANN ENTOMOL SOC AM 46: 43-48.

*Sieckert, E.E.; Ritenour, G.L.; Davis, H.G. 1974 An artificial medium and rearing techniques for the colonization of *Bactra verutana*. ENVIRON ENTOMOL 3: 196-197.

*Silhacek, D.L.; Miller, G.L. 1972 Growth and development of the Indian meal moth, *Plodia interpunctella* (Lepidoptera: Phycitidae), under laboratory mass-rearing conditions. ANN ENTOMOL SOC AM 65: 1084-1087.

**Silva, G.M. de 1970 Sterile-male technique for control of the olive fly. pp. 119-130. *in* Sterile-male Technique for Control of Fruit Flies. Proceedings of a Panel Organized by the Joint FAO/IAEA Division of Atomic Energy in Food and Agriculture.

*Silva, G.M. de 1973 The use of parafilm as an ovipositioning device. A contribution to the mass rearing of the olive fly (*Dacus oleae* Gmel.). AGRON LUSIT 35: 85-90.

Silverman, P.H.; Levinson, Z.H. 1954 Lipid requirements of the larvae of the housefly *Musca vicina* (Macq.) reared under non-aseptic conditions. BIOCHEM J 58: 291-294.

Silverman, P.H.; Silverman, L. 1953 Growth measurements of *Musca vicina* (Macq.) reared with a known bacterial flora. RIV PARASSITOL 14: 89-95.

**Simmonds, F.J. 1944 The propagation of insect parasites on unnatural hosts. BULL ENTOMOL RES 35: 219-226.

**Simmonds, F.J. 1964 Mass production of insect parasites and predators. BULL W H O 31: 511-512.

*Simon, J.E. 1969 Present stage of research into the eradication of the Mediterranean and South American fruit flies and the cotton stainer in Peru by the sterile-male technique. pp. 115-121. *in* Sterile-male Technique for Eradication or Control of Harmful Insects. Proceedings of a Panel Organized by the Joint FAO/IAEA Division of Atomic Energy in Food and Agriculture.

Singh, K.R.P. 1955a The nutrition of *Aedes aegypti* L. REP ENTOMOL SOC ONT 86: 17-19.

*Singh, K.R.P. 1955b Studies on nutrition of *Corcyra cephalonica* Staint (Galleridae, Lepidoptera) with special reference to carbohydrate requirements. SCI CULT 20: 339-340.

*Singh, K.R.P.; Brown, A.W.A. 1957 Nutritional requirements of *Aedes aegypti* L. J INSECT PHYSIOL 1: 199-220.

Singh, P. 1969 Effects of antimicrobial agents on dipterous parasites, *Agria affinis*, reared on chemically defined diets. PROC ENTOMOL SOC ONT 100: 238.

**Singh, P. 1972 Bibliography of Artificial Diets for Insects and Mites. N Z DEP SCI IND RES BULL 209: 75 pp.

*Singh, P. 1974a Aseptic rearing of *Epiphyas postvittana* (Lepidoptera: Tortricidae) on a meridic diet. N Z J ZOOL 1: 111-117.

*Singh, P. 1974b A chemically defined medium for rearing *Epiphyas postvittana* (Lepidoptera: Tortricidae). N Z J ZOOL 1: 241-243.

*Singh, P. 1974c Improvements in or relating to artificial dietary mixtures for insects. N Z Patent, Application No. 174449 Provisional Application 4th June, 1974 Complete Specification filed 4 September, 1975.

**Singh, P. 1974d Artificial Diets for Insects: a Compilation of References with Abstracts (1970-1972). N Z DEP SCI IND RES BULL 214: 96 pp.

*Singh, P. 1976a Artificial diet for rearing *Drosophila busckii* (Diptera: Drosophilidae). Unpublished.

*Singh, P. 1976b Improved chemically defined diet for rearing housefly larvae. Unpublished.

**Singh, P. 1976c Synthetic diets for insects and mites. in press. *in* Rechcigl, M. (Ed.), Handbook of Nutrition and Food. CRC Press, Cleveland, Ohio.

Singh, P.; Bucher, G.E. 1971 Efficacy of 'safe' levels of antimicrobial food additives to control microbial contaminants in a synthetic diet for *Agria affinis* larvae. ENTOMOL EXP APPL 14: 297-309.

**Singh, P.; Charles, J.G. 1975 A List of Laboratory Cultures and Rearing Methods of Terrestrial Arthropods in New Zealand. ENTOMOL SOC N Z BULL 3: 30 pp.

*Singh, P.; Charles, J.G. 1976 An artificial diet for the potato tuber moth (*Phthorimaea operculella* (Zeller)) (Lepidoptera: Gelechiidae). Unpublished.

Singh, P.; House, H.L. 1970a Antimicrobial agents: their detrimental effects on size of an insect, *Agria affinis*. CAN ENTOMOL 102: 1340-1344

Singh, P.; House, H.L. 1970b Effects of streptomycin and potassium sorbate levels in relation to nutrient levels on the larvae of *Agria affinis*. J ECON ENTOMOL 63: 449-454.

Singh, P.; House, H.L. 1970c Antimicrobials: 'safe' levels in a synthetic diet of an insect, *Agria affinis*. J INSECT PHYSIOL 16: 1769-1782.

*Singh, P.; Jerram, E.M. 1976 Rearing of housefly larvae in plastic bags. N Z J ZOOL 3: 57-58.

Singh, P.; Mabbett, F.E. 1976 A note on the life history of the magpie moth, *Nyctemera annulata* (Lepidoptera: Arctiidae). N Z J ZOOL 3: (in press).

*Singh, P.; Taylor, B.; Rodriguez, J.G. 1968 Bioassay of insecticides for control of *Drosophila melanogaster*. J ECON ENTOMOL 61: 336-337.

*Singh, Z.; Howe, W.L. 1971 Feeding, longevity, and fecundity of the adult western corn rootworm fed artificial diets. J ECON ENTOMOL 64: 1136-1137.

Sinha, M. 1974 Breeding the common house fly *Musca domestica* L. for physiological studies (Diptera: Muscidae). CURR SCI 43: 320.

Sinoir, Y. 1968 Étude de quelques facteurs conditionnant la prise de nourriture chez les larves du criquet migrateur, *Locusta migratoria migratorioides* (Orthoptera: Acrididae). I. Facteurs externes. ENTOMOL EXP APPL 11: 195-210. (In French with English summary)

*Sivaramakrishnan, R.; Sarma, P.S. 1957 Effect of soybean inhibitor and vitamin B_{12} on growth of rice moth larva (*Corcyra cephalonica* St.). CURR SCI 26: 283-284.

*Smilowitz, Z.; Smith, C.L. 1970 Distributions and frequencies of weight of cabbage looper larvae reared on artificial diet. J ECON ENTOMOL 63: 1106-1107.

*Smirnoff, W.A. 1958 An artificial diet for rearing coccinellid beetles. CAN ENTOMOL 90: 563-565.

*Smith, B.C. 1965 Effects of food on the longevity, fecundity, and development of adult coccinellids (Coleoptera: Coccinellidae). CAN ENTOMOL 97: 910-919.

**Smith, C.N. (Ed.) 1966 Insect Colonization and Mass Production. Academic Press, New York and London, 618 pp.

**Smith, C.N. 1967 Mass-breeding procedures. pp. 653-672. *in* Wright, J.W.; Pal, R. (Eds), Genetics of Insect Vectors of Disease. Elsevier Publishing Company, Amsterdam.

*Smith, H.R.; Richardson, E.G.; Yonce, C.E.; Jacklin, S.W. 1969 A method of rearing the peach tree borer on artificial diet. J ECON ENTOMOL 62: 961-962.

*Smith, J.J.B.; Friend, W.G. 1970 Feeding in *Rhodnius prolixus:* responses to artificial diets as revealed by changes in electrical resistance. J INSECT PHYSIOL 16: 1709-1720.

Smith, T.L. 1937 Breeding methods for *Galleria mellonella.* pp. 349-355. *in* Needham, J.G. *et al.* (Eds), Culture Methods for Invertebrate Animals. Comstock Publishing Company, Ithaca, New York (Dover Edition, 1959, unaltered and unabridged).

Soderstrom, E.L.; Lovitt, A.E. 1967 A volumetric dispenser for insect media. J ECON ENTOMOL 60: 1179-1180.

*Solomon, J.D. 1966 Artificial rearing of the carpenter-worm, *Prionoxystus robiniae* (Lepidoptera: Cossidae), and observations of its development. ANN ENTOMOL SOC AM 59: 1197-1200.

Solomon, J.D. 1967 Rearing the carpenterworm, *Prionoxystus robiniae,* in the forest (Lepidoptera: Cossidae). ANN ENTOMOL SOC AM 60: 283-285.

Solomon, M.E.; Cunnington, A.M. 1964 Rearing acaroid mites. ACAROLOGIA 6: 399-403.

**Sonleitner, F.J. 1964 The stress factor in insect colonies. BULL W H O 31: 545-549.

*Soo Hoo, C.F.; Fraenkel, G. 1964a The resistance of ferns to the feeding of *Prodenia eridania* larvae. ANN ENTOMOL SOC AM 57: 788-790.

Soo Hoo, C.F.; Fraenkel, G. 1964b A simplified laboratory method for rearing the southern armyworm, *Prodenia eridania,* for feeding experiments. ANN ENTOMOL SOC AM 57: 798-799.

Soo Hoo, C.F.; Fraenkel, G. 1966 The selection of food plants in a polyphagous insect, *Prodenia eridania* (Cramer). J INSECT PHYSIOL 12: 693-709.

**Spector, W.S. (Ed.) 1956 Handbook of Biological Data. W.B. Saunders Co., Philadelphia.

Speers, C.F.; Cody, M.G. 1975 Technique for rearing the pales weevil, *Hylobius pales* (Coleoptera: Curculionidae). J GA ENTOMOL SOC 10: 251-254.

Spencer, N.E. 1971 Sterilization of insect diet by *gamma* irradiation. J ECON ENTOMOL 64: 753-754.

Spencer, W.P. 1950 Collection and laboratory culture. pp. 535-590. *in* Demerec, M. (Ed.), Biology of Drosophila. Wiley, Chapman and Hall, New York and London.

Spiller, D. 1963 Procedure for rearing houseflies. NATURE LONDON 199: 405.

Spiller, D. 1964a Bacterial status of a housefly rearing medium. N Z ENTOMOL 3: 33-38.

**Spiller, D. 1964b Nutrition and diet of muscoid flies. BULL W H O 31: 551-554.

*Spiller, D. 1966 House flies. pp. 203-225. *in* Smith, C.N. (Ed.), Insect Colonization and Mass Production. Academic Press, New York and London.

*Splittstoesser, C.M.; McEwen, F.L. 1961 A bioassay technique for determining the insecticidal activity of preparations containing *Bacillus thuringiensis* Berliner. J INSECT PATHOL 3: 391-398.

*Srivastava, P.D.; Kumar, S.S.; Bhamburkar, M.W.; Pradhan, S. 1964 An artificial diet for rearing the desert locust. INDIAN J ENTOMOL 26: 352-356.

*Srivastava, P.N.; Auclair, J.L. 1970 Responses of the large milkweed bug, *Oncopeltus fasciatus* (Hemiptera: Lygaeidae) to chemically defined diets. ENTOMOL EXP APPL 13: 208-216.

*Srivastava, P.N.; Auclair, J.L. 1971a An improved chemically defined diet for the pea aphid, *Acyrthosiphon pisum.* ANN ENTOMOL SOC AM 64: 474-478.

Srivastava, P.N.; Auclair, J.L. 1971b Influence of sucrose concentration on diet uptake and performance by the pea aphid, *Acyrthosiphon pisum.* ANN ENTOMOL SOC AM 64: 739-743.

Srivastava, P.N.; Auclair, J.L. 1974 Effect of amino acid concentration on diet uptake and performance by the pea aphid, *Acyrthosiphon pisum* (Homoptera: Aphididae). CAN ENTOMOL 106: 149-156.

Srivastava, P.N.; Auclair, J.L. 1975 Role of single amino acids in phagostimulation, growth and survival of *Acyrthosiphon pisum.* J INSECT PHYSIOL 21: 1865-1871.

Stadelbacher, E.A.; Brewer, F.D. 1974 Dispensers of insect diets made from a stainless steel sprayer. J ECON ENTOMOL 67: 312-313.

*Städler, E. 1971 An improved mass-rearing method of the carrot rust fly, *Psila rosae* (Diptera: Psilidae). CAN ENTOMOL 103: 1033-1038.

*Stahler, N.; Terzian, L.A. 1953 Studies in the laboratory mating habits of *Anopheles quadrimaculatus* Say. J EXP ZOOL 124: 317-328.

Stairs, G.R. 1965 Dosage-mortality response of *Galleria mellonella* (Linnaeus) to a nuclear-polyhedrosis virus. J INVERTEBR PATHOL 7: 5-9.

*Standifer, L.N.; Haydak, M.H.; Mills, J.P.; Levin, M.D. 1973 Influence of pollen in artificial diets on food consumption and brood production in honey bee colonies. AM BEE J 113: 94-95.

*Standifer, L.N.; Poole, H.K.; Doull, K.M. 1971 Egg production, oviposition and survival of isolated queen honey bees fed experimental diets. ANN ENTOMOL SOC AM 64: 228-232.

*Steffan, W.A. 1966 A generic revision of the family Sciaridae (Diptera) of America north of Mexico. UNIV CALIF BERKELEY PUBL ENTOMOL 44: 77 pp.

*Steiner, L.F.; Mitchel, S. 1966 Tephritid fruit flies. pp. 555-583. *in* Smith, C.N. (Ed.), Insect Colonization and Mass Production. Academic Press, New York and London.

*Steinhaus, E.A. 1963 Insect Pathology- An Advance Treatise Vol I & II. Academic Press, New York & London.

*Sterling, W.L.; Adkisson, P.L. 1966 An artificial diet for laboratory culture of boll weevil larvae and adults. J ECON ENTOMOL 59: 1074-1077.

Sterling, W.L.; Wellso, S.G.; Adkisson, P.L.; Dorough, H.W. 1965 A cottonseed-meal diet for laboratory cultures of the boll weevil. J ECON ENTOMOL 58: 867-869.

*Steve, P.C. 1960 Biology and control of the little housefly *Fannia canicularis* in Massachusetts. J ECON ENTOMOL 53: 999-1004.

Stevens, W.C. 1971 Nutritional studies including quantitative mineral requirements of the almond moth, *Cadra (Ephestia) cautella* (Lepidoptera: Pyralidae). DISS ABSTR INT B: 32: 355.

Stewart, J.R.; Glover, D. Jr; Torries, W.J. 1972 Automated system for mass producing diet for boll weevil adults. J ECON ENTOMOL 65: 861-862.

Stewart, P.A.; Baker, A.P. Jr 1970 Rate of growth of larval tobacco hornworms reared on tobacco leaves and on artificial diets. J ECON ENTOMOL 63: 535-536.

Stimmann, M.W.; Pangaldan, R.; Schureman, B.S. 1972 Improved method of rearing the beet armyworm. J ECON ENTOMOL 65: 596-597.

*Stoffolano, J.G. Jr 1974 Influence of diapause and diet on the development of the gonads and accessory reproductive glands of the black blowfly, *Phormia regina* (Meigen). CAN J ZOOL 52: 981-988.

*Stones, L.C. 1972 The Stablefly. pp. 556-563. *in* University Federation of Animal Welfare (Ed.), The UFAW Handbook on the Care and Management of Laboratory Animals. Fourth Edition. Churchill Livingstone, Edinburgh and London.

*Storms, J.J.H.; Noordink, J.Ph.W. 1970 Nutritional requirements of the two-spotted spider mite, *Tetranychus urticae*, Acarina: Tetranychidae. 7 EUR MITE SYMP POL AKAD 29-39.

Street, H.R.; Palmer, L.S. 1935 Requirement of the flour beetle (*Tribolium confusum* Duval) for vitamins in the B group. PROC SOC EXP BIOL MED 32: 1500-1501.

Strickberger, M.W. 1962 Experiments in Genetics with *Drosophila*. Wiley, New York.

*Stride, G.O. 1953 On the nutrition of *Carpophilus hemipterus* L. (Coleoptera: Nitidulidae). TRANS R ENTOMOL SOC LONDON 104: 171-194.

*Strong, F.E.; Kruitwagen, E. 1969 Feeding and nutrition of *Lygus hesperus*. III. Limited growth and development on a meridic diet. ANN ENTOMOL SOC AM 62: 148-155.

*Strong, F.E.; Kruitwagen, E. 1970 Gustatory discrimination between meridic diets by the bug, *Lygus hesperus*. J INSECT PHYSIOL 16: 521-530.

*Strong, F.E.; Landes, D.A. 1965 Feeding and nutrition of *Lygus hesperus* (Hemiptera: Miridae). II. An estimation of normal feeding rates. ANN ENTOMOL SOC AM 58: 309-314.

*Strong, R.G.; Partida, G.J.; Warner, D.N. 1968 Rearing stored-product insects for laboratory studies: six species of moths. J ECON ENTOMOL 61: 1237-1249.

Sturtevant, A.H. 1937 Culture methods for *Drosophila*. pp. 437-446. *in* Needham, J.G. *et al.* (Eds), Culture Methods for Invertebrate Animals. Comstock Publishing Company, Ithaca, New York (Dover Edition 1959 Unaltered and Unabridged).

Subbarow, Y.; Trager, W. 1940 The chemical nature of growth factors required by mosquito larvae. II. Pantothenic acid and vitamin B_6. J GEN PHYSIOL 23: 561-568.

Sun, P.I. 1963 Studies on the development of a chemically defined diet for the two-spotted spider mite (*Tetranychus telarius* L.). Masters Thesis, Department of Entomology, University of Kentucky, Lexington, U.S.A.

Sun, P.I.; Seay, T.N.; Rodriguez, J.G. 1963 Studies on rearing the two-spotted spider mite, *Tetranychus telarius* L., on a chemically defined diet. PROC NORTH CENT BRANCH ENTOMOL SOC AM 18: 92.

Sundaramurthy, V.T.; Subramaniam, T.R. 1969 A laboratory technique for mass rearing of the tobacco caterpillar *Spodoptera littoralis* (Boisd.) *Prodenia litura* (F.) (Noctuidae: Lepidoptera) on synthetic diet. CURR SCI 38: 569-570.

Sundby, R.A. 1966 A comparative study of the efficiency of three predatory insects *Coccinella septempunctata* L. (Coleoptera, Coccinellidae) *Chrysopa carnea* St. (Neuroptera, Chrysopidae) and *Syrphus ribesii* L. (Diptera, Syrphidae) at two different temperatures. ENTOMOPHAGA 11: 395-404.

Sundby, R.A. 1967 Influence of food on the fecundity of *Chrysopa carnea* Stephens (Neuroptera: Chrysopidae). ENTOMOPHAGA 12: 475-479.

Sutcliffe, J.F.; McIver, S.B. 1975 Artificial feeding of simuliids (*Simulium venustum*): factors associated with probing and gorging. EXPERIENTIA 31: 694-695.

Sutherland, O.R.W.; Hillier, J.R. 1976 The influence of maltose and other carbohydrates on the feeding behaviour of *Heteronychus arator* (Scarabaeidae: Coleoptera). EXPERIENTIA 32: 701.

Sutherland, O.R.W.; Mittler, T.E. 1971 Influence of diet composition and crowding on wing production by the aphid *Myzus persicae*. J INSECT PHYSIOL 17: 321-328.

Suto, M.; Higuchi, Y.; Matsuda, M.; Kawasugi, S.; Fukuda, T.; Kurose, T.; Shishido, J.; Sugiyama, T. 1966 Rearing experiments of silkworms on artificial diet during early instars, followed on fresh leaves regenerated from mulberry trees met with the frost damage. J SERIC SCI JPN 35: 281-284.

*Sutter, G.R.; Kieckhefer, R.W. 1969 Performance of the English grain aphid on sterilized and unsterilized artificial diet. J ECON ENTOMOL 62: 254-255.

*Sutter, G.R.; Krysan, J.L.; Guss, P.L. 1971 Rearing the southern corn rootworm on artificial diet. J ECON ENTOMOL 64: 65-67.

*Sutter, G.R.; Miller, E. 1972 Rearing the army cutworm on an artificial diet. J ECON ENTOMOL 65: 717-718.

*Sutter, G.R.; Miller, E.; Calkins, C.O. 1972 Rearing the pale western cutworm on artificial diet. J ECON ENTOMOL 65: 1470-1471.

Swailes, G.E. 1967 A substrate for oviposition by the cabbage maggot in laboratory cultures. J ECON ENTOMOL 60: 619-620.

Sweetman, H.L. 1956 Rearing successive generations of the carpet beetle under controlled conditions. J ECON ENTOMOL 49: 277-278.

**La Symbiose des Insectes avec les Microorganisms. Colloques Union Internationales des Sciences Biologiques, Ser B 10, 174 pp., 1952.

**Symposium on Culture Procedures for Arthropod Vectors and their Biological Control Agents. 1964. BULL W H O 32: 433-622.

Szalay-Marzso', L. 1972 Alkalic inactivation of granulosis virus of codling moth (*Lespeyresia pomonella* L.) and some problems of rearing hygiene. NORENYVEDELEM 8: 402-408. (In Hungarian with English summary)

*Taguchi, T. 1963 Evaluation of dosage of various ingredients and the possible substitutes for carrot in the carrot-yeast medium in the larval culture of the oriental fruit fly, *Dacus dorsalis* Hendel. RES BULL PLANT PROT SERV JPN 38: 17-27.

*Taguchi, T. 1966 Influence of different diets upon the longevity of the adult of oriental fruit fly, *Dacus dorsalis* Hendel. RES BULL PLANT PROT SERV JPN 41: 16-19. (In Japanese with English summary)

*Tahori, A.S.; Hazan, A. 1970 Rearing of the black citrus aphid *Toxoptera aurantii* on chemically defined diets. J INSECT PHYSIOL 16: 1975-1981.

Takei, G.H.; Tamashiro, M. 1974 Axenic rearing of the lawn armyworm, *Spodoptera mauritia acronyctoides* (Guenée), and synxenic inoculation techniques. PROC HAWAII ENTOMOL SOC (1972) 21: 475-483.

*Tamaki, G.; Turner, J.E.; Wallis, R.L. 1972 Life tables for evaluating the rearing of the zebra caterpillar. J ECON ENTOMOL 65: 1024-1027.

*Tamaki, Y. 1959 Studies on nutrition and metabolism of the smaller tea tortrix, *Adoxophyes orana* Fischer von Röslerstamm. I. Aseptic rearing of the larvae on synthetic diets. JPN APPL ENTOMOL ZOOL 3: 286-289.

Tamaki, Y. 1961a Studies on nutrition and metabolism of the smaller tea tortrix, *Adoxophyes orana* (Fischer von Röslerstamm). II. An essential factor for adult emergence. JPN J APPL ENTOMOL ZOOL 5: 58-63.

Tamaki, Y. 1961b Studies on nutrition and metabolism of the smaller tea tortrix, *Adoxophyes orana* (Fischer von Röslerstamm). III. Nature of an unknown dietary factor for larval growth. JPN J APPL ENTOMOL ZOOL 5: 203-206.

Tamaki, Y. 1962 Utilization of antiseptics for rearing of the smaller tea tortrix (*Adoxophyes orana*) on synthetic food medium. JPN J APPL ENTOMOL ZOOL 6: 248-250. (In Japanese)

*Tamaki, Y. 1964 Studies on nutrition and metabolism of the smaller tea tortrix, *Adoxophyes orana* Fischer von Röslerstamm. IV. Some chemical properties of an unknown dietary factor responsible for larval growth. JPN J APPL ENTOMOL ZOOL 8: 55-61.

*Tamaki, Y. 1966a Rearing of *Adoxophyes orana* Fischer von Röslerstamm and *Homona magnanima* Diakonoff on simplified artificial food. JPN J APPL ENTOMOL ZOOL 10: 46-48. (In Japanese)

*Tamaki, Y. 1966b Mass rearing of the smaller tea tortrix, *Adoxophyes orana* Fischer von Röslerstamm, on a simplified artificial diet for successive generations (Lepidoptera: Tortricidae). JPN J APPL ENTOMOL ZOOL 10: 120-124.

Tanada, Y.; Chang, G.Y. 1968 Resistance of the alfalfa caterpillar *Colias eurytheme*, at high temperature to a cytoplasmic-polyhedrosis virus and thermal inactivation point of the virus. J INVERTEBR PATHOL 10: 79-83.

Tanaka, N.; Hart, R.A.; Okamoto, R.Y.; Steiner, L.F. 1972 Control of excessive metabolic heat produced in diet by a high density of larvae of the Mediterranean fruit fly. J ECON ENTOMOL 65: 866-867.

*Tanaka, N.; Okamoto, R.; Chambers, D.L. 1970 Methods of mass rearing the Mediterranean fruit fly currently used by the U.S. Department of Agriculture. pp. 19-23. *in* Sterile-male Technique for Control of Fruit Flies. Proceedings of a Panel Organized by the Joint FAO/IAEA Division of Atomic Energy in Food and Agriculture.

**Tanaka, N.; Steiner, L.F.; Ohinata, K.; Okamoto, R. 1969 Low-cost larval rearing medium for mass production of oriental and Mediterranean fruit flies. J ECON ENTOMOL 62: 967-968.

Tanton, M.T. 1965 Agar and chemostimulant concentrations and their effect on intake of synthetic food by larvae of the mustard beetle, *Phaedon cochleariae* Fab. ENTOMOL EXP APPL 8: 74-82.

Tatum, E.L. 1939 Nutritional requirements of *Drosophila melanogaster*. PROC NATL ACAD SCI U S A 25: 490-497.

Tatum, E.L. 1941 Vitamin B requirements of *Drosophila melanogaster*. PROC NATL ACAD SCI U S A 27: 193-197.

Taylor, M.W.; Medici, J.C. 1966 Amino acid requirements of grain beetles. J NUTR 88: 176-180.

*Tenquist, J.D. 1971 Rearing of *Lucilia sericata* (Diptera: Calliphoridae) as a modified *Musca domestica* medium. N Z ENTOMOL 5: 30-31.

*Terriere, L.C.; Grau, P.A. 1972 Dietary requirements and tissue levels of fatty acids in three Noctuidae. J INSECT PHYSIOL 18: 63[illegible] 647.

Terzian, L.A.; Stahler, N. 1954 The effect [illegible] age and sex ratio on the mating activity of *Anopheles quadrimaculatus* Say. J EXP ZOOL 12[illegible] 389-396.

*Theron, P.P.A. 1945 Rearing *Euxoa* on artifici[illegible] media. J ENTOMOL SOC SOUTH AFR 8: 165-166.

*Thomas, H.A. 1969 A meridic diet and rearing technique for the pales weevil larva. J ECON ENTOMOL 62: 1491-1494.

Thomas, H.A. 1971 Development of pales weevil larvae on an emulsified synthetic diet. U S FOR SERV RES NOTE SE-163, 3 pp.

*Thomas, W.P. 1968 An artificial diet for rearing the light brown apple moth *Epiphyas postvittana* (Walk.) (Lepidoptera: Tortricidae). N Z ENTOMOL 4: 31-32.

Thompson, C.G. 1965 Douglas-fir tussock moth virus tests. Phase III simulated field test to determine virus dosages and concentrations. U S FOR SERV PAC NORTHWEST FOR RANGE EXP STN PROG REP, 26 pp.

Thompson, E.G.; Bell, L.H. 1968 Laboratory studies on the biology of *Anopheles stephensi* Liston. MOSQ NEWS 28: 639-642.

*Thompson, S.N. 1975 Defined meridic and holidic diets and aseptic feeding procedures for artificially rearing the ectoparasitoid *Exeristes roborator* (Fabricius). ANN ENTOMOL SOC AM 68: 220-226.

Thorsteinson, A.J. 1953 The chemotactic responses that determine host specificity in an oligophagous insect (*Plutella maculipennis* (Cart.) Lepidoptera). CAN J ZOOL 31: 52-72.

Thorsteinson, A.J.; Nayar, J.K. 1963 Plant phospholipids as feeding stimulants for grasshoppers. CAN J ZOOL 41: 931-935.

*Ticheler, J. 1971 Rearing of the onion-fly, *Hylemya antiqua* (Meigen), with a view to release of sterilized insects. pp. 341-346. *in* Sterility Principle for Insect Control or Eradication. Proceedings of a Panel Organized by the Joint FAO/IAEA Division of Atomic Energy in Food and Agriculture.

Ticheler, J.; Noordink, J.Ph.W. 1968 Application of the sterile-male technique on the onion-fly *Hylemya antiqua* (Meig.), in the Netherlands. Progress report. pp. 111-115. *in* Radiation, Radioisotopes and Rearing Methods in Control of Insect Pests. Proceedings of a Panel Organized by the Joint FAO/IAEA Division of Atomic Energy in Food and Agriculture.

Tingle, F.C.; Lane, H.C.; King, E.E.; Lloyd, E.P. 1971 Influence of nutrients in the adult diet on diapause in the boll weevil. J ECON ENTOMOL 64: 812-814.

*Tischlinger, A. 1975 Rearing olive fly at Seibersdorf. pp. 137-139. *in* Controlling Fruit Flies by the Sterile-insect Technique. Proceedings of a Panel Organized by the Joint FAO/IAEA Division of Atomic Energy in Food and Agriculture.

Toba, H.H.; Kishaba, A.N. 1972 Modifications of larval diet for cabbage loopers. J ECON ENTOMOL 65: 127-129.

*Toba, H.H.; Kishaba, A.N.; Pangaldan, R. 1970 A rating system for evaluating the suitability of larval diets for cabbage loopers. J ECON ENTOMOL 63: 759-760.

*Toba, H.H.; Kishaba, A.N.; Pangaldan, R.; Riggs, S. 1969 Laboratory rearing of pepper weevils on artificial diets. J ECON ENTOMOL 62: 257-258.

*Trager, W. 1935a The culture of mosquito larvae free from living microorganisms. AM J HYG 22: 18-25.

Trager, W. 1935b On the nutritional requirements of mosquito larvae (*Aedes aegypti*). AM J HYG 22: 475-493.

Trager, W. 1936 The utilization of solutes by mosquito larvae. BIOL BULL WOODS HOLE MASS 71: 343-351.

Trager, W. 1937a The culture of mosquito larvae free from living micro-organisms. pp. 389-390. *in* Needham, J.G. *et al.* (Eds), Culture Methods for Invertebrate Animals. Comstock Publishing Company, Ithaca, New York (Dover Edition 1959).

Trager, W. 1937b A growth factor required by mosquito larvae. J EXP BIOL 14: 240-251.

*Trager, W. 1941 The nutrition of the invertebrates. PHYSIOL REV 21: 1-35.

*Trager, W. 1947 Insect nutrition. BIOL REV 22: 148-177.

*Trager, W. 1948 Biotin and fat-soluble materials with biotin activity in the nutrition of mosquito larvae. J BIOL CHEM 176: 1211-1223.

*Trager, W. 1953 Nutrition. pp. 350-386. *in* Roeder, K.D. (Ed.), Insect Physiology. John Wiley & Sons Inc, New York.

Travaglini, E.C.; Tartof, D. 1972 Instant *Drosophila:* a method for mass culturing large numbers of *Drosophila*. DROSOPH INF SERV 48: 157.

*Treat, T.L.; Halfhill, J.E. 1973 Rearing alfalfa loopers and celery loopers on an artificial diet. J ECON ENTOMOL 66: 569-570.

*Trimble, R.M.; Corbet, P.S. 1975 Laboratory colonization of *Toxorhynchites rutilus septentrionalis* (Diptera: Culicidae). ANN ENTOMOL SOC AM 68: 217-219.

*Tsitsipis, J.A. 1975 Mass rearing of the olive fruit fly, *Dacus oleae* (Gmelin), at "Democritos". pp. 93-100. *in* Controlling Fruit Flies by the Sterile-insect Technique. Proceedings of a Panel Organized by the Joint FAO/IAEA Division of Atomic Energy in Food and Agriculture.

Tsitsipis, J.A.; Mittler, T.E. 1970 Convenient lighting system for inducing the production of sexual forms of aphids feeding on artificial diets. ANN ENTOMOL SOC AM 63: 1665-1667.

*Tsukuda, R.; Kiyoku, M. 1966 Laboratory cultures of the tobacco-cutworms *Prodenia litura* Fabricius on the synthetic media. SCI REP FAC AGRIC OKAYAMA UNIV 28: 13-18. (In Japanese with English summary)

Tulisalo, U.; Korpela, S. 1973 Mass rearing of the green lacewing (*Chrysopa carnea* Steph.). ANN ENTOMOL FENN 39: 143-144.

*Turner, E.C. Jr; Hair, J. 1967 Effect of diet on longevity and fecundity of laboratory-reared face flies. J ECON ENTOMOL 60: 857-860.

Turner, R.B. 1971 Dietary amino acid requirements of the cotton aphid, *Aphis gossypii:* the sulphur-containing amino acids. J INSECT PHYSIOL 17: 2451-2456.

Turunen, S. 1973a Utilization of fatty acids by *Pieris brassicae* L. reared on artificial and natural diets. J INSECT PHYSIOL 19: 1999-2009.

Turunen, S. 1973b Role of labelled dietary fatty acids and acetate in phospholipids during the metamorphosis of *Pieris brassicae*. J INSECT PHYSIOL 19: 2327-2340.

Turunen, S. 1974a Lipid utilization in adult *Pieris brassicae* with special reference to the role of linolenic acid. J INSECT PHYSIOL 20: 1257-1269.

*Turunen, S. 1974b Polyunsaturated fatty acids in the nutrition of *Pieris brassicae* (Lepidoptera). ANN ZOOL FENN 11: 300-303.

**T-W-Fiennes, R.N. (Ed.) 1972 Biology of Nutrition. 18 International Encyclopaedia of Food and Nutrition, Pergamon Press, 681 pp.

*Twine, B.H. 1971 Cannibalistic behaviour of *Heliothis armigera* (Hübn.). QUEENSL J AGRIC ANIM SCI 28: 153-157.

Tzanakakis, M.E. 1967 Control of the olive fruit fly *Dacus oleae* (Gmelin) with radiation or chemical sterilization procedures. U S DEP AGRIC F-G-GR-102, 81 pp.

Tzanakakis, M.E. 1971 Rearing methods for the olive fruit fly *Dacus oleae* (Gmelin). ANN SCH AGRIC FOR UNIV THESSALONIKI 14: 293-326.

Tzanakakis, M.E.; Economopoulos, A.P.; Tsitsipis, J.A. 1966 Improved artificial food media for larvae of *Dacus oleae* (Gmelin) (Diptera: Tephritidae). Z ANGEW ENTOMOL 58: 373-383.

*Tzanakakis, M.E.; Economopoulos, A.P. 1967 Two efficient larval diets for continuous rearing of the olive fruit fly. J ECON ENTOMOL 60: 660-663.

Tzanakakis, M.E.; Economopoulos, A.P.; Tsitsipis, J.A. 1967 The importance of conditions during the adult stage in evaluating an artificial food larvae of *Dacus oleae* (Gmelin) (Diptera: Tephritidae). Z ANGEW ENTOMOL 59: 127-130.

Tzanakakis, M.E.; Economopoulos, A.P.; Tsitsipis, J.A. 1968 Artificial rearing of the olive fly. Progress report. pp. 123-130. *in* Radiation, Radioisotopes and Rearing Methods in Control of Insect Pests. Proceedings of a Panel Organized by the Joint FAO/IAEA Division of Atomic Energy in Food and Agriculture.

*Tzanakakis, M.E.; Economopoulos, A.P.; Tsitsipis, J.A. 1970 Rearing and nutrition of the olive fruit fly. I. Improved larval diet and simple containers. J ECON ENTOMOL 63: 317-318.

*Tzanakakis, M.E.; Phillips, J.H.H. 1969 Artificial diets for larvae of the oriental fruit moth. J ECON ENTOMOL 62: 879-882.

*Tzanakakis, M.E.; Savopoulou, M.C. 1973 Artificial diets for larvae of *Lobesia botrana* (Lepidoptera: Tortricidae). ANN ENTOMOL SOC AM 66: 470-471.

Tzanakakis, M.E.; Tsitsipis, J.A.; Steiner, L.F. 1967 Egg production of olive fruit fly fed solids or liquids containing protein hydrolysate. J ECON ENTOMOL 60: 352-354.

*Uberoi, N.K. 1960 Fatty acids in the lipid metabolism of the rice moth *Corcyra cephalonica* Stainton. J ZOOL SOC INDIA 8: 85-90.

Umebachi, Y.; Tsuchitani, K. 1955 The presence of xanthurenic acid in the fruit fly, *Drosophila melanogaster*. J BIOCHEM TOKYO 42: 817-824.

**Universities Federation for Animal Welfare (Ed.) 1972 The UFAW Handbook on the Care and Management of Laboratory Animals. Fourth Edition. Churchill Livingstone, Edinburgh and London. 624 pp.

Uspenskaya, N.V.; Khlistovskiy, E.D. 1967 A universal semi-synthetic medium for the rearing of phytophagous insects. pp. 66-69. *in* The Useful and Injurious Vertebrates of Uzbekistan. Tashkent Branch of the USSR Academy of Science. (In Russian).

*Uspenskaya, N.V.; Kozhaeva, K.I. 1974 A new semi-synthetic nutrient medium for rearing larvae of *Agrotis segetum* (Lepidoptera, Noctuidae). ZOOL ZH 53: 782-784. (In Russian with English summary)

**Uvarov, B.P. 1928 Insect nutrition and metabolism. A summary of the literature. TRANS ENTOMOL SOC LONDON 65: 255-343.

**Vago, C. 1972 Problèmes pathologiques concernant les élevages en masse d'invertébrés. ENTOMOPHAGA 17: 111-129. (In French with English summary)

*Vago, C.; Fosset, J. 1965 Élevage aseptique des gryllides. 90 CONGR SOC SAVANTES NICE II: 501-505. (In French).

**Vago, C.; Fosset, J.; Meynadier, G. 1961 Élevage aseptique d'insectes phytophages. REV PATHOL VEG ENTOMOL AGRIC FR 40: 111-129. (In French)

*Vail, P.V.; Anderson, S.J.; Jay, D.L. 1973 New procedures for rearing cabbage loopers and other lepidopterous larvae for propagation of nuclear polyhedrosis virus. ENVIRON ENTOMOL 2: 339-344.

Vail, P.V.; Henneberry, T.J.; Kishaba, A.N.; Arakawa, K.Y. 1968 Sodium hypochlorite and formalin as antiviral agents against nuclear-polyhedrosis virus in larvae of the cabbage looper. J INVERTEBR PATHOL 10: 84-93.

*Vail, P.V.; Henneberry, T.J.; Pengalden, R. 1967 An artificial diet for rearing the salt marsh caterpillar, *Estigmene acrea* (Lepidoptera: Arctiidae), with notes on the biology of the species. ANN ENTOMOL SOC AM 60: 134-138.

*Valder, S.M.; Hopkins, T.L. 1968 Effects of nutrition of egg development and survival in the face fly, *Musca autumnalis*. ANN ENTOMOL SOC AM 61: 827-834.

*Vandamme, P.; Angelini, A. 1968 Comparaison de trois milieus nutritifs artificiels pour l'élevage d'*Heliothis armigera* Hübn. COTON FIBRES TROP 23: 417-422. (In French with English summary)

*Vanderzant, E.S. 1957 Growth and reproduction of the pink bollworm on an amino acid medium. J ECON ENTOMOL 50: 219-221.

Vanderzant, E.S. 1958 The amino acid requirements of the pink bollworm. J ECON ENTOMOL 51: 309-311.

*Vanderzant, E.S. 1959 Inositol: an indispensable dietary requirement for the boll weevil. J ECON ENTOMOL 52: 1018-1019.

*Vanderzant, E.S. 1963a Nutrition of the adult boll weevil: oviposition on defined diets and amino acid requirements. J INSECT PHYSIOL 9: 683-691.

Vanderzant, E.S. 1963b Nutrition of the boll weevil larva. J ECON ENTOMOL 56: 357-362.

*Vanderzant, E.S. 1965 Axenic rearing of the boll weevil on defined diets: amino acid, carbohydrate, and mineral requirements. J INSECT PHYSIOL 11: 659-670.

**Vanderzant, E.S. 1966 Defined diets for phytophagous insects. pp. 273-303. *in* Smith, C.N. (Ed.), Insect Colonization and Mass Production. Academic Press, New York and London.

*Vanderzant, E.S. 1967a Wheat-germ diets for insects: rearing the boll weevil and the salt-marsh caterpillar. ANN ENTOMOL SOC AM 60: 1062-1066.

*Vanderzant, E.S. 1967b Rearing lygus bugs on artificial diets. J ECON ENTOMOL 60: 813-816.

*Vanderzant, E.S. 1968 Dietary requirements of the bollworm, *Heliothis zea* (Lepidoptera: Noctuidae), for lipids, choline, and inositol and the effect of fats and fatty acids on the composition of the body fat. ANN ENTOMOL SOC AM 61: 120-125.

**Vanderzant, E.S. 1969a Physical aspects of artificial diets. ENTOMOL EXP APPL 12: 642-650.

*Vanderzant, E.S. 1969b An artificial diet for larvae and adults of *Chrysopa carnea*, an insect predator of crop pests. J ECON ENTOMOL 62: 256-257.

Vanderzant, E.S. 1973a Axenic rearing of larvae and adults of the boll weevils on defined diets: additional tests with amino acids and vitamins. ANN ENTOMOL SOC AM 66: 1184-1186.

*Vanderzant, E.S. 1973b Improvements in the rearing diet for *Chrysopa carnea* and the amino acid requirements for growth. J ECON ENTOMOL 66: 336-338.

**Vanderzant, E.S. 1974 Development, significance and application of artificial diets for insects. ANNU REV ENTOMOL 19: 139-160.

Vanderzant, E.S. 1975 Effect of heat treatment on the ascorbic acid content of a diet and the effect on development of the tobacco budworm. J ECON ENTOMOL 68: 375-376.

Vanderzant, E.S.; Chremos, J.H. 1971 Dietary requirement of the boll weevil for arginine and the effect of arginine analogues on growth and on the composition of the body amino acids. ANN ENTOMOL SOC AM 64: 480-485.

*Vanderzant, E.S.; Davich, T.B. 1958 Laboratory rearing of the boll weevil: a satisfactory larval diet and oviposition studies. J ECON ENTOMOL 51: 288-291.

Vanderzant, E.S.; Davich, T.B. 1961 Artificial diets for the boll weevil and techniques for obtaining eggs. J ECON ENTOMOL 54: 923-928.

*Vanderzant, E.S.; Kerur, D.; Reiser, R. 1957 The role of dietary fatty acids in the development of the pink bollworm. J ECON ENTOMOL 50: 606-608.

*Vanderzant, E.S.; Pool, M.C.; Richardson, C.D. 1962 The role of ascorbic acid in the nutrition of three cotton insects. J INSECT PHYSIOL 8: 287-297.

*Vanderzant, E.S.; Reiser, R. 1956a Aseptic rearing of the pink bollworm on synthetic media. J ECON ENTOMOL 49: 7-10.

Vanderzant, E.S.; Reiser, R. 1956b Studies on the nutrition of the pink bollworm using purified casein media. J ECON ENTOMOL 49: 454-458.

Vanderzant, E.S.; Richardson, C.D. 1963 Ascorbic acid in the nutrition of plant-feeding insects. SCIENCE 140: 989-991.

Vanderzant, E.S.; Richardson, C.D. 1964 Nutrition of the adult boll weevil: lipid requirements. J INSECT PHYSIOL 10: 267-272.

*Vanderzant, E.S.; Richardson, C.D.; Davich, T.B. 1959 Feeding and oviposition by the boll weevil on artificial diets. J ECON ENTOMOL 52: 1138-1143.

*Vanderzant, E.S.; Richardson, C.D.; Fort, S.W. Jr 1962 Rearing of the bollworm on artificial diet. J ECON ENTOMOL 55: 140.

Van Emden, H.F. 1967 An increase in the longevity of adult *Aphis fabae* fed artificially through parafilm membranes on liquids under pressure. ENTOMOL EXP APPL 10: 166-170.

Van Emden, H.F.; Adams, J.B. 1972 The biological properties of aphids and their host plant relationships. pp. 48-104. *in* Van Emden, H.F. (Ed.), Aphid Technology. Academic Press, New York and London.

Van Herrewege, C. 1971 Consommation alimentaire chez les mâles adultes de *Blatella germanica* (L): influence de l'âge, de la nourriture larvaire et du jeune. ARCH SCI PHYSIOL 25: 401-413. (In French)

Van Herrewege, C. 1974 Régulation de la prise de nourriture, après un jeûne, chez les mâles de la blatte germanique dans différentes conditions alimentaires. ENTOMOL EXP APPL 17: 234-244. (In French with English summary)

*Van Herrewege, C.; David, J. 1969 Une nouvelle méthode pour mesurer la prise de nourriture et l'utilisation digestive des alimants chez *Blattella germanica* (L.). ANN NUTR ALIMENT 23: 253-268. (In French)

Van Herrewege, J. 1973a Importance de la qualité des protéines d'un milieu semi-synthétique pour la fécondité et la longévité de *Drosophila melanogaster*. ARCH INT PHYSIOL BIOCHIM 81: 57-66. (In French with English summary)

Van Herrewege, J. 1973b Physiologie des insectes. Concentration optimale de sucre pour la nutrition des drosophiles adultes: variation en fonction du milieu nutritif et du caractère etudie. C R ACAD SCI SER D 276: 2565-2568. (In French)

Van Herrewege, J. 1974 Nutritional requirements of adult *Drosophila melanogaster:* the influence of the casein concentration on the duration of life (Diptera: Drosophilidae). J EXP GERONTOL 9: 191-198.

Vargues, H. 1969a Action de divers antibiotiques sur le développement larvaire de la Mouche des fruits (*Ceratitis capitata* Wied). C R SEANCES SOC BIOL PARIS 163: 1915-1918. (In French)

Vargues, H. 1969b Action de divers antibiotiques sur la fécondité de la Mouche des fruits (*Ceratitis capitata* Wied). C R SEANCES SOC BIOL PARIS 163: 1933-1938. (In French)

*Varma, A.; Avasthy, P.N. 1973 An artificial diet for the rearing of stalk borer, *Chilo auricilius* Dudg. EXPERIENTIA 29: 1161-1162.

*Vasiljevic, L.J.; Injac, M. 1971 Artificial diet for the gypsy moth (*Lymantria dispar* L.). ZAST BILJA 22: 389-396.

*Vereecke, A.; Hertveldt, L. 1971 Laboratory rearing of the cabbage maggot. J ECON ENTOMOL 64: 670-673.

Villacorta, A. 1970 Influence of nutrition, light and temperature on the development of wheat stem saw fly, *Cephus cinctus* Norton. DISS ABSTR INT B 32: 1968.

*Villacorta, A.; Bell, R.A.; Callenbach, J.A. 1971 An artificial plant stem as an oviposition site for the wheat stem saw fly. J ECON ENTOMOL 64: 752-753.

*Villee, C.A.; Bissell, H.B. 1948 Nucleic acids as growth factors in *Drosophila*. J BIOL CHEM 172: 59-66.

Vinson, S.B. 1967 Effect of several nutrients on DDT resistance in tobacco budworm. J ECON ENTOMOL 60: 565-568.

*Wahid, A.; Akhtar, M. 1971 Nutritional requirements of sugar-cane top borer, *Scirpophaga nivella* Fab. PAK J ZOOL 3: 15-21.

*Waites, R.E.; Gothilf, S. 1969 Nutrition of the almond moth. I. Analysis and improvements of the experimental diet. J ECON ENTOMOL 62: 301-305.

*Waki, S.; Matsumoto, K. 1973a Studies on the environmental requirements for the breeding of the dust mite, *Dermatophagoides farinae* Hughes, 1961. Part 1. Observations on the mode of breeding under various temperature and humidity conditions. EISEI DOBUTSU 23: 159-163. (In Japanese with English summary)

Waki, S.; Matsumoto, K. 1973b Studies on the environmental requirements for the breeding of the dust mite, *Dermatophagoides farinae* Hughes, 1961. Part 2. Observations on the mode of breeding in various kinds of food. EISEI DOBUTSU 24: 117-121. (In Japanese with English summary)

*Wal, Y.C.; Agarwal, P.N. 1966 Nutritional requirements of woolly bear. Part I. A new synthetic medium for rapid growth and development of *Anthrenus flavipes* Lec. LABDEV PART B 4: 71.

*Wal, Y.C.; Agarwal, P.N. 1968 Nutritional requirements of the woolly bear. Part II. Some B-vitamins requirements of the larvae of *Anthrenus flavipes* Lec. LABDEV PART B 6: 51-52.

**Waldbauer, G.P. 1968 The consumption and utilization of foo by insects. ADV INSECT PHYSIOL 5: 229-288.

*Waldbauer, G.P.; Yamamoto, R.T.; Bowers, W.S. 1964 Laboratory rearing of the tobacco hornworm, *Protoparce sexta* (Johnson) (Lepidoptera: Sphingidae). J ECON ENTOMOL 57: 93-95.

Walker, D.W. 1968 Potential for control of the sugar-cane borer through radio-induced sterility. pp. 131-140. *in* Radiation, Radioisotopes and Rearing Methods in Control of Insect Pests. Proceedings of a Panel Organized by the Joint FAO/IAEA Division of Atomic Energy in Food and Agriculture.

*Walker, D.W.; Alemany, A.; Quintana, V.; Padovani, F.; Hagen, K.S. 1966 Improved xenic diets for rearing the sugarcane borer in Puerto Rico. J ECON ENTOMOL 59: 1-4.

Walling, M.V.; White, D.C.; Rodriguez, J.G. 1968 Characterization, distribution, catabolism, and synthesis of the fatty acids of the two-spotted spider mite, *Tetranychus urticae*. J INSECT PHYSIOL 14: 1445-1458.

*Wang, C.M. 1964 Laboratory observations on the life history and habits of the face fly, *Musca autumnalis* (Diptera: Muscidae). ANN ENTOMOL SOC AM 57: 563-569.

*Ward, A.G.; Pass, B.C. 1969 Rearing sod webworms on artificial diets. J ECON ENTOMOL 62: 510-511.

**Wardojo, S. 1967 Artificial diets as replacement of the host plant. MEDED RIJKSFAC LANDBOUWWET GENT 32: 357-364.

*Wardojo, S. 1969a Some factors relating to the larval growth of the Colorado potato beetle, *Leptinotarsa decemlineata* Say (Coleoptera: Chrysomelidae), on artificial diets. MEDED LANDBOUWHOGESCH WAGENINGEN 69-16: 75 pp.

*Wardojo, S. 1969b Artificial diet without crude plant material for two oligophagous leaf feeders. ENTOMOL EXP APPL 12: 698-702.

*Waterhouse, D.F. 1959 Axenic culture of wax moths for digestion studies. ANN N Y ACAD SCI 77: 283-289.

Wearing, C.H. 1968 Responses of aphids to pressure applied to liquid diet behind parafilm membrane. Longevity and larviposition of *Myzus persicae* (Sulz.) and *Brevicoryne brassicae* (L.) (Homoptera: Aphididae) feeding on sucrose and sinigrin solutions. N Z J SCI 11: 105-121.

*Weaver, N. 1964 A pollen substitute for honeybee colonies. GLEAN BEE CULT 92: 550-553.

Webber, L.G. 1958 Nutrition and reproduction in the Australian sheep blow-fly *Lucilia cuprina*. AUST J ZOOL 6: 139-144.

**Wehrstein, A. 1975 Adult olive fly diets: a comparison. pp. 140-144. *in* Controlling Fruit Flies by the Sterile-insect Technique. Proceedings of a Panel Organized by the Joint FAO/IAEA Division of Atomic Energy in Food and Agriculture.

*Weismann, L.; Podmanická, D. 1970 Einfluss der Nahrung und der Zuchttemperatur auf die länge der Larvalentwicklung von *Scotia segetum* Den. et Schiff. BIOLOGIA BRATISLAVA 25: 537-545. (In German with English summary)

*Wellington, E.F. 1949 Artificial media for rearing some phytophagous Lepidoptera. NATURE LONDON 163: 574.

*Wellso, S.G.; Connin, R.V.; Hoxie, R.P.; Cobb, D.L. 1970 Storage and behaviour of plant and diet-fed adult cereal leaf beetle, *Oulema melanopus* (Coleoptera: Chrysomelidae). MICH ENTOMOL 3: 103-107.

Wenk, P. 1963 Kopulation, Blutsaugen und Eiablage von *Boophthora erythrocephala* de Geer 1776 (Diptera, Simuliidae) im Laboratorium. NATURWISSENSCHAFTEN 11: 409-410. (In German)

Wensler, R.J.D. 1962 Mode of host selection by an aphid. NATURE LONDON 195: 830-831.

**West, L.S.; Peters, O.N. 1973 An Annotated Bibliography of *Musca domestica* Linnaeus. Folkstones. Dawsons of Pall Mall. 743 pp.

Wetzel, H.W.; Bauer, B. 1974 Influence of antibiotics on *Glossina moristans* Westwood. pp. 501-505. *in* Sterility Principle for Insect Control. Proceedings of a Panel Organized by the Joint FAO/IAEA Division of Atomic Energy in Food and Agriculture.

*Wheeler, M.R. 1972 Fruitflies. pp. 543-548. *in* Universities Federation for Animal Welfare (Ed.), The UFAW Handbook on the Care and Management of Laboratory Animals. Fourth Edition. Churchill Livingstone, Edinburgh and London.

*White, D. 1972 Effect of varying dietary amino acid and sucrose concentrations on production of apterous cabbage aphids. J INSECT PHYSIOL 18: 1241-1243.

White, G.F. 1931 Production of sterile maggots for surgical use. J PARASITOL 18: 133.

*White, G.F. 1937 Rearing maggots for sugrical use. pp. 418-427. *in* Needham, J.G. *et al.* (Eds), Culture Methods for Invertebrate Animals. Comstock Publishing Company, Ithaca, New York (Dover Edition 1959)

White, L.D. 1966 Desilking of codling moth pupae by use of cupriethylene diamine. BULL ENTOMOL SOC AM 12: 152.

Whitlock, V.H. 1973 Modifications to an existing laboratory rearing technique for *Heliothis armigera* (Hübn). J ENTOMOL SOC S AFR 36: 307-308.

*Whitlock, V.H. 1974 A substitute for casein in an artificial rearing medium for the American bollworm, *Heliothis armigera* (Hübn). PHYTOPHYLACTICA 6: 133-134.

*Wielgus, R.S. 1974 Artificial diet: the key to the mass rearing of *Megathymus* larvae. J RES LEPID 13: 271-277.

*Wientjens, W.H.J.M. 1972 De omzetting van β-sitosterol in cholesterol bij drie soorten kakkerlakken. MEDED GENT LAB DELFT 452: 35-46. (In Dutch)

**Wigglesworth, V.B. 1952 Symbiosis in blood sucking insects. TIJDSCHR ENTOMOL 95: 63-68.

**Wigglesworth, V.B. 1960 Nutrition and reproduction in insects. PROC NUTR SOC ENG SCOT 19: 18-23.

*Wightman, J.A. 1968 Rearing and feeding *Lygocoris pabulinus* (Heteroptera: Miridae). ANNU REP AGRIC HORT RES STN UNIV BRISTOL 157-160 pp.

*Wilkes, A.; Bucher, G.E.; Cameron, J.W.M.; West, A.S. Jr 1948 Studies on the housefly (*Musca domestica* L.). 1. The biology and large scale production of laboratory populations. CAN J RES D26: 8-25.

*Wilkinson, J.D.; Morrison, R.K.; Peters, P.K. 1972 Effect of Calco Oil Red N-1700 dye incorporated into a semiartificial diet of the imported cabbageworm, cornearworm and cabbage looper. J ECON ENTOMOL 65: 264-268.

*Wilkinson, R.S. 1966 The David-Gardiner method of feeding lepidopterous larvae on a semi-synthetic diet. MICH ENTOMOL 1: 26-28.

*Williams, K.L.; Nurmi, S.; Birt, L.M. 1974 Developmental and biochemical characteristics of sterile cultures of the blowfly, *Lucilia cuprina.* LAB ANIM 8: 177-187.

Wilson, G.G. 1974 The use of Fumidil B to suppress the microsoprian *Nosema fumiferanae* in stock cultures of the spruce budworm *Choristoneura fumiferana* (Lepidoptera: Tortricidae). CAN ENTOMOL 106: 995-996.

Wilson, L.P.; Birch, M.M. 1944 Tolerance of larvae of *Drosophila* for amino acids: arginine. GROWTH 8: 125-148.

Winchester, A.M. 1933 A method of increasing the yield of *Drosophila.* SCIENCE 78: 483-484.

*Wollerman, E.H.; Adams, C.; Heaton, G.C. 1969 Continuous laboratory culture of the locust borer, *Megacyllene robiniae.* ANN ENTOMOL SOC AM 62: 647-649.

*Wongsiri, T.; Randolph, N.M. 1962 A comparison of the biology of the sugarcane borer on artificial and natural diets. J ECON ENTOMOL 55: 472-473.

*Wood, D.M.; Davies, D.M. 1965 The rearing of simuliids (Diptera). PROC 12 INT CONGR ENTOMOL LONDON (1964): 821-823.

*Wood, J. 1970 Rearing *Wiseana* species in the laboratory. N Z ENTOMOL 4: 3-7.

*Woodring, J.P.; Cook, E.F. 1962 The biology of *Ceratozetes cisalpinus* Berlese, *Scheloribates laevigatus* Koch, and *Oppia neerlandica* Oudemans (Oribatei) with a description of all stages. ACAROLOGIA 4: 109-137.

Woodring, J.P.; Cutcher, J.J. 1968 Vital dye marking of tyroglyphid (Acaridae) mites. ANN ENTOMOL SOC AM 61: 1031-1032.

Woodruff, L.C. 1939 Linear growth ratios for *Blattella germanica* L. J EXP ZOOL 81: 287-298.

**Workshop on the Genetics of Insect Behaviour held at Gainesville, Florida, September 24-29, 1974.

**Wostmann, B.S.; Pleasants, J.R.; Reddy, B.S. 1967 Water-soluble, non-antigenic diets. pp. 187-205. *in* Proceedings of the Third International Symposium of the International Committee on Laboratory Animals, on the Husbandry of Laboratory Animals. September 1965, Ireland, Academic Press, New York and London.

Wressell, H.B. 1955 Rearing the European corn borer, *Pyrausta nubilalis* (Hbn.) (Lepidoptera: Pyralididae) on an artificial diet. REP ENTOMOL SOC ONT 86: 10-13.

Wright, J.E.; Oehler, D.D. 1971 Fatty acids in the larval diet and in the phorate adult of the horn fly, *Haematobia irritans.* J INSECT PHYSIOL 17: 1479-1488.

**Wyniger, R. 1974 Insektenzucht: Methoden der Zucht Haltung von Insekten und Milben im Laboratorium. Verlag Eugen Ulmer Stuttgart, 368 pp. (In German)

Yakimova, N.L. 1973 A method of mass rearing of the codling moth on artificial media. Conference on Progressive Methods of Pest Control in Agricultural Crops Moscow 146-148. (In Russian)

Yamada, H.; Kato, M. 1967 Oleic acid as potent growth factor for the silkworm. PROC JPN ACAD 43: 230-233.

Yamamoto, R.T. 1968 Mass rearing of the tobacco hornworm. I. Egg production. J ECON ENTOMOL 61: 170-174.

*Yamamoto, R.T. 1969 Mass rearing of the tobacco hornworm. II. Larval rearing and pupation. J ECON ENTOMOL 62: 1427-1431.

*Yamaya, K.; Tamaki, Y. 1972 Mass rearing method for Tortricidae. SHOKUBUTSU BOEKI 26: 165-168. (In Japanese)

Yanishevskaya, L.V. 1969 Influence of some antiseptics on diet sterility and development of the codling moth. ZAKHYST ROSL 4: 102-105. (In Ukranian).

*Yanishevskaya, L.V.; Pristavko, V.P. 1969 The problem of rearing winged insects on artificial nutrient media. ZAKHYST ROSL KIEV 10: 46-58. (In Ukranian)

Yatomi, Y.; Kaneko, T.; Fukazawa, N.; Ishii, S. 1951 Sterile culture of the dual-transformation (dimorphic) rice borer). J AGRIC HORTIC 26: 55-56. (In Japanese)

*Yazgan, S. 1972 A chemically defined synthetic diet and larval nutritional requirements of the endoparasitoid *Itoplectis conquisitor* (Hymenoptera). J INSECT PHYSIOL 18: 2123-2141.

Yazgan, S.; House, H.L. 1970 An hymenopterous insect, the parasitoid *Itoplectis conquisitor*, reared axenically on a chemically-defined synthetic diet. CAN ENTOMOL 102: 1304-1306.

*Yearian, W.C.; Gilbert, K.L.; Warren, L.O. 1966 Rearing the fall webworm, *Hyphantria cunea* (Drury) (Lepidoptera: Arctiidae) on a wheat germ medium. J KANS ENTOMOL SOC 39: 495-499.

*Yearian, W.C.; Wilkinson, R.C. 1963 An artificial rearing medium for *Ips calligraphus* Germ. FLA ENTOMOL 46: 319-320.

*Yearian, W.C.; Wilkinson, R.C. 1965 Two larval rearing media for *Ips* bark beetles. FLA ENTOMOL 48: 25-27.

Yen, D.F. 1972 Rearing studies on the rice stem borer, *Chilo suppressalis* Walker. JPN PESTIC INF 10: 127.

Yendol, W.G. 1970 Fatty acid composition of *Galleria* larvae, hemolymph, and diet (Lepidoptera: Galleriidae). ANN ENTOMOL SOC AM 63: 339-341.

*Yonce, C.E.; Gentry, C.R.; Pate, R.R. 1971 Artificial diets for rearing larvae of the plum curculio. J ECON ENTOMOL 64: 1111-1112.

*Yonce, C.E.; Gentry, C.R.; Pate, R.R. 1973 A complete artificial diet for rearing the plum curculio. J ECON ENTOMOL 66: 362-363.

*Yoshida, T.; Matsuoka, M.; Kimura, K. 1960 On the rearing of silkworm larvae with an artificial diet containing dried mulberry leaf powder as its basic material. BULL SERIC EXP STN JPN 15: 543-586. (In Japanese with English summary)

Young, R.C. 1961 The effects of dietary beeswax and wax components on the larvae of the greater wax moth, *Galleria mellonella* (L.). ANN ENTOMOL SOC AM 54: 657-659.

**Yushima, T. 1962 Rearing of insects on artificial diets - present and future. NOGYO GIJUTSU 17: 172-175, 212-215, 269-273, 314-317, 369-372, 419-422, 526-529, 581-586. (In Japanese)

**Yushima, T. 1968 Rearing methods of aphids on artificial diets. SHOKUBUTSU BOEKI 22: 306-310. (In Japanese)

**Yushima, T. 1970 The mass rearing of insects and its utilization in insect pest control. NOGYO OYOBI ENGEI 45: 1631-1636. (In Japanese)

Yushima, T.; Kamano, S. 1964 On the successive rearing of rice stem borer on the artificial diets under aseptic conditions. III. Relation between electrophoretic pattern of protein in larval haemolymph and dietary ascorbic acid level. JPN J APPL ENTOMOL ZOOL 8: 218-221. (In Japanese with English summary)

Żabiński, J. 1926 Observations sur l'élevage des cafards nourris avec des aliments artificiels. C R SOC BIOL 95: 545-548. (In French)

Żabiński, J. 1928a Élevage des Blattides soumis à une alimentation artificielle. C R SOC BIOL 98: 73-77. (In French)

Żabiński, J. 1928b Nouvelles recherches sur l'élevage des Blattides soumis à une alimentation artificielle. Influence de la composition du régime. C R SOC BIOL 98: 78-80. (In French)

*Żabiński, J. 1929 The growth of blackbeetles and of cockroaches on artificial and on incomplete diets. Part I. J EXP BIOL 6: 360-385.

*Zerillo, R.T.; Odell, T.M. 1973 White pine weevil: a rearing procedure and artificial medium. J ECON ENTOMOL 66: 593-594.

Zhlistovskiy, E.D. 1968 Semi-synthetic nutrient media and technique for the rearing of noctuid (Lepidoptera: Noctuidae) caterpillar on them. Thesis Uzbek Academy of Sciences Branch of USSR Academy of Sciences Tashkent: 1-23. (In Russian)

*Zlotin, A.Z.; Lymareva, M.A. 1966 Rearing insects in winter for biological tests of insecticides. ZOOL ZH 45: 1100-1102. (In Russian with English summary)

Zuska, J. 1975a Concentration of dietary glucose or sucrose determining longevity of adult *Piophila casei* (Diptera, Piophilidae). ACTA ENTOMOL BOHEMOSLOV 72: 80-86.

*Zuska, J. 1975b Simplified laboratory culture of the cheese skipper, *Piophila casei* (Diptera, Piophilidae). FOLIA PARASITOL 22: 140.

Author Index

(Italicized page numbers indicate references occurring in the bibliography.)

Species Index

(Italicized page numbers indicate references occurring in the bibliography.)

GPSR Compliance

The European Union's (EU) General Product Safety Regulation (GPSR) is a set of rules that requires consumer products to be safe and our obligations to ensure this.

If you have any concerns about our products, you can contact us on ProductSafety@springernature.com

In case Publisher is established outside the EU, the EU authorized representative is:

Springer Nature Customer Service Center GmbH
Europaplatz 3
69115 Heidelberg, Germany

Batch number: 10838523

Printed by Printforce, the Netherlands